MOLECULAR BIOLOGY INTELLIGENCE UNIT

Translation Mechanisms

Jacques Lapointe, Ph.D.

Professeur titulaire
Département de Biochimie et de Microbiologie
Faculté des Sciences et de Génie
Université Laval
Québec, Québec, Canada

Léa Brakier-Gingras, Ph.D.

Professeur titulaire
Département de Biochimie
Faculté de Médecine
Université de Montréal
Montréal, Québec, Canada

LANDES BIOSCIENCE / EUREKAH.COM
GEORGETOWN, TEXAS
U.S.A.

KLUWER ACADEMIC / PLENUM PUBLISHERS
NEW YORK, NEW YORK
U.S.A.

TRANSLATION MECHANISMS

Molecular Biology Intelligence Unit

Eurekah.com / Landes Bioscience
Kluwer Academic / Plenum Publishers

Printed in the U.S.A.

Kluwer Academic / Plenum Publishers, 233 Spring Street, New York, New York, U.S.A. 10013
http://www.wkap.nl/

Please address all inquiries to the Publishers:
Eurekah.com / Landes Bioscience, 810 South Church Street
Georgetown, Texas, U.S.A. 78626
Phone: 512/ 863 7762; FAX: 512/ 863 0081
www.Eurekah.com
www.landesbioscience.com

Translation Mechanisms edited by Jacques Lapointe and Léa Brakier-Gingras, Landes / Kluwer dual imprint / Landes series: Molecular Biology Intelligence Unit

ISBN: 0-306-47839-0

While the authors, editors and publisher believe that drug selection and dosage and the specifications and usage of equipment and devices, as set forth in this book, are in accord with current recommendations and practice at the time of publication, they make no warranty, expressed or implied, with respect to material described in this book. In view of the ongoing research, equipment development, changes in governmental regulations and the rapid accumulation of information relating to the biomedical sciences, the reader is urged to carefully review and evaluate the information provided herein.

Library of Congress Cataloging-in-Publication Data

Translation mechanisms / [edited by] Jacques Lapointe, Léa Brakier-Gingras.
      p. ; cm. -- (Molecular biology intelligence unit)
Includes bibliographical references.
      ISBN 0-306-47839-0
1. Genetic translation.
      [DNLM: 1. Translation, Genetic. QH 450.5 T77215 2003] I. Lapointe, Jacques, Ph.D. II. Brakier-Gingras, L. III. Series.
QH450.5.T719 2003
572′ .645--dc21

Dedication

To Yolande and Gabriel

CONTENTS

IV: Regulation of the Biosynthesis of the Translational Machinery

23. Control of Stable RNA Synthesis 370
Melanie M. Barker and Richard L. Gourse

24. Regulation of the Expression of Aminoacyl-tRNA Synthetases and Translation Factors 388
Harald Putzer and Soumaya Laalami

V: Inhibitors of Protein Synthesis

25. Inhibitors of Aminoacyl-tRNA Synthetases as Antibiotics and Tools for Structural and Mechanistic Studies 416
Robert Chênevert, Stéphane Bernier and Jacques Lapointe

26. Antibiotics as Indicators of the Functional Components of the Ribosome 429
Dominique Fourmy, Satoko Yoshizawa and Stephen Douthwaite

EDITORS

Jacques Lapointe, Ph.D.
Professeur titulaire
Département de Biochimie et de Microbiologie
Faculté des Sciences et de Génie
Université Laval
Québec, Québec, Canada
email: jacques.lapointe@bcm.ulaval.ca
Chapter 25

Léa Brakier-Gingras, Ph.D.
Professeur titulaire
Département de Biochimie
Faculté de Médecine
Université de Montréal
Montréal, Québec, Canada
email: lea.brakier.gingras@umontreal.ca
Chapter 16

CONTRIBUTORS

J. Christopher Anderson
Department of Chemistry
The Scripps Research Institute
La Jolla, California, U.S.A.
Chapter 6

John F. Atkins
Department of Human Genetics
University of Utah
Salt Lake City, Utah, U.S.A.
email: john.atkins@genetics.utah.edu
Chapter 22

Jamie M. Bacher
Institute for Cellular and Molecular
   Biology
College of Natural Sciences
University of Texas
Austin, Texas, U.S.A.
email: jbacher@mail.utexas.edu
Chapter 5

Kristian E. Baker
Department of Biochemistry
   and Molecular Biology
University of British Columbia
Vancouver BC, Canada
Chapter 9

Melanie M. Barker
Department of Bacteriology
University of Wisconsin
Madison, Wisconsin, U.S.A.
email: mmbarker@students.wisc.edu
Chapter 23

François Bélanger
Département de Biochimie
Université de Montréal
Montréal, Québec, Canada
email: belangf@magellan.umontreal.ca
Chapter 16

Rudolf K.Beran
Department of Microbiology,
    Immunology, and Molecular Genetics
University of California
Los Angeles, California, U.S.A.
Chapter 9

Stéphane Bernier
Département de Chimie
Université Laval
Québec, Canada
Chapter 25

Sylvain Blanquet
Laboratoire de Biochimie
Unité Mixte de Recherche n°7654
CNRS-École Polytechnique
Palaiseau Cedex, France
email: blanquet@botrytis.polytechnique.fr
Chapter 4

Richard Buckingham
UPR 9073 du CNRS
Institut de Biologie Physico-Chimique
Paris, France
email: Richard.Buckingham@ibpc.fr
Chapter 21

Robert Chênevert
Département de Chimie
Université Laval
Québec, Canada
email: robert.chenevert@chm.ulaval.ca
Chapter 25

Brian F.C. Clark
Institute of Molecular and Structural
    Biology
University of Aarhus
Aarhus C, Denmark
email: clark@biobase.dk
Chapter 20

Maarten H. de Smit
Department of Biochemistry
Leiden Institute of Chemistry
Leiden University
Leiden, The Netherlands
email: m.smit@chem.leidenuniv.nl
Chapter 19

Stephen Douthwaite
Department of Biochemistry
    and Molecular Biology
Odense University
Odense, Denmark
email: srd@molbiol.ou.dk
Chapter 26

Marc Dreyfus
Laboratoire de Génétique Moléculaire
École Normale Supérieure
Paris, France
email: mdreyfus@biologie.ens.fr
Chapter 10

Måns Ehrenberg
Department of Cell and Molecular
    Biology
Uppsala University
Uppsala, Sweden
ehrenberg@xray.bmc.uu.se
Chapter 21

Andrew D. Ellington
Department of Chemistry
    and Biochemistry
Institute for Cellular and Molecular
    Biology
College of Natural Sciences
University of Texas
Austin, Texas, U.S.A.
email: andy.ellington@mail.utexas.edu
Chapter 5

Dominique Fourmy
Institut de Chimie des Substances
    Naturelles, CNRS
Gif-sur-Yvette, France
email: dominique.fourmy@icsn.cnrs-gif.fr
Chapter 26

Catherine Florentz
Département 'Mécanismes
    et Macromolécules de la Synthèse
    Protéique et Cristallogenèse'
Institut de Biologie Moléculaire
    et Cellulaire du CNRS
Strasbourg Cedex, France
email: florentz@ibmc.u-strasbg.fr
Chapter 8

Magali Frugier
Départment 'Mécanismes
    et Macromolécules de la Synthèse
    Protéique et Cristallogenèse'
Institut de Biologie Moléculaire
    et Cellulaire du CNRS
Strasbourg Cedex, France
Chapter 1

Raymond F. Gesteland
Department of Human Genetics
University of Utah
Salt Lake City, Utah, U.S.A.
Chapter 22

Richard Giegé
Départment 'Mécanismes
    et Macromolécules de la Synthèse
    Protéique et Cristallogenèse'
Institut de Biologie Moléculaire
    et Cellulaire du CNRS
Strasbourg Cedex, France
email: r.giege@ibmc.u-strasbg.fr
Chapter 1

Richard L. Gourse
Department of Bacteriology
University of Wisconsin
Madison, Wisconsin, U.S.A.
email: rgourse@bact.wisc.edu
Chapter 23

Olga L. Gurvich
Department of Human Genetics
University of Utah
Salt Lake City, Utah, U.S.A.
email: ogurvich@howard.genetics.utah.edu
Chapter 22

Tamara L. Hendrickson
Department of Chemistry
Johns Hopkins University
Baltimore, Maryland, U.S.A.
email: Tamara.Hendrickson@jhu.edu
Chapter 3

Scott P. Hennelly
Division of Biological Sciences
University of Montana
Missoula, Montana, U.S.A.
email: nomistic@yahoo.com
Chapter 17

Brad Herberich
Department of Chemistry
The Scripps Research Institute
La Jolla, California, U.S.A.
Chapter 6

Michael Ibba
Department of Microbiology
The Ohio State University
Columbus, Ohio, U.S.A.
email: ibba.1@osu.edu
Chapter 2

Ivaylo P. Ivanov
Department of Human Genetics
University of Utah
Salt Lake City, Utah, U.S.A.
email: Ivaylo.ivanov@m.cc.utah.edu
Chapter 22

Susan Joyce
Department of Biology and Biochemistry
University of Bath
Bath, U.K.
Chapter 10

Rob Knight
Department of Molecular, Cellular
    and Developmental Biology
University of Colorado
Boulder, Colorado, U.S.A.
Chapter 7

Soumaya Laalami
Institut Jacques Monod
CNRS
Paris, France
email: Harald.Putzer@ibpc.fr
Chapter 24

Laura Landweber
Department of Ecology
    and Evolutionary Biology
Princeton University
Princeton, New Jersey, U.S.A.
Chapter 7

J. Stephen Lodmell
Division of Biological Sciences
University of Montana
Missoula, Montana, U.S.A.
email: lodmell@selway.umt.edu
Chapter 17

George A. Mackie
Department of Biochemistry
    and Molecular Biology
University of British Columbia
Vancouver BC, Canada
email: gamackie@interchange.ubc.ca
Chapter 9

Thomas J. Magliery
Department of Chemistry
University of California
Berkeley, California, U.S.A.
Chapter 6

Lynne E. Maquat
Department of Biochemistry
    and Biophysics
School of Medicine and Dentistry
University of Rochester
Rochester, New York, U.S.A.
email: lynne_maquat@urmc.rochester.edu
Chapter 13

Paulo E. Marujo
Institut de Biologie Physico-Chimique
CNRS-UPR 9073
Paris, France
Chapter 11

Yves Mechulam
Laboratoire de Biochimie
Unité Mixte de Recherche n°7654
CNRS-École Polytechnique
Palaiseau Cedex, France
Chapter 4

Eric Meggers
Department of Chemistry
The Scripps Research Institute
La Jolla, California, U.S.A.
Chapter 6

Xin Miao
Department of Biochemistry
    and Molecular Biology
University of British Columbia
Vancouver BC, Canada
Chapter 9

Philip Mitchell
Institute of Cell and Molecular Biology
University of Edinburgh
Edinburgh, U.K.
email: pmitch@holyrood.ed.ac.uk
Chapter 14

Poul Nissen
Institute of Molecular and Structural
    Biology
University of Aarhus
Aarhus C, Denmark
Chapter 20

Jens Nyborg
Institute of Molecular and Structural
   Biology
University of Aarhus
Aarhus C, Denmark
email: jnb@imsb.au.dk
Chapter 20

Michael O'Connor
School of Biological Sciences
University of Missouri
Kansas City, Missouri, U.S.A.
email: oconnormi@umkc.edu
Chapter 16

Miro Pastrnak
Department of Molecular and Cellular
   Biology
University of California
Berkeley, California, U.S.A.
Chapter 6

Francis Poulin
Department of Biochemistry
McGill University
Montréal, Québec, Canada
email: francis.poulin@mail.mcgill.ca
Chapter 18

Thomas Preiss
Gene Regulation Unit
Victor Chang Cardiac Research Institute
Sidney, Australia
email: t.preiss@victorchang.unsw.edu.au
Chapter 12

Annie Prud'homme-Généreux
Department of Biochemistry
   and Molecular Biology
University of British Columbia
Vancouver BC, Canada
Chapter 9

Harald Putzer
Institut de Biologie Physico-Chimique
CNRS-UPR 9073
Paris, France
email: Harald.Putzer@ibpc.fr
Chapter 24

Philippe Regnier
Institut de Biologie Physico-Chimique
CNRS-UPR 9073
Paris, France
email: Philippe.Regnier@ibpc.fr
Chapter 11

Stephen W. Santoro
Department of Chemistry
The Scripps Research Institute
La Jolla, California, U.S.A.
Chapter 6

Paul Schimmel
The Scripps Research Institute
The Skaggs Institute for Chemical
   Biology
La Jolla, California, U.S.A.
email: schimmel@scripps.edu
Chapter 3

Emmanuelle Schmitt
Laboratoire de Biochimie
Unité Mixte de Recherche n°7654
CNRS-École Polytechnique
Palaiseau Cedex, France
Chapter 4

Peter G. Schultz
Genomics Institute of the Novartis
   Research Foundation
San Diego, California, U.S.A.
email: schultz@scripps.edu.
Chapter 6

Robert W. Simons
Department of Microbiology,
    Immunology, and Molecular Genetics
University of California
Los Angeles, California, U.S.A.
email: bobs@microbio.ucla.edu
Chapter 9

Marie Sissler
Département 'Mécanismes
    et Macromolécules de la Synthèse
    Protéique et Cristallogenèse'
Institut de Biologie Moléculaire
    et Cellulaire du CNRS
Strasbourg Cedex, France
Chapter 8

Dieter Söll
Department of Molecular Biophysics
    and Biochemistry
Yale University
New Haven, Connecticut, U.S.A.
email: Soll@trna.chem.yale.edu
Chapter 2

Nahum Sonenberg
Department of Biochemistry
McGill University
Montréal, Québec, Canada
email: nsonen@med.mcgill.ca
Chapter 18

Jan van Duin
Department of Biochemistry
Leiden Institute of Chemistry
Leiden University
Leiden, The Netherlands
email: j.duin@chem.leidenuniv.nl
Chapter 19

Lionel Vial
Laboratoire de Biochimie
Unité Mixte de Recherche n°7654
CNRS-École Polytechnique
Palaiseau Cedex, France
Chapter 4

Lei Wang
Department of Chemistry
University of California
Berkeley, California, U.S.A.
Chapter 6

Brian T. Wimberly
Rib-X Pharmaceuticals
New Haven, Connecticut, U.S.A.
email: wimberly@rib-x.com
Chapter 15

Michael Yarus
Department of Molecular, Cellular
    and Developmental Biology
University of Colorado
Boulder, Colorado, U.S.A.
email: yarus@stripe.colorado.edu
Chapter 7

Satoko Yoshizawa
Institut de Chimie des Substances
    Naturelles, CNRS
Gif-sur-Yvette, France
email: satoko.yoshizawa@icsn.cnrs-gif.fr
Chapter 26

Ribosomes require correctly aminoacylated tRNAs to perform protein synthesis directed by intact mRNAs. This book presents a review of the recent explosive progress that revealed the high-resolution 3D-structure of several components of the translational machinery and provided more precise information on their structure/function relationships. It also reviews recent discoveries on tRNA charging with the correct amino acid or with an unnatural one, on the factor-mediated transfer of aminoacyl-tRNAs to the ribosome and on the control of messenger RNA stability. This wealth of information clearly and concisely presented in this book will be an invaluable tool for investigators and graduate students working on protein biosynthesis or on the regulation of gene expression in general. It will be useful for designing and testing models of yet misunderstood steps of translation involving interactions between well-characterized components of the translational machinery. Here are some highlights of the 26 chapters:

Section I deals with various facets of tRNA aminoacylation:

Chapter 1 presents new high-resolution tRNA structures, tRNA plasticity, and the paradigm of a structural framework displaying identity elements for the aminoacyl-tRNA synthetases (aaRS).

Chapter 2 reviews the major impact of complementary structural and phylogenetic studies of aaRSs on our understanding of the evolution of living cells.

Chapters 3 and 4 present molecular mechanisms of translational accuracy control and efficiency with tRNA-dependent amino acid discrimination by 9 of the 20 aaRSs and discuss the maturation of some misacylated tRNAs and the recycling of prematurely released peptidyl-tRNAs.

Chapters 5 and 6 review two complementary strategies for incorporating unnatural amino acids into proteins and their use in vitro and in vivo for protein engineering and for expanding the genetic code.

Chapter 7 goes back to the RNA world, and shows that amino acid-binding RNAs appear to recapture some assignments of the modern genetic code, thus revealing traces of the primordial stereochemical relationships at its origin.

Chapter 8 presents disease-related mutations in human mitochondrial tRNA genes, with perspectives for therapeutic approaches, and analyzes distinct structural features of these tRNAs.

Section II examines the complexity of the control of mRNA decay and analyzes the action of various enzymes involved in this control:

Chapter 9 presents the enzymes involved in the different steps of mRNA decay in bacteria, with the description of a novel molecular machine, the degradosome.

Chapter 10 analyzes the relationship between translation and mRNA degradation in prokaryotes.

Chapter 11 focuses on the implication of polyadenylation in the degradation of mRNA in *E. coli*.

Chapter 12 analyzes the importance of the poly(A) tail in the control of mRNA degradation in eukaryotes and the involvement of this tail in translation initiation through mRNA circularization in eukaryotes.

Chapter 13 analyzes the relationship between the mode of surveillance of mRNA with a premature stop codon and pre-mRNA splicing and presents a model for the pioneer translation contributing to this surveillance.

Chapter 14 details the pathways of mRNA decays in yeast and dissects the function of the exosome complex in one of these pathways.

Section III analyzes the high-resolution structure of the ribosome, supporting that it is a ribozyme, reviews the importance of mutagenesis in probing the role of this ribozyme and stresses the importance of the dynamics of the ribosome during the different steps of translation. It also examines recent progress concerning these steps, with an original view of translation initiation in bacteria and a review of the non-conventional use of the genetic code:

Chapters 15, 16 and 17 present high-resolution structures of bacterial ribosomes, free or complexed to various natural or unnatural ligands (tRNA or analogs, factors, antibiotics), together with a wealth of genetic and biochemical data on normal and mutated ribosomes, that reveal several conformations of the ribosome during the successive steps of protein biosynthesis, and provide information for designing novel antibiotics. This knowledge will be invaluable for studying the more complex eukaryotic ribosomes, with direct implications for human health and biotechnology.

Chapter 18 discusses the role of the numerous eukaryotic initiation factors and the subsequent mechanisms that enable eukaryotes to control the initiation step.

Chapter 19 examines how ribosomes succeed in initiating at sites that are only transiently exposed in bacteria.

Chapters 20 and 21 examine how elongation and release factors interact intimately with the functional centers of the ribosome.

Chapter 22 reviews the various ways that ribosomes use recoding to read the information encoded in the mRNAs.

Section IV covers the regulation of the genes encoding components of the translational apparatus:

Chapter 23 outlines proposed regulatory mechanisms that coordinate the rate of rRNA synthesis with the translational needs of the cell.

Chapter 24 describes the elaborate mechanisms (some supported by high-resolution structures) regulating the expression of the genes encoding several aaRSs and translation factors.

Section V (Chapters 25 and 26) describes the use of various inhibitors of the ribosome or of aminoacyl-tRNA synthetases as antibiotics and as tools for structural and mechanistic studies.

We thank all the authors for their remarkable reviews of various aspects of translation mechanisms and for having captured the pace and excitement of recent developments. In particular, we wish to thank Robert W. Simons for his suggestions in the planning of section II on mRNA stability. Finally, we thank Cynthia Conomos, Celeste Carlton, William Bown, Jesse Kelly-Landes and Ron Landes for their efficient support at various stages of our editorial work.

Jacques Lapointe, Ph.D.
Université Laval
Québec, Québec, Canada

Léa Brakier-Gingras, Ph.D.
Université de Montréal
Montréal, Québec, Canada

CHAPTER 1

Transfer RNA Structure and Identity

Richard Giegé and Magali Frugier

Abstract

The structure of tRNA and its relationship with the biological necessity of specific tRNA aminoacylation reactions, in other words with identity, is reviewed. New structural data show the typical L-shaped tRNA architecture in great detail and highlight how adequate rigidity and plasticity of the molecule is essential for interaction with its biological partners, in particular with aminoacyl-tRNA synthetases. Identity is ensured by a small number of nucleosides predominantly located at the two distal extremities of the tRNA molecule. In several crystallographic complexes, these residues have been shown in contact with amino acids from the synthetases. In most cases, the interaction is accompanied by a conformational change of the tRNA. Assuming that the structural framework of tRNA displays identity elements to synthetases implies that altered and/or simplified RNA architectures can fulfill this role provided they contain correctly located identity elements. This paradigm holds true in nature where atypical and tRNA-like domains have been selected by evolution. Rationale-based engineering or selection by artificial evolution of novel RNA molecules recognized and aminoacylated by synthetases also verified it.

Introduction

Transfer RNAs (RNAs) are ubiquitous molecules present in all forms of life. Unprecedented in biology, their necessity as the adapters translating the genetic code was predicted,[1] before their biochemical characterization.[2] Beside protein synthesis, specialized tRNAs and tRNA-like structures can participate in diverse metabolic pathways (reviewed in ref. 3). The universal occurrence of tRNA, its amino acid donor key role in protein synthesis, and its involvement in other cellular processes underline its ancient origin. Contemporary tRNAs participating in protein synthesis are structurally homogeneous but are versatile in function since during their functional cycle they interact with many enzymes (maturation and modification enzymes including RNase P, aminoacyl-tRNA synthetases), protein factors (for initiation, elongation and termination) as well as with mRNA and the ribosome. Specialized tRNA functions such as initiation of protein synthesis,[4,5] incorporation of selenocysteine into proteins,[6] are in general correlated with atypical structural features. In what follows, emphasis is given to the tRNAs directly involved in protein synthesis and to the problem of tRNA identity that underlies their capacity to be specifically aminoacylated by synthetases and consequently to be responsible for the correct translation of the genetic message from the DNA into the protein language.

Structure of tRNAs

Sequences and Modified Nucleosides

Since 1965, when yeast tRNA^{Ala} was sequenced,[7] sequences of more than 4000 tRNAs from more than 300 organisms have been included in the tRNA database (http://www.uni-bayreuth.de/departments/biochemie/trna/),[8] and these numbers increase steadily due to the many genome projects. A vast majority of tRNA sequences adopt the cloverleaf folding which highlights the structural and functional domains of the molecule (Fig. 1A).

Translation Mechanisms, edited by Jacques Lapointe and Léa Brakier-Gingras.
©2003 Eurekah.com and Kluwer Academic / Plenum Publishers.

Most information came from gene sequencing (3700 DNA versus 550 RNA sequences). The more difficult RNA sequencing revealed the presence of many modified nucleotides. To date, 81 such residues are in the tRNA modification database (http://medlib.med.utah.edu/RNAmods/).[9] Some are common to almost all tRNA species, such as dihydrouridine (D) in D-loops and ribothymidine (T) in T-loops (for abbreviations of modified nucleosides, see ref. 8). Others are characteristic of individual tRNA species, of groups of organisms or of a given living kingdom (archaea, prokarya including the derived organelles or eukarya). For instance, wybutosine, a hypermodified G-residue formerly called the Y-base, and its derivatives, are exclusively found at position 37 of tRNA[Phe] and queuosine derivatives, other G-analogs, are found at the first anticodon position of certain tRNA[Asn], tRNA[Asp], and tRNA[Tyr] species. The triad Gm18, s^2T54 and m^1A58 in the D- and T-loop characterizes tRNAs from thermophilic organisms, as found in tRNA[Asp] from *Thermus thermophilus*,[10] and likely contributes to their stability by increasing their melting temperature.[11] The methylated residue m^1A at position 9 is typically mitochondrial. As demonstrated in the case of mitochondrial human tRNA[Lys], it prevents misfolding of the molecule into an elongated hairpin by preventing formation of an alternative pairing with U64 from the T-stem,[12] and therefore can be considered as an internal chaperone for correct folding. Likewise, archeosine, the G-like 7-deazaguanosine analog with ring substituted at position 7 by a formamidino group is exclusively archaeal and is found at position 15 in D-loops where it forms an atypical R15-Y48 Levitt tertiary pair.[13]

Sequence data revealed soon the presence of conserved (U8, A14, A21, U33, G53, T54, Ψ55, C56, A58, C61, C74, C75 and A76) and semiconserved (Y11, R15, R24, Y32, R37, Y48, R57, and Y60) residues in tRNA. The steadily incoming sequences confirm their occurrence in almost all tRNAs, the main exceptions being organellar tRNAs (see e.g., website for mammalian mitochondrial tRNAs,[14] [http://mamit-tRNA.u-strasbg.fr]) that often lack canonical D- and T-loop organizations. As demonstrated by crystallography, many of the conserved and semiconserved residues participate in the tertiary interactions that govern the three-dimensional structure of tRNA. In contrast, those in the anticodon loop play a role in mRNA decoding[15] and sometimes in tRNA identity,[16] the conserved 3'-terminal CCA triplet being crucial for tRNA recognition in the active site of synthetases. Helical regions of tRNAs are rich in G-U wobble base-pairs and to a lesser extent in other non-Watson-Crick pairs. The G-U pairs have unique structural characteristics that can be crucial to tRNA function.[17,18]

Organellar tRNAs, like human mitochondrial tRNA[Ser] or mitochondrial tRNAs from nematode worms, with part(s) of the canonical structure lacking constitute remarkable exceptions to the conserved cloverleaf folding of cytoplasmic tRNAs.[19,20] However, sequence compensations allow these molecules to adopt L-shaped conformations.[21] Noticeable, the sequence features in canonical tRNAs deviating from consensus, like the −1 residue at the 5'-terminus of histidine specific tRNAs,[22] or the uncommon G15-G48 Levitt pair (instead of R15-Y48) in *Escherichia coli* tRNA[Cys] were shown to be identity elements in these tRNAs.[23,24]

Three-Dimensional Structures of Free tRNAs

As first proposed for yeast tRNA[Phe], the tRNA molecule has an L-shaped architecture that was established independently in three laboratories.[25-28] It is based on the cloverleaf folding and shows a symmetrical organization with two helical branches of similar size oriented perpendicularly (Fig. 1B). One branch is formed by the amino acid acceptor stem stacked over the T-arm and the dumbbell-like other branch by a stack of the anticodon- and D-arms. The same overall structural organization is found in yeast tRNA[Asp],[29,30] in *Escherichia coli* initiator tRNA[fMet],[31] as well as in yeast initiator tRNA[Met],[32] and in the recently solved structure of human tRNA[Lys3], the primer of HIV reverse-transcriptase.[33] An NMR investigation of the anticodon stem-loop of this human tRNA[Lys3] confirms the canonical U-turn structure of the anticodon loop,[34] as found in the crystal structure of the whole tRNA. This conclusion arising from solution data excludes that packing effects in crystal lattices trigger U-turn conformations. Interestingly, the NMR study shows further that the modified nucleosides (mnm^5s^2U34, t^6A37, and Ψ39) stabilize this structure.

The above cited X-ray structures are of medium resolution (2.5Å at best) (Table 1) and except for tRNA[Lys3], were refined with nonoptimized methods. For long, these structures constituted the only structural RNA database. They revealed the conserved tRNA conformation and gave an architectural

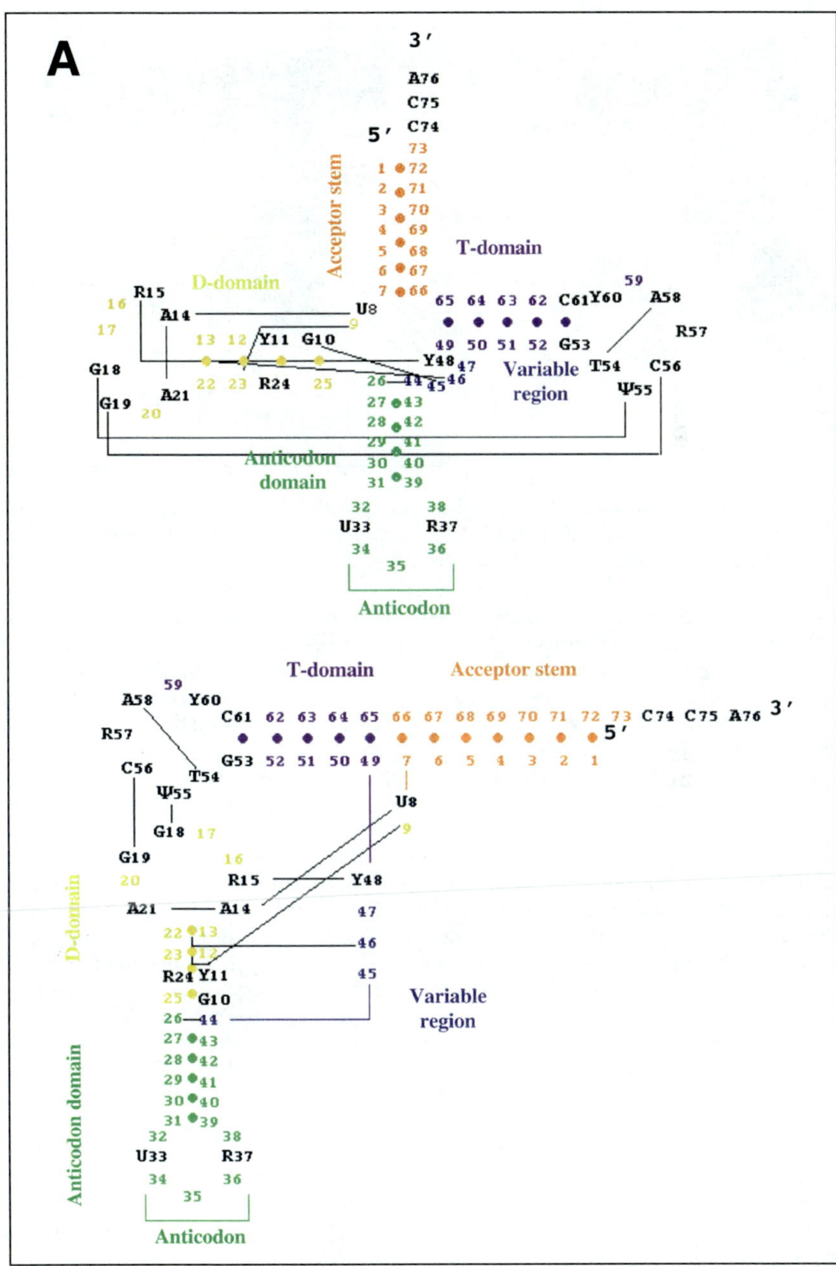

Figure 1A. Structure of cytoplasmic tRNA participating in protein synthesis. (top) Cloverleaf folding emphasizing the different domains of the tRNA molecule as well as the position of conserved and semiconserved residues. Length of tRNA sequences ranges from 72 to 95 nts; R = purine and Y = pyrimidine. Structural variability mainly originates from length-variations in the D-loop and variable region and from the location of the two conserved G18 and G19 residues in the D-loop. (bottom) L-shaped folding showing how the tRNA domains organize to form the three-dimensional structure stabilized by triple and other tertiary base-pairs. Continued on next page.

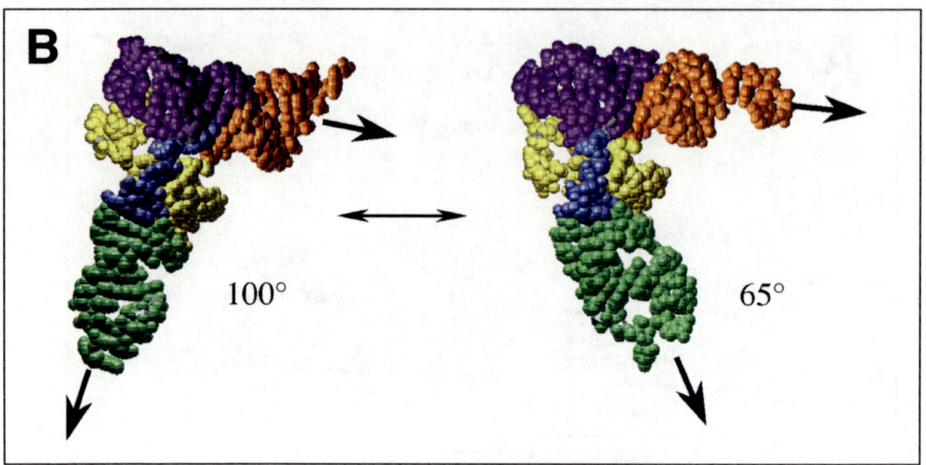

Figure 1B (continued from previous page). Example of two crystallographic conformers of tRNA emphasizing its structural plasticity: at the left, free tRNA[Asp] mimicking a tRNA interacting with mRNA and at the right, the same tRNA complexed with cognate AspRS.

significance to the conserved and semiconserved residues of tRNA that interconnect the D- and T-loops. They showed also novel structural features in RNA, like noncanonical pairings and metal binding sites, and suggested existence of faint conformational differences between tRNAs due to differences in semiconserved residues and in localization of the G18G19 dinucleotide in the D-loop and length variations in this loop and the variable region.[35] Existence of such differences was clearly shown by structural mapping of tRNA by chemical probes.[36]

The structure of yeast tRNA[Asp] is of particular type (Fig. 1B). It deviates from that of tRNA[Phe] by a significant opening of the angle between the two branches of the molecule and by changes in the D-/T-loop interaction with a disruption of the G19-C56 tertiary base-pair. These features result from crystal packing effects, but have biological meaning. Indeed, tRNA[Asp] molecules interact via GUC/GUC anticodon/anticodon pairing in the crystal and this pairing mimics a tRNA/mRNA interaction. Thus, the crystal structure of tRNA[Asp] would mimic the structure of a tRNA interacting with mRNA.[37] This interpretation was confirmed by solution data with tRNAs forming dimers, with in particular the demonstration of the opening of the G19-C56 pair by chemical probing.[37,38]

Recently, the structure of yeast tRNA[Phe] was revisited. New diffraction data were collected from monoclinic and orthorhombic crystals at synchrotron sources and structural models were refined with advanced techniques. As a result, high resolution electron density maps at ~2Å resolution were obtained,[39,40] even from 15-year-old crystals that yielded at best 2.5Å data in the past.[40] The new versions of the tRNA[Phe] structure overall confirm the previous structural information but reveal novel details. The triple and other tertiary base-pairs are now seen with high accuracy together with their hydration patterns and associated Mg^{2+} ions well defined (Fig. 2). Other striking improvements concern identification of new divalent cation binding sites and of many ordered water molecules.[39] Interestingly, in the case of the monoclinic crystals, four different cleavage sites, the major one in the D-loop, have been localized next to bound Mg^{2+} ions. The presence of such in situ Mg^{2+}-induced cleavages was suggested in the orthorhombic tRNA[Phe] crystals,[41] and clearly demonstrated in tRNA[Asp] crystals.[37] It is likely that this intrinsic chemical fragility explains many crystallization failures with free tRNA.

Polyamines are obligatory additives for the crystallization of free tRNAs.[42] Their role becomes clearly apparent in the high-resolution structure of tRNA[Phe] which shows a spermine molecule in the major groove of the T-arm that connects a symmetry-related tRNA molecule in the crystal lattice.[40]

Antibiotics from the aminoglycoside family are tRNA ligands, besides interacting with various other targets such as the ribosome (see Chapter 26 by D. Fourmy et al). The recently solved X-ray

Table 1. Summary of tRNA structures in their free state or in complexes with proteins

tRNAs	Ligand(s)		Resolution (Å)	Comments (e.g. differences with tRNA[Phe] structure)	Refs.
Free tRNAs					
Sc tRNA[Phe(GAA)]	-		1.93	is the reference conformation	25-28, 39-41
Ec tRNA[fMet]	-		3.5	C1-A72 mismatched pair; acceptor strand curled back; constant U33 unstacked	31
Sc tRNA[Meti]	-		3.0	tight interaction between D- and T-loop, different orientation of A20	32
Hs tRNA[Lys(SUU)]	-		3.3	overall structure as in tRNA[Phe]	33
Sc tRNA[Asp(GUC)]	tRNA[Asp]		3.0	more open overall conformation (110°)[a]; contact between D- and T-loop altered	29,30
Sc tRNA[Phe(GAA)]	neomycin B		2.6	unchanged structure, except displacement of metal ions by neomycin	43
tRNAs Complexed with Class I Synthetases					
Sc tRNA[Arg(ICG)]	Sc ArgRS	Ia	2.2	new AC-loop conformation ; CCA-strand folded back	60
Tt tRNA[Val(CAC)] (u.m.)	Tt ValRS**	Ia	2.9	conformation as for free tRNA[Phe] but G1-C72 base-pair disrupted	61
Ec tRNA[Ile(CAU)] (u.m.)	Sa IleRS**	Ia	2.2	no kinked CCA-strand (editing complex)	62
Tt tRNA[Glu(CUC)] *	Tt GluRS	Ib	2.4	AC-loop as in tRNA[Phe] ; no major distortion in tRNA conformation	63
Ec tRNA[Gln(CUG)]	Ec GlnRS	Ib	2.8	major alteration in AC-loop and CCA-strand folded back	51,56
Ec tRNA[Gln(CUG)] (u.m.)	Ec GlnRS	Ib	2.5	absence of certain bound water molecules	54
Ec tRNA[Gln(CUG)] (mutants)	Ec GlnRS**	Ib	2.6-3.0	altered geometry of the15-48 pair and its surrounding	58
tRNAs Complexed with Class II Synthetases					
Tt tRNA[Ser(GGA)]	Tt SerRS**	IIa	2.7	tRNA only partly visible; special architecture of the tRNA core	64,65
Ec tRNA[Thr(CGU)] *	Ec ThrRS	IIa	2.9	large deformation of the anticodon-loop with all bases splayed out; G35 and U36 are in H-bond interaction	66
Tt tRNA[Pro(CGG)]	Tt ProRS	IIa	2.85	major AC-loop distorsion; G47 not flipped out in solvent, but stacked against edge of A21 and G46	67
Sc tRNA[Asp(GUC)]	Sc AspRS	IIb	3.0	large conformational change of AC-arm starting at G30-C40 base pair	52
Ec tRNA[Asp(QUC)] *	Ec AspRS**	IIb	2.4	AC-loop as in other subclass IIb complexes; conformational change in acceptor stem due to G4-U68 wobble pair	68
Tt tRNA[Asp(XUC)] *	Tt AspRS	IIb	3.5	overall closed conformation (95°)[a]	69
Ec tRNA[Asp(QUC)] *	Tt AspRS	IIb	3.0	as in homologous Tt complex	69
Ec tRNA[Lys(U*UU)] *	Tt LysRS**	IIb	2.75	AC-loop as in other subclass IIb complexes	70
Tt tRNA[Lys(CUU)] (u.m.)	Tt LysRS	IIb	2.9	AC-loop as in other subclass IIb complexes	70
Tt tRNA[Phe(GAA)]	Tt PheRS	IIc	3.28	conformational changes in T- and D-loops; no major change in AC-loop	71

continued on next page

Table 1. Continued

tRNAs	Ligand(s)	Resolution (Å)	Comments (e.g. differences with tRNA[Phe] structure)	Refs.
tRNA Complexed with Other Proteins				
Sc phe-tRNA[Phe(GAA)]	Ta EF-Tu	2.7	conformation as for free tRNA[Phe], with slight changes in acceptor stem	72
Ec cys-tRNA[Cys(GCA)]	Ta EF-Tu	2.6	novel tertiary pairs: C16-C59, G15-G48, s^{4}U8-A14-A46	59
Ec fmet-tRNA[fMet] *	Ec transformylase**	2.8	C1-A72 pair is open and CCA-arm bend	73
tRNA on the Ribosome				
Ec tRNA[fMet] & [Lys], & ALS[Phe] fragment	Tt 70S ribosome	7.8	L-shape of tRNA can be seen in A-, P-, and E-sites	74
Ec tRNA[fMet], mRNA & ALS[Phe] fragment	Tt 70S ribosome	5.5	L-shape of tRNA in ribosome well defined	75

In human tRNA[Lys], S is the hypermodified A residue ms^{2}t^{6}A ; in yeast tRNA[Arg], I is inosine ; in *E. coli* tRNA[Asp], Q is queuosine; in *E. coli* tRNA[Lys], U* is mnm^{5}s^{2}U; u.m. means undermodified. ALS[Phe] is a anticodon-loop-stem fragment derived from tRNA[Phe]. [a]Angle between the two limbs of the tRNA. *Overproduced tRNA in *E. coli* cells (may be undermodified), **with adenylate analogs. Ec for *Escherichia coli*, Hs for *Homo sapiens*, Sa for *Staphylococcus aureus*, Sc for *Saccharomyces cerevisiae*, Ta for *Thermus aquaticus*, Tt for *Thermus thermophilus*.

structure at 2.6Å resolution of a complex between yeast tRNA[Phe] and neomycin B shows that the elongated antibiotic molecule is anchored in the tRNA core between residues G20, A44, and G45.[43] This binding site overlaps with known divalent metal ion binding sites and corresponds also to the location of major determinants for *E. coli* PheRS recognition. This suggests that binding of the antibiotic occurs by metal displacement and explains why neomycin and other structurally related aminoglycosides inhibit Pb^{2+} cleavage of tRNA[Phe] by hindering binding of the metal cation. It explains also that neomycin B inhibits aminoacylation of *E. coli* tRNA[Phe].[43]

The modular structure of tRNA in two domains connected by a network of tertiary interactions explains that the molecule can be dissected in individual parts with intrinsic structure and functional properties.[44] For instance, NMR[45] and crystallographic[46] investigations on a duplex RNA recapitulating the acceptor stem of *E. coli* tRNA[Ala] informed about the structural environment of the G3-U70 base-pair that determines alanine identity. Other biochemical and biophysical studies confirmed the existence of anticodon- and T-loop structures and highlighted the importance of modified residues for their local conformation (see e.g., refs. 34, 47-50).

The Structure of tRNA in Macromolecular Complexes

As examples, Figure 3A shows how *E. coli* tRNA[Gln] and yeast tRNA[Asp] bind to class I and class II synthetases.[51,52] In both tRNAs, binding to the synthetase triggers significant conformational changes in the anticodon loop. Bases become unstacked and point towards the recognition amino acids from the synthetases, and in the aspartate system the overall structure of complexed tRNA[Asp] becomes more closed than in the free molecule (Fig. 1B). In tRNA[Gln], the terminal base-pair is disrupted to facilitate bending of the 3'-terminal CCA into the active site of GlnRS. Such bending is not needed in tRNA[Asp] where the CCA-strand is in helical continuity with the acceptor stem to reach the catalytic site of AspRS. These structural differences in the acceptor stem of the two tRNAs rely to different binding modes in the catalytic site of class I and class II synthetases: in the minor groove in tRNA[Gln] (class I) and the major groove in tRNA[Asp] (class II). This differential binding has been verified by chemical probing of an unmodified tRNA[Asp] transcript chargeable by both class I ArgRS and class II AspRS (Fig. 3B).[53]

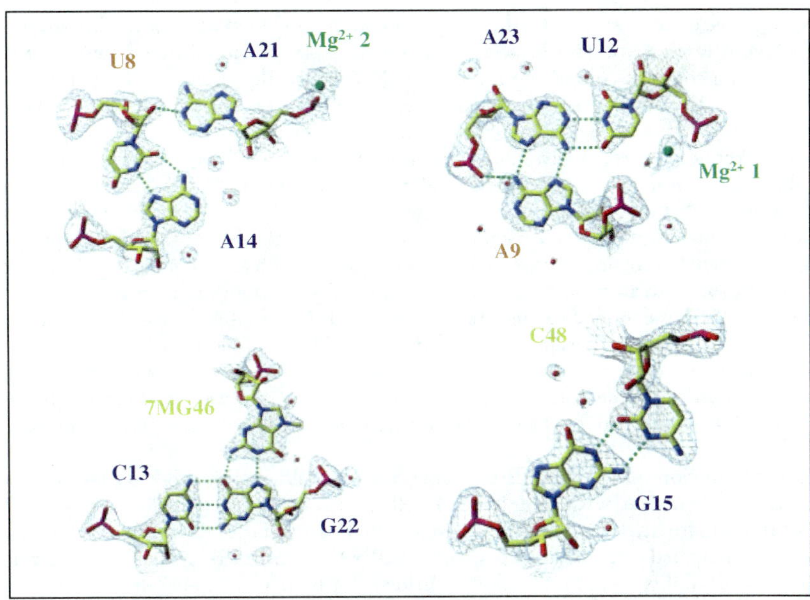

Figure 2. Details in the high-resolution structure of yeast tRNA^{Phe} as seen in the monoclinic crystal form.[40] Typical examples of tertiary interactions are shown: the U8-A14-A21, A9-[A23-U12], and [C13-G22]-m^{7}G46 triples and the G15-C48 Levitt pair. Notice the well-defined densities for bound water molecules and Mg^{2+} ions.

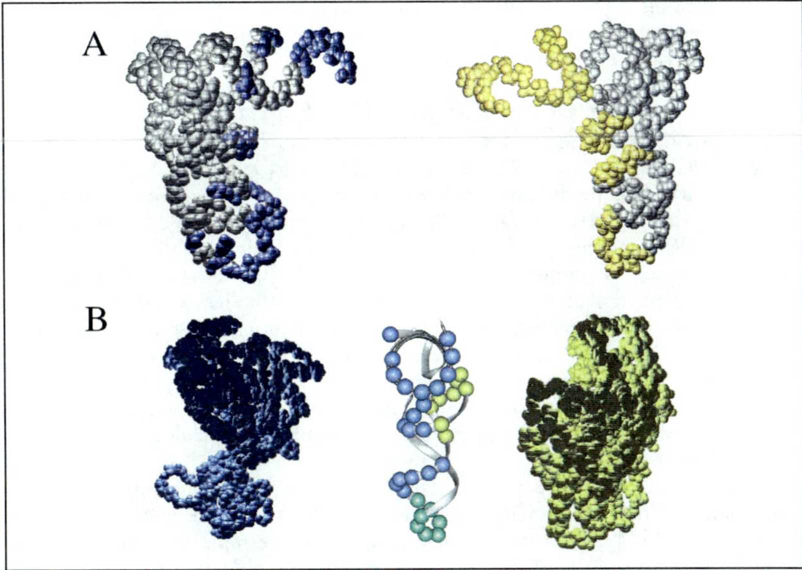

Figure 3. Comparison of tRNA binding to class I and class II synthetases. (A) Structure of tRNA^{Asp} (left) and tRNA^{Gln} (right) with residues in contact with AspRS and GlnRS shown in blue and yellow, respectively. (B) Mirror image recognition of a tRNA^{Asp} unmodified transcript aminoacylatable by either yeast AspRS or ArgRS with contact residues with the two synthetases. The orientation of the tRNA is turned 90° compared with panel (A) (adapted from ref. 53). The green color in the anticodon region indicates that this domain can be in contact with either AspRS (in blue) or ArgRS (in yellow).

In nature, tRNAs are modified and the obvious question is to know whether modifications affect binding to synthetases. A complex between unmodified tRNA[Gln] and GlnRS shows no structural differences with a complex comprising modified tRNA, except the absence of bound water molecules crosslinking the N5 atom of Ψ-residues to their 5'-phosphate. This finding suggests a possible role of pseudouridylation in tRNA stabilization through water-mediated binding of Ψ-residues to the tRNA backbone.[54] Interestingly, one of the three Ψ's in tRNA[Gln] is located in the anticodon loop at identity position 38,[55] and contacts GlnRS.[56] These data on tRNA[Gln] are in line with the general assumption that modified tRNAs have more rigid structures than unmodified transcripts.[48,57]

From a different viewpoint, two crystal structures of active tRNA[Gln] mutants in complexes with GlnRS and an adenylate analog were solved.[58] These mutants with G15-G48 atypical Levitt pairs, and for one of them with an additional change in sequence and length of the variable region, show structural perturbations confined to the core of the molecule near to the Levitt pair and the variable region. They consist among others in a *syn* conformation of the guanine ring of G48 with respect to the ribose moiety. This conformation differs from what observed in the native structure, where C48 is present as an *anti* conformer. Surprisingly, this *anti* conformation was also found in the crystal structure of a tRNA with a native G15-G48 Levitt pair, namely that of *E. coli* tRNA[Cys] complexed to EF-Tu.[59]

Further information on tRNA structure came from crystallographic work on four class I,[60-63] and nine class II,[64-71] tRNA/aaRS complexes, as well as from studies on complexes between tRNAs and other protein partners,[59,72,73] and recently even from X-ray studies of the complete ribosome,[74,75] (Table 1). As anticipated, the L-shaped structure of tRNA is confirmed, as well as the existence of conformational flexibility in the molecule (see below). Of particular interest are the structures of tRNA[Ser] and tRNA[Cys], the two tRNAs with uncommon sequence characteristics. For tRNA[Ser], the structure of the complex with SerRS identified novel tertiary interactions in the core of the molecule. Notably, it showed the role of conserved residue G20b in the D-loop that determines the orientation of the long variable arm at ~45° to the plane of the L-shaped tRNA molecule.[64] The case of *E. coli* tRNA[Cys] is noteworthy, since, to solve its structure, it was crystallized on purpose as a complex with elongation factor. Overall, the structure is canonical, but with a large interstem angle of ~100° and a geometry of its noncanonical G15-G48 Levitt pair that exposes the N2-N3 side of G15 and the O6-N7 side of G48 to the solvent.[59]

RNA Plasticity versus RNA Rigidity for tRNA Function

Indirect evidence has soon suggested the necessity of structural plasticity of tRNA during its life cycle. Reversible inactive forms of tRNA were described (see e.g., ref. 76) and, in the case of *E. coli* tRNA[Glu], it was shown recently that inactivation results from Mg^{2+}-dependent alternate folding of the tRNA.[77] Other changes were correlated with synthetase interaction.[78] Crystallography brought direct evidence for tRNA plasticity and comparison of data in Table 1 highlights major conformational differences between the canonical structure of yeast tRNA[Phe] and that of other tRNAs in their free or complexed forms. In many systems, but not in all, anticodon loops undergo drastic conformational changes upon interaction with synthetases. Other functional constraints determine overall structural effects on tRNA, as the opening of its interstem angle (Fig. 1B) that likely occurs on the ribosome. This global tRNA flexibility in solution can be quantitatively approached by transient electric birefringence experiments. The method examines the bending motions of the L-shaped tRNA architecture and, with tRNA[Phe], shows that the core of the molecule gains flexibility in the absence of Mg^{2+} ions.[79]

More subtle and local changes were seen by NMR spectroscopy. Studying microhelices recapitulating the acceptor stem of two *E. coli* tRNAs, it was shown that the nature of the discriminator base N73 influences in tRNA[Ala] the structure of this stem,[80] and determines in initiator tRNA[Met] the conformation of the 3'-terminal -N73CCA sequence which is stacked over the stem in the A73 wild-type molecule and folded back in the U73 variant.[81] A structural deformability of the acceptor stem has been invoked in the mechanism of alanine identity expression,[82] and a flexibility of the single-stranded acceptor end was shown to be a necessity for the editing function of class I IleRS,[62] and ValRS.[61]

At opposite, conformational rigidity in tRNA domains can have a functional role. This is the case of base modifications in the anticodon loop of certain tRNA species. For instance, in a minor tRNA[Leu] from *E. coli*, the conformational rigidity of Cm and cmnm^{5}Um in the first position of the anticodon guarantees correct codon reading.[83] From another viewpoint, conformational changes observed in the glutamine and aspartate class I and class II tRNA/synthetase complexes and thought to be class specific appear to be more system (or subclass) specific. So, in contrast with the opening of the first base-pair 1-72 in tRNA[Gln] interacting with class I GlnRS, this pair does not disrupt in tRNA[Val] interacting with class I ValRS.[84] Likewise, the anticodon loop in tRNA[Phe] interacting with class II PheRS keeps its native conformation,[71] and does not deform as in other tRNAs interacting with class II synthetases.

Aminoacylation and Identity of tRNAs

The Aminoacylation Reaction and the Concept of Identity

Aminoacylation of tRNA occurs in two step reactions: first activation of the amino acids by ATP to form enzyme-bound aminoacyl-adenylates and second transfer of the activated amino acids to the 3'-terminal adenosine of tRNA. Amino acid attachment to tRNA occurs by ester bond formation with hydroxyl groups of the terminal ribose, either 2'-OH for the reactions catalyzed by class I synthetases or 3'-OH in the case of class II enzymes (reviewed in ref. 85). The activation step is tRNA independent, except for ArgRS, GlnRS, and GluRS. Overall, aminoacylation reactions have to yield tRNAs correctly charged, otherwise wrong amino acids will be falsely incorporated into proteins.[86] This implies correct recognition of both amino acids and tRNAs by the synthetases. However, synthetases can misactivate amino acids, are able to recognize noncognate tRNAs and can catalyze mischarging reactions (reviewed in ref. 35). A first answer to the dilemma was brought when it was realized that fidelity in tRNA aminoacylation, and consequently in protein synthesis, mainly relies to highest kinetic efficiency of the synthetases for their cognate substrates and is mostly governed by the k$_{cat}$ of the tRNA charging reactions.[87] This answer was refined when proofreading (or editing) mechanisms were discovered (reviewed in ref. 85). Altogether, the phenomenological view of tRNA aminoacylation fidelity implies a strict correspondence between the charged amino acid and the codon read by the carrier tRNA according to the rules of the genetic code. This correspondence is mediated by identity determinants within tRNAs and is defined by the tRNA identity rules, which constitute a 'second' genetic code.

Major elements defining identity of all *E. coli* tRNAs have been deciphered and much is known about identity determinants of most yeast tRNAs and of a few tRNAs from other organisms.[16,88-91] In short, identity of a tRNA is determined by a small number of nucleosides, and more precisely by chemical groups carried by these nucleosides that often have been seen interacting with amino acids on the synthetases. In each tRNA, these nucleosides constitute the so-called 'identity set' that can be completed by structural elements of the nucleic acid. Negative elements that prevent a tRNA to be mischarged by noncognate synthetases can participate in identity. Some of the positive identity determinants can be considered as 'strong' since their mutation strongly reduces the aminoacylation capacity of the mutant tRNA, others are 'moderate' or 'weak'. As examples, Table 2 displays the strongest conserved identity determinants that have been characterized to date in prokaryotic tRNAs. At first glance, several features emerge: (i) Identity elements are mainly located at the two distal extremities of the tRNA. (ii) Except for glutamate and threonine identities, the discriminator base is a determinant, at least in *E. coli* tRNAs. (iii) Specific structural elements in tRNA often serve as identity determinants (i.e., the -1 residue in tRNA[His], the long extra-arm in tRNA[Ser], the G3-U70 wobble pair in tRNA[Ala]). (iv) For tRNAs specific for amino acids coded by more than four codons (leucine, serine, arginine), anticodon residues either do not participate in identity (leucine, serine) or, only for the middle C35 and semiconserved U/G36 positions (arginine).

In most systems, modified nucleosides do not participate in identity and thus are not recognized by the cognate synthetases. But they can play a major role in negative discrimination by preventing tRNAs to be recognized by noncognate synthetases, as was shown in the isoleucine and aspartate systems.[92,93] At present, the universal nature of the identity rules is rather well established and only faint differences distinguish identity sets for a given amino acid specificity along evolution. However, much has to be learned about the structural and evolutionary relationships between identity sets, the precise molecular mechanisms by which they are expressed in different organisms, and the exact nature of identities of atypical tRNAs, in particular those of mitochondrial origin.

Table 2. *Major identity determinants found in* E. coli *tRNAs and correlations with class-ranking of synthetases*

aaRSs		Amino Acid Acceptor Arm		Architectural Core	Anticodon Arm			
		Discriminator 73	aa Stem nt or bp(n°)*		ac Region nt or bp(n°)	34	35	36
Class I								
Ia	ValRS	A	2 bp(3,4)	---	---	---	**A**	C
	IleRS	A	1 bp(4)	yes	[37,38]**	**I/G**	A	**U**
	LeuRS	A	---	yes	---	**---**	**---**	**---**
	MetRS	A	2 bp(2,3)	---	**[32,33,37]**	**C**	A	**U**
	CysRS	U	2 bp(2,3)	yes	---	G	C	A
	ArgRS	**A / G**	---	yes	---	**---**	**C**	**U/G**
Ib	TyrRS	A	---	---	---	---	U	---
	TrpRS	**G**	3 bp(1-3)	---	---	**C**	**C**	A
Ic	GluRS	---	2 bp(2,3)	yes	[37]	---	s²U	G
	GlnRS	G	3 bp(1-3)	yes	[37,38]	C/U	U	G
Class II								
IIa	SerRS	**G**	3 bp(1-3)	**yes***	---	**---**	**---**	**---**
	ThrRS	---	2 bp(**1**,2)		---	G	G	U
	ProRS	A	(72)	yes	---	---	**G**	**G**
	GlyRS	U	3 bp(1,**2**,3)	---	---	---	C	C
	HisRS	C	**-1**	---	---	**G**	**U**	**G**
IIb	AspRS	**G**	1 bp(2)	yes	**[38]**	**G**	**U**	**C**
	AsnRS	G	---	---	---	U	U	G
	LysRS	A	---	---	---	U	U	U
IIc	AlaRS	A	3 bp(2,**3**,4)	yes	---	---	---	---
	PheRS	**A**	---	**yes**	2 bp(27,28)	**G**	**A**	**A**

x bp(n°)*, base-pair(s) with position of 5'-nucleotides; [n°]**, nucleotide(s) in anticodon loop; yes***, concerns the variable helical region; (---), position not involved in identity (if in bold, also the case in other organisms). Identity positions in bold, conserved in evolution (at least in two kingdoms); underlined, not conserved in evolution.

Establishment of tRNA Identities

Characterizing identity determinants constituted a challenge and was tackled by different methods. Determinants were searched both in vitro and in vivo by studying the aminoacylation capacity of tRNA variants or their effects on protein synthesis. Two breakthroughs were responsible for the wealth of results that accumulated in the last decade. First, use of mutant suppressor tRNAs in a genetic system based on expression of the reporter protein dihydrofolate reductase, with stop codon at position 10, and identification of the amino acid incorporated at the suppressed codon. This system was first employed in Abelson's laboratory to characterize leucine identity determinants in *E. coli* tRNA[Leu],[94] and soon later in McClain's and Schimmel's laboratories to identify the G3-U70 base-pair as the major alanine determinant in *E. coli* tRNA[Ala].[95,96] Second, the possibility to easily prepare tRNA variants by in vitro transcription of artificial genes by T7 RNA polymerase. This possibility was pioneered in Uhlenbeck's laboratory for deciphering the identity set of yeast tRNA[97] and was largely employed for understanding the yeast aspartate identity,[98,99] and many other identities (reviewed in ref. 16). In its original and simplest version, the in vitro approach considered as potential determinants those nucleotides where mutation lead to a drastic loss of aminoacylation capacity. The assumption underlying this statement implies that mutating an identity determinant does not affect the three-dimensional structure of the tRNA, which is not always true. Completeness of an identity set is verified if transplantation of the putative determinants in another tRNA confers the new identity to this molecule. Aminoacylation efficiency of the transplanted tRNA, however, is often not optimal, indicating that sequence context and/or architectural features play a role in identity expression. The fact that engineering the structural framework of the receiving tRNA can improve activity supports this conclusion.[16]

Each of the two approaches has advantages and drawbacks. The in vivo method does not check anticodon residues and yields only rough estimates of the effects of mutations on tRNA aminoacylation efficiency. However, and this is most useful, conclusions arising from in vivo experiments take into account the cellular environment and the competition between tRNAs and synthetases of diverse specificities. On the other hand, the in vitro method does not probe modified nucleosides, but yields quantitative kinetic parameters and approaches mechanistic aspects of the aminoacylation reactions. Finally, one has to keep in mind that identity sets were seldom determined by both in vitro and in vivo approaches, that for a given identity mutational analyses were done on only one isoaccepting tRNA species, and that identity swap experiments concerned only one or a few foreign tRNAs and in most cases were conducted in a qualitative manner. Novel results along these lines are awaited and will refine the view on identity.

Selected Examples of Identities

Four representative identities (alanine, phenylalanine, aspartate, and glutamine) are discussed below. The aim is to present hard facts, to raise points of debate, and to highlight more subtle aspects underlying identity expression in general. Among others, a refined view of tRNA identity has to distinguish between direct and indirect effects triggered by identity determinants and to take into account the functional role of architectural features in tRNA.

Alanine Identity

It is at present well accepted that the G3-U70 wobble base-pair is the major alanine identity determinant.[95,96] This pair is conserved in evolution,[96,100] and its transplantation in other tRNA frameworks, and even in minihelices recapitulating amino acid accepting arms,[44,101] confers to these molecules the capacity to be aminoacylated by AlaRSs, both in vitro and in vivo. The debate concerns the mechanisms by which this unique feature triggers specific charging. Does the recognition mechanism involve direct interaction of the synthetase with chemical groups on the G3-U70 identity pair or does it imply recognition of a shape created by this pair? In the absence of a crystallographic structure of a tRNA[Ala]/AlaRS complex, only indirect evidences can be invoked. In support to the first possibility are the so-called 'atomic mutagenesis' experiments, which were used to decipher the functional role of chemical groups in the G3-U70 pair. To this end, minimalist substrates were chemically synthesized including variants with G3 or U70 replaced by base analogs. Experiments led to the conclusion that the exocyclic 2-NH_2 group of G3 is essential for alanylation.[102] Other studies, with tRNAs having their ribose substituted by deoxynucleotides, indicated dramatic activity decreases for the mutants substituted at identity position 70 and at neighboring position 71,[103] suggesting disruption of contacts with AlaRS or perturbation of the solvation pattern around the G3-U70 pair. At the opposite, genetic investigations coupled with NMR analyses have shown that replacing the G3-U70 wobble pair with a C-C mispair preserves tRNA[Ala] aminoacylation in vivo. Likely, the C-C pair, as does a G-U pair, provides deformability in the acceptor stem which does not occur in a structurally more rigid stem with a G-C pair.[82] Altogether, these last data are better in line with an indirect recognition mechanism. What is then the real mechanism? Presumably, direct and indirect recognition mechanisms are not exclusive. In other words, alternate mechanisms can lead to the same level of specificity. We believe that structural adaptability may even confer functional advantages in maintaining a high degree of specificity when mutations or external circumstances (pH, salt, temperature changes...) perturb precise tuning of the tRNA/synthetase complex.

Related with these considerations are other data highlighting the importance of structural plasticity in tRNA alanylation systems and the existence of alternate mechanisms in identity expression. As to the first point, *E. coli* suppressor tRNA$^{Ala(CUA)}$ can accommodate substantial structural flexibility in its core, since 15 out of 16 possible nucleotide pairs at positions 15 and 48 are functional in vivo and therefore are likely alanylated by AlaRS.[104] Even the inactive variant in suppression, with the A15/A48 combination, is an efficient substrate of AlaRS in vitro, but in that case the mutation triggers a distal structural alteration in the anticodon loop that hampers its correct functioning in protein synthesis.[104] By tuning the plasticity of tRNA^{Ala} at the level of the 15-48 nucleotide pair evolution established species discrimination in alanine identity. Indeed, although a majority of tRNA^{Ala} sequences have R15-Y48 pairs, one finds all the 12 other combinations in cytoplasmic and mitochondrial tRNA^{Ala} species. Concerning the second point, the presence of a shifted G2-U71 alanine identity pair in a mitochondrial insect tRNA^{Ala} is noteworthy[105] and implies an alternate recognition mechanism by AlaRS. Other shifted determinants for expression of yeast arginine identity were discovered in anticodon loops within different tRNA frameworks.[106,107]

Phenylalanine Identity

The identity of yeast tRNA^{Phe} was the first deciphered by the in vitro method. It is given by five elements, namely four determinants located at the extremities of the L-shaped tRNA (A73, the discriminator base and G34 A35 A36, the three anticodon residues) and a fifth determinant located in the central core of the molecule (G20, in the D-loop).[97] Given the similarity of the phenylalanine identity sets in yeast and *T. thermophilus* and the crystal structure of the tRNA^{Phe}/PheRS complex from *T. thermophilus*,[71,108] it is likely that all five determinants in yeast tRNA^{Phe} contact amino acids in yeast PheRS. Mutation of individual determinants, that would remove a contact, have rather moderate effects on specificity (k_{cat}/K_M reduced at most by a factor of ~200 for an A35U mutation in the anticodon), but specificity is increased by the additive effects of the five determinants that act independently.[109] The surprise arose when transplanted molecules with poor phenylalanine acceptance but bearing the five phenylalanine determinants were discovered.[110] Indeed, specificity was found reduced by a factor of ~2000 for variants with insertion of a G2-C71 pair in acceptor stems within different tRNA frameworks, notably that of *E. coli* tRNA^{Ala} and even of cognate yeast tRNA^{Phe}. Surprisingly, the deleterious effect of the G2-C72 pair could be compensated by the insertion of a G3-U70 wobble pair, which by itself has no effect on phenylalanylation. From a mechanistic point of view, the G2-C72/G3-U70 combination is not a classical recognition element since its antideterminant effect can be compensated. Combinations that do not hinder aminoacylation are called "permissive". Presence of permissive elements implies that no nucleotide within a tRNA is of random nature but has been selected by evolution so that tRNAs can fulfill their functions efficiently. In agreement with this view are recent studies on cysteine and glutamine identities, showing that structural effects due to alteration of the Levitt pair and its surrounding in the core of the tRNA can be interpreted in terms of permissive and non-permissive elements.[58]

Aspartate Identity

Like for phenylalanine, aspartate identity of yeast tRNA^{Asp} is also triggered by determinants located at the distal extremities of the molecule and in its central core, namely G73, G34U35C36, C38, and G10-U25,[98,111] with strongest determinants being the discriminator G73 base and the GUC anticodon. Expression of aspartate identity, however, differs from what was observed in the phenylalanine case, since strong anti-cooperative effects were found between discriminator and anticodon determinants.[112] As a consequence, minihelices or microhelices with the single G73 aspartate determinant can be efficiently aspartylated.[113] Aspartate determinants make direct contacts with AspRS, mainly via chemical groups of the bases.[114] If some of these contacts are disrupted, structural adaptation of the tRNA on AspRS is not optimal and charging efficiency declines as concluded from footprinting and functional analyses.[115] Most intriguing was the finding of a close functional relationship between the sequences of yeast tRNA^{Asp} and yeast arginine accepting tRNAs, since the unmodified aspartate transcript can be efficiently charged by yeast ArgRS.[93] However, biological specificity is prevented in native tRNA^{Asp} by the m^1 methylation of residue G37 next to the anticodon that act as an antideterminant.[116] Presence of antideterminants in tRNA (modified nucleosides or other structural elements) is not restricted to this particular system and is becoming better documented (see e.g., refs. 16, 92, 105, 117). In fact, negative discrimination appears to be a general mechanism in nature for improving specificity of enzymatic reactions involving tRNAs (see e.g., refs. 118, 119).

Glutamine Identity

Identity of *E. coli* tRNA^{Gln} has been studied in depth both from the structural and functional viewpoints. In contrast to other identities, glutamine identity is specified by a large set of nucleotides.[120-122] It includes five residues in the anticodon region (the YUG anticodon itself and the two 3'-adjacent A37 and U38 residues) and five residues near the accepting end (the discriminator residue G73 and the second G2-C71 and third G3-C70 base-pair in the stem). As in the aspartate system,[98] mutations at identity positions essentially affect k_{cat}, but specificity constants (k_{cat}/K_M) in the glutamine system are decreased by greater factors (up to 10^5) than in the aspartate system (at most 530-fold). The tRNA determinants find their counterpart on the synthetase and it was shown that both nucleotides (e.g., U35) and amino acid (e.g., Arg341) determinants are coupled in establishing specificity during tRNA^{Gln}/GlnRS recognition.[123] The tRNA acts as a cofactor in the recognition process and enables correct positioning of its acceptor arm in the catalytic site of the synthetase, thus optimizing amino acid activation.[124] In this step, the 3'-terminal A of tRNA^{Gln} mediates the tRNA-dependent amino acid recognition by GlnRS.[125] This gives structural support to the well known fact that amino acid activation by GlnRS requires tRNA. The next question was to verify whether identity nucleotides determine also the cognate amino acid affinity of the synthetase. To this end, glutaminylation kinetics of tRNA^{Gln} mutants were conducted under saturating amino acid conditions.[126] Results indicate that the identity determinants can be subdivided in specificity determinants (exclusively in the acceptor stem of tRNA^{Gln}) and in elements responsible for binding (mainly the G10-C25 pair and the anticodon). All recognition events leading to tRNA charging imply conformational changes during the mutual adaptation of tRNA^{Gln} and GlnRS. Spectroscopic investigations suggest that only the cognate tRNA triggers the conformational changes that confer specific aminoacylation.[127] Finally, correct charging of tRNA^{Gln}, or its mischarging, is dependent on the balance (in vivo or in vitro) of tRNA^{Gln} with GlnRS and noncognate synthetases.[128]

Peculiar Implications of Synthetases and Other Proteins in Identity Expression

Aminoacyl-tRNA synthetases are modular proteins comprising catalytic and anticodon recognizing domains, and often additional domains not indispensable for the aminoacylation function. This is in particular the case of eukaryotic synthetases that present appended domains to the N- or C-terminal end of their core structure.[129] As examples we discuss the role in tRNA aminoacylation of two N-terminal appendices in yeast class I GlnRS[130], and class II AspRS.[131]

The demonstration of a functional role of the appendix in yeast GlnRS came from genetic experiments aimed to rescue a GlnRS-deficient yeast strain by GlnRS from *E. coli*.[130] In its first version, when the native *E. coli* enzyme was used, the experiment failed. Rescue occurred when the domain from yeast GlnRS was fused to *E. coli* GlnRS. In vitro experiments confirmed that the chimerical *E. coli* enzyme with the yeast appendix binds and glutaminylates better yeast tRNA^{Gln} than does native *E. coli* GlnRS.

The dimeric yeast AspRS is extremely sensitive to proteolytic cleavage leading to truncated species deprived of the first 20 to 70 residues but that retain enzymatic activity and dimeric structure. This N-terminal appendix is not seen in the crystallographic model of the tRNA^{Asp}/AspRS complex and thus was considered as not essential for aminoacylation. However, it participates in tRNA binding, a finding that could be generalized to eukaryotic class IIb synthetases.[131] Biochemical and mutagenesis experiments indeed showed that the extension connected to the anticodon binding module of AspRS contacts tRNA^{Asp} on the minor groove side of its anticodon stem and, as a consequence, lead to a stronger binding to AspRS. Sequence comparison of eukaryotic class IIb synthetases identified a lysine-rich sequence with the consensus xSKxxLKKxxK that is important for this binding.

Taken together, these two examples illustrate a more general phenomenon, namely the involvement of protein domains (as found also in the alanine system[132]) or even of specialized proteins like Arc1p,[133] and Trbp111,[134] that bind to tRNA and increase the global aminoacylation efficiency by chaperone-like effects. Along these lines, elongation factor EF-Tu that binds certain misacylated tRNAs more strongly (or weakly) than cognate aminoacyl-tRNAs may contribute to better identity expression and consequently to translational accuracy.[135]

Finally, mechanistic aspects of the tRNA aminoacylation reactions are certainly of crucial importance in expression of identities and need to be explored more thoroughly. For instance, the competitions between synthetases for interacting with a given tRNA, as studied for glutamine identity expression,[128] are likely of more general relevance. This is also true for effects mediated by pyrophosphate, a reaction product of amino acid activation, on the activity of variants with mutation at identity positions, as seen in tRNA^{Phe},[136] and in tRNA^{Asp} (Khvorova, Wolfson and Giegé, unpublished results).

Structure and Identity of Atypical tRNAs and of tRNA-Like Domains

Figure 4 gives examples of the folding of natural RNAs with structural and/or functional characteristics different from those of canonical tRNAs. Structural deviations in these molecules likely reflect their specialized functions (protein synthesis in mitochondria, donor of selenocysteine in protein synthesis, tagging of abnormal proteins on the ribosome for proteolysis in the case of tmRNA, replication of viral RNA genomes for plant virus tRNA-like structures). All these RNAs are recognized by aminoacyl-tRNA synthetases and thus should contain identity determinants mimicking the determinants present in canonical tRNAs.

Several bovine mitochondrial tRNAs have been studied by chemical probing, modeling and NMR methods. For tRNA$^{Ser(UCN)}$, the structure is close to that of classical tRNA, but the connector between acceptor stem and D-stem has only one nucleotide and the anticodon stem contains six base-pairs.[137] In tRNA$^{Ser(AGY)}$, the D-domain is missing and a novel pattern of tertiary interactions accounts for the three-dimensional structure of the molecule.[138] Interestingly, and in contrast with cytoplasmic tRNA^{Ser} species which have a large variable region, in the mitochondrial species, this region is always of small size. Further, because of the peculiar D- and T-loop sequences, interaction between these two loops is missing or strongly altered in most mitochondrial tRNAs. From the functional viewpoint, little is known about the identity determinants in these tRNAs, although it can be predicted from phylogenetic considerations and sequence comparisons that mitochondrial tRNAs contain identity elements found in *E. coli* tRNAs,[14] but their importance awaits to be verified experimentally. This was already done for aspartate identity in a mitochondrial tRNA^{Asp} from a marsupial, which like in prokaryotic tRNA^{Asp}, relies on anticodon, notably on the central U35 residue.[139] Of different outcome, however, were studies on serine identity of mitochondrial bovine tRNA$^{Ser(AGY)}$. Here, the importance of the T-loop and in particular of A58 is highlighted.[140-142] This differs from what found for serine identity of *E. coli* tRNA^{Ser} that is given by determinants in the acceptor stem and variable region.[143] For more details, see chapter 8 by C. Florentz and M. Sissler, on mitochondrial synthetases.[144]

Selenocysteine inserting tRNAs differ from canonical tRNAs by an extended amino acid branch made of 13 base-pairs and novel tertiary interactions. They are recognized by SerRS, likely as are canonical tRNA^{Ser} species. The specific properties of these atypical molecules rely on their potential to interact with specialized elongation factors and more generally to participate in the pathway of selenocysteine synthesis and incorporation into proteins (reviewed in ref. 145; see also chapter 4 by S. Blanquet et al, and chapter 22 by I.P. Ivanov et al). Another family of atypical tRNAs is constituted by the tmRNAs that have both mRNA and tRNA properties (see e.g., ref. 146 for recent structural aspects about the tRNA-like domain within *E. coli* tmRNA, and chapter 21 by R.H. Buckingham and M. Ehrenberg). These molecules are aminoacylated by AlaRS,[147] and, as anticipated have a G3-U70 identity pair.

Viral tRNA-like structures were discovered at the 3'-end of genomic RNAs of several genera of plant viral RNAs (for early literature see e.g., refs. 148-151). Three groups of mimics have been characterized to date on the basis of their aminoacylation identity (valine, histidine and tyrosine). Folding of these domains deviates markedly from the canonical tRNA cloverleaf. Closest sequence similarities with tRNA are found in the valine accepting structures in tymoviruses (e.g., TYMV). All the viral tRNA-like domains present a pseudoknotted acceptor stem. Recent advances in the field brought better understanding of the architectural features that actually mimic tRNA as well as of the rules that confer aminoacylation capacity to these molecules (reviewed in ref. 152). Because of the architectural properties of pseudoknotted stems,[153] all three families of tRNA-like domains (from TYMV, TMV, BMV) present in their acceptor arm a mimic of the -1 major histidine identity nucleotide and consequently were found histidylatable.[24] Interestingly, studies on tRNA showed

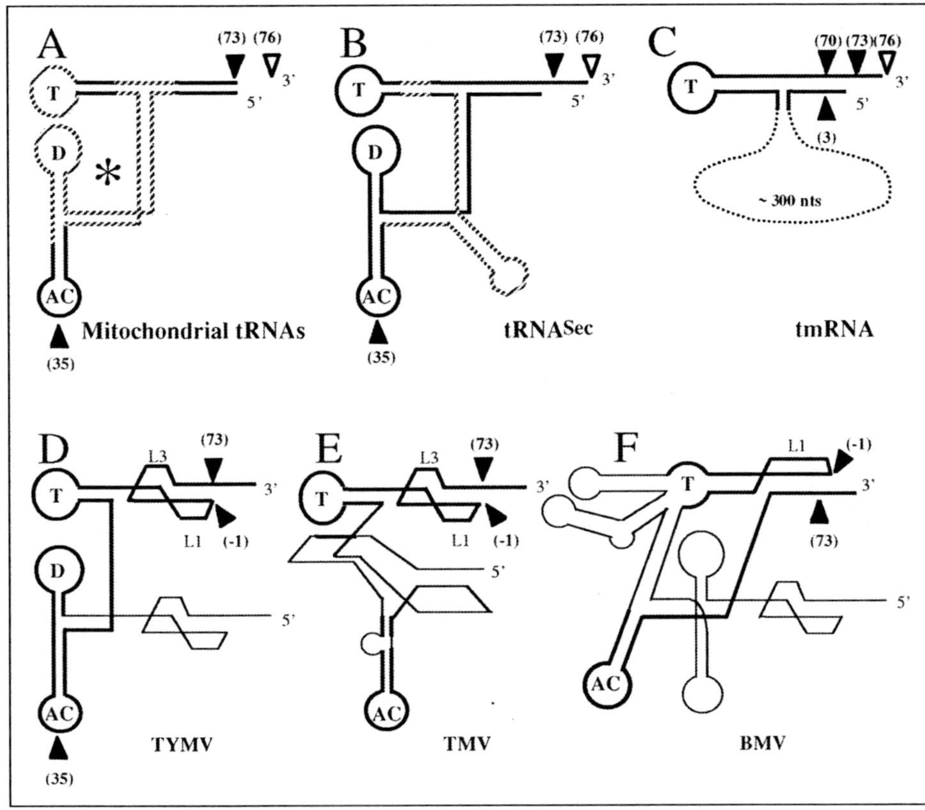

Figure 4. Schematized folding of atypical tRNAs emphasizing deviations from the canonical fold (dashed lines), domains mimicking tRNA features (in heavy lines), and location of identity determinants (full triangles) (adapted from ref. 161). (A) Mitochondrial tRNAs,[14,21] (*, indicates the importance of architectural features in identity). (B) Selenocysteine inserting tRNA.[145] (C) tmRNA.[147] (D-F) Viral tRNA-like domains,[152] (TYMV for turnip yellow mosaic virus, TMV for tobacco mosaic virus, and BMV for brome mosaic virus).

that the phosphate group of residue -1 is in fact the actual histidine identity determinant.[154] Strength of this determinant is strongest in the TMV tRNA-like structure and is much weaker in the context of TYMV and BMV RNAs, especially when compared to the dominant identity in these tRNA-like molecules (valine and tyrosine, respectively). It seems therefore unlikely that this property has in vivo implications in contemporary systems,[155] and we believe it is more likely that histidine identity in TYMV and BMV RNAs is a functional remnant of the evolutionary history of these molecules. As to valine identity of TYMV RNA, mutagenesis experiments pointed to the importance of discriminator and anticodon residues,[156,157] like in canonical tRNA^{Val}. A recent investigation based on in vitro selection of molecules derived from the TYMV tRNA-like structure randomized in the anticodon loop and in the pseudoknot confirmed the importance of the anticodon residues in valine identity.[158] It showed further that architectural rather than sequence features in the pseudoknot are important for efficient valylation. Finally, viruses belonging to the bromo-, cucumo- and hordeivirus genera possess tyrosine accepting tRNA-like structures at their 3'-extremity. All share a particularly intricate structure, as highlighted in the case of the BMV tRNA-like structure (Fig. 4F). In this structure, residues in the pseudoknotted acceptor arm,[159] functionally mimic the major tyrosine identity determinants found in yeast tRNA^{Tyr}, namely A73 and the first base-pair C1-G72 of the accepting stem.[160] Tyrosine identity in yeast, however, relies also on anticodon, but, strikingly, the

BMV tRNA-like structure possesses neither a canonical anticodon loop nor a tyrosine anticodon triplet GUA.[160] In fact, recent experiments showed that yeast TyrRS interacts with a hairpin, oriented perpendicularly to the acceptor branch. This hairpin anchors the tRNA-like structure on the synthetase and contributes to the efficiency of the tyrosylation reaction, which otherwise relies only on identity elements in the acceptor branch.[159]

Altogether, the conclusions arising from investigations on atypical tRNAs and tRNA-like domains favor the view of a strict conservation of the identity nucleotides for recognition of different types of RNA substrates by synthetases, whatever the structural scaffold in which these nucleotides are embedded.[16,161]

Engineering Structure and Identity of tRNA

The concept of a structural scaffold carrying identity elements at its extremities, as visualized in Figure 5, provides a rational basis for engineering structure and identity of tRNA. It accounts for understanding the output of identity transplantation experiments. If presentation is sub-optimal, aminoacylation efficiency is poor, but can be improved by engineering the scaffold of the receiving tRNA. The concept is illustrated by identity switch experiments between tRNAs aminoacylated by class I GlnRS and class II AspRS,[111] and by the design of a tRNA derived from yeast tRNA[Asp] having optimal phenylalanine identity,[162] after remodeling the frameworks of the tRNAs. A further example is a *E. coli* tRNA[Gln] variant that remained efficiently recognized by GlnRS upon transplantation of the large extra-arm of tRNA[Ser] (characteristic of class II tRNAs) in its architectural core.[163] Generalizing such engineering, it was possible to construct four distinct designs of the core of class II tRNAs that present stable folding and are efficiently aminoacylated by GlnRS.[164]

Along the same lines, a molecule with multiple identity having lost its original identity (aspartate) and acquired three new identities (alanine, phenylalanine, and valine) was designed.[165] Two of these identities have well-separated determinants (phenylalanine and alanine) and are therefore expressed as efficiently as in their original framework, while the third one (valine) has determinants overlapping with those of phenylalanine and consequently is expressed less efficiently. Similarly, a full-length and bimolecular (obtained by annealing of two fragments) versions of a tRNA with dual phenylalanine and alanine identity were obtained upon introduction of the G3-U70 alanine identity determinant in the framework of yeast tRNA[Phe].[103] Going one step further, the framework of atypical tRNAs was shown to be recognized by AspRS when it was realized that mutations in the D- and T-loops of yeast tRNA[Asp], compatible with aminoacylation activity,[166] created a core sharing structural resemblance with that of tRNA[Sec]. Direct identity swap experiments confirmed that the tRNA framework of tRNA[Sec] is indeed fully recognized by AspRS.[167]

In a more general perspective, the central core of the tRNA can be completely reorganized and again it can be anticipated that functional molecules can be obtained, provided that identity elements are properly located in the new scaffold. This engineering can be done either by combinatorial methods or by rationale based approaches (reviewed in ref. 168). Figure 6 shows examples of such engineered molecules that were derived from *E. coli* tRNA[Phe],[169] and yeast tRNA[Asp].[113,170] From a highly degenerate library of 63 nucleotides, several RNAs that tightly bind to *E. coli* PheRS were selected by a filter binding assay. Interestingly, all binders are missing canonical core features and contain an anticodon stem-loop of atypical sequence with a phenylalanine GAA anticodon.[169] But the presence of this triplet, part of the *E. coli* phenylalanine identity set (Table 2), is not sufficient to confer phenylalanylation capacity to these molecules. Presumably, their flexibility is restricted so that proper accommodation of the accepting 3'-end in the catalytic site of PheRS is prevented. This accommodation is possible for three types of tRNA[Asp]-mimics having deep alteration in their central cores. A first architecture, well aspartylatable, resembles metazoan mitochondrial tRNA[Ser] lacking the D-arm. The second one lacks both D- and T-arms and has its acceptor and anticodon helices joined by two connectors. This construct is a substrate of AspRS that behaves like a minihelix, since mutating the anticodon identity determinants does not affect aminoacylation. Removing the connector at the 5'-side provides more flexibility, and allows aminoacylation that is dependent on both G73 and anticodon identity elements.[170] Thus, neither a helical structure in the acceptor stem nor the presence of a D- or T-arm is mandatory for specific aspartylation. This conclusion is not restricted to the yeast aspartate system, since yeast PheRS is able to charge a tRNA[Phe] fragment encompassing a single-stranded acceptor domain.[171]

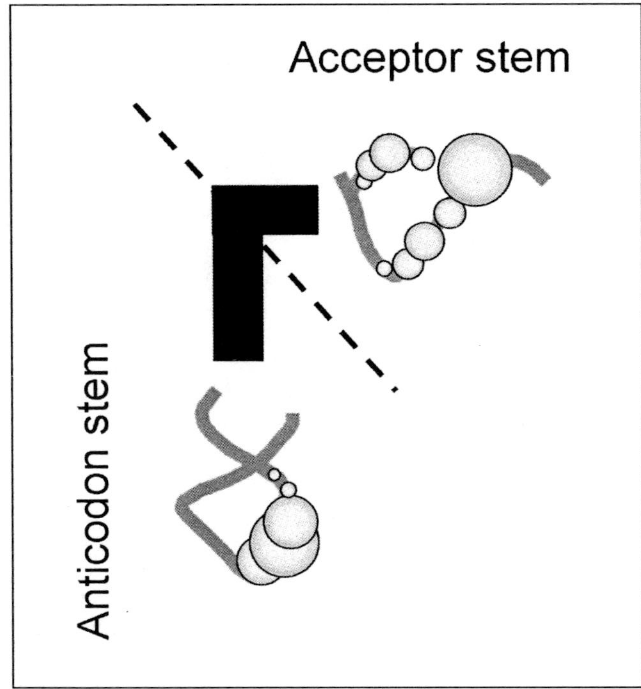

Figure 5. L-shaped RNA scaffold of tRNA and distribution of major identity residues at its distal extremities (adapted from ref. 161). Size of spheres is proportional to the occurrence of identity elements – note the predominant occurrence at discriminator N73 position and at anticodon positions. The dashed line symbolizes the symmetrical architecture of tRNA in two structurally independent domains.

A Few Remarks on Evolution

The structure of contemporary tRNA is the result of a long evolutionary history and it is most likely that the two domains of its modular L-shaped architecture have arisen independently with the acceptor branch appearing first. Likewise, modular structures, with well-defined catalytic cores, are found in synthetases. Here also the primitive versions of these enzymes would have been restricted to catalytic domains recognizing minimalist acceptor RNAs.[172,173] The implication is that the primordial signals defining aminoacylation of tRNA should be present in the minimal structural elements needed for function. As seen in Table 2, identity elements are indeed present in the acceptor arm of all tRNA species. On the other hand, the similar structure of all tRNAs and the structural similarities in the catalytic core of each of the two classes of synthetases, imply evolutionary relationships between the different tRNA aminoacylation systems. Again, this assumption finds support in the distribution and nature of tRNA identity elements (Table 2). Although a complete picture on determinants is presently only available for *E. coli* tRNAs and that information on tRNAs from other organisms except yeast,[16] is scarcer, it appears clearly that many determinants are conserved in evolution. This is true in many systems for the discriminator base and adjacent base-pairs in the acceptor stem. Exceptions to this trend are the nonconservations of the discriminator base in tRNA[Gly] and tRNA[His] and even its noninvolvement in the case of tRNA[Thr] identity. Interestingly, these idiosyncrasies seem to be compensated by conservation in evolution of determinants in the acceptor stem of these tRNAs recognized by class IIa synthetases.

In the context of evolution, the structural and functional relationships that were discovered in the glutamate/glutamine and aspartate/asparagine tRNA aminoacylation systems deserve special attention. In both pairs the synthetases are structurally related, belonging to subclass Ic for GluRS and GlnRS and subclass IIb for AspRS and AsnRS. Likewise the identity sets in the corresponding

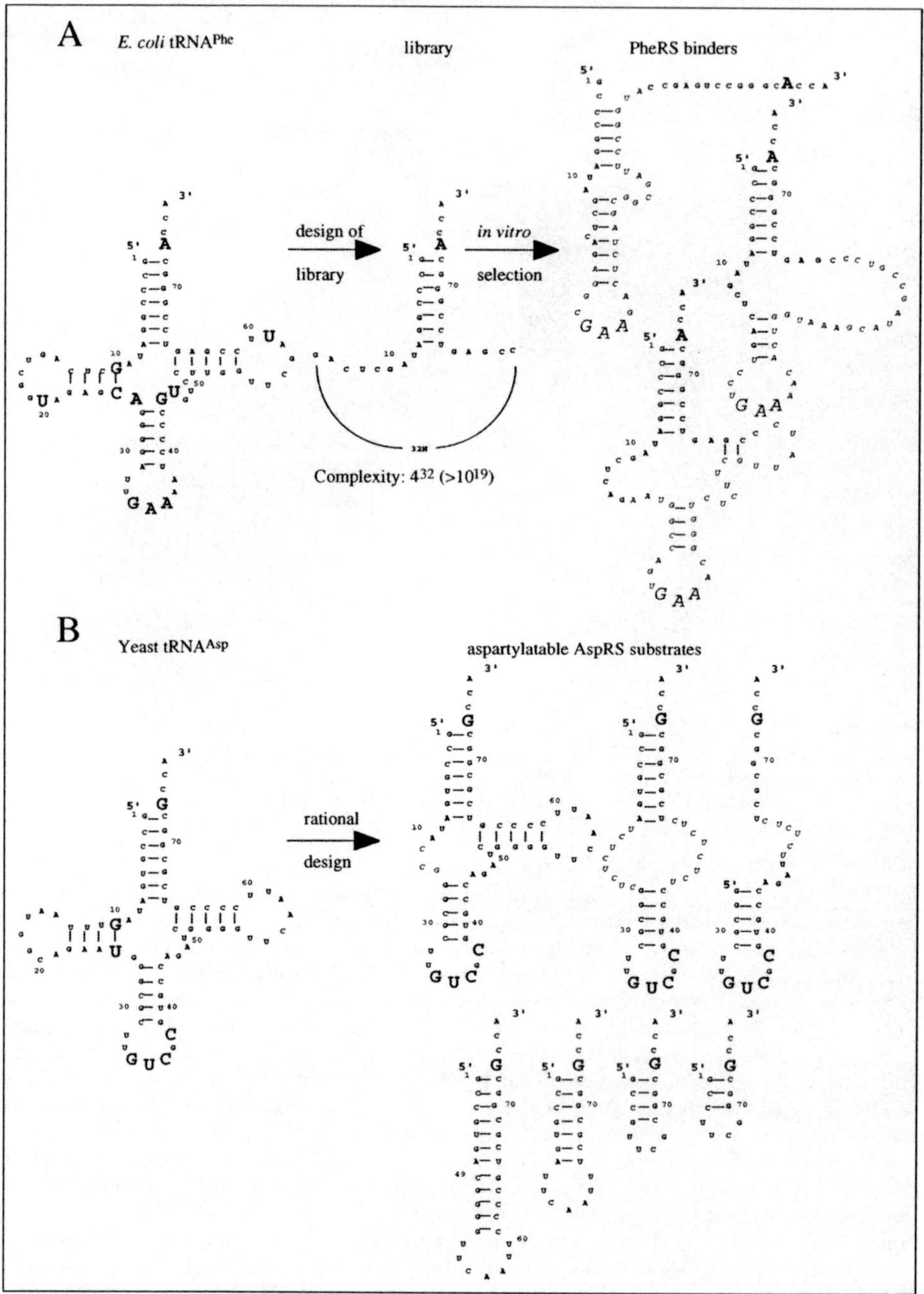

Figure 6. Engineered tRNA structures that interact with a synthetase or are aminoacylatable. (A) PheRS-binders selected from a library based on the acceptor stem of *E. coli* tRNA^Phe with the other parts of the tRNA molecule degenerated (adapted from ref. 169). (B) RNAs aspartylated by yeast AspRS designed by a rational approach (adapted from refs. 113, 170). Sequences that differ from wild-type are indicated in italics. Identity nucleotides are in enlarged letters.

tRNAs, specific for either glutamate and glutamine or aspartate and asparagine, overlap and share strong homologies. Altogether this suggests common evolutionary origins of the glutamate/glutamine and aspartate/asparagine pairs. The fact that in some organisms GlnRS or AsnRS is absent and that aminoacylation of tRNA^{Gln} or tRNA^{Asn} occurs via a two step process comprising tRNA mischarging by heterologous GluRS or AspRS followed by amidation of the mischarged amino acids by tRNA-dependent amidases (reviewed in ref. 174) supports this view. It suggests further that glutamine and asparagine identities originate from the glutamate and aspartate identities.

To conclude, we notice the idiosyncratic architectures of the anticodon-binding modules in synthetases that likely were appended to their catalytic cores late in the course of the primordial evolution of life. In most systems, these modules recognize identity determinants within the anticodon of the homologous tRNAs. Because of the universal nature of the genetic code, determinants in anticodons are obviously conserved in evolution. From that and other considerations, it can be suggested that the origin of tRNA aminoacylation systems is connected with the development of the genetic code (ref. 173 and references therein).

Acknowledgements

We thank C. Florentz for suggestions and critical reading of the manuscript. This work was supported by grants from the Centre National de la Recherche Scientifique (CNRS), Ministère de la Recherche (Programme PRFMMIP), Université Louis Pasteur, Strasbourg, and the European Community (BIO4-CT98-0189).

References

1. Crick FHC. On protein synthesis. Symp Soc Exp Biol 1958; 12:128-163.
2. Hoagland MB, Stephenson ML, Scott JF et al. A soluble ribonucleic acid intermediate in protein synthesis. J Biol Chem 1958; 231:241-257.
3. Söll D. Transfer RNA—an RNA for all seasons. In: Gesteland R, Atkins J, eds. The RNA World. Cold Spring Harbor, New York: Cold Spring Harbor Laboratory Press, 1993:157-184.
4. Kiesewetter S, Ott G, Sprinzl M. The role of modified purine 64 in initiator/elongator discrimination of tRNA^{Meti} from yeast and wheat germ. Nucleic Acids Res 1990; 18:4677-4682.
5. RajBhandary UL. Initiator transfer RNAs. J Bacteriol 1994; 176:547-552.
6. Hüttenhofer A, Böck A. RNA structures involved in selenoprotein synthesis. In: Simons RW, Grunberg-Manago M, eds. RNA structure and Function. Cold Spring Harbor: Cold Spring Harbor Laboratory Press, 1998:377-413.
7. Holley RW, Apgar J, Everett GA et al. Structure of a ribonucleic acid. Science 1965; 147:1462-1465.
8. Sprinzl M, Horn C, Brown M et al. Compilation of tRNA sequences and sequences of tRNA genes. Nucleic Acids Res 1998; 26:148-153.
9. Rozenski J, Crain PF, McCloskey JA. The RNA modification database: 1999 update. Nucleic Acids Res 1999; 27:196-197.
10. Keith G, Yusupov M, Brion C et al. Sequence of tRNA^{Asp} from *Thermus thermophilus* HB8. Nucleic Acids Res 1993; 21:4399.
11. Horie N, Hara-Yokoyama M, Yokoyama S et al. Two tRNA^{Ile1} species from an extreme thermophile, *Thermus thermophilus* HB8: Effect of 2-thiolation of ribothymidine on the thermostability of tRNA. Biochemistry 1985; 24:5711-5715.
12. Helm M, Giegé R, Florentz C. A Watson-Crick base-pair disrupting methyl group (m^{1}A9) is sufficient for cloverleaf folding of human mitochondrial tRNA^{Lys}. Biochemistry 1999; 38:13338-13346.
13. Gregson JM, Crain PF, Edmonds CG, et al. Structure of the archaeal transfer RNA nucleoside G*-15 (2-amino-4,7-dihydro-4-oxo-7-beta-D-ribofuranosyl-1H-pyrrolo[2,3-d]pyrimidine-5-carboximidamide (archaeosine)). J Biol Chem 1993; 268:10076-10086.
14. Helm M, Brulé H, Friede D et al. Search for characteristic structural features of mammalian mitochondrial tRNAs. RNA 2000; 6:1356-1379.
15. Yokoyama S, Nishimura S. Modified nucleosides and codon recognition. In: Söll D, RajBhandary U, eds. tRNA: Structure, Biosynthesis, and Function. Washington DC: Am Soc Microbiol Press, 1995:207-223.
16. Giegé R, Sissler M, Florentz C. Universal rules and idiosyncratic features in tRNA identity. Nucleic Acids Res 1998; 26:5017-5035.
17. Masquida B, Westhof E. On the wobble G-U and related pairs. RNA 2000; 6:9-15.
18. Varani G, MaClain WH. The G-U wobble base pair. A fundamental building block of RNA structure crucial to RNA function in diverse biological systems. EMBO Reports 2000; 1:18-23.
19. de Bruijn MH, Schreier PH, Eperon IC et al. A mammalian mitochondrial serine transfer RNA lacking the "dihydrouridine" loop and stem. Nucleic Acids Res 1980; 8:5213-5222.
20. Wolstenholme DR, Okimoto R, Mcfarlane JL. Nucleotide correlations that suggest tertiary interactions in the TV-replacement loop-containing mitochondrial tRNAs of the nematodes, *Caenorhabditis elegans* and *Ascaris suum*. Nucleic Acids Res 1994; 22:4300-4306.

21. Steinberg S, Leclerc F, Cedergren R. Structural rules and conformational compensations in the tRNA L-form. J Mol Biol 1997; 266:269-282.
22. Cooley L, Appel B, Söll D. Post-transcriptional nucleotide addition is responsible for the formation of the 5'-terminus of histidine tRNA. Proc Natl Acad Sci USA 1986; 79:6475-6479.
23. Hou Y-M, Westhof E, Giegé R. An unusual RNA tertiary interaction has a role for the specific aminoacylation of a transfer RNA. Proc Natl Acad Sci USA 1993; 90:6776-6780.
24. Rudinger J, Felden B, Florentz C et al. Strategy for RNA recognition by yeast histidyl-tRNA synthetase. Bioorg Med Chem 1997; 5:1001-1009.
25. Kim SH, Quigley GJ, Suddath FL et al. Three dimensional structure of yeast phenylalanine transfer RNA. Folding of the polynucleotide chain. Science 1973; 179:285-288.
26. Kim SH, Suddath FL, Quigley GJ et al. Three dimensional tertiary structure of yeast phenylalanine transfer RNA. Science 1974; 185:435-440.
27. Robertus JD, Ladner JE, Finch JT et al. Structure of yeast phenylalanine tRNA at 3Å resolution. Nature 1974; 250:546-551.
28. Stout CD, Mizuno H, Rubin J et al. Atomic coordinates and molecular conformation of yeast phenylalanyl tRNA. An independent investigation. Nucleic Acids Res 1976; 3:1111-1123.
29. Moras D, Comarmond MB, Fischer J et al. Crystal structure of yeast tRNA^{Asp}. Nature 1980; 288:669-674.
30. Westhof E, Dumas P, Moras D. Crystallographic refinement of yeast aspartic acid transfer RNA. J Mol Biol 1985; 184:119-145.
31. Woo NH, Roe BA, Rich A. Three-dimensional structure of *Escherichia coli* initiator tRNA^{fMet}. Nature 1980; 286:346-351.
32. Basavappa R, Sigler PB. The 3Å crystal structure of yeast initiator tRNA: Functional implications in initiator/elongator discrimination. EMBO J 1991; 10:3105-3111.
33. Bénas P, Bec G, Keith G et al. The crystal structure of HIV reverse-transcription primer tRNA$_3$^{Lys} shows a canonical anticodon loop. RNA 2000; 6:1347-1355.
34. Sundaram M, Durant PC, Davis DR. Hypermodified nucleosides in the anticodon of tRNA^{Lys} stabilize a canonical U-turn structure. Biochemistry 2000; 39:12575-12584.
35. Giegé R, Puglisi JD, Florentz C. tRNA structure and aminoacylation efficiency. Prog Nucleic Acid Res Mol Biol 1993; 45:129-206.
36. Giegé R, Helm M, Florentz C. Chemical and enzymatic probing of RNA structure. In: Söll D, Nishimura S, Moore P, eds. Prebiotic Chemistry, Molecular Fossils, Nucleosides, and RNA. Oxford: Pergamon, 1999:63-80.
37. Moras D, Dock AC, Dumas P et al. The structure of yeast tRNA^{Asp}. A model for tRNA interacting with mesenger RNA. J Biomol Struct Dyn 1985; 3:479-493.
38. Moras D, Dock A-C, Dumas P et al. Anticodon-anticodon interaction induces conformational changes in tRNA: Yeast tRNA^{Asp}, a model for tRNA-mRNA recognition. Proc Natl Acad Sci USA 1986; 83:932-936.
39. Shi HJ, Moore PB. The crystal structure of yeast phenylalanine tRNA at 1.93Å resolution: a classic structure revisited. RNA 2000; 6:1091-1105.
40. Jovine L, Djordjevic S, Rhodes D. The crystal structure of yeast phenylalanine tRNA at 2.0Å resolution: cleavage by Mg^{2+} in 15-year old crystals. J Mol Biol 2000; 301:401-414.
41. Sussman JL, Holbrook SR, Warrant RW et al. Crystal structure of yeast phenylalanine transfer RNA. I. Crystallographic refinement. J Mol Biol 1978; 123:607-630.
42. Dock-Bregeon A-C, Moras D, Giegé R. Nucleic acids and their complexes. In: Ducruix A, Giegé R, eds. Crystallization of Nucleic Acids and Proteins. A Practical Approach (second edition). Oxford: IRL Press, 1999:209-243.
43. Mikkelsen NE, Johansson K, Virtanen A et al. Aminoglycoside binding displaces a divalent metal ion in tRNA-neomycin B complex. Nature Struct Biol 2001; 8:510-514.
44. Martinis SA, Schimmel P. Small RNA oligonucleotide substrates for specific aminoacylations. In: Söll D, RajBhandary UL, eds. tRNA: Structure, Biosynthesis, and Function. Washington, DC: Am Soc Microbiol Press, 1995:349-370.
45. Ramos A, Varani G. Structure of the acceptor stem of *Escherichia coli* tRNA^{Ala}: role of the G3:U70 base pair in synthetase recognition. Nucleic Acids Res 1997; 25:2083-2090.
46. Mueller U, Schubel H, Sprinzl M et al. Crystal structure of acceptor stem of tRNA(Ala) from *Escherichia coli* shows unique G.U wobble base pair at 1.16 Å resolution. RNA 1999; 5:670-677.
47. Romby P, Carbon P, Westhof E et al. Importance of conserved residues for the conformation of the T-loop in tRNAs. J Biomol Struct Dyn 1987; 5:669-687.
48. Agris PF. The importance of being modified: Roles of modified nucleosides and Mg^{2+} in RNA structure and function. Prog Nucleic Acid Res Mol Biol 1996; 53:79-129.
49. Koshlap KM, Guenther R, Sochacka E et al. A distictive RNA fold: the solution structure of an analogue of the yeast tRNA^{Phe} TYC domain. *Biochemistry* 1999; 38:8647-8656.
50. Auffinger P, Westhof E. An extended structural signature for the tRNA anticodon loop. RNA 2001; 7:334-341.
51. Rould MA, Perona JJ, Söll D et al. Structure of E. *coli* glutaminyl-tRNA synthetase complexed with tRNA^{Gln} and ATP at 2.8Å resolution. Science 1989; 246:1135-1142.
52. Ruff M, Krishnaswamy S, Boeglin M et al. Class II aminoacyl transfer RNA synthetases: Crystal structure of yeast aspartyl-tRNA synthetase complexed with tRNA^{Asp}. Science 1991; 252:1682-1689.
53. Sissler M, Eriani G, Martin F et al. Mirror image alternative interaction patterns of the same tRNA with either class I arginyl-tRNA synthetase or class II aspartyl-tRNA synthetase. Nucleic Acids Res 1997; 25:4899-4906.

54. Arnez JG, Steitz TA. Crystal structure of unmodified tRNA^{Gln} complexed with glutaminyl-tRNA synthetase and ATP suggests a possible role for pseudo-uridines in stabilization of RNA structure. Biochemistry 1994; 33:7560-7567.

55. Hayase Y, Jahn M, Rogers MJ et al. Recognition of bases in *E. coli* tRNA^{Gln} by glutaminyl-tRNA synthetase: A complete identity set. EMBO J 1992; 11:4159-4165.

56. Rould MA, Perona JJ, Steitz TA. Structural basis of anticodon loop recognition by glutaminyl-tRNA synthetase. Nature 1991; 352:213-218.

57. Hall KB, Sampson JR, Uhlenbeck OC et al. Structure of an unmodified tRNA molecule. Biochemistry 1989; 28:5794-5801.

58. Sherlin LD, Bullock TL, Newberry KJ et al. Influence of transfer RNA tertiary structure on aminoacylation efficiency by glutaminyl and cysteinyl-tRNA synthetases. J Mol Biol 2000; 299:431-446.

59. Nissen P, Thirup S, Kjeldgaard M et al. The crystal structure of Cys-tRNA^{Cys}-EF-Tu-GDPNP reveals general and specific features in the ternary complex and in tRNA. Structure 1999; 7:143-156.

60. Delagoutte B, Moras D, Cavarelli J. tRNA aminoacylation by arginyl-tRNA synthetase: induced conformations during substrates binding. EMBO J 2000; 19:5599-5610.

61. Fukai S, Nureki O, Sekine S et al. Structural basis for double-sieve discrimination of L-valine from L-isoleucine and L-threonine by the complex of tRNA(Val) and valyl-tRNA synthetase. Cell 2001; 103:793-803.

62. Silvian LF, Wang J, Steitz TA. Insights into editing from an ile-tRNA synthetase structure with tRNA^{Ile} and mupirocin. Science 1999; 285:1074-1077.

63. Sekine S, Nureki O, Shimada A et al. Structural basis for anticodon recognition by discriminating glutamyl-tRNA synthetase. Nature Struct Biol 2001; 8:203-206.

64. Biou V, Yaremchuk A, Tukalo et al. The 2.9 Å crystal structure of *T. thermophilus* seryl-tRNA synthetase complexed with tRNA^{Ser}. Science 1994; 263:1404-1410.

65. Cusack S, Yaremchuck A, Tukalo M. The crystal structure of the ternary complex of *T. thermophilus* seryl-tRNA^{Ser} and a seryl-adenylate analogue reveals a conformational switch in the active site. EMBO J 1996; 15:2834-2842.

66. Sankaranarayanan R, Dock-Bregeon A-C, Romby P et al. The structure of threonyl-tRNA synthetase-tRNA^{Thr} complex enlightens its repressor activity and reveals an essential zinc ion in the active site. Cell 1999; 97:371-381.

67. Yaremchuk A, Cusack S, Tukalo M. Crystal structure of a eukaryote/archaeon-like prolyl-tRNA synthetase and its complex with tRNA(Pro)(CGG). EMBO J 2000; 19:4745-4758.

68. Eiler S, Dock-Bregeon AC, Moulinier L et al. Synthesis of aspartyl-tRNA^{Asp} in *Escherichia coli*-a snapshot of the second step. EMBO J 1999; 18:6532-6541.

69. Briand C, Poterszman A, Eiler S et al. An intermediate step in the recognition of tRNA^{Asp} by aspartyl-tRNA synthetase. J Mol Biol 2000; 299:1051-1060.

70. Cusack S, Yaremchuk A, Tukalo M. The crystal structures of *T. thermophilus* lysyl-tRNA synthetase complexed with *E. coli* tRNA^{Lys} and a *T. thermophilus* tRNA^{Lys} transcript: anticodon recognition and conformational changes upon binding of a lysyl-adenylate analogue. EMBO J 1996; 15:6321-6334.

71. Goldgur Y, Mosyak L, Reshetnikova L et al. The crystal structure of phenylalanyl-tRNA synthetase from *Thermus thermophilus* complexed with cognate tRNA^{Phe}. Structure 1997; 5:59-68.

72. Nissen P, Kjeldgaard M, Thirup S et al. Crystal structure of the ternary complex of Phe-tRNA^{Phe}, EF-Tu, and a GTP analog. Science 1995; 270:1464-1472.

73. Schmitt E, Panvert M, Blanquet S et al. Crystal structure of methionyl-tRNA^{fMet} transformylase complexed with the initiator formyl-methionyl-tRNA^{fMet}. EMBO J 1998; 17:6819-6826.

74. Cate JH, Yusupov MM, Yusupova GZ et al. X-ray crystal structures of 70S ribosome functional complexes. Science 1999; 285:2095-2104.

75. Yusupov MM, Yusupova GZ, Baucom A et al. Crystal structure of the ribosome at 5.5Å resolution. Science 2001; 292:883-896.

76. Streeck RE, Zachau HG. Characterization of the native and denatured conformations of tRNA^{Ser} and tRNA^{Phe} from yeast. Eur J Biochem 1972; 30:382-391.

77. Madore E, Florentz C, Giegé R et al. Magnesium-dependent alternate foldings of active and inactive *Escherichia coli* tRNA^{Glu} revealed by chemical probing. Nucleic Acids Res 1999; 27:3583-3588.

78. Krauss G, Pingoud A, Boehme D et al. Equivalent and non-equivalent binding sites for tRNA on aminoacyl-tRNA synthetases. Eur J Biochem 1975; 55:517-29.

79. Friederich MW, Vacano E, Hagerman PJ. Global flexibility of tertiary structure in RNA: yeast tRNA^{Phe} as a model system. Proc Natl Acad Sci USA 1998; 95:3572-3577.

80. Limmer S, Hofmann HP, Ott G et al. The 3'-terminal end (NCCA) of tRNA determines the structure and stability of the aminoacyl acceptor stem. Proc Natl Acad Sci USA 1993; 90:6199-6202.

81. Puglisi EV, Puglisi JD, Williamson JR et al. NMR analysis of tRNA acceptor stem microhelices: discriminator base change affects tRNA conformation at the 3'end. Proc Natl Acad Sci USA 1994; 91:11467-11471.

82. Chang KY, Varani G, Bhattacharya S et al. Correlation of deformability at a tRNA recognition site and aminoacylation specificity. Proc Natl Acad Sci USA 1999; 96:11764-11769.

83. Horie N, Yamaizumi Z, Kuchino Y et al. Modified nucleosides in the first positions of the anticodons of tRNA$_4$^{Leu} and tRNA$_5$^{Leu} from *Escherichia coli*. Biochemisrtry 1999; 38:207-217.

84. Liu M, Chu W-C, Liu JC-H et al. Role of acceptor stem conformation in tRNA^{Val} recognition by its cognate synthetase. Nucleic Acids Res 1997; 25:4883-4890.

85. First EA. Catalysis of tRNA aminoacylation by class I and class II aminoacyl-tRNA synthetases. In: Sinnott M, ed. Comprehensive Biological Catalysis. New-York: Academic Press, 1998:573-607.
86. Chapeville F, Lipman F, Ehrenstein GV et al. On the role of soluble ribonucleic acid in coding for amino acids. Proc Natl Acad Sci USA 1962; 48:1086-1092.
87. Ebel J-P, Giegé R, Bonnet J et al. Factors determining the specificity of the tRNA aminoacylation reaction. Biochimie 1973; 55:547-557.
88. Normanly J, Abelson J. tRNA identity. Annu. Rev. Biochem. 1989; 58:1029-1049.
89. Schulman LH. Recognition of tRNAs by aminoacyl-tRNA synthetases. Prog Nucleic Acid Res Mol Biol 1991; 41:23-87.
90. McClain WH. Rules that govern tRNA identity in protein synthesis. J Mol Biol 1993; 234:257-280.
91. Saks ME, Sampson JR, Abelson JN. The transfer RNA identity problem: A search for rules. Science 1994; 263:191-197.
92. Muramatsu T, Nishikawa K, Nemoto F et al. Codon and amino-acid specificities of a transfer RNA are both converted by a single post-transcriptional modification. Nature 1988; 336:179-181.
93. Perret V, Garcia A, Grosjean H et al. Relaxation of transfer RNA specificity by removal of modified nucleotides. Nature 1990; 344:787-789.
94. Normanly J, Ogden RC, Horvath SJ et al. Changing the identity of a transfer RNA. Nature 1986; 321:213-219.
95. Hou Y-M, Schimmel P. A simple structural feature is a major determinant of the identity of a transfer RNA. Nature 1988; 333:140-145.
96. McClain WH, Foss K. Changing the identity of a tRNA by introducing a G-U wobble pair near the 3' acceptor end. Science 1988; 240:793-796.
97. Sampson JR, DiRenzo AB, Behlen LS et al. Nucleotides in yeast tRNA^{Phe} required for the specific recognition by its cognate synthetase. Science 1989; 243:1363-1366.
98. Pütz J, Puglisi JD, Florentz C et al. Identity elements for specific aminoacylation of yeast tRNA^{Asp} by cognate aspartyl-tRNA synthetase. Science 1991; 252:1696-1699.
99. Giegé R, Florentz C, Kern D et al. Aspartate identity of transfer RNAs. Biochimie 1996; 78:605-623.
100. Hou YM, Schimmel P. Evidence that a major determinant for the identity of a transfer RNA is conserved in evolution. Biochemistry 1989; 28:6800-6804.
101. Francklyn C, Schimmel P. Aminoacylation of RNA minihelices with alanine. Nature 1989; 337:478-481.
102. Beuning PJ, Yang F, Schimmel P et al. Specific atomic groups and RNA helix geometry in acceptor stem recognition by a tRNA synthetase. Proc Natl Acad Sci USA 1997; 94:10150-10154.
103. Pleiss JA, Wolfson AD, Uhlenbeck OC. Mapping contacts between *Escherichia coli* alanyl-tRNA synthetase and 2' hydroxyls using a complete tRNA molecule. Biochemistry 2000; 39:8250-8258.
104. Hou Y-M, Sterner T, Jansen M. Permutation of a pair of tertiary nucleotides in a transfer RNA. Biochemistry 1995; 34:2978-2984.
105. Lovato MA, Chihade JW, Schimmel P. Translocation within the acceptor helix of a major tRNA identity determinant. EMBO J 2001; 20:4846-4853.
106. Sissler M, Giegé R, Florentz C. Arginine aminoacylation identity is context-dependent and ensured by alternate recognition sets in the anticodon loop of accepting tRNA transcripts. EMBO J 1996; 15:5069-5076.
107. Sissler M, Giegé R, Florentz C. The tRNA sequence context defines the mechanistic routes by which yeast arginyl-tRNA synthetase charges tRNA. RNA 1998; 4:647-657.
108. Moor NA, Ankilova VN, Lavrik OI. Recognition of tRNA^{Phe} by phenyalanyl-tRNA synthetase of *Thermus thermophilus*. Eur J Biochem 1995; 234:897-902.
109. Sampson JR, Behlen LS, DiRenzo AB et al. Recognition of yeast tRNA^{Phe} by its cognate yeast phenylalanyl-tRNA synthetase; an analysis of specificity. Biochemistry 1992; 31:4164-4167.
110. Frugier M, Helm M, Felden B et al. Sequences outside recognition sets are not neutral for tRNA aminoacylation: evidence for non-permissive combinations of nucleotides in the acceptor stem of yeast tRNA^{Phe}. J Biol Chem 1998; 273:11605-11610.
111. Frugier M, Söll D, Giegé R et al. Identity switches between tRNAs aminoacylated by class I glutaminyl- and class II aspartyl-tRNA synthetase. Biochemistry 1994; 33:9912-9921.
112. Pütz J, Puglisi JD, Florentz C et al. Additive, cooperative and anti-cooperative effects between identity nucleotides of a tRNA. EMBO J 1993; 12:2949-2957.
113. Frugier M, Florentz C, Giegé R. Efficient aminoacylation of resected RNA helices by class II aspartyl-tRNA synthetase dependent on a single nucleotide. EMBO J 1994; 13:2218-2226.
114. Cavarelli J, Rees B, Ruff M et al. Yeast tRNA^{Asp} recognition by its cognate class II aminoacyl-tRNA synthetase. Nature 1993; 362:181-184.
115. Rudinger J, Puglisi JD, Pütz J et al. Determinant nucleotides of yeast tRNA^{Asp} interact directly with aspartyl-tRNA synthetase. Proc Natl Acad Sci USA 1992; 89:5882-5886.
116. Pütz J, Florentz C, Benseler F et al. A single methyl group prevents the mischarging of a tRNA. Nature Struct Biol 1994; 1:580-582.
117. Ibba M, Losey HC, Kawarabayasi Y et al. Substrate recognition by class I lysyl-tRNA synthetase: A molecular basis for gene displacement. Proc Natl Acad Sci USA 1999; 96:418-423.
118. Rudinger J, Hillenbrandt R, Sprinzl M et al. Antideterminants present in minihelix^{Sec} hinder its recognition by prokaryotic elongation factor Tu. EMBO J 1996; 15:650-657.
119. Mohan A, Whyte S, Wang X et al. The 3' end CCA of mature tRNA is an antideterminant for eukaryotic 3'-tRNase. RNA 1999; 5:245-256.

120. Rogers MJ, Söll D. Discrimination between glutaminyl-tRNA synthetase and seryl-tRNA synthetase involves nucleotides in the acceptor helix of tRNA. Proc Natl Acad Sci USA 1988; 85:6627-6631.
121. Jahn M, Rogers MJ, Söll D. Anticodon and acceptor stem nucleotides in tRNA^{Gln} are major recognition elements for *E. coli* glutaminyl-tRNA synthetase. Nature 1991; 352:258-260.
122. Rogers MJ, Adachi T, Inokuchi H et al. Switching tRNA^{Gln} identity from glutamine to tryptophan. Proc Natl Acad Sci USA 1992; 89:3463-3467.
123. Sherman JM, Thomann H-U, Söll D. Functionnal connectivity between tRNA binding domains in glutaminyl-tRNA synthetase. J Mol Biol 1996; 256:818-828.
124. Hong K-W, Ibba M, Weygand-Durasevic I et al Transfer RNA-dependant cognate amino acid recognition by an aminoacyl-tRNA-synthetase. EMBO J 1996; 15:1983-1991.
125. Liu J, Ibba M, Hong K-W et al. The terminal adenosine of tRNA^{Gln} mediates tRNA-dependent amino acid recognition by glutaminyl-tRNA synthetase. Biochemistry 1998; 37:9836-9842.
126. Ibba M, Hong K-W, Sherman JM et al. Interactions between tRNA identity nucleotides and their recognition sites in glutaminyl-tRNA synthetase determine the cognate amino acid affinitiy of the enzyme. Proc Natl Acad Sci USA 1996; 93:6953-6958.
127. Mandal AK, Bhattacharyya A, Bhattacharyya S et al. A cognate tRNA specific conformational change in glutaminyl-tRNA synthetase and its implication for specificity. Protein Science 1998; 7:1046-1051.
128. Sherman JM, Rogers MJ, Söll D. Competition of aminoacyl-tRNA synthetases for tRNA ensures the accuracy of aminoacylation. Nucleic Acids Res 1992; 20:2847-2852.
129. Mirande M. Aminoacyl-tRNA synthetase family from prokaryotes and eukaryotes: Structural domains and their implications. Prog Nucleic Acid Res Mol Biol 1991; 40:95-142.
130. Whelihan EF, Schimmel P. Rescuing an essential enzyme-RNA complex with a non-essential appended domain. EMBO J 1997; 16:2968-2974.
131. Frugier M, Moulinier L, Giegé R. A domain in the N-terminal extension of class IIb eukaryotic aminoacyl-tRNA synthetases is important for tRNA binding. EMBO J 2000; 19:2371-2380.
132. Chihade JW, Schimmel P. Assembly of a catalytic unit for RNA microhelix aminoacylation using nonspecific RNA binding domains. Proc Natl Acad Sci USA 1999; 96:12316-12321.
133. Simos G, Sauer A, Fasiolo F et al. A conserved domain within Arc1p delivers tRNA to aminoacyl-tRNA synthetases. Mol Cell 1998; 1:235-242.
134. Nomanbhoy T, Morales AJ, Abraham AT et al. Simultaneous binding of two proteins to opposite sides of a single transfer RNA. Nature Struct Biol 2001; 8:344-348.
135. LaRiviere FJ, Wolfson AD, Uhlenbeck OC. Uniform binding of aminoacyl-tRNAs to elongation factor Tu by thermodynamic compensation. Science 2001; 294:165-168.
136. Khvorova AM, Motorin Y, Wolfson AD. Pyrophosphate mediates the effect of certain tRNA mutations on aminoacylation of yeast tRNA^{Phe}. Nucleic Acids Res 1999; 27:4451-4456.
137. Hayashi I, Kawai G, Watanabe K. Higher-order structure and thermal instability of bovine mitochondrial tRNA^{Ser}UGA investigated by proton NMR spectroscopy. J Mol Biol 1998; 284:57-69.
138. Hayashi I, Yokogawa T, Kawai G et al. Assignment of imino proton signals of G-C base pairs and magnesium ion binding: an NMR study of bovine mitochondrial tRNA^{SerGCU} lacking the entire D arm. J Biochem 1997; 121:1115-1122.
139. Börner GV, Mörl M, Janke A et al. RNA editing changes the identity of a mitochondrial tRNA in marsupials. EMBO J 1996; 15:5949-5957.
140. Ueda T, Yotsumoto Y, Ikeda K et al. The T-loop region of animal mitochondrial tRNA$^{Ser(AGY)}$ is a main recognition site for homologous seryl-tRNA synthetase. Nucleic Acids Res 1992; 20:2217-2222.
141. Yokogawa T, Shimada N, Takeuchi N et al. Characterization and tRNA recognition of mammalian mitochondrial seryl-tRNA synthetase. J Biol Chem 2000; 275:19913-19920.
142. Shimada N, Suzuki T, Watanabe K. Dual mode of recognition of two isoacceptor tRNAs by mammalian mitochondrial seryl-tRNA synthetase. J Biol Chem 2001; 276:46770-46778.
143. Saks ME, Sampson JR. Variant minihelix RNAs reveal sequence-specific recognition of the helical tRNA^{Ser} acceptor stem by *E. coli* seryl-tRNA synthetase. EMBO J 1996; 15:2843-2849.
144. Florentz C, Sissler M., chapter 8 of this book.
145. Commans S, Böck A. Selenocysteine inserting tRNAs: an overview. FEMS Microbiology Reviews 1999; 23:335-351.
146. Zwieb C, Guven SA, Wower IK et al. Three-dimensional folding of the tRNA-like domain of *Escherichia coli* tmRNA. Biochemistry 2001; 40:9587-9595.
147. Komine Y, Kitabatake M, Yokogawa T et al. A tRNA-like structure is present in 10Sa RNA, a small stable RNA from *Escherichia coli*. Proc Natl Acad Sci USA 1994; 91:9223-9277.
148. Pinck M, Yot P, Chapeville F et al. Enzymatic binding of valine to the 3' end of TYMV RNA. Nature 1970; 226:954-956.
149. Dreher TW, Bujarski JJ, Hall TC. Mutant viral RNAs synthesized in vitro show altered aminoacylation and replicase template activities. Nature 1984; 311:171-175.
150. Mans MW, Pleij CWA, Bosch L. tRNA-like structures. Structure, function and evolutionary significance. Eur J Biochem 1991; 201:303-324.
151. Florentz C, Giegé R. tRNA-like structures in viral RNAs. In: Söll D, RajBhandary UL, eds. tRNA: Structure, Biosynthesis, and Function. Washington, DC: Am Soc Microbiol Press, 1995:141-163.
152. Fechter P, Rudinger-Thirion J, Florentz C et al. Novel features in the tRNA-like world of plant viral RNAs. CMLS, Cellular Mol Life Sciences 2001; 58:1547-1561.

153. Pleij CWA, Rietveld K, Bosch L. A new principle of folding based on pseudoknotting. Nucleic Acids Res 1985; 13:1717-1731.
154. Fromant M, Plateau P, Blanquet S. Function of the extra 5'-phosphate carried by histidine tRNA. Biochemistry 2000; 39:4062-4067.
155. Dreher TW. Functions of 3'-untranslated regions of positive strand RNA viral genomes. Annu Rev Phytopathol 1999; 37:151-174.
156. Florentz C, Dreher TW, Rudinger J et al. Specific valylation identity of turnip yellow mosaic virus RNA by yeast valyl-tRNA synthetase is directed by the anticodon in a kinetic rather than affinity-based discrimination. Eur J Biochem 1991; 195:229-234.
157. Dreher TW, Tsai C-H, Florentz C et al. Specific valylation of turnip yellow mosaic virus RNA by wheat germ valyl-tRNA synthetase is determined by three anticodon loop nucleotides. Biochemistry 1992; 31:9183-9189.
158. Wientges J, Pütz J, Giegé R et al. Selection of viral RNA-derived tRNA-like structures with improved valylation activities. Biochemistry 2000; 39:6207-6218.
159. Fechter P, Giegé R, Rudinger-Thirion J. Specific tyrosylation of the bulky tRNA-like structure of Brome Mosaic Virus RNA relies solely on identity nucleotides present in its amino acid accepting stem. J Mol Biol 2001; 309:387-399.
160. Fechter P, Rudinger-Thirion J, Théobald-Dietrich A et al. Identity of tRNA for yeast tyrosyl-tRNA synthetase: Tyrosylation is more sensitive to identity nucleotides than to structural features. Biochemistry 2000; 39:1725-1733.
161. Giegé R, Frugier M, Rudinger J. tRNA mimics. Curr Opin Struct Biol 1998; 8:286-293.
162. Perret V, Florentz C, Puglisi JD et al. Effect of conformational features on the aminoacylation of tRNAs and consequences on the permutation of tRNA specificities. J Mol Biol 1992; 226:323-333.
163. Nissan TA, Oliphant B, Perona JJ. An engineered class I transfer RNA with a class II tertiary fold. RNA 1999; 5:434-445.
164. Nissan TA, Perona JJ. Alternative designs for construction of the class II transfer RNA tertiary core. RNA 2000; 6:1585-1596.
165. Frugier M, Florentz C, Schimmel P et al. Triple aminoacylation specificity of a chimerized transfer RNA. Biochemistry 1993; 32:14053-14061.
166. Puglisi JD, Pütz J, Florentz C et al. Influence of tRNA tertiary structure and stability on aminoacylation by yeast aspartyl-tRNA synthetase. Nucleic Acids Res 1993; 21:41-49.
167. Rudinger-Thirion J, Giegé R. The peculiar architectural framework of tRNA^{Sec} is fully recognized by yeast AspRS. RNA 1999; 5:495-502.
168. Vörtler S, Pütz J, Giegé R. Manipulation of tRNA properties by structure-based and combinatorial in vitro approaches. Prog Nucleic Acid Res Mol Biol 2001; 70:291-334.
169. Tinkle Peterson E, Pan T, Coleman J et al. In vitro selection of small RNAs that bind to *Escherichia coli* phenylalanyl-tRNA synthetase. J Mol Biol 1994; 242:186-192.
170. Wolfson AD, Khvorova AM, Sauter C et al. Mimics of yeast tRNA^{Asp} and their recognition by aspartyl-tRNA synthetase. Biochemistry 1999; 38:11926-11932.
171. Renaud M, Ehrlich R, Bonnet J et al. Lack of correlation between affinity of the tRNA for the aminoacyl-tRNA synthetase and aminoacylation capacity as studied with modified tRNA^{Phe}. Eur J Biochem 1979; 100:157-164.
172. Schimmel P, Giegé R, Moras D et al. An operational RNA code for amino acids and possible relationship to genetic code. Proc Natl Acad Sci USA 1993; 90:8763-8768.
173. Ribas de Pouplana L, Schimmel P. Aminoacyl-tRNA synthetases: potential markers of genetic code development. Trends Biochem Sci 2001; 26:591-596.
174. Ibba, M, Söll, D. Aminoacyl-tRNA synthesis. Annu Rev Biochem 2000; 69:617-650.

CHAPTER 2

Aminoacyl-tRNA Synthetase Structure and Evolution

Dieter Söll and Michael Ibba

Summary

The pairing of codons in mRNA with tRNA anticodons determines the order of amino acids in a protein. It is therefore imperative for accurate translation that tRNAs are only coupled to amino acids corresponding to the RNA anticodon. This is mostly, but not exclusively, achieved by the direct attachment of the appropriate amino acid to the 3'-end of the corresponding tRNA by the aminoacyl-tRNA synthetases. To ensure the accurate translation of genetic information, the aminoacyl-tRNA synthetases must display an extremely high level of substrate specificity. Despite this highly conserved function, recent studies arising from the analysis of whole genomes have shown a significant degree of evolutionary diversity in aminoacyl-tRNA synthesis. For example, noncanonical routes have been identified for the synthesis of Asn-tRNA, Cys-tRNA, Gln-tRNA and Lys-tRNA. Characterization of noncanonical aminoacyl-tRNA synthesis has revealed an unexpected level of evolutionary divergence and has also provided new insights into the possible precursors of contemporary aminoacyl-tRNA synthetases.

Introduction

The ribosomal synthesis of proteins from a messenger RNA (mRNA) template is one of the defining features of the central dogma of molecular biology.[1] Proteins are made by the sequential translation of codons into their corresponding amino acids, resulting in the synthesis of a polypeptide whose sequence corresponds to that defined in the respective mRNA. Amino acids are delivered for protein synthesis as aminoacyl-tRNA:translation factor complexes. The identity of an amino acid inserted at a particular position in a nascent polypeptide is determined by two key molecular recognition events: the interaction of the aminoacyl-tRNA anticodon with an appropriate codon in mRNA, and the correct pairing of amino acid and tRNA anticodon in the aminoacyl-tRNA (Fig. 1). As a result, the fidelity of protein synthesis is dependent on the presence in the cell of a complete set of correctly aminoacylated tRNAs.[2]

Aminoacyl-tRNAs are synthesized by esterification at the 3'-terminus of tRNAs with the appropriate amino acids in a two-step reaction catalyzed by a family of enzymes collectively known as the aminoacyl-tRNA synthetases. In the first step of the reaction, ATP (or rarely other NTPs[3]) and amino acid bind at the active site. The respective positioning of the α-phosphate of ATP and the α-carboxylate of the amino acid allow the latter to attack the former by an in-line nucleophilic displacement mechanism. This leads to the formation of an enzyme-bound mixed anhydride (aminoacyl-adenylate), and an inorganic pyrophosphate-leaving group, as summarized below:

$$AA + ATP + AARS \leftrightarrows AARS \cdot AA\text{-}AMP + PP_i \qquad (1)$$

(AA, amino acid; AARS, aminoacyl-tRNA synthetase; PP$_i$, inorganic pyrophosphate)

In the second step of the reaction, the 2'- or 3'-hydroxyl of the terminal adenosine of tRNA nucleophilically attacks the α-carbonyl of the aminoacyl-adenylate. This results in the 3'-esterification of the tRNA with the amino acid moiety and generation of AMP as the leaving group, followed by product release:

Translation Mechanisms, edited by Jacques Lapointe and Léa Brakier-Gingras.

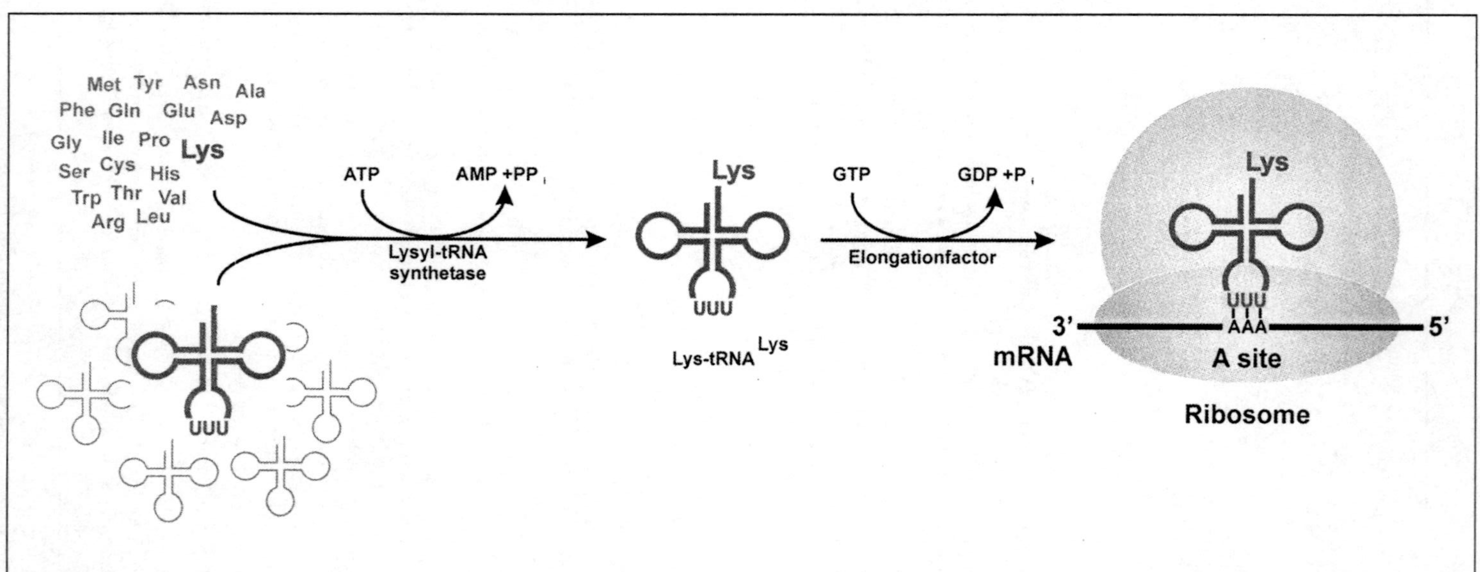

Figure 1. The cellular synthesis of an aminoacyl-tRNA and its role in protein synthesis. An uncharged tRNA and the corresponding amino acid are first selected from the cellular pools of similar molecules by the appropriate AARS. After synthesis and release from the AARS, the aminoacyl-tRNA is delivered to the ribosome, where its anticodon can then interact with the corresponding codon in mRNA. The example shown illustrates how this leads to the translation of the codon AAA as lysine during the elongation phase of protein synthesis.

$$\text{AARS} \cdot \text{AA-AMP} + \text{tRNA} \rightleftarrows \text{AARS} + \text{AA-tRNA} + \text{AMP} \qquad (2)$$

Numerous studies have shown that all AARSs catalyze essentially the same reaction using the scheme summarized above in equations 1 and 2.[4] Some differences have been observed with respect to the rate-determining step of the reaction, which is most commonly 3'-esterification of the tRNA but may also be aminoacyl-tRNA release. The other major difference observed concerns the requirement for tRNA binding prior to aminoacyl-adenylate synthesis in a small sub-group of AARSs, namely arginyl-tRNA synthetase, glutamyl-tRNA synthetase, glutaminyl-tRNA synthetase and the class I lysyl-tRNA synthetase[5] (LysRS1).

Aminoacyl-tRNA Synthetases

Class I and Class II Enzymes

The aminoacyl-tRNA synthetases are divided into two unrelated classes (I and II) based upon the existence of mutually exclusive sequence motifs that reflect distinct active site topologies (Table 1). Structural studies have shown that in class I synthetases the active site contains a Rossmann dinucleotide binding domain, whereas this fold is absent from the active site of class II enzymes which instead contain a novel anti-parallel β-fold. One result of this difference in active site structure is that class I enzymes bind ATP in an extended conformation, class II in a bent conformation.[4] The other major difference between the two AARS classes is in their binding of tRNA and the site at which the amino acid is subsequently attached to it. Class I enzymes approach the acceptor stem of tRNA from the minor groove side and acylate the 2'-hydroxyl of the terminal adenosine, whereas class II synthetases approach the major groove side and generally acylate the 3'-hydroxyl. Whether the different modes of tRNA binding are found for all members of each class remains unclear, although recent analyses indicate this to be the case.[6] Furthermore, extensive biochemical studies have suggested that alanyl-tRNA synthetase, a class II enzyme, recognizes the minor groove of the tRNA^{Ala} acceptor stem,[7] although this may be achieved while approaching the major groove as seen in the structure of the ThrRS:tRNA^{Thr} complex.[8] Generalizations about modes of tRNA binding in the AARSs are further complicated by the absence of common RNA-binding motifs in all but a few examples, described in more detail below.

Amino Acid Recognition and Catalysis

Presteady-state kinetic analyses have shown that the aminoacylation reaction does not involve covalent or acid-base catalysis, but instead uses the binding energy provided by enzyme-reactant interactions to stabilize the two transition states.[9] Mutagenesis of the class I signature motif KMSKS has shown that these residues are part of a mobile loop which serves to stabilize the transition state of the amino acid activation reaction using ATP binding energy.[10] It has also been shown that the other class I signature motif, HIGH, plays a similar role.[11] The roles of the KMSKS and HIGH motifs in ATP binding and transition state stabilization were found to be similar in GlnRS from structural studies,[12] despite a notable difference in the amino acid activation reaction of this enzyme. GlnRS (together with ArgRS, GluRS and LysRS1) requires the presence of tRNA for catalysis of amino acid activation, in contrast to all other synthetases which do not require tRNA. Taken together, these various studies provide a detailed understanding at the molecular level of aminoacyl-adenylate synthesis by the class I AARSs.

Structural analyses of the class II AARSs have also revealed several key aspects of the amino acid activation reaction.[4,13] The amino acid binding sites are rigid templates that bind the amino acid and ATP substrates in the optimal positions for transition state formation by in-line nucleophilic displacement. Within these binding sites, the specificity of recognition is determined by idiosyncratic contacts to the amino acid side-chain in the context of an active site shaped to accommodate the particular amino acid. The specificity of substrate recognition is further enhanced by an induced fit mechanism whereby ATP binding leads to ordering of the motif 2 loop. The importance of this substrate-induced ordering of the active site varies: for some class II AARSs such as AspRS and SerRS, the effect is mostly local whereas in others such as HisRS, LysRS2 and ProRS a more global change involving the relative positions of other domains is observed.[13]

A number of AARSs have also been observed to recognize noncognate amino acids,[14] sometimes with a frequency as high as 1 in 150 compared to the cognate amino acid (the recognition of valine

Table 1. The aminoacyl-tRNA synthetase class division and characteristic features of each class

	Class I	Class II
Members	Subclass Ia	Subclass IIa
	ArgRS, CysRS, IleRS, LeuRS, MetRS, ValRS	AlaRS, GlyRS, HisRS, ProRS, ThrRS, SerRS
	Subclass Ib	Subclass IIb
	GlnRS, GluRS, LysRS1	AsnRS, AspRS, LysRS2
	Subclass Ic	Subclass IIc
	TrpRS, TyrRS	PheRS
Sequence motifs	HIGH, KMSKS	Motifs 1, 2 and 3
Active site topology	Rossmann-fold	Anti-parallel β-fold
tRNA binding	Minor groove side of acceptor stem	Major groove side of acceptor stem
Aminoacylation site	2'-OH of tRNA	3'-OH of tRNA (except PheRS)

Subclasses based upon structural and sequence similarities are as previously defined.[9] Specific aminoacyl-tRNA synthetases are denoted by their three letter amino acid designation, e.g., AlaRS, alanyl-tRNA synthetase. Similarly, alanine tRNA or tRNA[Ala] denote uncharged tRNA specific for alanine; alanyl-tRNA or Ala-tRNA, tRNA aminoacylated with alanine.

versus isoleucine by IleRS). The inability to discriminate similar amino acids does not compromise the fidelity of translation as the respective AARSs have proofreading activities (for a review see chapter 3 by T. Hendrickson and P. Schimmel in this volume). For example, IleRS contains two distinct catalytic sites that present a double sieve during substrate selection,[15,16] ensuring that only about 1 in 3000 isoleucine codons are misread as valine during protein synthesis.[17]

tRNA Recognition by Aminoacyl-tRNA Synthetases

AARSs employ a wide range of strategies to ensure the accurate selection of their cognate tRNA substrates. Principal among these are the adoption of sequence-dependent alternative tRNA conformations, sequence-specific amino acid-tRNA interactions and the use of distinct domains to bind specific parts of the tRNA molecule. For example, when tRNA[Gln] binds GlnRS, two significant changes occur in its structure: the terminal U:A base pair of the acceptor stem is broken, facilitating the direction of the 3'-end of the tRNA into the active site, and the anticodon stem is extended by two nonWatson-Crick base pairs, resulting in the splaying out of the anticodon bases which enables their interaction with specific residues in GlnRS.[18] The need to destabilize the first base pair in the acceptor stem has also been deduced for the interactions of tRNA[Met] and tRNA[Ile] with MetRS and IleRS respectively, although comparison with other AARS:tRNA interactions suggests that this phenomenon is confined to certain class I AARSs while it remains an open possibility for some class II enzymes.[19] Structural re-arrangements in the tRNA anticodon have been more widely reported and provide a general means of optimizing recognition in both class I and class II AARSs (e.g., LysRS2[20]). One interesting exception is the case of SerRS which does not recognize the anticodon of tRNA[Ser], but instead recognizes the long extra arm found in all tRNA[Ser] isoacceptors by means of a C-terminal coiled-coil domain.[21]

The role of particular domains in tRNA recognition, as seen in SerRS, has also been observed in most other AARSs. Generally, the conserved catalytic domain contains regions proximal to the active site responsible for binding of the acceptor stem. In *Escherichia coli* GlnRS, the Rossmann fold contains an inserted acceptor-binding domain composed of three structural elements which specifically interact with a number of positions in the acceptor-stem of tRNA[Gln].[22,23] A similar role for a domain inserted in the Rossmann fold has been observed in TyrRS, where a short 39 amino acid peptide was found to specify species-specific recognition of the acceptor-stem.[24] Within this insertion, a single residue was found to strongly influence the rejection of noncognate tRNAs.[25] A distinct structural module appended to the "core" catalytic domain of AARSs generally achieves anticodon recognition. For example, comparison of structural and functional data for GlnRS and GluRS[26] has shown them to contain highly homologous N-terminal catalytic domains but unrelated C-terminal anticodon binding regions (a β-barrel structure in the case of GlnRS, an all α-helix cage in GluRS). The modularity of such anticodon recognition domains has been directly demonstrated in MetRS, the isolated C-terminal domain of which is functional in tRNA binding.[27] Considerable information is also available concerning anticodon recognition by class II AARSs, in particular for subclass II b enzymes (AsnRS, AspRS and LysRS2) which all employ an OB-fold in their N-terminal anticodon-binding domains.[28] Taken together, these various examples illustrate how the AARSs utilize a range of idiosyncratic structural modules to achieve specificity during recognition of their cognate tRNA substrates.

The sequence and structural elements that determine the identity of a specific tRNA during aminoacylation have been characterized in considerable detail both in vivo and in vitro, as described in detail elsewhere in this volume. The accuracy of cognate tRNA selection is not solely dependent on recognition, being enhanced in particular cases by the stabilization of the transition state for tRNA charging in cognate tRNA:AARS complexes,[29] tRNA sequence-specific effects on amino acid affinity[30] and the existence of anti-determinants in certain tRNAs which prevent interaction with noncognate AARSs.[31] The specificity of tRNA selection has also been found to be dependent on idiosyncratic N- and C-terminal extensions, which are primarily found in eukaryotic enzymes. For example, the C-terminal extension of hamster LysRS2 (which is absent from other LysRS2s) is critical for interactions with the acceptor stem of mammalian tRNA^{Lys},[32] and the N-terminal extension of yeast GlnRS has been implicated in determining the specificity of interactions with tRNA^{Gln}.[33]

Protein Complexes in Aminoacyl-tRNA Synthesis

In higher eukaryotes, a number of aminoacyl-tRNA synthetases have been identified as components of multi-enzyme complexes, a feature believed to be unique to metazoans.[34] To date, two such complexes have been characterized. One is a complex between ValRS and the translation elongation factor EF-1 delta subunit, where association between the two is mediated by the idiosyncratic N-terminal extension of the AARS.[35] This complex has also been shown to functionally interact with EF-1H.[36] The other, larger, complex is composed of ArgRS, AspRS, GlnRS, IleRS, LeuRS, LysRS2, MetRS, the multifunctional GluRS-ProRS protein and three non-AARS components designated p18, p38 and p43 (reviewed in ref. 37). Genetic, biochemical and stuctural approaches have provided a picture of the interactions between the various components of the larger complex and have implicated many of the idiosyncratic insertion and extension domains in complex formation (see for example 38 and references therein). The possible roles of the accessory proteins have also been clarified. The sequence of p18 suggests that it is responsible for the transient interaction of the complex with EF-1H (the "heavy" form of eukaryotic EF-1, composed of subunits alpha, beta, gamma, and delta);[39] p38 is important for the assembly of the complex;[40] p43, which is partly homologous to Arc1p, probably enhances tRNA binding by the complex.[41] Arc1p was first described in yeast where it was shown to bind GluRS and MetRS and improve tRNA binding by the latter synthetase.[42]

In addition to the proofreading mechanisms that target mischarged tRNAs as a means of quality control, the aminoacylation of tRNAs also may provide an important check-point for translational fidelity. In much the same way that tRNA maturation and transport are dependent on the recognition of specific conformations (e.g., refs. 43,44), a number of tRNAs are only recognized and exported from eukaryotic nuclei after aminoacylation.[45,46] Recent studies have shown that many of the AARSs present in mammalian nuclei are also part of a large multi-enzyme complex,[47] which might itself also facilitate tRNA export through association with elongation factor EF-1 alpha.[48] Accessory proteins functional in tRNA-binding have also been described in bacteria, suggesting that higher order aminoacylation complexes may exist outside the eukaryal kingdom.[49] While the exact functional significance of such higher order complexes is not known, it is possible that they may provide a means of optimizing aminoacyl-tRNA synthesis and its subsequent channeling to the ribosome.

Noncanonical Aminoacyl-tRNA Synthetases

Class I Lysyl-tRNA Synthetase

An aminoacyl-tRNA synthetase of particular substrate specificity will normally belong to either class I or class II regardless of its biological origin, reflecting the ancient evolution of this enzyme family.[50] The first exception to this rule was found among the lysyl-tRNA synthetases with the discovery of a class I enzyme in certain archaea,[51] all previously characterized members of this family belonging to class II. Subsequent work originating from analysis of whole genome sequences showed that the class I-type LysRS (LysRS1) is found in the majority of archaea and a scattering of bacteria, to the exclusion of the more common class II-type protein in all but one case (LysRS2, see refs. 52,53). To date, all known eukaryotic LysRSs (both cytoplasmic and organellar) are of the class II-type. Despite their lack of sequence similarity, LysRS1 and LysRS2 are able to recognize the same amino acid and tRNA substrates both in vitro and in vivo, providing an example of functional convergence by divergent enzymes.[5] Comparison of tRNA^{Lys} recognition by LysRS1 and LysRS2 proteins indicated that while they approach the acceptor stem from different sides, both recognize the same identity elements in tRNA^{Lys} namely the discriminator base, the acceptor stem and the anticodon. This, together with phylogenetic analyses of the class I LysRS family, suggests that tRNA^{Lys} may predate at least one of the LysRS families in the evolution of aminoacyl-tRNA synthesis.[5,54]

ProlylCysteinyl-tRNA Synthetase

The genome sequences of the thermophilic archaea *Methanococcus jannaschii* and *Methanobacterium thermoautotrophicum* do not contain any identifiable genes encoding CysRS proteins, in contrast to the genomes of more than 40 other organisms from all the three kingdoms, which encode canonical class I CysRS enzymes. This apparent discrepancy was resolved by biochemical and genetic studies that showed the enzyme responsible for the formation of Cys-tRNA^{Cys} to be a class II enzyme, prolyl-tRNA synthetase (ProRS).[55,56] To date, this is the only known example of a single AARS that can specify two different amino acids in translation. While no organisms outside of the archaea have yet been found to lack a gene encoding a canonical CysRS, the dual function ProRS is not confined to archaea. Molecular phylogenies of ProRS amino acid sequences suggested that the deep-rooted eukaryon *Giardia lamblia* might also contain a ProRS with CysRS activity. This possibility was subsequently confirmed experimentally, raising the possibility that ProCysRS enzymes may be present in other organisms.[57]

The basis of the dual substrate specificity of ProCysRS is related to differences in the mechanisms by which the two aminoacyl-tRNA products are synthesized. Although the structural basis for this activity is currently unclear, mutagenesis of active site residues suggests that the binding sites for cysteine and proline overlap.[58]

Nondiscriminating Aspartyl-tRNA and Glutamyl-tRNA Synthetases

Recent studies reveal that the most common divergence from canonical aminoacyl-tRNA synthesis is exhibited by the tRNA-dependent amino acid transformation pathways (eqn. 1-4). These two-step, indirect routes to glutaminyl-tRNA^{Gln} (Gln-tRNA^{Gln}) and asparaginyl-tRNA^{Asn} (Asn-tRNA^{Asn}) generally exist when glutaminyl-tRNA synthetase (GlnRS) or asparaginyl-tRNA synthetase (AsnRS), respectively, is absent. During Gln-tRNA^{Gln} synthesis (eqn. 1-2), tRNA^{Gln} is

$$\text{Glu} + \text{tRNA}^{Gln} + \text{ATP} \leftrightarrows \text{Glu-tRNA}^{Gln} + \text{AMP} + \text{PP}_i \qquad (1)$$
$$\text{Glu-tRNA}^{Gln} + \text{Gln} + \text{ATP} \leftrightarrows \text{Gln-tRNA}^{Gln} + \text{Glu} + \text{ADP} + \text{P}_i \qquad (2)$$

$$\text{Asp} + \text{tRNA}^{Asn} + \text{ATP} \leftrightarrows \text{Asp-tRNA}^{Asn} + \text{AMP} + \text{PP}_i \qquad (3)$$
$$\text{Asp-tRNA}^{Asn} + \text{Gln} + \text{ATP} \leftrightarrows \text{Asn-tRNA}^{Asn} + \text{Glu} + \text{ADP} + \text{P}_i \qquad (4)$$

first misaminoacylated with glutamate by a nondiscriminating (relaxed tRNA specificity) glutamyl-tRNA synthetase (GluRS), which, in addition to generating Glu-tRNA^{Glu}, can also synthesize Glu-tRNA^{Gln}. The resulting mischarged tRNA is then specifically recognized by glutamyl-tRNA^{Gln} amidotransferase (GluAdT[59]) and converted into Gln-tRNA^{Gln}. Similarly, Asn-tRNA^{Asn} is formed (eqn. 3-4) via a nondiscriminating aspartyl-tRNA synthetase (AspRS) and an aspartyl-tRNA^{Asn} amidotransferase (AspAdT[60]). For both GluAdT and AspAdT activity, bacteria and archaea use a single, heterotrimeric enzyme encoded by *gatCAB*.[61,62] In addition, archaea possess a unique GluAdT.[62] Thus, the amide aminoacyl-tRNA pathways of the three kingdoms (Bacteria, Archaea and Eukarya) use different enzymes and mechanisms. Similarly, kingdom-specific tRNA-dependent amino acid transformation pathways are also responsible for synthesizing selenocysteinyl-tRNA (from Ser-tRNA^{Sec}, ref. 63) and formylmethionyl-tRNA (from Met-tRNA$_i^{Met}$, ref. 64) (for a review, see Chapter 4 by S. Blanquet et al in this volume).

Characterization of the indirect synthetic pathways to Gln-tRNA^{Gln} and Asn-tRNA^{Asn} indicates that they evolved as distinct systems in the three kingdoms. All known examples of eukaryal cytoplasmic protein synthesis use exclusively GlnRS and AsnRS. In contrast, bacteria and eukaryal organelles use predominantly GluAdT and AsnRS (reviewed in ref. 62). Most of the exceptions to this rule seem to result from horizontal gene transfer (i.e., transfer of genes between different organisms). For example, GlnRS has been described only in some proteobacteria (reviewed in ref. 65), in the *Thermus/Deinococcus* group[61,66] and in the mitochondria of trypanosomatids.[67] The current lack of sequence data on this last group restricts speculation on the origin of its GlnRS. Phylogenetic analyses of bacterial GlnRS sequences consistently suggest a recent gene transfer from the eukarya. This, in turn, is suggestive of loss of the indirect pathway in some of these organisms, or for recruitment of GatCAB to Asn-tRNA^{Asn} formation as seen in *Deinococcus radiodurans* and *Thermus thermophilus*.[61,66] Most strikingly, the archaea use the indirect transamidation pathway almost exclusively,[68] as GlnRS activity and the corresponding gene have never been found in this kingdom. Only a few archaea have AsnRS genes (phylogenetic analysis again suggesting horizontal transfer[50]), whereas the vast majority use the AspAdT pathway characterized in *Haloferax volcanii*[62] and *M. thermoautotrophicum*.[62] While some archaea could potentially use GatCAB for both GluAdT and AspAdT function, each archaeal genome also encodes a second GluAdT enzyme. This heterodimeric enzyme, encoded by the *gatD* and *gatE* genes, is strictly archaeal and not found elsewhere. The purified *M. thermoautotrophicum* GatDE enzyme has GluAdT activity in vitro and is unable to form Asn-tRNA. The respective roles of the two archaeal GluAdT enzymes have yet to be determined. However, the existence of GatDE in every archaeal genome, even in the presence of GatCAB, suggests a critical function.

Evolution of Aminoacyl-tRNA Synthetases

The only AARSs for which clear evolutionary histories have been established are GlnRS and AsnRS. Phylogenetic and structural considerations leave little doubt that GlnRS and AsnRS were the last AARSs to emerge, resulting from duplication and diversification of ancestral GluRS and AspRS respectively (reviewed in ref. 50). Functional and structural considerations suggest a strikingly high degree of similarity between GluRS and LysRS1, although it is currently unclear whether this is also indicative of a comparatively recent origin for one of these AARSs.

While a consensus has arisen that aminoacyl-tRNA synthesis per se may have originated as an RNA-catalyzed process (e.g., refs. 69,70; see also Chapter 7 by Knight et al in this volume), the evolution of existing AARSs remains more contentious. Phylogenetic comparisons[71] and structural sub-groupings[6] indicate clear relationships between the AARSs, but, with few exceptions, do not reveal how they may have evolved from a presumably smaller number of ancestral proteins although recent studies are starting to clarify some potential mechanisms.[72,73] Until recently, the final transition in this scheme, from synthetases of broad to narrow substrate specificity, while suggested by AARS phylogenies, was not supported by any experimental findings. However, recent data suggest the validity of this proposed transition. For example, studies in archaea have shown that, at least the canonical cysteinyl-tRNA synthetase is not essential for viability,[74] suggesting that particular AARS enzymes may synthesize more than one aminoacyl-tRNA.

References

1. Crick F. Central dogma of molecular biology. Nature 1970; 227:561-563.
2. Ibba M, Söll D. Aminoacyl-tRNA synthesis. Annu Rev Biochem 2000; 69:617-650.
3. Fujiwara S, Lee SG, Haruki M et al. Unusual enzyme characteristics of aspartyl-tRNA synthetase from hyperthermophilic archaeon *Pyrococcus* sp. KOD1. FEBS Lett 1996; 394:66-70.
4. Arnez JG, Moras D. Structural and functional considerations of the aminoacylation reaction. Trends Biochem Sci 1997; 22:211-216.
5. Ibba M, Losey HC, Kawarabayasi Y et al. Substrate recognition by class I lysyl-tRNA synthetases: a molecular basis for gene displacement. Proc Natl Acad Sci USA 1999; 96:418-423.

6. Ribas de Pouplana L, Schimmel P. Two classes of tRNA synthetases suggested by sterically compatible dockings on tRNA acceptor stem. Cell 2001; 104:191-193.

7. Beuning PJ, Yang F, Schimmel P et al. Specific atomic groups and RNA helix geometry in acceptor stem recognition by a tRNA synthetase. Proc Natl Acad Sci USA 1997; 94:10150-10154.

8. Sankaranarayanan R, Dock-Bregeon AC, Romby P et al. The structure of threonyl-tRNA synthetase-tRNA^{Thr} complex enlightens its repressor activity and reveals an essential zinc ion in the active site. Cell 1999; 97:371-381.

9. First EA. Catalysis of tRNA aminoacylation by class I and class II aminoacyl-tRNA synthetases. In: Sinnot ML, ed. Comprehensive Biological Catalysis. London: Academic Press, 1998:573-607.

10. First EA, Fersht AR. Analysis of the role of the KMSKS loop in the catalytic mechanism of the tyrosyl-tRNA synthetase using multimutant cycles. Biochemistry 1995; 34:5030-5043.

11. Schmitt E, Panvert M, Blanquet S et al. Transition state stabilization by the 'high' motif of class I aminoacyl- tRNA synthetases: the case of *Escherichia coli* methionyl-tRNA synthetase. Nucleic Acids Res 1995; 23:4793-4798.

12. Perona JJ, Rould MA, Steitz TA. Structural basis for transfer RNA aminoacylation by *Escherichia coli* glutaminyl-tRNA synthetase. Biochemistry 1993; 32:8758-8771.

13. Cusack S. Aminoacyl-tRNA synthetases. Curr Opin Struct Biol 1997; 7:881-889.

14. Jakubowski H, Goldman E. Editing of errors in selection of amino acids for protein synthesis. Microbiol Rev 1992; 56:412-429.

15. Nureki O, Vassylyev DG, Tateno M et al. Enzyme structure with two catalytic sites for double-sieve selection of substrate. Science 1998; 280:578-582.

16. Silvian LF, Wang J, Steitz TA. Insights into editing from an ile-tRNA synthetase structure with tRNA^{Ile} and mupirocin. Science 1999; 285:1074-1077.

17. Loftfield RB, Vanderjagt D. The frequency of errors in protein biosynthesis. Biochem J 1972; 128:1353-1356.

18. Rould MA, Perona JJ, Steitz TA. Structural basis of anticodon loop recognition by glutaminyl-tRNA synthetase. Nature 1991; 352:213-218.

19. Alexander RW, Nordin BE, Schimmel P. Activation of microhelix charging by localized helix destabilization. Proc Natl Acad Sci USA 1998; 95:12214-12219.

20. Cusack S, Yaremchuk A, Tukalo M. The crystal structures of *T. thermophilus* lysyl-tRNA synthetase complexed with *E. coli* tRNA^{Lys} and a *T. thermophilus* tRNA^{Lys} transcript: anticodon recognition and conformational changes upon binding of a lysyl-adenylate analogue. EMBO J 1996; 15:6321-6334.

21. Biou V, Yaremchuk A, Tukalo M et al. The 2.9 Å crystal structure of *T. thermophilus* seryl-tRNA synthetase complexed with tRNA^{Ser}. Science 1994; 263:1404-1410.

22. Schwob E, Söll D. Selection of a 'minimal' glutaminyl-tRNA synthetase and the evolution of class I synthetases. EMBO J 1993; 12:5201-5208.

23. Arnez JG, Steitz TA. Crystal structures of three misacylating mutants of *Escherichia coli* glutaminyl-tRNA synthetase complexed with tRNA^{Gln} and ATP. Biochemistry 1996; 35:14725-14733.

24. Wakasugi K, Quinn CL, Tao N et al. Genetic code in evolution: switching species-specific aminoacylation with a peptide transplant. EMBO J 1998; 17:297-305.

25. Bedouelle H, Nageotte R. Macromolecular recognition through electrostatic repulsion. EMBO J 1995; 14:2945-2950.

26. Nureki O, Vassylyev DG, Katayanagi K et al. Architectures of class-defining and specific domains of glutamyl-tRNA synthetase. Science 1995; 267:1958-1965.

27. Gale AJ, Schimmel P. Isolated RNA binding domain of a class I tRNA synthetase. Biochemistry 1995; 34:8896-8903.

28. Berthet-Colominas C, Seignovert L, Hartlein M et al. The crystal structure of asparaginyl-tRNA synthetase from *Thermus thermophilus* and its complexes with ATP and asparaginyl-adenylate: the mechanism of discrimination between asparagine and aspartic acid. EMBO J 1998; 17:2947-2960.

29. Ibba M, Sever S, Praetorius-Ibba M et al. Transfer RNA identity contributes to transition state stabilization during aminoacyl-tRNA synthesis. Nucleic Acids Res 1999; 27:3631-3637.

30. Ibba M, Hong KW, Sherman JM et al. Interactions between tRNA identity nucleotides and their recognition sites in glutaminyl-tRNA synthetase determine the cognate amino acid affinity of the enzyme. Proc Natl Acad Sci USA 1996; 93:6953-6958.

31. Giegé R, Sissler M, Florentz C. Universal rules and idiosyncratic features in tRNA identity. Nucleic Acids Res 1998; 26:5017-5035.

32. Agou F, Quevillon S, Kerjan P et al. Functional replacement of hamster lysyl-tRNA synthetase by the yeast enzyme requires cognate amino acid sequences for proper tRNA recognition. Biochemistry 1996; 35:15322-15331.

33. Whelihan EF, Schimmel P. Rescuing an essential enzyme-RNA complex with a non-essential appended domain. EMBO J 1997; 16:2968-2974.

34. Kerjan P, Cerini C, Semeriva M et al. The multienzyme complex containing nine aminoacyl-tRNA synthetases is ubiquitous from *Drosophila* to mammals. Biochim Biophys Acta 1994; 1199:293-297.

35. Bec G, Kerjan P, Waller JP. Reconstitution in vitro of the valyl-tRNA synthetase-elongation factor (EF) 1 beta gamma delta complex. Essential roles of the NH2-terminal extension of valyl-tRNA synthetase and of the EF-1 delta subunit in complex formation. J Biol Chem 1994; 269:2086-2092.

36. Negrutskii BS, Shalak VF, Kerjan P et al. Functional interaction of mammalian valyl-tRNA synthetase with elongation factor EF-1alpha in the complex with EF-1H. J Biol Chem 1999; 274:4545-4550.

37. Yang DC. Mammalian aminoacyl-tRNA synthetases. Curr Top Cell Regul 1996; 34:101-136.

38. Robinson JC, Kerjan P, Mirande M. Macromolecular assemblage of aminoacyl-tRNA synthetases: quantitative analysis of protein-protein interactions and mechanism of complex assembly. J Mol Biol 2000; 304:983-994.

39. Quevillon S, Mirande M. The p18 component of the multisynthetase complex shares a protein motif with the beta and gamma subunits of eukaryotic elongation factor 1. FEBS Lett 1996; 395:63-67.
40. Quevillon S, Robinson JC, Berthonneau E et al. Macromolecular assemblage of aminoacyl-tRNA synthetases: identification of protein-protein interactions and characterization of a core protein. J Mol Biol 1999; 285:183-195.
41. Quevillon S, Agou F, Robinson JC et al. The p43 component of the mammalian multi-synthetase complex is likely to be the precursor of the endothelial monocyte-activating polypeptide II cytokine. J Biol Chem 1997; 272:32573-32579.
42. Simos G, Sauer A, Fasiolo F et al. A conserved domain within Arc1p delivers tRNA to aminoacyl-tRNA synthetases. Mol Cell 1998; 1:235-242.
43. McClain WH, Seidman JG. Genetic perturbations that reveal tertiary conformation of tRNA precursor molecules. Nature 1975; 257:106-110.
44. Lipowsky G, Bischoff FR, Izaurralde E et al. Coordination of tRNA nuclear export with processing of tRNA. RNA 1999; 5:539-549.
45. Lund E, Dahlberg JE. Proofreading and aminoacylation of tRNAs before export from the nucleus. Science 1998; 282:2082-2085.
46. Sarkar S, Azad AK, Hopper AK. Nuclear tRNA aminoacylation and its role in nuclear export of endogenous tRNAs in *Saccharomyces cerevisiae*. Proc Natl Acad Sci USA 1999; 96:14366-14371.
47. Nathanson L, Deutscher MP. Active aminoacyl-tRNA synthetases are present in nuclei as a high molecular weight multienzyme complex. J Biol Chem 2000; 275:31559-31562.
48. Grosshans H, Hurt E, Simos G. An aminoacylation-dependent nuclear tRNA export pathway in yeast. Genes Dev 2000; 14:830-840.
49. Swairjo MA, Morales AJ, Wang CC et al. Crystal structure of trbp111: a structure-specific tRNA-binding protein. EMBO J 2000; 19:6287-6298.
50. Woese CR, Olsen GJ, Ibba M et al. Aminoacyl-tRNA synthetases, the genetic code, and the evolutionary process. Microbiol Mol Biol Rev 2000; 64:202-236.
51. Ibba M, Morgan S, Curnow AW et al. A euryarchaeal lysyl-tRNA synthetase: resemblance to class I synthetases. Science 1997; 278:1119-1122.
52. Ibba M, Bono JL, Rosa PA et al. Archaeal-type lysyl-tRNA synthetase in the Lyme disease spirochete *Borrelia burgdorferi*. Proc Natl Acad Sci USA 1997; 94:14383-14388.
53. Söll D, Becker HD, Plateau P et al. Context-dependent anticodon recognition by class I lysyl-tRNA synthetases. Proc Natl Acad Sci USA 2000; 97:14224-14228.
54. Ribas de Pouplana L, Turner RJ, Steer BA et al. Genetic code origins: tRNAs older than their synthetases? Proc Natl Acad Sci USA 1998; 95:11295-11300.
55. Stathopoulos C, Li T, Longman R et al. One polypeptide with two aminoacyl-tRNA synthetase activities. Science 2000; 287:479-482.
56. Lipman RS, Sowers KR, Hou YM. Synthesis of cysteinyl-tRNA^{Cys} by a genome that lacks the normal cysteine-tRNA synthetase. Biochemistry 2000; 39:7792-7798.
57. Bunjun S, Stathopoulos C, Graham D et al. A dual-specificity aminoacyl-tRNA synthetase in the deep-rooted eukaryote *Giardia lamblia*. Proc Natl Acad Sci USA 2000; 97:12997-13002.
58. Stathopoulos C, Jacquin-Becker C, Becker HD et al. *Methanococcus jannaschii* prolyl-cysteinyl-tRNA synthetase possesses overlapping amino acid binding sites. Biochemistry 2001; 40:46-52.
59. Curnow AW, Hong KW, Yuan R et al. tRNA-dependent amino acid transformations. Nucleic Acids Symp Ser 1997; 36:2-4.
60. Curnow AW, Ibba M, Söll D. tRNA-dependent asparagine formation. Nature 1996; 382:589-590.
61. Curnow AW, Tumbula DL, Pelaschier JT et al. Glutamyl-tRNA^{Gln} amidotransferase in *Deinococcus radiodurans* may be confined to asparagine biosynthesis. Proc Natl Acad Sci USA 1998; 95:12838-12843.
62. Tumbula DL, Becker HD, Chang WZ et al. Domain-specific recruitment of amide amino acids for protein synthesis. Nature 2000; 407:106-110.
63. Commans S, Böck A. Selenocysteine inserting tRNAs: an overview. FEMS Microbiol Rev 1999; 23:335-351.
64. RajBhandary UL. Initiator transfer RNAs. J Bacteriol 1994; 176:547-552.
65. Brown JR, Doolittle WF. Gene descent, duplication, and horizontal transfer in the evolution of glutamyl- and glutaminyl-tRNA synthetases. J Mol Evol 1999; 49:485-495.
66. Becker HD, Kern D. *Thermus thermophilus*: a link in evolution of the tRNA-dependent amino acid amidation pathways. Proc Natl Acad Sci USA 1998; 95:12832-12837.
67. Nabholz CE, Hauser R, Schneider A. *Leishmania tarentolae* contains distinct cytosolic and mitochondrial glutaminyl-tRNA synthetase activities. Proc Natl Acad Sci USA 1997; 94:7903-7908.
68. Tumbula D, Vothknecht UC, Kim HS et al. Archaeal aminoacyl-tRNA synthesis: diversity replaces dogma. Genetics 1999; 152:1269-1276.
69. Illangasekare M, Yarus M. A tiny RNA that catalyzes both aminoacyl-RNA and peptidyl-RNA synthesis. RNA 1999; 5:1482-1489.
70. Saito H, Kourouklis D, Suga H. An in vitro evolved precursor tRNA with aminoacylation activity. EMBO J 2001; 20:1797-1806.
71. Nagel GM, Doolittle RF. Phylogenetic analysis of the aminoacyl-tRNA synthetases. J Mol Evol 1995; 40:487-498.
72. Ribas de Pouplana L, Schimmel P. Operational RNA code for amino acids in relation to genetic code in evolution. J Biol Chem 2001; 276:6881-6884.
73. Ribas de Pouplana L, Schimmel P. Aminoacyl-tRNA synthetases: potential markers of genetic code development. Trends Biochem Sci 2001; 26:591-596.
74. Stathopoulos C, Kim W, Li T et al. Cysteinyl-tRNA synthetase is not essential for viability of the archaeon *Methanococcus maripaludis*. Proc Natl Acad Sci USA 2001; 98:14292-14297.

CHAPTER 3

Transfer RNA-Dependent Amino Acid Discrimination by Aminoacyl-tRNA Synthetases

Tamara L. Hendrickson and Paul Schimmel

Abstract

Aminoacyl-tRNA synthetases (AARSs) form a direct connection between the trinucleotide codons of the genetic code and their corresponding amino acids. Each AARS catalyzes the biosynthesis of a specific, cognate set of AA-tRNA^{AA} isoacceptors. In some cases, the cognate amino acid is structurally similar to one or more encoded amino acids and/or other available metabolites, leading to misactivation and misacylation of non-cognate amino acids. To remedy these errors, many AARSs have tRNA-dependent hydrolytic editing activities against misactivated, non-cognate amino acids. In this manner, the accuracy of translation is maintained at a level that is higher than could be achieved by simple, one-step, side chain recognition. A resurgence in interest in tRNA-dependent editing mechanisms has occurred over the past decade. Proofreading mechanisms have now been identified in as many as nine different AARSs. Here, the role of tRNA-dependent editing in guaranteeing the accuracy of tRNA aminoacylation is summarized.

Introduction

Life on Earth thrives under remarkably diverse conditions.[1] Organisms have been identified that grow in environments of extreme salinity, pH (acid or base) and temperatures ranging from −17 °C[2] to 113°C [3,4] Despite this heterogeneity of viable climates, many metabolic pathways remain highly conserved. Most notably, enzymes that maintain and translate the genetic code are essential and conserved, offering a ready tool for delineating phylogenetic relationships and arguing that much of the genetic code was all but fixed in evolution prior to the emergence of modern-day life.

Translation is a dynamic process: information stored in DNA dictates sequences of proteins (and RNA); some of these proteins maintain and duplicate the code and are consequently responsible for accurate replication of their own genes. For this reason, the exquisite precision by which the code is converted into functional proteins is the most important aspect of protein translation. Even small errors in protein biosynthesis could introduce errors in DNA replication, leading to greater errors in protein biosynthesis, and a potentially catastrophic cycle.[5] Thus, to guarantee accuracy, many error-correcting mechanisms have evolved that uphold the fidelity of the various steps of translation. These proofreading steps are often highly conserved across evolution,[6-9] demonstrating their importance in the maintenance of the genetic code from generation to generation.

A series of well-integrated steps (Fig. 1) come together during protein translation. A crucial step is accomplished by the aminoacyl-tRNAs (aa-tRNAs), as these intermediates relay the information stored in mRNA codons directly into sequences of nascent polypeptide chains.[10] This decoding of genetic information takes place within the complex structure of the ribosome (Fig. 1) , where the anticodon of each AA-tRNA is aligned with its appropriate codon(s), positioning the aminoacylated end of the tRNA in the ribosome's peptidyl transferase center.[11] Importantly, by its very nature, the ribosome accepts and transfers all encoded amino acids (and many unnatural ones as well;[12] see also Chapters 5 and 6 by J.M. Bacher and A.D. Ellington, and by T.J. Magliery et al, respectively); its role is to decode genetic information, not proofread tRNA aminoacylation. For this reason,

Translation Mechanisms, edited by Jacques Lapointe and Léa Brakier-Gingras.
©2003 Eurekah.com and Kluwer Academic / Plenum Publishers.

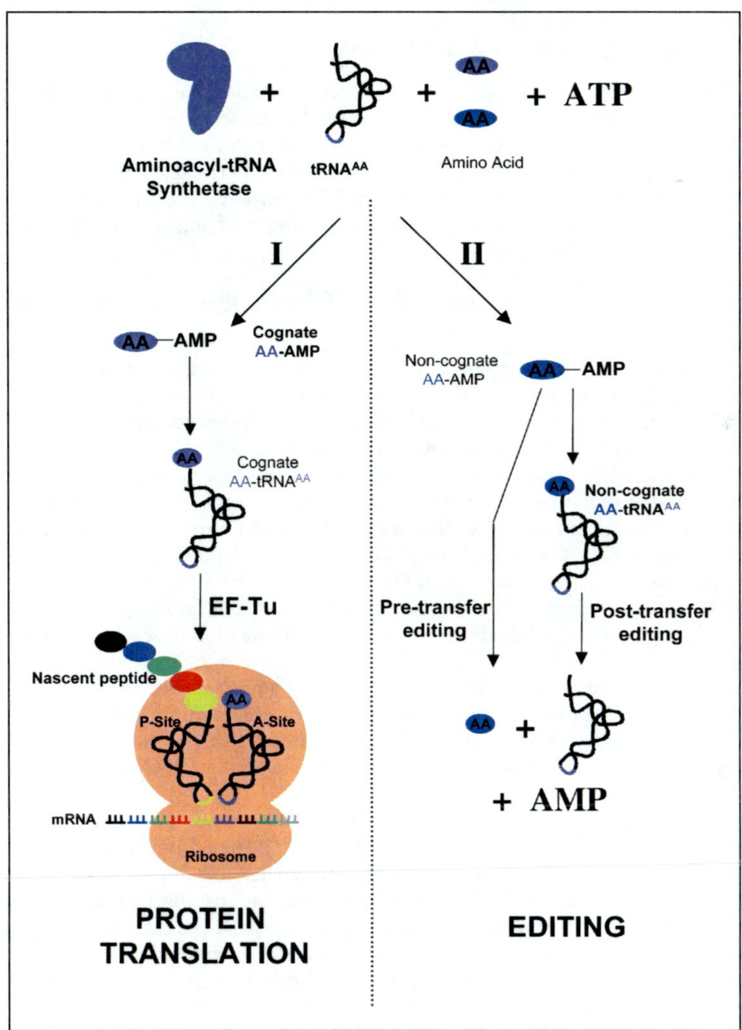

Figure 1. Accurate tRNA aminoacylation in protein biosynthesis. Pathway I: Each aminoacyl-tRNA synthetase transfers its cognate amino acid to the 3'-end of its cognate tRNA (AA-tRNA^{AA}), via the formation of an aminoacyl adenylate (AA-AMP). This AA-tRNA^{AA} is loaded onto the ribosome via the action of EF-Tu. Pathway II: Some aminoacyl-tRNA synthetases misactivate non-cognate amino acids to generate aminoacyl adenylates and even misacylated aminoacyl-tRNAs. The aminoacyl-tRNA synthetase recognizes these errors and corrects them, either via pre-transfer hydrolysis of the aminoacyl adenylate or by post-transfer hydrolysis of the aminoacyl-tRNA. In this way, the misacylated tRNA is eliminated before the errors can be transferred to the ribosome.

accuracy at the stage of AA-tRNA biosynthesis is the primary step where the relationship between cognate amino acid and cognate tRNA can be verified and corrected.

In most cases, AA-tRNAs are synthesized by aminoacyl-tRNA synthetases (AARSs, Fig. 1), with one AARS for each encoded amino acid.[10] Typically, synthesis occurs in two steps. First, the cognate amino acid is condensed with one molecule of ATP to form an aminoacyl adenylate (aa-AMP, Eq. 1) and inorganic pyrophosphate. Next, the amino acid is transferred to either the 2' or 3' hydroxyl on the 3'-end of the corresponding cognate tRNA, to generate the product AA-tRNA (Eq. 2).

$$AA + ATP + AARS \rightarrow AA\text{-}AMP + PPi \tag{1}$$
$$AA\text{-}AMP + tRNA^{AA} + AARS \rightarrow AA\text{-}tRNA^{AA} + AMP \tag{2}$$

Some AARSs cannot rigorously select only their cognate amino acid via a single recognition event (Eq. 1); non-cognate amino acids are occasionally misactivated and even inadvertently misacylated onto the wrong tRNA. Thus, to ensure accuracy in AA-tRNA biosynthesis and in protein translation, some AARSs have evolved corrective editing mechanisms that clear these errors, before they can be propagated into nascent proteins. The purpose of this review is to detail these tRNA-dependent proofreading reactions. Other proofreading mechanisms that contribute to accurate AA-tRNA biosynthesis also occur (see below).

Aminoacyl-tRNA Synthetases and the Challenge of Accurate Amino Acid Discrimination

The most conserved feature of protein translation is the invariant set of twenty amino acids that is used to construct proteins (Fig. 2). In all three domains of life, these amino acids are the basic building blocks. Remarkably, as a series, the encoded amino acids have many structural and chemical redundancies (e.g., isoleucine versus valine, threonine versus serine, aspartate versus glutamate), and yet they offer sufficient variety to generate the catalytic diversity of enzymes. This diversity is enhanced beyond what might be predicted because sometimes near-identical amino acids play dramatically different biological roles once they are incorporated into proteins. For example, serine proteases rely on an active site serine side chain hydroxyl to initiate hydrolysis of a peptide bond; threonine proteases are much less common and less reactive, presumably because of steric hindrance due to the β-branch of threonine.[13] Similarly, in eukaryotic cells, asparagine residues are often co-translationally glycosylated, while the side chain of glutamine (that differs by a single methylene group) is not glycosylated.[14]

Thus, despite the apparent redundancy of the encoded amino acids, the identity of a specific residue is often critical to a protein's cellular role, such that even a minor perturbation in protein translation would be intolerable. (Error rates in protein translation are estimated to be less than one in 3000[15] and perhaps as low as 1 in 38,000.[16]) The mechanisms by which each AARS guarantees accuracy are critical for the fidelity of protein translation.

In most cases, there is one AARS for each encoded amino acid. The twenty AARSs can be subdivided into two different classes (class I and class II),[17-20] based on conserved sequence and structural motifs. Each enzyme efficiently attaches only its amino acid to its cognate tRNA(s) (Eq. 1 and 2, above) and a single catalytic or "synthetic" active site catalyzes both of these reactions. The specificity of each AARS is determined by the overall architecture of the enzyme's active site. (There are only two known exceptions to this rule of one amino acid to one AARS: In organisms such as gram-positive bacteria and archaea, glutamyl- and aspartyl-tRNA synthetases aminoacylate tRNA^{Gln} and tRNA^{Asn}, in addition to their cognate tRNAs.[21] And, ProRS from *Methanococcus jannaschii* is reported to generate Cys-tRNA^{Cys} as well as Pro-tRNA^{Pro}).[22]

Precise substrate specificity is fundamental to the accuracy observed in translation because steps subsequent to aminoacylation are less specific for the amino acid. For some AARSs, the process of amino acid substrate selection is straightforward (e.g., tryptophanyl-[23] or arginyl-tRNA synthetase[24]), because the cognate amino acids are structurally dissimilar to other metabolites. A single active site activates the cognate amino acid while concomitantly rejecting all others. In some instances substrate selection by a single site is challenged by competition from other substrates that are closely related in structure (Table 1 gives some examples). Pauling noted that isoleucyl-tRNA synthetase (IleRS) must select isoleucine while rejecting the closely related non-cognate amino acid valine.[25] The isopropyl side chain of valine differs by one methylene from the isobutyl side chain of isoleucine. Pauling hypothesized that an enzyme active site designed to accommodate isoleucine would be unable to exclude valine (Fig. 3, top). He estimated that both amino acids would bind to the active site of IleRS with binding energies differing by only 1-2 kcal/mole. Thus, IleRS should misactivate valine and transfer it to tRNA^{Ile} with an efficiency that would lead to valine mistakenly replacing isoleucine in proteins at levels approaching 20% (or an error rate of 1 in 5).[25] This prediction, however, is in direct contradiction with the high levels of fidelity routinely observed in translated proteins.

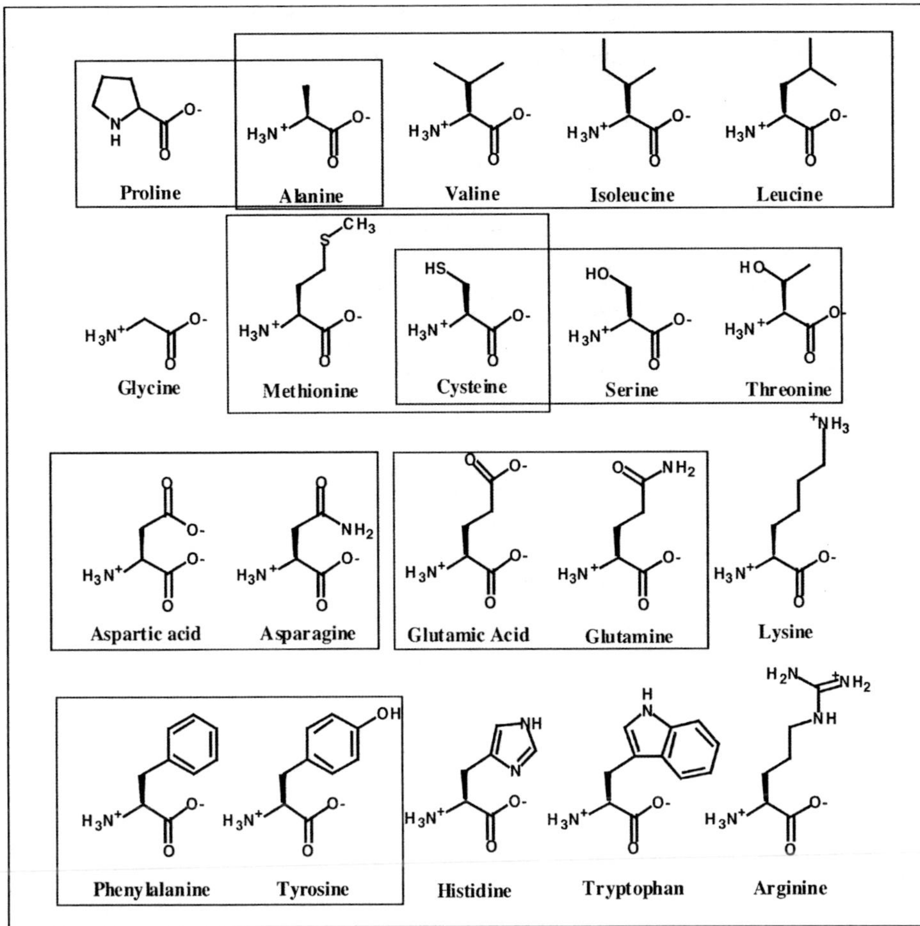

Figure 2. The structures of the twenty encoded amino acids. The boxes represent sets of amino acids that are structurally similar and therefore pose a challenge in molecular recognition for their respective aminoacyl-tRNA synthetases. In addition, challenges are posed by amino acids that are smaller than the one cognate to a particular synthetase. Smaller amino acids can fit into the binding pocket without steric constraints.

Consistent with Pauling's original prediction, IleRS misactivates Val-AMP.[26] Remarkably, however, misacylated tRNA[Ile] is never isolated. An explanation for this apparent contradiction was first offered by Baldwin and Berg in 1966[26] and later expanded by two different groups, Schimmel and coworkers[27,28] and Fersht and coworkers.[29-31] While Baldwin and Berg isolated a non-covalent complex of IleRS and Val-AMP, addition of tRNA[Ile] to this complex caused hydrolysis of Val-AMP to valine and AMP.[26]

At the time, it was not clear whether valine was transiently transferred to tRNA[Ile], followed by hydrolysis of Val-tRNA[Ile]. Using a special procedure, Val-tRNA[Ile] was synthesized and then challenged with IleRS. The mischarged tRNA[Ile] was rapidly hydrolyzed by the enzyme. At the same time, Schreier and Schimmel[28] showed that this deacylation activity was shared by many tRNA synthetases. Thus, deacylation of mischarged tRNA appeared to be a major mechanism by which mistakes are cleared. Indeed, Yarus and coworkers showed that PheRS rapidly deacylates Ile-tRNA[Phe].[32]

Table 1. *Examples of metabolites that could compete for the active site of a given AARS*

AARS	Potential Non-Cognate Competitors
Alanyl-	Glycine, serine, cysteine, α-aminobutyrate
Argininyl-	Lysine
Asparaginyl-	Aspartic acid, glutamine
Aspartyl-	Asparagine, glutamic acid
Cysteinyl-	Serine, alanine, α-aminobutyrate
Glutamyl-	Glutamine, aspartic acid
Glutaminyl-	Glutamic acid, asparagine
Glycyl-	None
Histidyl-	None
Isoleucyl-	Valine, leucine, alanine, threonine, O-methylthreonine
Leucyl-	Valine, alanine, norvaline, methionine, isoleucine
Lysyl-	Methionine, ornithine
Methionyl-	Homocysteine
Phenylalanyl-	Tyrosine
Prolyl-	Alanine
Seryl-	Cysteine, α-aminobutyric acid
Threonyl-	Valine, serine, cysteine, α-aminobutyric acid
Tryptophanyl-	None
Tyrosyl-	Phenylalanine
Valyl-	Threonine, serine, alanine, α-aminobutyric acid

Subsequently, Fersht and coworkers used rapid-quench stopped-flow kinetic methods to study editing of non-cognate amino acids. These investigations suggested that, in addition to the hydrolysis of Val-tRNA[Ile], the misactivated aminoacyl adenylate itself was hydrolyzed before transfer of the valyl moiety to the tRNA. Thus, editing was viewed as broken down to two reactions: pre-transfer and post-transfer (Fig. 3, middle).

Two questions remained. How can a single enzyme catalyze four reactions (amino acid activation, aminoacylation, pre-transfer and post-transfer editing)? And, how do the two editing reactions effectively discriminate between isoleucine and valine, so that Ile-AMP and Ile-tRNA[Ile] remain unskathed? Fersht suggested that this apparent complexity could be explained through the use of two distinct active sites: one for aminoacylation, and one for editing.[30] The synthetic active site (Fig. 3, top), or "coarse sieve", would generate Ile-AMP, Val-AMP, Ile-tRNA[Ile] and Val-tRNA[Ile]. This site would sterically exclude amino acids that are larger or significantly dissimilar to isoleucine, but smaller non-cognate amino acids would bind. The second active site (Fig. 3, bottom), or "fine sieve" would be structurally distinct from the synthetic active site.[30] The amino acid binding pocket would be smaller so that it would recognize the isopropyl side chain of valine but exclude the larger isobutyl group of isoleucine. Thus, the second site would hydrolyze both Val-AMP and Val-tRNA[Ile], while leaving Ile-AMP and Ile-tRNA[Ile] intact.[33] This proposal has been coined the "double-sieve" mechanism of amino acid discrimination.

Since these early reports, many features of tRNA-dependent editing have been elucidated, particularly in the case of IleRS. In addition to IleRS, editing reactions have now been identified and characterized to varying degrees in valyl-,[34,35] leucyl-,[36-38] methionyl-,[39-41] alanyl-,[42] lysyl-,[43,44] prolyl-,[45,46] phenylalanyl-,[47] and threonyl-tRNA synthetases[48] (Tables 2 and 3).

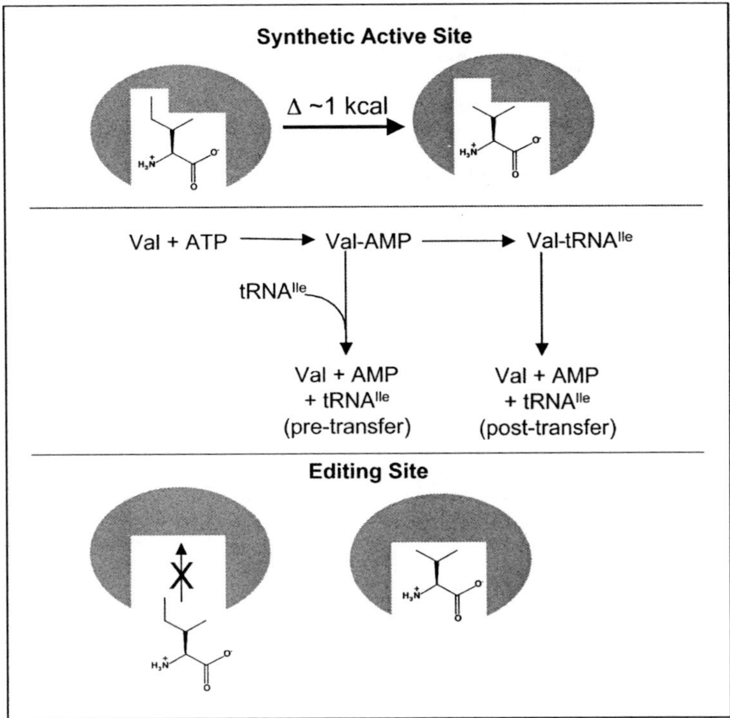

Figure 3. Substrate recognition by isoleucyl-tRNA synthetase. Top) The active site of IleRS can accommodate valine nearly as well as isoleucine. Middle) IleRS catalyzes two tRNA-dependent editing reactions to remove misactivated valine from Val-AMP and Val-tRNA[Ile]. Bottom) The editing site of IleRS sterically prohibits Ile-AMP and/or Ile-tRNA[Ile] binding, but can accommodate the smaller Val-AMP and Val-tRNA[Ile].

Transfer RNA-Dependent Editing of Noncognate Amino Acids

Interest in the role of tRNA-dependent editing in the maintenance of translational accuracy has seen a resurgence over the past five to ten years, particularly as the number of AARSs that are known to catalyze editing reactions has increased. Because much of this research focused on IleRS, it is used here as a case study, to demonstrate methods used to elucidate concepts surrounding tRNA-dependent editing.

Deacylation of Mischarged tRNAs

Of the two editing reactions, post-transfer editing is most understood. This reaction can be directly observed via a straightforward kinetic assay that was first developed for IleRS with Val-tRNA[Ile] as substrate.[27] Two different methods can be used to generate the misacylated tRNA[Ile]. Traditionally, Val-tRNA[Ile] was generated by incubating valyl-tRNA synthetase (ValRS) with tRNA[Ile], tritiated valine and ATP, in buffer containing 20% DMSO.[27,49] Addition of the organic solvent reduced the specificity of ValRS and led to misacylation of tRNA[Ile]. This method typically resulted in approximately 20-40% misacylated tRNA, due to the inefficiency of tRNA[Ile] as a substrate for ValRS. Recently, a new method was developed that utilizes an IleRS mutant (T242P, see below) to misacylate tRNA[Ile].[50,51] This mutant enzyme is defective in tRNA-dependent editing and therefore readily misacylates tRNA[Ile] with valine.[50] The use of T242P IleRS increases the efficiency of misacylation dramatically, resulting in virtually quantitative yields of Val-tRNA[Ile].[51] With this substrate in hand, post-transfer editing can be directly assessed by incubating (³H)Val-tRNA[Ile] with IleRS and monitoring the loss of radiolabelled valine that can be precipitated by trichloroacetic acid.

Table 2. *Class I aminoacyl-tRNA synthetases with known tRNA-dependent editing activities*

AARS	Edits	Pre-Transfer Editing	Post-Transfer Editing	References[1]
ArgRS	No			117
CysRS	No[2]			83
IleRS	Yes	Yes	Yes	26
GluRS	Unknown			
GlnRS	Unknown			
LeuRS	Yes		Yes	36
LysRSI	Unknown			
MetRS	Yes	Yes	Yes	118
TrpRS	Unknown			
TyrRS	No			119
ValRS	Yes		Yes	34

[1] An early or key reference is given for each enzyme that has known editing activity. See the text for detailed references.

[2] Cysteine can be converted to the cysteine thiolactone following activation by CysRS.[120] This could be considered an editing mechanism (see section VI.)

The ability to generate misacylated tRNA^{Ile} led Schmidt and Schimmel to explore the identity and location of the IleRS editing active site, relative to the active site of IleRS. In the absence of a crystal structure of IleRS, several mutations were designed, based on structural similarities between IleRS and other class I AARSs of known structure (e.g., MetRS). In particular, mutations that were predicted to be in the synthetic active site were examined in assays for amino acid activation (Ile-AMP) and for deacylation of Val-tRNA^{Ile}. Two mutations, F570S and G56P, disrupted amino acid activation by more than four orders of magnitude. In contrast, the deacylation rates evidenced by these enzymes, when confronted with Val-tRNA^{Ile}, were indistinguishable from that of the wild-type enzyme. These results demonstrated that the activities of amino acid activation and editing were functionally distinct and could be isolated by mutation.

To identify the location of the editing site, the amino termini of the amino acids in Ile-tRNA^{Ile} and Val-tRNA^{Ile} were each chemically modified with bromoacetic acid to form potential affinity labels, as BrAc-Ile-tRNA^{Ile} and BrAc-Val-tRNA^{Ile} (Fig. 4). Each BrAc-AA-tRNA^{Ile} was incubated with IleRS to allow crosslinking via loss of bromine. The crosslinking site(s) were identified by proteolysis of the modified IleRS and peptide sequencing. BrAc-Ile-tRNA^{Ile} crosslinked to a single peptide (Fig. 4, peptide 2), localized within the known synthetic active site of the enzyme. In contrast, Val-tRNA^{Ile} crosslinked to two different peptides: peptide 2 and a second peptide fragment (Fig. 4, Peptide 1), located in a large polypeptide insertion (CP1) that at the time was of unknown function. (This insertion had been previously identified as a feature of Class I tRNA synthetases.[52,53] It splits the Rossman nucleotide binding fold and was designated connective polypeptide I (CP1) because it connects one half of the fold with the other.) Site-directed mutagenesis of conserved residues in each of these peptides confirmed that peptide 1 was in the editing site of IleRS whereas peptide 2 was in the synthetic active site. These results offered the first experimental support of the hypothesis that the editing and active sites of IleRS were spatially distinct.[49]

Finally, CP1 was stably overexpressed and purified.[54] This ~250 amino acid polypeptide catalyzed the deacylation of Val-tRNA^{Ile}, confirming that the active site for post-transfer editing was within the CP1 domain and was distinct from the synthetic active site.

Table 3. Class II aminoacyl-tRNA synthetases and their tRNA-dependent editing activities

AARS	Total Editing (ATPase)	Pre-Transfer Editing	Post-Transfer Editing	References[3]
AlaRS	Yes	Yes	Yes	42
AspRS	Unknown			
AsnRS	Unknown			
GlyRS	Unknown			
HisRS	Unknown			
LysRSII	Yes[4]	Yes	?	43
PheRS	Yes		Yes	47
ProRS	Yes	Yes	Yes	45
SerRS	Unknown			
ThrRS	Yes		Yes	48

[3]An early or key reference is given for each enzyme that has known editing activity. See the text for detailed references.

[4]Some substrates for LysRS are observed to cyclize at the AA-AMP stage of activation. See section VI.

Pre-Transfer Editing

Pre-transfer editing, although conceptually described in the 1960's,[26] has proven to be more difficult to elucidate. One large hurdle has been an inability to directly assess the pre-transfer editing pathway. In addition to the post-transfer deacylation assay described above, total editing (pre- and post-transfer) can be monitored by observing the rate of ATP consumption, induced by the addition of valine and tRNA^{Ile} (Fig. 3, middle).[55] With this assay, a single molecule of ATP is hydrolyzed every time valine is misactivated, independent of whether or not the valine is edited by the pre- or post-transfer pathway.

Evidence in favor of pre-transfer hydrolysis was obtained by detailed kinetic analyses.[30] These experiments sought to quantitate the precise contribution of post-transfer editing to the observed editing and aminoacylation rates. When combined with a low steady-state concentration of Val-tRNA^{Ile} (0.8%), Fersht argued for pre-transfer editing as the dominant pathway used by IleRS, with post-transfer editing serving to cleanup errors that elude the pre-transfer pathway.[30]

The challenge of directly observing pre-transfer editing was reexamined by Hale and Schimmel, using DNA selection. A 61 nucleotide DNA, containing a sequence of 25 randomized nucleotides in the center, was subjected to multiple rounds of selection based on binding to an IleRS•Val-AMP complex.[56] (Note: A non-hydrolyzable analog of Val-AMP was used.) After seven rounds of selection, one aptamer emerged as the dominant sequence. This aptamer was termed DNA^{A}. Further evaluation of DNA^{A} demonstrated that it was not aminoacylated by IleRS (consistent with the lack of a CCA end and a non-tRNA-like predicted secondary structure). The aptamer stimulated ATP hydrolysis in the presence of valine but not isoleucine. Thus, DNA^{A} replaced tRNA^{Ile} and initiated only pre-transfer editing. Subsequently, the 3'-end of DNA^{A} was modified to contain 2'-deoxy-5-iodouridine. The modified aptamer was photocrosslinked to IleRS and the site of modification was localized to the same region of CP1 that had been identified with BrAc-Val-tRNA^{Ile},[57] as described above. Later, DNA^{A}-induced editing was shown to be sensitive to mutations in the editing active site and to mutations in the aptamer itself.[58] In total, these experiments offered the first direct observation of pre-transfer editing. They also demonstrated that the catalytic site(s) for pre-transfer editing was contained within the CP1 domain.

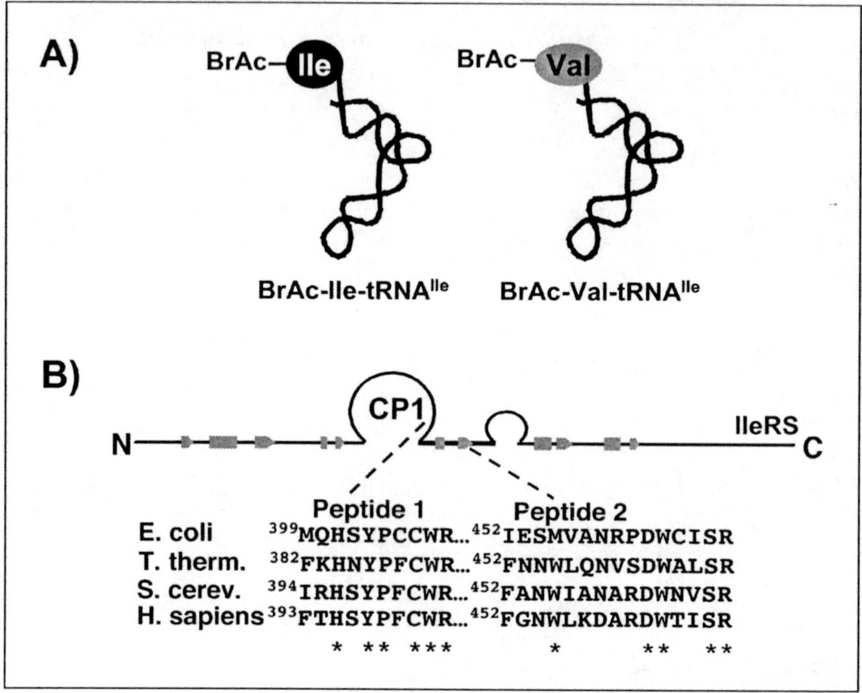

Figure 4. Identification of the IleRS editing site via crosslinking to aminoacyl-tRNAs.[49] A) Ile-tRNA^{Ile} and Val-tRNA^{Ile} were each modified with bromoacetic acid to generate the crosslinkers BrAc-Ile-tRNA^{Ile} and BrAc-Val-tRNA^{Ile}. B) The predicted Rossman fold of IleRS is shown as alternating pentagons and rectangles, representing the alternating α-helices and β-sheets that comprise the enzyme's active site. Both BrAc-Ile-tRNA^{Ile} and BrAc-Val-tRNA^{Ile} covalently crosslinked to peptide 2, within the enzyme's active site. In contrast, only BrAc-Val-tRNA^{Ile} crosslinked to peptide 1, within the CP1 insertion.

Nucleotide Determinants

Aminoacyl-tRNA synthetases recognize only their cognate tRNA(s), from a cellular pool containing all other tRNAs. Molecular recognition is accomplished by a series of identity determinants within the tRNA. (For a detailed review of identity determinants in tRNA aminoacylation, see Giegé et al[59]) Identity elements required for aminoacylation of tRNA^{Ile} have been reported (Fig. 5A, squares).[60] Significantly, tRNA^{Val} is a close relative of tRNA^{Ile} that binds competitively to IleRS but is not aminoacylated. Moreover, tRNA^{Val} does not stimulate editing by IleRS. The minimum set of alterations required to convert tRNA^{Val} into a tRNA that could stimulate IleRS-dependent editing of valylated-tRNAs[61] was thus investigated. A series of tRNA^{Val}/tRNA^{Ile} chimeras were constructed (Fig. 5B). Each tRNA$^{Val/Ile}$ chimera was evaluated with IleRS for its substrate capabilities in a standard aminoacylation reaction (with Ile) and as an inducer of tRNA-dependent editing of valine. Importantly, determinants for editing and aminoacylation were shown to be distinct. Each chimera (except tRNA^{Val}) contained the tRNA^{Ile} anticodon (GAU), which was sufficient to enable aminoacylation (Fig. 5B). In contrast, only those tRNAs that contained the sequence of the D-loop of tRNA^{Ile} were capable of initiating tRNA-dependent editing. A comparison of the similar D-loops of tRNA^{Ile} and tRNA^{Val} resulted in the identification of three nucleotides necessary for editing (Fig. 5A, circles).[61] Subsequently, each of these nucleotides was examined individually and shown to contribute to the efficiency of overall editing (ATPase assay).[62] The mechanism by which these identity elements are recognized and communicated to the IleRS editing site remains unclear, particularly in light of the co-crystal structure of IleRS•tRNA^{Ile}, which shows the D-loop of tRNA^{Ile} positioned away from the enzyme.[63]

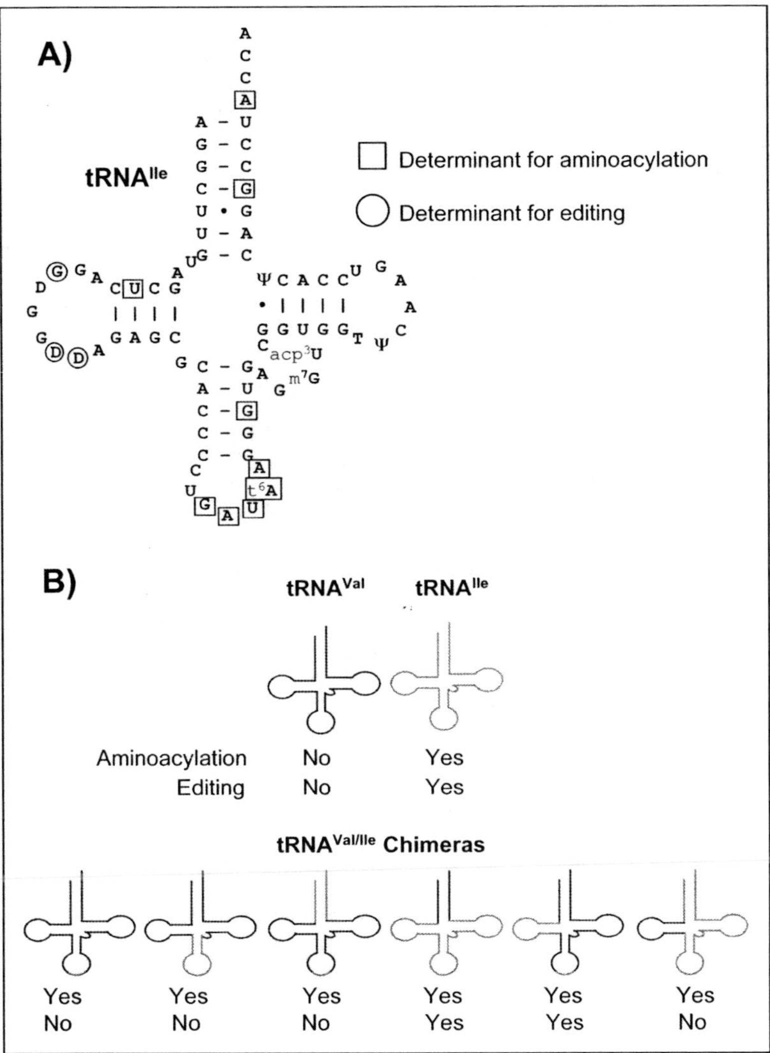

Figure 5. The tRNA identity determinants for IleRS-catalyzed tRNA-dependent total editing.[61,62] A) The secondary structure and post-translational modifications of tRNA^Ile are shown. Nucleotides which are known determinants for tRNA aminoacylation are boxed,[60] and those which are determinants for tRNA-dependent editing are encircled.[61,62] B) Chimeras, introducing features of tRNA^Ile into tRNA^Val, were used to determine the nucleotide identity elements for editing; tRNA^Ile is shown in black and tRNA^Val is in gray. Each tRNA was assayed for aminoacylation activity by IleRS (formation of Ile-tRNA) and for total editing with IleRS and valine, results are indicated below each tRNA.

Beyond the three identity determinants in the D-loop of tRNA^Ile, the full context of the tRNA is also required for efficient editing of misactivated valine. A small minihelix RNA, which reconstitutes the acceptor stem and TΨC-stem of tRNA^Ile (MH^Ile), was examined within the broad context of editing. MH^Ile was misacylated with valine and, unlike tRNA^Ile, the concentration of Val-MH^Ile built-up over time. Thus, MH^Ile cannot stimulate editing. Addition of an exogenous D-loop minihelix, containing all three editing identity elements of tRNA^Ile, did not induce editing of Val- MH^Ile. Therefore, the D-loop and the acceptor stem must be covalently connected for efficient editing.[64]

Translocation

The crystal structure of *Thermus thermophilus* IleRS offered the first three-dimensional picture of the relationship between the two catalytic sites (Fig. 6).[65] Not unexpectedly, CP1 is structurally distinct within the global IleRS structure (Fig. 6). Additionally, diffusion of isoleucine and of valine (separately) into the crystal clarified the substrate binding pockets and relative orientations of the two active sites. Isoleucine bound to the synthetic active site of the enzyme. In contrast, two molecules of valine were bound – one in the same orientation as Ile (within the synthetic active site), and a second in the editing site of the enzyme, within CP1 (Fig. 6). The positioning of this second valine within the editing site was verified by site-directed mutagenesis, leading to the first mutations in IleRS that were deficient in total editing (described below).[65]

The two sites are separated by ~30 Å, thus implying that the misactivated amino acid (as Val-AMP or Val-tRNA^{Ile}) has to move from one site to the other, in order for editing to occur. This movement could be via transient dimerization of normally monomeric IleRS or by translocation in cis from one site to the other. Dimerization was ruled out by direct experiments that showed that editing was not a quadratic function of enzyme concentration and that site-to-site transfer was by translocation.[66]

A fluorescence-based assay was developed that monitored translocation on the one hand, and total editing on the other. (Fig. 7).[66,67] The assay used the ATP analog N-methanthrinoyl-dATP (Mant-dATP, Fig. 7), which is a competitive inhibitor by binding to the ATP binding pocket within the synthetic active site (Mant-dATP cannot replace ATP as a substrate). Additionally, the Mant group introduces a fluorophore, with λ_{Ex} = 360 nm and λ_{Emit} = 440 nm. By excitation of the tryptophan residues in IleRS (at λ_{Ex} = 295 nm; λ_{Emit} = 330 nm), resonance energy transfer occurs when Mant-dATP is bound to the active site. Thus, the fluorescence of emission of Mant-dATP can be used to monitor clearance of the active site, after translocation. When the active site is cleared, Mant-dATP rebinds immediately, yielding emission at 490 nm.[66]

By using limiting concentrations of exogenously added Val-AMP, to give a single round of editing, the rate of translocation is observed as a transient fluorescent signal. Application of this approach showed that translocation (and not hydrolysis) is the rate-limiting step for editing.[66]

Like overall editing, the total rate of translocation is comprised of two components: translocation of Val-AMP and translocation of Val-tRNA^{Ile}. The co-crystal structure of IleRS•tRNA^{Ile} from *S. aureus* showed the acceptor stem of tRNA^{Ile} unwound so that the 3'-end extends into the CP1 editing site.[63] Thus, it seems probable that Val-tRNA^{Ile} is transferred to the editing active site by a structural rearrangement of the tRNA. Possibly, the D-loop editing identity determinants may participate in allowing this tRNA flexibility.

Recently, a specific residue within the CP1 domain of IleRS (D342) was identified that, when mutated, has a dramatic effect on the observed rate of translocation.[68] D342 was mutated, independently, to alanine (D342A), asparagine (D342N) and to glutamate (D342E). All three mutations were deficient in total editing, but to varying extents (>20 fold, >10 fold, and 2-3 fold respectively). Each mutant enzyme was examined to determine whether or not D342 played a role in translocation. In the fluorescence translocation assay (described above), the D342A and D342N IleRS variants were severely deficient in translocation, whereas D342E was only partially deficient. To examine whether or not these results arose from a translocation defect alone or by a defect in post-transfer editing that subsequently perturbed translocation, D342A and D342N were each cloned into the CP1 domain alone and examined for post-transfer editing. Neither of these mutant enzymes could catalyze the deacylation of Val-tRNA^{Ile}, demonstrating that each mutant is defective in post-transfer editing, even when isolated from the need for a translocation mechanism. Thus, D342 plays a role in translocation and loss of translocation by mutation of D342 results from a disruption of post-transfer editing.

Prior to the identification of D342, nothing was known about the role of IleRS in directing translocation of substrates for editing. D342 is a conserved residue in a region of the CP1 editing domain that is predicted to form a Val-tRNA^{Ile} binding pocket;[69] it does not lie along a path between the synthetic active site and the editing site. Kinetic studies demonstrated that D342E IleRS variant binds Val-tRNA^{Ile} in the editing reaction, with a somewhat reduced K$_{M-App}$ of 36 μM (compared to 8 μM for wild-type IleRS). Thus, this residue appears to be involved in binding of the post-transfer editing substrate Val-tRNA^{Ile}. Based on these and other results, the authors proposed a model for editing wherein a single round of post-transfer editing would occur to establish an

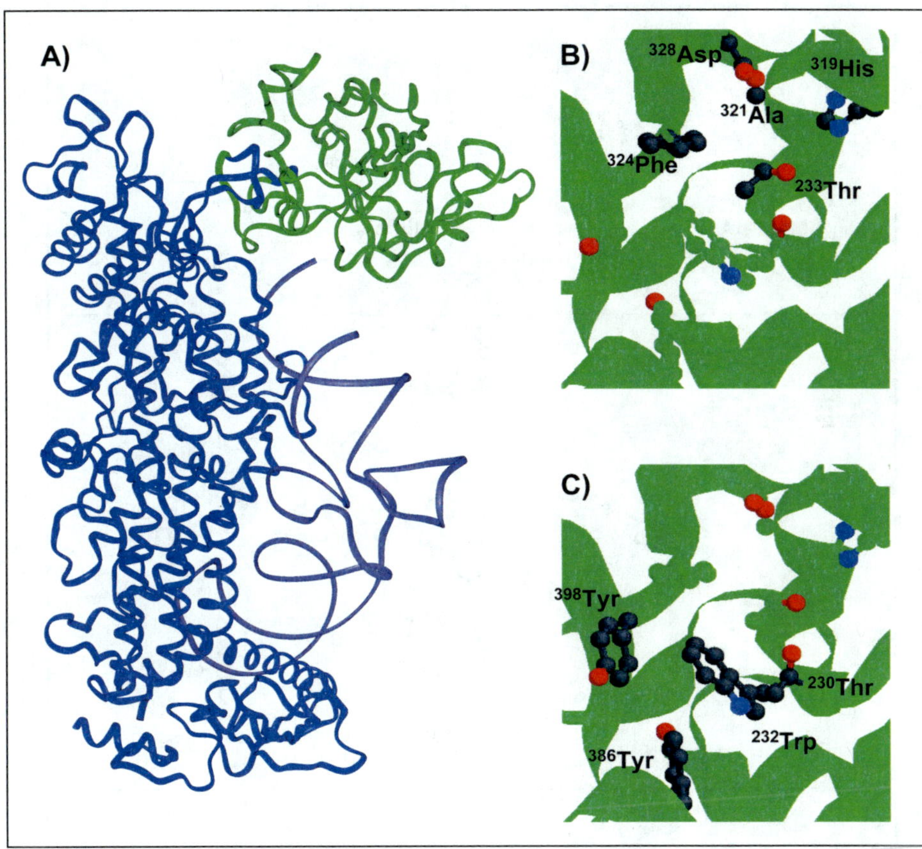

Figure 6. The structure of IleRS, complexed to tRNA^{Ile}; the editing site is divided into two binding pockets. A) *S. aureus* IleRS is shown in blue; the editing site is shown in green and tRNA^{Ile} is shown in magenta.[63] B) The pre-transfer binding pocket of *T. thermophilus* IleRS.[65] Residues that comprise the editing site are shown as ball-and-stick representations, with those predicted to form the pre-transfer editing pocket labeled and shown in gray (numbering system is that of *T. thermophilus* enzyme). C) The putative post-transfer binding pocket of IleRS. Residues predicted to form the post-transfer editing pocket are labeled and shown in grey (numbering system is that of *T. thermophilus* enzyme).

enzyme-tRNA^{Ile} complex that would be active for multiple rounds of pre-transfer editing. D342 could play a role in the assembly of this complex via the formation of a salt bridge with the amino terminus of Val-tRNA^{Ile}.

The mechanism by which Val-AMP is translocated to the editing site remains somewhat unclear. This substrate is apparently hydrolyzed by the CP1 editing domain, because a mutation like T242P disrupts total editing but not translocation of Val-AMP.[50] One possible mechanism would be the formation of a channel between the two IleRS active sites. This channel would require the formation of an IleRS•tRNA^{Ile} editing complex. No such channel is readily visible in the co-crystal structure of IleRS•tRNA^{Ile}, although the CP1 editing domain is not resolved in the structure.[63]

Mutagenesis of the Editing Center of IleRS

In addition to the experiments described above, the editing site of IleRS has been extensively examined by site-directed mutagenesis. Alignment of IleRS sequences reveals many conserved residues within the editing active site. These residues are conserved across all three domains of

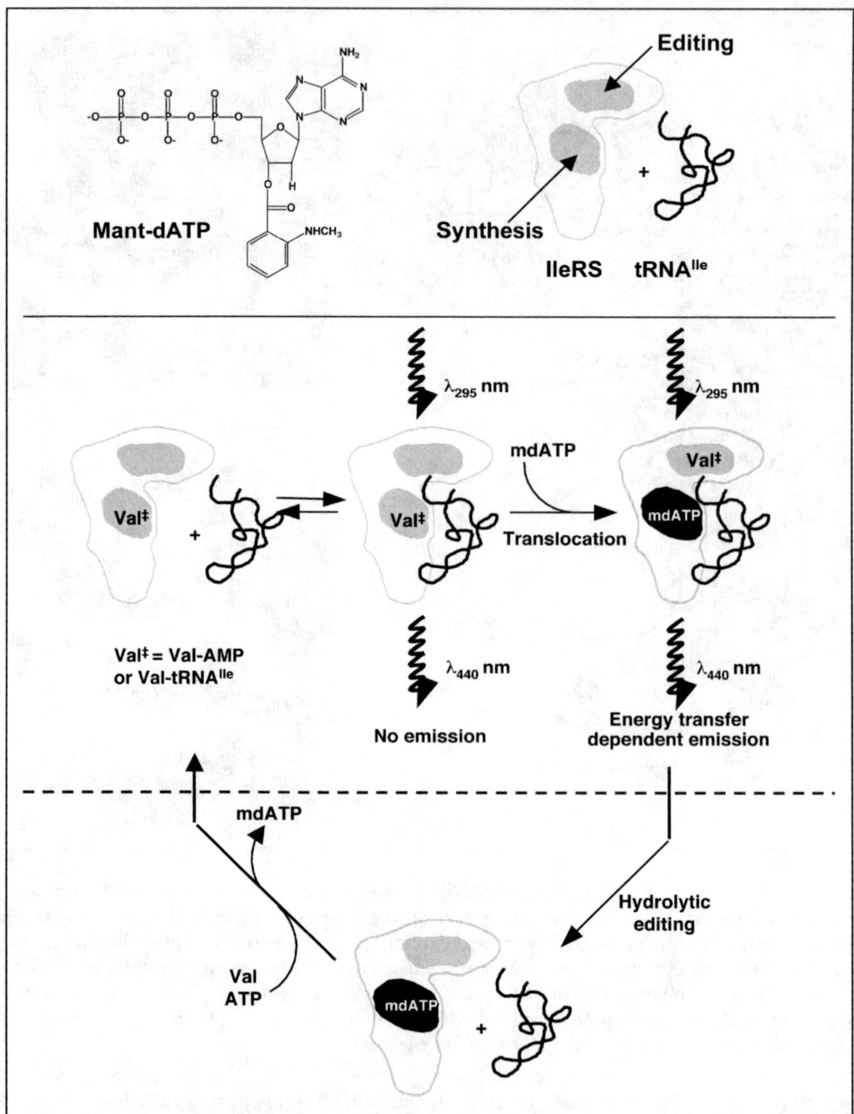

Figure 7. A fluorescence assay for editing.[66] Top) The chemical structure of Mant-dATP (left) and a schematic of the structure of IleRS, denoting the position of the synthetic active site and the editing site. Bottom) Prior to the addition of tRNA^Ile, misactivated valine occupies the synthetic active site of IleRS as Val-AMP (Val‡). Upon addition of tRNA^Ile, the misactivated valine is translocated to the editing site (either as Val-AMP or following transfer to the 3' end of tRNA^Ile – both substrates are denoted Val‡). This translocation event clears the synthetic active site. In the absence of excess valine and ATP, Mant-dATP binds to the synthetic active site. When bound, Mant-dATP fluorescence can be observed at 440 nm, following excitation of the IleRS tryptophans (at 295 nm) and fluorescence resonance energy transfer from tryptophan at the active site to Mant-dATP. The concentration of valine can be adjusted for multiple rounds of editing ([Val]>>[IleRS]) or for a single turnover that allows translocation to be observed specifically ([Val]≤[IleRS]).

life (Fig. 8, see page 51). Interestingly, site-directed mutagenesis demonstrated that many of these residues can be altered with little to no effect. The results of these mutagenesis studies are summarized in Table 4.

Mutations within the editing site fall into four categories, based on in vitro kinetic analyses of the resultant mutant enzymes: 1) Total editing is disrupted, 2) The specificity of post-transfer editing is altered, 3) Pre-transfer, but not post-transfer, editing is disrupted, or 4) no effect is detected.

Mutations That Disrupt Total tRNA-Dependent Editing

In the crystal structure of IleRS from *T. Thermophilus*, a cluster of possible hydrogen bonds forms between H333, N250 and either T242 or T243.[65] This cluster is adjacent to the carboxylate of the valine bound at the editing site (the carboxyl carbon of the valine moiety is the site of hydrolysis in both pre- and post-transfer editing). Thus, this H-bond relay could form the basis of catalytic activity for both of the editing reactions.[65]

To test this possibility, each of the residues in question was individually mutated to alanine and the resultant enzymes were evaluated for tRNA-dependent ATPase activity. T243A and H333A IleRS both showed wild-type ATPase activity. In contrast, T242A and N250A IleRS were both diminished (but not ablated) in total editing activity.[65] Furthermore, both of these mutant enzymes could generate low levels of Val-tRNA^{Ile}, in sharp contrast to wild-type IleRS, which does not catalyze the misacylation of its cognate tRNA.[50]

The effects of two additional mutations were investigated—T242P and the double mutant T242A/N250A.[50] Editing in each of these mutant enzymes was nearly eliminated. ATPase assays detected no total editing and neither mutant enzyme could deacylate exogenously supplied Val-tRNA^{Ile}. Consistent with the lack of activity, both of these mutant enzymes generated large amounts of Val-tRNA^{Ile}. T242P IleRS is now used to generate Val-tRNA^{Ile} for deacylation assays.[51] Thus, the functional side chain of T242 and N250 do not participate directly in the catalysis of tRNA-dependent editing. However, T242 and N250 are important for the structural integrity of the editing site and significant perturbations in this site can lead to a loss of editing.

Mutations That Affect Only Post-Transfer Editing

Several mutations have been identified in the editing site of *Escherichia coli* IleRS that perturb the specificity of post-transfer editing. To date, however, the identification of a mutation that is deficient only in post-transfer editing has remained elusive.

The first round of mutations was constructed based on results from crosslinking experiments with BrAc-Val-tRNA^{Ile}, described above. In these studies, three conserved residues, H401, Y403 and R408, were identified that, when mutated, perturbed the specificity of IleRS-catalyzed Val-tRNA^{Ile} deacylation.[49] H401Q, Y403F, R408A and R408Q IleRS were constructed. Each enzyme was evaluated for deacylation of Val-tRNA^{Ile} and with the cognate product Ile-tRNA^{Ile}. (Ile-tRNA^{Ile} is a poor substrate for post-transfer editing by wild-type IleRS.[27]) Deacylation of Val-tRNA^{Ile} by both H401Q and Y403F IleRS was faster than with the wild-type enzyme. Additionally, each of these enzymes only negligibly deacylates Ile-tRNA^{Ile}. Thus, each mutant enzyme was more than 35 times more efficient than wild-type IleRS at editing Val-tRNA^{Ile} versus Ile-tRNA^{Ile} [49]. (This number is based on the discrimination factor, or the ratio of the rates of deacylation of Val-tRNA^{Ile}/Ile-tRNA^{Ile}. The discrimination factor for wild-type IleRS is 32, while that of H401Q IleRS is >1200 and that of Y403F IleRS is 1100.) These results raise the question as to why H401 and Y403 are rigorously conserved.

Two mutations analyzed in this study, R408A and R408Q, conferred the opposite effect.[49] Each mutant enzyme was more efficient at deacylating Ile-tRNA^{Ile}, than was the wild-type enzyme (by 3.4- and 2.2-fold, respectively). Both enzymes are also diminished in their ability to deacylate Val-tRNA^{Ile}, by approximately 25%, compared to wild-type IleRS. Thus, these two mutations had discrimination factors of 4.6 (R408A) and 7.7 (R408Q).

In a separate study, H333A IleRS was shown to be incapable of discriminating between Val-tRNA^{Ile} and Ile-tRNA^{Ile}, deacylating both compounds with a similar efficiency. For this mutant enzyme, the discrimination factor is ~1.0. The H333A substitution most probably widens the editing site so that the valyl- and isoleucyl- moieties fit equally well into the pocket for editing.[51]

Table 4. Editing active site mutations in E. coli IleRS

Mutation	Consequences in vitro	Generates Val-tRNA[Ile]	Phenotype	References
Wild-type	NA	No	WT	26
T241A	No effect	No	WT	51
T242A	Diminished in total editing	Yes, weakly	WT	65,50
T242P	Strongly deficient in total editing	Yes, strongly	WT	
T243A	No	No	WT	121,65
T243R	Deficient in pre-transfer editing	Yes	Poor growth	51
W245E	No effect	No	WT	51
T246A	No effect	No	WT	51
N250A	Diminished in total editing	Yes, weakly	WT	65,50
T241A/ N250A	Deficient in total editing	Yes, strongly	WT	50
G328A	No effect	No	No	51
G328A/ G330A	No effect	No	No	51
H333A	Deacylates Ile-tRNA[Ile]	No	WT	51
D342A	Deficient in post-transfer editing and translocation	Yes, strongly	No	68
D342N	Deficient in post-transfer editing and translocation	Yes, strongly	No	68
D342E	Diminished in post-transfer editing and translocation	Yes, weakly	No	68
H401Q	Val-tRNA[Ile]/Ile-tRNA[Ile] > WT	No	ND	49
Y403F	Val-tRNA[Ile]/Ile-tRNA[Ile] > WT	No	ND	49
R408Q	Val-tRNA[Ile]/Ile-tRNA[Ile] < WT	No	ND	49
R408A	Val-tRNA[Ile]/Ile-tRNA[Ile] < WT	No	ND	49

WT – wild-type; ND – Not determined.
Note: all mutagenesis studies, to date, have been conducted on IleRS from *E. coli*. The amino acid numberings given above all refer to the *E. coli* sequence.

Mutations That Affect Pre-Transfer Editing

Yokoyama and coworkers recently compared the crystal structure of IleRS to that of the co-crystal of ValRS•tRNA[Val].[69] This comparison, in concert with detailed molecular modeling studies, led to the hypothesis that the editing site was comprised of two separate pockets – one for Val-AMP and one for Val-tRNA[Ile]. With this proposal, it seemed probable that these two subsites could be functionally separated by mutations in one or the other pocket. For example, H333 is in the proposed post-transfer editing cavity and the H333A mutation affects only post-transfer editing. Similarly, a T243R mutation disrupts only the pre-transfer editing site. This mutation introduced a positive charge into the proposed pre-transfer hydrophobic binding pocket. The total editing activity of T243R IleRS was diminished about six-fold. In contrast, T243R deacylates Val-tRNA[Ile] at the same rate as does wild-type IleRS. Thus, the T243R variant is diminished in pre-transfer but not post-transfer editing. Because overall editing is diminished six-fold even when deacylation of Val-tRNA[Ile] is unaffected, the kinetic data support the conclusion that pre-transfer editing is the dominant pathway used by IleRS.[51]

To evaluate the phenotypes of many of the mutations described above, a Δ*ileS* null strain was constructed by disrupting the chromosomal copy of the *ileS* gene in *E. coli* with the gene encoding kanamycin resistance. Growth of this strain was supported by a plasmid encoding wild-type *ileS*. P4 transduction was used to replace the wild-type gene with one containing the mutations of interest. In this way, the resultant strains were evaluated for viability, on plates or in liquid culture, in the absence of wild-type IleRS. T243R IleRS had a slow growth phenotype in liquid culture, relative to wild-type, T242P and H333A IleRS variants. As T243R IleRS is the only mutant with an apparently absolute deficit in pre-transfer editing, this toxicity argues that post-transfer editing alone is insufficient to maintain viability.[51]

Mutations That Have No Effect on tRNA-Dependent Editing

The high level of sequence conservation observed in the editing sites of IleRS sequences suggests that the composition of the editing site has been maintained by positive selective pressure. However, as stated above, many of these residues can be altered by site-directed mutagenesis without deleterious consequences on the in vitro activity of the enzyme. For example, T241A, T243A, W245E, T246A, G328A, and G330A IleRS variants all have wild-type editing activities (Schmidt and Schimmel, unpublished data and Hendrickson et al[51]). And, while H401Q and Y403F mutations perturb post-transfer editing, this effect is only positive, generating an enzyme that is more accurate.[49] Finally, mutations like T242P that virtually ablate editing activity show no phenotype in vivo.[51]

Why, then, is the sequence of the editing site so rigorously conserved across evolution? The answer to this question most likely reflects a mixture of factors. For example, the high level of conservation observed in CP1 may reflect the constraints on an active site that responds to a variety of different substrates that are not usually assessed by in vitro assays. Thus, the tRNA-dependent editing activity of IleRS is not limited to valine; IleRS also misactivates and hydrolytically edits leucine, alanine, and threonine and probably others.[70] In most editing assays, however, the enzyme's activity against valine is the only one to be examined. The high level of conservation observed in CP1 may reflect the constraints on an active site that responds to a variety of different substrates. Additionally or alternatively, the editing site may have evolved to include a high level of functional redundancy. Perhaps, if one residue (e.g., 241) is mutated, its role in editing is compensated by an adjacent residue (e.g., T242).

The Chemical Mechanism of tRNA-Dependent Editing

The chemical mechanism of the two hydrolytic editing reactions of IleRS remains unsolved. Also not known is whether the two substrates for editing, Val-AMP and Val-tRNA^{Ile}, are hydrolyzed via the same mechanism. The substrates are chemically different, with the mixed anhydride of the valyl adenylate being more susceptible to hydrolysis. Because extensive mutagenesis has failed to reveal a single residue that is essential for editing, the editing site probably positions the reactive valyl carboxylate of each substrate adjacent to an ordered water molecule. In fact, a water molecule is present in the editing site of IleRS.[65] This water could effect catalysis via nucleophilic activation. This kind of mechanism is an attractive choice based on its simplicity and has been proposed for editing by LeuRS.[36] Additionally, the contrast between the activities of the H333A and T243R enzymes (see above) supports the idea that the two substrates for editing bind to different subsites within CP1.[51] Within the scope of the proposed mechanism, the role of these two subsites may be to accurately position the two structurally distinct substrates adjacent to a single water molecule.

Other Aminoacyl-tRNA Synthetases with Known tRNA-Dependent Editing Reactions

Class I

In addition to IleRS, specific tRNA-dependent editing reactions have been identified in two other class I enzymes: leucyl- and valyl-tRNA synthetases (LeuRS and ValRS, respectively).[26,34-38] (MetRS and other AARSs edit homocysteine (Hcy), in a reaction that can be tRNA-dependent or independent[39] – see below). These two AARSs are closely related to each other and to IleRS, as seen in the high levels of sequence and structural similarities. Like IleRS, LeuRS and ValRS each contain a CP1 insertion that contains the editing active site.[52]

ValRS

The editing site of ValRS is contained in a large CP1 insertion that is similar in primary sequence and structure to the CP1 domain of IleRS (See Figs. 8 and 9).[52,69] The accuracy of ValRS is challenged by threonine, which is isosteric to valine.[34] Thus, the editing domain of ValRS must distinguish between valine and threonine, to only hydrolyze the more hydrophilic substrates Thr-AMP and Thr-tRNA^{Val}. Unlike the problem faced by IleRS, these substrates cannot be distinguished by simple steric occlusion.

The CP1 Editing Domain of ValRS

The CP1 insertion of ValRS (CP1^{Val}) was cloned, stably expressed, and purified. CP1^{Val} efficiently hydrolyzed Thr-tRNA^{Val} (post-transfer editing), demonstrating that the editing site of ValRS is contained within the CP1^{Val}.[35] Consistent with the double-sieve model, the editing active site of ValRS is too small to accommodate an isoleucyl moiety fused to tRNA.[71,54]

Further evidence that the synthetic active site is distinct from editing was obtained by site-directed mutagenesis.[35] Several residues within the ValRS synthetic active site (P47, G85 and D87) were mutated and evaluated for aminoacylation activity and for post-transfer editing. Two mutant enzymes -P47I and D87A ValRS- were deficient in amino acid activation and aminoacylation of tRNA^{Val} (with both valine and threonine). In contrast, these enzymes catalyzed the deacylation of Thr-tRNA^{Val} with wild-type efficiencies, demonstrating that the synthetic active site of ValRS can be disrupted without affecting the editing activity.[35] Thus, like IleRS, the synthetic and editing active sites of ValRS can be functionally separated.

An alignment of key regions of the IleRS editing domain with the same regions of several different ValRS orthologs is shown in Figure 8. This alignment, when combined with the crystal structure of ValRS:tRNA^{Val} (Fig. 9),[69] provides a clear measure of how ValRS can discriminate between a hydrophilic and a hydrophobic amino acid. Several conserved residues in the IleRS editing domain, which are near the bound valine in the crystal structure (e.g., T243 and W245),[65] are replaced in ValRS by conserved residues that are significantly more polar or charged (e.g., R223 and E225).[69] Thus, the double sieve of ValRS is comprised of a slightly indiscriminate active site, which misactivates threonine, and a hydrophilic editing site, designed to accommodate and hydrolyze threonine-containing substrates.

Crosslinking experiments led to the partial characterization of the editing domain of ValRS in vitro.[72] Valine, threonine, norleucine and phenylalanine were all synthetically modified to contain a bromomethyl ketone in place of their respective carboxylic acids (Fig. 10). Each of these unnatural amino acids was incubated independently with ValRS and each irreversibly inactivated the enzyme. Inactivation could be competitively protected by the addition of exogenous L-valine. The modification sites for each analog were identified by trypsin-digestion of the labelled proteins, followed by extensive analysis by MALDI-TOF mass spectrometry. All four amino acid analogs covalently modified residues in the synthetic active site (either H424, H433, or C829). In contrast, the three non-cognate analogs (threonine, norleucine and phenylalanine) also covalently modified residues in the CP1 editing active site (H266, C275 and H282).[72] Although none of the modified residues are rigorously conserved in ValRS, they are found in the editing pocket of the recent crystal structure.[69]

Distinguishing between Pre-and Post-Transfer Editing in ValRS

ValRS from *B. stearothermophilus* was first examined in detail by Fersht and Kaethner.[34] The kinetics for amino acid activation, ATP hydrolysis, and deacylation, were evaluated in detail using a quenched flow apparatus so that reaction intermediates could be detected on a millisecond time scale. Remarkably, and in sharp contrast to IleRS, ValRS deacylates Thr-tRNA^{Val} with a rate constant of 40 sec^{-1} (IleRS deacylates Val-tRNA^{Ile} with a rate of 10 sec^{-1}).[30] A model where 62% of the editing activity of ValRS was via the post-transfer deacylation of Thr-tRNA^{Val} was proposed.[34] Yeast ValRS has a similarly enhanced rate of deacylation (80 sec^{-1}).[73] Thus, unlike IleRS, post-transfer editing appears to be the dominant pathway used to reject threonine.

Pre-transfer editing has never been directly observed for ValRS, and remains somewhat controversial, particularly because the enzyme rapidly deacylates Thr-tRNA^{Val}.[34] Similar to IleRS, molecular modeling studies suggested that the editing site is divided into two different, but overlapping substrate recognition pockets, one which binds Thr-AMP and one which binds Thr-tRNA^{Val}.[69] (In

```
           238                323                    401
    EcI    VIWTTTPWTLPANR...VTLDAGTGAVHTAPGHGPDD...HSYPCCWRHKTPIIFRATPQWFV
    MjI    VIWTTTPWTLVANL...VTLDAGTGAVHTAPGHGEED...HSYPHCWRCKTPLLFRATEQWFL
    ScI    VAWTTTPWTLPSNL...VTSDSGTGIVHNAPAFGEED...HSYPFCWRSDTPLLYRSVPAWFV
    HsI    VAWTTTPWTLPSNL...VKEEEGTGVVHQAPYFGAED...HSYPFCWRSDTPLIYKAVPSWFV
           * ********* :*      *. : *** ** ** .* :*    **** *** .**:::::. **:

           218                267                    362
    EcV    VVATTRPETLLGDT...ADMEKGTGCVKITPAHDFND...LTVPYGDRGGVVIEPMLTDQWYV
    MjV    LIATTRPELMAACV...VEKEFGTGAVMVCTFGDKTD...QNVGVCWRCKTPIEIIVTEQWFV
    ScV    IIATTRPETIFGDT...VDMEFGTGAVKITPAHDQND...MTIPTCSRSGDIIEPLLKPQWWV
    HsV    VVATTRIETMLGDV...VDMDFGTGAVKITPAHDQND...MVVPLCNRSKDVVEPLLRPQWYV
           ::**** * : .    .:: ***. : . * .*   :    *    :* :: **:*
```

Figure 8. Alignment of portions of the editing sites of IleRS and ValRS. An alignment of portions of the CP1 domains of both IleRS and ValRS are shown, using representative sequences of each enzyme. Asterisks indicate residues that are universally conserved in more than 95% of all known sequences of a given enzyme. Residues highlighted in red are universally conserved (>95%) in both IleRS and ValRS. Residues marked in blue (IleRS) and green (ValRS) are universally conserved as different amino acids in the two enzymes (e. g., IleRS W240 is ValRS A220). Residue numbering corresponds to the amino acid numbering in the *E. coli* enzyme. Abbreviations: EcI, *E. coli* IleRS;[122] MjI, *Methanococcus janaschii* IleRS;[123] ScI, *Saccharomyces cerevisiae* IleRS;[124,125] HsI, *Homo sapiens* IleRS; EcV, *E. coli* ValRS;[126] MjV, *Methanococcus janaschii* ValRS;[123] ScV, *Saccharomyces cerevisiae* ValRS;[127] HsV, *Homo sapiens* ValRS.[127]

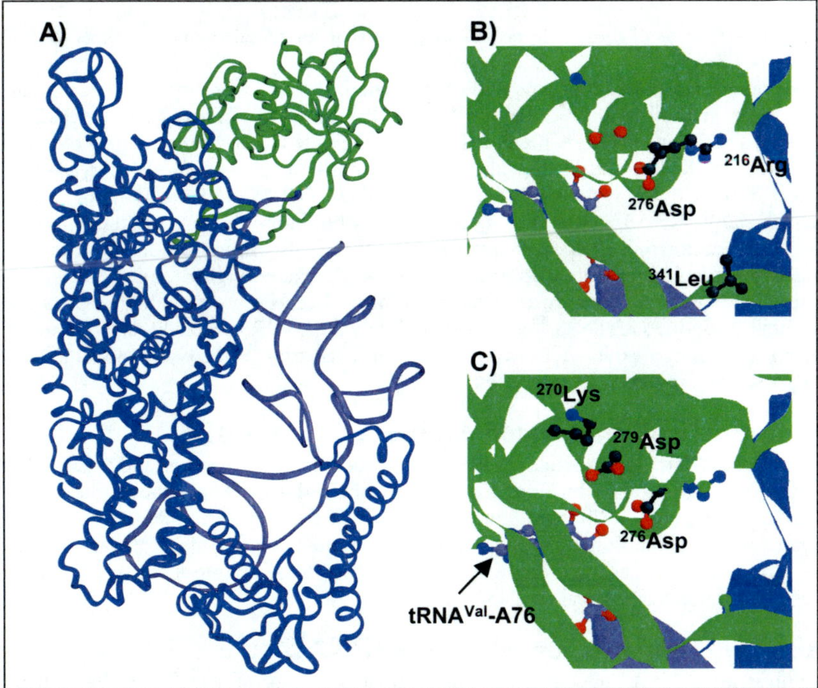

Figure 9. The structure of ValRS, complexed with tRNA^Val.[69] *T. thermophilus* ValRS is shown in blue; the editing site is shown in green and tRNA^Val is shown in magenta. B) The putative pre-transfer editing pocket; residues predicted to contact the threonine side chain in Thr-AMP are shown in gray as ball-and-stick representations. C) The putative post-transfer binding pocket of ValRS. Residues predicted to contact the threonine side chain in Thr-tRNA^Val are shown as space filling models in gray.

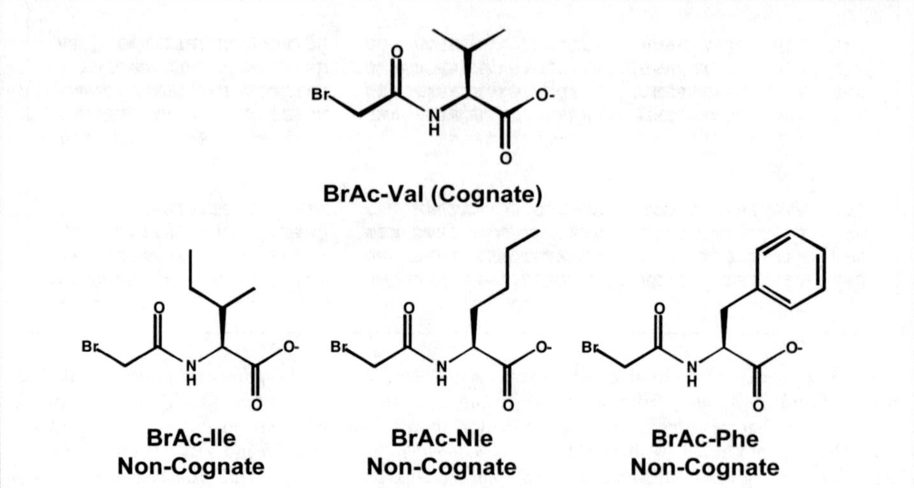

BrAc-Val (Cognate)

BrAc-Ile
Non-Cognate

BrAc-Nle
Non-Cognate

BrAc-Phe
Non-Cognate

Figure 10. Crosslinking reagents for the characterization of ValRS.[72] The structures of the bromoacetylated analogs of valine (Val), isoleucine (Ile), norleucine (Nle) and phenylalanine (Phe) are shown. Each of these compounds were incubated with ValRS and analyzed for the formation of crosslinks.

fact, it was this analysis of the ValRS crystal structure that suggested the two pockets in the IleRS editing site.)

Among others, a conserved aspartic acid (Asp276 in *T. thermophilus*, Asp283 in *E. coli* ValRS) is proposed to be in the binding pocket for Thr-AMP, whereas a conserved lysine (Lys270 in *T. thermophilus*, Lys277 in *E. coli*) is predicted to be in the post-transfer binding pocket.[69] Interestingly, the bromomethyl ketone analogs employed by Hountondji et al,[72] (Fig. 10) alkylated residues that appear to flank both proposed binding pockets; these analogs modified His266, Cys275 and His282 in *E. coli* ValRS. So, statistically, the analogs may bind to both the pre-and the post-transfer pockets. Thus, examination of the crystal structure of ValRS, after modification with the threonine bromomethyl ketone derivative, could demonstrate two substrate binding modes.

Because IleRS and ValRS are closely related, and the existence of two binding pockets has been experimentally supported for IleRS,[50] and suggested for ValRS, ValRS probably catalyzes pre-transfer editing, at least to minor extent. This hypothesis awaits evaluation of the putative ValRS Thr-AMP binding pocket by site-directed mutagenesis.

Translocation of Misactivated Amino Acids by ValRS

ValRS misactivation is promiscuous. ValRS misactivates threonine, cysteine and α-aminobutyrate (Abu), offering an opportunity for comparing the rate of translocation from the synthetic active site to the editing active site for different amino acids. Each of these three non-cognate amino acids is misactivated with different kinetic constants, with threonine being the most efficient non-cognate substrate. In sharp contrast, however, these three amino acids are all translocated between the two sites (>25 Å) with nearly identical rates (3.3 sec^{-1}).[67]

Identity Determinants for Editing in tRNA^{Val}

In contrast to tRNA^{Ile}, precise identity determinants for editing have yet to be evaluated for tRNA^{Val}. The 3'-terminal adenosine, however, is essential for editing. Replacement of this nucleoside with either a C or a U generated a mutant tRNA^{Val} that can be aminoacylated with valine or misacylated with threonine.[74] These two tRNA variants can also be misacylated with isoleucine, alanine, serine, and cysteine.[75] G76 tRNA^{Val} is active in aminoacylation with valine but is not misacylated with any of the above mentioned amino acids. Interestingly, C and U76 tRNA^{Val}

mutants are defective in post-transfer editing, offering an explanation for the misacylation results with C76 and U76. G76 tRNA^{Val} can't be mischarged, but correctly acylated Val-G76-tRNA^{Val} is immune to the low level of deacylation usually seen. Thus, all three N76 tRNA^{Val} variants are defective in post-transfer editing.[75] In total, these results suggest that the editing site of ValRS forms direct contacts with the terminal adenine in tRNA^{Val}, making formation of the correct 3'-end essential for accurate aminoacylation.

In Vivo Selection of ValRS Derivatives that Are Defective in tRNA-Dependent Editing

A selection scheme that generated ValRS variants that misacylated tRNA^{Val} in vivo was developed.[76] An essential gene (*thyA*) was mutated to replace a required codon for cysteine with one for valine. Thus, to survive under these conditions, bacteria were required to incorporate cysteine into thymidylate synthase at a position designated for valine. After mutagenesis of the entire chromosome, several mutant strains were isolated that grew in the presence of exogenous cysteine. Although mutagenesis could have occurred anywhere in the entire chromosome, all mutants had substitutions in the CP1 domain of ValRS (T222P, R223H, D230N, V276A, K277Q).

The strain carrying the T222P mutation was evaluated further[76] and shown to be sensitive to cysteine, threonine and Abu. These amino acids are close analogs of valine. As expected, the T222P enzyme generated Thr-tRNA^{Val}. In a strain bearing the T222P mutant allele, amino acid analysis and MALDI-mass spectrometry showed that 24% of the valine residues in *E. coli* proteins were replaced with Abu.[76] Mass spectrometry of tryptic digests of a sample protein showed that Abu was specifically inserted at the positions of valine codons. These data demonstrate the importance of the CP1 editing domain for the genetic code.

LeuRS

LeuRS misactivates homocysteine,[36,77,78] γ-hydroxyleucine,[36] δ-hydroxyleucine,[36] γ-hydroxyisoleucine,[36] δ-hydroxyisoleucine,[36] norvaline, norleucine,[79] methionine,[80] and isoleucine.[80] As understanding of the LeuRS editing site emerged, the strong similarities between LeuRS and IleRS and ValRS was apparent. The CP1 polypeptide houses the editing domain. Disruption of the CP1 domain by insertion of a 40 amino acid repeat generates a LeuRS variant that misacylates tRNA^{Leu} with methionine and with isoleucine.[80] A T252A mutation of a conserved threonine yields an enzyme that efficiently edits Leu-tRNA^{Leu}. The authors proposed that the novel editing of Leu-tRNA^{Leu} was due to a widening of the LeuRS editing active site.[38] In contrast, the same mutation in *E. coli* IleRS (T246A) is not diminished in total editing (Schmidt and Schimmel, unpublished results). Separately, A293I, F, Y, and R variants of LeuRS were each able to misacylate tRNA^{Leu} with either isoleucine or methionine.[37]

Finally, LeuRS misactivates Hcy, to generate Hcy-AMP [77] and Hcy-tRNA^{Leu}.[36] Both of these misactivated intermediates are removed via cyclization of Hcy to form the Hcy thiolactone (*vide infra*).

MetRS

By using bacterial strains that are auxotrophic for methionine, selenomethionine and/or telluromethionine can be incorporated into proteins to facilitate crystal structure interpretation.[81] Neither of these amino acids are natural metabolites and both are so similar to methionine that it is not surprising that they are misincorporated into proteins in place of methionine. Thus, selenomethionine and telluromethionine are not removed by an editing pathway. Similarly, MetRS can catalyze misincorporation of S-nitrosohomocysteine (NHcy) into proteins in vivo.[78] In contrast, unmodified Hcy is not misincorporated into proteins because it is removed via cyclization to the Hcy thiolactone (see below). But, unlike Hcy, the modified thiol side chain of NHcy cannot mediate cyclization. The lack of an editing response against NHcy, selenomethionine and telluromethionine is most likely because these novel amino acids are not metabolites in vivo.

In contrast, the amino acid Hcy meets all of the requirements for a non-cognate amino acid that is challenging to MetRS in vivo. Hcy is present in vivo because it is a side product of S-adenosylmethionine-mediated methyl transfer reactions. In contrast to the editing reactions of the other class I AARSs, Hcy can be edited via a tRNA-independent mechanism[82,39] (See below, Editing via cyclization).

CysRS

Although bacterial CysRSs are closely related to MetRS, IleRS, ValRS and LeuRS, they contain a much smaller CP1 insertion.[52] The kinetics for misactivation of serine, Abu and alanine were determined for *E. coli* CysRS, in order to predict whether the enzyme would require hydrolytic proofreading.[83] None of these three amino acids was efficiently activated. Abu was the best non-cognate substrate, but only one Abu was misactivated for every 34,000 cysteines. This rate is well within the levels of accuracy required for protein translation, demonstrating that an editing response would not be necessary in principle. Neither Cys-tRNA^{Cys} nor Ala-tRNA^{Cys} was deacylated by CysRS. The conclusion of this study was that *E. coli* CysRS does not possess appreciable editing activity – sufficient substrate selectivity is achieved during amino acid activation.[83]

Class II

That some class I AARSs use tRNA-dependent editing reactions to guarantee accuracy in their respective tRNA aminoacylation reactions has been recognized for more than 30 years.[26] In contrast, editing by class II AARSs have received less attention. Editing reactions have been identified in five class II AARSs: ProRS.[45,46] ThrRS.[48,84] PheRS,[47] AlaR,[42] and LysRS.[44] The characteristics of the editing reactions of these enzymes are distinct in some of the details from those of their class I counterparts.

ProRS

Unlike other class II enzymes, the phylogenetic distribution of ProRS does not conform to the canonical universal tree of life.[85] Class II enzymes are characterized by an eight-stranded β-structure with flanking α-helices. Three highly degenerate sequence motifs (motifs 1, 2, and 3) are imbedded in this structure and can be used to identify class II enzymes in sequence databases. Many bacterial ProRS sequences (including *E. coli*) are typified by the presence of a large polypeptide (~180 a.a.) inserted between motifs 1 and 2 in the enzyme's active site.[86,85] Based on size similarity between this polypeptide insertion and the CP1 domain of IleRS, Beuning and Musier-Forsyth suggested that this domain could serve as a second sieve for editing, possibly involved in hydrolytic proofreading of smaller amino acids like alanine.[45] To address this possibility, they evaluated *E. coli* ProRS for pre- and post-transfer editing activities against alanine and several other proline-related metabolites (Fig. 11).

ProRS has ATPase activity when challenged with alanine, azetidine-4-carboxylic acid, and cis- and trans-4-hydroxyproline.[45] In contrast to class I editing reactions, this hydrolytic editing activity was tRNA-independent – addition of tRNA to the assay reaction did not increase the rate of ATP hydrolysis. Thus, the authors concluded that this proofreading activity was solely due to pre-transfer editing.[45]

To evaluate post-transfer editing by *E. coli* ProRS, Beuning and Musier-Forsyth constructed a modified tRNA^{Pro}, containing crucial identity determinants for AlaRS (G1:C72 and U70). This chimera was misacylated with alanine by AlaRS to generate Ala-tRNA^{Pro}, which was then evaluated as a substrate for deacylation by ProRS.[45] Indeed, this misacylated tRNA was efficiently deacylated by *E. coli* ProRS. Neither Pro-tRNA^{Pro} nor Lys-tRNA^{Pro} were deacylated under the same reaction conditions. Thus, the post-transfer editing reaction of ProRS is consistent with the presence of a second active site.

A C443G mutation in motif 3 of *E. coli* ProRS dramatically diminished post-transfer editing. The mutation had no effect on ProRS aminoacylation kinetics or on the tRNA-independent pre-transfer editing. Possibly, C443 is important for translocation of misacylated tRNAs from the active site to the editing center that is speculated to be located in the insertion between motifs 2 and 3.[45]

The editing activities of *M. janaschii* ProRS are similar to those of its *E. coli* ortholog. *M. janaschii* ProRS catalyzes tRNA-independent pre-transfer hydrolysis of alanine, azetidine, cis- and trans-hydroxyproline and cysteine. Similarly, the enzyme catalyzes the deacylation of *M. janaschii* Ala-tRNA^{Pro}. Interestingly, *M. janaschii* ProRS does not contain the insertion polypeptide found in *E. coli* ProRS, suggesting that this domain does not contain the ProRS editing site or that the two orthologs edit Ala-tRNA^{Pro} by distinct mechanisms.[46] In contrast to the *M. janaschii* and *E. coli* ProRS orthologs, human ProRS does not catalyze either pre- or post-transfer editing against any of the amino acids evaluated.[46]

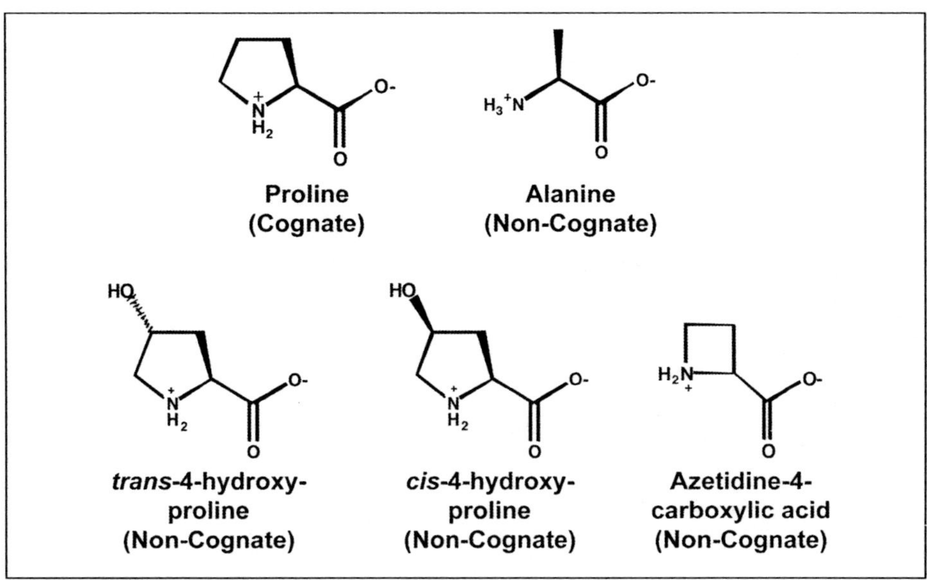

Figure 11. The chemical structure of proline and its close structural analogs. Each of the analogs shown was evaluated for misactivation and editing by ProRS.[45,46]

M. janaschii ProRS is reported to catalyze two different aminoacylation reactions, to generate Pro-tRNA[Pro] and, remarkably, to generate Cys-tRNA[Cys].[87,22,88] It has been designated a ProCysRS.[22] Amino acid discrimination is apparently accomplished via two separate but overlapping substrate binding pockets, one for proline and one for cysteine.[88] Separate amino acid recognition elements argue that the acceptor stems of tRNA[Pro] and of tRNA[Cys] may be differently directed into the enzyme's active site, offering an additional mechanism for accurate pairing of each tRNA to its amino acid.[89]

Accurate Pro-tRNA[Pro] biosynthesis by *M. janaschii* ProCysRS is only maintained by fully modified tRNA[Pro]. Unmodified tRNA[Pro] transcripts, but not modified tRNA[Pro], are misacylated by ProCysRS to generate Cys-tRNA[Pro].[89] (The reverse product, Pro-tRNA[Cys], is not observed.) Because ProCysRS edits Ala-tRNA[Pro],[45] misactivates cysteine, and catalyzes the formation of Cys-tRNA[Pro], editing of cysteine via either the pre- or post-transfer route was investigated. Pre-transfer editing of misactivated cysteine was tRNA-independent and detected at levels that were reduced compared to that for alanine. Unmodified Cys-tRNA[Pro] deacylation was not stimulated by the addition of ProCysRS.[89] Thus, these data argue that the post-transcriptional modifications in tRNA[Pro] are sufficient to guarantee accuracy in aminoacylation and are therefore sufficient to prevent misaminoacylation.[89]

ThrRS

ThrRS must distinguish between threonine and similar non-cognate amino acids like valine and serine. The structural similarities of these amino acids suggest an absolute requirement for a proofreading mechanism. Recent work by Moras and coworkers, including a the report of the crystal structure of *E. coli* ThrRS with a Thr-AMP analog bound to the enzyme's active site, has clarified in part how this enzyme guarantees accuracy in aminoacylation.[48,84]

The active site of the free enzyme contains a unique zinc ion tetracoordinated to Cys334, His385, His511 and a molecule of water.[90] Upon binding of threonine, coordination of the zinc is adjusted to a pentacoordinate ligand sphere, in which the water is displaced and the substrate threonine (as Thr-AMP) forms bidentate contacts with the zinc, through its side chain hydroxyl group and its α-amine (Fig. 12).[84] As the valine side chain does not contain the requisite hydroxyl group, it

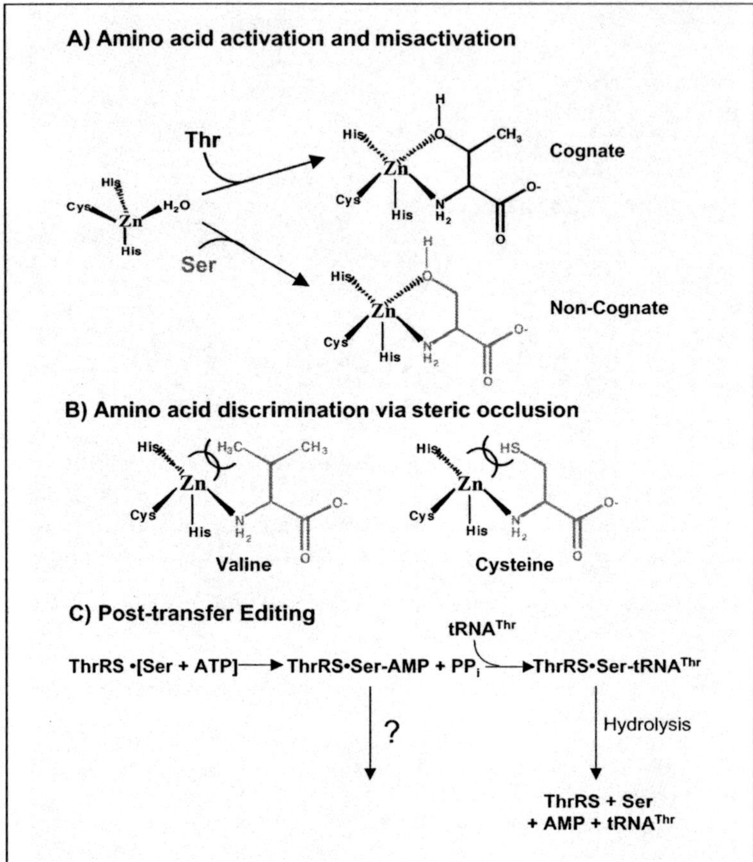

Figure 12. Amino acid misactivation and editing by threonyl-tRNA synthetase.[48,91] A) The active site of ThrRS contains a zinc ion that is tetrahedral, in the absence of substrate. Threonine and serine can occupy this active site, forming a bidentate coordination scheme with the zinc, which converts to a trigonal-bipyramidal geometry. B) Valine and cysteine are too bulky to occupy the active site of ThrRS in the correct orientation. C) ThrRS edits misactivated Ser via post-transfer hydrolysis of Ser-tRNA^Thr.[48]

cannot bind to the active site zinc ion and therefore is not misactivated by ThrRS.[91,84] Thus, in contrast to ValRS, which must edit the non-cognate threonine, ThrRS does not need to edit misactivated valine.

The non-cognate amino acid serine can form the bidentate coordination sphere (Fig. 12) and this amino acid is subsequently misacylated by ThrRS.[48] Ser-tRNA^Thr is rapidly deacylated by ThrRS. (It remains to be seen whether this enzyme catalyzes pre-transfer editing.) Like the class I enzymes, ThrRS contains an extra domain, in this instance appended to its N-terminus. This domain is unrelated to the CP1 domain found in some class I AARSs. Elimination of the N-terminal domain creates a ThrRS variant that is incapable of deacylating Ser-tRNA^Thr and subsequently efficiently misacylates tRNA^Thr.[48] Based on the crystal structure, site-directed mutagenesis was used to further probe the putative editing site, by introducing permutations into a cleft in the N-terminal domain. Both a D180A point mutation and a H73A-H77A double mutation generated ThrRS variants that could not deacylate Ser-tRNA^Thr. Both mutant enzymes also catalyzed the formation of Ser-tRNA^Thr. Thus, ThrRS has a distinct editing active site that is responsible for maintaining translational accuracy.

PheRS

Little is known about editing by phenylalanyl-tRNA synthetase. In vitro, yeast PheRS misactivates tyrosine, leucine and methionine.[92] *E. coli* PheRS misactivates isoleucine and generates Ile-tRNA[Phe] that is efficiently deacylated via post-transfer editing.[32,47] Treatment of Ile-tRNA[Phe] with nitrous acid to generate hydroxymethylvaleryl-tRNA[Phe] (replacing the α-amino group with a hydroxide) eliminated post-transfer editing.[32,47] Thus, the enzyme is sensitive to the presence of the α-amino group of the misactivated amino acid.

AlaRS

Tsui and Fersht evaluated AlaRS for the ability to misactivate serine, glycine, Abu, cysteine, norvaline and norleucine.[42] Only serine and glycine were misactivated at rates that were comparable to that of the cognate amino acid alanine (1/500 and 1/250, respectively). (Careful analysis established that these misactivations were not due to contaminating alanine in their amino acid stocks.) An ATPase activity was detected that was sensitive to addition of exogenous tRNA. Further work has established a post-transfer editing pathway that can be disrupted by specific mutations in the enzyme (Beebe and Schimmel, unpublished).

That serine is sterically larger than alanine is in direct contradiction to the double-sieve model of editing, wherein only smaller, non-cognate amino acids are recognized by an editing site. To address this contradiction, Fersht and Tsui suggested that the editing active sites of PheRS, AlaRS and ValRS were hydrophilic, to prohibit their more hydrophobic cognate substrates from binding.[42] The crystal structure of ValRS has recently confirmed this hypothesis through visualization of a hydrophilic site for editing.[69]

LysRS

With rare exceptions, Lysyl-tRNA synthetases (LysRS) are class II enzymes.[93] *E. coli* LysRS misactivates Hcy and ornithine.[43,44] Like MetRS and Hcy, these two amino acids are removed via cyclization to a thiolactone and a lactam, respectively (*vide infra*)

Editing via Cyclization – *Homocysteine* (Hcy), *Homoserine* (Hse) and *Ornithine* (Orn)

Hcy Cyclization to the Hcy Thiolactone (HcyT)

Hcy is a side product of methylation reactions in vivo. Its metabolic degradation is currently of intense interest because of a correlation between heart disease and serum concentrations of Hcy and the HcyT (Fig. 13).[94,95] The mechanism by which the HcyT leads to an increased risk in cardiovascular disease remains unknown. In human cell cultures, HcyT readily acylates various proteins and may, therefore, contribute to build-up of plaque.[96]

HcyT is generated via the action of various AARSs, all of which convert Hcy to the Hcy-AMP (reviewed by Jakubowski,[97] also see references therein). Hcy is misactivated in vitro by methionyl-, lysyl-, valyl-, isoleucyl-, and leucyl-tRNA synthetases, to generate Hcy-AMP. Subsequent to this misactivation, Hcy-AMP breaks down via an intramolecular cyclization to HcyT and AMP (Fig. 13).[39] This pattern of misactivation and cyclization has also been observed in vivo.[40,41,77,96] At least in some cases (e.g., LeuRS), Hcy is transferred to tRNA[Leu] and subsequently removed via a post-transfer mechanism.[36]

Intramolecular Cyclization of Homoserine and Ornithine

Cyclization reactions have been demonstrated for the non-coded amino acids homoserine and ornithine, after misactivation to their respective AA-AMPs. IleRS, ValRS and LysRS all misactivate homoserine, with release of the lactone.[97] LysRS misactivates ornithine, resulting in formation of the corresponding lactam.[44] These cyclization reactions are virtually identical to that for misactivated Hcy (Fig. 13). Possibly, these three amino acids were prohibited from entering the genetic code simply because their corresponding aminoacyl adenylates are susceptible to non-productive cyclization reactions.

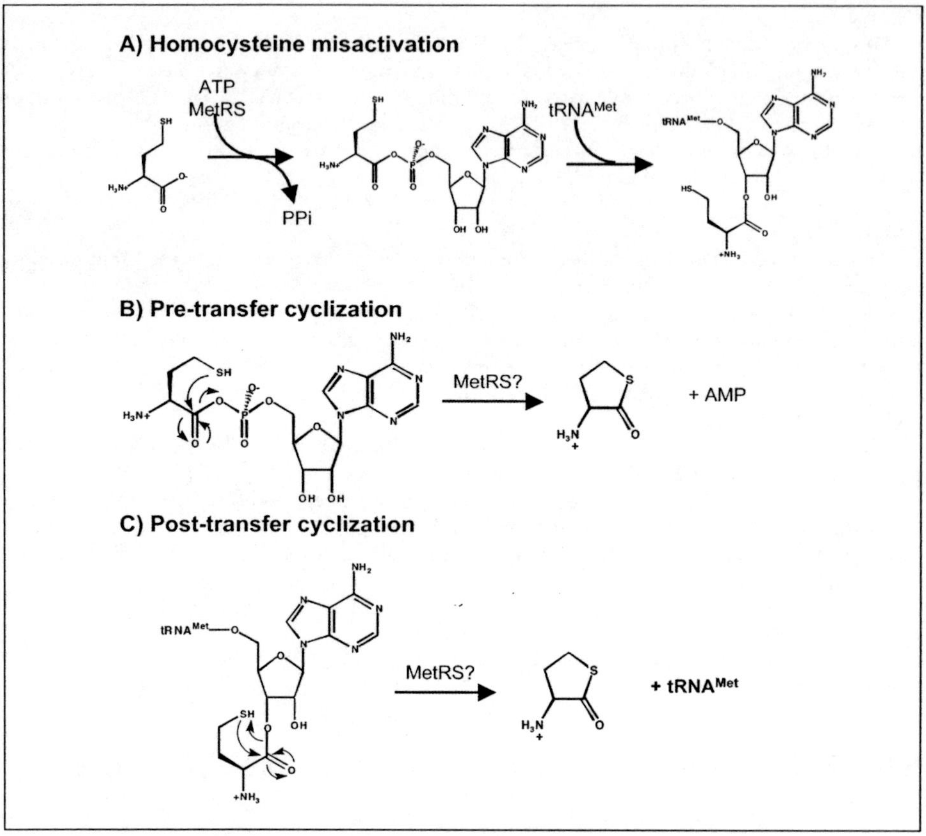

A) Homocysteine misactivation

B) Pre-transfer cyclization

C) Post-transfer cyclization

Figure 13. Homocysteine misactivation and transfer (see text for references). A) Homocysteine is misactivated by MetRS, to form Hcy-AMP and even Hcy-tRNA^{Met}. B) Pre-transfer cyclization converts homocysteine to the homocysteine thiolactone with release of AMP. C) The same cyclization reaction can occur once homocysteine has been transferred to the 3'-end of tRNA^{Met}.

Hcy and MetRS

Although little is known about homoserine and ornithine cyclization reactions, significant effort has gone into exploring the cyclization propensities of Hcy in the presence of various AARSs.[97] Hcy is misactivated by a wide-range of AARSs, from both class I and class II. Because of the close structural relationship between methionine and Hcy, MetRS misactivation of Hcy is a classic example of a challenge in amino acid recognition. In early work, HcyT was generated with MetRS in the absence of tRNA^{Met}. The rate of cyclization was doubled when tRNA was added.[39] Subsequently, Jakubowski expanded on these studies to demonstrate that misactivation is also mediated by several other AARSs in vitro and in vivo, including MetRS, IleRS, ValRS, LeuRS, and LysRS (Jakubowski[97] and references therein). A direct connection between these AARSs and heart disease has been suggested (reviewed in Jakubowski[97]).

A Putative Thiol-Binding Domain

Cyclization of Hcy-AMP to HcyT has been proposed to be an additional editing reaction. It is tRNA-*independent* and independent of the CP1 editing site of Class I enzymes.[98,99] Because Met-AMP cyclizes to the S-methylated HcyT[100], Jakubowski suggested that MetRS can accommodate AA-AMPs in two different binding pockets. The first is the active site's methionine-binding pocket, which is sterically restrictive and therefore prevents cyclization. The second is a proposed thiol-binding domain, which recognizes the free thiol side chain of Hcy, favorably positioning it for nucleophilic attack (Fig. 13).[97-99] The concept of two binding pockets is similar to the double-sieve mechanism proposed by Fersht.[71]

The presence of a thiol-binding domain was supported by deacylation of aminoacyl-tRNAs by exogenous thiol molecules, via a mechanism of trans-thioesterification.[43,98,99,101] If this putative thiol domain is distinct from the synthetic active site, then this additional sieve would offer a new target for the development of cardiovascular therapies. To date, this concept has not been supported by site-directed mutagenesis of the putative thiol-binding domain. However, mutagenesis of the active site of MetRS can perturb the extent to which Hcy is misactivated relative to methionine.[101] Additionally, AA-AMPs are reactive towards exogenous non-thiol based nucleophiles like hydroxylamine,[42,102] showing that small nucleophiles have access to the active sites of AARSs, without a requirement for a specific thiol-binding site. Indeed, a thiol on a flexible arm of the enzyme could swing in and out of the active site.

Possible Mechanisms for Hcy Cyclization

In the original report by Jakubowski and Fersht,[39] two possible mechanisms for Hcy cyclization were suggested. The first is that IleRS, ValRS, and MetRS participate directly in catalyzing the cyclization and breakdown of Hcy-AMP (and even Hcy-tRNA^{AA}). Within the scope of this mechanism, the role of the putative thiol-binding domain would be to position the Hcy side chain in such a way as to favor cyclization.[97] The second possibility is that cyclization could arise spontaneously from the activation of the Hcy carbonyl (either as the mixed anhydride (Hcy-AMP) or as the ester in Hcy-tRNA^{AA}), the favorability of five-membered ring formation reactions and the nucleophilicity of the thiol side chain of Hcy. Thus, Hcy editing may be fortuitous and not require direct enzyme participation.

For example, aminoacyl-tRNA synthetases stabilize each inherently reactive AA-AMP until tRNA aminoacylation is accomplished. Thus, an AARS active site could deter Hcy-AMP breakdown, by sterically confining the Hcy side chain such that it is less free for nucleophilic attack. MetRS releases Met-AMP at a measurable rate, in the absence of tRNA.[103] Perhaps the rate of Hcy-AMP release is much higher. Thus, Hcy editing may result from release of Hcy-AMP so that cyclization can spontaneously occur.

Post-transfer hydrolysis of Hcy-tRNA^{Leu}, in the presence and absence of LeuRS, has been directly compared.[36] An enzyme-catalyzed enhancement of rates was observed ($t_{1/2}$ = 14 minutes with LeuRS; $t_{1/2}$ = 4 hours without LeuRS). In light of the identification of the role of the CP1 domain as the site for editing, these results may reflect more traditional editing via a translocation event to CP1. Reports that demonstrate that LeuRS edited a variety of straight-chain amino acid substrates,[79] support a direct CP1-based hydrolysis of Hcy-tRNA^{Ile,} which seems reasonable.

Two experiments would shed light on the role of each AARS in catalyzing thiolactone formation. First, if one or more AARS directly catalyzes thiolactone cyclization, then this role should be proven by site-directed mutagenesis. No mutation has been identified that blocks thiolactone formation without at the same time disrupting Hcy-AMP synthesis. Second, a comparison of the rates of Hcy-AMP cyclization in the presence and absence of catalytic amounts of each AARS would be informative.

EF-Tu Discrimination of Misacylated-tRNAs as a Third Sieve

The focus of this chapter has been on the mechanisms by which AARSs ensure accuracy in aminoacylation of their cognate tRNAs. Several advances suggest additional mechanisms to prohibit misacylated tRNAs from accessing the ribosome. Following aminoacylation, AA-tRNAs are bound to EF-Tu or another elongation factor (such as SelB, see Chapter 4 by S. Blanquet et al) and then loaded onto the ribosome.[104] EF-Tu has broad tRNA specificities, binding most AA-tRNAs with similar affinities.[105,106] Consistent with this observation, crystal structures of EF-Tu complexed with Phe-tRNA^{Phe} or Cys-tRNA^{Cys} [107,108] demonstrate that primary contacts are between the protein and the phosphate backbone of each tRNA. At least in these crystal structures, EF-Tu does not recognize a specific nucleotide base. In the acceptor stem of a tRNA, several 2'-OH groups have been identified that thermodynamically contribute to the formation of a stable EF-Tu:AA-tRNA complex.

Because EF-Tu loads aminoacylated tRNAs onto the ribosome, it is an obvious target for a final proofreading step. And, despite the diversity of AA-tRNAs recognized by EF-Tu, EF-Tu discriminates between some mischarged AA-tRNAs, effectively prohibiting them from binding to the ribosome and thus from being misincorporated into proteins.[109-112,106] The structure of the amino acid side chain can stabilize the expected EF-Tu: AA-tRNA complex.[109,110] For example, Phe-tRNA^{Phe} and Tyr-tRNA^{Phe} bind EF-Tu with similar efficiencies. However, misacylated Phe-tRNA^{Lys} binds to EF-Tu more strongly than does Lys-tRNA^{Lys}.[110]

That EF-Tu can discriminate against some misacylated tRNAs was demonstrated clearly from a study of EF-Tu from the chloroplast *Pisum sativum*.[111] This organism lacks glutaminyl-tRNA synthetase (GlnRS) and instead indirectly forms Gln-tRNA^{Gln} via a misacylated Glu-tRNA^{Gln} intermediate.[113,114] (The Glu-tRNA^{Gln} is subsequently repaired by a glutamine-dependent amidotransferase, Glu-Adt.[115]) Stanzel et al[111] demonstrated that *P. sativum* EF-Tu bound Gln-tRNA^{Gln}, but not Glu-tRNA^{Gln}. A similar circumstance was evaluated in the thermophilic bacterium *T. thermophilus*, which generates Asp-tRNA^{Asn} as part of a novel mechanism for asparagine biosynthesis.[112] *T. thermophilus* EF-Tu bound Asn-tRNA^{Asn}, but not Asp-tRNA^{Asn}. Thus, EF-Tu plays a direct role in guaranteeing the accuracy of translation at asparagine and glutamine codons, by rejecting both Asp-tRNA^{Asn} and Glu-tRNA^{Gln}.[112]

LaRiviere et al. provided evidence that EF-Tu thermodynamically recognizes AA-tRNAs by favoring certain amino acid and tRNA pairs.[116] A series of accurately aminoacylated and intentionally misacylated tRNAs were systematically evaluated for complex formation with EF-Tu. Surprisingly, although each cognate AA-tRNA^{AA} bound with roughly the same affinity (K_d's ranged from 4.4 to 36 nM), the equilibrium dissociation constants for the misacylated tRNAs were spread over nearly four orders of magnitude, ranging from 0.05 nM to 260 nM. Thus, within the context of an AA-tRNA, both the identity of the amino acid and of the tRNA thermodynamically contribute to formation of an EF-Tu: AA-tRNA complex.[116] Some amino acids and tRNAs are poor binders while others are strong binders. The right combination offers a mechanism for compensation. For example, Gln-tRNA^{Gln} bound to EF-Tu with a K_d of 4.4 nM; Ala-tRNA^{Gln}, K_d = 260 nM; Ala-tRNA^{Ala}, K_d = 6.2 nM; and Gln-tRNA^{Ala}, K_d = 0.05 nM.[116] Thus, while alanine and tRNA^{Gln} interact less favorably with EF-Tu, tRNA^{Ala} and glutamine are favorable. Other results showed that glutamate was a weak binder, providing a mechanism where Glu-tRNA^{Gln} is disfavored relative to Gln-tRNA^{Gln}. The rejection of some misacylated tRNAs by EF-Tu can be considered as a third sieve, by which accuracy in aminoacylation is confirmed.

Conclusions

Transfer RNA-dependent repair of misactivated amino acids was first postulated more than 30 years ago. Over the past decade, an understanding of AARS-catalyzed tRNA-dependent editing reactions has expanded significantly, including identification of editing active site domains in several AARSs and generation of editing-deficient mutations that have deleterious phenotypes in vivo (T243R IleRS and T222P ValRS). Many questions remain however. For example, the mechanism of the hydrolytic reaction is not known. In spite of extensive mutagenesis, a critical catalytic residue has not been identified. Most likely, activated water has a role in catalysis. The mechanism of translocation from the synthetic to the editing site is also not solved. Essential to this process are the structural determinants within the tRNA, whose detailed role is not clear. Finally, as for the mechanism of thermodynamic compensation employed by EF-Tu in rejecting mischarged tRNAs, more mischarged tRNA:EF-Tu complexes need to be investigated to test the generality of the concept. Also, structural information on EF-Tu-misacylated-tRNA complexes would provide clarification.

Acknowledgements

The authors would like to thank Professor Ya-Ming Hou for providing the copy of a manuscript prior to publication and Dan Grilley for help with Figures 6 and 9. This work was supported by grants GM15539 and GM23562 from the National Institutes of Health and by a fellowship from the National Foundation for Cancer Research.

References

1. Madigan MT, Marrs BL. Extremophiles. Sci Am 1997; 276:82-87.
2. Carpenter EJ, Lin S, Capone DG. Bacterial activity in South Pole snow. Appl Environ Microbiol 2000; 66(10):4514-7.
3. Blochl E, Rachel R, Burggraf S et al. *Pyrolobus fumarii*, gen. and *sp. nov.*, represents a novel group of archaea, extending the upper temperature limit for life to 113 degrees C. Extremophiles 1997; 1(1):14-21.
4. Stetter KO. Extremophiles and their adaptation to hot environments. FEBS Lett 1999; 452(1-2):22-5.
5. Blomberg C, Johansson J, Liljenstrom H. Error propagation in *E. coli* protein synthesis. J Theor Biol 1985; 113(3):407-23.
6. Ibba M, Soll D. Quality control mechanisms during translation. Science 1999; 286(5446):1893-7.
7. Jager J, Pata JD. Getting a grip: polymerases and their substrate complexes. Curr Opin Struct Biol 1999; 9(1):21-8.
8. Kunkel TA, Bebenek K. DNA replication fidelity. Annu Rev Biochem 2000; 69:497-529.
9. Rodnina MV, Wintermeyer W. Ribosome fidelity: tRNA discrimination, proofreading and induced fit. Trends Biochem Sci 2001; 26(2):124-30.
10. Ibba M, Soll D. Aminoacyl-tRNA Synthesis. Annu Rev Biochem 2000; 69:617-650.
11. Green R, Noller HF. Ribosomes and translation. Annu Rev Biochem 1997; 66:679-716.
12. Ellman J, Mendel D, Anthony-Cahill A et al. Biosynthetic Method for Introducing Unnatural amino acids Site-Specifically into Proteins. Methods Enzymol 1991; 202:301-336.
13. Seemuller E, Lupas A, Stock D et al. Proteasome from *Thermoplasma acidophilum*: a threonine protease. Science 1995; 268(5210):579-82.
14. Welply JK, Shenbagamurthi P, Lennarz WJ et al. Substrate recognition by oligosaccharyltransferase. Studies on glycosylation of modified Asn-X-Thr/Ser tripeptides. J Biol Chem 1983; 258(19):11856-63.
15. Loftfield RB, Vanderjagt D. The frequency of errors in protein biosynthesis. Biochem J 1972; 128:1353-1356.
16. Freist W, Sternbach H, Cramer F. Isoleucyl-tRNA synthetase from baker's yeast and from *Escherichia coli MRE 600*. Discrimination of 20 amino acids in aminoacylation of tRNA(Ile)-C-C-A. Eur J Biochem 1988; 173(1):27-34.
17. Webster T, Tsai H, Kula M et al. Specific sequence homology and three-dimensional structure of an aminoacyl transfer RNA synthetase. Science 1984; 226(4680):1315-7.
18. Ludmerer SW, Schimmel P. Gene for yeast glutamine tRNA synthetase encodes a large amino-terminal extension and provides a strong confirmation of the signature sequence for a group of the aminoacyl-tRNA synthetases. J Biol Chem 1987; 262(22):10801-6.
19. Cusack S, Berthet-Colominas C, Hartlein M et al. A second class of synthetase structure revealed by X-ray analysis of *Escherichia coli* seryl-tRNA synthetase at 2.5 A. Nature 1990; 347(6290):249-55.
20. Eriani G, Delarue M, Poch O et al. Partition of tRNA synthetases into two classes based on mutually exclusive sets of sequence motifs. Nature 1990; 347(6289):203-6.
21. Ibba M, Curnow AW, Soll D. Aminoacyl-tRNA synthesis: divergent routes to a common goal. Trends Biochem Sci 1997; 22(2):39-42.
22. Stathopoulos C, Li T, Longman R et al. One polypeptide with two aminoacyl-tRNA synthetase activities [see comments]. Science 2000; 287(5452):479-82.
23. Praetorius-Ibba M, Stange-Thomann N, Kitabatake M et al. Ancient adaptation of the active site of tryptophanyl-tRNA synthetase for tryptophan binding. Biochemistry 2000; 39(43):13136-43.
24. Thiebe R. Arginyl-tRNA synthetase from brewer's yeast. Purification, properties, and steady-state mechanism. Eur J Biochem 1983; 130(3):517-24.
25. Pauling L. In: Festschrift fuer Prof. Dr. Arthur Stoll Siebzigsten. Basel: Birkhayser-Verlag, 1958:597-602.
26. Baldwin AN, Berg P. Transfer Ribonucleic Acid-induced Hydrolysis of Valyladenylate Bound to Isoleucyl Ribonucleic Acid Synthetase. J Biol Chem 1966; 241(4):839-845.
27. Eldred EW, Schimmel PR. Rapid deacylation by isoleucyl transfer ribonucleic acid synthetase of isoleucine-specific transfer ribonucleic acid aminoacylated with valine. J Biol Chem 1972; 247(9):2961-4.
28. Schreier AA, Schimmel PR. Transfer ribonucleic acid synthetase catalyzed deacylation of aminoacyl transfer ribonucleic acid in the absence of adenosine monophosphate and pyrophosphate. Biochemistry 1972; 11(9):1582-9.
29. Fersht AR, Kaethner MM. Mechanism of aminoacylation of tRNA. Proof of the aminoacyl adenylate pathway for the isoleucyl- and tyrosyl-tRNA synthetases from *Escherichia coli K12*. Biochemistry 1976; 15(4):818-23.
30. Fersht AR. Editing mechanisms in protein synthesis. Rejection of valine by the isoleucyl-tRNA synthetase. Biochemistry 1977; 16(5):1025-30.
31. Moe JG, Piszkiewicz D. Isoleucyl transfer ribonucleic acid synthetase. Steady-state kinetic analysis. Biochemistry 1979; 18(13):2804-10.
32. Yarus M. Verification of misacylated tRNA^{phe} is apparently carried out only by phenylalanyl-tRNA synthetase. Nat New Biol 1973; 245(140):5-6.
33. Fersht AR. Sieves in sequence [comment]. Science 1998; 280(5363):541.
34. Fersht AR, Kaethner MM. Enzyme hyperspecificity. Rejection of threonine by the valyl-tRNA synthetase by misacylation and hydrolytic editing. Biochemistry 1976; 15(15):3342-6.
35. Lin L, Schimmel P. Mutational analysis suggests the same design for editing activities of two tRNA synthetases. Biochemistry 1996; 35(17):5596-601.

36. Englisch S, Englisch U, von der Haar F et al. The proofreading of hydroxy analogues of leucine and isoleucine by leucyl-tRNA synthetases from *E. coli* and yeast. Nucleic Acids Res 1986; 14(19):7529-39.
37. Chen JF, Li T, Wang ED et al. Effect of alanine-293 replacement on the activity, ATP binding, and editing of *Escherichia coli* leucyl-tRNA synthetase. Biochemistry 2001; 40(5):1144-9.
38. Mursinna RS, Lincecum TL, Jr., Martinis SA. A conserved threonine within *Escherichia coli* leucyl-tRNA synthetase prevents hydrolytic editing of leucyl-tRNA^{Leu}. Biochemistry 2001; 40(18):5376-81.
39. Jakubowski H, Fersht AR. Alternative pathways for editing non-cognate amino acids by aminoacyl- tRNA synthetases. Nucleic Acids Res 1981; 9(13):3105-17.
40. Jakubowski H. Proofreading in vivo: editing of homocysteine by methionyl-tRNA synthetase in *Escherichia coli*. Proc Natl Acad Sci USA 1990; 87(12):4504-8.
41. Jakubowski H. Proofreading in vivo: editing of homocysteine by methionyl-tRNA synthetase in the yeast *Saccharomyces cerevisiae*. EMBO J 1991; 10(3):593-8.
42. Tsui WC, Fersht AR. Probing the principles of amino acid selection using the alanyl-tRNA synthetase from *Escherichia coli*. Nucleic Acids Res 1981; 9(18):4627-37.
43. Jakubowski H. Aminoacyl thioester chemistry of class II aminoacyl-tRNA synthetases. Biochemistry 1997; 36(37):11077-85.
44. Jakubowski H. Misacylation of tRNA^{Lys} with noncognate amino acids by lysyl-tRNA synthetase. Biochemistry 1999; 38(25):8088-93.
45. Beuning PJ, Musier-Forsyth K. Hydrolytic editing by a class II aminoacyl-tRNA synthetase. Proc Natl Acad Sci USA 2000; 97(16):8916-20.
46. Beuning PJ, Musier-Forsyth K. Species-specific differences in amino acid editing by class II prolyl- tRNA synthetase. J Biol Chem 2001; 276(33):30779-85.
47. Yarus M. Phenylalanyl-tRNA synthetase and isoleucyl-tRNA^{Phe}: a possible verification mechanism for aminoacyl-tRNA. Proc Natl Acad Sci USA 1972; 69(7):1915-9.
48. Dock-Bregeon A, Sankaranarayanan R, Romby P et al. Transfer RNA-mediated editing in threonyl-tRNA synthetase. The class II solution to the double discrimination problem [In Process Citation]. Cell 2000; 103(6):877-84.
49. Schmidt E, Schimmel P. Residues in a class I tRNA synthetase which determine selectivity of amino acid recognition in the context of tRNA. Biochemistry 1995; 34(35):11204-10.
50. Hendrickson TL, Nomanbhoy TK, Schimmel P. Errors from selective disruption of the editing center in a tRNA synthetase. Biochemistry 2000; 39(28):8180-6.
51. Hendrickson TL, Nomanbhoy TK, de Crecy-Lagard V et al. Mutational Separation of Two Pathways for editing by a Class I tRNA Synthetase. Mol Cell 2002.
52. Starzyk RM, Webster TA, Schimmel P. Evidence for dispensable sequences inserted into a nucleotide fold. Science 1987; 237(4822):1614-8.
53. Hou YM, Shiba K, Mottes C et al. Sequence determination and modeling of structural motifs for the smallest monomeric aminoacyl-tRNA synthetase. Proc Natl Acad Sci USA 1991; 88(3):976-80.
54. Lin L, Hale SP, Schimmel Pf. Aminoacylation error correction [letter]. Nature 1996; 384(6604):33-4.
55. Schmidt E, Schimmel P. Mutational isolation of a sieve for editing in a transfer RNA synthetase. Science 1994; 264(5156):265-7.
56. Hale SP, Schimmel P. Protein synthesis editing by a DNA aptamer. Proc Natl Acad Sci USA 1996; 93(7):2755-8.
57. Hale SP, Schimmel P. DNA Aptamer Targets Translational Editing Motif in a tRNA Synthetase. Tetrahedron 1997; 53(35):11985-11994.
58. Farrow MA, Schimmel P. Editing by a tRNA synthetase: DNA aptamer-induced translocation and hydrolysis of a misactivated amino acid. Biochemistry 2001; 40(14):4478-83.
59. Giege R, Sissler M, Florentz C. Universal rules and idiosyncratic features in tRNA identity. Nucleic Acids Res 1998; 26(22):5017-35.
60. Nureki O, Niimi T, Muramatsu T et al. Molecular recognition of the identity-determinant set of isoleucine transfer RNA from *Escherichia coli*. J Mol Biol 1994; 236(3):710-24.
61. Hale SP, Auld DS, Schmidt E et al. Discrete determinants in transfer RNA for editing and aminoacylation. Science 1997; 276(5316):1250-2.
62. Farrow MA, Nordin BE, Schimmel P. Nucleotide determinants for tRNA-dependent amino acid discrimination by a class I tRNA synthetase. Biochemistry 1999; 38(51):16898-903.
63. Silvian LF, Wang J, Steitz TA. Insights into editing from an ile-tRNA synthetase structure with tRNA^{Ile} and Mupirocin. Science 1999; 285(5430):1074-7.
64. Nordin BE, Schimmel P. RNA determinants for translational editing. Mischarging a minihelix substrate by a tRNA synthetase. J Biol Chem 1999; 274(11):6835-8.
65. Nureki O, Vassylyev DG, Tateno M et al. Enzyme structure with two catalytic sites for double-sieve selection of substrate. Science 1998; 280(5363):578-82.
66. Nomanbhoy TK, Hendrickson TL, Schimmel P. Transfer RNA-Dependent Translocation ofMisactivated Amino acids to Prevent Errors in Protein synthesis. Mol Cell 1999; 4:519-528.
67. Nomanbhoy TK, Schimmel PR. Misactivated amino acids translocate at similar rates across surface of a tRNA synthetase. Proc Natl Acad Sci USA 2000; 97(10):5119-22.
68. Bishop AC, Nomanbhoy TK, Schimmel P. Blocking site-to-site translocation of a misactivated amino acid by mutation of a class I tRNA synthetase. Proc Natl Acad Sci USA 2002; in press.
69. Fukai S, Nureki O, Sekine S et al. Structural basis for double-sieve discrimination of L-valine from L-isoleucine and L-threonine by the complex of tRNA(Val) and valyl-tRNA synthetase [In Process Citation]. Cell 2000; 103(5):793-803.

70. Jakubowski H. Amino acid selectivity in the aminoacylation of coenzyme A and RNA minihelices by aminoacyl-tRNA synthetases. J Biol Chem 2000; 275(45):34845-8.
71. Fersht AR, Dingwall C. Evidence for the double-sieve editing mechanism in protein synthesis. Steric exclusion of isoleucine by valyl-tRNA synthetases. Biochemistry 1979; 18(12):2627-31.
72. Hountondji C, Beauvallet C, Dessen P et al. Valyl-tRNA synthetase from *Escherichia coli* MALDI-MS identification of the binding sites for L-valine or for noncognate amino acids upon qualitative comparative labeling with reactive amino-acid analogs. Eur J Biochem 2000; 267(15):4789-98.
73. Fersht AR, Dingwall C. Establishing the misacylation/deacylation of the tRNA pathway for the editing mechanism of prokaryotic and eukaryotic valyl-tRNA synthetases. Biochemistry 1979; 18(7):1238-45.
74. Tamura K, Nameki N, Hasegawa T et al. Role of the CCA terminal sequence of tRNA(Val) in aminoacylation with valyl-tRNA synthetase. J Biol Chem 1994; 269(35):22173-7.
75. Tardif KD, Liu M, Vitseva O et al. Misacylation and editing by *Escherichia coli* valyl-tRNA synthetase: evidence for two tRNA binding sites. Biochemistry 2001; 40(27):8118-25.
76. Doring V, Mootz HD, Nangle LA et al. Enlarging the amino acid set of *Escherichia coli* by infiltration of the valine coding pathway. Science 2001; 292(5516):501-4.
77. Jakubowski H. Proofreading in vivo. Editing of homocysteine by aminoacyl-tRNA synthetases in *Escherichia coli*. J Biol Chem 1995; 270(30):17672-3.
78. Jakubowski H. Translational incorporation of S-nitrosohomocysteine into protein. J Biol Chem 2000; 275(29):21813-6.
79. Martinis SA, Fox GE. Non-standard amino acid recognition by *Escherichia coli* leucyl-tRNA synthetase. Nucleic Acids Symp Ser 1997; 36:125-8.
80. Chen JF, Guo NN, Li T et al. CP1 domain in *Escherichia coli* leucyl-tRNA synthetase is crucial for its editing function. Biochemistry 2000; 39(22):6726-31.
81. Besse D, Budisa N, Karnbrock W et al. Chalcogen-analogs of amino acids. Their use in X-ray crystallographic and folding studies of peptides and proteins. Biol Chem 1997; 378(3-4):211-8.
82. Fersht AR, Dingwall C. An editing mechanism for the methionyl-tRNA synthetase in the selection of amino acids in protein synthesis. Biochemistry 1979; 18(7):1250-6.
83. Fersht AR, Dingwall C. Cysteinyl-tRNA synthetase from *Escherichia coli* does not need an editing mechanism to reject serine and alanine. High binding energy of small groups in specific molecular interactions. Biochemistry 1979; 18(7):1245-9.
84. Sankaranarayanan R, Dock-Bregeon AC, Rees B et al. Zinc ion mediated amino acid discrimination by threonyl-tRNA synthetase [see comments]. Nat Struct Biol 2000; 7(6):461-5.
85. Stehlin C, Burke B, Yang F et al. Species-specific differences in the operational RNA code for aminoacylation of tRNA^{Pro}. Biochemistry 1998; 37(23):8605-13.
86. Cusack S, Yaremchuk A, Krikliviy I et al. tRNA(Pro) anticodon recognition by *Thermus thermophilus* prolyl-tRNA synthetase. Structure 1998; 6(1):101-8.
87. Lipman RS, Sowers KR, Hou YM. Synthesis of cysteinyl-tRNA(Cys) by a genome that lacks the normal cysteine-tRNA synthetase. Biochemistry 2000; 39(26):7792-8.
88. Stathopoulos C, Jacquin-Becker C, Becker HD et al. *Methanococcus jannaschii* Prolyl-Cysteinyl-tRNA synthetase Possesses Overlapping Amino acid Binding Sites. Biochemistry 2001; 40(1):46-52.
89. Lipman RS, Wang J, Sowers KR et al. Prevention of mis-aminoacylation of a dual-specificity aminoacyl-tRNA synthetase. J Mol Biol 2002; 315(5):943-9.
90. Sankaranarayanan R, Dock-Bregeon AC, Romby P et al. The structure of threonyl-tRNA synthetase-tRNA(Thr) complex enlightens its repressor activity and reveals an essential zinc ion in the active site. Cell 1999; 97(3):371-81.
91. Musier-Forsyth K, Beuning PJ. Role of zinc ion in translational accuracy becomes crystal clear. Nat Struct Biol 2000; 7(6):435-6.
92. Igloi GL, von der Haar F, Cramer F. Aminoacyl-tRNA synthetases from yeast: generality of chemical proofreading in the prevention of misaminoacylation of tRNA. Biochemistry 1978; 17(17):3459-68.
93. Ibba M, Morgan S, Curnow AW et al. A euryarchaeal lysyl-tRNA synthetase: resemblance to class I synthetases. Science 1997; 278(5340):1119-22.
94. McCully KS, Vezeridis MP. Homocysteine thiolactone in arteriosclerosis and cancer. Res Commun Chem Pathol Pharmacol 1988; 59(1):107-19.
95. Wood D. Established and emerging cardiovascular risk factors. Am Heart J 2001; 141(2 Suppl):S49-57.
96. Jakubowski H. Metabolism of homocysteine thiolactone in human cell cultures. Possible mechanism for pathological consequences of elevated homocysteine levels. J Biol Chem 1997; 272(3):1935-42.
97. Jakubowski H. Translational accuracy of Aminoacyl-tRNA Synthetases: Implications for Atherosclerosis. J Nutr 2001; 131(11):2983S-7S.
98. Jakubowski H. Proofreading in trans by an aminoacyl-tRNA synthetase: a model for single site editing by isoleucyl-tRNA synthetase. Nucleic Acids Res 1996; 24(13):2505-10.
99. Jakubowski H. The synthetic/editing active site of an aminoacyl-tRNA synthetase: evidence for binding of thiols in the editing subsite. Biochemistry 1996; 35(25):8252-9.
100. Jakubowski H. Proofreading and the evolution of a methyl donor function. Cyclization of methionine to S-methyl homocysteine thiolactone by *Escherichia coli* methionyl-tRNA synthetase. J Biol Chem 1993; 268(9):6549-53.
101. Kim HY, Ghosh G, Schulman LH et al. The relationship between synthetic and editing functions of the active site of an aminoacyl-tRNA synthetase. Proc Natl Acad Sci USA 1993; 90(24):11553-7.
102. Loftfield RB, Eigner EA. The specificity of enzymic reactions. Aminoacyl-soluble RNA ligases. Biochim Biophys Acta 1966; 130(2):426-48.

103. Gillet S, Hountondji C, Schmitter JM et al. Covalent methionylation of *Escherichia coli* methionyl-tRNA synthethase: identification of the labeled amino acid residues by matrix-assisted laser desorption-ionization mass spectrometry. Protein Sci 1997; 6(11):2426-35.

104. Clark BF, Nyborg J. The ternary complex of EF-Tu and its role in protein biosynthesis. Curr Opin Struct Biol 1997; 7(1):110-6.

105. Ott G, Schiesswohl M, Kiesewetter S et al. Ternary complexes of *Escherichia coli* aminoacyl-tRNAs with the elongation factor Tu and GTP: thermodynamic and structural studies. Biochim Biophys Acta 1990; 1050(1-3):222-5.

106. Pleiss JA, Uhlenbeck OC. Identification of thermodynamically relevant interactions between EF-Tu and backbone elements of tRNA. J Mol Biol 2001; 308(5):895-905.

107. Nissen P, Kjeldgaard M, Thirup S et al. Crystal structure of the ternary complex of Phe-tRNA^{Phe}, EF-Tu, and a GTP analog. Science 1995; 270(5241):1464-72.

108. Nissen P, Thirup S, Kjeldgaard M et al. The crystal structure of Cys-tRNA^{Cys}-EF-Tu-GDPNP reveals general and specific features in the ternary complex and in tRNA. Structure Fold Des 1999; 7(2):143-56.

109. Knowlton RG, Yarus M. Discrimination between aminoacyl groups on su+ 7 tRNA by elongation factor Tu. J Mol Biol 1980; 139(4):721-32.

110. Wagner T, Sprinzl M. The complex formation between *Escherichia coli* aminoacyl-tRNA, elongation factor Tu and GTP. The effect of the side-chain of the amino acid linked to tRNA. Eur J Biochem 1980; 108(1):213-21.

111. Stanzel M, Schon A, Sprinzl M. Discrimination against misacylated tRNA by chloroplast elongation factor Tu. Eur J Biochem 1994; 219(1-2):435-9.

112. Becker HD, Kern D. *Thermus thermophilus*: a link in evolution of the tRNA-dependent amino acid amidation pathways. Proc Natl Acad Sci USA 1998; 95(22):12832-7.

113. Schon A, Kannangara CG, Gough S et al. Protein biosynthesis in organelles requires misaminoacylation of tRNA. Nature 1988; 331(6152):187-90.

114. Jahn D, Kim YC, Ishino Y et al. Purification and functional characterization of the Glu-tRNA(Gln) amidotransferase from *Chlamydomonas reinhardtii*. J Biol Chem 1990; 265(14):8059-64.

115. Curnow AW, Hong K, Yuan R et al. Glu-tRNA^{Gln} amidotransferase: a novel heterotrimeric enzyme required for correct decoding of glutamine codons during translation [see comments]. Proc Natl Acad Sci USA 1997; 94(22):11819-26.

116. LaRiviere FJ, Wolfson AD, Uhlenbeck OC. Uniform binding of aminoacyl-tRNAs to elongation factor Tu by thermodynamic compensation. Science 2001; 294(5540):165-8.

117. Freist W, Sternbach H, Cramer F. Arginyl-tRNA synthetase from yeast. Discrimination between 20 amino acids in aminoacylation of tRNA(Arg)-C-C-A and tRNA(Arg)-C-C-A(3'NH2). Eur J Biochem 1989; 186(3):535-41.

118. Fersht AR, Dingwall C. An editing mechanism for the methionyl-tRNA synthetase in the selection of amino acids in protein synthesis. Biochemistry 1979; 18(7):1250-6.

119. Fersht AR, Shindler JS, Tsui WC. Probing the limits of protein-amino acid side chain recognition with the aminoacyl-tRNA synthetases. Discrimination against phenylalanine by tyrosyl-tRNA synthetases. Biochemistry 1980; 19(24):5520-4.

120. Jakubowski H. Editing function of *Escherichia coli* cysteinyl-tRNA synthetase: cyclization of cysteine to cysteine thiolactone. Nucleic Acids Res 1994; 22(7):1155-60.

121. Nureki O, Vassylyev DG, Tateno M et al. Proofreading by Isoleucyl-tRNA synthetase: Response. Science 1998; 283:459.

122. Blattner FR, Plunkett G, 3rd, Bloch CA et al. The complete genome sequence of *Escherichia coli* K-12. Science 1997; 277(5331):1453-74.

123. Bult CJ, White O, Olsen GJ et al. Complete genome sequence of the methanogenic archaeon, *Methanococcus jannaschii*. Science 1996; 273(5278):1058-73.

124. Shiba K, Suzuki N, Shigesada K et al. Human cytoplasmic isoleucyl-tRNA synthetase: selective divergence of the anticodon-binding domain and acquisition of a new structural unit. Proc Natl Acad Sci USA 1994; 91(16):7435-9.

125. Goffeau A, Barrell BG, Bussey H et al. Life with 6000 genes. Science 1996; 274(5287):546, 563-7.

126. Hartlein M, Frank R, Madern D. Nucleotide sequence of *Escherichia coli* valyl-tRNA synthetase gene valS. Nucleic Acids Res 1987; 15(21):9081-2.

127. Tettelin H, Agostoni Carbone ML, Albermann K et al. The nucleotide sequence of *Saccharomyces cerevisiae* chromosome VII. Nature 1997; 387(6632 Suppl):81-4.

CHAPTER 4

Aminoacyl-Transfer RNA Maturation

Sylvain Blanquet, Yves Mechulam, Emmanuelle Schmitt and Lionel Vial

Summary

Some aminoacylated tRNAs are subject to specific maturation. These maturations can be necessary either to produce the correct aminoacylated species for the ribosome or to regenerate toxic or nonproductive tRNAs produced after errors of the translational machinery. This chapter will first focus on the transamidation pathways for asparaginylation and glutaminylation of tRNAs, on the production of selenocysteinyl-tRNA and on the formylation of the initiator Met-tRNA in bacteria and organelles. The second part will deal with the recycling of peptidyl-tRNAs produced from premature polypeptide chain termination and with the hydrolysis of D-aminoacylated-tRNAs. Finally, the specificity of EF-Tu will be discussed in relation with the above processes.

Introduction: Aminoacyl-tRNA Processing in the Context of Translation

Any free functional tRNA is intended to be recognized as substrate by a given aminoacyl-tRNA synthetase. This correspondence results in the esterification of each tRNA with the cognate amino acid. After aminoacylation, most aminoacyl-tRNAs are ready for use in translation. They are directed towards the ribosomal A site where they participate in polypeptide chain elongation. To achieve this step, interaction of aminoacylated tRNA with elongation factor, EF-Tu in eubacteria and organelles, and eEF1A in archaea and the cytoplasm of eukaryotes, is required.

However, a few aminoacylated tRNAs species are not ready-to-use substrates of protein biosynthesis. To reach the ribosome, they must undergo maturation and therefore follow specific routes (Fig. 1). These routes imply recognition by specific proteins as well as protection against misappropriation by the canonical elongation factor.

Besides the dogma of an aminoacyl-tRNA synthetase for each of the 20 amino acids, there are cases where an aminoacyl-tRNA synthetase is lacking. In these cases, correct aminoacylation can be obtained through adequate maturation of a precursor amino acid attached to a tRNA. In many organisms, Asn-tRNA^{Asn} and/or Gln-tRNA^{Gln} derive from Asp-tRNA^{Asn} or Glu-tRNA^{Gln}. Furthermore, according to its ubiquitous character, selenocysteine may be considered as the twenty first amino acid although it is very rarely used in protein. Aminoacylation of tRNA^{Sec} is achieved through modification of Ser-tRNA^{Sec} precursor.

In eubacterial protein biosynthesis, as well as in mitochondria and chloroplasts, prior to its recruitment towards the ribosomal P site, the initiator tRNA, called tRNA$_f^{Met}$, is modified by formylation of the esterified methionine. In *E. coli*, it was demonstrated that this step is crucial in the acquisition by the initiator tRNA of its initiator identity. Indeed, formylation enables specific interaction with the initiation factor, IF2. Formylation also impairs binding of the initiator tRNA to EF-Tu.

Finally, following an accidental process, some tRNAs become incorrectly esterified and must not be associated with elongation factor to avoid poisoning of the protein biosynthesis machinery. One example is the release of peptidyl-tRNAs in the cytoplasm upon premature termination of translation. Recycling of such peptidyl-tRNAs is insured by a peptidyl-tRNA hydrolase, and can be also considered as a maturation. It should be underlined that the eubacterial enzyme is specific enough to

Translation Mechanisms, edited by Jacques Lapointe and Léa Brakier-Gingras.
©2003 Eurekah.com and Kluwer Academic / Plenum Publishers.

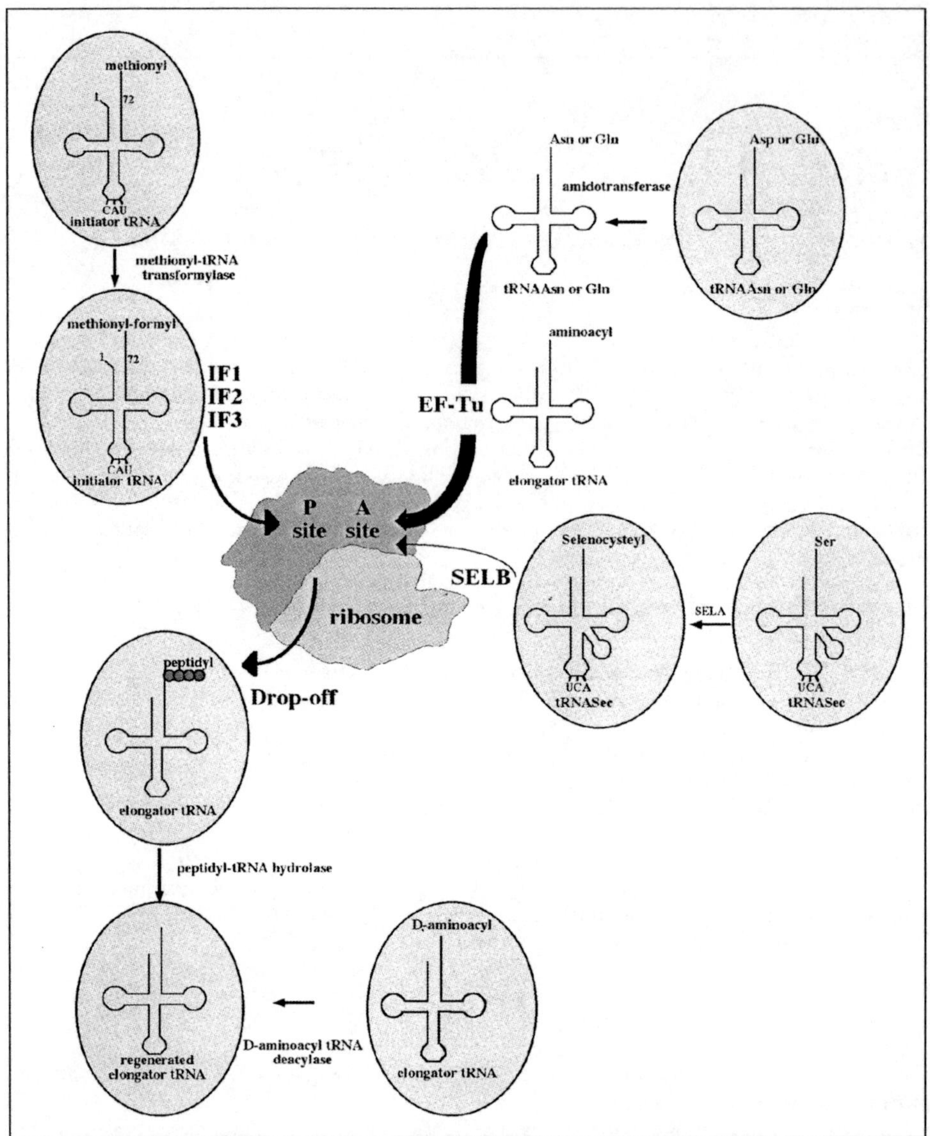

Figure 1. Different pathways for an aminoacyl-tRNA in eubacterial cells. tRNA species schematized on a grey background must be protected from EF-Tu binding.

distinguish between an elongator tRNA carrying a peptidyl moiety and initiator tRNA esterified with an N-blocked methionine.

Other incorrect aminoacyl-tRNAs result from aminoacylation errors. Such errors appear to be often corrected by the aminoacyl-tRNA synthetases themselves. Immediate correction by the synthetase precludes binding by the elongation factor. Corresponding editing-hydrolysis reactions are beyond the scope of this chapter, and are discussed by T Hendrickson and P. Schimmel in Chapter 3. However, we shall consider here the case of the misacylation of tRNA with a D amino acid. Indeed, maturation of D-aa-tRNA into free tRNA involves a specific D-aa-tRNA deacylase.

The goal of this chapter is to describe the above aminoacyl-tRNA maturation processes, and to discuss their implications in the cellular context. The chapter will focus on the documented case of eubacteria, but, whenever possible, the cases of other organisms will also be considered.

Processing of Elongator tRNAs

Transamidation Pathways for the Production of Asn-tRNA or Gln-tRNA

Beyond the canonical process to produce the aminoacylated tRNAs, there are indirect routes leading to correctly aminoacylated species. Such pathways often occur for two aminoacylated tRNAs, Asn-tRNA^{Asn} and Gln-tRNA^{Gln}, which can be produced through a two-step mechanism (Fig. 2).

In the first step, tRNA^{Asn} or tRNA^{Gln} are esterified with aspartate or glutamate, respectively, by a nondiscriminative aspartyl- or glutamyl-tRNA synthetase. In the second step, an amidotransferase catalyzes the transamidation reaction from an amide-nitrogen donor (L-asparagine or L-glutamine) to the mischarged tRNA to give the right species. These enzymes were named glutamyl-tRNA^{Gln} amidotransferase (GluAdT) and aspartyl-tRNA^{Asn} amidotransferase (AspAdT). The availability of complete genome sequences of various organisms has recently allowed to envisage the transamidation process in the different kingdoms.

The transamidation reaction to give Gln-tRNA^{Gln} was first evidenced in some gram-positive bacteria,[1-3] in organelles and in cyanobacteria.[4] This process was further evidenced in the gram-negative bacteria *Rhizobium meliloti*[5] and in some archaea.[6-8] The indirect pathway also exists in the case of asparagine in archaea and in some bacteria.[7,9] It is usually related to the absence in the considered organism of the corresponding glutaminyl-tRNA synthetase (GlnRS) or asparaginyl-tRNA synthetase (AsnRS) and to the presence of an amidotransferase.

The first GluAdT was identified in *Bacillus subtilis*.[10,11] This enzyme is a heterotrimeric protein expressed from three genes, *gatC, gatA, gatB*. Orthologs of these genes were further identified and characterized in *Deinococcus radiodurans*.[12] Sequence alignements allow the prediction of the presence of a GluAdT in many bacteria. The putative presence of this protein is systematically associated with the lack of a gene coding for a Gln-tRNA synthetase and with the presence of a nondiscriminative Glu-tRNA synthetase.[3,13,14] The transamidation reaction is therefore the only way to produce Gln-tRNA^{Gln} in these procaryotic species. In agreement with this conclusion, inactivation of the *gatCAB* operon in *B. subtilis* is lethal.[11] Biochemical studies showed that the amidotransferase is specific for the misaminoacylated species Glu-tRNA^{Gln}. In particular, Glu-tRNA^{Glu} is not a substrate.[11] The transamidation reaction can be described as follows.[1,2]. First, the amido donor, Asn or Gln, is hydrolyzed to give an NH_3 group sequestered by the enzyme. Second, the γ-carboxyl group of the glutamyl moiety of Glu-tRNA^{Gln} is activated by ATP to give a γ-phospho-Glu-tRNA^{Gln} intermediate. Finally, the activated intermediate is aminolyzed by the sequestered NH_3 group to form the final product. The three above activities of Glu-AdT are tightly coupled, reflecting communication between putatively distant active sites.[15] The GatB subunit might be involved in the recognition of the tRNA part of the substrate, whereas the GatA subunit shares amino acid sequence similarities with the amidase family.[11] The GatC subunit possibly keeps the three subunits correctly assembled. In vitro, GluAdT from *D. radiodurans*, *Thermus thermophilus* or *B. subtilis* also converts Asp-tRNA^{Asn} into Asn-tRNA^{Asn}.[11,12,16] In vivo, expression of a dual Asp/GluAdT activity depends of the cellular context. For instance, *B. subtilis* contains a nondiscriminating glutamyl-tRNA synthetase (GluRS) able to form Glu-tRNA^{Gln} and a discriminative aspartyl-tRNA synthetase (AspRS), unable to generate the Asp-tRNA^{Asn} intermediate. Therefore, in this organism, the transamidation pathway is restricted to Gln-tRNA^{Gln} formation. Conversely, in *D. radiodurans* and in *T. thermophilus*, the GluRS is discriminative whereas there exists two different AspRS of which one is able to generate Asp-tRNA^{Asn}.[12,17,18] Therefore in these two species, the Asp/Glu-AdT is an Asp-AdT. In *Acidithiobacillus ferrooxidans*, AsnRS and GlnRS are both lacking and the amidotransferase catalyzes either Asn-tRNA^{Asn} or Gln-tRNA^{Gln} formation.[19] Finally, it is noticeable that some bacteria can use both direct and indirect routes to catalyze Asn-tRNA formation.[9,12] In such cases, the coexistence of the two pathways can be related to the absence of a tRNA-independent asparagine synthase. Therefore, in these cases, the indirect route which converts Asp into Asn would be dedicated to the biosynthesis of asparagine, not to tRNA aminoacylation per se.

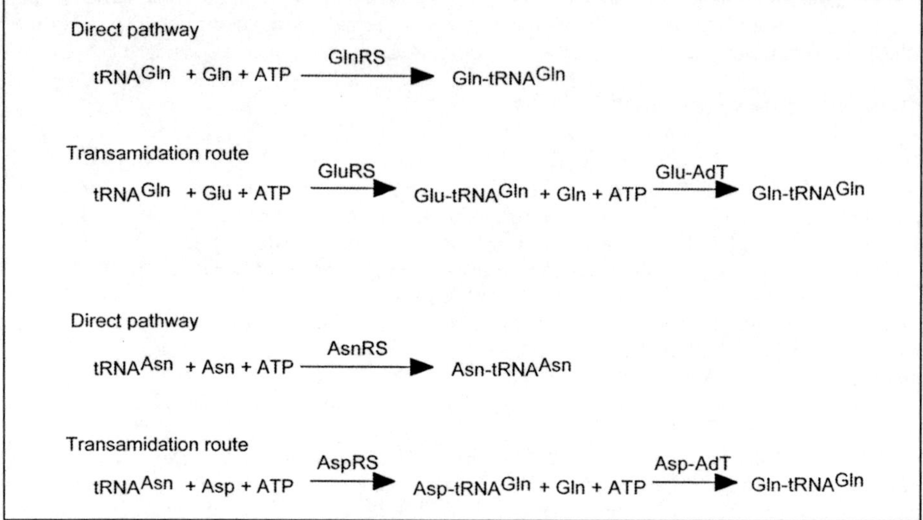

Figure 2. Schemes for the direct and indirect pathways for glutaminylation and asparaginylation.

In archaebacteria, recent sequenced genomes showed the systematic presence of a new heterodimeric amidotransferase (encoded by *gatD* and *gatE*). This new archaeal amidotransferase is specific for Gln-tRNA formation.[8] In contrast, Asn-tRNA synthesis is divergent. Some archaea use asparaginyl-tRNA synthetase whereas others do not possess AsnRS and convert Asp-tRNA^{Asn} into Asn-tRNA^{Asn} using a heterotrimeric amidotransferase, GatCAB.[8] Nevertheless, some cases remain intriguing. For example, the *Pyrococcus* species likely expresses a nondiscriminative AspRS.[18,20] However, an AsnRS plus an asparagine synthase are also present. Moreover, the *gatCAB* genes are missing and a GatDE amidotransferase could be evidenced. Whether the nondiscriminative AspRS produces Asp-tRNA^{Asn} in vivo and whether this aminoacyl-tRNA can be converted by an amidotransferase remains an open question.

In the light of the recent results, one picture emerges where Asp and Glu are early amino acids. In a second step of evolution, occurrence of a tRNA-dependent amino acid biosynthesis would have favoured the emergence of glutamine and asparagine. Such a link between amino acid metabolism and protein biosynthesis is possibly illustrated by the strong resemblance between the three-dimensional structure of asparagine synthase and that of aspartyl-tRNA synthetase.[21] The case of histidine biosynthesis where a paralog of histidyl-tRNA synthetase is involved, may also reflect co-evolution of the translation apparatus with the amino acid biosynthetic routes.[22]

The Selenocysteyl-tRNA Pathway

In all three kingdoms of life, some proteins contain selenocysteine. Most of the selenocysteine containing proteins are oxidoreductases. These enzymes take advantage of the lower redox potential of the selenol group as compared to that of the thiol group of cysteine. The occurrence of this unusual amino acid in a protein was first identified in glycine reductase from *Clostridium sticklandii*.[23] The presence of TGA stop codons interrupting the reading frames of genes encoding selenocysteine containing enzymes (mouse glutathione peroxidase,[24] and *Escherichia coli* formate dehydrogenase[25]) gave the first evidences for co-translational insertion of selenocysteine in response to UGA codons in a special mRNA context. The *E. coli* system responsible of selenocysteine incorporation has been extensively studied by Böck and co-workers (reviewed in refs. 26 and 27). The key feature is the existence of a specialized tRNA^{Sec}, encoded by the *selC* gene, carrying a UCA anticodon.[28] This tRNA is first aminoacylated by seryl-tRNA synthetase. The seryl group esterified on tRNA^{Sec} is then modified into a selenocysteyl group by a selenocysteine synthase, encoded by *selA*. Phosphoselenate, which serves as selenium donor in this reaction, is produced by an enzyme encoded by the *selD* gene.

The UGA codons encoding selenocysteine are distinguishable from the UGA stop codons by the presence on the mRNA of a specific stem-loop structure (the SECIS element) comprising about 40 nucleotides immediately 3' to the UGA Sec codon. Decoding of the UGA Sec codons is then achieved with the help of a specialized elongation factor, the product of *selB*.[29] This factor is made of two domains, one of which is homologous to EF-Tu.[30,31] The second domain is involved in the recognition of the SECIS element, thereby accounting for the specificity of the decoding process.[32]

In eukaryotes, the selenocysteine incorporation pathway mostly resembles the bacterial one. The main difference is that the SECIS element is located in the 3' untranslated region of the mRNA, and can be as far as several kilobases downstream from the UGA Sec codon.[33] Moreover, two polypeptides are necessary for the recruitment of Sec-tRNA^{Sec} on the ribosome at the level of UGA Sec codons.[34-36] The first polypeptide is a specific elongation factor, unable to bind to the SECIS element, whereas the second polypeptide is able to bind the SECIS hairpin with high affinity. The situation in Archaea appears similar to that in eukaryotes.[37-39]

tRNA^{Sec} possesses several specific features which allow it to follow its unusual route. This tRNA is usually larger than other tRNAs. For instance, *E. coli* tRNA^{Sec} contains 95 nucleotides. This length is first accounted for by a long extra arm, which allows recognition by seryl-tRNA synthetase. Second, all tRNAs^{Sec} possess a 13 base pair acceptor helix (i.e., the acceptor stem plus the T stem). Bacterial tRNAs^{Sec} possess eight base pairs in the acceptor stem, instead of seven in canonical tRNAs, and a standard five base pair T stem. In eukaryotes, it is likely that the acceptor stem contains nine base pairs, and the T stem four base pairs.[40] In addition, eubacterial tRNAs^{Sec} possess a purine-pyrimidine pair at position 11-24, a characteristic that is only shared by initiator tRNAs. Finally, tRNAs^{Sec} have a D stem extended to six base pairs, and a D loop reduced to four nucleotides. This, together with unusual nucleotides at positions involved in tRNA folding, might result in a reduced number of tertiary interactions in tRNA^{Sec}.[27,41,42] The relationships between these features and the biological properties of tRNA^{Sec} remain to be fully understood. It is however clear that the 13 base pair acceptor helix is a key element in the selenocysteine identity of tRNA^{Sec}. Indeed, deletion of one base pair in the acceptor stem of *E. coli* tRNA^{Sec} impairs recognition by the specific elongation factor SELB and permits appropriation by the general elongation factor EF-Tu.[43] Moreover, the mutated tRNA becomes a poor substrate of selenocysteine synthase (SELA).

Selenocysteine synthase activity has been evidenced in *E. coli*[44,45] as well as in eukaryotic cells.[46,47] In all cases, the selenium donor is selenophosphate. However, only in eubacteria has the enzyme been biochemically characterized.[45,48] This enzyme is a 500 kDa homodecamer able to bind one tRNA per two 50 kDa subunits.[45,49] Accordingly, electron microscopy experiments have highlighted a five-fold symmetry, with subunits arranged in two rings and five tRNA molecules bound.[50] Selenocysteine synthase belongs to the alpha-gamma superfamily of pyridoxal-5'-phosphate-dependent enzymes. The prosthetic group is held by a Lys residue.[49] The reaction mechanism involves the formation of a Schiff base between the formyl group of pyridoxal-5'-phosphate and the α-amino group of Ser-tRNA^{Sec}.[48] After water elimination, the OH group of the Ser side chain is removed, and a double bond is formed between the α carbon and the CH_2 side chain. The aminoacrylyl-tRNA intermediate then reacts with selenophosphate to produce selenocysteyl-tRNA.

The reasons why such a complex pathway for the incorporation of selenocysteine has emerged and was conserved in living cells remain to be fully understood.

Formylation of Initiator tRNA in Bacteria and Organelles

In eubacteria and in organelles (mitochondria and chloroplasts), orientation of methionylated initiator tRNA towards the initiation of translation requires an N-formylation reaction. This reaction is catalyzed by methionyl-tRNA$_f^{Met}$ transformylase (FMT) using N-10 formyltetrahydrofolate (FTHF) as the formyl donor. In *E. coli*, inactivation of the *fmt* gene severely impairs bacterial growth at 37°C and is lethal at 42°C.[51] In contrast, disruption of the gene for mitochondrial formylase in *S. cerevisiae* does not affect respiratory growth.[52] Formylation is highly specific. Indeed, FMT is able to select the initiator tRNA charged with methionine among all aminoacylated elongator tRNAs. Otherwise, N-blocking of an elongator aminoacylated tRNA would be detrimental to the elongation of proteins by the ribosome. In *E. coli*, nucleotidic determinants important for recognition by FMT are concentrated within the acceptor stem of the initiator tRNA, although the peculiar Pu11-Py24 pair in the D-stem also plays a role. The most important feature is the lack of strong base pairing at

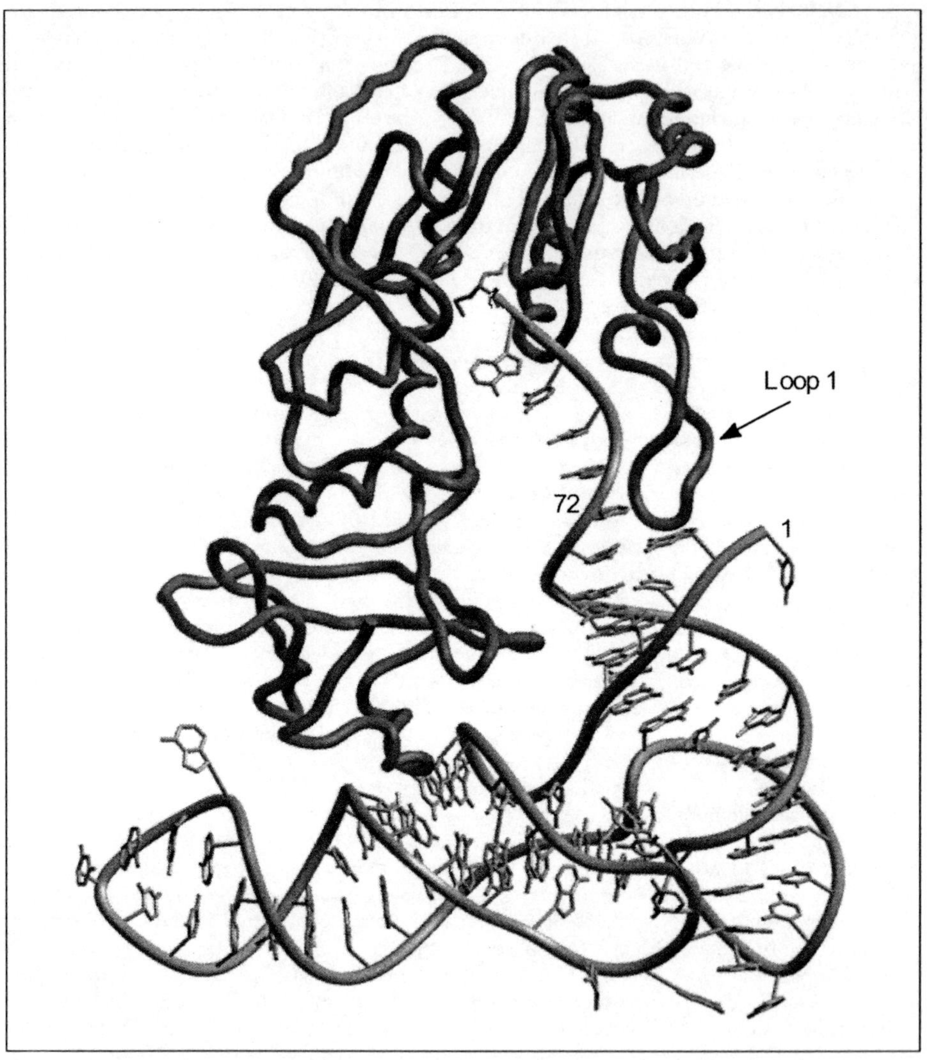

Figure 3. Schematic representation of the *E. coli* formylase complexed to f-met-tRNA$_f^{Met}$.[58] Loop 1 of formylase and bases 1-72 of tRNA are labeled. The figure was drawn with Setor.[97]

position 1-72. The A73 discriminator base and base pairs 2:71, 3:70 and 4:69 also contribute to the formylation reaction.[53,54] Finally, the enzyme can recognize initiator tRNAs artificially misacylated with various amino acids. However, methionine is much preferred with respect to the efficiency of the reaction.[54-56] The crystal structure of the *E. coli* FMT free or complexed to the formyl-methionyl-tRNA$_f^{Met}$ was solved (Fig. 3).[57,58] The enzyme shows an elongated shape with two domains connected by an extended peptide. The catalytic N-terminal domain contains a Rossmann fold and is highly homologous to the glycinamide ribonucleotide transfomylase (GARF) which also uses FTHF as formyl donor. The C-terminal domain is built around a β-barrel. In the structure of the complex, both domains of FMT as well as the linker peptide have contacts inside the L-shaped tRNA molecule. tRNA binds the enzyme through the D-stem side and the major groove of the acceptor stem. In the D-stem, the peculiar A11-U24 base pair is held by the C-terminal domain of the enzyme. In the N-terminal domain, a large loop (loop 1) inserted within the Rossmann fold becomes structured upon tRNA binding and is wedged in the major groove of the acceptor helix (Fig. 3). As a result, the C1⁻ A72 mismatch in the tRNA structure is split and the 3'-end bends inside the active site crevice.[58] Many interactions occur between the acceptor stem and the crucial loop 1 and are at the basis of the specificity of the formylase. The functional importance of these contacts could be further established by studying mutated FMT and tRNA molecules.[58-61]

Sequences of available formylases were aligned (Fig. 4). In formylases from eubacteria as well as from the plants *Arabidopsis thaliana* and *Brassica napus*, the sequence of the crucial loop 1 necessary for disruption of the 1:72 pair appears well conserved. This suggests an identical mechanism for the specificity of these formylases. In support to this idea, it is notable that all known initiator tRNAs from eubacteria, plant mitochondria or plant chloroplasts have a mismatched 1-72 pair. The case of mitochondria from single cell eukaryotes appears more diverse since initiator tRNAs can have various base pairs at position 1-72 including G-C. Hopefully, future comparison of the corresponding formylase sequences will help to clarify apparent differences in these organelles. In the case of animal mitochondria, the mechanism of discrimination between initiator and elongator tRNA[Met] is particular. Indeed, a single mitochondrial tRNA[Met] species with a A1-U72 base pair has yet been detected. This tRNA is thought to be used for both initiation and elongation of translation, depending on whether or not it has been formylated. tRNA[Met] identity elements governing formylation by bovine mitochondrial transformylase have been searched for.[62] The conclusion was that mitochondrial formylase preferentially recognizes the methionyl moiety of its tRNA substrate. This behaviour is consistent with the existence of a single tRNA[Met]. Indeed, the methionyl group is sufficient now to distinguish the correct substrate of formylation from any other aminoacylated tRNA. It should be moreover underlined that the acceptor stem of many elongator tRNAs in animal mitochondria with A-U or U-A base pair at position 1-72 may resemble the 3' end of tRNA[Met]. Therefore, a relatively small weight of the tRNA acceptor stem in the recognition process appears well adapted to the context of organelles. In agreement with this biochemical analysis, sequence examination indicates the absence of loop1 in mitochondrial formylases of animal origin (Fig. 4).[62]

The Recycling of Incorrectly Esterified tRNAs

Peptidyl-tRNAs and Peptidyl-tRNA Hydrolase

In the course of the elongation of a protein, premature dissociation of peptidyl-tRNAs from the ribosome causes erroneous polypeptide chain termination.[63] Therefore, peptidyl-tRNAs may accumulate in the cytoplasm and were shown to be toxic for the cell.[63-65] Activities which can recycle peptidyl-tRNAs into tRNAs have been characterized in eubacteria,[66-68] in yeast[67,69] and in higher eukaryotes.[70,71] In bacteria and in yeast, this activity, carried by peptidyl-tRNA hydrolase (PTH), is an esterase one. In the case of rabbit reticulocytes, another mechanism was reported, with peptidyl-AMP and tRNA-CC being the products of the reaction.[70] However, the recent availability of several mammalian genomes enables the recognition of proteins homologous to bacterial PTH. Therefore, which mechanism mammals actually use to recycle peptidyl-tRNAs is still an open question.

In eubacteria, the PTH activity is essential for cell viability.[72-75] Any N-acyl-aminoacyl-tRNA except formyl-methionyl-tRNA$_f^{Met}$ can be a substrate of the *E. coli* enzyme.[76,77] Resistance to hydrolysis of the formylated tRNA is required to allow the participation of the initiator tRNA to translation. The main structural characteristic explaining the resistance of the initiator tRNA is the absence of a strong base pairing at position 1-72.[76,77] Thus, the same structural feature which allows selection by the formylase is used to promote exclusion by PTH.[53,54] Moreover, the rate of PTH-catalyzed hydrolysis depends on the presence of a 5' phosphate group in the nucleotidic moiety of the substrate. The crystallographic structure of PTH was solved.[75] This enzyme is formed by a single globular domain built around a mixed β-sheet (Fig. 5). Discrimination by PTH between elongator tRNAs and initiator tRNA is reminiscent to that of EF-Tu.[75,78] Indeed, like EF-Tu, PTH recognizes elongator tRNA molecules possessing a full base pair at position 1-72. In the two systems, selection of elongator tRNAs appears mediated through recognition of a correctly located phosphate group at the 5' top of the polynucleotide. EF-Tu possesses two basic residues forming a binding clamp for the tRNA 5' phosphate group. In *E. coli* PTH, screening of the role of basic residues located at the surface of the enzyme led to identification of two residues, K105 and R133, necessary to the recognition and anchoring of the tRNA 5' phosphate.[78] The lack of pairing at position 1-72

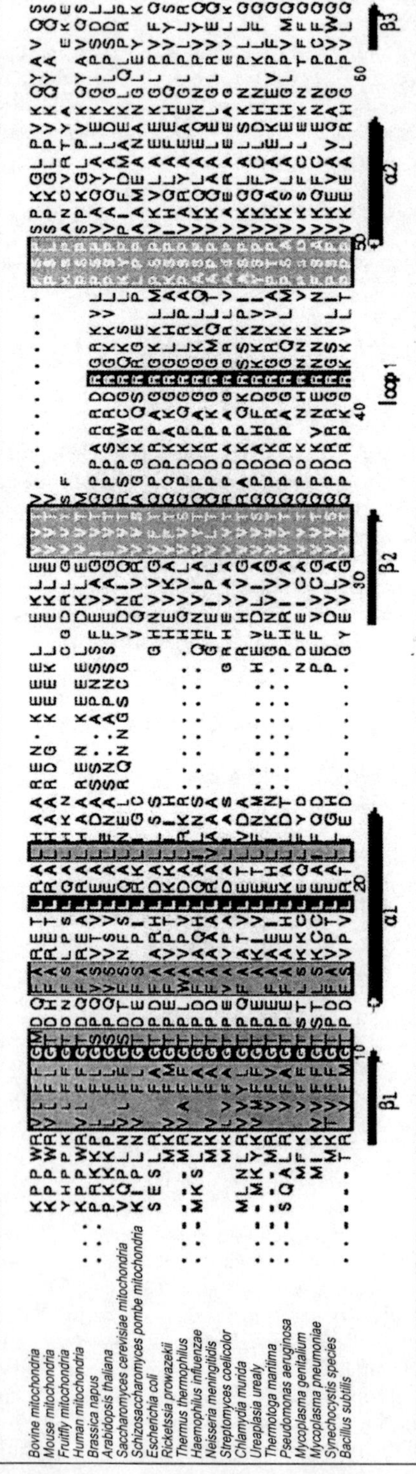

Figure 4. Alignment of methionyl-tRNA$_f^{Met}$ formylase sequences from various sources in the region of loop 1. Amino acid sequences were obtained from the SwissProt database (http://www.expasy.ch/sprot) and aligned using the Clustal X program.[98] Numbering is according to the *E. coli* sequence. The secondary structures are those of *E. coli* formylase 3D structure. Strictly conserved residues are boxed in black with white letters. The crucial R42 residue of *E. coli* formylase, strictly conserved in all formylases having a loop 1, is also boxed in black. Conservative replacements are shown on a grey background with black letters. Two conserved regions bordering loop 1 are boxed in grey with white letters. The figure was drawn with Alscript.[99]

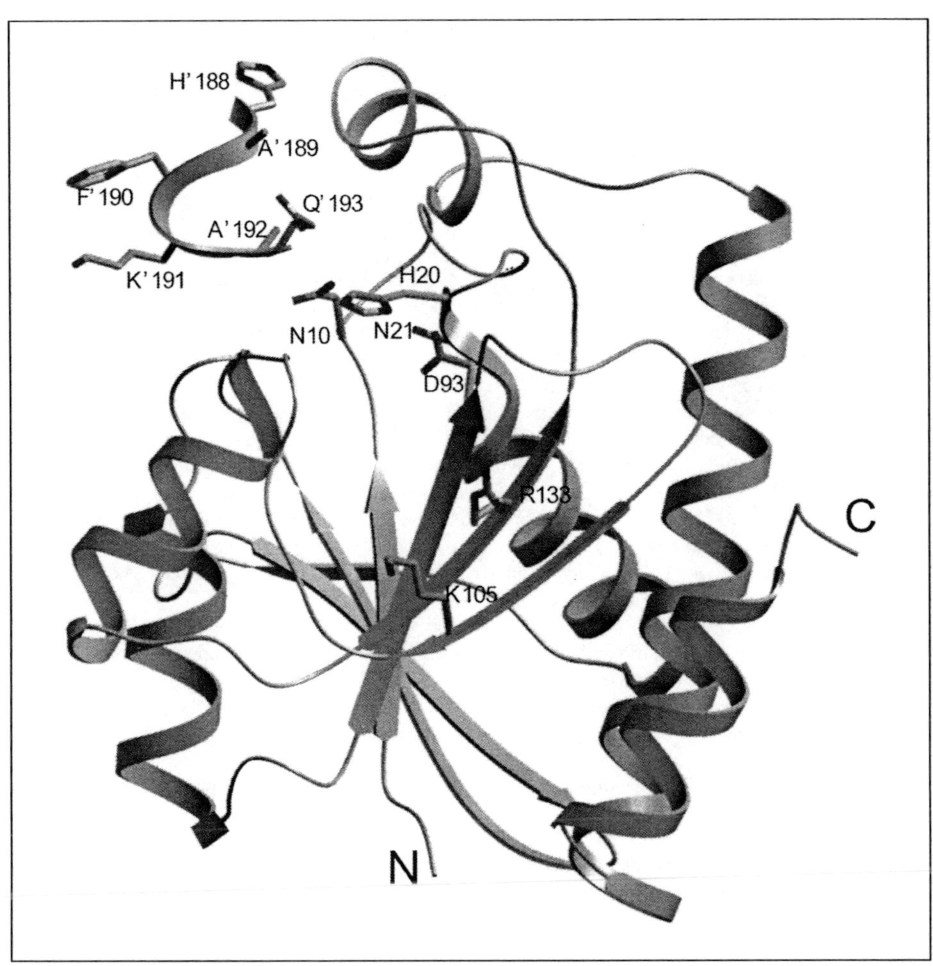

Figure 5. Schematic representation the 3D structure of *E. coli* peptidyl-tRNA hydrolase (193 residues).[75] In the crystal packing, the hydrolase molecules establish close contacts. In particular, the C-terminus of a neighbouring molecule enters inside the active cleft of the hydrolase. This contact is thought to mimic the peptidyl part of the substrate of the enzyme. Residues belonging to this peptide are labelled with a prime. In the hydrolase structure, the side chains of residues N10, H20, N21 and D93 which belong to the active site are drawn with sticks. K105 and R133 form the electropositive clamp recognizing the 5' phosphate of the peptidyl-tRNA substrate.[78] The figure was drawn with Setor.[97]

of the initiator tRNA would result in an abnormal mobility of the 5' phosphate and would consequently impair the docking to either the PTH enzyme or the elongation factor.

Hydrolysis of D-Aminoacyl-tRNAs by D-Aminoacyl-tRNA Deacylase

Aminoacyl-tRNA synthetases are believed to discriminate against the D-form of amino acids. However, it was previously observed that purified *E. coli* and *Bacillus subtilis* tyrosyl-tRNA synthetases were able to transfer D-tyrosine onto tRNA.[79,80] Later, extracts of *E. coli*, yeast, rabbit reticulocytes or rat liver were shown to contain an enzyme activity, called D-Tyr-tRNA^{Tyr} deacylase, capable of accelerating the hydrolysis of the ester linkage of D-Tyr-tRNA^{Tyr} in the production of free tRNA and D-tyrosine.[80] L-Tyr-tRNA^{Tyr} was left intact by this activity.

Recently, two other *E. coli* aminoacyl-tRNA synthetases, tryptophanyl- and aspartyl-tRNA synthetases, were also shown in vitro to transfer the D-form of their cognate amino acid onto the cognate tRNAs.[81] The D-Trp- and D-Asp-tRNA products behave in vitro as substrates of purified D-Tyr-tRNA deacylase.[82]

E. coli and *Saccharomyces cerevisiae* D-Tyr-tRNA deacylases are encoded by the *dtd* and DTD1 genes, respectively.[83] Orthologs of these genes occur in many bacteria as well as in *Caenorhabditis elegans, Arabidopsis thaliana,* mice and humans. Involvement of the deacylase in the protection of *E. coli* against in vivo produced D-Tyr, D-Trp- or D-Asp-tRNA is supported by the exacerbation of the toxicity of each of these three D-amino acids in response to the inactivation of the *dtd* gene.[81,83] The toxic effects of D-Gln and D-Ser also respond to the absence of the *dtd* product, therefore suggesting that Gln- and Ser-tRNA synthetases from *E. coli* produce D-Gln-tRNA and D-Ser-tRNA, respectively, in vivo.[81] In the case of yeast, inactivation of the DTD1 gene leads to the toxicity of D-Tyr and D-Leu.[82] In agreement with this observation, tyrosyl-tRNA synthetase from *S. cerevisiae* could be shown to transfer D-Tyr onto tRNA^{Tyr}.

Taken together, the above data indicate that the specificity of D-Tyr-tRNA deacylase is large enough to accommodate any D-aminoacyl-tRNA substrate. Indeed, D-Tyr-, D-Trp- and D-Asp-tRNA were shown to be substrates in vitro. Calendar and Berg noted that partially purified *E. coli* D-Tyr-tRNA deacylase also cleaved D-Phe-tRNA and Gly-tRNA.[80] The in vivo data strongly suggest that D-Gln- and D-Ser-tRNA are recognized by the *E. coli* deacylase [81] and that D-Leu-tRNA is a substrate of the yeast deacylase.[81] In contrast with such a broad specificity, the deacylase rejects all substrates containing an L-amino acid moiety. N-blocked D-aminoacylated tRNAs also fully resist the action of D-Tyr-tRNA deacylase.[83]

However, the deacylase appears not to simply behave as a D-amino acid esterase. Calendar and Berg showed that D-Tyr-adenosine was not hydrolyzed by the enzyme. In contrast, a D-Tyr-esterified oligonucleotide produced by RNAse T1 digestion of D-Tyr-tRNA was a substrate.[80] In the derived 19-mer oligonucleotide, the deacylase possibly recognizes a feature common to all tRNAs, for instance the CCA triplet at the end of the acceptor stem of the polynucleotide. On the side of the amino acid moiety, the deacylase would only distinguish the stereoisomeric character of the Cα.

The deacylase of *E. coli* has recently been characterized as a 2x16 kDa homodimer. No sequence similarity could be evidenced between the deacylase and the editing domain of aminoacyl-tRNA synthetases. The crystallographic structure of dimeric *Escherichia coli* D-Tyr-tRNA^{Tyr} deacylase at 1.55 Å resolution indicates a β-barrel closed on one side by a β-sheet lid (Fig. 6).[84] This barrel results from the assembly of the two subunits. Analysis of the structure in relation with sequence homologies in the orthologous family suggests location of two symmetrical active sites at the carboxy end of the β-strands. The solved 3-D structure markedly differs from those of all other documented tRNA-dependent hydrolases. Further structural and biochemical studies are required to understand the basis of the broad specificicity of D-Tyr-tRNA^{Tyr} deacylase.

Conclusion: Rejection by Eubacterial EF-Tu of Aminoacyl-tRNA Intermediates

In eubacteria, the fate of a canonical aminoacyl-tRNA is to bind the elongation factor EF-Tu.GTP complex and to be thereafter directed towards the ribosomal A site in order to participate to peptide chain elongation. However, as described in this chapter, some aminoacyl-tRNAs are not the final products used for translation. They are reaction intermediates which need further modification before being able to fulfill their role. Such intermediates, which include Ser-tRNA^{Sec}, Asp-tRNA^{Asn} and Glu-tRNA^{Gln}, as well as Met-tRNA$_f^{Met}$ in eubacteria and in organelles, must therefore be prevented from interacting with EF-Tu.GTP. This view can be extended either to Sec-tRNA^{Sec} and formyl-Met-tRNA$_f^{Met}$ which are tRNA products following a route distinct from that of EF-Tu but can be used by the ribosome, or to toxic peptidyl-tRNAs and D-aminoacyl-tRNAs which must not be appropriated by the elongation factor.

Biochemical and structural studies have established that bacterial EF-Tu closely recognizes the amino group of the esterified amino acid.[85,86] Consequently, abnormal 3D positioning of the amino group with respect to the ester bond in a D-aminocylated tRNA explains exclusion by EF-Tu.[87] This property also accounts straightforwardly for the protection against EF-Tu binding of

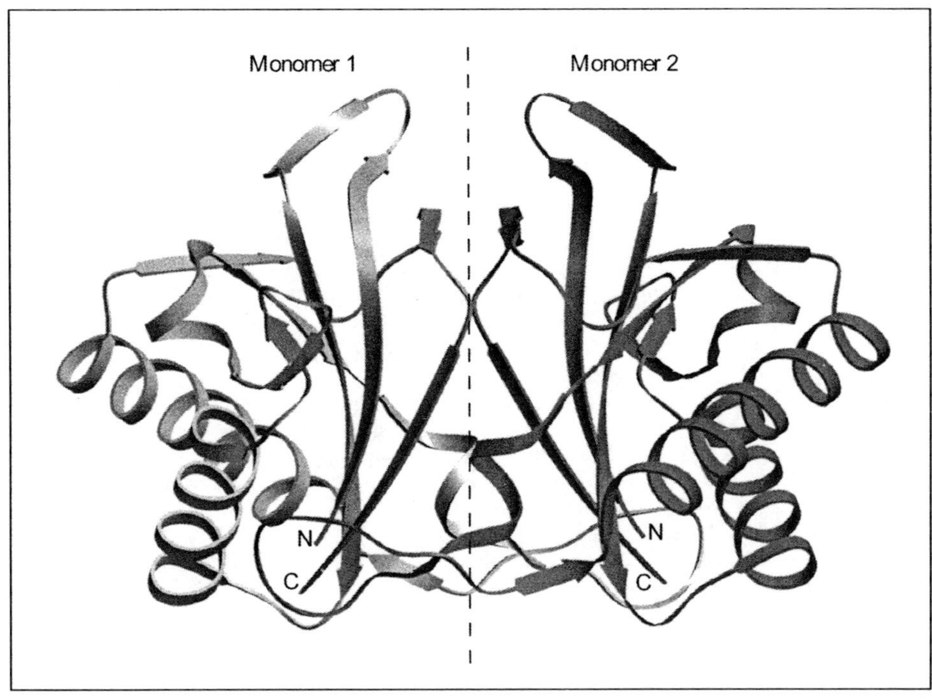

Figure 6. Schematic representation of the 3D structure of *E. coli* D-tyr-tRNA deacylase dimer.[84] The two-fold axis corresponding to the dimeric assembly is represented. The figure was drawn with Setor.[97]

formyl-Met-tRNA$_f^{Met}$ and peptidyl-tRNAs. Accordingly, although formylation strongly contributes to the initiator identity of a tRNA by favouring its binding to initiation factor IF2, it is also important to impair its misappropriation by EF-Tu.[88-91]

In vivo studies in *E. coli* cells have brought evidence for the occurrence of a competition between EF-Tu.GTP and methionyl-tRNA$_f^{Met}$ transformylase (FMT) in the uptake of Met-tRNA$_f^{Met}$.[90,91] In vitro studies indicate that the factor EF-Tu.GTP has ten times less affinity towards Met-tRNA$_f^{Met}$ than towards any canonical aminoacylated elongator tRNA.[85] Such a discrepancy can be explained on the basis of the 1-72 mismatch in the initiator tRNA. Indeed, as described above, EF-Tu weakly recognizes the 5' phosphate of the initiator tRNA because of the 1-72 mismatch and the resulting mobility of the 5' phosphate.[86] However, the intracellular concentration of EF-Tu (100 μM) is much larger than that of FMT (1μM). Consequently, despite its relatively low affinity, the initiator tRNA charged with methionine is likely to be at least partly saturated by EF-Tu. This view led to the proposal that FMT might have the capacity to bind and process a Met-tRNA$_f^{Met}$ molecule that has been already misappropriated by EF-Tu, without the requirement of prior full dissociation of the erroneous complex.[58] In the model, a ternary complex containing the polynucleotide and the two proteins was imagined. Such a double association can be imagined because tRNA is docked to EF-Tu on the T-stem side while the initiator tRNA is bound by FMT on the opposite D-stem side. Such a pathway would decrease the negative impact of the competition for Met-tRNA$_f^{Met}$ between FMT and EF-Tu in eubacterial cells.

Finally, animal mitochondria contain a single tRNA^{Met} species (see above). It is assumed that upon methionylation, this tRNA^{Met} molecule can either bind EF-Tu and further participate to chain elongation, or undergo formylation through the action of mitochondrial FMT and then be directed to the initiation machinery. Thus, competition between EF-Tu and FMT for the binding of Met-tRNA^{Met} would, in this case, ensure the correct balance between the elongator and initiator activities of mitochondrial tRNA^{Met}.

Both seryl-tRNA[Sec] and selenocysteyl-tRNA[Sec] are prevented from binding to the canonical elongation factor.[92] In the eubacterial system, this behaviour was shown to result from a specific nucleotide combination in the acceptor helix of tRNA[Sec].[93] This combination, named "antideterminant box", corresponds to the 8th, 9th and 10th base pairs in the acceptor helix, i.e., to the last base pair at the bottom of the acceptor stem followed by the two first base pairs in the T-stem (C_7G_{66}-$G_{49}U_{65}$-$C_{50}G_{64}$). To fulfil its role, the antideterminant box must be correctly positioned with respect to the 3' end, since deletion of the 6th base pair in the acceptor stem of tRNA[Sec] results in a tRNA able to bind EF-Tu.GTP.[43] Finally, in order to prevent translational incorporation of serine instead of selenocysteine, the specific elongation factor SELB must be able to distinguish between Sec-tRNA[Sec] and Ser-tRNA[Sec]. Characterization of *E. coli* SELB has demonstrated that the factor is indeed able to recognize the side chain of the amino acid esterified to its tRNA ligand, and to exclusively bind Sec-tRNA[Sec].[29] A structural model of SELB was proposed on the basis of the crystal structure of EF-Tu.[30] It suggests the existence on the SELB factor of a binding pocket specific for selenocysteine.

In the case of those organisms which have adopted the indirect route for the synthesis of Asn-tRNA[Asn] or Gln-tRNA[Gln], the canonical EF-Tu must also be able to discriminate against Asp-tRNA[Asn] and/or Glu-tRNA[Gln] (see above). Such a discrimination was first evidenced in the case of chloroplasts from *Pisum sativum*.[94] The authors observed that chloroplastic Glu-tRNA[Glu] efficiently binds the chloroplastic elongation factor, while Glu-tRNA[Gln] cannot. In contrast, both chloroplastic Glu-tRNA[Glu] and Glu-tRNA[Gln] are bound by EF-Tu from *E. coli*. In the case of *T. thermophilus*, both direct and indirect routes are used for the synthesis of Asn-tRNA[Asn].[9] The corresponding EF-Tu binds efficiently Asn-tRNA[Asn], while it rejects the Asp-tRNA[Asn] intermediate.[9] Further studies are required to understand the bases of the side-chain discriminating properties of EF-Tu in all these organisms.

To conclude, protection against EF-Tu binding of those aminoacyl-tRNAs which have to follow a special route is of prime importance for faithful interpretation of the genetic code in eubacteria. Similar protections against elongation factor binding are likely to exist in archaeal and eukaryal cells. This problem is for instance crucial for the initiator tRNA, since it is not formylated in these kingdoms. In yeast cells, a bulky modification at base 64 was demonstrated to exclude cytoplasmic initiator tRNA from elongation factor binding.[95,96] Beyond this interesting observation, the problem of the protection against elongation factor binding in archaea and in eukaryotes remains widely open.

References

1. Wilcox M, Nirenberg M. Transfer RNA as a cofactor coupling amino acid synthesis with that of protein. Proc Natl Acad Sci USA 1968; 61:229-236.
2. Wilcox M. γ-Glutamyl phosphate attached to Glutamine-specific tRNA. Eur J Biochem 1969; 11:405-412.
3. Schön A, Hottinger H, Söll D. Misaminoacylation and transamidation are required for protein biosynthesis in *Lactobacillus bulgaricus*. Biochimie 1988; 70:391-394.
4. Schön A, Kannangara CG, Gough S et al. Protein biosynthesis in organelles requires misaminoacylation of tRNA. Nature 1988; 331:187-190.
5. Gagnon Y, Lacoste L, Champagne N et al. Widespread use of the Glu-tRNA[Gln] transamidation pathway among bacteria. J Biol Chem 1996; 271:14856-14863.
6. White BN, Bayley ST. Further codon assignments in an extremely halophilic bacterium using a cell-free protein-synthesizing system and a ribosomal binding assay. Can J Biochem 1972; 50:600-609.
7. Curnow AW, Ibba M, Söll D. tRNA-dependant asparagine formation. Nature 1996; 382:589-590.
8. Tumbula DL, Becker H, Chang W-Z et al. Domain-specific recruitment of amide amino acids for protein synthesis. Nature 2000; 407:106-110.
9. Becker HD, Kern D. *Thermus thermophilus*: a link in evolution of the tRNA-dependent amino acid amidation pathways. Proc Natl Acad Sci USA 1998; 95:12832-12837.
10. Strauch MA, Zalkin H, Aronso AI. Characterization of the glutamyl-tRNA[Gln] to glutaminyl-tRNA[Gln] amidotrasferase reaction of *Bacillus subtilis*. J Bact 1988; 170:916-920.
11. Curnow AW, Hong K-W, Yuan R et al. Glu-tRNAGln amidotransferase: a novel heterotrimeric enzyme required for correct decoding of glutamine codons during translation. Proc Natl Acad Sci USA 1997; 94:11819-11826.
12. Curnow AW, Tumbula DL, Pelaschier JT et al. Glutamyl-tRNA[Gln] amidotransferase in *Deinococcus radiodurans* may be confined to asparagine biosynthesis. Proc Natl Acad Sci USA 1998; 95:12838-12843.
13. Lapointe J, Duplain L, Proulx M. A single glutamyl-tRNA synthetase aminoacylates tRNA[Glu] and tRNA[Gln] in *Bacillus subtilis* and efficiently misacylates *Escherichia coli* tRNA$_1$[Gln] in vitro. J Bacteriol 1986; 165:88-93.
14. Chen M-W, Jahn D, Schön A et al. Purification and characterization of *Chlamydomonas reinhardtii* chloroplast glutamyl-tRNA synthetase, a natural misacylating enzyme. J Biol Chem 1990; 265:4054-4057.

15. Horiuchi KY, Harpel MR, Shen L et al. Mechanistic studies of reaction coupling in Glu-tRNAGln amidotransferase. Biochemistry 2001; 40:6450-6457.
16. Becker HD, Bokkee M, Jacobi C et al. The heterotrimeric *Thermus thermophilus* Asp-tRNA^{Asn} amidotransferase can also generate Gln-tRNA^{Gln}. FEBS Lett 2000; 476:140-144.
17. Becker HD, Reinbolt J, Kreutzer R et al. Existence of two distinct aspartyl-tRNA synthetase in *Thermus thermophilus*. Structural and biochemical properties of the two enzymes. Biochemistry 1997; 36:8785-8797.
18. Becker HD, Roy H, Moulinier L et al. *Thermus thermophilus* contains an eubacterial and an archaebacterial aspartyl-tRNA synthetase. Biochemistry 2000; 39:3216-3230.
19. Salazar J, Zuniga R, Raczniak G et al. A dual-specific Glu-tRNA^{Gln} and Asp-tRNA^{Asn} amidotransferase is involved in decoding glutamine and asparagine codons in *Acidithiobacillus ferrooxidans*. FEBS Lett 2001; 500:129-131.
20. Schmitt E, Moulinier L, Fujiwara S et al. Crystal structure of aspartyl-tRNA synthetase from *Pyrococcus kodakaraensis* KOD: archaeon specificity and catalytic mechanism of adenylate formation. EMBO J 1998; 17:5227-5237.
21. Nakatsu T, Kato H, Oda J. Crystal structure of asparagine synthetase reveals a close evolutionary relationship to class II aminoacyl-tRNA synthetase. Nature Struct Biol 1998; 5:15-19.
22. Sissler M, Delorme C, Bond J et al. An aminoacyl-tRNA synthetase paralog with a catalytic role in histidine biosynthesis. Proc Natl Acad Sci USA 1999; 96:8985-8990.
23. Cone JE, Martin del Rio R, Davis JN et al. Chemical characterisation of the selenoprotein component of clostridial glycine reductase: identification of selenocysteine as the organoselenium moiety. Proc Natl Acad Sci USA 1976; 73:2659-2663.
24. Chambers I, Frampton J, Goldfarb P et al. The structure of the mouse glutathione peroxidase gene: the selenocysteine in the active site is encoded by the 'termination' codon, TGA. EMBO J 1986; 5:1221-1227.
25. Zinoni F, Birkmann A, Stadtman TC et al. Nucleotide sequence and expression of the selenocysteine-containing polypeptide of formate dehydrogenase (formate-hydrogen-lyase-linked) from *Escherichia coli*. Proc Natl Acad Sci USA 1986; 83:4650-4654.
26. Böck A. Biosynthesis of selenoproteins—an overview. Biofactors 2000; 11:77-78.
27. Commans S, Böck A. Selenocysteine inserting tRNAs: an overview. FEMS Microbiol Rev 1999; 23:335-351.
28. Leinfelder W, Zehelein E, Mandrand-Berthelot MA et al. Gene for a novel tRNA species that accepts L-serine and cotranslationally inserts selenocysteine. Nature 1988; 331:723-725.
29. Forchhammer K, Leinfelder W, Böck A. Identification of a novel translation factor necessary for the incorporation of selenocysteine into protein. Nature 1989; 342:453-456.
30. Hilgenfeld R, Böck A, Wilting R. Structural model for the selenocysteine-specific elongation factor SelB. Biochimie 1996; 78:971-978.
31. Böck A, Hilgenfeld R, Tormay P et al. Domain structure of the selenocysteine-specific translation factor SelB in prokaryotes. Biomed Environ Sci 1997; 10:125-128.
32. Kromayer M, Wilting R, Tormay P et al. Domain structure of the prokaryotic selenocysteine-specific elongation factor SelB. J Mol Biol 1996; 262:413-420.
33. Berry MJ, Banu L, Chen YY et al. Recognition of UGA as a selenocysteine codon in type I deiodinase requires sequences in the 3' untranslated region. Nature 1991; 353:273-276.
34. Copeland PR, Fletcher JE, Carlson BA et al. A novel RNA binding protein, SBP2, is required for the translation of mammalian selenoprotein mRNAs. EMBO J 2000; 19:306-314.
35. Fagegaltier D, Hubert N, Yamada K et al. Characterization of mSelB, a novel mammalian elongation factor for selenoprotein translation. EMBO J 2000; 19:4796-4805.
36. Tujebajeva RM, Copeland PR, Xu XM et al. Decoding apparatus for eukaryotic selenocysteine insertion. EMBO Rep 2000; 1:158-163.
37. Rother M, Resch A, Gardner WL et al. Heterologous expression of archaeal selenoprotein genes directed by the SECIS element located in the 3' non-translated region. Mol Microbiol 2001; 40:900-908.
38. Rother M, Wilting R, Commans S et al. Identification and characterisation of the selenocysteine-specific translation factor SelB from the archaeon *Methanococcus jannaschii*. J Mol Biol 2000; 299:351-358.
39. Wilting R, Schorling S, Persson BC et al. Selenoprotein synthesis in archaea: identification of an mRNA element of *Methanococcus jannaschii* probably directing selenocysteine insertion. J Mol Biol 1997; 266:637-641.
40. Hubert N, Sturchler C, Westhof E et al. The 9/4 secondary structure of eukaryotic selenocysteine tRNA: more pieces of evidence. RNA 1998; 4:1029-1033.
41. Baron C, Westhof E, Böck A et al. Solution structure of selenocysteine-inserting tRNA from *Escherichia coli*. Comparison with canonical tRNA^{Ser}. J Mol Biol 1993; 231:274-292.
42. Sturchler C, Westhof E, Carbon P et al. Unique secondary and tertiary structural features of the eucaryotic selenocysteine tRNA(Sec). Nucleic Acids Res 1993; 21:1073-1079.
43. Baron C, Böck A. The length of the aminoacyl-acceptor stem of selenocysteine-specific tRNA of *Escherichia coli* is the determinant for binding to elongation factors SELB or EF-Tu. J Biol Chem 1991; 266:20375-20379.
44. Leinfelder W, Stadtman TC, Böck A. Occurrence in vivo of selenocysteyl-tRNA$^{Ser}_{UCA}$ in *Escherichia coli*. J Biol Chem 1989; 264:9720-9723.
45. Forchhammer K, Leinfelder W, Boesmiller K et al. Selenocysteine synthase from *Escherichia coli*. Nucleotide sequence of the gene (*selA*) and purification of the protein. J Biol Chem 1991; 266:6318-6323.
46. Mizutani T, Kurata H, Yamada K et al. Some properties of murine selenocysteine synthase. Biochem J 1992; 284:827-834.

47. Amberg R, Mizutani T, Wu XQ et al. Selenocysteine synthesis in mammalia: an identity switch from tRNA(Ser) to tRNA(Sec). J Mol Biol 1996; 263:8-19.
48. Forchhammer K, Böck A. Selenocysteine synthase from *Escherichia coli*: analysis of the reaction sequence. J Biol Chem 1991; 266:6324-6328.
49. Tormay P, Wilting R, Lottspeich F et al. Bacterial selenocysteine synthase: structural and functional properties. Eur J Biochem 1998; 254:655-661.
50. Engelhardt H, Forchhammer K, Muller S et al. Structure of selenocysteine synthase from *Escherichia coli* and location of tRNA in the seryl-tRNA(sec)-enzyme complex. Mol Microbiol 1992; 6:3461-3467.
51. Guillon JM, Mechulam Y, Schmitter JM et al. Disruption of the gene for Met-tRNA$_f^{Met}$ formyltransferase severely impairs growth of *Escherichia coli*. J Bacteriol 1992; 174:4294-4301.
52. Li Y, Holmes WB, Appling DR et al. Initiation of protein synthesis in *Saccharomyces cerevisiae* mitochondria without formylation of the initiator tRNA. J Bacteriol 2000; 182:2886-2892.
53. Lee CP, Seong BL, RalBhandary UL. Structural and sequence elements important for recognition of *Escherichia coli* formylmethionine tRNA by methionyl-tRNA transformylase are clustered in the acceptor stem. J Biol Chem 1991; 266:18012-18017.
54. Guillon JM, Meinnel T, Mechulam Y et al. Nucleotides of tRNA governing the specificity of *Escherichia coli* methionyl-tRNA$^{Met}_f$ formyltransferase. J Mol Biol 1992; 224:359-367.
55. Giegé R, Ebel JP, Clark BFC. Formylation of mischarged *E. coli* tRNA$^{Met}_f$. FEBS Lett 1973; 30:291-295.
56. Li S, Kumar NV, Varsney U et al. Important role of the amino acid attached to tRNA in formylation and in initiation of protein synthesis in *Escherichia coli*. J Biol Chem 1996; 271:1022-1028.
57. Schmitt E, Blanquet S, Mechulam Y. Structure of crystalline *Escherichia coli* methionyl-tRNA$_f^{Met}$ formyltransferase: comparison with glycinamide ribonucleotide formyltransferase. EMBO J 1996; 15:4749-4758.
58. Schmitt E, Panvert M, Blanquet S et al. Crystal structure of methionyl-tRNAfMet transformylase complexed with the initiator formyl-methionyl-tRNAfMet. EMBO J 1998; 17:6819-6826.
59. Ramesh V, Gite S, Li Y et al. Suppressor mutations in *Escherichia coli* methionyl-tRNA formyltransferase: role of a 16-amino acid insertion module in initiator tRNA recognition. Proc Natl Acad Sci USA 1997; 94:13524-13529.
60. Ramesh V, Gite S, RajBhandary UL. Functional interaction of an arginine conserved in the sixteen amino acid insertion module of *Escherichia coli* methionyl-tRNA formyltransferase with determinants for formylation in the initiator tRNA. Biochemistry 1998; 37:15925-15932.
61. Ramesh V, Mayer C, Dyson MR et al. Induced fit of a peptide loop of methionyl-tRNA formyltransferase triggered by the initiator tRNA substrate. Proc Natl Acad Sci USA 1999; 96:875-880.
62. Takeuchi N, Vial L, Panvert M et al. Recognition of tRNAs by methionyl-tRNA transformylase from mammalian mitochondria. J Biol Chem 2001; 276:20064-20068.
63. Menninger JR. Peptidyl-transfer RNA dissociates during protein synthesis from ribosomes of *Escherichia coli*. J Biol Chem 1976; 251:3392-3398.
64. Atherly AG. Peptidyl-transfer RNA hydrolase prevents inhibition of protein synthesis initiation. Nature 1978; 275:769.
65. Chapeville F, Yot P, Paulin D. Enzymatic hydrolysis of N-acylaminoacyl transfer RNAs. Cold Spring Harbor Symp Quant Biol 1969; 34:493-498.
66. Cuzin F, Kretchmer N, Greenberg RE et al. Enzymatic hydrolysis of N-substituted aminoacyl-tRNA. Proc Natl Acad Sci USA 1967; 58:2079-2086.
67. Kössel H, RajBhandary UL. Studies on polynucleotides. LXXXVI. Enzymic hydrolysis of N-acylaminoacyl-transfer RNA. 1968; 35:539-560.
68. Kössel H. Purification and properties of peptidyl-tRNA hydrolase from *Escherichia coli*. Biochim Biophys Acta 1969; 204:191-202.
69. Jost JP, Böck RM. Enzymatic hydrolysis of N-substituted aminoacyl transfer ribonucleic acid in yeast. J Biol Chem 1969; 244:5866-5873.
70. Gross M, Crow P, White J. The site of hydrolysis by rabbit reticulocyte peptidyl-tRNA hydrolase is the 3'-AMP terminus of susceptible tRNA substrates. J Biol Chem 1992; 267:2080-2086.
71. Gross M, Starn TK, Rundquist C et al. Purification and initial characterization of peptidyl-tRNA hydrolase from rabbit reticulocytes. J Biol Chem 1992; 267:2073-2079.
72. Atherly AG, Menninger JR. Mutant *E. coli* strain with temperature sensitive peptidyl-transfer RNA hydrolase. Nature New Biol 1972; 240:245-246.
73. Garcia-Villegas MR, De La Vega FM, Galindo JM et al. Peptidyl-tRNA hydrolase is involved in λ inhibition of host protein synthesis. EMBO J 1991; 10:3549-3555.
74. Menninger JR, Coleman RA. Lincosamide antibiotics stimulate dissociation of peptidyl-tRNA from ribosomes. Antimicrob Agents Chemother 1993; 37:2027-2029.
75. Schmitt E, Mechulam Y, Fromant M et al. Crystal structure at 1.2 A resolution and active site mapping of *Escherichia coli* peptidyl-tRNA hydrolase. EMBO J 1997; 16:4760-4769.
76. Schulman LH, Pelka H. The structural basis for the resistance of *Escherichia coli* formylmethionyl transfer ribonucleic acid to cleavage by *Escherichia coli* peptidyl transfer ribonucleic acid hydrolase. J Biol Chem 1975; 250:542-547.
77. Dutka S, Meinnel T, Lazennec C et al. Role of the 1-72 base pair in tRNAs for the activity of *Escherichia coli* peptidyl-tRNA hydrolase. Nucleic Acids Research 1993; 21:4025-4030.
78. Fromant M, Plateau P, Schmitt E et al. Receptor site for the 5'-phosphate of elongator tRNAs governs substrate selection by peptidyl-tRNA hydrolase. Biochemistry 1999; 38:4982-4987.

79. Calendar R, Berg P. The catalytic properties of tyrosyl ribonucleic acid synthetases from *Escherichia coli* and *Bacillus subtilis*. Biochemistry 1966; 5:1690-1695.
80. Calendar R, Berg P. D-tyrosyl RNA: formation, hydrolysis and utilization for protein synthesis. J Mol Biol 1967; 26:39-54.
81. Soutourina J, Plateau P, Blanquet S. Metabolism of D-aminoacyl-tRNAs in *Escherichia coli* and *Saccharomyces cerevisiae* cells. J Biol Chem 2000; 275:32535-32542.
82. Soutourina J, Blanquet S, Plateau P. D-tyrosyl-tRNA^{Tyr} metabolism in *Saccharomyces cerevisiae*. J Biol Chem 2000; 275:11626-11630.
83. Soutourina J, Plateau P, Delort F et al. Functional characterization of the D-tyr-tRNA^{Tyr} deacylase from *Escherichia coli*. J Biol Chem 1999; 274:19109-19114.
84. Ferri-Fioni M-L, Schmitt E, Soutourina J et al. Structure of crystalline D-Tyr-tRNATyr deacylase: a representative of a new class of tRNA-dependent hydrolase. J Biol Chem 2001; 276:47285-47290.
85. Janiak F, Dell VA, Abrahamson JK et al. Fluorescence characterization of the interaction of various transfer RNA species with elongation factor Tu.GTP: evidence of a new functional role for elongation factor Tu in protein biosynthesis. Biochemistry 1990; 29:4268-4277.
86. Nissen P, Kjeldgaard M, Thirup S et al. Crystal structure of the ternary complex of Phe-tRNA^{Phe}, EF-Tu, and a GTP analog. Science 1995; 270:1464-1472.
87. Yamane T, Miller DL, Hopfield JJ. Discrimination between D- and L-tyrosyl transfer ribonucleic acids in peptide chain elongation. Biochemistry 1981; 20:7059-7064.
88. Petersen HU, Røll T, Grunberg-Manago M et al. Specific interaction of initiator factor IF$_2$ of *E. coli* with formylmethionyl-tRNA^{Met}$_f$. Biochem Byophys Res Commun 1979; 91:1068-1074.
89. Sundari R, Stringer L, Schulman L et al. Interaction of initiation factor 2 with initiator tRNA. J Biol Chem 1976; 251:3338-3345.
90. Guillon JM, Mechulam Y, Blanquet S et al. Importance of formylability and anticodon stem sequence to give tRNA^{Met} an initiator identity in *Escherichia coli*. J Bacteriol 1993; 175:4507-4514.
91. Guillon JM, Heiss S, Suturina J et al. Interplay of methionine tRNAs with translation elongation factor Tu and translation initiation factor 2 in *Escherichia coli*. J Biol Chem 1996; 271:22321-22325.
92. Forster C, Ott G, Forchhammer K et al. Interaction of a selenocysteine-incorporating tRNA with elongation factor Tu from *E.coli*. Nucleic Acids Res 1990; 18:487-491.
93. Rudinger J, Hillenbrandt R, Sprinzl M et al. Antideterminants present in minihelix(Sec) hinder its recognition by prokaryotic elongation factor Tu. EMBO J 1996; 15:650-657.
94. Stanzel M, Schon A, Sprinzl M. Discrimination against misacylated tRNA by chloroplast elongation factor Tu. Eur J Biochem 1994; 219:435-439.
95. Kiesewetter S, Ott G, Sprinzl M. The role of modified purine 64 in initiator/elongator discrimination of tRNA(iMet) from yeast and wheat germ. Nucleic Acids Res 1990; 18:4677-4682.
96. Astrom SU, Bystrom AS. Rit1, a tRNA backbone-modifying enzyme that mediates initiator and elongator tRNA discrimination. Cell 1994; 79:535-546.
97. Evans SV. Setor: hardware lighted three-dimensional solid model representation of macromolecules. J Mol Graphics 1993; 11:134-138.
98. Thompson JD, Gibson TJ, Plewniak F et al. The CLUSTAL_X windows interface: flexible strategies for multiple sequence alignment aided by quality analysis tools. Nucleic Acids Res 1997; 25:4876-4882.
99. Barton GJ. ALSCRIPT: a tool to format multiple sequence alignments. Prot Eng 1993; 6:37-40.

CHAPTER 5

The Directed Evolution of Organismal Chemistry: Unnatural Amino Acid Incorporation

Jamie M. Bacher and Andrew D. Ellington

Abstract

Life on earth is tremendously diverse. Organisms that have adapted to extreme environments are surprisingly ubiquitous. However, all known extremophiles have adapted to unusual physicochemical environments using the same underlying biochemistry. Although small perturbations in the genetic code are known, these are clearly deviations from the genetic code that presumably became universal at the time of the last universal common ancestor. No known genetic code encodes an amino acid other than the canonical twenty; the so-called twenty-first amino acid, selenocysteine, requires the activity of extra, specialized gene products. With this framework in mind, the one remaining boundary to life, new biochemistries, has begun to be explored experimentally. Here we discuss the different approaches used to alter the underlying biochemistry of life.

Introduction

Life on earth has adapted to an extremely wide variety of environments, including temperatures over 100°C, pH values as low as 0 and as high as 11.5, salinity of up to several molar NaCl, extreme aridity, and other seemingly inhospitable physical or chemical settings (reviewed in ref. 1). In general, since living systems strive to maintain homeostasis, these extreme environments have yielded adaptations that maintain an intracellular milieu which is similar to that of organisms from more moderate environments.[1]

At the molecular level, adaptations to extreme environments frequently involve alterations of the sequences of proteins. A comparison of the characteristics of proteins from organisms isolated from a range of temperatures has enhanced our understanding of how such sequence changes contribute to conflicting requirements of stability and flexibility (reviewed in refs. 2, 3). Essentially, thousands of individual stabilizing interactions are counterbalanced by a number of destabilizing forces, and the overall stability of a given protein depends on the energy of just a few hydrogen bonds or other interactions. Temperature affects both stabilizing and destabilizing factors, but proteins that have evolved in either thermophiles or psychrophiles (heat-loving and cold-loving organisms, respectively) nonetheless maintain an appropriate balance of stability and flexibility. As a broad generalization, thermophilic proteins tend to substitute nonpolar uncharged residues for polar residues, and tend to be smaller than their mesophilic homologues. Cold-adapted proteins tend to have substitutions with the opposite physical characteristics, and are larger.

Piezophilic organisms (reviewed in ref. 4) are those that have optimal growth rates at pressures above ~0.1 MPa. High pressure has the effect of decreasing volume and primarily affects the relative level of protein hydration and the interactions between amino acid side chains. Ultimately, piezophilic organisms buffer proteins against high pressure by substituting hydrophobic amino acids for charged amino acids. Solvation of buried hydrophobic side chains accompanies a positive volume change, and hydrophobic substitutions thus counteract the negative volume change induced by increased pressure (which typically leads to denaturation), thus maintaining the protein in a folded state.

Halophilic organisms have adapted to high salt concentration in two different ways: eubacteria maintain osmolarity by synthesizing various osmotic solutes, whereas archaebacteria just maintain high intracellular salt concentrations. Adaptation to salinity in turn requires protein sequence adaptations (reviewed in refs. 5, 6). Increased salt concentrations promote protein denaturation (the purification of proteins by 'salting out' at high ammonium sulfate concentrations is another example). To compensate, halophilic proteins tend to have more acidic amino acids than halosensitive homologues.

Other extremophiles have adapted to potentially deadly environments by having redundant systems. *Deinococcus radiodurans* is highly resistant to radiation, dessication, and starvation in part because of a plethora of DNA repair mechanisms.[7,8] Similarly, *Ralstonia sp* CH34 has several operons dedicated to the export of heavy metals and can grow at metal concentrations several-fold above those that are inhibitory to *E. coli.*[9]

Taken together, the extraordinary environments exploited by organisms can be considered boundary conditions for life. Since any organismal phenotype is at root a function of biosynthetic chemistry, it is remarkable that such a wide range of environments have been conquered by living systems that all have much the same basic biochemistry. As these vignettes demonstrate, although extremophiles are adapted to exceptional environments, they do not use any special chemistries to do so. They are subject to the same physicochemical rules as other organisms. However, the combinatorial nature of protein sequences, lipid compositions, and metabolic pathways has provided adequate fodder for natural selection to widely diversify the environmental niches and habitats of organisms.

From a different vantage, it is remarkable that of all the boundary conditions that have been explored by life, the one boundary that seems to have remained relatively sacrosanct is organismal chemistry itself. This could of course be because the various other adaptations require an internal homeostasis in chemistry, such as a constant set of amino acids and a standard genetic code. Conversely, it could be that the bounds on organismal chemistry are a result of history rather than function. That is, if it is evolutionarily simple to adopt amino acid changes, and evolutionarily difficult to adopt new amino acids, then the former strategy will be embraced long before the latter.

In fact, there are experimental indications that organismal chemistry can in fact be significantly perturbed. Early experiments were facilitated by the ready availability of unnatural isotopes following the development of domestic nuclear weapons and energy programs. The growth of *E. coli* is drastically inhibited in media containing deuterated water and lactose as a carbon source.[10] This is likely an aggregate result of the differences in the activities of deuterated catalysts and metabolites. In support of this hypothesis, fully deuterated alkaline phosphatase shows decreased activity.[11] Nevertheless, an organism capable of growth in the presence of unusual isotopes has been described. *Chlorella vulgaris* can grow under conditions of 99.7% D_2O.[12] Furthermore, this organism can be grown on D_2O and $^{13}CO_2$, the combination of which surprisingly diminishes abnormalities seen when grown on D_2O alone.[13]

Thus, while chemical diversity does not appear to have been fully embraced by natural organisms, it may be a biotechnological resource to be exploited. Natural evolution with conventional chemistries can obviously yield amazing new functionality, but may do so only rarely and certainly slowly; this is in part because of homeostasis and the fact that protein function is generally maintained regardless of environment. For this reason, it can be argued that that new organismal chemistries might facilitate the evolution of entirely new organismal functions, or at least foster more streamlined evolutionary solutions to the environmental problems that have already been solved by extremophiles.

Chemical Homeostasis and Variety During Organismal Evolution

Unaccountably, organisms are phenotypically diverse and chemically uniform. We will first examine several modifications to organismal chemistry that nature has introduced, in particular the adoption of amino acids and genetic codes, in order to gain better insights into whether large scale changes in organismal chemistry may be experimentally accessible.

The Biochemical Uniformity of Life

All forms of life discovered thus far have the same canonical nucleotides (with the exception of some bacteriophage[14]) and amino acids. Moreover, all forms of life have similar biochemistry, from the virtually universal pathways for metabolite biosynthesis to the highly conserved machinery for protein biosynthesis. The similarity in biochemistry is likely a result of history, of descent from a single 'last common ancestor.'[15] Moreover, the fact that both the ribosome and various essential cofactors are derived primarily from ribonucleotides, rather than proteins or amino acids, suggests that the last common ancestor was preceded by a phylogenetic spectrum of organisms whose primary catalysts may have been ribozymes rather than protein enzymes, as predicted by the so-called RNA world hypothesis.[15]

The fact that the modern protein world apparently displaced the RNA world is an indication of the magnitude of functional change that is required to engender large changes in organismal biochemistry. Based on this comparison, it should come as no surprise that the protein world and its underlying molecular traits, the complement of amino acids, genetic code, and translation machinery, are virtually fixed throughout life. However, there are several indications from natural organisms that 'upheavals' in the protein synthesis apparatus may be possible. First, unnatural amino acids can be found in organisms; second, the genetic code is not universal; and finally, directed evolution experiments have shown that the translation machinery is malleable.

Unnatural Amino Acids in Organisms

There are a variety of unnatural amino acids found in organisms. For example, the post-translational modification of proteins is a common mechanism for enhancing the functionality available to proteins. Commonly known protein modifications include phosphorylation, acetylation, and ribosylation. Of course, the glycosylation of proteins is one of the most ubiquitous forms of post-translational modification (reviewed in ref. 16). A relatively straightforward example of this are the lantibiotics, a group of bacteriocins produced by gram-positive bacteria that are defined by their post-translational modification to include lanthionine (Lan) and similar derivatives (Fig. 1; reviewed in ref. 17). Specific proteins are responsible for these post-translational modifications: LanB dehydrates serine or threonine to form 2,3-didehydroalanine (Dha) or 2,3-didehydrobutyrine (Dhb), respectively. LanC then forms a thioether bond between Dha or Dhb and Cys, forming Lan and 3-methyllanthionine (MeLan), respectively. Certain lantibiotics are also decarboxylated by LanD. These thioether post-translational modifications assist certain lantibiotics, such as cinnamycin, in maintaining an elongated form, while others, such as nisin, are maintained in a globular form. This is but a single example of unnatural amino acid incorporation into peptides; there are numerous other examples distributed throughout phylogeny, including the direct synthesis of modified peptide antibiotics such as gramicidin S by special nonribosomal peptide synthetases.

Sometimes amino acids undergo chemical modification to enhance protein function (reviewed in ref. 18). For example, relative to other amino acids, tyrosine residues are commonly modified to form proteinaceous cofactors. Copper-induced autocatalysis of tyrosine can lead to the formation of 2,4,5-trihydrophenylalanine quinone (Tpq, Fig. 1) within the peptide –TXXNY(D/E)Y- (in which the first tyrosine is converted to Tpq). This modified Tyr residue can then act as a Schiff base in amine oxidases. Other modifications of Tyr include a Cys-Tyr crosslink that has been shown to be important in the coordination of copper in fungal galactose oxidase by promoting one-electron chemistry. Similarly, a Tyr-His crosslink has been shown to be important in the coordination of heme-oxygen and copper in cytochrome c oxidase. His-Tyr crosslinks are also found in catalase HPII from *E. coli*. Finally, the triad of SYG forms the chromophore of Green Fluorescent Protein (Fig. 1) as the serine and glycine autocatalytically cyclize (reviewed in ref. 19). This list of tyrosine modifications is far from exhaustive, and there are similar functional modifications of other residues.

The Evolution of Nonstandard Codes

While the genetic code is extremely highly conserved, it is not universal. *Micrococcus luteus* lacks several codons ending in A,[20] and certain bacterial, protozoan, ciliate and archaebacterial translational systems or subcellular compartments have altered genetic codes.[21] The presence of such altered codes is either the result of history or function or both (reviewed in ref. 22). In this respect, it is interesting to note that the canonical code may largely be the result of function. A well-adapted code would be one that minimizes damage done by either mutations or mistakes in translation. In fact, a recent analysis[23] comparing the universal genetic code to a million random variants of the code found a single genetic code that was superior to the universal code when considering only the minimization of mutational effects. However, when translational errors were also considered, the universal genetic code was found to minimize errors most efficiently. Problems with historical

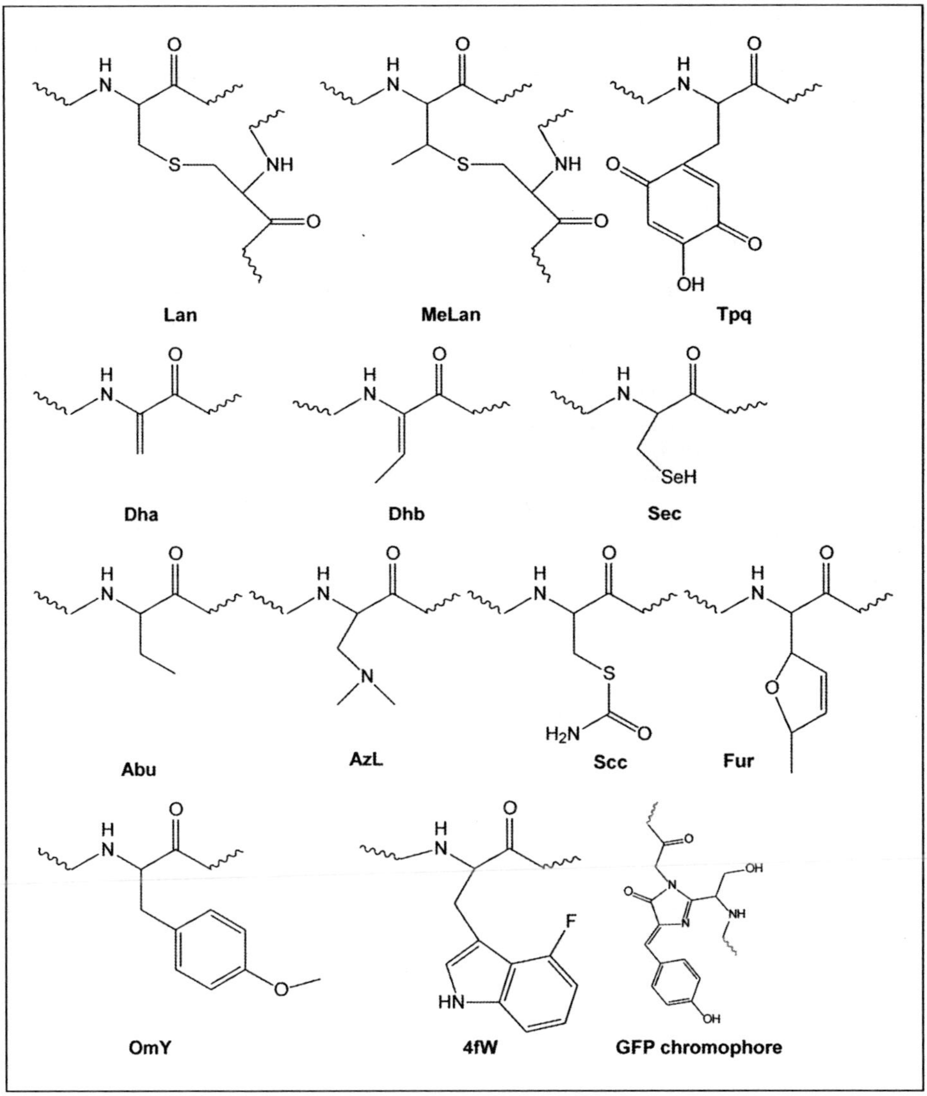

Figure 1. Structures of unnatural amino acids. The structures of unnatural amino acids discussed in the chapter are shown. Wavy bonds represent linkages to other amino acids. Lan: lanthionine. MeLan: methyllanthionine. Tpq: 2,4,5-trihydrophenylalanine quinone. Dha: 2,3-dihydroalanine. Dhb: 2,3-dihydrobutyrine. Fur: furanomycin. Sec: selenocysteine. Abu: α-aminobutyrate. AzL: azaleucine. Scc: S-carbamoyl-L-cysteine. OmY: *O*-methyl-L-tyrosine. 4fW: 4-fluorotryptophan. The GFP chromophore is formed by a cyclization of the triad SYG.

explanations for the origins of the genetic code[24,25] also make function the most likely explanation for codon assignment.

The balance between the robustness and malleability of the universal genetic code has recently been explored.[26] Robustness minimizes the effect of mutation and mistranslation, while malleability allows for the exploration of new phenotypes, and possibly allows for the adaptation to new environments. Overall, these statistical arguments validate previous chemical-based reasoning that the genetic code has remained experimentally and naturally mutable.[27] Thus, the nonstandard codes can potentially be viewed as functional adaptations, rather than as historical accidents. This is not to say that they impart greater or different function to their resultant proteomes; indeed, altered codes in organelles can to some extent be viewed as a defense against genomic dissipation by gene transfer. The alternative genetic codes of ciliates may similarly be a defense against viral parasitism of their translational apparatuses; no viruses of ciliates have yet been identified.

Nonstandard genetic codes are typically relatively small perturbations relative to the universal genetic code,[21] as might be expected of a system that optimized both robustness and malleability. One example of genetic code reassignment is the capture of stop codons by amino acids. For example, several lineages of ciliates encode glutamine using UAA and UAG codons, while other lineages of ciliates use UGA to encode either cysteine or tryptophan. A comparison of ciliate phylogenies based on 28S rRNA with the distribution of stop codons in the α-tubulin and phosphoglycerate kinase genes suggested that noncanonical but identical codes may have arisen independently.[28] Stop codon capture appears to be facilitated by mutations in eukaryotic release factor 1 (ref. 29). Modified genetic codes are certainly not limited to ciliates; archaea, bacteria, fungi, protozoa, mitochondria and chloroplasts are all known to have divergent codes. In addition, stop codon capture is certainly not the only deviation from the canonical code: vertebrate mitochondria are known to have at least two reassignments, in addition to an alternative use of a stop codon.[21]

Despite the widespread variation in the genetic code, few explanations for mechanism are satisfying. One possible mechanism for the reassignment of codons is through a genomic bias towards AT or GC,[30-32] in which codons are lost from the genetic code through a drift or directional bias, and then recovered, possibly coding for a new amino acid, when the bias drifts or is selected to reverse. A survey of *M. luteus* codons revealed that, in correspondence with its high GC content, certain codons ending in A did not appear at all.[20] Furthermore, translational systems extracted from these organisms resulted in truncated peptides attached to ribosomes when translation of mRNA containing AGA or AUA codons was tested. This suggests that these codons were unassigned and did not cause an interaction with translation release factors. Similar results were obtained with *Mycoplasma capricolum* when tested with the codon CGG.[33] Furthermore, the *M. luteus* cell-free translation system was later used to allow the in vitro suppression of AGA with chemically acylated tRNA.[34] Organisms such as *M. luteus* and *M. capricolum* may represent intermediate stages between the universal genetic code and divergent genetic codes.

However, while exceptions to the canonical genetic code exist, it remains true that no organism studied to date encodes an amino acid other than the canonical twenty. This is interesting, given the range and number of post-translational modifications that occur. However, while there is not a canonical encoding of a new amino acid, there is one example of a specialized mechanism of encoding a 21st amino acid, selenocysteine. The stop codon UGA encodes selenocysteine (Sec), but only in the presence of specific *cis*-acting signals and *trans*-acting factors (reviewed in refs. 35, 36, Fig. 2). The tRNA^{Sec} is a UGA suppressor. tRNA^{Sec} is charged by seryl-tRNA synthetase with serine. This seryl-tRNA^{Sec} is then modified by selenocysteine synthase (the product of *selA*) to form aminoacrylyl-tRNA^{Sec}. The reactive selenium donor, selenium phosphate, then donates selenium to form selenocysteyl-tRNA^{Sec}. Selenium phosphate is formed by the product of *selD*, selenophosphate synthetase, from selenide and ATP. The Sec-tRNA^{Sec} is bound by the protein SELB, which acts instead of EF-Tu to bring the charged tRNA to the ribosome. In fact, SELB associates ~100-fold more tightly with Ser-tRNA^{Sec} than does EF-Tu.[37] This discrimination may be due to the fact that tRNA^{Sec} is unusual among tRNAs in two ways: it has an eight-base acceptor stem, compared with the usual seven, and it has an unusually long variable loop. However, in order to suppress only the correct UGA codons, the SELB protein also recognizes a *cis*-acting element in the mRNA, known as the selenocysteine insertion sequence (SECIS), which occurs following the UGA to be suppressed. The minimal version of the SECIS is a 17 bp stem-loop.[38] This process has been shown to be rather inefficient using an *fdhF – lacZ* gene fusion as an assay for selenocysteine incorporation.[39] The mechanism for insertion in eukaryotes is thought to be similar, but is not entirely known (see Chapter 4 by S. Blanquet et al).

While this mechanism of incorporation is cumbersome, the functional importance of Sec is clear: it provides a reductive power that is unobtainable by Cys. To illustrate this effect, the selenoprotein formate dehydrogenase H of *E. coli* has a k_{cat}/K_M about 100-fold greater than a

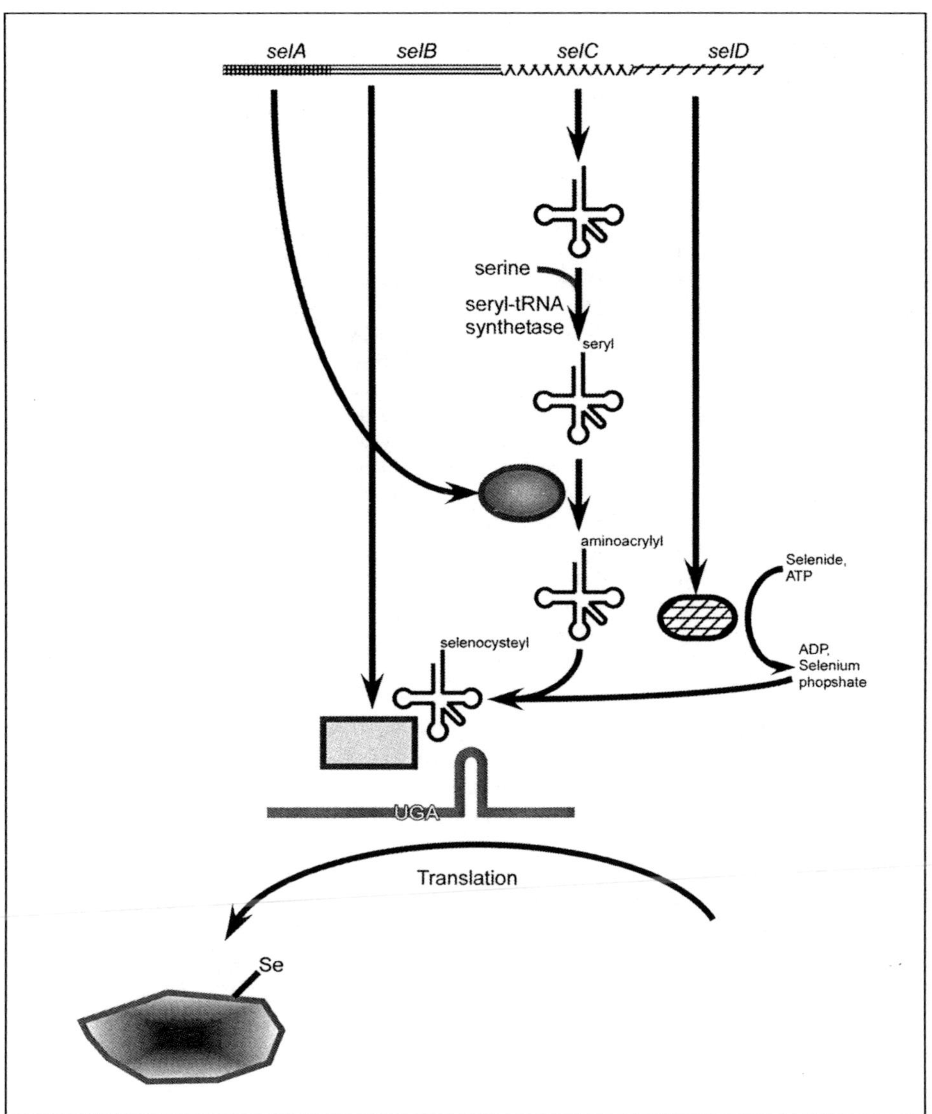

Figure 2. Selenocysteine translation. Selenocysteine incorporation depends on the function of four genes, *selABCD*. The product of *selC* is tRNA^{Sec}. This tRNA is charged with serine by SerRS, but converted to aminoacrylyl-tRNA^{Sec} by selenocysteine syntherase, the product of *selA*. ATP and selenide are converted to ADP and selenium phosphate, which then reacts with the aminoacrylyl-tRNA^{Sec} to form Sec-tRNA^{Sec}. This tRNA is bound by SELB and inserted into a growing peptide chain when encoded by the UGA codon and when the SECIS is present. Note that gene positions are shown for clarity; in fact, they do not form an operon on the *E. coli* chromosome.

cysteine-substituted version of the same protein.[40] The use of this unusual amino acid thus allows bacteria chemical options that were previously unavailable to them. This finding bodes well for the inclusion of other chemically diverse amino acids in the genetic code; with sufficient selection pressure, nature is capable of inventing methods for their incorporation.

Engineering Organismal Chemistry

An examination of natural systems suggests that the adoption of novel organismal chemistries is entirely possible, but is generally not the first evolutionary adaptation that is explored. Thus, it should be possible to use directed evolution to generate selection pressures sufficient to force the incorporation of new chemistries, in particular novel amino acids.

Strategies for the Incorporation of Unnatural Amino Acids

There are two general strategies for the incorporation of unnatural amino acids into organisms that can be loosely termed 'top down' strategies and 'bottom up' strategies. 'Top down' strategies treat the organism as the unit of selection, and typically involve feeding an unnatural amino acid to an auxotroph in hopes of having the auxotroph tolerate or incorporate the analogue. 'Bottom up' approaches assume that our knowledge of organismal biochemistry is sufficient to engineer single or multiple incorporations of unnatural amino acids into particular proteins or potentially throughout the proteome. In general, 'bottom up' approaches involve engineering aminoacyl-tRNA synthetase:tRNA pairs to direct the incorporation of unnatural amino acids at particular sites, such as stop codons.

Assuming that engineering organismal chemistry is the desired goal, it is likely that 'top down' strategies will be more successful than 'bottom up' strategies, since it is not only the translation machinery that must accommodate a new amino acid but the entire proteome. This has in fact proven to be the case: the only known example of complete substitution of an unnatural amino acid for a natural amino acid involved a 'top down' strategy. However, the sheer scale of sequence space that must be explored during organismal evolution argues strongly that 'top down' strategies should be melded with 'bottom up' strategies, in order that rare evolutionary steps in aminoacyl-tRNA synthetases and tRNAs that might otherwise never occur can actually serve as starting points for the accommodation of new amino acid chemistries into an organismal proteome.

Directed Evolution of Unnatural Amino Acid Incorporation

The fact that the genetic code has proven to be malleable during the course of natural evolution is consistent with the 'top down' hypothesis that organisms can be evolved to tolerate and incorporate unnatural amino acids. In fact, there is a body of literature that examines the ability of organisms to incorporate unnatural amino acids on a limited basis. For example, heavy atom analogues of methionine, such as selenomethionine and telluromethionine, are routinely incorporated into proteins for use in multiwavelength anomalous diffraction for protein crystallography. This is typically done by growing the organism in limiting amounts of methionine and excess analogue, and inducing protein expression when the natural amino acid has been exhausted.[41] Another interesting example is the accidental charging of furanomycin onto tRNA^{Ile} by IleRS. Although furanomycin is unlike isoleucine, it is nevertheless effectively charged onto tRNA and incorporated into proteins.[42] Finally, 4-fluorotryptophan (4fW) has been extensively studied as a tryptophan substitute. Although *E. coli* tryptophan auxotrophs are known to be incapable of survival for more than a few generations, the effects of these analogues on cellular growth are known.[43,44] The effect of 4fW incorporation on various proteins has also been determined. Many proteins are inhibited when translated in the presence of 4fW; β-galactosidase exhibits only 28% of its wild-type activity,[44] while lactose permease induction is diminished by 65%.[43] Conversely, arginyl-tRNA synthetase is essentially unaffected by 4fW incorporation,[45] while the specific activity of lactate dehydrogenase actually increases 2-fold.[43]

Since organisms show limited growth in the presence of 4fW and since the effects of this analogue on protein function are relatively modest, it is reasonable to assume that an organism could evolve to accommodate the unnatural amino acid. Unlike *E. coli*, tryptophan auxotrophs of *Bacillus subtilis* are capable of repeatable growth in the presence of 4fW. Jeffrey Wong exploited this phenotype to select organisms that could more efficiently incorporate 4fW into their proteome.[46] Wong plated a thick spread of bacteria on rich, but defined, media with a plentiful supply of 4fW and no W, and picked a large colony (Fig. 3). He repeated this positive selection a second time and obtained a clone with roughly equivalent growth characteristics on 4fW and W. Additional rounds of selection included a strong negative selection: following N-methyl-N'-nitro-N-nitrosoguanidine (MNNG) mutagenesis, cells were grown on plates with plentiful W and extremely limiting 4fW. A clone that formed small colonies, suggesting it had poor growth capabilities on W, was picked. This clone was

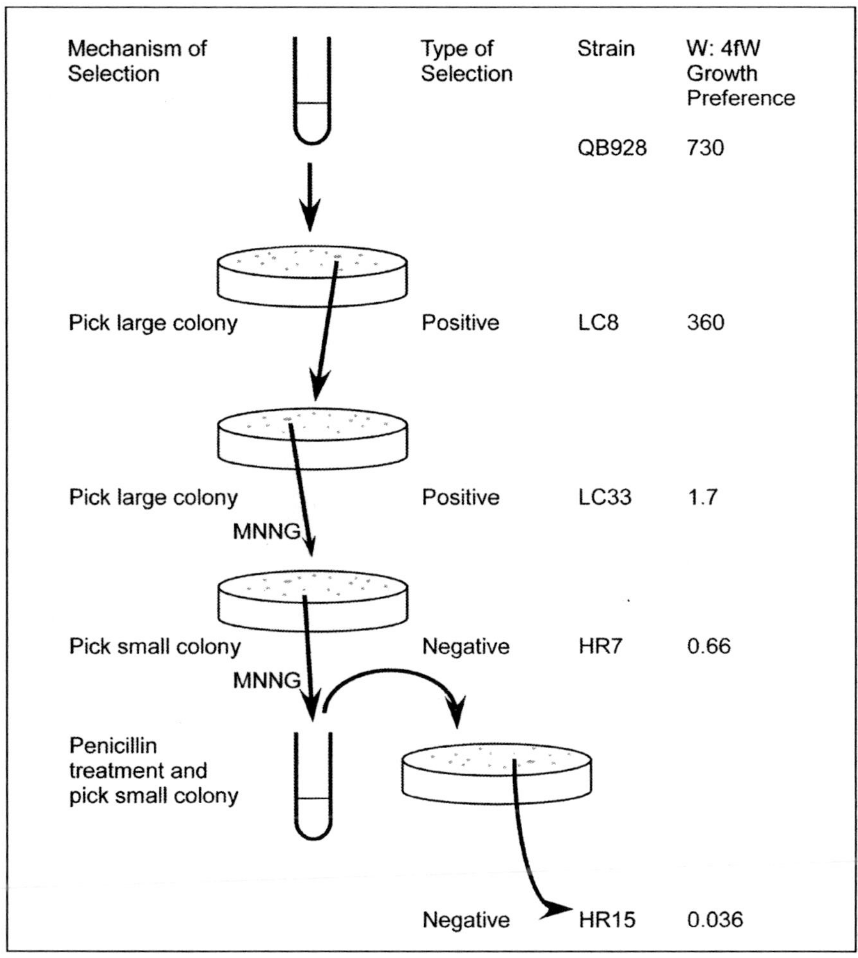

Figure 3. Strategy of selection to isolate a *B. subtilis* strain which prefers 4fW over W. In order to select variants of *B. subtilis* that could grow on 4fW, Wong[46] used a combination of positive and negative selections by picking large colonies on 4fW media, or small colonies on media with W and trace amounts of 4fW. The final strain achieved a switch in growth on W:4fW of over 2×10^4. (Data adapted from ref. 46)

then grown up and once again subjected to mutagenesis. Two negative selection steps were applied. First, the population was supplied solely with W in the presence of penicillin. Cells capable of growth with W in the absence of 4fW were killed by the action of penicillin. The surviving bacteria were then plated on W with limiting 4fW, and small colonies were again picked. This strong combination of positive and negative selections ultimately produced a strain of bacteria, HR15, which showed a 2×10^4-fold switch in its preference for 4fW relative to W. Amino acid analyses of whole-cell protein extracted from this strain clearly showed incorporation of 4fW. HR23, a derivative of HR15 (F. Mat and J. T. Wong, unpubl.), shows an extraordinary preference for 4fW: it grows well on 4fW, slightly on W and not at all on eight other tryptophan analogues (J. M. B. and A. D. E., unpubl.). While it is always difficult to claim that there are no remaining tryptophan residues in an organismal proteome, especially since many tryptophan analogues are contaminated to a small extent with tryptophan (see, for example, ref 47), the fact that HR15 and HR23 prefer the unnatural amino acid analogue strongly supports the hypothesis that incorporation is complete.

More recently, we have attempted to recreate this experiment,[47] this time using an *E. coli* tryptophan auxotroph. While Wong chose *B. subtilis* for its slight ability to grow on 4fW, the *E. coli* strain C600p (which is Δ *trpE*) initially showed absolutely no growth on 4fW. However, like other auxotrophs that have been examined, C600p grew well on a mixture of 95% 4fW:5% W, but not at all when only 1% of the total available W was the natural amino acid. A serial dilution technique was used to enrich for variants that could grow on progressively higher ratios of 4fW:W. Ultimately, bacteria were isolated that could grow on 99.97% 4fW (0.03% of the commercial 4fW preparation was tryptophan). However, these 'winners' did not display the same preference for 4fW that was exhibited by the *B. subtilis* strain HR15. In fact, the selected *E. coli* strain maintained a marked preference for W, and grew extremely slowly on 4fW. Nevertheless, mass spectrometry data of individual proteins clearly showed the incorporation of 4fW. Genetic characterizations of the evolved strain revealed mutations in proteins involved in the incorporation of tryptophan: the aromatic amino acid permease, *aroP* (three amino acid substitutions), tryptophanyl-tRNA synthetase, *trpS* (one amino acid substitution) and the tyrosine repressor, *tyrR* (nonsense mutation). The retained preference for tryptophan coupled with the fact that the isolated tryptophanyl-tRNA synthetase enzyme appeared to discriminate against 4fW suggested that the selected *E. coli* were scavenging trace amounts of W for growth while nonetheless tolerating proteomic substitution of 4fW, a feat the ancestral strain was incapable of.

It is interesting to speculate on why such very different courses of evolution were adopted by HR15 and the 4fW-adapted *E. coli*. Two possible reasons for the different results are the different media used and the types of selection employed. The experiment by Wong used a rich media, which removed or reduced selective constraints on perhaps one third to one half of the *B. subtilis* genome.[48] In fact, it has now been shown that methionine is an absolute requirement for HR23 (F. Mat and J.T. Wong, pers. comm.). This suggests that one or more proteins in this biosynthetic pathway is nonfunctional when synthesized with 4fW. In addition, the use of both positive and negative selections may have favored an 'incorporation' strategy over a 'scavenger' strategy. The techniques used by Wong directly selected against the utilization of W, while the serial dilution technique we adopted did not select against this option.

The lesson here is in part a reiteration of the adage "you get what you select for," but it may also indicate that both evolutionary pathways are surprisingly accessible. *B. subtilis* was poised to adopt an 'incorporation' strategy because it had an initial ability to grow continuously on high concentrations of 4fW. The strong positive selection demanded that variants grow on limited or no W, and the negative selection removed from the population those variants that maintained a preference for growth on W. Thus, it is possible that if the *E. coli* selection was performed in a rich media that variants capable of growth on 4fW would have been selected. Conversely, the metabolic load on *B. subtilis* grown in a minimal media may have been too great to allow the ready incorporation of 4fW and the concomitant degradation of biosynthetic enzyme activities. If the 'incorporation' and 'scavenger' pathways are in fact in equipoise, at least in terms of mutational probabilities, then this bodes well for the generation of organisms that incorporate unnatural amino acids with the proviso that strong negative selections against the incorporation of natural amino acids may always need to be included.

So far we have only considered 'all or none' approaches in which an unnatural amino acid is substituted for a natural one. However, it is also possible that genetic code may be become chemically degenerate, and that a given codon or codon set can specify more than one type of amino acid. Along these lines, the Marlière group showed that misincorporation of azaleucine by leucyl-tRNA synthetase could suppress the lethal phenotype of the thymidylate synthase (*thyA*) mutant R126L mutation. Selection for azaleucine-dependent growth resulted in the recovery of clones that had only the leucine codons at this position.[49] This clever genetic selection was similarly applied to force the misincorporation of cysteine for isoleucine.[50] The function of *thyA* is absolutely dependant on C146. The codon at position 146 was altered to the isoleucine/methionine block AUN and these four codons were assayed with four cysteine tRNAs with NAU anticodons. When position 146 was encoded by either the isoleucine codons AUU or AUC, the cysteine GAU anticodon restored nearly wild-type levels of growth. A long-term serial dilution experiment showed that under selective conditions the altered tRNA, and presumably the misincorporation of Cys for Ile, was maintained. This experiment effectively demonstrated that the forced inclusion of an unnatural amino acid at a critical site could continually suppress a lethal phenotype.

Taken together these experiments provided the necessary background for taking advantage of misincorporation not only by tRNA binding, but also by tRNA charging. In a collaboration between the Marlière and Schimmel groups, both the isoleucine (AUN) and valine (GUN) codon groups were substituted into *thyA* at position 146. It was found that abundant Cys in the medium could suppress the otherwise lethal C146V mutation in *thyA*.[51] To enhance this suppression, two selections were separately performed to enhance incorporation of cysteine across from a valine codon; both selections attempted to limit the oxidation of cysteine in the media. First, cells were grown on the cysteine analogue S-carbamoyl-cysteine (Scc). This one-step screen found four winning mutants. The second selection was a serial dilution experiment in which cells were propagated anaerobically for 100 generations in the presence of limiting Cys. One winner of this selection was characterized. All winning strains were found to cotransduce the Cys-suppression phenotype with a selectable marker, which was near to the ValRS. In fact, all five strains were found to carry mutations in ValRS; specifically in the editing domain of this protein, which hydrolyses mischarged amino acids. ValRS normally mischarges tRNA^{Val} with Thr at a level of about 1:200 to 1:400, but only ~1:3000 valine residues is substituted by Thr in proteins due to the editing domain of ValRS, CP1.[52] This editing domain of ValRS is highly conserved, and all mutations were found to be in absolutely conserved residues of the protein, presumably interdicting function.[51,52] Kinetic analysis of one of the mutants showed that it charged tRNA^{Val} with Thr, whereas the wild-type protein had no charging capacity for Thr. Presumably the editing mutant allows enough misincorporation of cysteine to promote growth of the *thyA* mutant strain.

The misincorporation mutants described above presumably do not only operate on thymidylate synthase, but throughout the proteome. Thus, it is interesting to note that when the editing mutant described above was grown in the presence of α-aminobutyrate (Abu), a structural analogue of valine, the unnatural amino acid was incorporated throughout the proteome at 24%. It is surprising that the organism could survive this level of incorporation, although these results are consistent with the previous body of work with fluorotryptophans and other unnatural amino acids. These results provide an indication of a third potential pathway for introducing unnatural amino acids into the genetic code, beyond the 'incorporation' and 'scavenging' mechanisms described above. An unnatural amino acid could potentially initially be partially misincorporated via a natural codon or codon set, and in the longer term could 'take over' a portion of a codon set for itself. Indeed, the technique of enforced incorporation championed by Marlière makes this scenario all the more possible. The researchers designed an intelligent system that paired a lethal phenotype with an analogue of the encoded amino acid. The use of cysteine as an analogue of valine ultimately allowed the substitution of valine with an unselected amino acid, α-aminobutyrate. Thus, it may be possible to first relax the coding restrictions inherent in the genetic code and in the translation apparatus as a whole, and then upon reoptimization of the proteome to a new amino acid set, to re-establish new coding relationships. In fact, it has been postulated that it is precisely this scenario that led to the alternative codes seen in mitochondria and elsewhere.[53-55]

The Role of Genome Size

While it has proven possible to adapt organisms to utilize or tolerate unnatural amino acids, it is unclear how many mutational steps may generally be required to achieve these feats. The partial incorporation of Abu required one mutation, while the complete incorporation of 4fW apparently required at least five in *E. coli* and likely a commensurate number in *B. subtilis*. Nonetheless, it might be suspected that more audacious chemical substitutions will require a correspondingly larger number of changes, spread throughout the organismal proteome.

In this respect, it can be hypothesized that there should be an inverse correlation between the size of a genome and its chemical evolvability, its malleability with respect to new chemical information such as unnatural amino acids. The larger the genome, the more proteins that must evolve, the harder it is expected to be to achieve complete substitution. To assess this hypothesis, we have recently adapted the small bacteriophage Qβ to grow in the presence of the unnatural amino acid 6-fluorotryptophan (6fW; J. M. B., J. J. Bull and A. D. E., submitted). The phage were in fact adapted to growth on tryptophan auxotrophs that grew on a 95:5 mixture of 6fW:W. When the entire phage genome was sequenced, phage populations with superior growth properties contained, on average, 4.5 coding mutations and 2.5 noncoding mutations; in no case did a mutation eliminate a tryptophan codon.

These results are somewhat surprising, in that the number of mutations in phage and in the much larger bacterial genomes so far appear to be equivalent. This is likely just due to the fact that complete information about the evolved bacterial genomes is unavailable, but may also reflect the hellish selective pressure and evolutionary optimization that guides bacteriophage evolution.

Evolutionary Engineering of Aminoacyl-tRNA Synthetases

An alternate approach to the incorporation of unnatural amino acids into organismal proteomes is to directly alter the specificities of the codon:tRNA:aminoacyl-tRNA synthetase sets that direct incorporation. We have seen that directed evolution of organismal chemistries can in fact lead to alterations in aminoacyl-tRNA synthetase specificities; in turn, the direct manipulation of the tRNA:synthetase pairs should greatly potentiate changes in organismal chemistry. There have been two mechanisms used in tandem for manipulating aminoacyl-tRNA synthetase specificities: rational design, and directed protein evolution.

Schultz and his co-workers have championed the incorporation of so-called 'orthogonal' tRNA:aminoacyl-tRNA synthetase pairs into organisms; these approaches are also discussed extensively in Chapter 6 by Magliery et al. An orthogonal pair refers to a tRNA that is not charged by any of the host synthetases, and a synthetase which does not charge any host tRNA. Orthogonal pairs typically use stop codon suppressors for site-specific incorporation of unnatural amino acids; tRNAs used as part of an orthogonal pair are typically mutated to recognize the UAG codon. It may also benefit this approach to examine mutants of release factors to enhance the use of particular codons for suppression.[29] Orthogonal pairs can be taken from any species; bacteria have served as a source for tRNA:aminoacyl-tRNA synthetase pairs for other bacteria, yeast have donated $tRNA^{Gln}$/GlnRS and $tRNA^{Tyr}$/TyrRS to *E. coli*, and a human initiator tRNA has been paired with an *E. coli* GlnRS as a donor for yeast.[56-59] However, imported pairs are not always orthogonal. In fact, even after several rounds of mutagenesis and screening, the *E. coli* GlnRS retained activity towards the natural $tRNA^{Gln}$ in *E. coli*.[58]

Nonetheless, Schultz and coworkers ultimately succeeded in evolving a completely orthogonal tRNA:aminoacyl-tRNA synthetase pair. They first described an orthogonal yeast GlnRS/$tRNA^{Gln}$ in *E. coli*.[60] This selection was successful due to the development of a combined positive and negative selection (Fig. 4). The selection depended on the suppression of AUG codons. The positive step selected for synthetases capable of charging AUG-suppressing tRNAs. An AUG codon was substituted for an inessential position in β-lactamase that preceded the active site. Growth on ampicillin indicated charging of the orthogonal tRNA by the synthetase. Selection for an orthogonal aminoacyl-tRNA synthetase was achieved by coordinate expression of Barnase with two or three stop codons in the absence of the unnatural amino acid of interest. If charging of the tRNA occurs, then the stop codons are suppressed and functional Barnase is produced. Conversely, if orthogonality is achieved, then none of the natural amino acids serve as a substrate for the aminoacyl-tRNA synthetase, the suppressor tRNA is not charged, and the cell survives.

The combination of these techniques has proven quite successful in generating useful orthogonal pairs. For example, the TyrRS/$tRNA^{Tyr}$ pair from *Methanococcus janaschii* has been adapted for use as an orthogonal pair in *E. coli*. The *M. janaschii* TyrRS was chosen because of its ability to effectively charge its own tRNA in *E. coli*; human and yeast TyrRS/$tRNA^{Tyr}$ pairs were found to be unsuitable.[56] Next, they selected tRNA mutants that were less able to be charged by native *E. coli* aminoacyl-tRNA synthetases, while retaining efficient charging by the *M. janaschii* TyrRS.[61] Finally, they selected mutants of the TyrRS that could recognize and effectively charge these mutant tRNAs with *o*-methyl-L-tyrosine and effectively demonstrated the complete replacement of Tyr for the analogue in a single, purified protein (the analogue was coded for by the stop codon UAG).[62] The resultant orthogonal pairs should prove to be extremely effective for the site-specific incorporation of unnatural amino acids; Schultz and co-workers have proposed the incorporation of keto-containing amino acids, fluorescent or photocleavable amino acids, as well as α-hydroxy acids and β-amino acids.[58,60]

It has also been shown that effective suppression and orthogonality can be achieved by simply importing an aminoacyl-tRNA synthetase, and appropriately modifying a native or closely related tRNA to serve as the orthogonal receptor of the pair.[59] The *E. coli* GlnRS was imported to yeast, but used a human initiator tRNA as the isoacceptor; the yeast TyrRS was imported to *E. coli*, and used a modified *E. coli* $tRNA^{fMet}$ as the isoacceptor. This was the first example of the use of an orthogonal

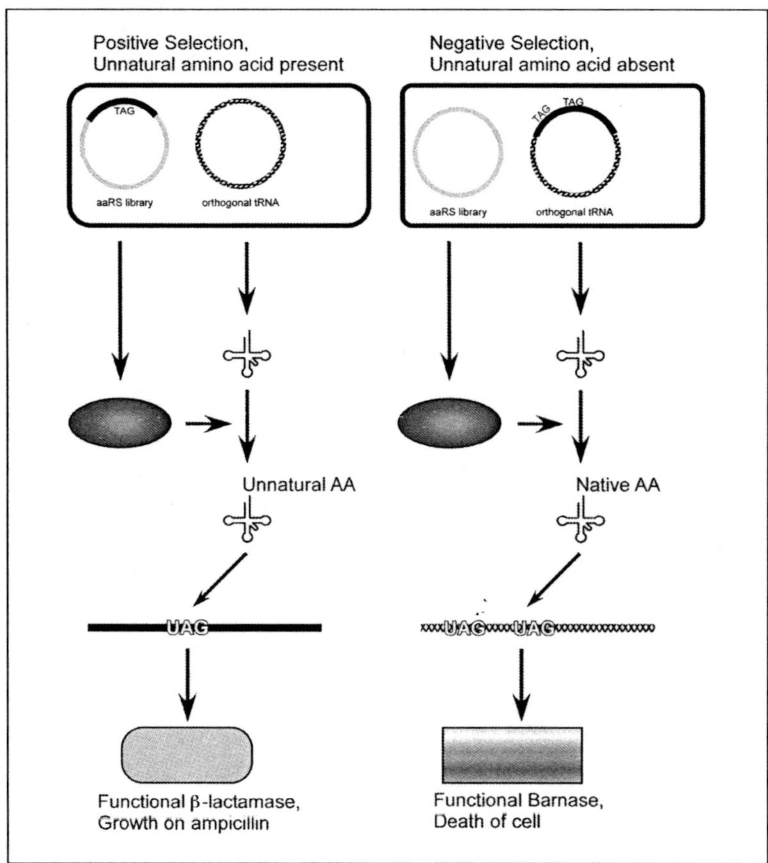

Figure 4. Strategy of selection to isolate an orthogonal aminoacyl-tRNA synthetase. This strategy uses both positive and negative selections to achieve a fully orthogonal tRNA:aminoacyl-tRNA synthetase pair.[60] A positive selection, performed in the presence of the unnatural amino acid of interest, selects for charging of the orthogonal tRNA by suppression of stop codons in β-lactamase. Growth on ampicillin indicates that the orthogonal tRNA was indeed charged. Positive clones are then subjected to a negative selection, which selects for the orthogonality of the aminoacyl-tRNA synthetase. The unnatural amino acid of interest is not included in the growth media. If the orthogonal tRNA is charged with a natural amino acid, then the stop codons in the Barnase gene are suppressed, and the bacteria are killed. This protocol can be modified to select for orthogonal tRNA.[61]

pair in eukaryotes. It also demonstrated a potential pitfall in this approach: although the yeast TyrRS was orthogonal with regards to *E. coli* tRNA^{Tyr}, it mischarged tRNA^{Pro}. Even selected mutants of TyrRS still mischarged tRNA^{Pro} to a limited extent. These results underscore the difficulty that researchers have in attempting to import proteins and tRNAs to foreign species without full knowledge of the 'second genetic code' and all of its phylogenetic variations.[63]

Another approach to engineering unnatural amino acid incorporation is just to alter the specificity of an endogenous aminoacyl-tRNA synthetase:tRNA pair by rational design or directed evolution. A residue in phenylalanyl-tRNA synthetase from *E. coli* was identified that was important for discrimination against *para*-position modifications of Phe.[64,65] Further mutagenesis of this position identified a mutant which more readily allowed the incorporation of *p*-F-phenylalanine, *p*-Cl-Phe and *p*-Br-Phe.[64,66-68] Hamano-Takaku et al generated a randomized pool of TyrRS mutants and screened variants for their ability to incorporate radioactive azatyrosine.[69] Based on this screen, they found a single mutant of TyrRS, F130S, which could incorporate azatyrosine with roughly twice the incorporation of the wild-type protein. Of course, in the absence of an orthogonal pair, engineering endogenous aminoacyl-tRNA synthetases will of necessity require a misincorporation strategy similar to that pursued by Marlière and Schimmel.[51,52]

Prospects for the Alteration and Expansion of the Genetic Code and Organismal Chemistry in General

The directed evolution of the genetic code is a resource awaiting exploitation. Extremophiles have been the source of a number of thermophilic enzymes that are used in industrial processes today.[70] The utility of unnatural organisms is unknown and has yet to be plumbed. This may be in part because it looks as though large-scale adaptation of new organismal chemistries may turn out to require extraordinarily complex evolutionary mechanisms. The incorporation of selenocysteine is cumbersome, requiring four genes, a complex mechanism of charging the tRNA, and the formation of a complex at the point of Sec insertion (Sec-tRNA^{Sec}, SELB and the SECIS). More generally, the genetic code maintains a certain level of changeability as a requirement for continued survival,[26] and the bias of AT or GC content of a genome may lead to the eventual reassignment of codons.[30] *M. luteus* is an example of an organism that may be experiencing a transition over geological time scales.

In contrast, laboratory experiments all but prove that organisms can adopt new chemistries, given the right selection pressures. It may be that the experiments that have so far been attempted are mere bagatelles, and that something like the proteome-wide incorporation of fluorotryptophan is far simpler than the protein-specific incorporation of selenocysteine. Nonetheless, the similarity between the results of genetic code alterations in natural organisms and genetic code alterations in engineered organisms argues that both nature and biotechnology should favor further interventions. For example, should an orthogonal tRNA and aminoacyl-tRNA synthetase be introduced into *M. luteus*, an organism that already contains a transitional code, this may well serve as a starting point for the expansion of the genetic code.[50]

Ultimately, the generation of unusual genetic codes should also require a synthesis of the techniques that have so far been brought to bear on the problem. The orthogonal tRNA:aminoacyl-tRNA synthetase approach[60,62] will prove indispensable simply because it is a requirement of an expanded genetic code. The use of enforced incorporation[49] denies to organisms the evolutionary strategy of avoiding amino acid analogues, as demonstrated by the successful selection for α-aminobutyrate-incorporating organisms.[51] This could be a critical factor in determining the success or failure of an otherwise precariously poised selection for genomic adaptation to an unnatural amino acid, as demonstrated by the differing results obtained in such selections by Wong and by our group.[46,47] And, finally, since the organism and proteome are ultimately the units of selection, rather than any single synthetase:tRNA pair or site of misincorporation, it will likely be the 'top down' approach of Wong[46] that will allow the canonical genetic code to be altered or even expanded upon.

Note

Since initial submittal of this chapter, it has been shown that a modified form of lysine, pyrrolysine, is incorporated by an unknown mechanism using a stop-codon suppressing tRNA in methyltransferase proteins of *Methanosarcina barkeri*.[71,72]

Acknowledgements

We thank F. Mat and J. T. Wong for sharing their unpublished data. We also thank J. T. Wong for comments on the mansucript and for interesting discussions on this topic. This work was supported by grants to A. D. E. from the Dreyfus Foundation, the Defense Advanced Research Projects Agency and the NASA Astrobiology Institute. J. M. B. is supported by a Harrington Dissertation Fellowship.

References

1. Rothschild LJ, Mancinelli RL. Life in extreme environments. Nature 2001; 409(6823):1092-101.
2. Fields PA. Review: Protein function at thermal extremes: balancing stability and flexibility. Comp Biochem Physiol A Mol Integr Physiol 2001; 129(2-3):417-31.

3. Russell NJ. Toward a molecular understanding of cold activity of enzymes from psychrophiles. Extremophiles 2000; 4(2):83-90.
4. Abe F, Horikoshi K. The biotechnological potential of piezophiles. Trends Biotechnol 2001; 19(3):102-8.
5. Madern D, Ebel C, Zaccai G. Halophilic adaptation of enzymes. Extremophiles 2000; 4(2):91-8.
6. Margesin R, Schinner F. Potential of halotolerant and halophilic microorganisms for biotechnology. Extremophiles 2001; 5(2):73-83.
7. White O, Eisen JA, Heidelberg JF et al. Genome sequence of the radioresistant bacterium *Deinococcus radiodurans* R1. Science 1999; 286(5444):1571-7.
8. Makarova KS, Aravind L, Daly MJ et al. Specific expansion of protein families in the radioresistant bacterium *Deinococcus radiodurans*. Genetica 2000; 108(1):25-34.
9. Nies DH. Heavy metal-resistant bacteria as extremophiles: molecular physiology and biotechnological use of *Ralstonia sp* CH34. Extremophiles 2000; 4(2):77-82.
10. Henderson TR, Dacus JM, Crespi HL et al. Deuterium isotope effects on beta-galactosidase formation by *Escherichia coli*. Arch Biochem Biophys 1967; 120(2):316-21.
11. Rokop S, Gajda L, Parmerter S et al. Purification and characterization of fully deuterated enzymes. Biochim Biophys Acta 1969; 191(3):707-15.
12. Uphaus RA, Flaumenhaft E, Katz JJ. A living organism of unusual isotopic composition. Sequential and cumulative replacement of stable isotopes in *Chlorella vulgaris*. Biochim Biophys Acta 1967; 141(3):625-32.
13. Flaumenhaft E, Uphaus RA, Katz JJ. Isotope biology of ^{13}C. Extensive incorporation of highly enriched ^{13}C in the alga *Chlorella vulgaris*. Biochim Biophys Acta 1970; 215(3):421-9.
14. Khudyakov IY, Kirnos MD, Alexandrushkina NI et al. Cyanophage S-2L contains DNA with 2,6-diaminopurine substituted for adenine. Virology 1978; 88(1):8-18.
15. Jeffares DC, Poole AM, Penny D. Relics from the RNA world. J Mol Evol 1998; 46(1):18-36.
16. Bertozzi CR, Kiessling LL. Chemical glycobiology. Science 2001; 291(5512):2357-64.
17. Guder A, Wiedemann I, Sahl HG. Posttranslationally modified bacteriocins—the lantibiotics. Biopolymers 2000; 55(1):62-73.
18. Okeley NM, van der Donk WA. Novel cofactors via post-translational modifications of enzyme active sites. Chem Biol 2000; 7(7):R159-71.
19. Tsien RY. The green fluorescent protein. Annu Rev Biochem 1998; 67:509-44.
20. Kano A, Ohama T, Abe R et al. Unassigned or nonsense codons in *Micrococcus luteus*. J Mol Biol 1993; 230(1):51-6.
21. Fox TD. Natural variation in the genetic code. Annu Rev Genet 1987; 21:67-91.
22. Knight RD, Freeland SJ, Landweber LF. Selection, history and chemistry: the three faces of the genetic code. Trends Biochem Sci 1999; 24(6):241-7.
23. Freeland SJ, Hurst LD. The genetic code is one in a million. J Mol Evol 1998; 47(3):238-48.
24. Ellington AD, Khrapov M, Shaw CA. The scene of a frozen accident. RNA 2000; 6(4):485-98.
25. Ronneberg TA, Landweber LF, Freeland SJ. Testing a biosynthetic theory of the genetic code: fact or artifact? Proc Natl Acad Sci USA 2000; 97(25):13690-5.
26. Maeshiro T, Kimura M. The role of robustness and changeability on the origin and evolution of genetic codes. Proc Natl Acad Sci USA 1998; 95(9):5088-93.
27. Wong JT. Evolution of the genetic code. Microbiol Sci 1988; 5(6):174-81.
28. Tourancheau AB, Tsao N, Klobutcher LA et al. Genetic code deviations in the ciliates: evidence for multiple and independent events. EMBO J 1995; 14(13):3262-7.
29. Lozupone CA, Knight RD, Landweber LF. The molecular basis of nuclear genetic code change in ciliates. Curr Biol 2001; 11(2):65-74.
30. Osawa S, Jukes TH. Codon reassignment (codon capture) in evolution. J Mol Evol 1989; 28(4):271-8.
31. Knight RD, Freeland SJ, Landweber LF. A simple model based on mutation and selection explains trends in codon and amino-acid usage and GC composition within and across genomes. Genome Biol 2001; 2(4):RESEARCH0010.1-0010.13.
32. Ehara M, Inagaki Y, Watanabe KI et al. Phylogenetic analysis of diatom coxI genes and implications of a fluctuating GC content on mitochondrial genetic code evolution. Curr Genet 2000; 37(1):29-33.
33. Oba T, Andachi Y, Muto A et al. CGG: an unassigned or nonsense codon in *Mycoplasma capricolum*. Proc Natl Acad Sci USA 1991; 88(3):921-5.
34. Kowal AK, Oliver JS. Exploiting unassigned codons in *Micrococcus luteus* for tRNA-based amino acid mutagenesis. Nucleic Acids Res 1997; 25(22):4685-9.
35. Patching SG, Gardiner PH. Recent developments in selenium metabolism and chemical speciation: a review. J Trace Elem Med Biol 1999; 13(4):193-214.
36. Stadtman TC. Selenocysteine. Annu Rev Biochem 1996; 65:83-100.
37. Forster C, Ott G, Forchhammer K et al. Interaction of a selenocysteine-incorporating tRNA with elongation factor Tu from *E.coli*. Nucleic Acids Res 1990; 18(3):487-91.
38. Liu Z, Reches M, Groisman I et al. The nature of the minimal 'selenocysteine insertion sequence' (SECIS) in *Escherichia coli*. Nucleic Acids Res 1998; 26(4):896-902.
39. Suppmann S, Persson BC, Bock A. Dynamics and efficiency in vivo of UGA-directed selenocysteine insertion at the ribosome. EMBO J 1999; 18(8):2284-93.
40. Axley MJ, Bock A, Stadtman TC. Catalytic properties of an *Escherichia coli* formate dehydrogenase mutant in which sulfur replaces selenium. Proc Natl Acad Sci USA 1991; 88(19):8450-4.

41. Budisa N, Steipe B, Demange P et al. High-level biosynthetic substitution of methionine in proteins by its analogs 2-aminohexanoic acid, selenomethionine, telluromethionine and ethionine in *Escherichia coli*. Eur J Biochem 1995; 230(2):788-96.
42. Kohno T, Kohda D, Haruki M et al. Nonprotein amino acid furanomycin, unlike isoleucine in chemical structure, is charged to isoleucine tRNA by isoleucyl-tRNA synthetase and incorporated into protein. J Biol Chem 1990; 265(12):6931-5.
43. Pratt EA, Ho C. Incorporation of fluorotryptophans into proteins of *Escherichia coli*. Biochemistry 1975; 14(13):3035-40.
44. Browne DR, Kenyon GL, Hegeman GD. Incorporation of monoflurotryptophans into protein during the growth of *Escherichia coli*. Biochem Biophys Res Commun 1970; 39(1):13-9.
45. Zhang QS, Shen L, Wang ED et al. Biosynthesis and characterization of 4-fluorotryptophan-labeled *Escherichia coli* arginyl-tRNA synthetase. J Protein Chem 1999; 18(2):187-92.
46. Wong JT. Membership mutation of the genetic code: loss of fitness by tryptophan. Proc Natl Acad Sci USA 1983; 80(20):6303-6.
47. Bacher JM, Ellington AD. Selection and Characterization of *Escherichia coli* Variants Capable of Growth on an Otherwise Toxic Tryptophan Analogue. J Bacteriol 2001; 183(18):5414-25.
48. Kunst F, Ogasawara N, Moszer I et al. The complete genome sequence of the gram-positive bacterium *Bacillus subtilis*. Nature 1997; 390(6657):249-56.
49. Lemeignan B, Sonigo P, Marlière P. Phenotypic suppression by incorporation of an alien amino acid. J Mol Biol 1993; 231(2):161-6.
50. Döring V, Marlière P. Reassigning cysteine in the genetic code of *Escherichia coli*. Genetics 1998; 150(2):543-51.
51. Döring V, Mootz HD, Nangle LA et al. Enlarging the amino acid set of *Escherichia coli* by infiltration of the valine coding pathway. Science 2001; 292(5516):501-4.
52. Lin L, Schimmel P. Mutational analysis suggests the same design for editing activities of two tRNA synthetases. Biochemistry 1996; 35(17):5596-601.
53. Schultz DW, Yarus M. On malleability in the genetic code. J Mol Evol 1996; 42(5):597-601.
54. Schultz DW, Yarus M. Transfer RNA mutation and the malleability of the genetic code. J Mol Biol 1994; 235(5):1377-80.
55. Yarus M, Schultz DW. Further comments on codon reassignment. Response. J Mol Evol 1997; 45(1):3-6.
56. Wang L, Magliery TJ, Liu DR et al. A new functional suppressor tRNA/aminoacyl-tRNA synthetase pair for the in vivo incorporation of unnatural amino acids into proteins. J Am Chem Soc 2000; 122:5010-1.
57. Ohno S, Yokogawa T, Fujii I et al. Co-expression of yeast amber suppressor tRNATyr and tyrosyl-tRNA synthetase in *Escherichia coli*: possibility to expand the genetic code. J Biochem (Tokyo) 1998; 124(6):1065-8.
58. Liu DR, Magliery TJ, Pastrnak M et al. Engineering a tRNA and aminoacyl-tRNA synthetase for the site-specific incorporation of unnatural amino acids into proteins in vivo. Proc Natl Acad Sci USA 1997; 94(19):10092-7.
59. Kowal AK, Köhrer C, RajBhandary UL. Twenty-first aminoacyl-tRNA synthetase-suppressor tRNA pairs for possible use in site-specific incorporation of amino acid analogues into proteins in eukaryotes and in eubacteria. Proc Natl Acad Sci USA 2001; 98(5):2268-73.
60. Liu DR, Schultz PG. Progress toward the evolution of an organism with an expanded genetic code. Proc Natl Acad Sci USA 1999; 96(9):4780-5.
61. Wang L, Schultz PG. A general approach for the generation of orthogonal tRNAs. Chem Biol 2001; 127:1-8.
62. Wang L, Brock A, Herberich B et al. Expanding the genetic code of *Escherichia coli*. Science 2001; 292(5516):498-500.
63. Giege R, Sissler M, Florentz C. Universal rules and idiosyncratic features in tRNA identity. Nucleic Acids Res 1998; 26(22):5017-35.
64. Kast P, Hennecke H. Amino acid substrate specificity of *Escherichia coli* phenylalanyl-tRNA synthetase altered by distinct mutations. J Mol Biol 1991; 222(1):99-124.
65. Hennecke H, Bock A. Altered alpha subunits in phenylalanyl-tRNA synthetases from p-fluorophenylalanine-resistant strains of *Escherichia coli*. Eur J Biochem 1975; 55(2):431-7.
66. Sharma N, Furter R, Kast P et al. Efficient introduction of aryl bromide functionality into proteins in vivo. FEBS Lett 2000; 467(1):37-40.
67. Ibba M, Kast P, Hennecke H. Substrate specificity is determined by amino acid binding pocket size in *Escherichia coli* phenylalanyl-tRNA synthetase. Biochemistry 1994; 33(23):7107-12.
68. Ibba M, Hennecke H. Relaxing the substrate specificity of an aminoacyl-tRNA synthetase allows in vitro and in vivo synthesis of proteins containing unnatural amino acids. FEBS Lett 1995; 364(3):272-5.
69. Hamano-Takaku F, Iwama T, Saito-Yano S et al. A mutant *Escherichia coli* tyrosyl-tRNA synthetase utilizes the unnatural amino acid azatyrosine more efficiently than tyrosine. J Biol Chem 2000; 275(51):40324-8.
70. Niehaus F, Bertoldo C, Kahler M et al. Extremophiles as a source of novel enzymes for industrial application. Appl Microbiol Biotechnol 1999; 51(6):711-29.
71. Srinivasan G, James CM, Krzycki JA. Pyrrolysine encoded by UAG in Archaea: charging of a UAG-decoding specialized tRNA. Science 2002; 296(5572):1459-62.
72. Hao B, Gong W, Ferguson TK et al. A new UAG-encoded residue in the structure of a methanogen methyltransferase. Science 2002; 296(5572):1462-6.

CHAPTER 6

In Vitro Tools and in Vivo Engineering:
Incorporation of Unnatural Amino Acids into Proteins

Thomas J. Magliery, Miro Pastrnak, J. Christopher Anderson,
Stephen W. Santoro, Brad Herberich, Eric Meggers, Lei Wang
and Peter G. Schultz

Summary

Unnatural protein mutagenesis has dramatically enhanced our ability to probe the basis of structure and function in protein biochemistry and has enabled us to create proteins with entirely novel properties. Recent advances in in vitro unnatural amino acid incorporation have expanded the scope of protein engineering and allow larger proteins in higher yields. The in vitro biosynthetic incorporation of unnatural amino acids has become especially robust for generating proteins of virtually any size, and novel insertion sequences like four-base codons and unnatural codons have proven useful. The first inroads have been made into engineering living cells to insert unnatural amino acids into proteins, effectively expanding the genetic code. Methods for equipping cells with the essential biomolecules for genetic code expansion—"orthogonal" tRNAs and aminoacyl-tRNA synthetases (aaRSs)—are discussed here in detail. Robust screens and selections have been designed to alter the amino acid specificity of aaRSs. These methods have yielded the first bacteria with an expanded genetic code, capable of site-selectively inserting O-methyltyrosine in response to the amber stop codon, and preliminary results suggest that a wide array of functionally-diverse tyrosine derivatives can be inserted using these methods. Future advances in these technologies will yield other cell types, such as human cell lines, capable of using unnatural amino acids, and will allow unprecedented advances in cell biology.

Introduction

Understanding biological structure and function at virtually any level requires an understanding of proteins, biopolymers of remarkable diversity composed of only twenty encoded amino acids. Advances in structure determination have contributed greatly to this understanding. One of the most powerful tools in elucidating how proteins work has been site-directed mutagenesis, a process by which residues in a protein can be swapped for other naturally-occurring amino acids.[1,2] Site-directed mutagenesis has allowed exploration of the parameters that control protein folding and stability, identified residues critical to binding and catalytic function, and enabled protein engineering for the dissection of signaling pathways and regulatory mechanisms.[3] However, with the limitation of the natural repertoire of amino acids, it is impossible to make the kinds of subtle changes to steric or electronic properties that have allowed physical organic chemists to carefully explore the mechanisms of organic reactions. Likewise, it is impossible to introduce side chains with novel physical and chemical properties, such as biophysical probes, photoactivatable moieties and unique reactive groups. These limitations have motivated a variety of approaches to generate full-length proteins containing "unnatural" amino acids.

Modern chemical methods have made direct synthesis of large peptides possible in vitro, and small proteins are now accessible through chemical and enzymatic ligation strategies. Combining the power of chemistry and biology, in vitro biosynthetic methods now allow the production of

Translation Mechanisms, edited by Jacques Lapointe and Léa Brakier-Gingras.
©2003 Eurekah.com and Kluwer Academic / Plenum Publishers.

proteins containing unnatural amino acids at specific sites in sufficient yields for characterization by techniques such as circular dichroism, X-ray crystallography and NMR. Also, the natural permissivity of the translational machinery has begun to be exploited to produce proteins in which natural amino acids have been substituted with unnatural ones. The culmination of these efforts has been the engineering of living organisms fully capable of incorporating unnatural amino acids in a site-specific manner and with high fidelity.

Chemical and in Vitro Biosynthetic Approaches

Chemical Synthesis and Ligation

Chemical synthesis is a straightforward strategy to incorporate unnatural amino acids into peptides.[4] Unfortunately, routine stepwise solid-phase synthesis is limited to producing peptides of about 50 amino acids or 6 kD, corresponding to the smallest proteins or domains known. A methodological breakthrough in the synthesis of larger peptides and proteins has been the recent development of chemical ligation of unprotected peptide fragments to construct full-length proteins.

The total synthesis of proteins via chemical ligation has been used to prepare a number of proteins up to 120 residues and including homodimers and heterodimers.[5-7] The proteins have been synthesized on 50-100 mg scale and for the most part fold correctly and maintain their biological function. Chemical ligation was initially accomplished through a nucleophilic substitution reaction between a thioester and an alkyl bromide to form a thioester linkage.[8] Further development led to native chemical ligation, which results in proteins with all native amide bonds.[9] In this procedure, the two peptide fragments are joined by an initial transthioesterification between a thioester and an N-terminal cysteine residue, which is followed by rearrangement to give the native peptide (Fig. 1). This method has been much improved recently by adapting it to the solid-phase.[10]

A number of proteins containing noncoded amino acids have been constructed by chemical synthesis.[11,12] For example, a backbone modification to an ester linkage was introduced in OMTKY3 to probe the strength of hydrogen bonds. From this study, a loss of 1.5 kcal mol^{-1} in stability was observed when an ester bond was substituted for an amide bond.[13] Unnatural amino acids have been inserted into proteins to probe stability, to perturb electronic properties of metals in metalloproteins, and to introduce fluorescent markers. Love et al substituted different side chains (Val26Abu, Ile30Nva) in the hydrophobic core of ubiquitin and evaluated their influence on α-helix stability.[14] In an iron-sulfur protein, Low and Hill investigated the perturbation of the electronics of the iron by substituting analogs for a neighboring tyrosine residue.[15] Using native chemical ligation, donor and acceptor dyes have been incorporated into chymotrypsin inhibitor 2 to study its folding properties.[16]

Recently, bacterially expressed proteins have been used in thioester-mediated native chemical ligation. Natural protein splicing involves the excision of an intein with the intermediacy of a thioester. A mutant version of the splicing domain has been generated that traps the thioester, and this has been exploited commercially as a means of protein purification, wherein the intein is linked to a chitin-binding domain and the recombinant protein is purified over chitin and released with DTT.[17-19] Muir and coworkers instead released the trapped thioester from the resin with a synthetic peptide bearing an N-terminal cysteine, which resulted in rearrangement to yield the native amide bond (Fig. 1).[20,21] Expressed protein ligation expands the scope of chemical protein synthesis to include large proteins that are not accessible by total synthesis techniques.[22]

An alternative approach for ligation of peptide segments is the use of modified proteases.[23] In 1987, Kaiser et al investigated a previously constructed serine protease, thiolsubtilisin, whose protease activity is decreased by converting its catalytic serine residue to cysteine. By using p-chlorophenyl esters as acylating agents at the C-terminus of one peptide, ligation was accomplished to the N-terminal amine of a second peptide.[24] Wells and co-workers engineered the protein further and positioned the catalytic cysteine residue at a more optimum distance from the peptide substrate.[25] Using this new subtilisin, termed subtiligase, synthetic peptides have been coupled to proteins containing certain N-terminal sequences. Affinity handles such as biotin, isotopic labels and heavy atom derivatives for X-ray crystallography have been introduced by this method.[26] In addition, subtiligase can be used for the preparation of complete proteins; for example, variants of ribonuclease A containing unnatural residues in the catalytic site were synthesized.[27]

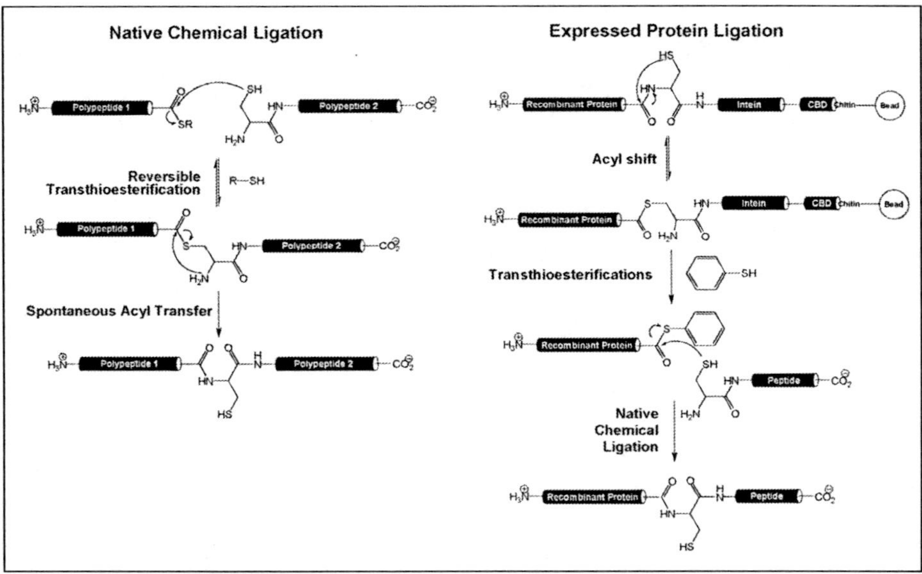

Figure 1. Native chemical ligation and expressed protein ligation. Schematic representation of techniques for asembling synthetic and semi-synthetic proteins from peptide fragments.

Modified peptides or proteins can also be obtained by chemical post-translational modification.[28-30] The sulfhydryl group of cysteine is usually the most reactive functional group in proteins and can be easily labeled. A common strategy involves the introduction of a unique cysteine via site-directed mutagenesis, which is then modified either by forming a disulfide bond (with a methanethiosulfonate or pyridyl disulfide reagent) or by reaction with an N-substituted maleimide or α-iodoacetamide compound. One example is the design of a semisynthetic substrate-linked DNA-cleaving enzyme by derivatizing a staphylococcal nuclease with a deoxyoligonucleotide.[31] The linkage was created between a unique cysteine near the active site of the enzyme and a thiol of a modified deoxyoligonucleotide through disulfide bond exchange. Another useful reactive group in proteins is the ε-amine of lysine. Lysine amines are reasonably good nucleophiles above pH 8 and readily form amide bonds with *N*-hydroxysuccinimide esters. Lysine side chains also selectively react with aldehydes to form Schiff bases, which can be reduced with sodium cyanoborohydride to give a stable alkylamine bond. Polyfunctional aldehydes are extensively used for intermolecular and intramolecular crosslinking of proteins.[32]

A selective method to modify peptides and proteins at the N-terminus is the very mild oxidative cleavage with periodate of N-terminal serine or threonine residues (generated synthetically or with proteolysis of a precursor protein)[33], yielding a reactive aldehyde functionality which then can be reductively aminated.[34] This method can be extended to peptide bond formation.[35] Other amino acids are more difficult to label selectively. Exceptions are activated side chains in active sites of enzymes. For example, trypsin was chemically converted into selenotrypsin by replacement of the nucleophilic Ser195 oxygen with an oxidatively labile Se atom, yielding glutathione peroxidase activity.[36] The two-step protocol includes the selective activation of the serine residue with phenylmethanesulfonyl fluoride and subsequent treatment of the sulfonylated enzyme with hydrogen selenide.[37]

In Vitro Biosynthesis

For over a decade, in vitro biosynthetic methods have been employed to make proteins containing over 100 different unnatural amino acids varying widely in structure, including near-analogs of

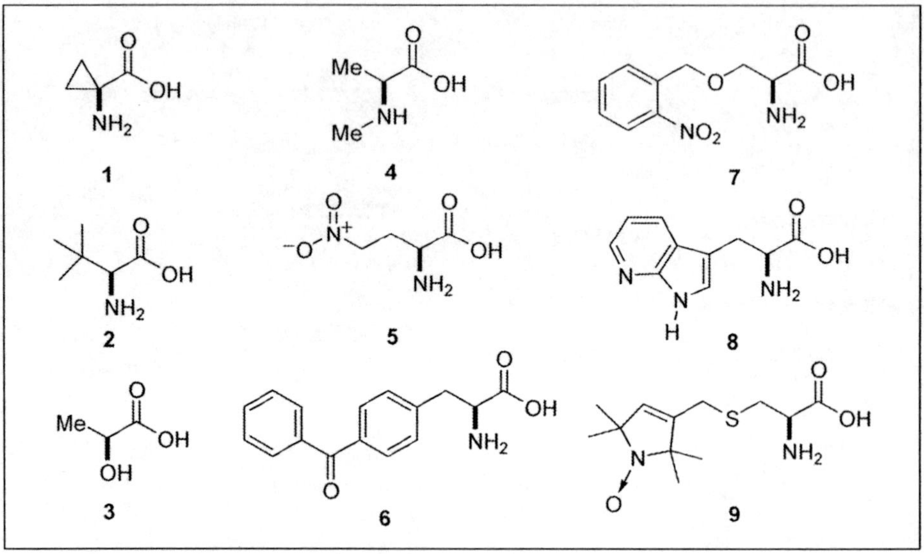

Figure 2. Some unnatural amino acids inserted with the Schultz in vitro biosynthetic method. Amino acids have been used to probe conformational restrictions (1 and 2), effects of alternate backbone linkages (3 and 4), electronic properties (5), protein-protein interactions with affinity labels (6), enzyme mechanism with "caged" amino acids (7) and biophysical properties with fluorescence (8) and spin (9) probes.

natural amino acids, fluorophores, photoreactive moieties, and α-hydroxy acids (Fig. 2).[38-42] A gene encoding the target polypeptide is mutated to contain an amber stop codon (TAG) at the site targeted for replacement with an unnatural amino acid. The DNA is then added to an in vitro transcription-translation system supplemented with an amber suppressor tRNA chemically acylated with the amino acid of choice. In the presence of the suppressor tRNA, the ribosome inserts the unnatural amino acid in response to the stop codon and resumes translation to produce the full-length product (Fig. 3). The suppressor tRNA must not only robustly suppress amber stop codons, but it must also be "orthogonal" to the aminoacyl-tRNA synthetases in the translation system to prevent insertion of natural amino acids at the UAG codon. This problem has been solved principally by employing tRNAs derived from species of different kingdoms. In the case of the *E. coli* S30 system (see below), a yeast tRNA^Phe modified with a CUA anticodon is competent for translation but is not misacylated by *E. coli* synthetases.[43]

To date, three different translation systems have been frequently employed. The most robust system is a supplemented S30 extract from *E. coli*, containing all the necessary components for transcription and translation.[44] Typically, transcription from the T7 promoter is driven in situ by the addition of T7 RNA polymerase to the extract. Alternative systems include a rabbit reticulocyte expression system and a *Xenopus* oocyte system that has an added advantage of allowing the synthesis of integral membrane proteins with unnatural amino acids (see below). Unfortunately, the method is technically demanding, and yields of mutant protein rarely exceed 100 μg mL^{-1} of suppression reaction. Testing different suppressor tRNAs, Schultz and coworkers found that an amber-suppressing derivative of *E. coli* tRNA^Asn provided significantly higher yields in the S30 extract system for some amino acids.[45] Changing the composition of the translation system can give rise to significantly higher yields.[46] Also, preparing the S30 extract from an *E. coli* strain with a temperature sensitive isoform of release factor 1 (RF1, the termination factor that acts at amber stop codons) can improve efficiency.[47,48] The most difficult aspect of the technology has been the chemical misacylation of tRNA.[49,50] The chemistry necessary to produce these molecules is outside the realm of expertise of most biochemistry labs.

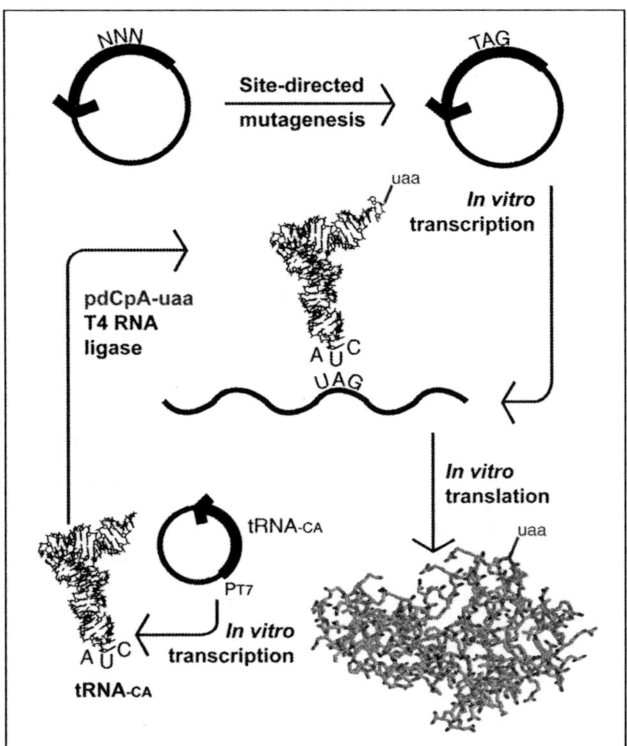

Figure 3. A biosynthetic approach to site-specific unnatural protein mutagenesis. Site-directed mutagenesis is first used to mutate the codon for the residue of choice to the amber stop codon. This is added to an in vitro transcription-translation mixture with amber suppressor tRNA chemically aminoacylated with unnatural amino acid to generate full-length protein bearing the unnatural amino acid. The amber suppressor tRNA is in vitro transcribed without its final pCpA-3', and a synthetic pdCpA acylated with unnatural amino acid is ligated to the tRNA$_{CA}$ with T4 RNA ligase.

Use of Alternate Codons

The use of the amber stop codon as an insertion signal limits us to a single unnatural amino acid per suppression reaction. To further expand the genetic code, it has been possible to introduce unnatural amino acids in response to other nonsense codons, rare codons, or codons with unnatural bases. Missense suppression of the rare arginine codon AGG has proven to be highly efficient; however, insertion of arginine competes with insertion of unnatural amino acid and produces a protein of the same length.[51] Benner and coworkers used an unnatural base to generate a novel (iso-dC)AG codon, which was decoded in an in vitro translation reaction containing a chemically misacylated tRNA bearing a CU(iso-dG) anticodon. Although the suppression was efficient, the unnatural tRNA and mRNA had to be chemically synthesized.[52,53] For such a process to be practical, it will first be necessary to develop means to produce the suppressor tRNA and mRNA enzymatically.

Significant advances have been made in developing novel base pairs to expand the genetic alphabet. Schultz, Romesberg and coworkers have synthesized unnatural hydrophobic bases capable of forming stable DNA base pairs in duplex DNA. Rather than using hydrogen bonding patterns to elicit specificity, these novel bases rely upon hydrophobic interactions. For example, the isocarbostyril derivative PICS was found to interact very stably and specifically as a self base pair in duplex DNA and has been replicated enzymatically.[54,55] Another interesting approach has been the development of metallobase pairs, ligand bases that form a stable pair upon binding of a metal ion.[56] Before such

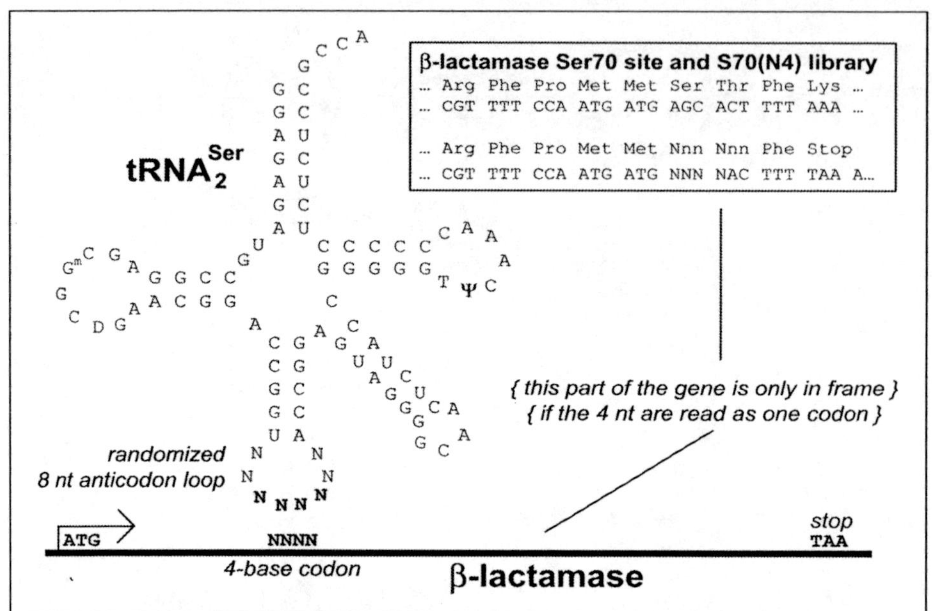

Figure 4. Selection scheme for suppressors of 4-base codons. *E. coli* were co-transformed with vectors expressing a β-lactamase reporter and a tRNA₂^Ser derivative and grown on media containing ampicillin. Full-length β-lactamase is produced only if the 4 nucleotides that replace the wild-type codon (here, AGC at Ser70) are read as a single codon (see inset for wild-type and reporter +1 frameshift sequences). This tRNA scaffold was chosen because SerRS will acylate tRNA^Ser regardless of the anticodon loop sequence. Libraries of tRNAs with randomized 8 and 9 nt anticodon loops were examined.

bases can be used for in vitro translation, it will be necessary to develop RNA polymerases capable of faithfully replicating unnatural base pairs without chain termination.[57]

Another promising approach is the use of four-base codons.[58] Moore et al have shown that a suppressor tRNA derived from a tRNA^Leu amber suppressor can efficiently decode the four-base codon UAGA with good specificity and efficiency in vivo.[59] Recently, we examined all possible four-base codons by replacing a codon in the gene for β-lactamase with the sequence NNNN where N is a mixture of all four bases. The resulting 256 reporter genes thereby contain a +1 frameshift and no longer encode the full-length product if read exclusively three bases at a time. In the presence of an appropriate frameshift suppressor tRNA, the four-base sequence is translated as a single codon and translation proceeds to generate full-length β-lactamase, conferring resistance to ampicillin. The suppressor tRNAs employed were derived from the tRNA₂^ser of *E. coli* and were modified by replacing the seven base anticodon loop with eight random bases. Using this selection (Fig. 4), several four-base codons—including AGGA, CUAG, UAGA, and CCCU—and suppressor tRNAs were identified that decoded only the canonical four-base codon with high efficiency.[60] We have subsequently examined suppression of 2-base through 6-base codons with tRNAs bearing anticodon loops with six to 10 nucleotides.[61]

Several groups have demonstrated the use of four-base codons with in vitro amino acid mutagenesis.[62] Perhaps the most promising application of four-base codons is the incorporation of multiple unnatural amino acids into the same polypeptide. Two unique four-base codons, CGGG and AGGT, were inserted into the gene for streptavidin. In the presence of two different chemically aminoacylated suppressor tRNAs, it was possible to incorporate 2-naphthylalanine and an NBD (fluorophore) derivative of lysine into the same polypeptide.[63-65]

Applications of the in Vitro Approach

Protein Stability and Function

Unnatural amino acid mutagenesis has a distinct advantage over conventional mutagenesis for the study of protein stability and function. Since almost any amino acid side chain can be incorporated into proteins, a series of highly related amino acids can be used to examine linear free energy relationships. Thorson et al took advantage of this ability to examine the strength of a tyrosine-glutamate hydrogen bond in staphylococcal nuclease (Fig. 5). The aromatic ring of Tyr27 was replaced by various polysubstituted fluorine derivatives of tyrosine. By comparing the pK_a values of the phenolic hydroxyl to the stability of the unnatural amino acid-containing proteins, a nearly equal sharing of the proton between donor and accepter was derived from the free-energy equation.[66,67] The effects of backbone perturbation have also been examined with conformationally restricted amino acids and ester linkages from α-hydroxy acids.[68,69]

Biophysical Probes

A popular method of examining conformational changes in proteins has been the use of intrinsic tryptophan fluorescence. Unfortunately, tryptophan is a weak fluorophore, and such experiments are often complicated by the presence of multiple tryptophan residues in the target protein. Small fluorophores such as azatryptophan can be readily incorporated.[47,70,71] Also, spin labels have been incorporated into proteins.[72] Photofunctional crosslinking groups like *p*-benzoylphenylalanine have been good substrates for translation. Such amino acids may be of great use to identify residues involved in protein-protein interactions.

Key residues in proteins have been replaced by amino acids protected with photolabile groups. The resulting proteins, unable to perform their usual function while protected, are restored upon irradiation.[73] Schultz and coworkers have implemented this strategy to study protein-protein interactions. A critical residue, Asp38, of the protein p21^{ras} was replaced by the o-nitrobenzyl ester of aspartic acid. Upon binding GTP, the ras-GTP complex can bind to and effect the activity of downstream enzymes. GAP proteins accelerate the rate of hydrolysis of GTP by the ras proteins thereby regulating their function. When Asp38 was caged, p120-GAP was no longer able to bind to and activate hydrolysis of the ras-GTP complex. Upon irradiation with 355 nm light, the o-nitrobenzyl protecting group was removed and p120-GAP binding was restored.[74] Similar success has been observed for *p*-phenylazophenylalanine. This amino acid switches from a *cis* to *trans* configuration in response to irradiation. One could imagine using such an amino acid to turn a protein on or off with light.

Reactive Handles

Many important proteins contain modifications that most likely cannot be introduced during translation due to their bulk. Most extracellular proteins in mammalian systems are heavily glycosylated by enzymes post-translationally. Such modifications are not performed in *E. coli*, making the recombinant production of these proteins difficult if not impossible. One strategy to generate these chemical modifications would be to use unnatural amino acid mutagenesis to incorporate reactive handles into the protein. Several chemistries are available and sufficiently specific to allow such methodology. Our group has introduced a ketone-modified tyrosine into T4 lysozyme at a surface exposed site, Ala82. The reaction of ketones with hydrazides and hydroxyl amines leads to stable linkages (hydrazones and oximes). It was possible to specifically label the ketones introduced into T4 lysozyme with a fluorescein hydrazide.[75] Clark et al have employed an alternative chemistry using olefin metathesis. In the presence of the Grubbs ruthenium carbene catalyst, it was possible to crosslink olefin-containing peptides and assemble a peptide cylinder.[76] Bertozzi and coworkers have been successful in the introduction of azides into sugars on the surface of cultured mammalian cells. The azides could be selectively reacted with a specifically engineered triarylphosphine ester moiety.[77]

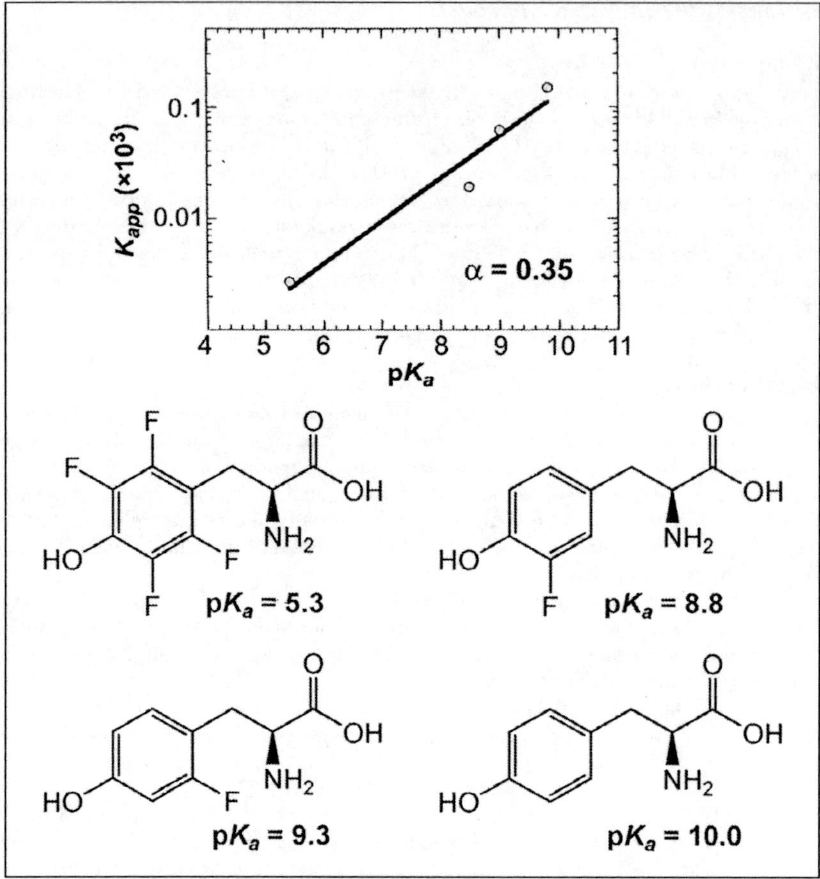

Figure 5. Linear free energy analysis of hydrogen bonding in proteins. The tyrosine in the Tyr27-Glu10 hydrogen bond of SNase was replaced with various fluorinated derivatives, and the stability of the resulting proteins was measured. The apparent equilibrium constants were fit against the pK_a of the residues in the equation $\log K_{app} = \alpha(pK_a) + C$, where α reflects the degree of proton transfer between Tyr27 and Glu10.

In Vivo Approaches

Chemically Acylated tRNAs in Vivo

Dougherty and coworkers have extended the in vitro biosynthetic method of unnatural protein mutagenesis for use in vivo by transferring chemically acylated tRNAs and the desired mRNA into living cells.[42] This in vivo method allows the study of proteins that cannot be easily studied in vitro, such as integral membrane proteins, as well as the use of relatively sensitive cell-based assays like patch-clamping, circumventing the need to produce large quantities of protein.[78] *Xenopus* oocytes have been the cells of choice, as they are amenable to microinjection. The method also requires a tRNA that is "orthogonal" within *Xenopus* oocytes. Dougherty and coworkers solved this problem by employing an amber suppressor tRNA derived from *Tetrahymena thermophila*, which is not recognized by the endogenous synthetases of the *Xenopus* oocyte and which efficiently incorporates unnatural amino acids in the oocyte system.[79]

Dougherty's group has largely exploited this method to study the nicotinic acetylcholine receptor, which recapitulates activity in the membrane of the oocyte. Applications of this method include use of leucine isosteres to probe hydrophobic interactions,[80] fluorinated tryptophan residues to probe cation-π interactions,[81] a biotin-containing amino acid to identify surface-exposed residues,[82] a photoactive amino acid to identify the role of a critical loop via photochemical proteolysis,[83] a fluorescent amino acid to measure distances using fluorescence resonance energy transfer,[84] caged tyrosine residues to probe conformational processes in real time,[85] and α-hydroxy amino acids to probe backbone hydrogen bonding interactions.[86] Recently, our group applied this technique to study the role of backbone carbonyls in an inwardly-rectifying potassium channel.[87] One clear limitation of this technology is that it is only amenable to use in microinjectable cells, excluding desired cell types, such as mammalian neurons. Moreover, poor suppression efficiencies and low protein yields require the use of very sensitive techniques to assay the synthesized protein.

Multisite Amino Acid Substitution

Apart from the in vivo suppression methodology discussed above, efforts to incorporate unnatural amino acids into proteins in vivo have focused primarily on multisite substitution strategies.[88] In contrast to site-specific methods that rely on amber suppression, multisite substitution results in partial to complete replacement of a particular natural amino acid with a structurally analogous unnatural one at all coded sites in proteins. These methods tend to be technically simple, requiring only addition of unnatural amino acid under appropriate culture conditions, and can be used to generate large quantities of protein.

The discovery that unnatural amino acids can be made to be incorporated into proteins in vivo was made long ago.[89,90] In recent years, several advances have dramatically improved the efficiency of amino acid substitution and made possible the incorporation of a variety of unnatural amino acids. These developments include: auxotrophic host strains in which a relevant metabolic pathway supplying the cell with a natural amino acid is switched off; powerful and inducible protein expression systems; and, in cases in which a desired unnatural amino acid is not accepted by the host's endogenous aminoacyl-tRNA synthetases, mutant synthetases with relaxed substrate specificities. Upon growth of an auxotrophic host strain in minimal media, to which an unnatural amino acid is added following induction of the expression system, unnatural amino acid-containing protein is expressed.

Multisite substitution has now been used to incorporate a variety of unnatural amino acids with efficiencies that approach quantitative. Examples of widespread replacement without changes in the cellular translation machinery include the incorporation of heavy atom-containing unnatural amino acids to facilitate protein structure determination by X-ray crystallography;[91,92] fluorinated functional groups to increase protein stability, improve protein assembly, strengthen ligand-receptor interactions or facilitate spectroscopic analysis;[93-96] and unsaturated functional groups to allow modification via bromination, hydroxylation and potentially other chemistries.[97-100]

Because this method relies on the aminoacylation activity of the endogenous synthetases of the host cell, it is limited by the fidelity of these enzymes. To expand the scope of the method, synthetase mutations have been identified that relax substrate specificity to allow the incorporation of a greater variety of amino acids. A single mutation within *E. coli* phenylalanyl-tRNA synthetase was found to allow the incorporation of *p*-Cl- or *p*-Br-phenylalanine.[101,102] Ibba has since shown that this mutant also is likely to accept benzofuranylalanine, which is photoreactive.[103] Similarly, a single point mutation within *E. coli* tyrosyl-tRNA synthetase was found to allow azatyrosine to be incorporated more efficiently than tyrosine.[104] A slightly different approach allowed the identification of mutations within the editing domain of *E. coli* valyl-tRNA synthetase that allowed the enzyme to accept both the natural amino acid cysteine and the unnatural (but sterically similar) aminobutyrate.[105] Again, the major limitation of this scheme is that throughout proteins all sites corresponding to a particular natural amino acid are replaced partially or completely with the analog.

Strategy for All-in Vivo Site-Specific Approaches

What the protein biochemist ultimately desires is an in vivo method that allows site-selective insertion of unnatural amino acids without affecting the normal translation of the common 20 amino acids—in essence, a means to expand the number of genetically coded amino acids. In vivo methods have the significant advantages of high yield of protein and easy scale-up, technical ease, and the potential to observe the altered proteins in the living cell—that is, to do "unnatural cell biology" with, one envisions, caged proteins, affinity labeled proteins or proteins bearing biophysical probes or moieties to expand their functionality.

As with the in vitro site-selective method, an in vivo scheme requires a unique signal for encoding the amino acid (i.e., a unique codon) and an "orthogonal" translation-competent suppressor tRNA (i.e., one that is not recognized by the cell's endogenous synthetases). However, there are at least three additional design considerations for an in vivo method.

1. Since the "orthogonal" tRNA must be acylated in vivo, an aminoacyl-tRNA synthetase is required that uniquely acylates the orthogonal tRNA and no other tRNAs in the cell.
2. This "orthogonal" synthetase must be capable of acylating the orthogonal tRNA with an unnatural amino acid while rejecting any other naturally occurring amino acids as substrates.
3. The unnatural amino acid must be uptaken by the cell or produced by it, and it cannot be inserted by any other synthetase or, more generally, be markedly toxic to the cell.

For its genetic tractability and high transformation efficiency, *E. coli* was chosen as the initial host organism. The amber stop codon was selected as the insertion signal due to the excellent suppression that is possible with suppressor tRNAs known in *E. coli*. However, attempts to make orthogonal tRNA/synthetase pairs using four-base codons are underway, in light of our identification of easily suppressible extended codons (J.C.A. & P.G.S., unpublished work). Initially, we adopted an engineering approach to the generation of an orthogonal tRNA/synthetase pair, starting with the extremely well-characterized *E. coli* glutaminyl pair. However, we and others have found it more advantageous to engineer amber suppressor tRNAs and aminoacyl-tRNA synthetases imported from other organisms (heterologous pairs). Of course, these orthogonal pairs insert a natural amino acid, and so robust screens and selections have been developed to find variants of these synthetases from carefully designed libraries of mutagenized enzymes capable of acylating tRNA with unnatural amino acids. Using these principles, our group and others have seen the first successes in engineering living cells that are capable of site-selectively inserting unnatural amino acids into proteins.

Development of Orthogonal Pairs

The first orthogonal tRNA developed for the purpose of in vivo site selective delivery of unnatural amino acids was derived from *E. coli* tRNA^{Gln}. Glutaminyl-tRNA synthetase (GlnRS) was known to acylate the amber-suppressing derivative of its tRNAs and biochemical and X-ray crystal structural information defined the nature of the interaction between tRNA and synthetase.[106-108] Three sites at which mutations were expected to modulate the ability of GlnRS to acylate the tRNA ("knobs") were selected, and tRNAs bearing mutations at each site were generated. These mutations were found to interact in complicated, nonadditive ways both with respect to aminoacylation by GlnRS and performance as tRNAs for delivery of amino acids at the level of translation. Based on in vitro aminoacylation with GlnRS and in vitro suppression studies, the amber suppressor tRNA with all three knob mutations, O-tRNA^{Gln}(CUA), was found to meet the criteria for an orthogonal tRNA in *E. coli*: it was not a substrate for endogenous synthetases but was competent to act in translation.[109]

This tRNA was also characterized in vivo by expression in an *E. coli* strain with an amber mutation in the gene for β-galactosidase (*lacZ*). The cells were incapable of surviving on lactose minimal medium due to the fact that the O-tRNA^{Gln}(CUA) was not appreciably acylated to produce full-length LacZ. This fact was used as the basis for a selection for a GlnRS mutant capable of aminoacylating the O-tRNA^{Gln}(CUA). The gene for GlnRS (*glnS*) was mutagenized by PCR and DNA shuffling, and co-transformed into the selection strain with the O-tRNA. After seven rounds of mutagenesis and selection, a mutant GlnRS was found that acylated the wild-type tRNA^{Gln} substrate only nine-fold better than the O-tRNA^{Gln}(CUA) and was down only 250-fold with respect to acylation of tRNA^{Gln} by wild-type GlnRS. This enzyme with overall 1,500-fold change in specificity was capable of acylating the O-tRNA^{Gln}(CUA) sufficiently to observe by Western blot full-length protein produced from an amber mutant of the gene for *E. coli* surface protein LamB. No full-length protein was observed with the O-tRNA^{Gln}(CUA) alone.[110] Despite the remarkable change in activity, this mutant GlnRS was still not ideal, since it acylated the wild-type tRNA^{Gln} about as well as the orthogonal substrate. This could cause toxicity through insertion of unnatural amino acid at other Gln sites in *E. coli* proteins, if this enzyme could be mutated to deliver such a substrate.

Recently, Schimmel et al showed that *E. coli* GlnRS (*Ec*GlnRS) does not acylate *Saccharomyces cerevisiae* tRNA^{Gln} (*Sc*tRNA^{Gln}) due to the lack of an N-terminal RNA-binding domain that *S. cerevisiae* GlnRS (*Sc*GlnRS) possesses.[111,112] Liu and Schultz showed that the amber suppressing derivative of *Sc*tRNA^{Gln} (O-*Sc*tRNA^{Gln}(CUA)) and *Sc*GlnRS constitute an orthogonal tRNA/synthetase pair in *E. coli*. The O-*Sc*tRNA^{Gln}(CUA) was shown to be orthogonal and translationally competent by in vitro aminoacylation studies and in vitro suppression. Likewise, *Sc*GlnRS was shown to acylate O-*Sc*tRNA^{Gln}(CUA) in vitro, but not *E. coli* tRNAs. This pair was also characterized in vivo by co-expression of the *Sc*GlnRS, O-*Sc*tRNA^{Gln}(CUA), and an amber mutant of the gene for β-lactamase (*amp*). This amber mutation occurs at a permissive site (Ala184), so that insertion of virtually any amino acid confers resistance to ampicillin.[113] With an inactive mutant of *Sc*GlnRS, these cells exhibited an IC$_{50}$ of about 20 μg ml^{-1} ampicillin, indicating virtually no acylation by endogenous synthetases. With an active *Sc*GlnRS, cells exhibit an IC$_{50}$ of about 500 μg ml^{-1} ampicillin, indicating that the *Sc*GlnRS acylates the O-*Sc*tRNA^{Gln}(CUA) in *E. coli*.[114]

More recently, the amber suppressing derivative of *Methanococcus jannaschii* tRNA^{Tyr} and *M. jannaschii* TyrRS were shown to be an orthogonal pair in *E. coli*.[115] *Mj*TyrRS recognizes a C1:G72 pair in the tRNA^{Tyr} acceptor stem, while *Ec*TyrRS strongly favors G1:C72. In fact, *Mj*TyrRS was shown to acylate crude tRNA from yeast (whose tRNA^{Tyr} has C1:G72) but not crude *E. coli* tRNA in vitro.[116,117] Using the ampicillin resistance test for amber suppression, O-*Mj*tRNA^{Tyr}(CUA) expression alone confers an IC$_{50}$ of about 55 μg ml^{-1}, but co-expression with the MjTyrRS confers resistance to an IC$_{50}$ of about 1,200 μg ml^{-1}. This indicates both that the O-*Mj*tRNA^{Tyr}(CUA) is slightly less orthogonal to endogenous *E. coli* synthetases than O-*Sc*tRNA^{Gln}(CUA) and that the *Mj*TyrRS is much more active than the *Sc*GlnRS under the expression conditions examined. Since the TyrRS lacks proofreading activity and has an active site that accommodates a relatively large, hydrophobic amino acid, it is a suitable starting point for attempts to acylate with interesting unnatural hydrophobic amino acids such as fluorophores or affinity labels.

We set out to improve the orthogonality of this *Mj*tRNA^{Tyr}(CUA) with a selection strategy. Here, eleven nucleotides were identified in the tRNA that were thought not to interact directly with the *Mj*TyrRS. These nucleotides were randomized and the resulting library was first passed through a negative selection, wherein acylation of the amber suppressor tRNA resulted in translation of a toxic gene product, barnase, through suppression of two or three amber codons. This step removed tRNA variants that can be acylated by endogenous *E. coli* tRNAs. The products of this selection were then passed through a positive selection step in the presence of the *Mj*TyrRS; here, survival of the cells, grown in the presence of antibiotic, required that the MjTyrRS acylate the tRNA variant to support translation of an amber mutant antibiotic resistance gene. The resulting O-*Mj*tRNA^{Tyr}(CUA)* supported survival on ampicillin in the β-lactamase suppression assay at an IC$_{50}$ of only 12.4 μg mL^{-1}, making it about four-fold more orthogonal than the unmodified *Mj*tRNA^{Tyr}(CUA). Nevertheless, the modified tRNA was still acylated sufficiently by *Mj*TyrRS to support survival at an IC$_{50}$ of 436 μg mL^{-1} ampicillin, down about three-fold from the unmodified suppressor.[118,119]

RajBhandhary and coworkers found that yeast TyrRS aminoacylated *E. coli* tRNA^{Pro}. Since this misacylation is lethal to *E. coli*, this was used as the basis for a negative selection for a mutant *Sc*TyrRS incapable of acylating *Ec*tRNA^{Pro}. It was also shown that an amber suppressing mutant of *E. coli* initiator tRNA^{fMet} was not acylated in *E. coli* but was acylated by *Sc*TyrRS due to the C1:G72 recognition element in this O-*Ec*tRNA^{fMet}(CUA). Thus, by co-expressing this tRNA and a library of mutant ScTyrRSs in an *E. coli* strain with an amber mutation in the gene for chloramphenicol acetyltransferase (CAT), survival on chloramphenicol demanded a synthetase both capable of acylating the O-*Ec*tRNA^{fMet}(CUA) and incapable of acylating *Ec*tRNA^{Pro}. Some such mutant ScTyrRSs were isolated, one with a specificity factor for O-*Ec*tRNA^{fMet}(CUA) 15-fold greater than that for *Ec*tRNA^{Pro}. It was also shown that *E. coli* GlnRS and an amber suppressing derivative of human initiator tRNA^{fMet} constitute an orthogonal pair in yeast cells, the first such pair demonstrated in cells other than *E. coli*.[120] Interestingly, RajBhandary and coworkers had previously demonstrated that a variant of the *E. coli* tRNA^{Gln}(CUA) is expressed and orthogonal in COS-1 and CV-1 cells, but that co-expression of *E. coli* GlnRS in these cells results in amber suppression.[121] This may constitute an orthogonal pair in mammalian cells, although it is not known to what degree *Ec*GlnRS acylates mammalian tRNAs. Although protein engineering in eukaryotic cells is difficult due to poor transformation efficiencies, it is possible that an unnatural amino acid-inserting active site from a bacterial selection system could "transplanted" into a synthetase to be used in mammalian cells, due to the high homology of synthetase active sites.

Due to the fact that the anticodon of tRNA^{Asp} is a critical recognition element of AspRS and that yeast tRNA^{Asp} is not acylated by *E. coli* synthetases, we speculated that an amber suppressing derivative of *Sc*tRNA^{Asp} might be orthogonal in *E. coli*.[122-124] Moreover, it was known that the Asp93‡Lys mutant of *E. coli* AspRS was able to acylate the amber suppressing derivative of *Ec*tRNA^{Asp}.[125] With the related Asp188→Lys mutation in *Sc*AspRS, this enzyme and O-*Sc*tRNA^{Asp}(CUA) constituted an orthogonal pair in *E. coli*, albeit with weak activity. (In fact, the activity is similar to the amount of background acylation of the O-*Mj*tRNA^{Tyr}(CUA) by endogenous *E. coli* synthetases.) However, to make this pair useful, expression levels of the synthetase and tRNA were increased and an RF1 deficient strain of *E. coli* was employed, substantially increasing the amount of amber-suppression mediated ampicillin or chloramphenicol resistance upon co-expression of the O-*Sc*AspRS with the O-*Sc*tRNA^{Asp}(CUA) (554 μg mL^{-1} ampicillin) versus the tRNA alone (135 μg mL^{-1} ampicillin).[126]

Selections and Screens for Novel Amino Acid Specificity

The method for isolation of a mutant aminoacyl-tRNA synthetase (aaRS) that is specific for an unnatural amino acid from a large pool of aaRS mutants must be: sensitive (since the mutants from the initial rounds could have very low levels of activity), applicable to a wide range of amino acids, applicable to large libraries of synthetase variants, capable of excluding mutants with broader substrate specificities, and tunable, since the ability to control the stringency of the selection from one round to the next may be important. A number of in vitro selection methods that fulfill many of these criteria can be envisaged, including panning of synthetase mutants displayed on phage against immobilized substrates, or a direct high-throughput enzyme assay. Most in vivo selection methods are based on producing a distinct phenotype upon suppression of nonsense, missense or frameshift codons in a reporter gene. Here we describe several approaches recently developed in our laboratory to evolve the active sites of aminoacyl-tRNA synthetases.

General Selection

In a general selection (Fig. 6), the library is passed through a positive selection in the presence of the unnatural amino acid (resulting in a pool of all active synthetases), followed by a negative selection in the absence of the unnatural amino acid (eliminating all mutants active towards natural amino acid substrates). One example of a general selection involves a positive selection based on suppression of an amber stop codon in a β-lactamase gene (conferring survival on ampicillin) and a negative selection based on amber suppression in a barnase gene (suppression results in RNA degradation and cell death).[114] This selection has been further improved by replacing the β-lactamase reporter with the chloramphenicol acetyltransferase (CAT) gene, which has better selection properties due to the bacteriostatic nature of chloramphenicol.[126,127] Antibiotic resistance genes have two significant advantages over other potential selection markers: they allow one to tune the stringency of the positive selection by altering the concentration of the antibiotic in the growth medium, and the phenotype (antibiotic resistance) is easily quantifiable and related to the amount of suppression. However, the readout of the antibiotic selection (i.e., growth rate) is indirect, and the selection results can be easily skewed by anything that affects growth rates, such as the pharmacology of the unnatural amino acid. Nonetheless, this general selection scheme was recently used to isolate a TyrRS mutant which efficiently acylates *O*-methyltyrosine onto the engineered orthogonal tyrosyl amber suppressor tRNA (see below).[118]

General Screen

A variation on the general, double-sieve selection has been introduced that uses fluorescence-activated cell sorting (FACS) and a variant of green fluorescent protein (GFP) as a reporter. In a first step, cells containing the gene for T7 RNA polymerase with multiple amber mutations, GFP under the control of the T7 promoter, orthogonal tRNA(CUA) and a library of variants of orthogonal aaRS are grown in the presence of unnatural amino acids. These cells are then examined for fluorescence, either by FACS or visually on plates with long-wave UV irradiation. Fluorescent cells are diluted and grown in the absence of unnatural amino acids. Here, cells that fail to fluoresce must contain a synthetase that is able to reject natural amino acids but is known to be

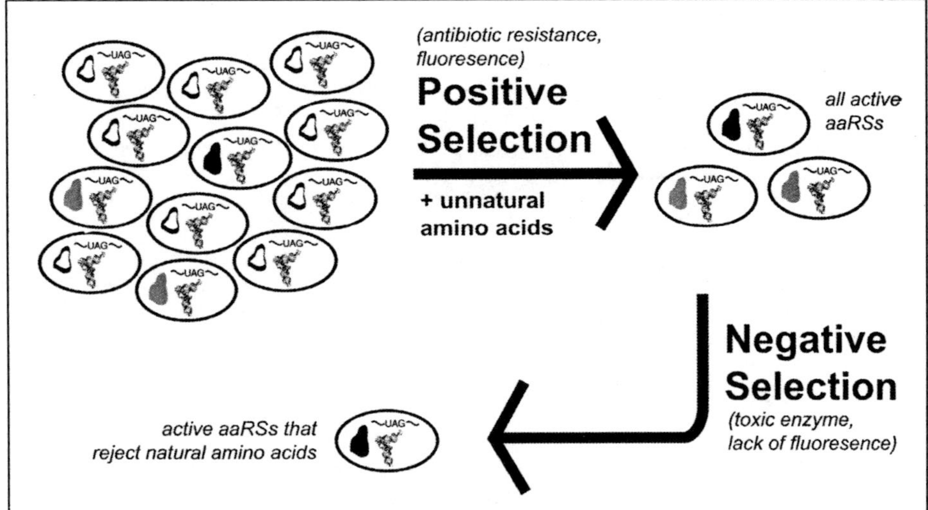

Figure 6. A general, double-sieve scheme for selecting aminoacyl-tRNA systhetases with unnatural amino acid specificity. A library of orthogonal aaRS variants is co-transformed into bacteria with an orthogonal amber-suppressing tRNA. First, a selection or screen for aaRS activity in the presence of unnatural amino acids is applied, where amber suppression confers survival or some marker that can be scored. Cells with aaRSs that acylate with natural (grey) or unnatural (black) amino acids survive, but those with very weak activity are not amplified. Second, a selection or screen against aaRS activity in the absence of amino acids is applied, where undesired amber suppression leads to expression of toxic gene product or a marker that can be scored. Survivors of both selections must, overall, contain synthetases capable of acylating the orthogonal tRNA but also capable of rejecting natural amino acids as substrates.

active toward the unnatural amino acid substrate from the first screen. This system has been shown to be useful for the yeast glutaminyl and *M. jannaschii* tyrosyl pairs, and is being used as a method for both the screening of libraries and the characterization of selectants from antibiotic selections. FACS offers the additional advantage that one can select the appropriate level of fluorescence as a cut-off for sorting, thereby altering the stringency of either the negative or positive step (S.W.S. & P.G.S., unpublished results). Modern FACS is capable of sorting a billion bacterial cells a day, which is comparable to the largest libraries that can be conveniently generated in *E. coli*.

Direct Positive Selection

In a direct selection, the unnatural amino acid side-chain would itself confer a selectable phenotype upon the cell. A simple example would be a genetic selection, in which incorporation of the unnatural amino acid into an essential protein is required for the survival of the cell. However, due to the fact that the target amino acids are by definition unnatural, a selection of this type is difficult to construct, in practice. For example, an unnatural side chain that mimics the function of one of the natural side-chains could confer a selectable phenotype upon the cell. A nonhydrolyzable analogue of a phosphorylated amino acid, such as *L*-2-amino-4-phosphonobutyric acid (phosphonoserine), might be able to substitute for phosphoserine in a protein involved in signal transduction. However, even for these two close analogues, the pK_a values differ by 1.2 units, which means that their protonation states will differ at physiological pH.

We have designed a selection that takes advantage of monoclonal antibodies specific for an unnatural amino acid presented in the context of a synthetic immunogenic peptide, the poliovirus C3 epitope (Fig. 7).[128] A C3 peptide was fused to the N-terminus of VCSM13 phage coat protein pIII, such that phage production requires suppression of the amber nonsense codon in the middle of the coding region for the C3 peptide. Cells were then transformed both with a phagemid encoding the synthetase library and the orthogonal tRNA, followed by induction of synthetase expression and

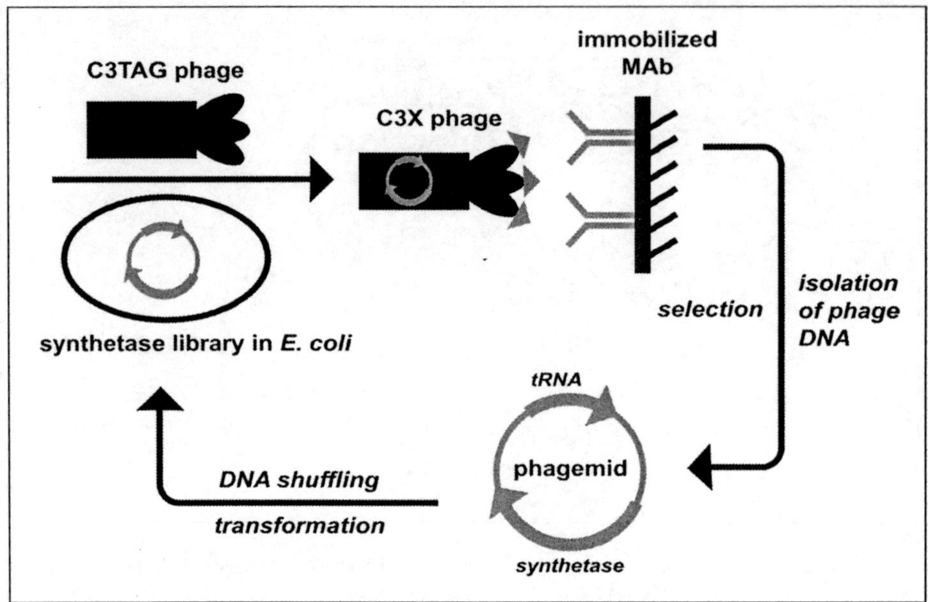

Figure 7. Phage-based selection for the incorporation of unnatural amino acids into a displayed epitope. The phage with a stop codon in the displayed fusion peptide gene is used to infect the selection strain carrying the synthetase library. The phagemid DNA coding for active synthetase mutants is packaged into phage, synthetase substrates are displayed on the phage surface, and the phage displaying the desired amino acids are isolated with immobilized antibodies. The enriched pool of mutant synsthetases is subjected to further mutagenesis and passed through additional rounds of selection.

infection with the C3TAG phage. Even a small amount of synthetase activity results in suppression of C3TAG and display of the amino acid substrate on the phage surface. Moreover, each phagemid carrying the synthetase gene is preferentially packaged in the same phage that displays the amino acid, since VCSM13 phage DNA does not have an intact M13 intergenic region necessary for efficient packaging. Subsequently, the phage pool representing all of the active synthetase genes in the library is incubated with immobilized monoclonal antibodies directed against the unnatural amino acid, in order to isolate only the phage carrying the synthetase with the desired amino acid specificity. An enrichment of up to 300-fold per round of selection was observed under model conditions that approximate a real selection.

Selection for Binding

In addition to in vivo selections for synthetases with novel amino acid specificities, we are also developing a number of in vitro screening methods. One such method relies on panning of phage-displayed synthetase libraries against immobilized sulfamoyl analogues of the aminoacyl adenylate intermediates. For example, after selection on a tyrosyl adenylate column, a monovalent display of phagemid-encoded *M. jannaschii* TyrRS fused to the pIII coat protein of M13 phage is enriched 50-fold over a control phage displaying an unrelated antibody. The actual enrichment value is probably much higher (perhaps 5,000 to 50,000), due to the fact that only 0.1 to 1 % of the starting TyrRS phage population displays the TyrRS protein, since the M13 helper phage dominates the starting pool. It is very encouraging that a K_i of 5 μM for the tyrosyl analogue translates into such a significant enrichment. However, this method might not be suitable for some synthetases: the K_i of the aspartyl analogue for the yeast orthogonal AspRS is in the low millimolar range, probably because the linker interferes with the synthetase binding. (E.M., B.H., M.P. & P.G.S, unpublished results)

The First "Unnatural" Organisms

Taking advantage of the fact that mutants of PheRS are known that reject *p*-F-Phe, Furter was partially successful in engineering a bacterium capable of inserting p-F-Phe in a site-selective manner. When yeast PheRS and tRNA[Phe](CUA G37A) were expressed in a *p*-F-Phe-resistant, Phe-auxotrophic strain of *E. coli*, growth in the presence of high levels of *p*-F-Phe resulted in largely site-specific insertion at amber mutations. Under optimal conditions, about 75% of the amber-encoded site in DHFR was occupied by *p*-F-Phe, while 20% was Phe and 5% was Lys. This indicates both that the O-tRNA[Phe](CUA) is being promiscuously acylated by *Ec*LysRS and that the ScPheRS inserts Phe in addition to *p*-F-Phe. Also, when the same site in DHFR was replaced with a Phe codon, 93% of this site was occupied by Phe, but 7% was p-F-Phe, indicating that the endogenous mutant *Ec*PheRS incorporates some *p*-F-Phe in addition to Phe. The yield of DHFR was high, however (about 10 mg L^{-1} culture), and since the natural amino acids are silent in ^{19}F NMR, this system may be useful for this application. Besides the lack of translational fidelity, the most significant limitation with this approach is that it is not general with respect to amino acid.[129]

Recently, we have developed a general strategy to incorporate unnatural amino acids site-specifically into proteins. Specifically, a directed library strategy combined with a selection based on chloramphenicol resistance was used to find a mutant of *Mj*TyrRS capable of acylating the selected O-*Mj*tRNA[Tyr](CUA)* (see above) specifically with the methyl ether of tyrosine (*O*-Me-Tyr). A library was constructed by randomization of five of the residues of *Mj*TyrRS proximal to the phenolic oxygen of the ligand tyrosine (Fig. 8A). This library was subjected to positive, chloramphenicol-based selection in the presence of *para*-substituted analogs of tyrosine (Fig. 8B) and negative screening in the absence of the unnatural amino acids. Selectants were pooled, subjected to DNA shuffling and reselected twice. Only four amino acid substitutions were found in the final selectant (Tyr32→Gln, Glu107→Thr, Asp158→Ala, Leu162→Pro); their relative importance has yet to be determined. The specificity of this mutant *Mj*OMeTyrRS is remarkable. When the tRNA and synthetase are present with an amber mutant of the gene coding for dihydrofolate reductase, no DHFR can be detected unless *O*-Me-Tyr is added to the medium. Moreover, using Fourier Transform Ion Cyclotron Resonance Mass Spectrometry, the DHFR isolated from growth in the presence of the unnatural amino acid was confirmed to carry the *O*-Me-Tyr at a single position. The yield of protein from cells grown in minimal media supplemented with *O*-Me-Tyr was 2 mg L^{-1}, the same as for wild-type DHFR. Not only is this mutant capable of adenylating the *O*-Me-Tyr about 8-fold faster than tyrosine at saturation (by pyrophosphate exchange assay), the K_M for tyrosine is about 13-fold higher than for *O*-Me-Tyr.[118]

Efforts are currently underway to "transplant" this active site into TyrRSs from other organisms, in order to transfect mammalian cells with a heterologous aaRS/tRNA(CUA) pair capable of specifically inserting *O*-Me-Tyr (T.J.M., S.W.S & P.G.S., unpublished). In addition, preliminary results suggest that other mutants of *Mj*TyrRS are capable of activating a number of other tyrosine analogs (S.W.S., L.W. & P.G.S., unpublished). Finally, similar selection strategies are being applied to other orthogonal tRNA/synthetase pairs, including pairs with tRNAs that decode four-base codons (J.C.A. & P.G.S., unpublished).

Conclusion

Modern methods of in vitro production of proteins allow us to study macromolecules with changes as small as a single atom or as significant as the addition of multiple biophysical probes. This exquisite ability to control protein composition is entering a new era with methods for the site-selective insertion of unnatural amino acids into proteins in living cells, effectively expanding the genetic code. A new array of tools, requiring no facility with synthetic chemistry, will give biochemists and cell biologists access to large quantities of proteins with novel physical and chemical properties. These methods will not only allow us to engineer proteins with interesting new properties, they will allow us to analyze the functions of these mutant proteins in living cells.

The major challenge that lies ahead is the application of selections for novel unnatural amino acid specificity toward side chains significantly different from the natural set to make useful tools for cell biology. As these new enzymes are generated, novel ways of inserting these unnatural amino acids site-specifically will have to be demonstrated, with, for example, four-base codons and unnatural codons.

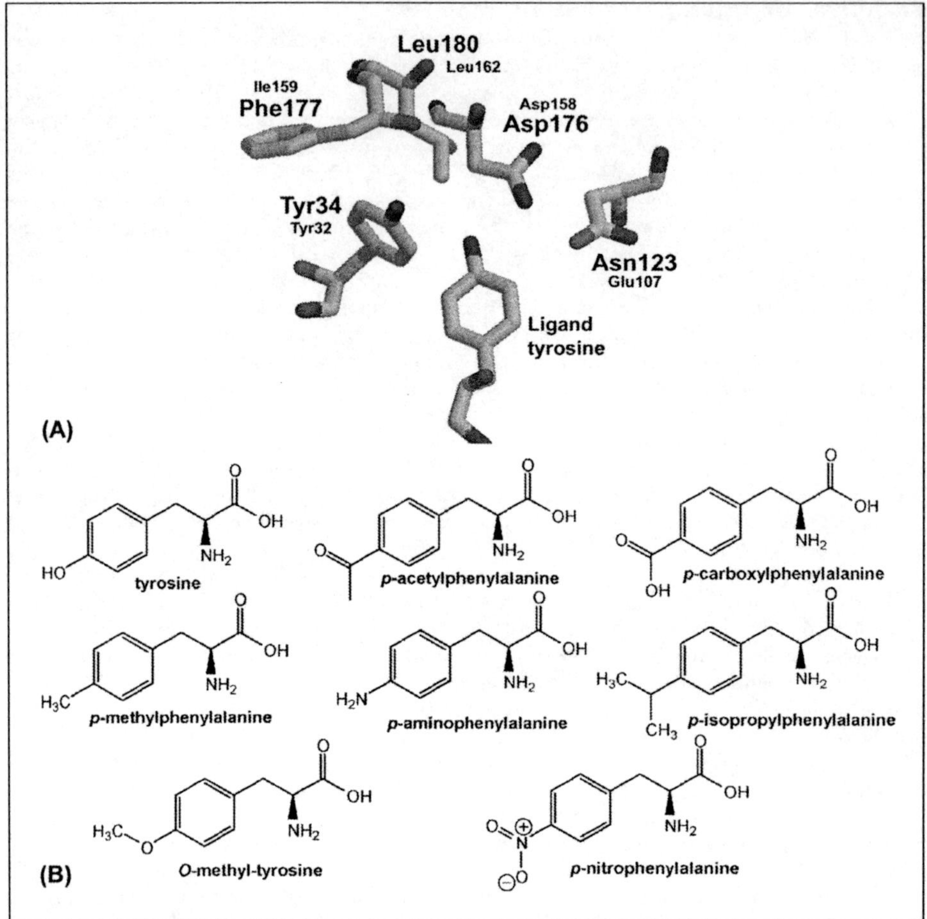

Figure 8. TyrRS *para*-targeted library and tyrosine *para*-substituted analogs. A) The indicated residues (smaller font) were randomized in *M. jannaschii* TyrRS due to proximity to the phenolic oxygen of the ligand tyrosine as evidenced in the co-crystal structure of *B. stearothermophilus* TyrRS with an analog of Tyr-AMP. The homologous resides identified from sequence alignment are indicated (larger font). B) *Para*-substituted analogs of tyrosine used in the selection with the library. (A) was created with RasMol 2.6 from the PDB file 3ts1.

Finally, transitioning these systems into eukaryotes like yeast and mammalian cells will allow us to study proteins involved in human disease with remarkable precision. Significant inroads have been made into each of these daunting problems to show that they will be surmountable in the years ahead.

Acknowledgments

T.J.M. and J.C.A. are National Science Foundation predoctoral fellows. S.W.S. is a postdoctoral fellow of The Jane Coffin Childs Memorial Fund for Medical Research. B.H. is a postdoctoral fellow of the National Institutes of Health. E.M. is a postdoctoral fellow of the German Research Foundation (DFG). Financial support was provided by the Office of Naval Research, Department of the Army and the Skaggs Research Fund.

References

1. Zoller MJ, Smith M. Oligonucleotide-directed mutagenesis of DNA fragments cloned into M13 vectors. Methods Enzymol 1983; 100(11):468-500.
2. In vitro mutagenesis protocols. In: Trower MK, ed. Methods in Molecular Biology. Vol 57. Totowa: Humana Press, 1996.
3. Knowles JR. Tinkering with enzymes: what are we learning? Science 1987; 236(4806):1252-1258.
4. Merrifield B. Solid phase synthesis. Science 1986; 232(4748):341-347.
5. Muir TW, Dawson PE, Kent SB. Protein synthesis by chemical ligation of unprotected peptides in aqueous solution. Methods Enzymol 1997; 289(11):266-298.
6. Wilken J, Kent SBH. Chemical protein synthesis. Curr Opin Biotechnol 1998; 9(4):412-426.
7. Dawson PE, Kent SB. Synthesis of native proteins by chemical ligation. Annu Rev Biochem 2000; 69 923-960.
8. Schnölzer, M, Kent SB. Constructing proteins by dovetailing unprotected synthetic peptides: backbone-engineered HIV protease. Science 1992; 256(5054):221-225.
9. Dawson PE, Muir TW, Clarklewis I et al. Synthesis of Proteins By Native Chemical Ligation. Science 1994; 266(5186):776-779.
10. Canne LE, Botti P, Simon RJ et al. Chemical protein synthesis by solid phase ligation of unprotected peptide segments. J Am Chem Soc 1999; 121(38):8720-8727.
11. Baca M, Kent SBH. Catalytic Contribution of Flap-Substrate Hydrogen Bonds in Hiv-1 Protease Explored By Chemical Synthesis. Proc Nat Acad Sci USA 1993; 90(24):11638-11642.
12. Cotton GJ, Muir TW. Peptide ligation and its application to protein engineering. Chem Biol 1999; 6(9):R247-R256.
13. Lu WY, Qasim MA, Laskowski M et al. Probing intermolecular main chain hydrogen bonding in serine proteinase-protein inhibitor complexes: Chemical synthesis of backbone-engineered turkey ovomucoid third domain. Biochemistry 1997; 36(4):673-679.
14. Love SG, Muir TW, Ramage R et al. Synthetic, structural and biological studies of the ubiquitin system: Synthesis and crystal structure of an analogue containing unnatural amino acids. Biochem J 1997; 323(PT3):727-734.
15. Low DW, Hill MG. Rational fine-tuning of the redox potentials in chemically synthesized rubredoxins. J Am Chem Soc 1998; 120(44):11536-11537.
16. Deniz AA, Laurence TA, Beligere GS et al. Single-molecule protein folding: Diffusion fluorescence resonance energy transfer studies of the denaturation of chymotrypsin inhibitor 2. Proc Nat Acad Sci USA 2000; 97(10):5179-5184.
17. Xu MQ, Perler FB. The Mechanism of Protein Splicing and Its Modulation By Mutation. EMBO J 1996; 15(19):5146-5153.
18. Chong SR, Mersha FB, Comb DG et al. Single-column purification of free recombinant proteins using a self-cleavable affinity tag derived from a protein splicing element. Gene 1997; 192(2):271-281.
19. Evans TC, Xu MQ. Intein-mediated protein ligation: Harnessing natures' escape artists. Biopolymers 1999; 51(5):333-342.
20. Severinov K, Muir TW. Expressed protein ligation, a novel method for studying protein-protein interactions in transcription. J Biol Chem 1998; 273(26):16205-16209.
21. Muir TW, Sondhi D, Cole PA. Expressed protein ligation: A general method for protein engineering. Proc Nat Acad Sci USA 1998; 95(12):6705-6710.
22. Ayers B, Blaschke UK, Camarero JA et al. Introduction of unnatural amino acids into proteins using expressed protein ligation. Biopolymers 1999; 51(5):343-354.
23. Braisted AC, Judice JK, Wells JA. Synthesis of proteins by subtiligase. Methods Enzymol 1997; 289(4):298-313.
24. Nakatsuka T, Sasaki T, Kaiser ET. Peptide segment coupling catalyzed by the semisynthetic enzyme thiosubtilisin. J Am Chem Soc 1987; 109 3808-3810.
25. Abrahmsen L, Tom J, Burnier J et al. Engineering Subtilisin and Its Substrates For Efficient Ligation of Peptide Bonds in Aqueous Solution. Biochemistry 1991; 30(17):4151-4159.
26. Chang TK, Jackson DY, Burnier JP et al. Subtiligase—a Tool For Semisynthesis of Proteins. Proc Nat Acad Sci USA 1994; 91(26):12544-12548.
27. Jackson DY, Burnier J, Quan C et al. A Designed Peptide Ligase For Total Synthesis of Ribonuclease a With Unnatural Catalytic Residues. Science 1994; 266(5183):243-247.
28. Lundblad RL. Chemical reagents for protein modification. 2nd ed. Boca Raton: CRC Press, 1991.
29. Brinkley MA. Brief Survey of Methods For Preparing Protein Conjugates With Dyes, Haptens, and Cross-Linking Reagents. Bioconj Chem 1992; 3(1):2-13.
30. Hermanson GT. Bioconjugate techniques. San Diego: Academic Press, 1996.
31. Corey DR, Schultz PG. Generation of a hybrid sequence-specific single-stranded deoxyribonuclease. Science 1987; 238(4832):1401-1403.
32. Stclair NL, Navia MA. Cross-Linked Enzyme Crystals As Robust Biocatalysts. J Am Chem Soc 1992; 114(18):7314-7316.
33. Gaertner HF, Rose K, Cotton R et al. Construction of Protein Analogues By Site-Specific Condensation of Unprotected Fragments. Bioconj Chem 1992; 3(3):262-268.
34. Geoghegan KF, Stroh JG. Site-Directed Conjugation of Nonpeptide Groups to Peptides and Proteins Via Periodate Oxidation of a 2-Amino Alcohol—Application to Modification At N-Terminal Serine. Bioconj Chem 1992; 3(2):138-146.

35. Kawakami T, Akaji K, Aimoto S. Peptide bond formation mediated by 4,5-dimethoxy-2-mercaptobenzylamine after periodate oxidation of the N-terminal serine residue. Org Lett 2001; 3(9):1403-1405.
36. Liu JQ, Jiang MS, Luo GM et al. Conversion of trypsin into a selenium containing enzyme by using chemical mutation. Biotechnol Lett 1998; 20(7):693-696.
37. Wu Z-P, Hilvert D. Conversion of a protease into an acyl transferase: selenolsubtilisin. J Am Chem Soc 1989; 111 4513-4514.
38. Noren CJ, Anthony-Cahill SJ, Griffith MC et al. A general method for site-specific incorporation of unnatural amino acids into proteins. Science 1989; 244(4901):182-188.
39. Bain JD, Glabe CG, Dix TA et al. Biosynthetic Site-Specific Incorporation of a Non-Natural Amikno Acid into a Polypeptide. J Am Chem Soc 1989; 111 8013-8014.
40. Cornish VW, Mendel D, Schultz PG. Probing Protein Structure and Function with an Expanded Genetic Code. Angew Chem Int Ed Engl 1995; 34 621-633.
41. Gilmore MA, Steward LE, Chamberlin AR. Incorporation of Noncoded Amino acids by In Vitro Protein Biosynthesis. Topics Curr Chem 1999; 202 77-99.
42. Dougherty DA. Unnatural amino acids as probes of protein structure and function. Curr Opin Chem Biol 2000; 4(6):645-652.
43. Ellman J, Mendel D, Anthony-Cahill S et al. Biosynthetic method for introducing unnatural amino acids site-specifically into proteins. Methods Enzymol 1991; 202 301-336.
44. Thorson JS, Cornish VW, Barrett JE et al. A biosynthetic approach for the incorporation of unnatural amino acids into proteins. Methods Mol Biol 1998; 77 43-73.
45. Cload ST, Liu DR, Froland WA et al. Development of improved tRNAs for in vitro biosynthesis of proteins containing unnatural amino acids. Chem Biol 1996; 3(12):1033-1038.
46. Kim D, Kigawa T, Choi C et al. A highly efficient cell-free protein synthesis sytem from *Escherichia coli*. Eur J Biochem 1996; 239 881-886.
47. Steward LE, Collins CS, Gilmore MA et al. In vitro site-specific incorporation of fluorescent probes into beta-galactosidase. J Am Chem Soc 1997; 119(1):6-11.
48. Short GF, Golovine SY, Hecht SM. Effects of Release Factor 1 on in Vitro Protein Translation and the Elaboration of Proteins Containing Unnatural amino acids. Biochemistry 1999; 38:8808-8819.
49. Robertson SA, Noren CJ, Anthonycahill SJ et al. The Use of 5'-Phospho-2 Deoxyribocytidylyl-riboadenosine As a Facile Route to Chemical Aminoacylation of Transfer RNA. Nucl Acids Res 1989; 17(23):9649-9660.
50. Robertson SA, Ellman JA, Schultz PG. A General and Efficient Route For Chemical Aminoacylation of Transfer RNAs. J Am Chem Soc 1991; 113(7):2722-2729.
51. Ma C, Kudlicki W, Odom OW et al. In vitro protein engineering using synthetic tRNA(Ala) with different anticodons. Biochemistry 1993; 32(31):7939-7945.
52. Piccirilli JA, Krauch T, Moroney S.E et al. Enzymatic incorporation of a new base pair into DNA and RNA extends the genetic alphabet. [see comments]. Nature 1990; 343(6253):33-37.
53. Bain JD, Switzer C, Chamberlin AR et al. Ribosome-mediated incorporation of a non-standard amino acid into a peptide through expansion of the genetic code. Nature 1992; 356(6369):537-539.
54. McMinn DL, Ogawa AK, Wu YQ et al. Efforts toward expansion of the genetic alphabet: DNA polymerase recognition of a highly stable, self-fairing hydrophobic base. J Am Chem Soc 1999; 121(49):11585-11586.
55. Ogawa AK, Wu YQ, McMinn DL et al. Efforts toward the expansion of the genetic alphabet: Information storage and replication with unnatural hydrophobic base pairs. J Am Chem Soc 2000; 122(14):3274-3287.
56. Meggers E, Holland PL, Tolman WB et al. A novel copper-mediated DNA base pair. J Am Chem Soc 2000; 122(43):10714-10715.
57. Kool ET. Synthetically modified DNAs as substrates for polymerases. Curr Opin Chem Biol 2000; 4(6):602-608.
58. Roth JR. Frameshift suppression. Cell 1981; 24(3):601-602.
59. Moore B, Persson BC, Nelson CC et al. Quadruplet codons: implications for code expansion and the specification of translation step size. J Mol Biol 2000; 298(2):195-209.
60. Magliery TJ, Anderson JC, Schultz PG. Expanding the genetic code: selection of efficient suppressors of four-base codons and identification of "shifty" four-base codons with a library approach in *Escherichia coli*. J Mol Biol 2001; 307(3):755-769.
61. Anderson JC, Magliery TJ, Schultz PG. Exploring the limits of codon and anticodon size. Chem Biol 2001; in press.
62. Kramer G, Kudlicki W, Hardesty B. In vitro engineering using synthetic tRNAs with altered anticodons including four-nucleotide anticodons. Methods Mol Biol 1998; 77(11):105-116.
63. Hohsaka T, Ashizuka Y, Murakami H et al. Incorporation of Nonnatural Amino acids Into Streptavidin Through in Vitro Frame-Shift Suppression. J Am Chem Soc 1996; 118(40):9778-9779.
64. Hohsaka T, Ashizuka Y, Sasaki H et al. Incorporation of two different nonnatural amino acids independently into a single protein through extension of the genetic code. J Am Chem Soc 1999; 121(51):12194-12195.
65. Hohsaka T, Ashizuka Y, Sisido M. Incorporation of two nonnatural amino acids into proteins through extension of the genetic code. Nucleic Acids Symp Ser 1999; 32(42):79-80.
66. Thorson JS, Chapman E, Murphy EC et al. Linear Free Energy Analysis of Hydrogen Bonding in Proteins. J Am Chem Soc 1995; 117 1157-1158.

67. Thorson JS, Chapman E, Schultz PG. Analysis of Hydrogen Bonding Strength in Proteins Using Unnatural amino acids. J Am Chem Soc 1995; 117 9361-9362.
68. Ellman JA, Mendel D, Schultz PG. Site-specific incorporation of novel backbone structures into proteins. Science 1992; 255(5041):197-200.
69. Mendel D, Ellman JA, Chang Z et al. Probing protein stability with unnatural amino acids. Science 1992; 256(5065):1798-1802.
70. Hohsaka T, Sato K, Sisido M et al. Site-specific incorporation of photofunctional nonnatural amino acids into a polypeptide through in vitro protein biosynthesis. FEBS Lett 1994; 344(2-3):171-174.
71. Hohsaka T, Kajihara D, Ashizuka Y et al. Efficient incorporation of nonnatural amino acids with large aromatic groups into streptavidin in in vitro protein synthesizing systems. J Am Chem Soc 1999; 121(1):34-40.
72. Cornish VW, Benson DR, Altenbach C.A et al. Site-specific incorporation of biophysical probes into proteins. Proc Nat Acad Sci USA 1994; 91(8):2910.
73. Mendel D, Ellman JA, Schultz PG. Construction of a Light-Activated Protein by Unnatural amino acid Mutagenesis. J Am Chem Soc 1991; 113 2758-2760.
74. Pollitt SK, Schultz PG. A Photochemical Switch for Controlling Protein-Protein Interactions. Angew Chem Int Ed Engl 1998; 37(15):2104-2107.
75. Cornish VW, Hahn KM, Schultz PG. Site-Specific Protein Modification Using a Ketone Handle. J Am Chem Soc 1996; 118 8150-8151.
76. Clark TD, Ghadiri MR. Supramolecular Design by Covalent Capture. Design of a Peptide Cylinder via Hydrogen-Bond-Promoted Intermolecular Olefin Metathesis. J Am Chem Soc 1995; 117 12364-12365.
77. Saxon E, Bertozzi CR. Cell Surface Engineering by a Modified Staudinger Reaction. Science 2000; 287 2007-2010.
78. Nowak MW, Kearney PC, Sampson JR et al. Nicotinic Receptor Binding Site Probed With Unnatural amino acid Incorporation in Intact Cells. Science 1995; 268(5209):439-442.
79. Saks ME, Sampson JR, Nowak MW et al. An Engineered *Tetrahymena* tRNA(Gln) For in Vivo Incorporation of Unnatural amino acids Into Proteins By Nonsense Suppression. J Biol Chem 1996; 271(38):23169-23175.
80. Kearney PC, Zhang HY, Zhong WG et al. Determinants of nicotinic receptor gating in natural and unnatural side chain structures at the M2 9' position. Neuron 1996; 17(6):1221-1229.
81. Zhong WG, Gallivan JP, Zhang YN et al. From ab initio quantum mechanics to molecular neurobiology: A cation-pi binding site in the nicotinic receptor. Proc Nat Acad Sci USA 1998; 95(21):12088-12093.
82. Gallivan JP, Lester HA, Dougherty DA. Site-specific incorporation of biotinylated amino acids to identify surface-exposed residues in integral membrane proteins. Chem Biol 1997; 4(10):739-749.
83. England PM, Lester HA, Davidson N et al. Site-specific, photochemical proteolysis applied to ion channels in vivo. Proc Nat Acad Sci USA 1997; 94(20):11025-11030.
84. Turcatti G, Nemeth K, Edgerton MD et al. Probing the Structure and Function of the Tachykinin Neurokinin-2 Receptor Through Biosynthetic Incorporation of Fluorescent Amino acids At Specific Sites. J Biol Chem 1996; 271(33):19991-19998.
85. Miller JC, Silverman SK, England PM et al. Flash decaging of tyrosine sidechains in an ion channel. Neuron 1998; 20(4):619-624.
86. England PM, Zhang YN, Dougherty DA et al. Backbone mutations in transmembrane domains of a ligand-gated ion channel: Implications for the mechanism of gating. Cell 1999; 96(1):89-98.
87. Lu T, Ting AY, Mainland J et al. Probing ion permeation and gating in a K+ channel with backbone mutations in the selectivity filter. Nature Neuroscience 2001; 4(3):239-246.
88. Budisa N, Minks C, Alefelder S et al. Toward the experimental codon reassignment in vivo: protein building with an expanded amino acid repertoire. FASEB J 1999; 13(1):41-51.
89. Cowie DB, Cohen GN. Biosynthesis by *Escherichia coli* of Active Altered Proteins Containing Selenium Instead of Sulphur. Biochim Biophys Acta 1957; 26 252-261.
90. Richmond MH. The Effect of Amino acid analogues on Growth and Protein synthesis in Microorganisms. Bacteriol Rev 1962; 26 398-420.
91. Yang W, Hendrickson WA, Crouch RJ et al. Structure of Ribonuclease-H Phased At 2-a Resolution By Mad Analysis of the Selenomethionyl Protein. Science 1990; 249(4975):1398-1405.
92. Budisa N, Karnbrock W, Steinbacher S et al. Bioincorporation of telluromethionine into proteins: A promising new approach for X-ray structure analysis of proteins. J Mol Biol 1997; 270(4):616-623.
93. Minks C, Huber R, Moroder L et al. Noninvasive tracing of recombinant proteins with "fluorophenylalanine-fingers". Anal Biochem 2000; 284(1):29-34.
94. Tang Y, Ghirlanda G, Vaidehi N et al. Stabilization of coiled-coil peptide domains by introduction of trifluoroleucine. Biochemistry 2001; 40(9):2790-2796.
95. Tang Y, Ghirlanda G, Petka WA et al. Fluorinated coiled-coil proteins prepared in vivo display enhanced thermal and chemical stability. Angew Chem Int Ed Engl 2001; 40(8):1494-1496,1331.
96. Renner C, Alefelder S, Bae JH et al. Fluoroprolines as tools for protein design and engineering. Angew Chem Int Ed Engl 2001; 40(5):923-925.
97. Deming TJ, Fournier MJ, Mason TL et al. Biosynthetic incorporation and chemical modification of alkene functionality in genetically engineered polymers. J Macromol Sci Pure Appl Chem 1997; A34(10):2143-2150.
98. vanHest JCM, Tirrell DA. Efficient introduction of alkene functionality into proteins in vivo. FEBS Lett 1998; 428(1-2):68-70.

99. Kiick KL, Tirrell DA. Protein engineering by in vivo incorporation of non-natural amino acids: Control of incorporation of methionine analogues by methionyl-tRNA synthetase. Tetrahedron 2000; 56(48):9487-9493.
100. van Hest JCM, Kiick KL, Tirrell DA. Efficient incorporation of unsaturated methionine analogues into proteins in vivo. J Am Chem Soc 2000; 122(7):1282-1288.
101. Ibba M, Hennecke H. Relaxing the Substrate Specificity of an Aminoacyl-tRNA synthetase Allows in Vitro and in Vivo Synthesis of Proteins Containing Unnatural amino acids. FEBS Lett 1995; 364(3):272-275.
102. Sharma N, Furter R, Kast P et al. Efficient introduction of aryl bromide functionality into proteins in vivo. FEBS Lett 2000; 467(1):37-40.
103. Behrens C, Nielsen JN, Fan XJ et al. Development of strategies for the site-specific in vivo incorporation of photoreactive amino acids: p-azidophenylalanine, p-acetylphenylalanine and benzofuranylalanine. Tetrahedron 2000; 56(48):9443-9449.
104. Hamano-Takaku F, Iwama T, Saito-Yano S et al. A mutant *Escherichia coli* tyrosyl-tRNA synthetase utilizes the unnatural amino acid azatyrosine more efficiently than tyrosine. J Biol Chem 2000; 275(51):40324-40328.
105. Doring V, Mootz HD, Nangle LA et al. Enlarging the amino acid set of *Escherichia coli* by infiltration of the valine coding pathway. Science 2001; 292(5516):501-504.
106. Freist W, Gauss DH, Ibba M et al. Glutaminyl-tRNA synthetase. Biol Chem 1997; 378(10):1103-1117.
107. Rath VL, Silvian LF, Beijer B et al. How glutaminyl-tRNA synthetase selects glutamine. Structure 1998; 6(4):439-449.
108. Ibba M, Soll D. Aminoacyl-tRNA synthesis. Annu Rev Biochem 2000; 69 617-650.
109. Liu DR, Magliery TJ, Schultz PG. Characterization of an 'orthogonal' suppressor tRNA derived from E-coli tRNA(2)(Gln). Chem Biol 1997; 4(9):685-691.
110. Liu DR, Magliery TJ, Pastrnak M et al. Engineering a tRNA and aminoacyl-tRNA synthetase for the site-specific incorporation of unnatural amino acids into proteins in vivo. Proc Nat Acad Sci USA 1997; 94(19):10092-10097.
111. Wang CC, Schimmel P. Species barrier to RNA recognition overcome with nonspecific RNA binding domains. J Biol Chem 1999; 274(23):16508-16512.
112. Whelihan EF, Schimmel P. Rescuing an essential enzyme RNA complex with a non-essential appended domain. EMBO J 1997; 16(10):2968-2974.
113. Huang WZ, Petrosino J, Hirsch M et al. Amino acid Sequence Determinants of Beta-Lactamase Structure and Activity. J Mol Biol 1996; 258(4):688-703.
114. Liu DR, Schultz PG. Progress toward the evolution of an organism with an expanded genetic code. Proc Nat Acad Sci USA 1999; 96(9):4780-4785.
115. Wang L, Magliery TJ, Liu DR et al. A new functional suppressor tRNA/aminoacyl-tRNA synthetase pair for the in vivo incorporation of unnatural amino acids into proteins. J Am Chem Soc 2000; 122(20):5010-5011.
116. Steer BA, Schimmel P. Major anticodon-binding region missing from an archaebacterial tRNA synthetase. J Biol Chem 1999; 274(50):35601-35606.
117. Steer BA, Schimmel P. Domain-domain communication in a miniature archaebacterial tRNA synthetase. Proc Nat Acad Sci USA 1999; 96(24):13644-13649.
118. Wang L, Brock A, Herberich B et al. Expanding the genetic code of *Escherichia coli*. Science 2001; 292(5516):498-500.
119. Wang L, Schultz PG. A general approach for the generation of orthogonal tRNAs. Chem Biol 2001; 8(9):883-890.
120. Kowal AK, Kohrer C, RajBhandary UL. Twenty-first aminoacyl-tRNA synthetase-suppressor tRNA pairs for possible use in site-specific incorporation of amino acid analogues into proteins in eukaryotes and in eubacteria. Proc Natl Acad Sci USA 2001; 98(5):2268-2273.
121. Drabkin HJ, Park HJ, Rajbhandary UL. Amber suppression in Mammalian Cells Dependent Upon Expression of an Escherichia coli Aminoacyl-tRNA synthetase Gene. Mol Cell Biol 1996; 16(3):907-913.
122. Doctor BP, Mudd JA. Species specificity of amino acid accepter ribonucleic acid and aminoacyl soluble ribonucleic acid synthetases. J Biol Chem 1963; 238(11):3677-3681.
123. Kwok Y, Wong JT. Evolutionary relationship between *Halobacterium cutirubrum* and eukaryotes determined by use of aminoacyl-tRNA synthetases as phylogenetic probes. Can J Biochem 1980; 58(3):213-218.
124. Giege R, Florentz C, Kern D et al. Aspartate Identity of Transfer RNAs. Biochimie 1996; 78(7):605-623.
125. Martin F. Thesis. Universite Louis Pasteur: Strasbourg, France, 1995.
126. Pastrnak M, Magliery TJ, Schultz PG. A new orthogonal suppressor tRNA/aminoacyl-tRNA synthetase pair for evolving an organism with an expanded genetic code. Helv Chim Acta 2000; 83(9):2277-2286.
127. Magliery TJ. Thesis in Chemistry. University of California: Berkeley, 2001.
128. Pastrnak M, Schultz PG. Phage selection for site-specific incorporation of unnatural amino acids into proteins in vivo. Bioorg Med Chem 2001; 9(9):2373-2379.
129. Furter R. Expansion of the genetic code: Site-directed p-fluoro-phenylalanine incorporation in Escherichia coli. Protein Sci 1998; 7(2):419-426.

CHAPTER 7

Tests of a Stereochemical Genetic Code

Rob Knight, Laura Landweber and Michael Yarus

Abstract

Does the genetic code assign similar codons to similar amino acids because of chemical interactions between them? Unlike adaptive explanations, which can only explain the relative positions of amino acids in the code, stereochemical explanations could tie codon assignments to absolute, verifiable rules. However, modern translation encodes amino acid sequences without direct codon/amino acid interaction. If there is a relationship between RNA sequences with intrinsic affinity for amino acids and the modern genetic code, we must therefore explain a historical transition in which direct interactions were abandoned. We review the literature and find no evidence that interactions between short sequences (mono-, di- or trinucleotides) and amino acids are strong or specific enough to originate genetic coding. Instead, interactions between amino acids and longer nucleic acid sequences appear to recapture some assignments of the modern code. For example, real codons are concentrated in newly selected amino acid binding sites to a greater extent than codons from similar, but randomized, codes. This implies that some initial coding assignments were made by interaction with macromolecular RNA-like molecules, and have survived. Thus, subsequent selection, such as selection to minimize coding errors, has not erased all primordial chemical relationships. Retention of initial stereochemical codon assignments for three of six amino acids (arginine, isoleucine, and tyrosine, but not glutamine, leucine or phenylalanine) is strongly supported.

Combining data for the six amino acids, significant stereochemical relationships are of more than one type—codons and anticodons are each concentrated in some binding sites. Further work will be required to catalog the relationships between amino acids and binding site sequences, especially if, as now appears, more than one type of interaction has been transmitted to the modern code.

The Codon Correspondence Hypothesis

The codon correspondence hypothesis, tested in any stereochemical theory of the origin of the genetic code, may be stated:

For each amino acid, there is a coding sequence for which it has the greatest association. The association between these sequences and amino acids influenced the form and content of the genetic code.

The codon-correspondence hypothesis is compatible with establishment of the genetic code either before or during the RNA world. A direct association between mono-, di- or trinucleotides and their cognate amino acids would suggest that the code arose before complex RNA catalysts, since trinucleotides would likely occur before the reproducible synthesis of longer oligonucleotides. Alternatively, an association between trinucleotides and their cognate amino acids that requires RNA tertiary structure would suggest that the genetic code arose in the RNA world (the earliest evolutionary time at which long RNA-like molecules were available). Larger RNAs loosen the constraint on the role of the coding sequences, which could then support the amino acid binding site but need not comprise it entirely. Amino acid/RNA complexes might have functioned in translation from the beginning, but alternatives abound. Their original functions may have been varied: as coenzyme sites for ribozymes,[1] to stabilize RNA double helices,[2] or to label tRNA-like genomic tags.[3,4]

Translation Mechanisms, edited by Jacques Lapointe and Léa Brakier-Gingras.

Chemical Associations: A Historical Perspective

The idea that the genetic code might be stereochemically determined predates the elucidation of the code. Gamow's 'diamond code', in which amino acids would fit specific pockets bounded by four DNA bases, relied on direct interaction between amino acids and nucleic acids.[5] More abstruse possibilities exist: mathematical (and even numerological) schemes for solving the coding problem abounded before the actual codon assignments were fully uncovered (reviewed in refs. 6, 7).

The structure of the code showed clear patterns. Chemical explanations for such order were sought by two routes. Physicochemical theorists[8,9] hoped to measure interaction between bases and amino acids. This might have resulted in chromatographic co-partitioning on the early earth, which would be reproducible today by chemical techniques. In contrast, stereochemical theorists[10,11] assumed that molecular modeling could reveal molecular complementarities between amino acids and coding triplets.

Stereochemistry/Molecular Models: The first chemical investigations of codon assignments were via molecular modeling. Molecular models have been said to prove that the genetic code was established in quite varied ways. For example, amino acids might pair with codons[12] or anticodons[10,13] in the tRNA. Codonic mononucleotides and α-helical homopolymeric amino acids may bind each other specifically (this model "correctly predicts the glycine codon GGG", although it unfortunately fails to predict any other).[14] Free glycine and free nucleotides[15] may have affinity, or free amino acids may intercalate into adjacent bases in the anticodon doublet through H-bonding between methylene groups and the π-electrons of the bases.[16] Specific 2' aminoacylation of the second position anticodon base may have been mediated by the first position anticodon base.[17] Amino acids may be able to intercalate between first and second position bases in double-stranded RNA molecules.[18] Cavities caused by removal of the second-position codon bases in B-DNA may accept amino acids.[19] Perhaps amino acids nestle into a pentanucleotide cup with the anticodon in the center.[20] Pairing between amino acid side-chains and cavities in a complex of four nucleotides (C4N) on the acceptor stem of tRNA might occur. Or perhaps amino acids can bind their codons transposed 3'-5'.[22,23] A double-stranded complex of the codon and anticodon has also been suggested.[18,24]

The modeling approach was tarnished early on when a claimed association between codons and amino acids[12] relied on models that had been built backwards, 3' to 5'.[25] Nevertheless, even the idea that there is a relationship between reversed codons and amino acids has been defended.[22,23]

Clearly, modeling methods used thus far are not sufficiently constrained. As a result, they allow too many solutions. Additionally, these approaches tend to assume that the entire code was uniquely determined by stereochemical fit (and even that modern variant codes reflect fits induced by different environments).[26] If amino acids were added to the code over time and for different reasons, as seems probable,[27,28] such explanations are overstatements that may prevent confirmation even if the basic hypothesis is true.

Physicochemical Effects/Chromatography: A second line of evidence comes from chromatography. Because chromatographic properties of amino acids show regular variation in the genetic code, any mechanism for the code's origin must account for this organization. Various studies have shown that the code conserves certain properties, such as polarity. The polar requirement of amino acids (the ratio of the log relative mobility to the log mole fraction water in a water-pyridine mixture) orders coding assignments impressively. Amino acids with U in the second position of their codon are hydrophobic while those with A are hydrophilic; those with C are intermediate, and those with G are mixed. Furthermore, codons that share a doublet have almost identical polar requirements even if not otherwise related (e.g., His and Gln; possibly Cys and Trp). Thus the code is ordered with respect to amino acid properties, but such evidence cannot tell us whether the code was optimized to minimize errors due to mutation or established by direct chemical interactions.[28]

Nor does such chemical order suggest a mechanism for actual codon assignments. Partitioning of amino acids and nucleotides between aqueous and organic phases, as in a primordial oil slick, might have associated AAA codons with Lys and UUU codons with Phe.[30] However, none of these molecules are produced in prebiotic syntheses[31] and a further hypothesis is required to bring chromatographic partitioning to bear on codon assignment. Analysis with two further chromatographic systems, water/micellar sodium dodecanoate and hexane/ dodecylammonium propionate-trapped water, confirmed the previous hydrophobicity scales in a context closer to prebiotic conditions.[32] The relative hydrophobicity of the homocodonic amino acids (Phe UUU, Pro CCC, Lys AAA, Gly GGG) and the four nucleotides in an ammonium acetate/ammonium sulfate system showed an anticodonic association, and for dinucleoside monophosphates the association was also with the anticodon, rather than the codon, doublets.[33] Multivariate analysis of the properties of dinucleoside monophosphates and amino acids, focusing on hydrophobicity, revealed many strong (p < 0.001) correlations between anticodons and amino acids, but not between codons and amino acids.[34]

Thus, chromatographic data suggest anticodonic, rather than codonic interactions (note the underlying assumption that molecules with similar properties interact). However, although chemical partitioning on the early earth could conceivably have led to specific cofractionation between particular nucleotides (or oligonucleotides) and prebiotic amino acids, there do not seem to be consistent correlations. Chromatographic separation on various plausibly prebiotic surfaces (silicates, clays, hydroxyapatite, calcium carbonate, etc.) showed that, on a silica surface under an aqueous solution of $MgCl_2$ and $(NH_4)H_2PO_4$, Ala comigrates with CMP and Gly comigrates with GMP.[35] Ala is assigned the GCN codon class, while Gly has the GGN codon class. However, there was no strong separation between GMP and UMP or between AMP and CMP even on silica, and many prebiotic amino acids (Pro, Ile, Leu, Val) fell well outside the range of the nucleotides. The situation was even worse on other surfaces, which did not provide any amino acid-nucleotide concordances. Thus, the data do not support the conclusion that copartitioning of nucleotides and amino acids led to the genetic code,[35] especially in the absence of a plausible mechanism for transforming a copartition into modern codon assignments.

Physicochemical Effects/Direct Interactions: The third type of evidence comes from tests for direct interaction between nucleotides and amino acids. Mononucleotides show nonspecific but charge-dependent interactions with polyamino acid chains, as measured by the change in turbidity of the cosolution.[14] Affinity chromatography, which tested retardation of the four nucleotide monophosphates by each of nine amino acids (Gly, Lys, Pro, Met, Arg, His, Phe, Trp, Tyr) immobilized by their carboxyl groups, showed no association between binding strength and codon or anticodon assignments.[36] Interactions between free amino acids and poly(A), as measured by the chemical shift of the C_2 and C_8 protons of A, are also "not easily reconcilable with the genetic code".[37] Further affinity chromatography and NMR experiments on the interaction between amino acids and mono-, di-, and trinucleotides showed that amino acids did selectively interact with specific bases,[38] although the interactions did not parallel the genetic code. Imidazole-activated amino acids esterify the 2'-OH groups of RNA homopolymers with some specificity.[39] However, since the two amino acids tested, phenylalanine and glycine, much preferred poly(U) over any other polynucleotide, the results do not support the authors' contention that this mechanism led to the present codon assignments.

The dissociation constants of AMP complexes with the methyl esters of amino acids also show selectivity, ranging about seven-fold from Trp (120 mM) to Ser (850 mM).[40] However, neither Trp (UGG) nor Ser (CUN, AGY) have particularly many or few A residues in their codons or anticodons, while the amino acids that do (Lys AAR, Phe UUY) have intermediate dissociation constants (320 and 196 mM respectively). These data did show a strong negative correlation between the association constant ($1/K_D$) and amino acid hydrophobicity. There are positive correlations between the dissociation constant and the number of codons assigned to the amino acid, and to frequency of the amino acid in proteins.[40] Condensation of dipeptides of the form Gly-X in the presence of AMP, CMP, poly(A) and poly(U) was mainly enhanced by the anticodonic nucleotides, where a pattern was apparent.[41] Different amino acids differ in their ability to stabilize poly(A)-poly(U) and poly(I)-poly(C) double helices, although the order is similar in each case and so cannot have contributed to the establishment of the genetic code. Finally, D-ribose adenosine biases esters with L-Phe but not D-Phe towards the 3'-OH (the pattern is reversed with L-ribose adenosine). Thus, single nucleotides moderately regio- and stereo- selectively aminoacylate themselves.[42]

Recent evidence also suggests that self-assembly of purine monolayers differentially affects adsorption of amino acids. The spacing between residues is consistent with peptide bond distances: such self-assembly might have formed a primordial code, although apparently one very different from the **modern** genetic code.[43-45]

Summary: Two comprehensive reviews of these and other data[46,47] suggested that if the genetic code were established by interactions between simple molecules (not more complicated than dipeptides or trinucleotides) and amino acids, then the greatest specific interaction was between amino acids and their anticodon nucleotides. However, individual experiments were equivocal or correlated with both anticodons and occasionally codons, so no strong direction is evident in the data.

The absence of obvious, strong or reproducible correlations from these highly varied approaches, considered alone or especially in sum, weakens the hypothesis that the code rests on the chemistry of trinucleotide-amino acid interactions. We suggest instead a later origin for the code, involving larger RNAs.

Adaptors and Adaptation

Perhaps the simplest explanation for the observed order in the genetic code[11,48-50] is that codon assignments were determined by stereochemical association between oligonucleotides and amino acids.[8-10,12] This mechanism would assign similar amino acids to similar codons because of intrinsic affinity, rather than as a result of natural selection among alternative codes. Although the resulting codon assignments might appear *adaptive*, in that they reduce various errors relative to other possible codes, they would not be an *adaptation*.

Stereochemical Pairing: Several such stereochemical schemes are conceivable. Thus, the primordial sequences with which pairing occurred can either be the actual codons, or some simple transform thereof.[9] As detailed in the above section entitled "Chemical Associations: A Historical Perspective", interactions have been proposed between amino acids and codons,[12] anticodons,[10,13] codons read 3'- 5' instead of 5'-3',[22,23] a complex of four nucleotides (C4N) formed by the three 5' nucleotides of tRNA with the fourth nucleotide from the 3' end,21 and a double-stranded complex of the codon and anticodon.[18,24]

A fundamental problem that all stereochemical models share is that codons and amino acids are never stereochemically linked in modern translation. Thus an implied evolutionary shift has occurred in which direct associations were lost, but their logic was nevertheless transmitted to the present. Such a conservative transition, required to make a stereochemical origin observable, is supported by a strong argument from continuity. The shift to indirect associations must occur in a translation apparatus that is making useful peptides (otherwise the translation apparatus itself could not have been selected). Thus the logic of the older direct interactions must be preserved or the altered translation apparatus will be of no use. After consideration of the evidence, we discuss this transition to indirect coding again.

The existence of adaptors, tRNAs and aminoacyl-tRNA synthetases, in the modern system allows codon assignments to be readily shuffled among amino acids.[51] Accordingly, adaptive evolution can erase primordial codon assignments. Thus we would only expect *some* amino acids to show codon/site associations, especially if others were added to the code later. Consequently, it is remarkable that any associations persist to the present.[52]

Amino Acid-Binding RNA: Most attention to sequence/binding site associations initially focused on arginine, since arginine binds specifically to two completely distinct classes of natural RNA molecules. The first class is the guanosine-binding site of self-splicing group I introns, which binds arginine as a competitive inhibitor. The guanidinium side-chain of arginine is similar in structure to the Watson-Crick face of G.[53] A conserved Arg codon confers this activity, and the binding site is almost invariably composed of several Arg codons in close juxtaposition.[54,55] The second class has been extensively studied because of potential medical importance: free arginine can mimic the natural interaction of HIV Tat peptides with TAR RNA.[56] In this case, however, no Arg codons are conserved at the binding site.[57]

Natural amino acid-binding RNAs are few; more significantly, they can provide only anecdotal evidence for codon/binding site interactions because they are almost certainly under strong selection for properties other than binding to the free amino acids. However, SELEX or selection-amplification, a technique for directed molecular evolution,[58-60] makes it possible to select those RNA molecules that perform a desired catalytic or binding function from large random pools (see ref. 61 for review). This technological advance makes it possible to find out whether RNA molecules that bind to particular amino acids share any characteristic motifs at their binding sites.

Aptamers have now been isolated from a variety of amino acids (Table 1), including hydrophobic amino acids such as valine,[62] phenylalanine/tyrosine,[63] isoleucine,[64] tyrosine,[65] leucine (I. Majerfeld and M. Yarus, unpublished data), and phenylalanine,[65a] and hydrophilic amino acids such as glutamine (G. Tocchini-Valentini, unpublished data) and citrulline, which is not normally found in proteins.[66]

Table 1. Natural and artificial amino acid-binding RNA

Amino Acid	Kd	Comments	Reference
Arg	400μM	Group I intron: naturally binds G	86
Arg	4mM	TAR: Naturally binds Tat peptide in HIV	56
Arg	1mM	3 families selected; no structures available	67
Arg	**4mM**	**Selected against GMP binding**	**68**
Arg	**2-4mM**	**Selected by salt elution to mimic TAR**	**70**
Arg	**60μM**	**Derived from citrulline binder by mutagenesis/reselection; NMR structure available**	**66**
Arg	**330nM**	**Intensive selection with heat-denaturation; only one sequence structurally characterized, though many selected**	**72**
Val	12mM	No structural data	62
Ile	**200-500μM**	**Only one family survived selection**	**64**
Phe/ Tyr	2-25mM	No structural data	63
Trp	18μM	Binds D-Trp-agarose, not free L-Trp; no structural data	87
Tyr	**35μM**	**Also binds Trp; evolved from L-DOPA binder**	**65**
Phe	**<1mM**	**Some clones bind only Phe-agarose**	**65a**
Leu	**~1mM**		**Majerfeld & Yarus, unpublished data**
Gln	**18-20mM**		**Mannironi et al, unpublished data**

Entries in **bold** are those with sufficient structural information to define binding site nucleotides, used to test for statistical association between binding sites and triplet motifs. Natural RNA sequences that bind arginine were excluded from the analysis, because they are probably under selection for other properties.

However, RNA aptamers for arginine are most abundant in the literature, and have been independently isolated in several different experiments.[66-73] Since structural information is available for many of these sequences, it becomes possible to ask whether particular sequences are overrepresented at recently selected binding sites, and, if so, whether these sequences have any relationship to the modern genetic code.

Statistical Evidence for Triplet/Binding Site Associations

The theory that the code arose by stereochemical means is both specific and unique; its predictions are explicit and different from other prevalent theories. Coevolution theories (that coding was extended along biosynthetic pathways[74]) are typically agnostic about which trinucleotide-amino acid pairing established the initial codon assignments, but predict that such pairings, if they exist at all, can account for only a small part of the codon catalog. Optimization theories (that coding minimizes errors in expression[75]) predict no correspondence at all between trinucleotides and amino acid binding sites.

Evolution of Binding Triplets: Assuming that original amino acid binding sites were RNA-like, they could have evolved into any of the components of modern translation: tRNA, rRNA, mRNA, or primitive aminoacyl-tRNA synthetases (subsequently replaced by protein enzymes). Depending on which modern translation component descended from ancient amino acid interactions, we predict different associations between coding nucleotides and amino acids. If binding sites evolved into tRNAs, for instance, the anticodons should be overrepresented in amino acid binding sites, whereas if they evolved into mRNA the codons should be overrepresented.[76]

The selection of RNA molecules (aptamers) that bind amino acid ligands has made such conjectures testable (Table 1). Because in vitro selection searches a large space of possible sequences for optimal or near-optimal "solutions" to particular binding problems, such directed evolution might be able to recapitulate primordial interactions between amino acids and short RNA sequences. If amino acids interact favorably with coding RNA sequences, this relation might be observed, or even proven. Since aptamers can be selected for each amino acid, and since the specific nucleotides important to binding can be determined, standard statistical tests for association (such as the χ^2 or G tests) will reveal any consistent relation between binding-site nucleotides and nucleotides in coding sequences.[77]

Such a search for motifs faces predictable difficulties. RNA is more versatile than might have once been thought, and many oligomers often bind an amino acid. The diversity of RNAs that bind arginine, for example, shows that efforts to emulate a unique primordial RNA for each amino acid would be futile.[57] Recurrence of specific sequence motifs in amino acid aptamers, such as codons or anticodons, cannot prove that similar interactions led to the establishment of present codon assignments. However, suppose that coding sequences embody such general interactions that they will still be detectable in the most probable modern binding sites. Proof of *any* specific pairings at all would show that the specificity existed to originate a genetic code. If specific pairings detected with in vitro selection actually match present codon assignments, then similar processes in ancient translation are supported. If there are frequent, strong associations between present codons or anticodons and amino acids, their involvement in the origin of the code is the only plausible explanation.

Binding Site Preferences: That any codon/binding site associations could survive to the present has been questioned.[78] However, the association between arginine and its binding sites is exceptionally strong, and has proven remarkably robust to statistical methodology, choice of binding sites, and choice of sequences from selected pools.[52,76-78] In particular, arginine binding sites show strong associations with arginine codons (Table 2), but not anticodons (Table 3), codon or anticodon sets for other amino acids, other groups of 4+2 codons incorporating a family box plus a doublet, or other short motifs. The relationship remains highly significant even under many modifications: sequences where the binding site overlaps the constant regions can be excluded, the binding sites can be corrected for biased base composition and alternative sequences can be chosen from reported pools, without altering the conclusion.

Arginine may be unique: it acts as a nucleotide mimic,[53] perhaps more so than other amino acids. However, significant associations between Tyr aptamer binding sites and codons have been reported,[52] and Ile aptamers contain conserved Ile codons at their active sites.[64] Data from several other amino acids have become available, allowing a more general test of generality for the association between binding sites and *codons*. We now extend the analysis to all available amino acids (Table 1) and reassess hypotheses about specific associations.

Testing Triplet/Site Associations: Codons occur more often in binding sites than expected for each of the six amino acids for which data are available, an improbable outcome itself ($P = (0.5)^6 = 0.016$). Individually, the arginine aptamers showed a significant codon/site association only. Tyrosine and isoleucine aptamers showed significant associations between both codons and anticodons: except for the association between tyrosine and its codons, these relationships persist even when corrected for six multiple comparisons ($P < 0.01$). Glutamine, leucine and phenylalanine have no significant tendency to locate codons or anticodons in their binding sites (when corrected for multiple comparisons). The most sensitive tests combine all data; then we observe highly significant associations overall with *both* codons *and* anticodons, even when the single most influential amino acid is excluded from the analysis ($P < 10^{-6}$ in all cases). Thus there is reason to believe that codons and anticodons are associated with binding sites, and this conclusion does not depend on any one selection or set of binding sites.

Table 2. **Tests for association between amino acid binding sites and their cognate codons**

Codons	Arg	Tyr	Ile	Gln	Phe	Leu
Ter	0.05	1.28	-5.02	-4.19	*15.86*	*2.65*
Ala	0.09	-16.95	-11.97	-11.57	-18.51	-0.38
Cys	-16.97	-0.66	-8.42	-3.32	-4.79	0.04
Asp	0.15	3.96	-3.45	2.89	-1.82	-1.08
Glu	3.44	-3.17	-1.79	-0.01	*6.81*	*1.47*
Phe	-3.38	-2.38	-8.42	1.26	**3.73**	-2.00
Gly	0.35	0.25	*31.57*	*8.94*	2.25	0.00
His	-1.04	-1.87	-6.14	-0.02	-3.69	-0.68
Ile	2.86	*9.18*	**10.35**	*3.43*	0.01	-4.60
Lys	1.34	-14.86	0.00	1.74	1.39	0.62
Leu	-19.92	-4.16	8.14	-10.60	-7.57	**0.83**
Met	-5.60	3.06	0.00	-0.02	-0.15	-1.35
Asn	5.46	-0.04	-1.79	*3.25*	1.04	0.01
Pro	0.00	-2.30	-11.17	-9.55	-8.26	-0.15
Gln	0.27	2.30	-2.85	**2.98**	2.00	0.62
Arg	**29.11**	0.24	-25.10	1.66	0.17	-0.78
Ser	-6.07	-4.95	-15.73	-7.54	-11.32	*5.65*
Thr	-0.10	0.57	-16.54	1.94	*-7.32*	*2.61*
Val	-0.13	4.45	-0.04	-0.38	*5.53*	*2.82*
Trp	-7.26	0.04	*42.58*	-1.14	-2.52	0.28
Tyr	-3.38	**6.69**	*10.90*	-0.33	0.03	-0.12
Rank	**1**	**2**	**4**	**4**	**4**	**6**

Rows: codon sets for each amino acid; columns: amino acids for which aptamers with known structures have been reported. **Bold** values indicate the cognate codon sets for each amino acid aptamer; values in *italics* indicate codon sets with at least as strong an association as the actual codon set. Tabulated numbers are G values for association between codons and binding sites, with the Williams correction;[88] negative values indicate codon sets that are found less frequently at binding sites than would be expected by chance. 'Rank' indicates the rank order of the cognate amino acid's codon set. Binding sites for this table and all others are taken from ref. 52 where applicable (Arg, Ile, Tyr), or otherwise from personal communications from the specific aptamer laboratories. See ref. 76 for discussion of the effects of different choices of binding site.

On the other hand, controls show that this method can rule out certain possibilities. There was no significant association for any amino acid, or for the set as a whole, with the codons reversed 3' to 5', indicating that this hypothesis can be clearly rejected.

It is possible that the 21 codon (or anticodon) sets are an unfair comparison class, since they range in size from 1 to 6 codons. A less precise, but perhaps more robust, test is to see whether there is a significant association between the amino acid binding sites and the codon (or anticodon) that contains the cognate doublet: this reflects the intuitively plausible idea that the primitive code may have assigned amino acids only to family boxes. However, doublet analysis (Table 4) does not greatly change the outcome. Significant associations are observed for both doublets and codons/anticodons. Thus, again, the results to date suggest both associations between codons and anticodons.

Table 3. ***Test for association between binding sites and the cognate codons, anticodons, and codons reversed 3' to 5'***

Codons	n	+b+c	+b-c	-b+c	-b-c	G	P
Arg	5	36	16	38	106	29.1	*3.4E-08*
Tyr	3	12	71	9	179	6.7	4.8E-03
Ile	5	15	25	30	181	10.4	*6.5E-04*
Gln	3	6	36	6	108	3.0	4.2E-02
Leu	2	16	46	19	78	0.8	1.8E-01
Phe	8	11	74	35	504	3.7	2.7E-02
Total	**26**	**96**	**268**	**137**	**1156**	**51.6**	***3.5E-13***
Total—Arg	**21**	**60**	**252**	**99**	**1050**	**25.1**	***2.7E-07***

Anticodons	# seq	+b+c	+b-c	-b+c	-b-c	G	P
Arg	5	20	32	37	107	2.9	4.5E-02
Tyr	3	18	65	6	182	21.7	1.6E-06
Ile	5	16	24	23	188	17.1	*1.7E-05*
Gln	3	1	41	17	97	-5.9	9.9E-01
Leu	2	27	35	23	74	6.7	4.7E-03
Phe	8	12	73	40	499	3.7	2.8E-02
Total	**26**	**94**	**270**	**146**	**1147**	**43.1**	***2.6E-11***
Total—Tyr	**21**	**74**	**238**	**109**	**1040**	**39.6**	***1.6E-10***

Rev. Codons	# seq	+b+c	+b-c	-b+c	-b-c	G	P
Arg	5	16	36	42	102	0.05	8.3E-01
Tyr	3	3	80	6	182	0.03	8.6E-01
Ile	1	10	30	25	186	4.10	4.3E-02
Gln	3	7	35	11	103	1.34	2.5E-01
Leu	2	12	50	29	68	-2.22	1.4E-01
Phe	8	2	83	43	496	-4.33	3.7E-02
Total	**22**	**50**	**314**	**156**	**1137**	**0.71**	**4.0E-01**

Column headings: n, number of sequences; +b+c, number of bases both in codons and in binding sites; +b-c, number of bases in binding sites but not in codons; -b+c, number of bases in codons but not in binding sites; -b-c, number of bases neither in codons nor in binding sites; G, the G test for association in a 2 x 2 table, with the Williams correction; P, 1-tailed test for independence with 1 degree of freedom. Values in italics are significant to $P < 0.01$ after correcting for 6 comparisons. There are significant associations between some amino acid binding sites and both codons and anticodons, even when the single most significant association is removed. However, there is no association at all between amino acid binding sites and the reversed codons.

Table 4. Test for association between binding sites and codon doublets (XYN) or anticodon doublets (NY'X'), where X and Y are specified and N is any base

Codon Doublets	# seq	+b+c	+b-c	-b+c	-b-c	G	P
Arg	5	24	28	24	120	16.4	*2.5E-05*
Tyr	3	22	61	20	168	10.2	*7.1E-04*
Ile	5	15	25	30	181	10.4	*6.5E-04*
Gln	3	9	33	15	99	1.5	1.1E-01
Leu	2	7	55	21	76	-2.9	9.6E-01
Phe	8	17	68	96	443	0.2	3.2E-01
Total	**26**	**94**	**270**	**206**	**1087**	**17.5**	***1.4E-05***
Total—Arg	**21**	**70**	**242**	**182**	**967**	**7.1**	**3.9E-03**

Anticodon Doublets	# seq	+b+c	+b-c	-b+c	-b-c	G	P
Arg	5	11	41	27	117	0.1	3.6E-01
Tyr	3	23	60	19	169	12.5	*2.1E-04*
Ile	5	8	32	45	166	0.0	5.7E-01
Gln	3	5	37	46	68	-12.6	1.0E+00
Leu	2	22	40	16	81	7.2	3.6E-03
Phe	8	27	58	72	467	15.6	*3.8E-05*
Total	**26**	**96**	**268**	**225**	**1068**	**13.8**	***1.0E-04***
Total—Phe	18	85	227	198	951	14.8	*6.1E-05*

For example, the codon doublet for Phe is UUN within a binding site, and the anticodon doublet is NAA within a site. Again, the specific associations hold for both codons and anticodons overall, although few of the results are individually significant. Italics indicate significant values after correction for 6 comparisons.

We can carry these conclusions a step further by freeing them of the assumptions required even for standard statistical tests. If there is an association between the triplets found at amino acid binding sites and the modern genetic code, it should be found only with the actual genetic code and not with randomized versions of it. Accordingly, we generated many alternative codes, and tested for codon/binding site associations. This preserves important aspects of the experimental results, such as the spatial correlations within binding sites (they occur in specific sections of the molecule), and the influence of the occurrence of each triplet on the probability of finding others. In order to eliminate stependence on any particular method for generating variant codes, we used several quite different permutation methods.

An ISO C program randomized the code according to the following schemes:

1. Codon permutation: a codon can randomly and independently take on any identity (including its real one). This keeps the number of codons per amino acid constant, but usually completely disrupts the fine structure of the code (such as wobble relations). This potentially generates $64! = 1.2 \times 10^{89}$ possible codes.

2. Amino acid permutation: any amino acid can randomly and independently take any existing coding block(s), including those of stop codons. This preserves the structure of the code entirely (the number and size of blocks for codons are preserved, and their relative positions are preserved within the coding table), but amino acids can be given different numbers of codons. At one extreme, Arg, which normally has 6 codons split into a 4-block and a 2-block, might end up with Trp's single codon. This potentially generates $21! = 5.1 \times 10^{19}$ possible codes.

3. Codon block permutation. Keeping the structure of the code constant, we randomly assorted amino acid identities among groups of codons of the same size. For example, the CGN block assigned to Arg might be swapped with the CCN block normally assigned to Pro, but could not swap with the single UGA codon assigned to Trp. Treating the three Ile codons as a 2-block and a 1-block, this leads to $8! \times 14! \times 4! = 8.4 \times 10^{16}$ codes with 8 4-blocks, 14 2-blocks, and 4 1-blocks. This "n-block" scheme completely preserves the degeneracy of the code, and also conserves the number of codons assigned to each amino acid. Compared to the other randomization schemes, amino acids are far more likely to retain some of their actual codons.

5. Base identity permutation: in addition to the block permutation of method 3, this method randomizes the meaning of the first and second position base . This partially disrupts the code's structure (so that, for example, the UGN codon block need not be split into blocks of 2, 1, and 1), but preserves the degeneracy across a row and down a column. This multiplies the number of codes from method 3 by a factor of $(4! \times 4!)/2$ for a total of 2.4×10^{19} codes, and dramatically reduces retention of fragments of the present code.

6. Codon doublet permutation: like method 4, except that any codon doublet independently takes on the meaning of any other codon doublet. This leads to $16!/(8! \times 6! \times 2!) = 360360$ times as many codes as method 3, for a total of 3.0×10^{22} possible codes. Both this and method 4 preserve the number of codons assigned to each amino acid and their block structure (e.g., Arg will always have a 4-block and a 2-block), but this method does not preserve the relation between blocks of particular sizes as does method 4.

We generated 10 million randomized codes for each of the 5 schemes listed above, and compared codon/site associations in observed amino acid binding sites with those found in the actual code (Fig. 1). The "n-block" model (#3) is uniquely right-skewed, because some of the codons can only swap with a few partners under this model (e.g., there are only 4 blocks containing one codon) so that some of the present structure of the code will often be preserved. Even under this highly constrained model, however, only 0.8% of randomized codes give apparent associations between codons and binding sites better than the actual code. For the other, more completely scrambled models, between 0.11% (method 2) and 0.04% (methods 4 and 5) of all random codes do better than the actual code. Said another way, real codons are more associated with real binding sites than in 99.2 to 99.96% of all randomized codes, even though randomized codes include fragments of the actual code. Using Fisher's method for independent probabilities rather than performing a G test on the summed counts gave similar results (data not shown). Thus, our result is general and not sensitive to choice of alternative codes or sensitive to statistical methodology. It is highly unlikely that we would see as significant an association between codons and binding sites for a genetic code picked at random as that actually seen with the real code. Randomization of anticodon assignments gives similar results, but slightly less significant than for codons. Randomized anticodons are less associated with binding sites than real ones in 99.2 to 99.5% of all codes. This small difference in significance appears also in the statistical tests (Table 3).

These controls argue strongly that the most probable modern RNA-amino acid binding sites capture something of the essential nature of the code. In particular, a stereochemical process involving macromolecular RNA-like binding sites containing codons, and perhaps anticodons, gave rise to the present genetic code. Considering individual amino acids, primordial RNA-like binding sites were probably relevant to the assignment of codons for at least three of six amino acids for which we have data.

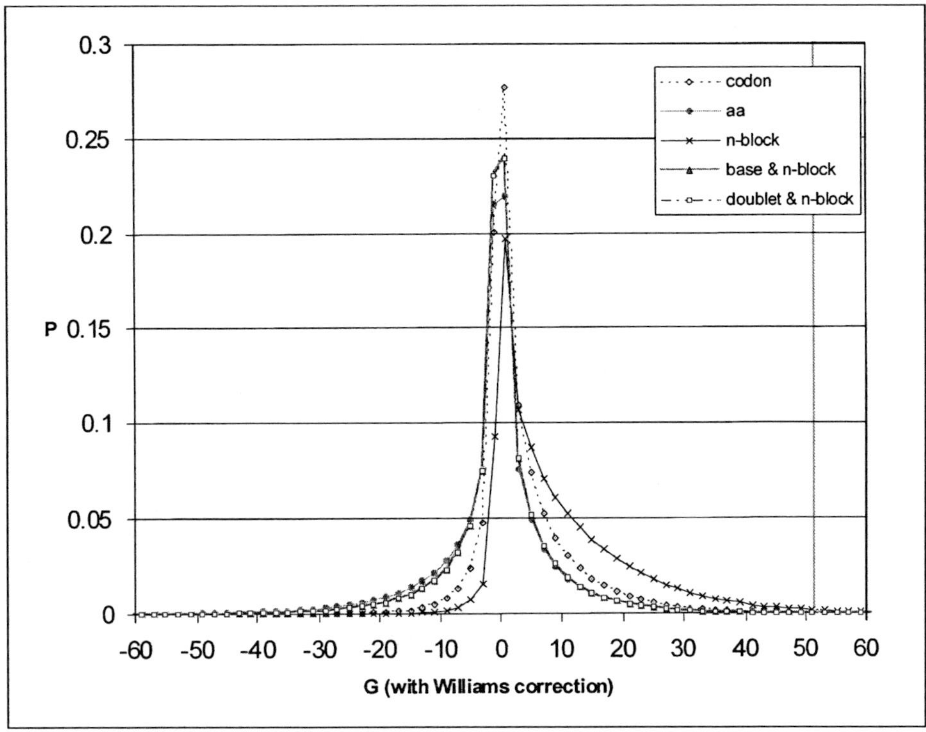

Figure 1. Distribution of likelihood for randomized genetic codes. The lines correspond to the different models for random codes described in the section on "statistical evidence for triplet/binding site associations." The gray line at the right (G = 51.5) gives the position of the actual code: very few randomized codes give a higher association between 'codons' and binding sites, making it highly unlikely that the observed association for the real code is due to chance. The "n-block" line (x) is skewed strongly to the right, because some codons can occupy relatively few blocks under this model. Thus n-block randomization preserves many similarities to the real code.

Concluding Remarks

We now return to the direct to indirect coding transition implied by every stereochemical model. RNA-amino acid binding sites contain sequences likely to be relevant to the appearance of the code. Thus the logically predicted transition from direct to indirect coding rests first on the ability of coding sequences to serve as structural elements in amino acid binding sites, and then to subsequently serve in normal base pairing. Triplets that became codons might begin as essential elements in binding sites (indirect coding), and later pair with primordial tRNAs (direct coding). Triplets that became anticodons might begin within binding sites (indirect), then employ their more well-known base-pairing activity when they begin to act as anticodons (direct coding). The conservative logic of the direct to indirect transition, required by argument from continuity, is implicit as soon as it is known that nucleotide triplets can be essential elements of amino acid binding sites (compare the DRT theory[57]).

Descendants of the original amino acid-binding sites could play four possible roles: as tRNAs, mRNAs, ribosomes, or aminoacyl-tRNA synthetases. All these activities are known to be possible activities for RNA,[79-85] because they exist in modern selected parallels. With present data, it appears that arginine may have been bound in primordial sites containing sequences that became codons in mRNA. We found no strong evidence for association between glutamine, leucine and phenylalanine and their coding sequences. These are negative results based on limited data; however, these codons may have been assigned by other means during later code evolution. Tyrosine and isoleucine present a case we had not anticipated, in which both codons and anticodons are overrepresented (though not because they are paired in the molecules). We cannot confidently specify the descent of the coding sequences for these amino acids. Their binding sequences could have become both tRNA-like and mRNA-like molecules, or these data may be the first indication of the need for a new, more comprehensive theory.

Ideally, with a large sample of independently derived families of aptamers that bind each of the amino acids, it should be possible to test associations between binding sites and individual trinucleotides. If there are, as now appears, to be several classes of amino acids with different relations to coding sequences, such high resolution may be required. It is possible that high-throughput techniques for aptamer isolation will achieve this in the future, but, for the moment, isolating aptamers and determining binding sites is a time-consuming process. Consequently, it may be several years before site/triplet associations are maximally resolved.

However, it is clearly not true that each aptamer binds its target amino acid using only the cognate codons. Amino acid binding sites always require other nucleotides for their construction. Where structures are known, the coding sequences can be in contact with the amino acid or providing less central support for the site—in some cases they are in both places.[52] The fact that binding sites with detectable affinities are far more complex than single trinucleotides strongly suggests that the code probably began in an RNA world, after complex RNA molecules were prevalent. Assuming that the RNA world biota were our immediate antecedents, translation was also probably devised in the RNA world.[89] An economical interpretation is therefore that coding assignments arose predominantly during initial selection for templated peptide synthesis, rather than via other activities.

These techniques have substantial potential for further analysis. It may be possible to discover why some amino acids have the actual codon assignments they do, and perhaps why some amino acids were incorporated into the code while others, available on the early earth or as metabolic intermediates, were excluded. Furthermore, with complete data in hand it may be possible to define a minimal, stereochemically determined code, and therefore to estimate the relative roles of chemistry and selection in shaping modern codon assignments.

References

1. Szathmáry E. Coding coenzyme handles: A hypothesis for the origin of the genetic code. Proc Natl Acad Sci USA 1993; 90:9916-9920.
2. Porschke D. Differential effect of amino acid residues on the stability of double helices formed from polyribonucleotides and its possible relation to the evolution of the genetic code. J Mol Evol 1985; 21:192-198.
3. Maizels N, Weiner AM. Peptide-specific ribosomes, genomic tags, and the origin of the genetic code. Cold Spring Harb Symp Quant Biol 1987; LII:743-749.
4. Maizels N, Weiner AM. The genomic tag hypothesis: modern viruses as molecular fossils of ancient strategies for genomic replication. In: Gesteland RF and Atkins JF, eds. The RNA world. New York: Cold Spring Harbor Laboratory Press, 1993:577-602.
5. Gamow G. Possible mathematical relation between deoxyribonucleic acid and protein. Kgl Dansk Videnskab Selskab Biol Medd 1954; 22:1-13.
6. Woese CR. The genetic code: the molecular basis for genetic expression. New York: Harper & Row, 1967.
7. Ycas M. The biological code. In: Neuberger A, Tatum EL, eds. North-Holland Research Monographs: Frontiers of Biology. Vol. 12. Amsterdam: North-Holland Publishing Company, 1969.
8. Woese CR, Dugre DH, Dugre SA et al. On the fundamental nature and evolution of the genetic code. Cold Spring Harb Symp Quant Biol 1966; 31:723-736.
9. Woese CR, Dugre DH, Saxinger WC et al. The molecular basis for the genetic code. Proc Natl Acad Sci USA, 1966; 55:966-974.
10. Dunnill P. Triplet nucleotide-amino acid pairing: a stereochemical basis for the division between protein and nonprotein amino acids. Nature 1966; 210:1267-1268.
11. Pelc SR. Correlation between coding triplets and amino acids. Nature 1965; 207:597-599.
12. Pelc SR, Welton MGE. Stereochemical relationship between coding triplets and amino-acids. Nature 1966; 209:868-872.
13. Ralph RK. A suggestion on the origin of the genetic code. Biochem Biophys Res Comm 1968; 33:213-218.
14. Lacey Jr JC, Pruitt KM. Origin of the genetic code. Nature 1969; 223:799-804.
15. Rendell MS, Harlos JP, Rein R. Specificity in the genetic code: the role of nucleotide base-amino acid interaction. Biopolymers 1971; 10:2083-2094.
16. Melcher G. Stereospecificity of the genetic code. J Mol Evol 1974; 3:121-140.
17. Nelsesteuen GL. Amino acid-directed nucleic acid synthesis. J Mol Evol 1978; 11:109-120.

18. Hendry LB, Whitham FH. Stereochemical recognition in nucleic acid-amino acid interactions and its implications in biological coding: a model approach. Perspect Biol Med 1979; 22:333-345.
19. Hendry LB, Bransome Jr ED, Hutson MS et al. First approximation of a stereochemical rationale for the genetic code based on the topography and physichemical properties of "cavities" constructed from models of DNA. Proc Natl Acad Sci USA 1981; 78:7440-7444.
20. Balasubramanian R. Origin of life: a hypothesis for the origin of adaptor-mediated ordered synthesis of proteins and an explanation for the choice of terminating codons in the genetic code. Bio Systems 1982; 15:99-104.
21. Shimizu M. Molecular basis for the genetic code. J Mol Evol 1982; 18:297-303.
22. Root-Bernstein RS. Amino acid pairing. J theor Biol 1982; 94:885-894.
23. Root-Bernstein RS. On the origin of the genetic code. J theor Biol 1982; 94:895-904.
24. Alberti S. The origin of the genetic code and protein synthesis. J Mol Evol 1997; 45:352-358.
25. Crick FHC. An error in model building. Nature 1967; 213:798.
26. Mellersh A. A model for the prebiotic synthesis of peptides and the genetic code. Orig Life Evol Biosph 1993; 23:261-274.
27. Crick FHC. The origin of the genetic code. J Mol Biol 1968; 38:367-379.
28. Knight RD, Freeland SJ, Landweber LF. Selection, history and chemistry: the three faces of the genetic code. Trends Biochem Sci 1999; 24:241-7.
29. Woese CR. Evolution of the genetic code. Naturwissenschaften 1973; 60:447-59.
30. Nagyvary J, Fendler JH. Origin of the genetic code: a physical-chemical model of primitive codon assignments. Orig Life 1974; 5:357-362.
31. Miller SL. Which organic compounds could have occurred on the prebiotic earth? Cold Spring Harb Symp Quant Biol 1987; LII:17-27.
32. Fendler JH, Nome F, Nagyvary J. compartmentalization of amino acids in surfactant aggregates. J Mol Evol 1975; 6:215-232.
33. Weber AL, Lacey Jr JC. Genetic code correlations: amino acids and their anticodon nucleotides. J Mol Evol 1978; 11:199-210.
34. Jungck JR. The genetic code as a periodic table. J Mol Evol 1978; 11:211-224.
35. Lehmann U. Chromatographic separation as selection process for prebiotic evolution and the origin of the genetic code. Bio Systems 1985; 17:193-208.
36. Saxinger C, Ponnamperuma C. Experimental investigation on the origin of the genetic code. J Mol Evol 1971; 1:63-73.
37. Raszka M, Mandel M. Is there a physical chemical basis for the present genetic code? J Mol Evol 1972; 2:38-43.
38. Saxinger C, Ponnamperuma C. Interactions between amino acids and nucleotides in the prebiotic milieu. Orig Life 1974; 5:189-200.
39. Lacey Jr JC, Weber AL, White Jr WE. A model for the coevolution of the genetic code and the process of protein synthesis: review and assessment. Orig Life 1975; 6:273-283.
40. Reuben J, Polk FE. Nucleotide-amino acid interactions and their relation to the genetic code. J Mol Evol 1980; 15:103-112.
41. Podder SK, Basu HS. Specificity of protein-nucleic acid interaction and the biochemcial evolution. Orig Life 1984; 14:477-484.
42. Lacey Jr JC, Wickramasinghe NSMD, Cook GW et al. Couplings of character and of chirality in the origin of the genetic system. J Mol Evol 1993; 37:233-239.
43. Sowerby SJ, Cohn CA, Heckl WM et al. Differential adsorption of nucleic acid bases: relevance to the origin of life. Proc Natl Acad Sci USA 2001; 98:820-822.
44. Sowerby SJ, Heckl WM. The role of self-assembled monolayers of the purine and pyrimidine bases in the emergence of life. Orig Life Evol Biosph 1998; 28:283-310.
45. Sowerby SJ, Stockwell PA, Heckl WM et al. Self-programmable, self-assembling two-dimensional genetic matter. Orig Life Evol Biosph 2000; 30:81-99.
46. Lacey Jr JC, Mullins Jr DW. Experimental studies related to the origin of the genetic code and the process of protein synthesis-a review. Orig Life 1983; 13:3-42.
47. Lacey Jr JC. Experimental studies on the origin of the genetic code and the process of protein synthesis: a review update. Orig Life Evol Biosph 1992; 22:243-275.
48. Epstein CJ. Role of the amino-acid 'code' and of selection for conformation in the evolution of proteins. Nature 1966; 210:25-28.
49. Volkenstein MV. Coding of polar and non-polar amino acids. Nature, 1965; 207:294-295.
50. Woese CR. Order in the genetic code. Proc Natl Acad Sci USA 1965; 54:71-75.
51. Saks ME, Sampson JR, Abelson J. Evolution of a transfer RNA gene through a point mutation in the anticodon. Science 1998; 279:1665-1670.
52. Yarus M. RNA-ligand chemistry: a testable source for the genetic code. RNA 2000; 6:475-484.
53. Yarus M. A specific amino acid binding site composed of RNA. Science 1988; 240:1751-1758.
54. Yarus M. Specificity of arginine binding by the Tetrahymena intron. Biochemistry 1989; 28:980-988.
55. Yarus M. An RNA-amino acid complex and the origin of the genetic code. New Biologist 1991; 3:183-189.
56. Tao J, Frankel AD. Specific binding of arginine to TAR RNA. Proc Natl Acad Sci USA 1992; 89:2723-2726.
57. Yarus M. Amino acids as RNA Ligands: a Direct-RNA-Template Theory for the Code's Origin. J Mol Evol 1998; 47:109-117.

58. Ellington AD, Szostak JW. In vitro selection of RNA molecules that bind specific ligands. Nature 1990; 346:818-822.
59. Tuerk C, Gold L. Systematic evolution of ligands by exponential enrichment: RNA ligands to bacteriophage T4 DNA polymerase. Science 1990; 249:505-510.
60. Robertson DL, Joyce GF. Selection in vitro of an RNA enzyme that specifically cleaves single- stranded DNA. Nature 1990; 344:467-468.
61. Ciesiolka J, Illangasekare M, Majerfeld I et al. Affinity selection-amplification from randomized ribooligonucleotide pools. Meth Enzymol 1996; 267:315-335.
62. Majerfeld I, Yarus M. An RNA pocket for an aliphatic hydrophobe. Nat Struct Biol 1994; 1:287-292.
63. Zinnen S, Yarus M. An RNA pocket for the planar aromatic side chains of phenylalanine and tryptophane. Nucl Acid Symp Ser 1995; 33:148-151.
64. Majerfeld I, Yarus M. Isoleucine:RNA sites with essential coding sequences. RNA 1998; 4:471-478.
65. Mannironi C, Scerch C, Fruscoloni P et al. Molecular recognition of amino acids by RNA aptamers: the evolution into an L-tyrosine binder of a dopamine-binding RNA motif. RNA 2000; 6:520-527.
65a. Illangasekare M, Yarus M. Phenylalanine-binding RNAs and genetic code evolution. J Mol Evol 2002; 54:298-311.
66. Famulok M. Molecular recognition of amino acids by RNA-aptamers: an L-citrulline binding RNA motif and its evolution into an L-arginine binder. J Am Chem Soc 1994; 116:1698-1706.
67. Connell GJ, Illangsekare M, Yarus M. Three small ribooligonucleotides with specific arginine sites. Biochemistry 1993; 32:5497-5502.
68. Connell GJ, Yarus M. RNAs with dual specificity and dual RNAs with similar specificity. Science 1994; 264:1137-1141.
69. Yarus M. An RNA-amino acid affinity, in The RNA World. Gesteland RF, Atkins JF, eds. New York: Cold Spring Harbor Laboratory Press, 1993:205-217.
70. Tao J, Frankel AD. Arginine-binding RNAs resembling TAR identified by in vitro selection. Biochemistry 1996; 35:2229-2238.
71. Burgstaller P, Kochoyan M, Famulok M. Structural probing and damage selection of citrulline- and arginine-specific RNA aptamers identify base positions required for binding. Nucl Acid Res 1995; 23:4769-4776.
72. Geiger A, Burgstaller P, von der Eltz H et al. RNA aptamers that bind L-arginine with sub-micromolar dissociation constants and high enantioselectivity. Nucl Acid Res 1996; 24:1029-1036.
73. Yang Y, Kochoyan M, Burgstaller P et al. Structural basis of ligand discrimination by two related RNA aptamers resolved by NMR spectroscopy. Science 1996; 272:1343-1346.
74. Wong JT-F. A co-evolution theory of the genetic code. Proc Natl Acad Sci USA 1975; 72:1909-1912.
75. Sonneborn TM. Degeneracy of the genetic code: extent, nature, and genetic implications. In: Bryson V and Vogel HJ, eds. Evolving Genes and Proteins. New York: Academic Press, 1965:377-297.
76. Knight RD, Landweber LF. Guilt by association: the arginine case revisited. RNA, 2000; 6:499-510.
77. Knight RD, Landweber LF. Rhyme or reason: RNA-arginine interactions and the genetic code. Chem Biol 1998; 5:R215-R220.
78. Ellington AD, Khrapov M, Shaw CA. The scene of a frozen accident. RNA 2000; 6:485-498.
79. Illangasekare M, Sanchez G, Nickles T et al. Aminoacyl-RNA synthesis catalyzed by an RNA. Science 1995; 267:643-647.
80. Illangasekare M, Yarus M. Specific, rapid synthesis of Phe-RNA by RNA. Proc Natl Acad Sci U S A 1999; 96:5470-5475.
81. Illangasekare M, Yarus M. A tiny RNA that catalyzes both aminoacyl-RNA and peptidyl-RNA synthesis. RNA 1999; 5:1482-1489.
82. Welch M, Majerfeld I, Yarus M. 23S rRNA similarity from selection for peptidyl transferase mimicry. Biochemistry 1997; 36:6614-6623.
83. Nissen P, Hansen J, Ban N et al. The structural basis of ribosome activity in peptide bond synthesis. Science 2000; 289:920-930.
84. Yarus M, Welch M. Peptidyl transferase: ancient and exiguous. Chem Biol 2000; 7:R187-R190.
85. Kumar RK, Yarus M. RNA-catalyzed amino acid activation. Biochemistry 2001; 40:6998-7004.
86. Yarus M, Majerfield I. Co-optimization of ribozyme substrate stacking and L-arginine binding. J Mol Biol 1992; 225:945-949.
87. Famulok M, Szostak JW. Stereospecific recognition of tryptophan agarose by in vitro selected RNA. J Am Chem Soc 1992; 114:3990-3991.
88. Sokal RR, Rohlf FJ, Biometry: The Principles and Practice of Statistics in Biological Research. 3rd ed. New York: W. H. Freeman and Company 1995.
89. Yarus, M. On translation by RNAs alone. Cold Spring Harb Symp Quant Biol 2001; 66:207-215.

CHAPTER 8

Mitochondrial tRNA Aminoacylation and Human Diseases

Catherine Florentz and Marie Sissler

Abstract

The human mitochondrial (mt) genome encodes for only 13 proteins which are all subunits of transmembranar respiratory chain complexes. These complexes contribute to a major mt functions namely the synthesis of energy in the way of ATP. Translation of the mRNAs is performed by a set of 22 tRNAs, also encoded by the mt genome, and aminoacylated by nuclear encoded aminoacyl-tRNA synthetases imported into the mitochondria. More and more point mutations affecting mt tRNA genes are reported as correlated to severe neurodegenerative disorders. Since these mutations generally lead to decreased mt protein synthesis, understanding the genotype/phenotype relationships is primarily based on investigation of the possible impacts of individual mutations on various aspects of tRNA structural and functional properties. Here, the present knowledge on human mt aminoacylation systems, as well as the strategies developed to investigate mt aminoacylation, and the effects of point mutations in tRNAs on this process are reviewed. The diversity in the effects observed so far, for a same mutation as well as for various mutations, highlights the ongoing technical limitations in studying human mt aminoacylation. They also suggest that aminoacylation may be a focus for therapeutic strategies for some mutations, while the impact of other mutations needs to be searched as well at other levels of the tRNA structure/ function relationship as at unforeseen levels in mitochondria.

Introduction

Mitochondria are the center of a number of metabolic pathways including oxidative phosphorylation and ATP synthesis, fatty acid degradation, tricarboxylic acid cycle, urea cycle and others.[1] To fulfill these numerous tasks, mitochondria mainly import several hundreds of nuclear encoded proteins but possess also a small genome, a circular DNA of 16569 bp, which allows for synthesis of further 13 proteins. These proteins are subunits of transmembranar respiratory chain complexes required for the oxidative phosphorylation pathway, and are complemented by about 70 nuclear encoded proteins to fulfill activity. The mitochondrial (mt) translation machinery is of dual origin with a set of 22 transfer RNAs (tRNAs) and two ribosomal RNAs (rRNAs) encoded by the mt DNA. All other factors are encoded by the nuclear genome and imported to mitochondria from the cytosol. These include at least aminoacyl-tRNA synthetases (aaRSs), the enzymes which charge tRNAs with their cognate amino acid, translation initiation-, elongation- and termination factors, as well as the numerous ribosomal proteins. So far, detailed knowledge on human, or more generally mammalian mt translation machineries, remains scarce. For example, the sequences of the 22 tRNAs and two rRNAs are known but only 5 tRNAs have been sequenced at the RNA level, their secondary structures and especially their tertiary structures remain incompletely understood. Only a few aaRS genes have been cloned, few of the synthetases have been purified by classical biochemical means from tissues.[2] Ribosomes are under investigation with progressive assignment of proteins.[3] They are smaller (55S, with a large 39S subunit and a small 28S subunit) than cytosolic or prokaryotic ribosomes, and have a higher ratio of proteins over RNA. A single initiation factor and several elongation factors have been purified and characterized.[4] Additional proteins, allowing for example the

Translation Mechanisms, edited by Jacques Lapointe and Léa Brakier-Gingras.
©2003 Eurekah.com and Kluwer Academic / Plenum Publishers.

integration of the newly synthesized proteins into the hydrophobic mt membranes, as is the case in yeast mitochondria,[5] still need to be found.

The limited knowledge on human mt translation is a drawback at a time where a number of human disorders become correlated to mutations in the mt genome. Within the 115 mutations reported since 1989, some occur in the protein coding regions (38%), a few have been detected in the rRNA sequences (4%), but most are present in the tRNA genes (58%).[6-10] The molecular mechanisms underlying the diseases remain mostly unknown. Due to a general negative effect (although not systematic) of tRNA mutations on mt translation, different aspects of tRNA structure/function relationships have been analyzed. Aminoacylation of tRNAs by aaRSs being the most specific and central step in translation, and likely the most sensitive to tRNA alterations, it has been the main focus of initial investigations.

Here, the still emerging fundamental knowledge on human mt aminoacylation systems is first summarized. The growing number of point mutations in tRNA genes as well as the variety of human disorders to which they are correlated will then highlight the complexity of the task towards understanding the genotype/phenotype relationships. After reviewing the technical strategies adapted to investigate aminoacylation of human mt tRNAs, specific cases analyzed will be reported. While aminoacylation was indeed found dramatically affected for some mutations, this is not the case for others. Most of the mutations however, have not been analyzed, leaving open a large field of research.

Aminoacyl-tRNA Synthetases and tRNAs in Human mt Translation

Special Needs for the Synthesis of Only 13 Proteins

Mammalian mitochondria need a complete translational apparatus for the synthesis of only 13 proteins. These proteins are all sub-units of the respiratory chain complexes and are thus actors in electron transfer and ATP synthesis processes. They consist of seven subunits of NADH-ubiquinone oxidoreductase (complex I), the cytochrome b apoprotein of ubiquinone-cytochrome c oxidoreductase (complex III), three subunits of cytochrome c oxidase (complex IV) and two subunits of ATP synthase (complex V). These proteins are strongly hydrophobic, a property which may account for maintenance of the corresponding genes in the mt genome, while other genes have been transferred to the nucleus along evolution.[1] Figure 1 illustrates the amino acid contents of the 13 proteins and shows the high proportion of aliphatic and aromatic amino acids (2279 over a total of 3833, i.e., 59.4%). Several amino acids are present with parsimony (Cys, Arg, Asp, Glu, Gln, Lys present at less than 3% each) while others are present at high levels (Ser, Ile and Thr are present at more than 7% and leucine is by far the winner with 14,4%). From these numbers, it can be anticipated that the requirements of each of the aminoacylated tRNAs for mt translation is not equivalent, although the total mt protein synthesis activity is not only dependent on the amino acid content of each protein, but also on the copy number, efficiency of translation and relative stability of the individual mRNAs. It has been estimated that the number of each mt mRNA per cell varies from less than 50 to about 1200, that the relative efficiency of their translation ranges from 1 to 8-fold and that the relative abundance of the respiratory chain complexes varies from 1 to 7.[11,12] However, since the amino acid content is about the same for each of the 13 proteins (Fig. 1), the total need of each aminoacylated tRNA to achieve translation can be considered as roughly proportional to the amino acid distribution. Thus, some aminoacyl-tRNAs (e.g., Ser-tRNA^{Ser}, Ile-tRNA^{Ile}, Thr-tRNA^{Thr} and Leu-tRNA^{Leu}) are required in situ to much higher steady-state levels than others (e.g., Cys-tRNA^{Cys}).

Mitochondrial tRNAs

The 22 mt tRNA genes are distributed all along the mt genome in alternance with protein coding genes or ribosomal RNA genes. After synthesis of two large primary transcripts covering the full length of each of the DNA strands and a smaller transcript covering the two rRNAs and two tRNA genes, the tRNA precursors are excised, leading concomitantly to mRNAs. Thus, tRNAs are considered as punctuations of the mt genome from a transcriptional point of view.[13] After processing at the 5' end by RNAse P[14,15] and processing (if necessary) at the 3'-end by a 3' tRNAse,[16] the nonencoded CCA 3'-terminal sequences are incorporated by a mt nucleotidyl-transferase.[17-19] Post-transcriptional modifications do take place. So far, direct sequencing of 5 human mt tRNAs

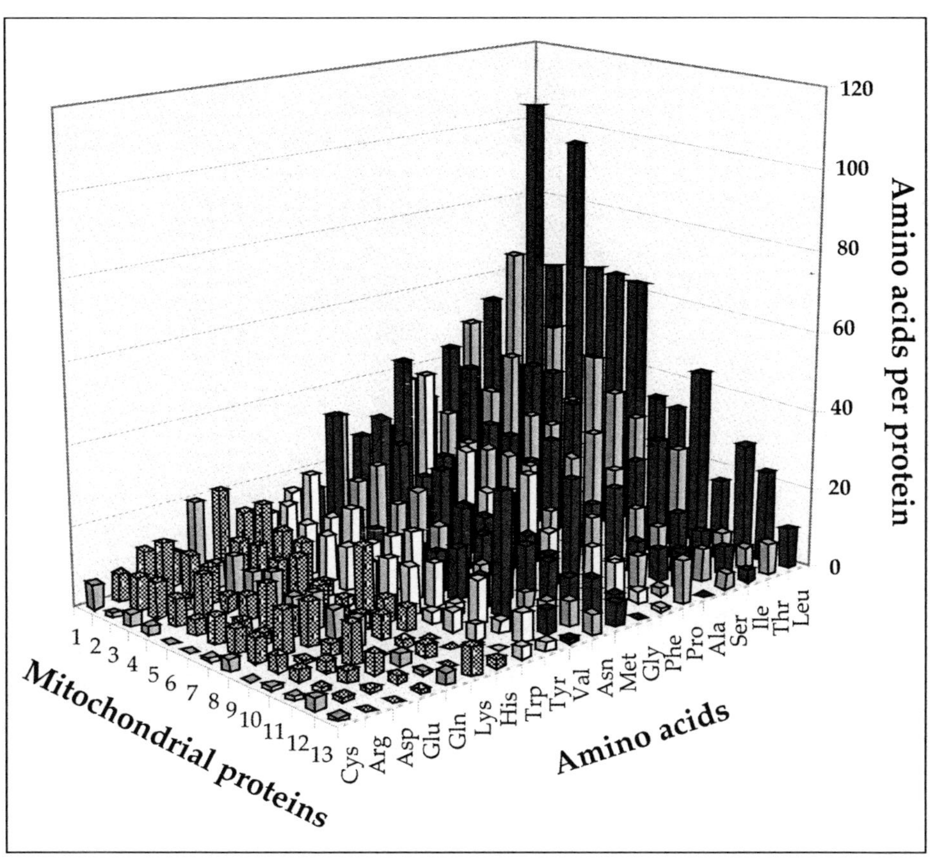

Figure 1. Hydrophobic character of human mt encoded proteins. Amino acid content and amino acid distribution in the 13 human mt encoded proteins. Amino acids are given by their 3 letters abbreviation code and are classified according to their total content in all 13 proteins. Histograms are shaded according to the physico-chemical properties of the different amino acids: black for hydrophobic aliphatic residues (Ala, Gly, Ile, Leu, Met, Val), white for hydrophobic aromatic residues (Phe, Trp and Tyr), grey for neutral residues (Asn, Cys, Gln, Pro, Ser and Thr), dark dots for hydrophilic basic residues (Arg, Lys and His) and light dots for hydrophilic acid residues (Asp and Glu). Proteins are classified by size. 1: ND5 (603 aa), subunit 5 of NADH ubiquinone oxidoreductase (complex I); 2: COX1 (513 aa), subunit 1 of cytochrome *c* oxidase (complex IV); 3: ND4 (459 aa), subunit 4 of complex I; 4: cytb (380 aa), apoprotein of cytochrome *b* ubiquinone oxidoreductase (complex III); 5: ND2 (347 aa), subunit 2 of complex I; 6: ND1 (318 aa), subunit 1 of complex I; 7: COX3 (278 aa), subunit 3 of complex IV; 8: COX2 (251 aa), subunit 2 of complex IV; 9: ATP6 (226 aa), subunit 6 of ATP synthase (complex V); 10: ND6 (174 aa), subunit 6 of complex I; 11: ND3 (115 aa), subunit 3 of complex I; 12: ND4L (101 aa), subunit 4L of complex I; 13: ATP8 (68 aa), subunit 8 of complex V.

revealed 11 types of modifications, most of these being conventional (m^1A, thymidine, pseudouridine, m^5C, dihydrouridine) but sometimes they are present at unexpected positions. Methylation of an adenine at position 9 is only seen in mt tRNAs.[20,21] Two new modifications have been found at uridine 34 of tRNA^{Lys} and tRNA$^{Leu(UUR)}$. Both tRNAs have a taurinomethyl group at position 5 of the base, with tRNA^{Lys} having in addition a thio group at position 2.[22] The lower number of post-transcriptional modifications in mt tRNAs as compared to "classical" tRNAs (tRNAs from prokaryotes or from eukaryotic cytosols), is in favor of a higher importance of these. Indeed, m^1A in tRNA^{Lys} is responsible for correct folding of the tRNA[20,23] and taurine-derived anticodon modifications in tRNA[Lys] and tRNA[Leu(UUR)] are involved in specific codon-anticodon recognition.[22] Mt tRNA biosynthesis in vivo undergoes a number of regulatory events (in particular, down regulations of over-expressed primary transcripts) so that the steady-state levels of each of them ends up in a same range (relative amounts of individual species vary only by a factor of one to fourfold).[24,25]

The 22 mt tRNAs correspond to one tRNA per amino acid, with the exception of leucine and serine for which two isoacceptors are present. One leucine isoacceptor tRNA[Leu(UUR)] (R for purines) reads UUA and UUG codons while the other isoacceptor tRNA[Leu(CUN)] reads CUN codons (N for any of the four nucleotides). The two categories of serine specific codons (AGY, with Y for pyrimidines, and UCN) are read by the corresponding isoacceptors tRNA[Ser(AGY)] and tRNA[Ser(UCN)]. Interestingly, for both of these tRNA families, there is a disproportion in codon usage with more than 5-fold less UUR leucine codons (89) than CUN leucine codons (553) and serine AGY codons (56) than UCN codons (237) in the 13 mt protein genes, showing that the respective contribution of each isoacceptor to translation is unequal. While 14 tRNAs are transcribed from the heavy mt DNA strand and are thus "light" (G-poor) tRNAs (Table 1), 8 are transcribed from the light strand and form the "heavy" (G-rich) tRNAs.[26] The nucleotide content of the two families of tRNAs leads to interesting characteristics. Light tRNAs are very rich in A, and as a consequence in pyrimidine-A stretches (about 13 per molecule) known as sensitive sites for RNA degradation,[27] while heavy tRNAs contain only about 8 such sites per molecule. With regard to secondary structure elements, light tRNAs are rich in A-U base-pairs and contain mismatches, while heavy tRNAs contain numerous G-U pairs but no mismatches (Table 1).

The 2D structure (theoretical or experimental) of each mt tRNA deviates more or less strongly from classical tRNAs[28] with the extreme situation of tRNA[Ser(AGY)] which completely misses the D-arm. The 21 other tRNAs do fold into theoretical cloverleaf structures with however some size variations. Deviations in the T-loop size from the classical seven nucleotides are most striking (Table 1 and Fig. 2). Interestingly also, none of the human mt tRNAs has a large variable region. Conserved nucleotides at strategic positions of the 2D structure are partially present with most tRNAs having U8, A9, A14, A15, A21, A26, U33, R38, Y48 and G53, C61. However, only four tRNAs do have residues G18 and G19 in the D-loop and U54, U55, C56, (to become TΨC after post-transcriptional modification), R57 and A58 in the T-loop, suggesting that tertiary interactions between both loops are either absent or at least unconventional in the majority of these tRNAs. These characteristics are mainly conserved all over mammalian mt tRNAs.[29] So far, experimental structures of only four mammalian mt tRNAs are available, namely bovine tRNA[Ser(AGY)] and tRNA[Ser(UCN)],[30-35] bovine tRNA[Phe],[36] and human tRNA[Lys].[23,37] Much larger structural deviations from canonical tRNAs such as complete absence of the T-arm or extensions of the anticodon arm can be found for mt tRNAs from other organisms.[38-42]

In summary, our present knowledge on human mt tRNAs points towards a number of features which distinguish them from classical tRNAs. In particular, their nucleotide content leads to richness in weak base-pairs in cloverleaf structures (1.5 to 1.8 times more A-U, G-U and mismatches than G-C pairs, Table 1). In addition, the absence of classical tertiary interactions suggests strongly that these tRNAs are structurally and chemically less stable than classical tRNAs. While post-transcriptional modifications may well compensate some of these features by stabilizing effects,[43] it can be understood why experimentation with the mt tRNAs may be technically difficult (see below).

Human mt Aminoacyl-tRNA Synthetases

While most of their aminoacylation activities were detected long ago (i.e., refs 24,44-46), mammalian mt aaRSs and in particular human mt aaRSs gained full attention only recently. The genes for human mt GlyRS[47] and IleRS[48] were the first to be cloned followed by a putative HisRS gene.[49] The genes for PheRS,[50] LeuRS,[51] LysRS[52] and TrpRS[53] have been recognized and cloned over the last two years. In addition, one other mammalian aaRS was cloned, namely bovine SerRS.[54] Since the mt genome codes for only a restricted number of proteins, aaRSs are necessarily of nuclear origin, synthesized in the cytoplasm and targeted towards mitochondria. An imminent step towards knowledge on genes encoding human mt aaRSs will be indexation of putative sequences in the human genome. The evolutionary closeness between aaRSs might facilitate this search by taking advantage of sequence similarities with aaRSs from other organisms. The present knowledge of genes encoding for human aaRSs with cytosolic location and the enzyme of same specificity with mt

Table 1. Striking features of human mitochondrial tRNAs

Specificity	Total	Primary Sequence Features				Py-A Seq.	Secondary Structure Features				
		Nucleotide Content					G-C	A-U	G-U	Mism.	Weak/Strong
"Light" tRNAs		A	C	G	U						
Phe	74	25	21	10	18	15	8	10	0	1	11/8
Val	72	24	20	11	17	13	8	9	1	2	12/8
Leu(UUR)	78	26	18	13	21	13	7	11	1	2	14/7
Ile	72	28	11	11	22	11	6	13	0	1	14/6
Met	71	21	20	13	17	12	12	6	0	3	9/12
Trp	71	26	15	11	19	12	7	10	1	1	12/8
Asp	71	29	11	7	24	15	5	15	1	0	12/7
Lys	73	28	16	9	20	14	8	13	0	0	13/8
Gly	71	28	13	7	23	14	5	15	1	0	16/5
Arg	68	28	10	7	23	14	5	14	0	1	15/5
His	72	26	17	10	19	13	7	13	0	1	14/7
Ser(AGY)	62	20	20	9	13	10	8	8	0	2	10/8
Leu(CUN)	74	27	16	11	20	12	7	13	0	1	14/7
Thr	69	24	16	11	18	11	9	9	0	1	10/9
Total	998	360	224	140	274	179	102	159	5	16	180/102
Mean/tRNA	**71.3**	**25.7**	**16**	**10**	**19.6**	**12.8**	**7.3**	**11.4**	**0.4**	**1.1**	**1.8**
"Heavy" tRNAs											
Pro	71	21	9	16	25	8	6	10	4	0	14/6
Glu	72	17	11	18	26	9	8	8	5	0	13/8
Ser(UCN)	72	14	14	22	22	4	9	8	5	0	13/9
Tyr	69	21	12	18	18	9	8	10	2	0	12/8
Cys	69	19	16	16	18	6	10	6	3	0	9/10
Asn	76	17	11	21	27	10	8	7	5	0	12/8
Ala	72	16	11	21	24	8	8	9	4	0	13/8
Gln	75	16	11	26	22	8	8	9	3	1	13/8
Total	576	141	95	158	182	62	65	67	31	1	99/65
Mean/tRNA	**72**	**17.6**	**11.8**	**19.8**	**22.7**	**7.8**	**8.2**	**8.4**	**3.9**	**0.1**	**1.5**

tRNAs are classified according to their location in the mitochondrial genome. Total nucleotide content includes a 3'-CCA end, although this sequence is not encoded. Py-A streches in primary sequences are known as weak phosphodiester bonds, prone to degradation. Secondary structure features refer to theoretical cloverleaf structures. Mismatches (Mism.) as well as G-U and A-U pairs are considered as weak pairs while G-C are strong pairs.

location, allows to distinguish specific features with regard to genomic origin, structural characteristics and aminoacylation properties and will be briefly summarized below. This has been extensively reviewed in ref. 2.

Mitochondria are remnant prokaryotes engulfed by primitive eukaryotic cells[55,56] so that the present-day organelle genomes result from the progressive transfer to the nucleus of endosymbiont genes. The genes for aaRSs were either duplicated or not, to give a separate set (or not) of genes for the cytosolic forms of the enzymes.[57,58] As a result, two different gene combinations are encountered for the production of cytosolic and mt aaRSs (Fig. 3). These are either the existence of two different genes leading to cytosolic and mt enzymes, or the existence of a same gene that leads to the two enzymes. In six of the so far known cases (HisRS, IleRS, LeuRS, PheRS, SerRS and TrpRS), cytosolic and mt targeted enzymes are encoded by distinct genes.[48-51,53,54] For four of them, the sequence similarities between both versions of a same enzyme are lower than those observed when comparing

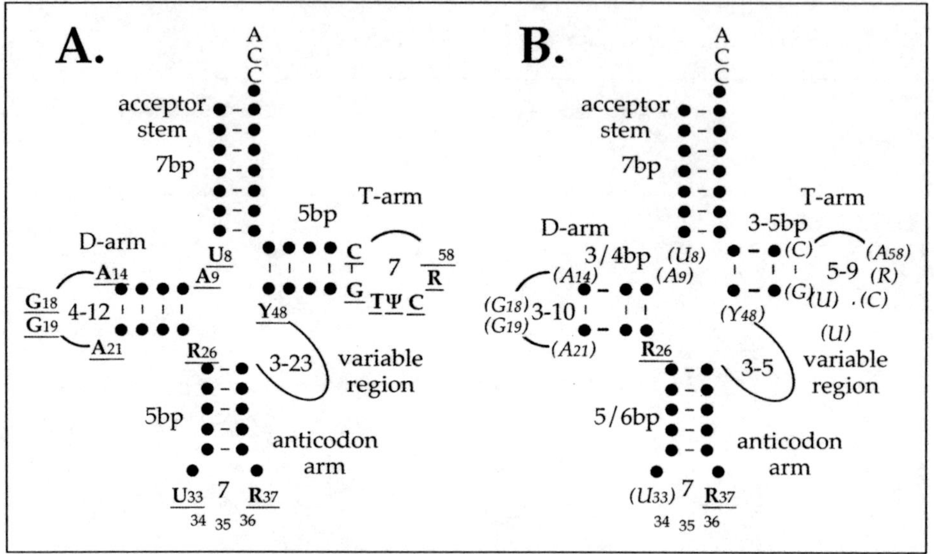

Figure 2. Typical secondary structure features of human mt tRNAs. A) Typical cloverleaf structure of canonical (prokaryotic and eukaryotic cytosolic) tRNAs. B) Typical secondary structure of human mt tRNAs (with the exception of tRNA$^{Ser(AGY)}$ in which the D-arm is replaced by a 4 nucleotides connector between the acceptor and the anticodon arms). Dots correspond to nucleotides, lines to variable numbers of nucleotides. Sizes of each structural domain are squared. To be noticed in B, the large size variability of the T-stem and T-loop and the restricted size of the variable region. Conserved nucleotides are indicated in bold underlined characters. R stands for purines, Y for pyrimidines, T for thymine and Ψ for pseudouridine. Since most of the mt tRNAs have not been sequenced at the RNA level, unmodified nucleotides are considered here. Classically conserved nucleotides are only found in a subset of human mt tRNAs. They are indicated between parentheses. Only R26 and R37 are present in each of the 22 tRNAs.

mt and the corresponding prokaryotic enzymes. Exceptions concern the putative mt HisRS gene which has more than 70% identity with the cytosolic enzyme,[49] and the mt PheRS gene, which resembles partially to that coding for the cytosolic enzyme.[50] For the two remaining enzymes, mt and cytosolic versions are encoded by a same gene. For GlyRS, two initiation sites for translation lead to two enzymes which differ only by a mt targeting signal. Once in the organelle matrix, this signal is removed, making the mt enzyme identical to the cytosolic one. The case of LysRS is somewhat different since alternate splicing allows the production of two aaRSs from a single gene. After cleavage of the signal peptide, the mature mt enzyme differs from the cytoplasmic by about 30 amino acids at its N-terminal.[52]

With regard to the classification of aaRS[59,60] gene, sequence analyses show that mt aaRSs possess the same typical signature motifs than other synthetases with the same specificity and thus belong to the same synthetase class. The oligomeric structures of the mt aaRSs are mostly identical to those of known prokaryotic or eukaryotic enzymes of the same identities with the exceptions of GlyRS and PheRS. Human mt GlyRS is dimeric (α_2) as is the case for other mt and cytosolic GlyRS while prokaryotic GlyRSs are heterotetrameric ($\alpha_2\beta_2$). More striking is the divergence between prokaryotic and eukaryotic tetrameric cytosolic PheRSs ($\alpha_2\beta_2$), and the monomeric (α) version of the human mt enzyme.[50] However, the same difference was found between yeast mt and cytosolic PheRSs.[61]

Only partial information on aminoacylation properties of mammalian mt aaRSs are so far available. The coexistence of two protein synthesizing compartments within eukaryotic cells lead to the question of possible cross-reactivities (aminoacylation of cytosolic tRNA by mt aaRSs, or of mt tRNA by cytosolic aaRSs). Cross-aminoacylation experiments performed on crude rat liver cytosolic and mt extracts,[62] and on HeLa cell cytosolic and mt extracts[24] revealed the general trend of organelle aaRSs to be able to charge cytosolic tRNAs while cytosolic enzymes are only rarely able to

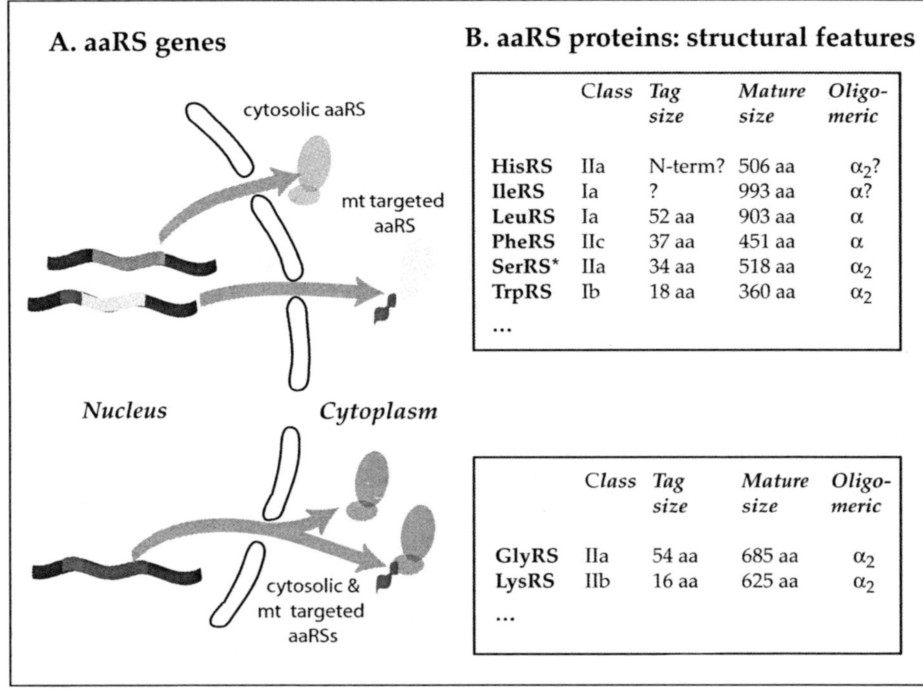

	Class	Tag size	Mature size	Oligo- meric
HisRS	IIa	N-term?	506 aa	α_2?
IleRS	Ia	?	993 aa	α?
LeuRS	Ia	52 aa	903 aa	α
PheRS	IIc	37 aa	451 aa	α
SerRS*	IIa	34 aa	518 aa	α_2
TrpRS	Ib	18 aa	360 aa	α_2
...				

	Class	Tag size	Mature size	Oligo- meric
GlyRS	IIa	54 aa	685 aa	α_2
LysRS	IIb	16 aa	625 aa	α_2
...				

Figure 3. Human mt aminoacyl-tRNA synthetases. A. Schematic representation of the nuclear gene distribution encoding for both cytosolic and mt aaRSs. Two situations have been highlighted so far namely (top) cytosolic and mt aaRSs of a same specificity are encoded by two distinct genes and (bottom) the 2 enzymes are encoded by a single gene. B. Structural features of so far investigated human mt aaRSs. The tag allows import to mitochondria and is removed by a specific peptidase. * stands for bovine mt aaRS.

aminoacylate mt tRNAs. Experiments performed with purified mt aaRSs, confirm these findings since mt GlysRS, IleRS, LeuRS, LysRS and PheRS are able to charge the corresponding cytoplasmic tRNAs. This suggests sequence or structural similarities between the two enzymes able to recognize the same substrate, as verified in the case for GlyRS and LysRS for which the two isoforms (cytosolic and mt enzymes) are encoded by a unique gene. At opposite, cross-reactivity is not detectable for TrpRS and SerRS, in agreement with the existence of two unrelated genes for the mt and the cytosolic enzymes. Because of the endosymbiotic origin of mitochondria, possible cross-species aminoacylation properties have been searched for, and in particular possible charging of *E. coli* tRNAs by mt enzymes.[45,46,63] The general trend is towards an efficient eubacterial tRNA aminoacylation by mt extracts whereas the prokaryotic enzymes are unable to charge bovine mt tRNAs.

Establishment of kinetic parameters of aminoacylation of cognate tRNA by mammalian mt aaRSs revealed slower enzymatic activities when compared to prokaryotic or cytoplasmic homologues. Cloned LeuRS, TrpRS and PheRS display significantly reduced efficiency of amino acid activation (100 to 250-fold lower) when compared to *E. coli* enzymes.[50,51,53] In addition, mt aaRSs have 20 to 400-fold lower specific activities than the corresponding *E. coli* or cytosolic homologues.[50,51] Specific tRNA recognition by homologous aaRSs, governed by identity elements (e.g., refs. 64,65) have been barely studied within mammalian mitochondria, a situation likely due to technical difficulties (see below). Only serine specific identity elements have been searched in a rational systematic way.[54,66,67] Interestingly, the same mt SerRS recognizes two tRNAs sharing neither consensus sequence elements nor the same structural architecture (tRNA$^{Ser(AGY)}$ misses a D-arm and tRNA$^{Ser(CUN)}$ has a short connector between the acceptor and the D stems and has an additional base-pair in the anticodon stem). A thorough analysis of in vitro transcribed tRNA[Ser] variants, combined to footprinting experiments, leads to the conclusion that recognition elements are within the T-stem in both tRNAs. Some hints as to the possible identity elements in mt tRNAs can be gained from the general conservation of major identity elements over evolution[64] and the endosymbiotic origin of mitochondria. The theoretical presence of major *E. coli* identity determinants within human mt tRNA sequences has been verified.[29] However, such extrapolations need to be made with caution because of the large structural variations observed in mt tRNAs which may change or influence aminoacylation rules of these particular tRNAs.

Human Diseases Correlated to Point Mutations in mt tRNA Genes

Human Disorders and Potential Molecular Impacts of Point Mutations

Since the first description of a mutation in the tRNA[Leu(UUR)] gene related to a mt encephalopathy with lactic acidosis and stroke-like episodes (MELAS),[68] about 70 additional point mutations in tRNAs have been reported as correlated to a variety of maternally inherited diseases.[69,70] These disorders include phenotypes such as myopathies, encephalomyopathies, cardiopathies, diabetes, ophtalmoplegia, deafness, etc, or combination of these (Table 2). So far, mutations do affect 20 out of the 22 tRNA genes, with the genes for tRNA[Leu(UUR)], tRNA[Ile] and tRNA[Lys] as hot spots and no mutations found so far in tRNA[Arg] and tRNA[His]. While mutations have been reported at the gene level, Figure 4 shows them in tRNA cloverleaf structures to highlight the variety in locations within a given tRNA (gene or gene product).

The relationships between the genotypes and the phenotypes are perplexing. Indeed, the same mutation can lead to different pathologies while the same pathology can be correlated to different individual mutations. In addition, different degrees of severity of a given disease can be observed for the same mutation, going from mild to lethal. Some of these observations can be explained by mitochondria-specific features such as mitotic segregation of mutation-carrying mitochondria and threshold effects in heteroplasmy (both wild-type and mutated versions of the mt genome are present in a single mitochondria or a single cell).[6-8,71,72] From a biochemical point of view, disease-related mitochondria display a decreased respiration capacity and decreased activities of individual respiratory chain complexes. Considering the central role of tRNA in protein biosynthesis and especially in the synthesis of mitochondria-encoded subunits of respiratory chain complexes, it is generally assumed that the decreased respiratory chain activities are due to the lower abundance of mt encoded subunits or to the presence of nonfunctional subunits.[6-8,71,72]

At a molecular level, numerous types of mutation impacts have been considered.[6-8,71,72] They concern both tRNA biosynthesis and tRNA functions. Aminoacylation of tRNAs by their cognate aaRS being a central event in protein synthesis, possible impacts on this process have been considered dominantly and experiments have been set up to test this possibility. On a theoretical level, a point mutation in a mt tRNA can hinder aminoacylation by the cognate enzyme ("loss of function" hypothesis) by removing an identity element, introducing an antideterminant against the cognate synthetase or by affecting the structure of the tRNA (folding or flexibility), so that the cognate enzyme becomes less efficient at the catalytic level. At the opposite, a mutation could also favor aminoacylation by a noncognate enzyme ("gain of function" hypothesis) by introducing an identity element corresponding to another identity specificity, or by removal of an antideterminant.[88] An overview of our present understanding of the impact of mutations on aminoacylation will be presented in what follows. However, we will first recall technical limitations in tRNA handling and measurement of aminoacylation capacities.

Experimental Strategies for the Investigation of Human mt tRNA Aminoacylation

Investigation of mt aminoacylation systems is complicated not only by the fact that access to human mt tRNA -and especially mutated tRNAs- is difficult, but also because of intrinsic properties of mitochondria, such as heteroplasmy. A series of strategies has been adapted to allow insight into mt aminoacylation, both in vivo and in vitro.[73] Measurement of in vivo steady-state levels of aminoacylated tRNAs can be performed by separation of aminoacylated and non aminoacylated

Table 2. Human disorders correlated to point mutations in mitochondrial tRNA genes

Acronym	Phenotype	Genotype
DCM	Dilated CardioMyopathy	**Leu(CUN)** (T12297C)
ECM	EncephaloCardioMyopathy	**Ile** (C4320T)
FICP	Fatal Infantile Cardiomyopathy Plus Melas-associated	**Ile** (A4269G, A4317G)
MHCM	Maternally inherited Hypertrophic CardioMyopathy	**Gly** (T9997C)
MICM	Maternal Inherited CardioMyopathy	**Lys** (G8363A)
MMC	Maternal Myopathy and Cardiomyopathy	**Leu(UUR)** (A3260G, C3303T)
A	Ataxia	**Trp** (G5549A)
AMDF	Ataxia, Mental deterioration, DeaFness	**Val** (G1606A)
CIPO	Chronic Intestinal PseudoObstruction with myopathy	**Gly** (A10006G), **Ser(AGY)** (C12246A)
EI	Exercise Intolerance	**Leu(UUR)** (T3258C, T3273C), **Tyr** (A5874G)
EM	EncephaloMyopathy	**Cys** (A5814G), **Gln** (C4332T), **Glu** (A14709G), **Gly** (T10010C, A10044G), **Ile** (A4269G), **Lys** (G8328A)
LIMM	Lethal Infantile Mitochondrial Myopathy	**Thr** (A15923G)
LW	Limb Weakness	**Tyr** (A5874G)
MERRF	Myoclonic Epilepsy and Ragged Red Fibers	**Lys** (A8296G, A8344G, T8356C, G8363A)
MM	Mitochondrial Myopathy	**Asn** (C5703T), **Gln** (insT4370), **Glu** (A14709G), **Leu(CUN)** (A12320G), **Leu(UUR)** (A3243T, T3250C, A3251G, C3254G, A3280G, A3288G, A3302G), **Met** (T4409C, G4450A), **Phe** (T618C), **Pro** (G15990A), **Ser(UCN)** (C7497T), **Thr** (G15915A, delT15940), **Trp** (G5521A)
MS	Myoclonic Seizures	**Asp** (A7543G), **Lys** (G8342A)
PEM	Progressive EncephaloMyopathy	**Leu(UUR)** (A3243T, delT3272), **Ser(UCN)** (C7497T, insG7472, A7512G), **Trp** G5540A
SM	Skeletal Myopathy	**Lys** (T8355C, T8362G)
DEMCHO	DEMentia, CHOrea	**Trp** (G5549A)
LA	Lactic Acidosis	**Leu(UUR)** (T3258C), **Ser(UCN)** (C7497T)
LS	Leigh Syndrome	**Lys** (G8363A), **Trp** (insT5537), **Val** (G1644T)
MELAS	Mitochondrial Encephalomyopathy, Lactis Acidose, Stroke-like episodes	**Cys** (A5814G), **Leu(UUR)** (A3243G, A3252G, C3256T, T3271C, T3291C), **Lys** (T8316C), **Phe** (G583A), **Val** (G1642A)
MNGIE	Mitochondrial NeuroGastroIntestinal Encephalomyopathy	**Lys** (G8313A)
CPEO	Chronic Progressive External Ophtalmoplegia	**Ala** (A5628G), **Asn** (A5692G, C5703T), **Ile** (T4274C, G4298A, G4309A), **Leu(CUN)** (G12315A)
D	Diabete	**Gln** (C4332T), **Glu** (A14709G), **Trp** (G5549A)
DEAF	Maternally inherited DEAFness	**Lys** (G8363A), **Ser(UCN)** (A7511G)

continued on next page

Table 2. Continued

Acronym	Phenotype	Genotype
DM	Diabetes Mellitus	**Leu(UUR)** (T3264C, T3271C)
DMDF	Diabetes Mellitus, DeaFness	**Leu(UUR)** (A3243G), **Lys** (A8296G), **Ser(AGY)** (C12258A)
KS	Kearns-Sayre syndrom	**Leu(UUR)** (G3249A)
LHON	Leber Hereditary Optic Neuropathy	**Leu(UUR)** (C3275A)
O	Ophtalmoplegia	**Leu(UUR)** (T3273C), **Pro** (G15990A),
PEO	Progressive External Ophtalmoplegia	**Asn** (C5698T), **Cys** (A5814G), **Ile** (T4285C), **Lys** (G8342A, T8355C)
SNHL	SensoriNeural Hearing Loss	**Ser(UCN)** (A7511G)
ADPD	Alzeimer's Disease and Parkinsons's Disease	**Gln** (A4336G), **Pro** (T15965C), **Thr** (G15950A)
AISA	Acquired Idiopathic Sideroblastic Anemia	**Leu(CUN)** (G12301A)
CD	Cox Deficiency	**Gln** (insT4370), **Trp** (G5540A), **Tyr** (A5874G)
M	Myoglobinuria	**Phe** (A606G)

tRNAs on acidic gels.[74] Acidity ensures stability of the ester bond between the amino acid and tRNA for charged tRNAs. Application of this approach to human mt tRNAs required the development of specific cell lines allowing to overcome the problem of heteroplasmy (mixture of both wild-type and mutated tRNA versions in the same cell). Cybrid cell lines were prepared, homoplasmic for either wild-type mt genome or for a genome with a single point mutation in a given tRNA gene (e.g., refs. 75-77). After cell growth, mitochondria are purified, total tRNA is extracted under acid conditions, and specific tRNAs are analyzed by northern blotting after separation on acidic gels.[78,79] While valid for several amino acids, this approach is however not of general application because it is not possible to separate aminoacylated from non aminoacylated tRNAs in some specific cases.[73] The "OXOCIRC" assay (oxidation followed by circularization of the tRNA) is also based on extraction of mt tRNA under acidic conditions.[80] tRNAs are extracted from heteroplasmic biopsies or from cybrid cell cultures (containing both wild-type and mutated tRNAs from a given amino acid specificity) and submitted, or not, to deacylation. After an oxidation step allowing destruction of the 3'-terminal ribose of non-aminoacylated tRNAs and circularization of the non-oxidized tRNAs (i.e., those aminoacylated), RT-PCR and sequencing allows us to distinguish qualitatively and quantitatively wild-type and mutated tRNAs.

An alternative to the in vivo measurements consists in purification of sufficient amounts of native tRNA and measurement of aminoacylation properties in vitro. The strategy of tRNA purification, based on hybridization of a given tRNA to a specific complementary oligonucleotide,[63] is not sensitive enough to distinguish wild-type and mutant tRNA with the same specificity. Thus, it requires also homoplasmic cell lines. This approach, although very powerful remains very time consuming (long term cell culture) and has a poor yield. Since mt aaRSs are able to cross-aminoacylate cytosolic tRNAs, trace contamination by these tRNAs (which are present in an excess of 130/1 in HeLa cells)[25] needs to be eliminated. The advantage of this approach is access to tRNAs which are representative of the cell situation, i.e., wild-type tRNAs optimal in structure and post-transcriptional modifications, and mutant tRNAs which have undergone (with more or less success) all steps of biosynthesis. This strategy has only been applied successfully once so far[22] in what can be considered as a *tour de force*.

In vitro transcription of cloned wild-type and mutated tRNA genes was thought to be the approach of choice leading to an easy access to any human mt tRNA, as was the case for many prokaryotic or eukaryotic tRNAs.[64] However, numerous unexpected difficulties arose for human mt tRNAs. In the specific case of human mt tRNA^{Lys}, the wild-type transcript was found unable to fold into a

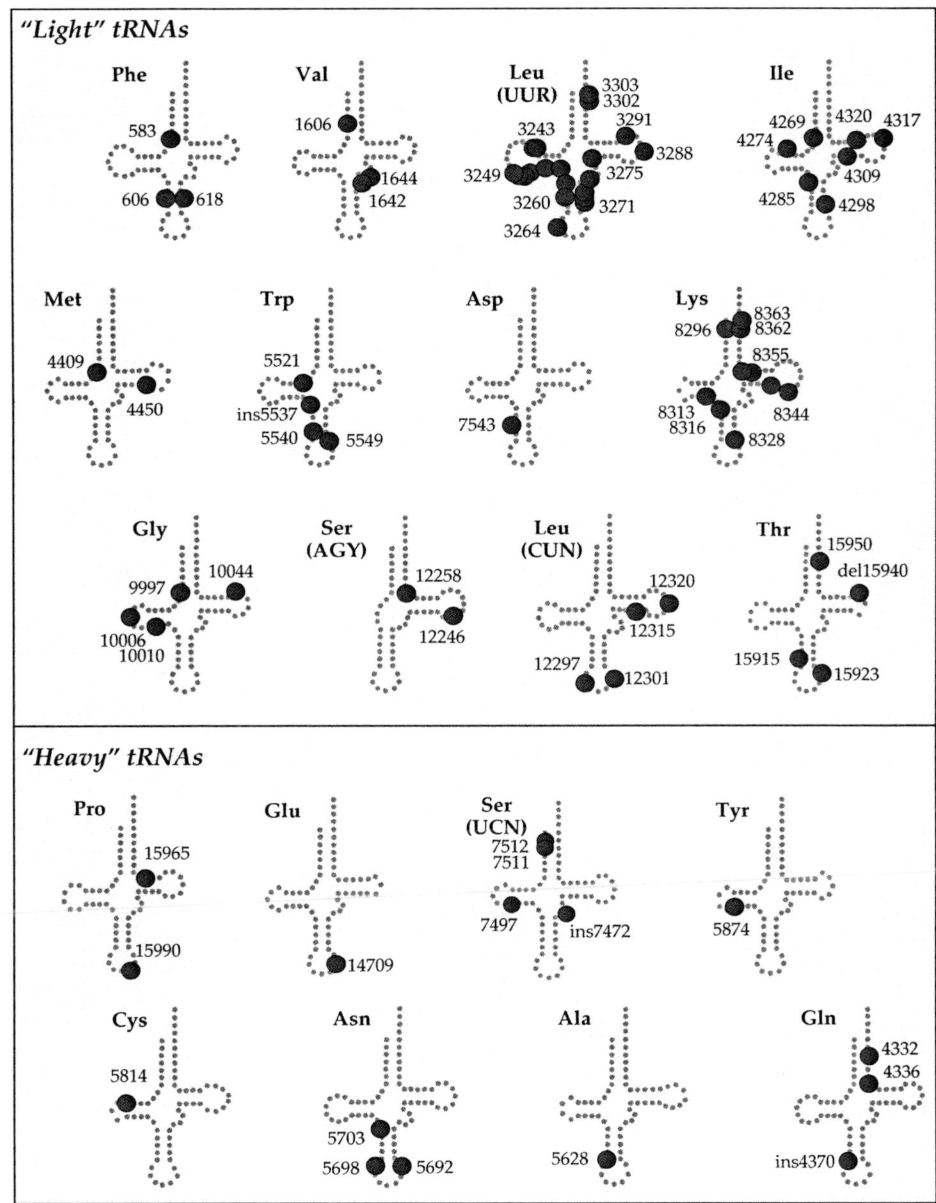

Figure 4. Disease-correlated mutations in human mt tRNA genes. Mutations are presented in the tRNA cloverleaf structures. "Light" tRNAs are those transcribed from the DNA heavy strand and "heavy" tRNAs those transcribed from the DNA light strand.[26] tRNAs are ordered according to their position on the genome. Nucleotide positions are those in the mt genome. Mutations are according to refs. [69,70]

cloverleaf.[20] A post-transcriptionally added methyl group is necessary and sufficient to hinder this hairpin fold and allows for the cloverleaf structure. Thus, the presence of a post-transcriptional modification is a prerequisite for correct folding of the tRNA.[23] Not only tRNA^{Lys}, but also several other mt tRNAs prepared by in vitro transcription turned out to be barely chargeable with an amino acid (our unpublished results, and refs. 50,52). The difficulties reported in aminoacylation of transcripts are probably primarily linked to the particular structural features or these tRNAs. Their nucleotide composition, secondary structural features and "loose" tertiary structures, allow for numerous alternative conformations and generally unstable structures. The presence of post-transcriptional modifications in natural tRNAs likely restricts the folding space and stabilizes functional conformations. In addition, they may correspond to positive identity elements required for recognition by the cognate aaRS. Despite these difficulties, it was possible to use the in vitro transcription approach to prepare functional tRNA^{Lys} and tRNA^{Ile}.

Mutations and Aminoacylation

The technical difficulties reported above have slowed down the investigation of possible impacts of pathology-related mutations on tRNA aminoacylation. Less than two handful of mutations have been considered, affecting tRNA^{Lys}, tRNA$^{Leu(UUR)}$, tRNA^{Asn} and tRNA^{Ile}. Much emphasis was put on the MERRF (myoclonic epilepsy with ragged red fibers) mutation A8344G in tRNA^{Lys} and the MELAS (mt myopathy, encephalopathy with lactic acidosis and stroke-like episodes) mutation A3243G in tRNA$^{Leu(UUR)}$. Both mutations are most often encountered in patients and are correlated to typical phenotypes. The analysis of different aspects of the molecular impacts of these two mutations parallels the technical breakthroughs in the field to allow for both in vivo and in vitro investigations. tRNA^{Ile} was the first human mt tRNA that could be aminoacylated when produced by in vitro transcription.

In Vivo Aminoacylation of Mutated tRNAs

The MERRF mutation A8344G is located in the T-loop of tRNA^{Lys}, at position 55 (conventional tRNA numbering) where classical tRNAs show the T54Ψ55C56 sequence important for tertiary interactions with the D-loop (Fig. 5). Neither in wild-type nor in mutant tRNA^{Lys}, the T-loop shows these conventional sequence elements so that the impact level of the mutation at a structural level cannot be foreseen. Pioneering experiments on the effect of the mutation on aminoacylation have been performed in vivo by the acidic gel approach, on osteosarcoma-derived cybrid cell lines.[78] It has been shown that (i) the steady-state level of mutated tRNA^{Lys} is decreased by 16% to 33% as compared to that of wild-type tRNA^{Lys} present in the sibling wild-type cells, likely due to higher instability of the mutated tRNA, (ii) the steady-state level of aminoacylated mutated tRNA^{Lys} is decreased by 37% to 49% as compared to the level of aminoacylated wild-type tRNA^{Lys}, (iii) the combination of both effects leads to a decrease of 50 to 60% of available Lys-tRNA^{Lys} in the disease-carrying cells as compared to the healthy cells. From control experiments, it was estimated that the mutated tRNA is not mischarged and that there is no lysine incorporation at inappropriate positions. Thus, the mutation does not affect codon reading nor interaction with elongation factor but infers translation by the insufficient available steady-state levels of aminoacylated tRNA^{Lys} leading to frameshifting and abortive termination events.[78] From this work, the primary effect of the mutation was interpreted to occur at the level of interaction with LysRS and lysylation.

The OXOCIRC approach was used to evaluate the effect of the MERRF mutation on aminoacylation of tRNA^{Lys} in biopsies. Intriguingly, it was found that a series of muscular biopsies, as well as fibroblasts in culture, do have the same steady-state levels of lysyl-tRNA^{Lys} than controls.[80] Thus, here aminoacylation is considered not affected by the mutation. However, when the osteosarcoma cell lines are analyzed by the OXOCIRC approach, the steady-state level of Lys-tRNA^{Lys} is found decreased.[80] The variation in results between cybrid cell lines and biopsies is likely linked to the different nuclear backgrounds and/or to the sensitivity of the two different approaches.

Mutation G5703A in tRNA^{Asn} gene is linked to a mt myopathy.[81] When considered in the tRNA cloverleaf structure, this mutation converts a C-G pair on top of the anticodon stem into a U-G pair (Fig. 5). Cybrid cell lines, homoplasmic for this mutation, have a large decrease in respiration capacity (over 90% reduction of oxygen consumption). Mt translation is also strongly reduced in all proteins synthesized. At the molecular level, a marked reduction in steady-state levels of tRNA^{Asn} was observed. Experimentation favors the possibility that this is not due to ineffective transcription or processing events, but rather to conformational alterations of the mature tRNA which may impair aminoacylation. Direct influence of the mutation on aminoacylation proved however difficult

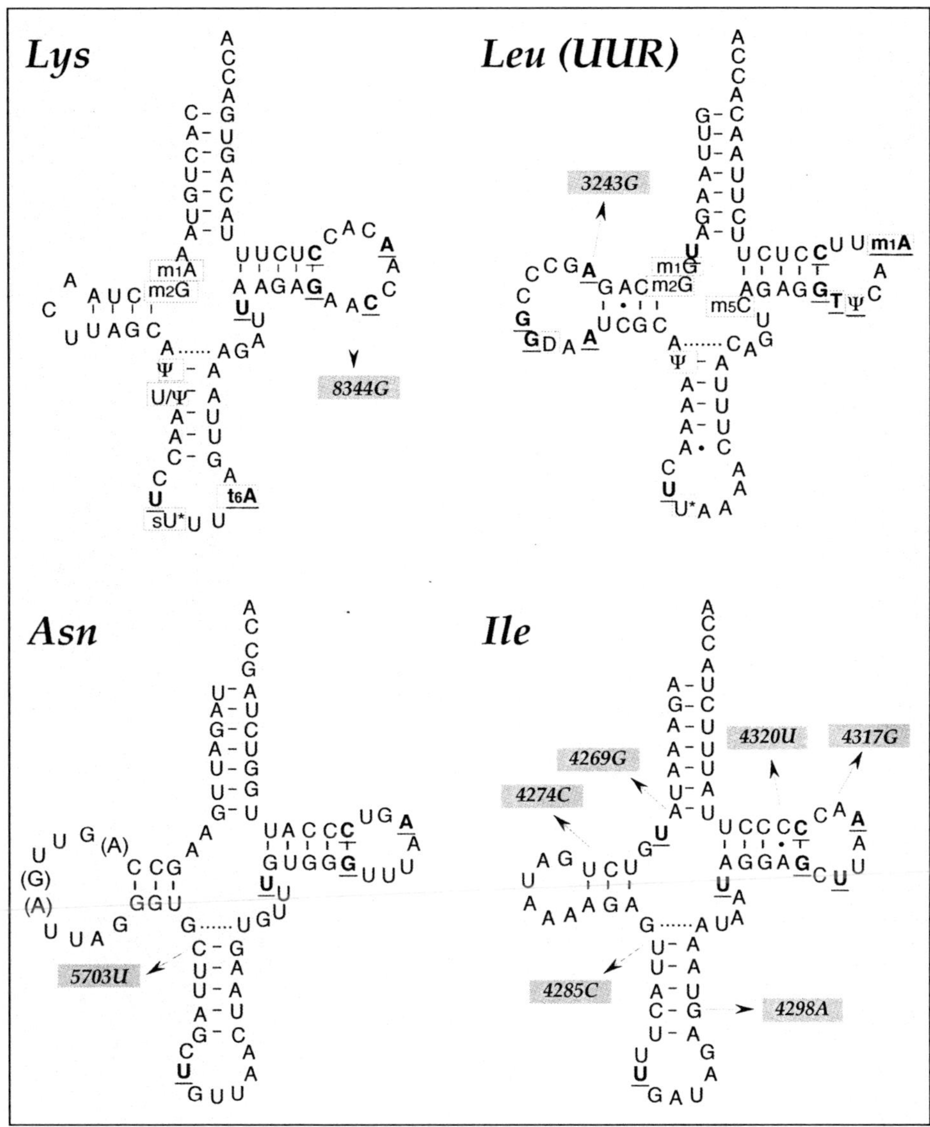

Figure 5. Mutations investigated for their possibe effect on aminoacylation. Both tRNA^{Lys} and tRNA$^{Leu(UUR)}$ have been sequenced at the RNA level so that their post-transcriptional modifications are indicated.[22,96] For tRNA^{Asn} and tRNA^{Ile}, only the gene sequences are known so far. Nucleotides conserved in all tRNAs (see Fig. 1) are bold and underlined. Mutations investigated are indicated with arrows and numbered as in the human mt genome.

to analyze in vivo, since asparaginylated and non charged tRNA^{Asn} did migrate similarly on acidic gels: a tentative conclusion is that this mutation does not affect aminoacylation.[81]

Mutation A3243G in tRNA$^{Leu(UUR)}$ is the most prevalent mutation found in patients. This mutation is typical of the MELAS syndrome, and is located in the D-loop of the tRNA, at position 14 (Fig. 5). tRNA$^{Leu(UUR)}$ has a predicted secondary structure very close to that of classical tRNAs

(Fig. 5) with conserved nucleotides in both D- and T-loops. It is thus likely that this tRNA has a "classical" tertiary structure. If so, mutation 3243 may interfere both with the tRNA folding (interference with the tertiary interaction U8-A14) and with aminoacylation since A14 is an identity element in *E. coli* leucylation.[82] The potential effect of this mutation on aminoacylation has been investigated both by the acidic gel approach on cybrid cell lines[83] and the OXOCIRC method on biopsies.[80] In cybrid cell lines, both the steady-state levels of mutated tRNA[Leu(UUR)] and of aminoacylated mutant tRNA[Leu(UUR)] are decreased as compared to wild-type tRNA, leading to an available amount of aminoacylated tRNA decreased by 75 %. In biopsy samples, the mutation was also found to affect aminoacylation. Interestingly, the same A3243G mutation, but associated to diabetes causes severe mt dysfunction without a strong decrease in protein synthesis rate, suggesting that in this case, aminoacylation is also not affected.[84]

In Vitro Aminoacylation Properties of Mutated tRNAs

In vitro aminoacylation assays allow to evaluate the effect of single point mutations from a kinetic point of view. Sufficient amounts of both wild-type and mutated tRNA[Lys] have been purified from cybrids derived from HeLa cells to perform in vitro aminoacylation assays and establish kinetic parameters in the presence of purified bovine mt LysRS.[22] This lead to very close K_M and V_{max} values for aminoacylation of both wild-type and mutated tRNAs, showing that the efficiency of aminoacylation is about the same (relative V_{max}/K_M values of 1 for the wild-type tRNA and of 0.32 for the mutated tRNA). Our own experiments, performed with human cytosolic LysRS and in vitro transcribed pseudo wild-type tRNA[Lys] (a transcript that folds correctly into the cloverleaf) and the same pseudo-tRNA[Lys] with the 8344 MERRF mutation, reveal the absence of interference of the mutation on the aminoacylation kinetic parameters (Sissler et al, in preparation). In vitro aminoacylation studies on the MELAS mutation have not yet been reported, either on native tRNAs nor on in vitro transcripts. The wild-type in vitro transcript was shown to be a poor substrate for LeuRS.[51]

A breakthrough came from in vitro transcribed tRNA[Ile]. The aminoacylation capacity of wild-type transcript has first been compared to that of native tRNA[Ile] extracted from human placental mitochondria, with a mt enzymatic crude extract and proved to be only fourfold less efficiently charged.[85] This made valid the in vitro tRNA transcript approach and allowed investigation on the effects of mutations. Mutations A42569G and A4317G (located in the tRNA T-loop and acceptor stem respectively, Fig. 5) are linked to fatal cardiopathies. While mutation A4269G does not affect aminoacylation, mutation A4317G leads to a 4-fold decrease in aminoacylation.[85] Three ophtalmoplegia-related mutations, replacing A-U or G-C pairs in stems by A•C mismatches, have been investigated using cloned and overproduced mt IleRS.[86] Mutation T4274C is located in the D-stem while mutations T4285C and G4298A are located in the anticodon stem (Fig. 5). For each mutation, aminoacylation was severely attenuated. The intensity of the effects is variable, varying from 25 to 50-fold decreases to total loss of aminoacylation for mutation G4298A. These were the first strong effects on aminoacylation reported for disease-related mutations. The negative impact of all mutations in stems could be overcome by compensatory mutations. Thus, the mutations induce conformational changes that interfere with recognition by the synthetase rather than affecting direct identity elements.[86,87] The mutation located in the T-loop was also proposed to induce structural changes in the tRNA rather than to affect recognition by the aaRS in a direct way.[85]

Interestingly, in addition to the effect of the mutation on the tRNA itself, it has been shown that the mutated tRNAs become strong inhibitors of the aminoacylation of wild-type tRNA.[86] When extrapolated to the in vivo situation of heteroplasmic tissues where wild-type and mutated tRNAs coexist, the effects of a single mutation will be at least at two levels. Indeed, not only will there be less aminoacylated tRNA[Ile] (the mutated tRNA being less well or even not aminoacylated), but there will also be restricted aminoacylation of the wild-type tRNA because of inhibition by the mutant molecules.

Unexpectedly, mutation C4320U in the T-stem, replacing a C•A mismatch by a U-A base-pair, a cardiopathy-related mutation, leads to improved aminoacylation capacity of the mutated tRNA.[87] While aminoacylation efficiency is actually increased by stabilization of a tRNA stem region, it remains striking that improved aminoacylation capacity can lead to a pathology. However, this structural stabilization may become restrictive for optimal activity of the tRNA at a post-aminoacylation step of translation such as interaction with the mito-ribosome which may require flexible structures.[87]

Variability in Effects of Mutations on Aminoacylation

Although only a limited number of pathology-related mutations have been analyzed so far with regard to their possible impact on tRNA aminoacylation, a variety of results was observed. While some mutations strongly hinder aminoacylation, others have no effects, and at least one has a stimulating effect. Worse, the same mutation lead to different effects according to the nuclear background of the investigated cell or cell line and/or to the experimental approach. While the first type of variability could be expected on the basis of the location of mutations within the tRNA structural domains (involved or not in aminoacylation), or on the basis of foreseen structural perturbations, the second type is challenging. A provocative comment entitled "The np 3243 MELAS mutation: Damned if you aminoacylate, damned if you don't" recalled however nicely that while aminoacylation may well be the case of primary effects of mutations, nuclear genetic and perhaps developmental background, as well as mt haplotypes (differences in mt DNA sequences between different individuals) and epigenetic processes, may induce a variety of compensatory mechanisms, leading to distinct outcomes.[88] In vivo aminoacylation analysis may allow to distinguish those mutations affecting aminoacylation from those not affecting this step of the tRNA life only if performed in a sensitive systematic comparative way, in a same nuclear background.[88] Studies on the aminoacylation properties of in vitro transcribed tRNAs would allow to establish kinetic parameters in addition to aminoacylation capacity of the affected tRNAs and would thus allow insight into detailed quantitative effects. This approach may however only become efficient once our knowledge on optimal folding and stabilization of these tRNAs, deprived of post-transcriptional modifications, has improved. It may well be that not yet deciphered mt factors may contribute to the tRNA functions in protecting them against degradation, maintaining and stabilizing their structure (chaperones) and/or playing the role of co-factors to activity. It is expected that the combination of both integrated biology and in vitro molecular biology approaches will allow a better comprehension of the effects of point mutations in human mt tRNA genes at the aminoacylation level.

Perspectives

Human mt tRNAs and aaRSs gained much interest recently, in parallel to the recognition of the possible contribution of defective versions to severe human disorders. Despite pioneering explorations in mammals, our knowledge on both partners of the aminoacylation reaction is still far behind that of prokaryotic or eukaryotic cytosolic systems with most important unsolved questions concerning the structures of the mt tRNAs. As much as 17 out of 22 tRNAs remain to be sequenced at the RNA level, a step required for identifying the post-transcriptional modifications which are likely to be important both for tRNA structure[23] and function.[22] The 3D structures of these tRNAs, although expected of L-shape to fit the ribosomal sites, remain undeciphered. In regard of aaRSs, the genes of 12 of them remain to be found. This should be possible in a near-future, taking advantage of the annotation of different eukaryotic genomes. Aminoacylation studies per se, will remain complicated due to difficulties with tRNA production and handling.

Mt tRNA genes are very sensitive to mutations so that the discovery of new pathogenic mutations is very likely.[9] Mutations affect statistically all positions in the tRNA structure[10] so that it is likely that not all of them will affect aminoacylation, but that various aspects of the tRNA life cycle can be perturbed. Indeed, abnormal transcription,[89-92] incomplete maturation,[93-95] incomplete post-transcriptional modification,[96-98] and lower stability[78,81,83,99] were observed for various mutated tRNAs as compared to corresponding wild-type references. From a functional point of view, disabled codon reading,[22] frameshifting[78,100] as well as slower polysome formation[83] have also been observed. It is well possible that all of these effects, which may individually be mild, combine and cumulate for a given mutation, leading to deleterious mt protein synthesis. Due to the link between mt and nuclear encoded proteins involved in the formation of the respiratory chain complexes, deleterious effects can further spread to the cytosol and its components and contribute to the disease status as well. Comparative proteomic investigations on healthy and disease-carrying mitochondria should help to distinguish protein partners involved in these long-range processes.[101]

Searching for primary effects of mutations in human mt tRNAs at the level of the aminoacylation process remains of major importance in the frame of therapeutic means. Indeed, aaRSs which are nuclear encoded proteins, appear as ideal tools at least at two levels. Transformation of affected cells with the corresponding genes would allow for higher steady-state levels of enzyme in mitochondria, a situation which may enhance aminoacylation of the defective tRNA, as was shown in *E. coli.*[102] Alternatively, aaRSs may become used as shuttles for import of wild-type tRNA which may compensate for the deleterious mutated molecules. Promising experiments along this line have already been reported.[103]

Acknowledgements

We thank Richard Giegé for permanent support and constructive comments on the manuscript. Our investigations on mitochondrial tRNAs are supported by Centre National de la Recherche Scientifique (CNRS), Université Louis Pasteur Strasbourg (ULP), Association Française contre les Myopathies (AFM) and European Comunity grant ALGA-CT-1999-00660 in the frame of the 5th Program.

References

1. Scheffler I. Mitochondria. New-York, Chichester, Weinheim, Brisbane, Singapore, Toronto: John Wiley & Sons, INC., Publication, 1999.
2. Sissler M, Pütz J, Fasiolo F et al. Mitochondrial aminoacyl-tRNA synthetases. In: Ibba M, Francklyn C, Cusack S, eds. The aminoacyl-tRNA synthetases. Georgetown: Landes Biosciences, 2003: in press.
3. Koc EC, Burkhart W, Blackburn K et al. A Proteomics approach to the identification of mammalian mitochondrial small subunit ribosomal proteins. J Biol Chem 2000; 275:32585-32591.
4. Schwarzbach C, Farwell M, KLiao H-X et al. Methods Enzymol 1996; 264:248-261.
5. Hell K, Neupert W, Stuart R. Oxa1p acts as a general membrane insertion machinery for proteins encoded by mitochondrial DNA. EMBO J 2001; 20:1281-1288.
6. DiMauro S, Moraes CT. Mitochondrial encephalomyopathies. Arch Neurol 1993; 50:1197-1208.
7. Schon EA, Bonilla E, DiMauro S. Mitochondrial DNA mutations and pathogenesis. J Bioenerg Biomemb 1997; 29:131-149.
8. Wallace DC. Mitochondrial diseases in man and mouse. Science 1999; 283:1482-1488.
9. DiMauro S, Andreu A. Mutations in mtDNA: are we scraping the bottom of the barrel? Brain Pathol 2000; 10:431-441.
10. Florentz C, Sissler M. Disease-related versus polymorphic mutations in human mitochondrial tRNAs: where is the difference? EMBO Reports 2001; 2:481-486.
11. Attardi G. The elucidation of the human mitochondrial genome. In: Benzi G, ed. Advances in myochemistry. John Libbey Eurotext, 1986:157-171.
12. Attardi G, Schatz G. Biogenesis of mitochondria. Annu Rev Cell Biol 1988; 4:289-333.
13. Ojala D, Montoya J, Attardi G. tRNA punctuation model of RNA processing in human mitochondria. Nature 1981; 290:470-474.
14. Rossmanith W, Tullo A, Potuschak T et al. Human mitochondrial tRNA processing. J Biol Chem 1995; 270:12885-12891.
15. Puranam RS, Attardi G. The RNase P associated with HeLa cell mitochondria contains an essantial RNA component identical in sequence to that of the nuclear RNase P. Mol Cell Biol 2001; 21:548-561.
16. Mohan A, Whyte S, Wang X et al. The 3' end CCA of mature tRNA is an antideterminant for eukaryotic 3'-tRNase. RNA 1999; 5:245-256.
17. Nagaike T, Suzuki T, Tomari Y et al. Identification and characterization of mammalian mitochondrial tRNA nucleotidyltransferase. J Biol Chem 2001; 276:40041-40049.
18. Reichert A, Rothbauer U, Mörl M. Processing and editing of overlapping tRNAs in human mitochondria. J Biol Chem 1998; 273:31977-31984.
19. Reichert A, Thurlow D, Mörl M. A eubacterial origin for the human tRNA nucleotidyltransferase? Biol Chem 2001; 382:1431-1438.
20. Helm M, Brulé H, Degoul F et al. The presence of modified nucleotides is required for cloverleaf folding of a human mitochondrial tRNA. Nucleic Acids Res 1998; 26:1636-1643.
21. Sprinzl M, Horn C, Brown M et al. Compilation of tRNA sequences and sequences of tRNA genes. Nucleic Acids Res 1998; 26:148-153.
22. Yasukawa T, Suzuki T, Ishii N et al. Wobble modification defect in tRNA disturbs codon-anticodon interaction in a mitochondrial disease. EMBO J 2001; 20:4794-4802.
23. Helm M, Giegé R, Florentz C. A Watson-Crick base-pair disrupting methyl group (m1A9) is sufficient for cloverleaf folding of human mitochondrial tRNA^Lys. Biochem 1999; 38:13338-13346.
24. Lynch DC, Attardi G. Amino acid specificity of the transfer RNA species coded for by HeLa cell mitochondrial DNA. J Mol Biol 1976; 102:125-141.
25. King MP, Attardi G. Post-transcriptional regulation of the steady-state levels of mitochondrial tRNAs in HeLa cells. J Biol Chem 1993; 268:10228-10237.
26. Anderson S, Bankier AT, Barrel BG et al. Sequence and organization of the human mitochondrial genome. Nature 1981; 290:457-465.
27. Dock-Bregeon A-C, Moras D. Conformational changes and dynamics of tRNAs: Evidence from hydrolysis patterns. Cold Spring Harbor, Symp Quant Biol 1987; 52:113-121.
28. Dirheimer G, Keith G, Dumas P et al. Primary, secondary and tertiary structures of tRNAs. In: Söll D, RajBhandary UL, eds. tRNA: Structure, Biosynthesis, and Function. Washington, DC: Am Soc Microbiol Press, 1995:93-126.

29. Helm M, Brulé H, Friede D et al. Search for characteristic structural features of mammalian mitochondrial tRNAs. RNA 2000; 6:1356-1379.

30. de Bruijn MH, Schreier PH, Eperon IC et al. A mammalian mitochondrial serine transfer RNA lacking the "dihydrouridine" loop and stem. Nucleic Acids Res 1980; 8:5213-5222.

31. de Bruijn MHL, Klug A. A model for the tertiary structure of mammalian mitochondrial transfer RNAs lacking the entire 'dihydrouridine' loop and stem. EMBO J 1983; 2:1309-1321.

32. Ueda T, Watanabe K, Ohta T. Structural analysis of bovine mitochondrial $tRNA^{Ser(AGY)}$. Nucleic Acids Res 1983; Symposium Series 12:141-144.

33. Watanabe Y-I, Kawai G, Yokogawa T et al. Higher-order structure of bovine mitochondrial $tRNA^{Ser}_{UGA}$: Chemical modification and computer modeling. Nucleic Acids Res 1994; 22:5378-5384.

34. Hayashi I, Yokogawa T, Kawai G et al. Assignment of imino proton signals of G-C base pairs and magnesium ion binding: an NMR study of bovine mitochondrial $tRNA^{Ser}_{GCU}$ lacking the entire D arm. J Biochem 1997; 121:1115-1122.

35. Hayashi I, Kawai G, Watanabe K. Higher-order structure and thermal instability of bovine mitochondrial $tRNA^{Ser}_{UGA}$ investigated by proton NMR spectroscopy. J Mol Biol 1998; 284:57-69.

36. Wakita K, Watanabe Y-I, Yokogawa T et al. Higher-order structure of bovine mitochondrial $tRNA^{Phe}$ lacking the 'conserved' GG and TΨCG sequences as inferred by enzymatic and chemical probing. Nucleic Acids Res 1994; 22:347-353.

37. Leehey MA, Squassoni CA, Friederich MW et al. A noncanonical tertiary conformation of a human mitochondrial transfer RNA. Biochem 1995; 34:16235-16239.

38. Wolstenholme DR, Macfarlane JL, Okimoto R et al. Bizarre tRNAs inferred from DNA sequences of mitochondrial genomes of nematode worms. Proc Natl Acad Sci USA 1987; 84:1324-1328.

39. Wolstenholme DR, Okimoto R, Mcfarlane JL. Nucleotide correlations that suggest tertiary interactions in the TV-replacement loop-containing mitochondrial tRNAs of the nematodes, Caenorhabditis elegans and Ascaris suum. Nucleic Acids Res 1994; 22:4300-4306.

40. Steinberg S, Gautheret D, Cedergren R. Fitting the structurally diverse animal mitochondrial $tRNAs^{Ser}$ to common three-dimensional constraints. J Mol Biol 1994; 236:982-989.

41. Steinberg S, Cedergren R. Structural compensation in atypical mitochondrial tRNAs. Nature Struct Biol 1994; 1:507-510.

42. Steinberg S, Leclerc F, Cedergren R. Structural rules and conformational compensations in the tRNA L-form. J Mol Biol 1997; 266:269-282.

43. Agris PF. The importance of being modified: Roles of modified nucleosides and Mg2+ in RNA structure and function. Prog Nucleic Acid Res Mol Biol 1996; 53:79-129.

44. Nass MMK, Buck CA. Studies on mitochondrial tRNA from animal cells. II. Hybridization of aminoacyl-tRNA from rat liver mitochondria with heavy and light complementary strands of mitochondrial DNA. J Mol Biol 1970; 54:187-198.

45. Kumazawa Y, Yokogawa T, Hasegawa E et al. The aminoacylation of structurally variant phenylalanine tRNAs from mitochondria and various nonmitochondrial sources by bovine mitochondrial phenylalanyl-tRNA synthetase. J Biol Chem 1989; 264:13005-13011.

46. Kumazawa Y, Himeno H, Miura K-I et al. Unilateral aminoacylation specificity between bovine mitochondria and eubacteria. J Biochem 1991; 109:421-427.

47. Shiba K, Schimmel P, Motegi H et al. Human glycyl-tRNA synthetase. Wide divergence of primary structure from bacterial counterpart and species-specific aminoacylation. J Biol Chem 1994; 269:30049-30055.

48. Shiba K, Suzuki N, Shigesada K et al. Human cytoplasmic isoleucyl-tRNA synthetase: Selective divergence of the anticodon-binding domain and acquisition of a new structural unit. Proc Natl Acad Sci USA 1994; 91:7435-7439.

49. O'Hanlon TP, Raben N, Miller FW. A novel gene oriented in a head-to-head configuration with the human histidyl-tRNA synthetase (HRS) gene encodes an mRNA that predicts a polypeptide homologous to HRS. Biochem Biophys Res Commun 1995; 210:556-566.

50. Bullard J, Cai Y-C, Demeler B et al. Expression and characterization of a human mitochondrial phenylalanyl-tRNA synthetase. J Mol Biol 1999; 288:567-577.

51. Bullard J, Cai Y-C, Spremulli L. Expression and characterization of the human mitochondrial leucyl-tRNA synthetase. Biochim Biophys Acta 2000; 1490:245-258.

52. Tolkunova E, Park H, Xia J et al. The human lysyl-tRNA synthetase gene encodes both the cytoplasmic and mitochondrial enzymes by means of an unusual splicing of the primary transcript. J Biol Chem 2000; 275:35063-35069.

53. Jörgensen R, Sögarrd MM, Rossing AB et al. Identification and characterization of human mitochondrial tryptophanyl-tRNA synthetase. J Biol Chem 2000; 275:16820-16826.

54. Yokogawa T, Shimada N, Takeuchi N et al. Characterization and tRNA recognition of mammalian mitochondrial seryl-tRNA synthetase. J Biol Chem 2000; 275:19913-19920.

55. Cavalier-Smith T. The simultaneous symbiotic origin of mitochondria, chloroplasts, and microbodies. Ann NY Acad Sci 1987; 503:55-71.

56. Gray MW. Origin and evolution of mitochondrial DNA. Annu Rev Cell Biol 1989; 5:25-50.

57. Ribas de Pouplana L, Schimmel P. Visions & Reflections. A view into the origin of life: aminoacyl-tRNA synthetases. Cell Mol Life Sci 2000; 57:865-870.

58. Chihade JW, Brown JR, Schimmel P et al. Origin of mitochondria in relation to evolutionary history of eukaryotic alanyl-tRNA synthetase. Proc Natl Acad Sci USA 2000; 97:12153-12157.

59. Eriani G, Delarue M, Poch O et al. Partition of tRNA synthetases into two classes based on mutually exclusive sets of sequence motifs. Nature 1990; 347:203-206.
60. Cusack S, Berthet-Colominas C, Härtlein M et al. A second class of synthetase structure revealed by X-ray analysis of *Escherichia coli* seryl-tRNA synthetase. Nature 1990; 347:249-255.
61. Sanni A, Walter P, Boulanger Y et al. Evolution of aminoacyl-tRNA synthetase quaternary structure and activity: *Saccharomyces cerevisiae* mitochondrial phenylalanyl-tRNA synthetase. Proc Natl Acad Sci USA 1991; 88:8387-8391.
62. Buck CA, Nass MMK. Studies on mitochondrial tRNA from animal cells. I. A comparison of mitochondrial and cytoplasmic tRNA and aminoacyl-tRNA synthetases. J Mol Biol 1969; 41:67-82.
63. Kumazawa Y, Yokogawa T, Miura K et al. Bovine mitochondrial tRNA^{Phe}, tRNA$^{Ser(AGY)}$ and tRNA$^{Ser(UCN)}$: preparation using a new detection method and their properties in aminoacylation. Nucleic Acids Symp Ser 1988; 19:97-100.
64. Giegé R, Sissler M, Florentz C. Universal rules and idiosyncratic features in tRNA identity. Nucleic Acids Res 1998; 26:5017-5035.
65. Giegé R, Frugier M. Transfer RNA structure and identity. In: Lapointe J, Brakier-Gringas L, eds. Translation Mechanisms. Georgetown: Landes Sciences, 2003: in press.
66. Ueda T, Yotsumoto Y, Ikeda K et al. The T-loop region of animal mitochondrial tRNA$^{Ser(AGY)}$ is a main recognition site for homologous seryl-tRNA synthetase. Nucleic Acids Res 1992; 20:2217-2222.
67. Shimada N, Suzuki T, Watanabe K. Dual mode of recognition of two isoacceptor tRNAs by mammalian mitochondrial seryl-tRNA synthetase. J Biol Chem 2001; 276:46770-46778.
68. Goto Y, Nonaka I, Horai S. A mutation in the tRNA$^{Leu(UUR)}$ gene associated with the MELAS subgroup of mitochondrial encephalomyopathies. Nature 1990; 348:651-653.
69. Kogelnik AM, Lott MT, Brown MD et al. MITOMAP: a human mitochondrial genome database -1998 update. Nucleic Acids Res 1998; 26:112-115.
70. Servidei S. Mitochondrial encephalomyopathies: gene mutations. Neuromuscul Disord 2001; 11:508-513.
71. Larsson N-G, Clayton DA. Molecular genetic aspects of human mitochondrial disorders. Annu Rev Genetics 1995; 29:151-178.
72. Schon E. Mitochondrial genetics and disease. Trends Biochem Sci 2000; 25:555-560.
73. Enriquez J, Attardi G. Analysis of aminoacylation of human mitochondrial tRNAs. Methods Enzymol 1996; 264:183-196.
74. Varshney U, Lee CP, RajBhandary UL. Direct analysis of aminoacylation levels of tRNAs in vivo. Application to studying recognition of E. coli initiator tRNA mutants by glutaminyl-tRNA synthetase. J Biol Chem 1991; 266:24712-24718.
75. King MP, Attardi G. Human cells lacking mtDNA: repopulation with exogenous mitochondria by complementation. Science 1989; 246: 500-503.
76. Chomyn A, Meola G, Bresolin N et al. In vitro genetic transfer of protein synthesis and respiration defects to mitochondrial DNA-less cells with myopathy-patient mitochondria. Mol Cell Biol 1991; 11:2236-2244.
77. Chomyn A, Lai S, Shakeley R et al. Platelet-mediated transformation of mtDNA-less human cells: analysis of phenotypic variability among clones from normal individuals-and complementation behavior of the tRNA^{Lys} mutation causing myoclonic epilepsy and ragged red fibers. Am J Hum Genet 1994; 54:966-974.
78. Enriquez JA, Chomyn A, Attardi G. MtDNA mutation in MERRF syndrome causes defective aminoacylation of tRNA^{Lys} and premature translation termination. Nature Gen 1995; 10:47-55.
79. Enriquez JA, Attardi G. Evidence for aminoacylation-induced conformational changes in human mitochondrial tRNAs. Proc Natl Acad Sci USA 1996; 93:8300-8305.
80. Börner GV, Zeviani M, Tiranti V et al. Decreased aminoacylation of mutant tRNAs in MELAS but not in MERRF patients. Hum Mol Genet 2000; 9:467-475.
81. Hao H, Moraes CT. A disease-associated G5703A mutation in human mitochondrial DNA causes a conformational change and a marked decrease in steady-state levels of mitochondrial tRNA^{Asn}. Mol Cel Biol 1997; 17:6831-6837.
82. Asahara H, Himeno H, Tamura K et al. Recognition nucleotides of *Escherichia coli* transfer RNALeu and its elements facilitating discrimination from transfer RNASer and transfer RNATyr. J Mol Biol 1993; 231:219-229.
83. Chomyn A, Enriquez JA, Micol V et al. The mitochondrial myopathy, encephalopathy, lactic acidosis, and stroke-like episode syndrome-associated human mitochondrial tRNA$^{Leu(UUR)}$ mutation causes aminoacylation deficiency and concomitant reduced association of mRNA with ribosomes. J Biol Chem 2000; 275:19198-19209.
84. Janssen G, Maassen J, van den Ouweland J. The diabetis-associated 3243 mutation in the mitochondrial tRNA$^{Leu(UUR)}$ gene causes severe mitochondrial dysfunction without a strong decrease in protein synthesis rate. J Biol Chem 1999; 274:29744-29748.
85. Degoul F, Brulé H, Cepanec C et al. Isoleucylation properties of native human mitochondrial tRNA^{Ile} and tRNA^{Ile} transcripts. Implications for cardiomyopathy-related point mutations (4269, 4317) in the tRNA^{Ile} gene. Hum Mol Gen 1998; 7:347-354.
86. Kelley S, Steinberg S, Schimmel P. Functional defects of pathogenic human mitochondrial tRNAs related to structural fragility. Nature Struc Biol 2000; 7:862-865.
87. Kelley SO, Steinberg SV, Schimmel P. Fragile T-stem in disease-associated human mitochondrial tRNA sensitizes structure to local and distant mutations. J Biol Chem 2001; 276:10607-106 11.
88. Jacobs HT, Holt IJ. The np 3243 MELAS mutation: damned if you aminoacylate, damned if you don't. Hum. Mol. Genet. 2000; 9:463-465.

89. Bindoff LA, Howell N, Poulton J et al. Abnormal RNA processing associated with a novel tRNA mutation in mitochondrial DNA. A potential disease mechanism. J Biol Chem 1993; 268:19559-19564.

90. King MP, Koga Y, Davidson M et al. Defects in mitochondrial protein synthesis and respiratory chain activity segregate with the tRNA$^{Leu(UUR)}$ mutation associated with mitochondrial myopathy, encephalopathy, lactic acidosis and strokelike episodes. Mol Cell Biol 1992; 12:480-490.

91. Koga Y, Davidson M, Schon EA et al. Analysis of cybrids harboring MELAS mutations in the mitochondrial tRNA$^{Leu(UUR)}$ gene. Muscle & Nerve 1995; Suppl 3:S119-S123.

92. Schon EA, Koga Y, Davidson M et al. The mitochondrial tRNA$^{Leu(UUR)}$ mutation in MELAS: A model for pathogenesis. Biochim Biophys Acta 1992; 1101:206-209.

93. Flierl A, Reichmann H, Seibel P. Pathophysiology of the MELAS 3243 transition mutation. J Biol Chem 1997; 272:27189-27196.

94. Rossmanith W, Karwan R. Impairment of tRNA processing by point mutations in mitochondrial tRNA$^{Leu(UUR)}$ associated with mitochondrial diseases. FEBS Lett 1998; 433:269-274.

95. Levinger L, Jacobs O, James M. In vitro 3' end endonucleolytic processing defect in a human mitochondrial tRNA$^{Ser(UCN)}$ precursor with the U7445C substitution, which causes nonsyndromic deafness. Nucleic Acids Res 2001; 29:4334-4340.

96. Helm M, Florentz C, Chomyn A et al. Search for differences in post-transcriptional modification patterns of mitochondrial DNA-encoded wild-type and mutant human tRNA^{Lys} and tRNA$^{Leu(UUR)}$. Nucleic Acids Res 1999; 27:756-763.

97. Yasukawa T, Suzuki T, Suzuki T et al. Modification defect at anticodon wobble nucleotide of mitochondrial tRNAs$^{Leu(UUR)}$ with pathogenic mutations of mitochondrial myopathy, encephalopathy, lactic acidosis and stroke-like episodes. J Biol Chem 2000; 275:4251-4257.

98. Yasukawa T, Suzuki T, Ishii N et al. Defect in modification at the anticodon wobble nucleotide of mitochondrial tRNA^{Lys} with the MERRF encephalomyopathy pathogenic mutation. FEBS Lett 2000; 467:175-178.

99. Yasukawa T, Hino N, Suzuki T et al. A pathogenic point mutation reduces stability of mitochondrial mutant tRNA^{Ile}. Nucleic Acids Res 2000; 28:3779-3784.

100. Masucci JP, Davidson M, Koga Y et al. In vitro analysis of mutations causing myoclonus epilepsy with ragged-red fibers in the mitochondrial tRNA^{Lys} gene : two genotypes produce similar phenotypes. Mol Cell Biol 1995; 15:2872-2881.

101. Rabilloud T, Strub JM, Carte N et al. Comparative proteomics as a new tool for exploring human mitochondrial tRNA disorders. Biochem 2001; 41:144-150.

102. Sherman JM, Rogers MJ, Söll D. Competition of aminoacyl-tRNA synthetases for tRNA ensures the accuracy of aminoacylation. Nucleic Acids Res 1992; 20:2847-2852.

103. Kolesnikova O, Entelis N, Mireau H et al. Suppression of mutations in mitochondrial DNA by tRNAs imported from the cytoplasm. Science 2000; 289:1931-1933.

mRNA decay in *Escherichia coli*: Enzymes, Mechanisms and Adaptation

Rudolf K. Beran, Annie Prud'homme-Généreux, Kristian E. Baker,
Xin Miao, Robert W. Simons and George A. Mackie

Abstract

The well recognized metabolic lability of mRNA places serious constraints on the yield of polypeptides which can be translated from a single mRNA. This property can also be exploited as an important mode of post-transcriptional regulation of gene expression. This chapter presents a model for mRNA decay in *Escherichia coli* that embraces several concepts initially recognized by David Apirion. Progress of the last decade has permitted a significant elaboration of his model to encompass the currently identified enzymes of mRNA decay, including a macromolecular complex, the RNA degradosome, which is specialized for degrading highly structured RNA fragments. Emphasis is placed on the finding that the single-strand-specific endoribonuclease, RNase E, widely believed to initiate the decay process on most mRNAs, displays 5'-end dependence which can rationalize several hitherto puzzling aspect of mRNA degradation. Three 3'-exoribonucleases, polynucleotide phosphorylase, RNase II and oligoribonuclease, act during later stages of the decay pathway to remove the initial products of endonucleolytic cleavage. In addition, the modification of mRNAs by polyadenylation facilitates the turnover of structured RNA fragments as well as misfolded RNAs. Finally, recent findings have documented the potential of the mRNA decay machinery to respond to environmental stresses, notably to cold shock.

Importance of mRNA Decay in Prokaryotes

Metabolic instability is a widely shared property of mRNAs.[1-3] Generally, mRNA lifetimes are short relative to cellular doubling times whereas stable RNAs such as tRNA and rRNA remain intact and functional for at least several generations. Thus in *Escherichia coli*, the prokaryote in which the process of mRNA decay is best characterized, many mRNAs exhibit half-lives in the order of 60-120 sec, although a few mRNAs (e.g., *ompA* mRNA) are significantly more stable, with half-lives of 15 min or more.[4] Why should mRNAs be so unstable? The reasons are both physical and functional. Stable RNAs in *E. coli* and other bacteria are tightly associated with proteins and/or are extensively folded into stable structures either or both of which effectively preclude access by the RNases implicated in mRNA turnover (see below). Even tRNA, which is not permanently liganded to a binding protein, associates with aminoacyl tRNA synthetases, EF-Tu, or the ribosome, thereby protecting itself. In contrast, although mRNAs can and do assume stable secondary structures, their coding sequences, their principal *raison d'être*, largely function in single-stranded form. Consequently, mRNAs are intrinsically vulnerable to endonuclease attack. Arguably, the most important functional consequence of the instability of mRNAs is the potential for rapid repression of gene expression in response to changes in environment. This property permits the amplification of negative regulatory signals.[5-7] Indeed, mRNA decay can account for the differential expression of gene products in some polycistronic mRNAs.[7] Regulation of gene expression by natural antisense RNAs (e.g., in IS10 or colE1 replicons) frequently relies on selective mRNA decay.[8-12] Interestingly, newly identified regulatory mechanisms in eukaryotes such as mRNA surveillance (including "nonsense-mediated

Translation Mechanisms, edited by Jacques Lapointe and Léa Brakier-Gingras.

decay") and RNA interference ("RNAi") provoke targeted mRNA degradation.[13-15] Finally, the constant turnover of mRNAs permits the recycling of ribonucleotides. This may be a significant contribution to energy metabolism in *E. coli*, as roughly 60% of the nascent RNA chains in vivo correspond to mRNAs.[16]

Despite the importance of mRNA decay in post-transcriptional regulation of gene expression, our knowledge of the process is still incomplete, particularly in organisms other than *E. coli* or *S. cerevisiae*. The rapidly increasing number of sequenced microbial genomes will help fill this gap, at least in part, by clarifying which enzymes of mRNA decay are conserved across species boundaries. This information will permit assessment of the extent to which the overall process is also conserved. Nonetheless, "the devil is in the details", and much remains to be learned about how individual mRNAs are recognized during mRNA decay and how processing/decay events influence the expression of the protein products of individual mRNAs. Much of the progress of recent years in *E. coli* is due to the choice of small, tractable mRNAs as models, either in vivo or in vitro. There is, consequently, a dearth of information regarding more complex, polycistronic mRNAs, particularly in vitro.

mRNA Decay in *Escherichia coli*: A Current Model

mRNA decay in *E. coli* has been reviewed to various degrees of depth by several authors in the past four years.[17-20] The following provides a synopsis of models of mRNA decay and a brief comparison with those applicable to other organisms. Until recently, mRNA decay in *E. coli* was best described by a model elaborated by David Apirion,[21] based on ideas and data accumulated by him and by others including Kennell, Kepes, Schlessinger and Yanofsky dating to the late 1960s.[17-24] This model (see Fig. 1A) predicted that the initial step in mRNA decay was an endonucleolytic cleavage followed by scavenging of the newly generated 3' ends by 3' to 5' exonucleases.[21] Moreover, successive endonucleolytic cleavages would result in a net 5' to 3' disappearance of mRNA (see below). Several features of this model have proven valid. Nonetheless, it suffered from major gaps, most crucially, in the identity of the initiating endonuclease, but also in explaining the large differences in stability exhibited among different mRNA species. A major accomplishment of the past decade has been the identification of many of the enzymes and factors involved in mRNA decay in *E. coli* (see Table 1), the cloning of their genes, the characterization of cleavage sites, the characterization of degradative intermediates in a number of mRNAs, the reconstitution of decay of a natural mRNA from purified components,[25-26] and the demonstration of a mechanism for 5'→3' vectorial mRNA decay.[27-29] This new information has permitted significant elaboration of Apirion's model, summarized in the "best current model" shown in Figure 1B and discussed in the following sections. Remarkably, several fundamental features of his model have remained substantially intact.

Table 1 summarizes the RNases and other enzymes known to play a role in RNA processing and decay in *E. coli*. Collectively, the 15 documented RNases can account for all RNA processing in *E. coli* except for the final 5'-maturation of 5S rRNA and the 3'-maturation of 16S rRNA. There is extensive functional redundancy and only three enzymes, RNase E, RNase P and oligo-RNase are essential. The list of nonnucleolytic "factors" which can modulate mRNA decay is undoubtedly incomplete and others will be elucidated as the metabolism of specific RNAs is investigated.

The Initiating Step in mRNA Decay

Work from numerous groups during the past decade has shown that the initiating event in mRNA decay is almost always catalyzed by the endoribonuclease, RNase E (see Fig. 1B and Table 1).[17-20,22-24] This enzyme was first characterized by D. Apirion and colleagues as an essential processing enzyme required for the penultimate steps in the maturation of 5S rRNA.[30] RNase E is now known to be involved as well in processing the 5'-spacer region of 16S rRNA[31-32] and most, if not all, tRNA precursors.[31a,31b] These functions probably account for it being an essential enzyme. Two mutations in RNase E, *rne-1* (G66S; formerly called *ams-1*) and *rne-3071* (L68F), render its activity thermolabile[33-35] and, more significantly, reduce the rate of decay of most mRNAs in vivo.[36-41] These results point to a central role for RNase E in both mRNA decay and stable RNA processing. Characterization of RNase E was hampered by difficulties in its purification and in cloning its gene.[17] Its large size, 1061 aa, was unexpected and hinted at its capacity for interactions with other macromolecules. In fact, the large Rne polypeptide bears at least three distinguishable domains: an

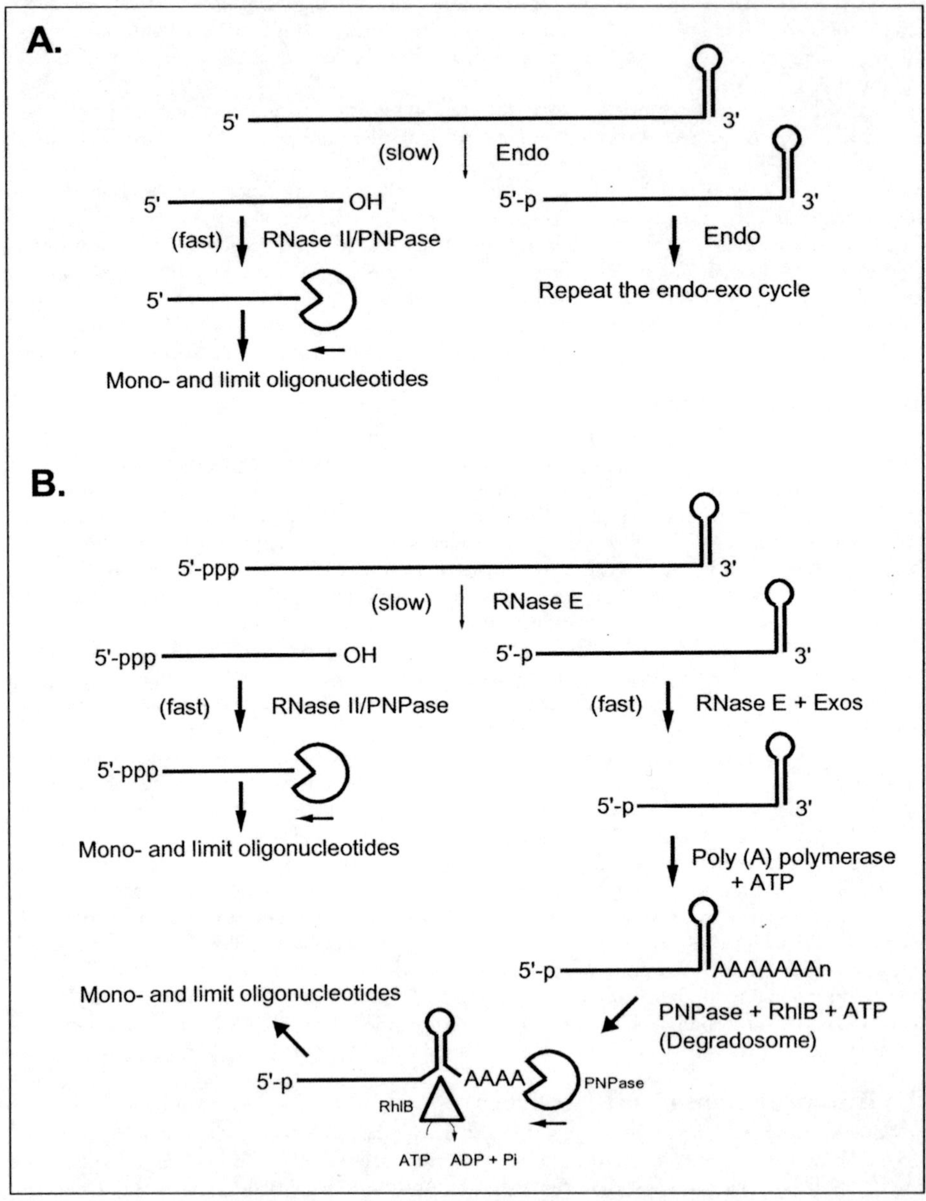

Figure 1. Models for mRNA decay. Panel A. Apirion's original model (adapted from ref. 21). Degradation of mRNA is initiated by an endoribonuclease; the initial cleavage products are attacked by 3'→5' exoribonucleases (PNPase and RNase II). Panel B. A current model for mRNA decay. RNase E initiates the decay process with a slow endonucleolytic cleavage dependent on the 5'-triphosphate of the substrate RNA; subsequent cleavages are accelerated by the 5'-monophosphate group on the initial cleavage product.[27] A terminal stem-loop is resistant to the exoribonucleases, but can be "edited" by poly(A) polymerase I, making its susceptible to the action of PNPase, facilitated by an ATP-dependent RNA helicase.

Table 1. Ribonucleases and mRNA decay factors in Escherichia coli

	Activity	Function	Comments
Endo-RNases	RNase I	scavenger	periplasmic
	RNase III	**mRNA decay &** rRNA processing	double-stranded RNA
	RNase E	**mRNA decay &** rRNA processing	major endonuclease; single-strand specific (†)
	RNase G	16S rRNA 5' end	homolog of RNase E
	RNase H	DNA replication	removal of RNA primers
	RNase HII	?	
	RNase P	tRNA 5' ends	ribozyme (†)
Exo-RNases	RNase II	**mRNA decay**	3'-exo; hydrolytic (‡, ●)
	RNase BN	tRNA 3' ends	3'-exo; hydrolytic
	RNase D	tRNA 3' ends	3'-exo; hydrolytic (●)
	RNase PH	tRNA 3' ends	3'-exo; phosphorolytic (●)
	RNase R	?	3'-exo; hydrolytic (§)
	RNase T	tRNA, 5S rRNA	3'-exo; hydrolytic
	PNPase	**mRNA decay**	3'-exo; phosphorolytic (‡)
	Oligo-RNase	**mRNA decay**	3'-exo; hydrolytic (†)
Factors	PAP I	**mRNA decay**	poly(A) polymerase
	RhlB	**mRNA decay**	DEAD-box RNA helicase
	DeaD/CsdA	cold shock	DEAD-box RNA helicase
	Hfq	several	ss RNA binding protein
	FinO	FinP RNA decay	RNA binding protein

Notes: † Function is essential in *E. coli*
‡ Functionally redundant; a double mutant is conditionally lethal.
§ Virulence factor homologous to RNase II; *aka* VacB
● Homolog found in the eukaryotic exosome

essential catalytic domain encompassing the N-terminal 500 residues, an apparently dispensable arginine-rich RNA binding domain and a degradosome assembly domain (platform) at the C-terminus (see Fig. 2 and below). Although RNase E is single-strand specific and frequently cleaves RNA 5' to AUu/c trinucleotides, its sequence specificity, if any, is still unclear.[17] RNase E can also catalyze the shortening of poly(A) or poly(U) tails in vitro, but this is simply an additional manifestation of its endonucleolytic activity.[42,43]

Despite the documented importance of RNase E in mRNA metabolism, the stability of a few mRNAs is not affected by *rne* mutations;[44] in these cases, some other endonuclease must initiate decay. RNase G, a close homolog of RNase E, is one such candidate although its primary function appears to be maturation of the 5'-end of 16S rRNA.[31-32,45-46] The recent demonstration that inactivation of RNase G can stabilize the *adhE* mRNA encoding alcohol dehydrogenase by over 2-fold reinforces this supposition.[47] Two groups have also postulated the existence of a poly(A)-dependent endoribonuclease although firm evidence for such an activity is lacking.[48-49] Nevertheless, it would be prudent to maintain an open mind regarding the existence of additional endoribonucleases in *E. coli*. In this regard, the remarkable processing of the *daaABCDPE* mRNA which requires translation of a short open reading frame, *daaP*,[50] cannot readily be explained by any of the activities listed in Table 1. In a few well-documented cases, RNase III, a double-strand-specific endonuclease, can initiate the degradative process.[51-53] However, functional inactivation of the gene encoding RNase III does not exert a general phenotype on mRNA decay.[54] Nonetheless, in one case, RNase III cleavage of its own mRNA renders the transcript sensitive to attack by RNase E (Matsunaga J, Simons RW, unpublished data). In another instance, RNase III cleavage actually stabilizes the cIII mRNA of bacteriophage lambda.[55]

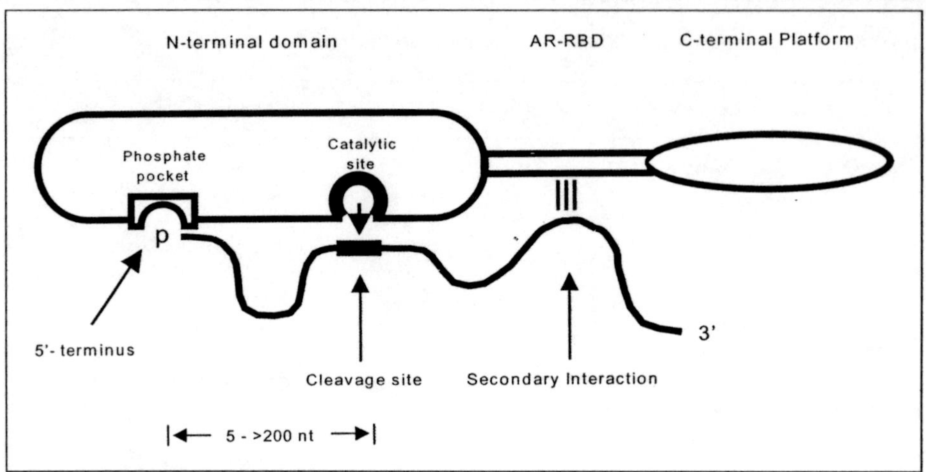

Figure 2. Simplified domain structure of RNase E. The three domains of RNase E encompass the following residues: N-terminal (catalytic) domain, residues 1-498;[60] AR-RBD (Arg-rich RNA binding domain), residues 500-700; C-terminal platform (binding site for RhlB, enolase and PNPase), residues 700-1061.[105] In the speculative model shown here, the putative phosphate binding pocket is located distant from the actual catalytic site in the N-terminal domain. Refer to Figure 3 for an alternative model. The 5' end of the RNA substrate would be engaged by the phosphate binding pocket, mono-phosphates more efficiently than triphosphates (see the text).[27,29] Additional interactions with the substrate may be provided by the ARRBD.

RNA binding proteins add a layer of complexity as they can modulate the action of RNase E. In the simplest case, the FinO protein masks an RNase E cleavage site in the *finP* RNA and stabilizes it, amplifying its antisense effects.[56] In a second example, Hfq appears to confer growth-rate-sensitive decay to the *ompA* mRNA, by reducing translational efficiency at low growth rates, thereby improving the ability of RNase E to access cleavage sites in the *ompA* leader.[57-58] We predict that as more mRNAs are investigated, additional examples of modulation of RNase E cleavage by bound proteins will emerge. Finally, ribosomes can play a major role, directly or indirectly, in controlling the stability of mRNAs. Chapter 10 (by M. Dreyfus and S. Joyce) in this volume provides a detailed review of the fascinating interplay between translation and mRNA stability.

Propagation of the Initiating Cleavage

A single endonucleolytic cleavage is insufficient to inactivate most mRNAs and especially those which are polycistronic. Apirion postulated that additional, distal endonucleolytic cleavages would follow the initial event.[21] The rapid removal of the 5' product of each endonucleolytic cleavage by 3' to 5' exonucleases would account for the apparent 5'-3' directionality of mRNA decay suggested by kinetic experiments. Indeed, the processing of the phage f1 mRNAs illustrates a "wave" of 5' to 3' RNase E cleavages exceptionally well.[19] The underlying mechanism for this phenomenon remained enigmatic until the surprising finding that RNase E is a 5'-end-dependent RNase, the first such example.[27] The basis for this conclusion relied on three lines of evidence. First, circular RNAs containing known sites of cleavage are relatively resistant to the action of purified RNase E, being cleaved 6-fold more slowly both in vitro and in vivo.[27,28] Second, RNA heteroduplexes between a substrate and a 13- residue DNA oligonucleotide are 5-7-fold more resistant to RNase E if the DNA oligonucleotide is flush with the 5'-end of the substrate compared to being annealed internally.[27] Finally, 5'-triphosphorylated (ppp) RNA substrates are cleaved up to 20-25-fold more slowly than otherwise identical 5'-monophosphorylated (p) substrates.[27,29] Extrapolating these in vitro data to the process of mRNA decay in vivo, the rate-limiting initiating cleavage by RNase E would depend on a single-stranded 5'-triphosphate (ppp) terminus to provide the initial, but weak, contact between an mRNA and this enzyme. The subsequent cleavage would occur at a nearby or moderately distant site to yield a 3' product with a 5'-monophosphate (p) terminus. This 5'-p mRNA fragment would then tether the substrate to RNase E relatively tightly resulting in ≥20-fold more rapid cleavage of the p-mRNA degradative intermediate than on the initial ppp-mRNA (Fig. 1B). Others have since confirmed the 5'-end dependence of RNase E, have mapped this property to its N-terminal 498 residues (which contains the catalytic site), and have extended it to the homologous enzyme, RNase G.[46,59]

Figure 2 shows a simple representation of RNase E which can explain its 5'-end-dependence. In this elementary model, a phosphate binding pocket is located in the N-terminal catalytic domain of RNase E, distinct from the actual catalytic site.[29,46,59,60] The two points of contact between RNase E and a substrate would enhance the enzyme's affinity for a susceptible RNA. Additional contacts have been documented between substrates and residues C-terminal to the catalytic domain in RNase E[59,62,63] and are shown in Figure 2. A more sophisticated model has been proposed in which RNase E is at least a dimer, and the catalytic sites on each subunit alternate between cleavage and product retention.[17] Thus, the phosphate binding pocket would simply be the catalytic site in another guise. An updated version of this model is shown in Figure 3. Regardless of the detailed mechanism, the 5'-end-dependence of RNase E rationalizes several fundamental observations about mRNA decay in vivo. First, it explains the "all-or-none" phenomenon[24,61] whereby the initial rate-limiting cleavage of an mRNA is followed by rapid disappearance of the cleavage products. This ensures that the decay apparatus completes the degradation of inactivated mRNAs rather than randomly nicking additional mRNAs. Second, the 5'- end-dependence of RNase E reinforces and accelerates the 5'→3' vectorial nature of mRNA decay.[19,21] This is particularly true for longer mRNAs where each successive RNase E cleavage provides a new entry point for the 3'-exonucleases.[19] Finally, this property can explain the longevity of mRNAs whose secondary structure renders the 5'- terminus inaccessible to RNase E.

Some mRNAs (e.g., *ompA* mRNA) possess stable 5'-terminal stem-loops which retard, but do not abolish, decay.[64-68] Likewise, circular forms of the *rpsT* mRNA are more stable than their linear counterparts, in vivo and in vitro.[27,28] In these cases, the initiating cleavage must occur in a slower, end-independent fashion. In this <u>secondary</u> pathway, the "internal entry model", shown in Figure 4, RNase E bypasses the normal 5'-end contact, albeit inefficiently, engaging a cleavage site directly. This slow initial cleavage then triggers a shift to the 5'-end-dependent mode of decay. Internal entry is highly sensitive to translational efficiency[28,58,66,68a] perhaps because ribosomes mask cleavage sites transiently during their passage along an mRNA. Supporting this idea, the pausing of ribosomes at the termination codon of the *rpsO* mRNA has been shown to mask access of RNase E to the rate-limiting M2 site in this mRNA.[69] Alternatively, internal entry could involve prior nonspecific binding of RNase E to a target RNA ("docking") followed by a search for a cleavage site by "gliding" or "exchange" (not unlike the interaction of RNA polymerase or Lac repressor with DNA). Translating ribosomes could compete for the docking event or could displace RNase E from the mRNA, retarding its direct search for a cleavage site. Interestingly, the best evidence for a docking event is provided by the autoregulatory cleavage of the *rne* mRNA itself where a conserved stem-loop complex in the 5-untranslated region is essential for autoregulation.[70]

Exonucleases and 3'-End Scavenging

Following the initial endonucleolytic cleavage of an mRNA, a new 3'-end is generated on the 5'-cleavage product (Fig. 1).[21,71,72] This and subsequently formed 3'-ends are rapidly attacked by one or more 3'-exonucleases. Among the eight such enzymes in *E. coli* (Table 1), RNase II accounts for 90% of the activity directed against large mRNA fragments, with polynucleotide phosphorylase (PNPase) accounting for the remainder.[73] In contrast, phosphorylytic exonuclease activity predominates in *Bacillus subtilis*.[73] Both RNase II and PNPase have been known for many years; indeed, PNPase was an essential tool in the elucidation of the Genetic Code.[74] RNase II which acts hydrolytically has been thought to be a monomer in solution,[75] but recent biophysical data suggest that it is a dimer or tetramer (Mosimann S, personal communication). In contrast, PNPase is unambiguously trimeric[76] and acts phosphorylytically, using inorganic phosphate as the nucleophile for phosphodiester bond cleavage, liberating nucleoside 5'-diphosphates.[71] Otherwise, the two enzymes appear to act almost interchangeably and are largely indistinguishable in their action on synthetic RNA substrates containing single-stranded extensions of at least 6-9 residues 3' to a stable stem-loop

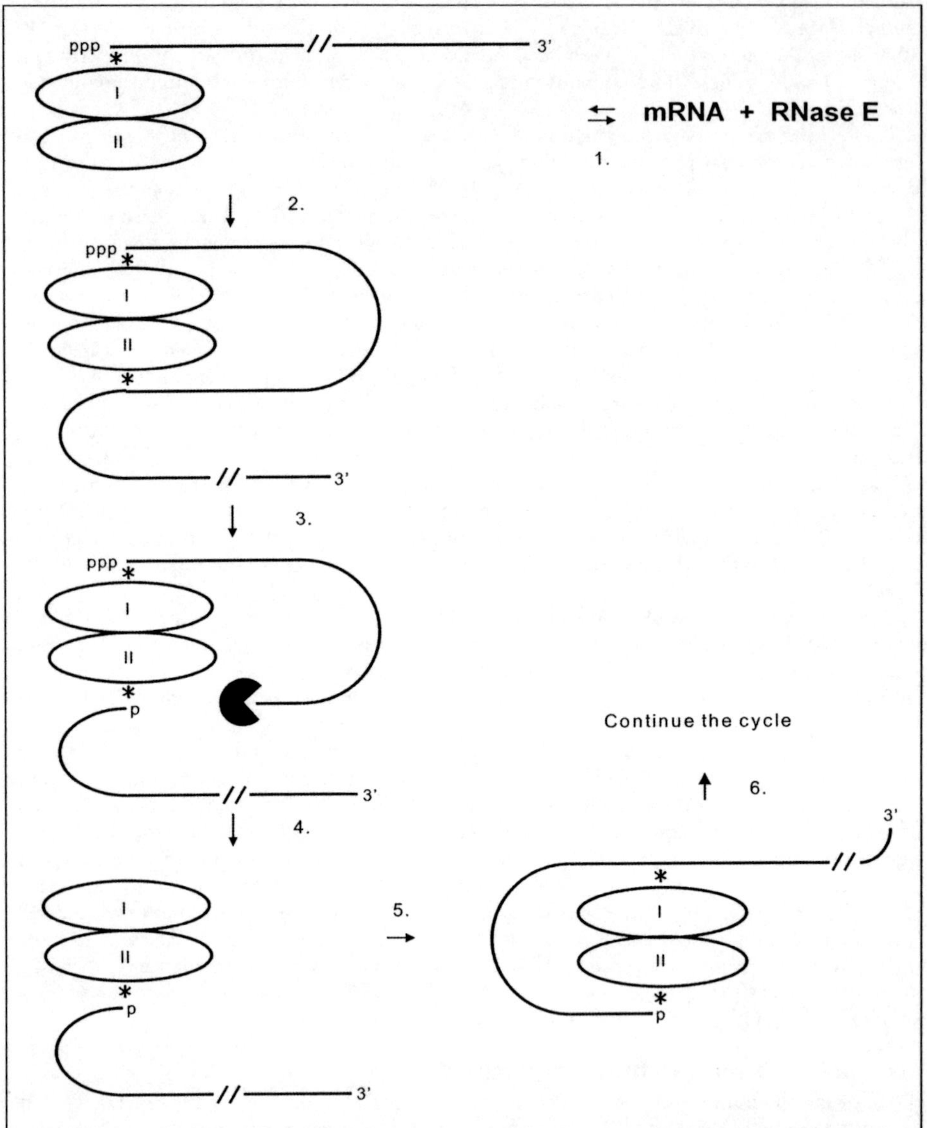

Figure 3. Alternating sites model for RNase E action. This model was originally proposed by Coburn and Mackie[17] and is presented here in slightly revised form to take into account newer data. In this model, the phosphate binding pocket and the active site are physically identical, but function alternatively. Subunit I in an RNase E dimer (or higher oligomer) initially contacts the 5'-end of the substrate RNA (step 1). The second subunit contacts a distant cleavage site by looping out the intervening RNA (step 2), stabilizing the mRNA-RNase E complex. Cleavage occurs exposing a new 3'- terminus which is attacked by a 3'-exonuclease (step 3). The removal of the 5'- fragment exposes the active site on subunit I while subunit II remains tightly bound to the 5'-P terminus revealed by the prior cleavage (step 4). Subunit I now contacts the next accessible cleavage site by looping out the intervening RNA (step 5). This cycle of binding, cleavage, and looping continues. Asterisks denote binding interactions.

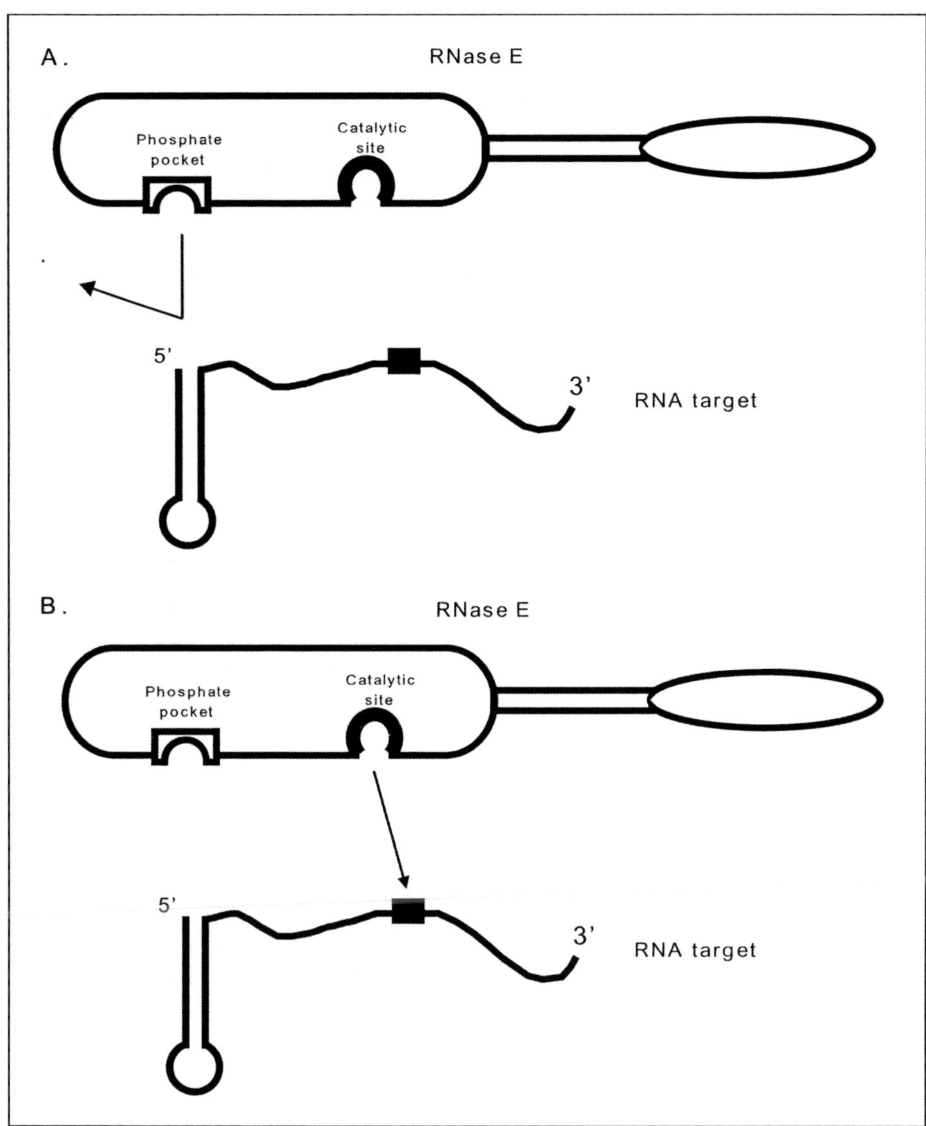

Figure 4. 5'-protection against RNase E action. 5'-stem loops structures strongly inhibit RNase E action by precluding the initial interaction between the 5'-end of the mRNA and the putative phosphate binding pocket. This results in stabilization of mRNAs in vivo (Panel A). Nonetheless, such mRNAs ultimately decay in an RNase E-dependent manner. This can be explained by an inefficient internal entry model in which RNase E engages a cleavage site without contacting the 5'-end (Panel B). The N-terminal domain of RNase E is represented by an oval with the same conventions for the phosphate binding pocket and active site as in Figure 2. The thickened region in the RNA substrate represents a cleavage site.

structure.[77] Importantly, both enzymes are processive: they remain bound to their substrates during the process of degradation until the substrates are shortened to less than ~15 nt or until they encounter a structural barrier such as a stem-loop of ≥7 bp at which point they dissociate from their substrate.[71,77] Moreover, although *E. coli* can tolerate loss-of-function mutations in either enzyme, simultaneous loss of RNase II and PNPase is lethal, further supporting the notion that these enzymes are functionally redundant.[72] These exonucleases do differ in one important regard: PNPase activity can be coupled to the action of one or more RNA helicases whereas RNase II activity cannot (see below).[26]

The enzymatic or structural basis for the preference of PNPase and RNase II for single-stranded RNA, for the mechanism of processivity, and their catalytic mechanisms remain unknown. Cannistraro and Kennell have proposed an elaborate model for the processivity of RNase II, primarily based on kinetic considerations.[78] In their model, single-stranded RNA is anchored to the enzyme at a site 15-25 residues from the 3'-end thereby preventing the enzyme-substrate complex from dissociating. Structural information, ideally for an enzyme-inhibitor complex at atomic resolution, would provide a critical test of their model. Interestingly, RNase II, like PNPase, RNase E and RNase G, contains an S1 domain, a putative single-stranded RNA binding motif (Fig. 5).[79] No information is yet available to suggest a function for this domain in RNase II. It could play a key role in binding of single-stranded RNA and providing processivity. Small crystals of RNase II from *E. coli* have been obtained and efforts are underway to solve its structure (Ling P, Mackie GA, Mosimann S, unpublished results).

Structural investigations of PNPase are more advanced as a crystal structure has been determined for the PNPase from *Streptomyces antibioticus* to 2.6 Å resolution.[80] Several important insights have emerged from this structure. First, the enzyme is indeed trimeric with a central channel. Second, each subunit consists of a duplicated core of two RNase PH-like domains[81,82] (residues 1-210 and 312-541, respectively, in the *E. coli* enzyme; see Fig. 5) joined by a linker. Such organization can be detected in most known PNPases by sequence comparison alone.[82] The second of the "PH-domains" contains the active site (Miao X and Mackie GA, unpublished data). Two known RNA binding motifs, a KH domain (residues 551-591) and an S1 domain (residues 620-690), follow the internal duplication (Fig. 5). Although the functional significance of these domains remains unknown, the *pnp*-71 mutation in a conserved glycine residue in the KH domain leads to a loss of autoregulation of PNPase and other functional deficiencies.[83] The presence of tandem RNA binding motifs may provide an extended RNA binding surface on PNPase as has been proposed for the NusA transcription factor from *Thermotoga maritima*.[84] This surface could play a key role in conferring processivity to PNPase. Finally, the crystal structure has permitted the identification of putative active site aspartate residues and a phosphate binding pocket (occupied by tungstate in the crystal).[80] Investigation of the role of these residues, and ideally obtaining the structure of a co-crystal of enzyme and substrate or inhibitor, should greatly assist efforts to understand the physical basis for the properties of PNPase.

The action of RNase E and the 3'-exonucleases would result in the accumulation of limit oligonucleotides of 12-20 residues. These fragments can be degraded to nucleoside-5'-monophosphates by the hydrolytic action of oligoribonuclease, a widely distributed 3'-exonuclease which is apparently an essential enzyme in *E. coli*.[85,86]

Polyadenylation and the Problem of Terminal Secondary Structures

Fragments of mRNA containing stable 3'-terminal stem-loops or < 9 unpaired residues at their 3'-ends pose a formidable problem since all the 3'-exonucleases are single-strand-specific. Moreover, both RNase II and PNPase require a single-stranded 3'-extension of ≥6-9 residues to bind to an RNA substrate.[71,77] Such structural barriers can be overcome in two ways. First, the 3'-termini of all RNAs in *E. coli* can be "edited" by the addition of poly(A) tails catalyzed by poly(A) polymerase I (PAP I), the product of the *pcnB* gene (see Fig. 1B).[87-91] Addition of an unstructured 3'- poly(A) tail to potential substrates permits the binding of 3'-exonucleases[77,91,92] and stimulates the activity of both RNase II and PNPase in vitro (Fig. 1B).[87-88] It has been suggested that polyadenylation is the initiating step in the degradation of some or even all mRNAs.[48] With the exception of RNA I, for which the evidence for a role for PAP I in its functional decay is strong,[88] we believe that this is unlikely in general. Strains in which *pcnB* is deleted are viable and exhibit only a mild growth defect.[93-95] Moreover, the initial rate of decay of several mRNAs which have been tested is unaffected in such strains and is reduced only in multiple RNase-deficient mutants.[25,90] Finally, both the length of the poly(A) tract and the fraction of a given mRNA species containing poly(A) is limited (see Chapter 11 by P. Régnier and P. E. Marujo).[96-97] Recent data suggest that the action of PAPI on

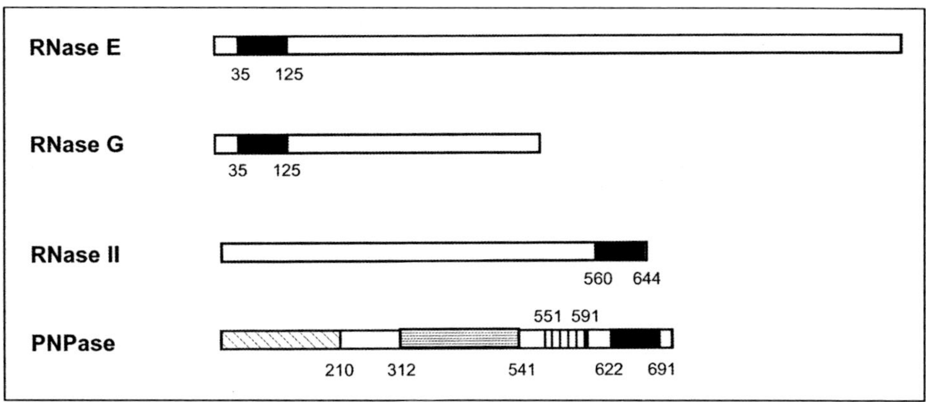

Figure 5. Domain structure of key ribonucleases. The positions of various domains recognizable by sequence alignment or structural determination are shown schematically as follows: S1 domain (solid block), PH-like (PH') domain (diagonal stripes), PH domain (stippled), KH domain (vertical stripes).

RNA I is 5'-end dependent and is stimulated by a 5'- monophosphate group.[98] The stimulation of poly(A)-dependent degradation of the *rpsO* mRNA by prior RNase E cleavage is consistent with this observation.[99] If 5'-end dependence of PAP I can be substantiated for other RNAs, it would imply that this enzyme acts efficiently only on RNAs which have previously undergone at least one endonucleolytic cleavage and that PAPI could not participate in initiating mRNA decay. Alternatively, the preference of PAP I for a 5'-monophosphate could reflect some idiosyncrasy of the structure of RNA I and tRNAs (e.g., the close proximity of 5' and 3'- termini).

The rediscovery of poly(A) tracts in bacteria prompted consideration of potential parallels between prokaryotic and eukaryotic mRNA metabolism.[48] We believe, however, that polyadenylation of bacterial mRNAs is akin to ubiquitination of proteins in eukaryotes:[100] it stimulates the removal of structured mRNA fragments[25,26,92,101] and effectively "marks" damaged, misfolded or partially degraded RNAs for decay.[101a] The basis for this form of "RNA surveillance" is the affinity of RNase II and PNPase for single-stranded RNA and their inability to degrade stem-loop structures exceeding 6 bp.[77,92] The addition of a poly(A) tract to the 3'-end of an RNA permits these 3'-exonucleases to "test" the integrity of adjacent secondary structures while removing the added "tail". Misfolded RNAs will be susceptible to exonuclease attack while correctly folded RNAs will be stable and thus resistant. Moreover, RNA binding proteins (e.g., ribosomal proteins; aminoacyl tRNA synthetases; EF-Tu) will select for RNAs presenting correctly folded structures and can also shield otherwise accessible 3'-ends from the action of both PAP I and 3'- exonucleases.

An alternative process for removal of 3'-terminal hairpin structures relies on a DEAD-box RNA helicase complexed to RNase E and PNPase in a "macromolecular machine", the RNA degradosome (see below and Fig. 1B).[102-104] The interaction of the RNA helicase, RhlB, with RNase E activates the former's ATP-dependent helicase activity so that it can (presumably) unwind otherwise highly stable stem-loops.[25,26,103,105] In some circumstances, an alternative helicase, DeaD [also called CsdA], may substitute for RhlB (see below). The action of the RNA helicase exposes a sufficient number of single-stranded residues to PNPase, also bound to RNase E, to permit it to initiate processive 3'-5' degradation of the substrate. Prior oligoadenylation of 3'-ends facilitates the action of RhlB helicase, possibly to facilitate its binding to the RNA.[26,101] The existence of the degradosome and the ATP requirement for mRNA turnover were not predicted by Apirion's original model[21] and represent significant recent surprises to the field (Fig. 1B).

The Enigma of the RNA Degradosome

Perhaps the most striking feature of the enzymes of RNA processing in *E. coli* is the association of RNase E, PNPase, RhlB and, curiously, enolase in the RNA degradosome (Table 1 and Fig. 6).[103-104] Other proteins and activities, including polyphosphate kinase and DnaK, copurify with

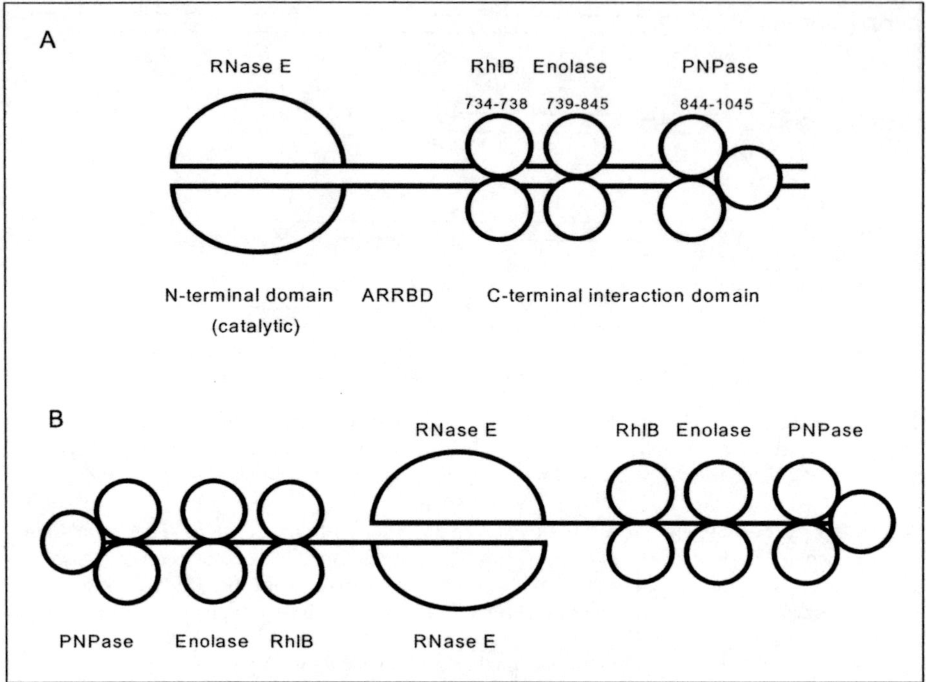

Figure 6. Models for the organization of the RNA degradosome. In the model in Panel A, taken from Vanzo et al,[105] RNase E is a head-to-head dimer. The C-terminal interaction domain (residues 700-1061) interacts with a presumed dimer of RhlB, a dimer of enolase, and a trimer of PNPase. The coordinates for the boundaries of the interaction sites are shown. In the model in Panel B, RNase E forms a head-to-tail dimer. The two models are distinguished by the larger size of the degradosome in Panel B, their different symmetries and by the stoichiometries of PNPase, RhlB and enolase relative to RNase E.

the degradosome,[104,106] but some of these may be contaminants. As mentioned above, the association of the first three is critical to the degradation of highly structured 3'-terminal RNA fragments because RNase E is required to activate RhlB.[26,105] A function for enolase in the degradosome is not obvious, however. The discovery of the degradosome was highly satisfying; moreover, its purification by several groups,[25,102-104] the mapping of specific interactions among its components (see Fig. 6),[26,105] its reconstitution,[26] and the co-localization of its components by immuno-electron microscopy[107] strongly suggest that it exists as a multi-subunit complex in vivo. The relative stoichiometry of the components and their spatial organization remain to be determined. Figure 6 illustrates two possible arrangements of the components of the degradosome consistent with existing data.

Despite the body of evidence in favour of an RNA degradosome, several lines of evidence have cast doubt upon its functional importance, if not its existence. First, Kido et al discovered a set of viable *rne* deletions whose translated products would lack over 450 residues at the C-terminus.[108] The best characterized of their alleles, *rne*-131, which encodes residues 1-584 of RNase E, displays only mild temperature-sensitivity, is fully competent for processing pre-5S rRNA, and suffers from a defect in the decay of some mRNAs, most notably the *rne* mRNA itself and a T7-*lacZ* chimeric mRNA.[59,109] Ow et al have conducted a more extensive deletion analysis of the 3' end of the *rne* gene and have found that only the N-terminal 426 residues of the Rne protein are required for viability.[110] Although their deletions do not affect the catalytic activity of RNase E, they lead to the concomitant loss of interactions with RhlB, enolase and PNPase. These findings clearly imply that RNA processing and the decay of some, but not all, mRNAs can occur at nearly normal rates in the absence of an RNA degradosome. In contrast, larger *rne* deletions, while viable, do result in significantly retarded rates of decay for a number of mRNAs.[110] A second line of evidence casting doubt on the existence of the degradosome arises from the fact that its components do not appear to be coordinately synthesized or regulated. Rather, PNPase accumulates in excess of RNase E, perhaps by as much as 10-fold, and enolase by 50-fold.[107] Moreover, loss-of-function mutations in *pnp* do not affect bulk mRNA stability or the cleavage of the *rpsT* mRNA by RNase E.[72,111] Finally, the C-terminal 640 residues of RNase E which contain the binding sites for RhlB, enolase and PNPase (see Fig. 6), are poorly conserved, if at all, among RNase E homologs from different bacterial species (Briant DJ, Mackie GA, unpublished data).[112] Collectively, these findings raise the possibility that RNase E and PNPase may function independently of each other. Indeed, the only likely occasion when RNase E and PNPase must function in concert occurs when RhlB is required for the degradation of tightly folded RNA hairpins (see above). Reconstitution experiments demonstrate that RNase II cannot substitute for PNPase in promoting RhlB-dependent degradation in vitro.[26] Thus it remains mysterious how stem-loop structures would be degraded in *pnp-7* or *rne-131* strains.

Adaptation of the RNA Decay Machinery

Since mRNA turnover is an important mechanism of post-transcriptional regulation of gene expression, the mRNA decay apparatus could participate in global responses to stresses such as heat shock, cold shock, anaerobiosis or pathogenesis. The evidence to support this postulate is, however, limited. An investigation into the effect of heat shock on mRNA decay failed to find a significant alteration in mRNA decay.[113] During anaerobiosis, the steady-state levels of the *bla* and *ompA* mRNAs are maintained relative to aerobic growth.[114] This is achieved because the rates of transcription of the corresponding genes are reduced while the mature mRNAs are significantly stabilized. These interesting observations have not, to our knowledge, been extended to a broader range of mRNAs. In contrast, cold shock of a bacterial culture (a rapid shift to 15°C), results in an instantaneous stabilization of many cellular RNAs, followed by the alteration of the mRNA decay apparatus during the adaptation period.[115-119] Following adaptation, the mRNAs induced by cold shock decay in a process requiring PNPase, PAP I, and DeaD.[120] Only the portion of RNase E expressed in the *rne-131* mutant is required for the typical expression profile during cold shock.[120]

In this context, the absence of a substantial phenotype for most *rne* mutations may reflect a limited survey of conditions, almost invariably LB medium at 37° C. Interestingly, *rneΔ610*, a mutant lacking a distal portion of the catalytic domain and both the AR-RBD and platform domains[110] (see Fig. 2), is not viable at 15° C (RKB and RWS; unpublished data). Such is not the case for the *rneΔ508* mutant,[110] in which RNase E retains the full catalytic domain but still lacks the AR-RBD and platform. Since both the *rneΔ508* and *rneΔ610* mutants are viable at 37°C (although the latter is severely defective for auto-regulation and slightly defective in pre-5S rRNA processing),[110] these results show that full activity of the RNase E catalytic domain is essential for growth at low temperatures, whereas the degradosome is not (*cf* ref. 120).

Nevertheless, other data suggest that some form of the degradosome may play an important role in mRNA decay at low temperatures. As mentioned above, mRNA stability increases dramatically and immediately upon a shift to low temperatures, but decreases somewhat over the following several hours, as an apparently new mRNA decay mechanism develops.[115-120] PNPase levels are induced during cold-temperature adaptation,[115,117-119] and RNase E and PNPase remain able to form a complex at cold temperatures (RKB, APG, GAM, and RWS; unpublished data). Interestingly, the DeaD protein, which is also induced during cold adaptation,[121,122] is homologous to the RhlB RNA-helicase,[123] suggesting that DeaD may replace RhlB in the degradosome, possibly providing an mRNA-degradative machine adapted for growth under such conditions. Indeed, DeaD can fully substitute for RhlB in an in vitro assay for degradosome function, and does so in an ATP- and RNase E-dependent manner (RKB, APG, GAM, and RWS, unpublished data). Preliminary evidence suggests that DeaD may associate with RNase E and/or PNPase in vitro (APG, RKB, RWS and GAM; unpublished data). Moreover, it is also reported to associate with PAP I,[123] supporting the possibility that a DeaD-containing degradosome complex may form in the cell. However, whether such a specialized degradosome plays a role in mRNA decay at cold temperatures remains to be established, and we can only speculate on whether additional forms of the degradosome assemble under other conditions.

Both PNPase and DeaD are required for cell growth at cold temperatures, but not at 30° C or higher.[117,122] However, since the degradosome is not essential under any of these conditions, PNPase and DeaD must be fulfilling some other critical role at low temperatures. Preliminary evidence suggests that PNPase is not essential, per se; rather, it appears to be required for the expression and/ or activity of RNase E, whose (possibly enhanced) function is crucial for cell viability at low temperatures (RKB and RWS; unpublished data). The basis for the cold-temperature essentiality of DeaD remains to be established.

Concluding Remarks

Our understanding of the mechanism of mRNA decay in *E. coli* has advanced remarkably during the past decade, a function of improved methodologies for the detection of RNAs and a determined effort to unravel the enzymology of the mRNA decay apparatus. A number of surprises have emerged. First, mRNA decay is a specialized case of mRNA processing in general, involving many of the same enzymes. Second, mRNA decay displays a previously unsuspected ATP requirement, particularly for dealing with stable secondary structures. Third, some of the enzymes of mRNA decay may be organized into a macromolecular complex, the RNA degradosome. Although the evidence for the latter's functional importance is not yet unequivocal, we believe that the degradosome exists in vivo and suggest that it constitutes a molecular machine in which the gears may be changed to adapt it and the RNA processing machinery to different conditions. However, as a machine, it represents an "extra cost option" for *E. coli* and is apparently dispensable under laboratory conditions.

Acknowledgements

We thank members of our laboratories for comments, particularly Doug Briant and Janet Hankins. GAM gratefully acknowledges the support of MRC/CIHR. RWS gratefully acknowledges research support from the NIH and NSF.

References

1. Astrachan L, Volkin E. Properties of ribonucleic acid turnover in T2-infected *Escherichia coli*. Biochim Biophys Acta 1958; 29:3-11.
2. Brenner S, Jacob F, Meselson M. An unstable intermediate carrying information from genes to ribosomes for protein synthesis. Nature 1961; 190:576-581.
3. Gros F, Hiatt H, Gilbert W et al. Unstable ribonucleic acid revealed by pulse labelling of *Escherichia coli*. Nature 1961; 190:73-77.
4. Belasco JG, Higgins CF. Mechanisms of mRNA decay in bacteria: a perspective. Gene 1988; 72:15-33.
5. Singer P, Nomura M. Stability of ribosomal protein mRNA and translational feedback regulation in *Escherichia coli*. Mol Gen Genet 1985; 199:543-546.
6. Greenberg ME, Ziff EB. Stimulation of 3T3 cells induces transcription of the *c-fos* proto-oncogene. Nature 1984; 311:433-438.
7. Newbury SF, Smith NH, Robinson EC et al. Stabilization of translationally active mRNA by prokaryotic REP sequences. Cell 1987; 48:297-310.
8. Simons RW, Kleckner N. Translational control of IS10 transposition. Cell 1983; 34:683-691.
9. Tomizawa J, Itoh T, Selzer G et al. Inhibition of Col E1 primer formation by a plasmid-specified small RNA. Proc Natl Acad Sci USA 1981; 78:1421-1425.
10. Inouye M. Antisense RNA: its functions and applications in gene expression—a review. Gene 1988; 72:25-34.
11. Simons RW, Kleckner, N. Biological regulation by antisense RNA in prokaryotes. Annu Rev Genet 1988; 22:567-600.
12. Eguchi Y, Itoh T, Tomizawa J. Antisense RNA. Annu Rev Biochem 1991; 60:631-652.
13. Maquat LE, Carmichael GG. Quality control of mRNA function. Cell 2001; 104:173-176.
14. Hilleren P, Parker R. Mechanisms of mRNA surveillance in eukaryotes. Annu Rev Genet 1999; 33:229-260.
15. Bass BL. Double-stranded RNA as a template for gene silencing. Cell 2000; 101:235-238.
16. Gausing K. Regulation of ribosome biogenesis in *E. coli*. In: Chambliss G, Craven GR, Davies J et al, ed. Ribosomes: structure, function and genetics. Baltimore: University Park Press, 1980:693-718.
17. Coburn GA, Mackie GA. Degradation of mRNA decay in bacteria: an old problem with some new twists. Prog Nucl Acid Res Mol Biol 1999; 62:55-108.
18. Grunberg-Manago M. Messenger RNA stability and its role in control of gene expression in bacteria and phages. Annu Rev Genet 1999; 33:193-227.
19. Steege DA. Emerging features of mRNA decay in bacteria. RNA 2000; 6:1079- 1090.
20. Régnier P, Arraiano CM. Degradation of mRNA in bacteria: emergence of ubiquitous features. Bioessays 2000; 22:235-244.
21. Apirion D. Degradation of RNA in Escherichia coli: a hypothesis. Mol Gen Genet 1972; 122:313-322.

22. Belasco JG. mRNA degradation in prokaryotic cells: an overview. In: Belasco JG, Brawerman G, eds. Control of messenger RNA stability. San Diego: Academic Press, 1993:3-12.

23. Higgins CF, Peltz SW, Jacobson A. Turnover of mRNA in prokaryotes and lower eukaryotes. Curr Opin Genet Devel 1992; 2:739-747.

24. Nierlich DP, Murakawa GJ. The decay of bacterial messenger RNA. Prog Nucl Acid Res Mol Biol 1996; 52:153-216.

25. Coburn GA, Mackie GA. Reconstitution of the degradation of the mRNA for ribosomal protein S20 with purified enzymes. J Mol Biol 1998; 279:1061-1074.

26. Coburn GA, Miao X, Briant DJ et al. Reconstitution of a minimal RNA degradosome demonstrates a functional coordination between a 3' exonuclease and a DEAD-box RNA helicase. Genes Devel 1999; 13:2594-2603.

27. Mackie GA. RNase E is a 5'-end-dependent endonuclease. Nature 1998; 395:720-724.

28. Mackie GA. Stabilization of circular *rpsT* mRNA demonstrates the 5'-end-dependence of RNase E action in vivo. J Biol Chem 2000; 275:25069-25072.

29. Spickler C, Stronge V, Mackie GA. Preferential cleavage of degradative intermediates of the *rpsT* mRNA by the *Escherichia coli* RNA degradosome. J Bacteriol 2001; 183:1106-1109.

30. Misra TK, Apirion D. RNase E, an RNA processing enzyme from *Escherichia coli*. J Biol Chem 1979; 254:11154-11159.

31. Li ZW, Pandit S, Deutscher MP. RNase G (CafA protein) and RNase E are both required for the 5' maturation of 16S ribosomal RNA. EMBO J 1999; 18:2878-2885.

31a. Li Z, Deutscher MP. RNase E plays an essential role in the maturation of *Escherichia coli* tRNA precursors. RNA 2002; 8:97-109.

31b. Ow MC, Kushner SR. Initiation of tRNA maturation by RNase E is essential for cell viability in *E. coli*. Genes Devel 2002; 16:1102-1115.

32. Wachi M, Umitsuki G, Shimizu M et al. *Escherichia coli cafA* gene encodes a novel RNase, designated as RNase G, involved in processing of the 5' end of 16S rRNA. Biochem Biophys Res Comm 1999; 259:483-488.

33. Ono M, Kuwano M. A conditional lethal mutation in an *Escherichia coli* strain with a longer chemical lifetime of mRNA. J Mol Biol 1979; 129:343-357.

34. McDowall KJ, Hernandez RG, Lin-Chao S et al. The *ams-1* and *rne-3071* temperature-sensitive mutations in the *ams* gene are in close proximity to each other and cause substitutions within a domain that resembles a product of the *Escherichia coli mre* locus. J Bacteriol 1993; 175:4245-4249.

35. Ghora BK, Apirion D. Structural analysis and in vitro processing to p5S rRNA of a 9S RNA molecule isolated from an *rne* mutant of *Escherichia coli*. Cell 1978; 15:1055-1066.

36. Mudd EA, Carpousis AJ, Krisch HM. *Escherichia coli* RNase E has a role in the decay of bacteriophage T4 mRNA. Genes Devel 1990; 4:873-881.

37. Mudd EA, Krisch HM, Higgins CF. RNase E, an endoribonuclease, has a general role in the chemical decay of *Escherichia coli* rRNA: evidence that *rne* and *ams* are the same genetic locus. Mol Microbiol 1990 4:2127-2135.

38. Babitzke P, Kushner SR. The *ams* (altered mRNA stability) protein and RNase E are encoded by the same structural gene of *Escherichia coli*. Proc Nat Acad Sci USA 1991; 88:1-5.

39. Taraseviciene L, Miczak A, Apirion D. The gene specifying RNase E (*rne*) and a gene affecting mRNA stability (*ams*) are the same gene. Mol Microbiol 1991; 5:851-855.

40. Melefors Ö, von Gabain A. Genetic studies of cleavage-initiated mRNA decay and processing of ribosomal 9S RNA show that the *Escherichia coli ams* and *rne loci* are the same. Mol Microbiol 1991; 5:857-864.

41. Mackie GA. Specific endonucleolytic cleavage of the mRNA for ribosomal protein S20 of *Escherichia coli* requires the product of the *ams* gene in vivo and in vitro. J Bacteriol 1991; 171:2488-2497.

42. Huang H, Liao J, Cohen SN. Poly(A) and poly(U)-specific RNA 3' tail shortening by *E. coli* ribonuclease E. Nature 1998; 391:99-102.

43. Walsh AP, Tock MR, Mallen MH et al. Cleavage of poly(A) tails on the 3'-end of RNA by ribonuclease E of *Escherichia coli*. Nucl Acids Res 2001; 29:1864-1871.

44. Bilge SS, Apostol JM jr, Aldape MA et al. mRNA processing independent of RNase III and RNase E in the expression of the F1845 fimbrial adhesin of *Escherichia coli*. Proc Nat Acad Sci USA 1993 90:1455-1459.

45. Wachi M, Umitsuki G, Nagai K. Functional relationship between *Escherichia coli* RNase E and the CafA protein. Mol Gen Gene 1997; 253:515-519.

46. Tock MR, Walsh AP, Carroll G et al. The CafA protein required for 5'- maturation of 16S rRNA is a 5'-end-dependent ribonuclease that has context-dependent broad sequence specificity. J Biol Chem 2000; 275:8726-8732.

47. Umitsuki G, Wachi M, Takada A et al. Involvement of RNase G in in vivo mRNA metabolism in *Escherichia coli*. Genes to Cells 2001; 6:403-410.

48. Kushner SR. mRNA decay. In: Neidhardt FC, ed. *Escherichia coli* and *Salmonella*, 2nd ed. Washington DC: ASM Press, 1996:849-860.

49. Hajnsdorf E, BraunF, Haugel-Nielsen J et al. Multiple degradation pathways of the *rpsO* mRNA of *Escherichia coli*. RNase E interacts with the 5' and 3' extremities of the primary transcript. Biochimie 1996; 78:416-424.

50. Loomis WP, Koo JT, Cheung TP et al. A tripeptide sequence within the nascent DaaP protein is required for mRNA processing of a fimbrial operon in *Escherichia coli*. Mol Microbiol. 2001; 39:693-707.

51. Portier C, Dondon L, Grunberg-Manago M et al. The first step in the functional inactivation of the *Escherichia coli* polynucleotide phosphorylase messenger is a ribonuclease III processing at the 5' end. EMBO J 1987; 6:2165-2170.
52. Bardwell JC, Régnier P, Chen SM et al. Autoregulation of RNase III operon by mRNA processing. EMBO J 1989; 8:3401-3407.
53. Régnier P, Grunberg-Manago M. RNase III cleavages in non-coding leaders of *Escherichia coli* transcripts control mRNA stability and genetic expression. Biochimie 1990; 72:825-834.
54. Babitzke P, Granger L, Olszewski J et al. Analysis of mRNA decay and rRNA processing in *Escherichia coli* multiple mutants carrying a deletion in RNase III. J Bacteriol 1993; 175:229-239.
55. Altuvia S, Kornitzer D, Kobi S et al. Functional and structural elements of the mRNA of the cIII gene of bacteriophage lambda. J Mol Biol 1991; 218:723-733.
56. Jerome LJ, van Biesen T, Frost LS. Degradation of FinP antisense RNA from F-like plasmids: the RNA-binding protein, FinO, protects FinP from ribonuclease E. J Mol Biol 1999; 285:1457-1473.
57. Vytvytska O, Jacobsen JS, Balcunaite G et al. Host factor I, Hfq, binds to *Escherichia coli ompA* mRNA in a growth rate-dependent fashion and regulates its stability. Proc Nat Acad Sci USA 1998; 95:14118-14123.
58. Vytvytska O, Moll I, Kaberdin VR et al. Hfq (HF1) stimulates *ompA* mRNA decay by interfering with ribosome binding. Genes Devel 2000; 14:1109-1118.
59. Jiang X, Diwa A, Belasco JG. Regions of RNase E important for 5'-end-dependent RNA cleavage and autoregulated synthesis. J Bacteriol 2000; 182:2468-2475.
60. McDowall KJ, Cohen SN. The N-terminal domain of the *rne* gene product has RNase E activity and is non-overlapping with the arginine-rich RNA-binding site. J Mol Biol 1996; 255:349-355.
61. von Gabain A, Belasco JG, Schottel JL et al. Decay of mRNA in *Escherichia coli*: investigation of the fate of specific segments of transcripts. Proc Nat Acad Sci USA 1983; 80:653-657.
62. Cormack RS, Genereaux, JL, Mackie GA. RNase E activity is conferred by a single polypeptide: overexpression, purification and properties of the *ams/rne/hmp1* gene product. Proc Nat Acad Sci USA 1993; 90:9006-9010.
63. Kaberdin VR, Walsh AP, Jakobsen T et al. Enhanced cleavage of RNA mediated by an interaction between substrates and the arginine-rich domain of *E. coli* ribonuclease E. J Mol Biol 2000; 301:257-264.
64. Chen L-H, S. Emory SA, Bricker AL et al. Structure and function of a bacterial mRNA stabilizer: analysis of the 5' untranslated region of *ompA* mRNA. J Bacteriol 1991; 173:4578-4586.
65. Emory SA, Bouvet P, Belasco JG. A 5'-terminal stem-loop structure can stabilize mRNA in *Escherichia coli*. Genes Devel 1992; 6:135-148.
66. Bouvet P, Belasco JG. Control of RNase E-mediated RNA degradation by 5'-terminal base pairing in *E. coli*. Nature 1992; 360:488-491.
67. Hansen MJ, Chen L-H, Fejzo MLS et al. The *ompA* 5' untranslated region impedes a major pathway for mRNA degradation in *Escherichia coli*. Mol Microbiol 1994; 12:707-716.
68. Arnold TE, Yu J, Belasco JG. mRNA stabilization by the *ompA* 5'-untranslated region: two protective elements hinder distinct pathways for mRNA degradation. RNA 1998; 4:319-330.
68a. Baker KE, Mackie GA. Ectopic RNase E sites promote bypass of 5'-end-dependent mRNA decay in *Escherichia coli*. Molec Microbiol 2002; in press.
69. Braun F, Le Derout J, Régnier P. Ribosomes inhibit an RNase E cleavage which induces the decay of the *rpsO* mRNA of *Escherichia coli*. EMBO J 1998; 17:4790-4797.
70. Diwa A, Bricker AL, Jain C et al. An evolutionarily conserved RNA stem-loop functions as a sensor that directs feedback regulation of RNase E gene expresssion. Genes Devel 2000; 14:1249-1260.
71. Littauer UZ, Soreq H. Polynucleotide phosphorylase. In: Boyer PD, ed. The Enzymes XV, part B. New York: Academic Press Inc, 1982:517-553.
72. Donovan WP, Kushner SR. Polynucleotide phosphorylase and ribonuclease II are required for cell viability and mRNA turnover in Escherichia coli K-12. Proc Nat Acad Sci USA 1986; 83:120-124.
73. Deutscher MP, Reuven NB. Enzymatic basis for hydrolytic versus phosphorylytic mRNA degradation in *Escherichia coli* and *Bacillus subtilis*. Proc Nat Acad Sci USA 1991; 88:3277-3280.
74. Grunberg-Manago M. Recollections on studies of polynucleotide phosphorylase: a commentary. Biochim Biophys Acta 1989; 1000:59-64.
75. Gupta RS, Kasai T, Schlessinger D. Purification and some novel properties of *Escherichia coli* RNase II. J Biol Chem 1977; 252:8945-8949.
76. Portier C. Quaternary structure of *Escherichia coli* polynucleotide phosphorylase: new evidence for a trimeric structure. FEBS Lett 1975; 50:79- 81.
77. Spickler C, Mackie GA. Action of RNase II and polynucleotide phosphorylase against RNAs containing stem-loops of defined structure. J Bacteriol 2000; 182:2422-2427.
78. Cannistraro V, Kennell D. The reaction mechanism of ribonuclease II and its interaction with nucleic acid secondary structures. Biochim Biophys Acta 1999; 1433:170-187.
79. Bycroft M, Hubbard TJP, Proctor M et al. The solution structure of the S1 RNA binding domain: a member of an ancient nucleic acid binding fold. Cell 1997; 88:235-242.
80. Symmons MF, Jones GH, Luisi BF. A duplicated fold is the structural basis for polynucleotide phosphorylase activity, processivity and regulation. Structure 2000; 8:1215-1226.
81. Mian IS. Comparative sequence analysis of ribonucleases HII, III, II, PH and D. Nucl Acids Res 1997; 25:3187-3195.

82. Yuo Z, Deutscher MP. Exoribonuclease superfamilies: structural analysis and phylogenetic distribution. Nucl Acids Res 2001; 29:1017-1026.

83. Garcia-Mena J, Das A, Sanchez-Trujillo A et al. A novel mutation in the KH domain of polynucleotide phosphorylase affects autoregulation and mRNA decay in *Escherichia coli*. Mol Microbiol 1999; 33:235-248.

84. Worbs M, Bourenkov GP, Bartunik HD et al. An extended RNA binding surface through arrayed S1 and KH domains in transcription factor NusA. Molec Cell 2000; 7:1177-1189.

85. Zhang X, Zhu L, Deutscher MP. Oligoribonuclease is encoded by a highly conserved gene in the 3'-5' exonuclease family. J Bacteriol 1998; 180:2779- 2781.

86. Ghosh S, Deutscher MP. Oligoribonuclease is an essential component of the mRNA decay pathway. Proc Nat Acad Sci USA 1999; 96:4372-4377.

87. Cao GJ, Sarkar N. Identification of the gene for *Escherichia coli* poly (A) polymerase. Proc Nat Acad Sci USA 1992; 89:10380-10384.

88. Xu F, Lin-Chao S, Cohen SN. The *Escherichia coli pcnB* gene promotes adenylylation of antisense RNA I of colE1-type plasmids in vivo and degradation of RNA I decay intermediates. Proc Nat Acad Sci USA 1993; 90:6756-6760.

89. O'Hara EB, Chekanova JA, Ingle CA et al. Polyadenylation helps regulate mRNA decay in *Escherichia coli*. Proc Nat Acad Sci USA 1995; 92:1807-1811.

90. Hajnsdorf E, Braun F, Haugel-Nielsen J et al. Polyadenylation destabilizes the *rpsO* mRNA of *Escherichia coli*. Proc Nat Acad Sci USA 1995; 92:3973-3977.

91. Xu F, Cohen SN. RNA degradation in *Escherichia coli* regulated by 3' adenylation and 5' phosphorylation. Nature 1995; 374:180-183.

92. Coburn GA, Mackie GA. Differential sensitivities of portions of the mRNA for ribosomal protein S20 to 3'-exonucleases dependent on oligoadenylation and RNA secondary structure. J Biol Chem 1996; 271:15776-15781.

93. Lopilato J, Bortner S, Beckwith J. Mutations in a new chromosomal gene of *Escherichia coli, pcnB,* reduce plasmid copy number of pBR322 and its derivatives. Mol Gen Genet 1985; 205:285-290.

94. He L, Söderbom F, Wagner EGH et al. PcnB is required for the rapid degradation of RNA I, the antisense RNA that controls the copy numbers of ColE1-related plasmids. Mol Microbiol 1993; 9:1131-1142.

95. Masters M, Colloms MD, Oliver IR et al. The *pcnB* gene of *Escherichia coli,* which is required for ColE1 copy number maintenance, is dispensable. J Bacteriol 1993; 175:4405-4413.

96. Marujo PE, Hajnsdorf E, Le Derout J et al. RNase II removes the oligo(A) tails that destabilize the *rpsO* mRNA of *Escherichia coli.* RNA 2000; 6:1185-1193.

97. Mohanty BK, Kushner SR. Analysis of the function of *Escherichia coli* poly (A) polymerase I in RNA metabolism. Mol Microbiol 1999; 34:1094-1106.

98. Feng Y, Cohen SN. Unpaired terminal nucleotides and 5'-monophosphorylation govern 3' polyadenylation by *Escherichia coli* poly (A) polymerase I. Proc Nat Acad Sci USA 2000; 97:6415-6420.

99. Hajnsdorf E, Régnier, P. *E. coli rpsO* mRNA decay: RNase E processing at the beginning of the coding sequence stimulates poly(A)-dependent degradation of the mRNA. J Mol Biol 1999; 286:1033-1043.

100. Ciechanover A. The ubiquitin-proteasome proteolytic pathway. Cell 1994; 79:13-21.

101. Blum E, Carpousis AJ, Higgins CF. Polyadenylation promotes degradation of 3'- structured RNA by the *Escherichia coli* mRNA degradosome in vitro. J Biol Chem 1999; 274:4009-4016.

101a. Li Z, Reimers S, Pandit S et al. RNA quality control: degradation of defective transfer RNA. EMBO J 2002; 21:1132-1138.

102. Carpousis AJ, Van Houwe G, Ehretsmann C et al. Copurification of *E. coli* RNase E and PNPase: evidence for a specific association between two enzymes important in RNA processing and degradation. Cell 1994; 76:889-900.

103. Py B, Higgins CF, Krisch HM et al. A DEAD-box RNA helicase in the *Escherichia coli* degradosome. Nature 1996; 381:169-172.

104. Miczak A, Kaberdin VR, Wei C-L et al. Proteins associated with RNase E in a multicomponent ribonucleolytic complex. Proc Nat Acad Sci USA 1996; 93:3865-3869.

105. Vanzo NF, Li YS, Py B et al. Ribonuclease E organizes the protein interactions in the *Escherichia coli* RNA degradosome. Genes Devel 1998; 12:2770-2781.

106. Blum E, Py B, Carpousis AJ et al. Polyphosphate kinase is a component of the *Escherichia coli* RNA degradosome. Mol Microbiol 1997; 26:387-398.

107. Liou G-G, Jane WN, Cohen SN et al. RNA degradosomes exist in vivo in *Escherichia coli* as multicomponent complexes associated with the cytoplasmic membrane via the N-terminal region of ribonuclease E. Proc Nat Acad Sci USA 2001; 98:63-68.

108. Kido M, Yamanaka K, Mitani T et al. RNase E polypeptides lacking a carboxyl-terminal half suppress a *mukB* mutation in *Escherichia coli.* J Bacteriol 178:3917-3925.

109. Lopez PJ, Marchand I, Joyce SA et al. The C-terminal half of RNase E, which organizes the *Escherichia coli* degradosome, participates in mRNA degradation but not rRNA processing in vivo. Mol Microbiol 1999; 33:189-199.

110. Ow MC, Liu Q, Kushner SR. Analysis of mRNA decay and rRNA processing in *Escherichia coli* in the absence of RNase E-based degradosome assembly. Mol Microbiol 2000; 38:854-866.

111. Mackie GA. Stabilization of the 3' one-third of *Escherichia coli* ribosomal protein S20 mRNA in mutants lacking polynucleotide phosphorylase. J Bacteriol 1989; 171:4112-4120.

112. Kaberdin VR, Miczak A, Jakobsen JS et al. The endoribonucleolytic N-terminal half of *Escherichia coli* RNase E is evolutionarily conserved in *Synechocystis* sp. and other bacteria but not the C-terminal half, which is sufficient for degradosome assembly. Proc Nat Acad Sci USA 1998; 95:11637-11642.
113. Henry MD, Yancey SD, Kushner SR. Role of the heat response in stability of mRNA in *Escherichia coli* K-12. J Bacteriol 1992; 174:743-748.
114. Georgellis D, Barlow T, Arvidson S et al. Retarded mRNA turnover in *Escherichia coli*: a means of maintaining gene expression during anaerobiosis. Mol Microbiol 1993; 9:375-381.
115. Jones PG, Van Bogelen RA, Neidhardt FC. Induction of proteins in response to low temperature in *Escherichia coli*. J Bacteriol 1987; 169:2092-2095.
116. Goldenberg D, Azar I, Oppenheim AB. Differential mRNA stability of the *cspA* gene in the cold-shock response of *Escherichia coli*. Mol Microbiol 1996; 19:241-248.
117. Beran RK, Simons RW. Cold-temperature induction of *Escherichia coli* polynucleotide phosphorylase occurs by reversal of its autoregulation. Mol Microbiol 2001; 39:112-125.
118. Mathy N, Jarrige AC, Robert le Meur M et al. Increased expression of *Escherichia coli* polynucleotide phosphorylase at low temperatures is linked to a decrease in the efficiency of autocontrol. J Bacteriol 2001; 183:3848-3854.
119. Zangrossi S, Briani F, Ghisotti D et al. Transcriptional and post-transcriptional control of polynucleotide phosphorylase during cold acclimatization in *Escherichia coli*. Mol Microbiol 2000; 36:1470-1480.
120. Yamanaka K, Inouye M. Selective mRNA degradation by polynucleotide phosphorylase in cold shock adaptation in *Escherichia coli*. J Bacteriol 2001; 189:2808-2816.
121. Toone WM, Rudd KE, Friesen JD. *deaD*, a new *Escherichia coli* gene encoding a presumed ATP-dependent RNA helicase, can suppress a mutation in *rpsB*, the gene encoding ribosomal protein S2. J Bacteriol 1991; 173:3291- 3302.
122. Jones PG, Mitta M, Kim Y et al. Cold shock induces a major ribosomal-associated protein that unwinds double-stranded RNA in *Escherichia coli*. Proc Natl Acad Sci USA 1996; 93:76-80.
123. Raynal LC, Carpousis AJ. Poly(A) polymerase I of *Escherichia coli*: characterization of the catalytic domain, an RNA binding site and regions for the interaction with proteins involved in mRNA degradation. Mol Microbiol 1999; 32:765-75.

CHAPTER 10

The Interplay between Translation and mRNA Decay in Prokaryotes:
A Discussion on Current Paradigms

Marc Dreyfus and Susan Joyce

Abstract

We discuss here the different aspects of the translation-mRNA degradation relationship in bacteria. Whereas mRNAs are usually protected against endonucleases by the presence of ribosomes, protection does not invariably require direct shielding of the cleavage sites. Possible mechanisms underlying this 'protection at a distance', which is best documented in *Bacilli* but also occurs in *Escherichia coli*, are discussed. It is also increasingly clear that besides their protective effects, ribosomes occasionally participate actively in degradation. This 'killing' activity appears to be mediated by a ribosome-associated endonuclease that might be activated by ribosome stalling. We also discuss the well-known phenomenon whereby a global inhibition of translation causes bulk mRNA stabilization. We show that this effect cannot be explained by changes in ribosome packing over individual mRNAs but reflects an apparent inhibition of the degradation machinery itself. After a translational block, rRNA becomes unstable; we propose that it permanently titrates the degradation machinery, explaining mRNA stabilization. Finally, we discuss the relative contributions of transcription, translation, and mechanism of action of endo- and exonucleases, to the $5' \rightarrow 3'$ directionality often observed during mRNA decay.

Introduction

Since early studies on *E. coli* have revealed the metabolic instability of mRNA,[1] the possibility that mRNA decay is influenced by translation has always been considered. However the nature of this influence was initially obscure, and progress in this field has been largely dependent upon our understanding of mRNA decay itself. After having long remained primitive, this understanding has expanded spectacularly over the last 10-15 years. This is particularly true with the model organisms *Escherichia coli* and *Saccharomyces cerevisiae* for which the major degradation pathways are now thought to be understood[2,3] (see Chapters 9, 11, 12 and 14). Strikingly, these pathways are not conserved in evolution. Whereas in yeast and higher cells mRNA decay is usually controlled by exonucleases—the initial and rate-limiting step being deadenylation—in *E. coli* it is largely controlled by a single endonuclease, RNase E. Because of this difference, translation should impact mRNA degradation very differently in *S. cerevisiae* (or other eucaryotes) and in *E. coli*. Deadenylation has no obvious reason of being sensitive to the presence of ribosomes over the body of the message whereas endonucleolytic cleavage has, particularly if it occurs within coding sequences. Translation is therefore expected to stabilize procaryotic but not necessarily eucaryotic mRNAs. The reality is however far more complex, partly because alternate decay mechanisms exist in addition to canonical ones. Thus, whereas the stability of some eucaryotic mRNAs is indeed independent of their translation,[4] for others the rate of deadenylation is coupled at a distance to the flow of ribosomes via specific mechanisms.[5] Moreover mRNAs carrying premature stop codons are identified and degraded by a dedicated mechanism that bypasses deadenylation (Nonsense-mediated decay; see Chapter 13). In bacteria, numerous mRNAs (or mRNA regions) do not show the usual positive correlation between

Translation Mechanisms, edited by Jacques Lapointe and Léa Brakier-Gingras.

translation and stability; in particular, some of them are quite stable whether translated or not. While in many cases this stability may reflect an absence of target for endonucleases (e.g., RNase E), there is also evidence, particularly in *Bacilli*, that mRNA regions can be stabilized at a distance by ribosomes located elsewhere on the mRNA. In a few emerging cases, the presence of ribosomes even destabilizes rather than protects mRNAs, a situation which has been documented for some time in eucaryotes. Finally, it is clear that some mRNAs undergo non-nucleolytic rather than nucleolytic inactivation, i.e., they loose their ability to support translation before degradation begins. In these cases, the presence or absence of ribosomes cannot affect degradation directly.

From these brief considerations, it should be clear that, in its widest sense, the relation between translation and degradation is quite intricate with few general rules. Yet it is also an essential and inescapable facet of gene expression. Aspects of the translation-degradation interplay in eucaryotes are reviewed elsewhere in this book (Chapters 12 and 13). Here, we summarize our understanding of the field in bacteria. The last review on this topic goes back to 1993, when C. Petersen published a thoughtful monograph entitled "Translation and mRNA stability in bacteria: a complex relationship".[6] Since then, many more examples of interference between translation and mRNA degradation have appeared, and our understanding of mRNA decay itself has greatly expanded. We shall focus particularly on the following three aspects. First, we emphasize some conceptual and technical points which are relevant to the translation-degradation relationship and when overlooked can create (and have created) much confusion. Second, we attempt to rationalize the data available for individual mRNAs in the light of our current understanding of mRNA decay. Finally, we discuss the mechanism whereby mRNAs are stabilized in bulk when translation is blocked. We stress that this latter effect cannot be reduced to the effect of inhibiting the translation of individual mRNAs, but reflects an inhibition of the mRNA degradation machinery itself, the mechanism of which is discussed.

Translation-Degradation Interplay: Definitions and Practical Considerations

Functional Versus Physical Lifetime

The "functional" lifetime of an mRNA is the time during which it can support protein synthesis, whereas its "physical" or "chemical" lifetime (often simply called "stability" here) is the time during which it remains physically intact, i.e., undegraded. In many cases, functional inactivation is a consequence of a physical degradation: this is the case, for instance, when an actively translated mRNA is cleaved within its coding sequence,[7] or within elements essential for translation (e.g., the Shine-Dalgarno sequence[8]). In these cases, the two definitions of the mRNA lifetime obviously coincide. However, mRNAs can also loose their ability to support protein synthesis because of non-nucleolytic events. Examples of such events are the binding of a translational repressor provided it is virtually irreversible,[9] the formation of a secondary structure element that irreversibly sequesters the Ribosome Binding Site (RBS),[10] or even the binding of the RNase that will subsequently degrade the mRNA (e.g., RNase E), provided this binding is distinguishable from the cleavage step and can on its own prevent ribosome loading (for discussion see ref. 11). If this 'non-nucleolytic inactivation' occurs faster than physical degradation, the functional lifetime will be shorter than the physical lifetime, although in practice the two lifetimes may be similar if the mRNA becomes prone to degradation after translation has ceased. This latter situation can be distinguished from bona fide nucleolytic inactivation by inhibiting the RNases involved in decay: in this case the physical lifetime will be increased whereas the functional lifetime will not. Non-nucleolytic inactivation has been recognized early in *E. coli*,[12-14] and it presumably holds in other bacteria as well (e.g., in *Bacillus subtilis*)[15,16].

From the viewpoint of gene expression, the distinction between nucleolytic and non-nucleolytic inactivation is essential: in the former case, mRNA degradation is a direct player in gene expression—the higher the stability, the higher the expression—whereas in the latter it is just a scavenging process without bearing on expression. The distinction is also important, albeit on a more subtle register, for the interpretation of the translation-degradation interplay. In the case of nucleolytic inactivation, any correlation between translation and stability reflects a direct effect of the presence of ribosomes upon the process of mRNA cleavage. In contrast, in the case of non-nucleolytic inactivation, such correlation only indicates that translation can accelerate or slow down the inactivation process, which itself preceeds degradation. Consequently, it may shed light on the mechanism of inactivation, but not necessarily on the mechanism of the subsequent degradation. It is therefore important to estimate the fraction of mRNAs that is inactivated non-nucleolytically. In *E. coli*, mutations that affect bulk mRNA stability can provide a clue. The conditional rne-1 or rne-3071 mutations, which at the non-permissive temperature of 43°C result in the inactivation of RNase E, stabilize bulk mRNA physically without a proportionate increase in functional lifetime.[17] This observation led to the perception that most mRNAs that decay via an RNase E-controlled process are inactivated non-nucleolytically rather than nucleolytically. However, the above mutations severely impair growth, so that the above conditions do not correspond to steady-state culture conditions. In contrast, the rne-131 mutation (formally called smbB131[18]) does not impair steady-state growth at 37°C, yet it stabilizes bulk mRNA significantly. In the presence of this mutation, the functional and physical lifetimes of bulk mRNA increase by the same factor.[19] We therefore believe that nucleolytic inactivation is the rule rather than the exception, and that in most cases the observed relationship between translation and degradation reflects the effect of the presence of ribosomes on the decay process itself.

The Interpretation of mRNA Stability Measurements

Direct or Translation-Mediated Changes of mRNA Stability

For years, the interpretation of cis-acting mutations that alter mRNA stability (or of any other experimental changes that affect this stability) has been confused by the translation-degradation interplay. Do these mutations affect stability directly (e.g., by introducing or removing endonuclease cleavage sites), or indirectly, via an effect on translation? At first sight, it seems straightforward to decide whether translation is affected or not, by measuring the «frequency of translation initiation» (FTI).[20] This parameter is usually defined as the ratio between the rate of synthesis of a protein and the steady-state concentration of its mRNA (or, more correctly, of the fraction of this mRNA that is functionally active[6]). At the microscopic level, the FTI reflects the spacing of ribosomes on the mRNA: if ribosome movement can be regarded as uniform, it is inversely proportional to this spacing. Now, in the case of a very tight coupling between translation and degradation, any delay in loading the next ribosome will cause the onset of mRNA decay (Fig. 1). The mRNA will therefore remain closely packed with ribosomes even if translation initiation is impaired by mutation: simply, its lifetime will then be shorter. This situation is illustrated in Figure 1B, where the inefficiently translated mRNA starts decaying after only a few ribosomes have been loaded. Formally, it is impossible to distinguish this situation from the opposite one in which the mutation affects primarily mRNA stability rather than translation: in both cases, ribosome packing (i.e., FTI) remains invariant and only stability decreases (compare Fig. 1B and Fig. 1C).

In practice, many cases are not really ambiguous. For instance, the introduction of a cleavage site for a well-defined endonuclease (e.g., RNase III[21]) is likely to destabilize the mRNA directly, rather than indirectly by adventitiously depressing translation. Likewise, the 5' ends of many mRNAs carry hairpins which in vitro are known to impede the entry of RNase E or RNase G (see '5' tethering pathway' below). The destabilization often observed upon removing these hairpins in vivo[22] is therefore likely to reflect a facilitated attack by these endonucleases, rather than a decrease in efficiency of translation. These situations correspond to Figure 1C. Conversely mutations which affect Shine-Dalgarno sequences or initiation codons or remove or introduce stop codons, and in the meantime affect mRNA stability,[20,22-28] are likely to do so via their effect on translation, i.e., they clearly correspond to the situation of Figure 1B. However, in several other cases, the above ambiguity was recognized but not fixed. Cho and Yanofsky have characterized mutations in the 5'UTR of the trpE gene that affected mRNA stability, but could not decide whether the primarily affected parameter was stability or translation.[29] Studying mutations downstream of the lacZ RBS that affected mRNA stability, Petersen first favored a direct effect on stability,[30,31] but subsequently considered the possibility that at least in some cases the primary target was, in fact, translation.[6] Lundberg et al observed that the stability of the ompA mRNA increases with growth rate whereas the FTI does not, and concluded that these stability changes were unrelated to translation.[32] However, in revisiting this question recently using an in vitro approach, Vytvytska et al reached the opposite conclusion that the parameter affected by growth rate is in fact translation, with stability being affected secondarily

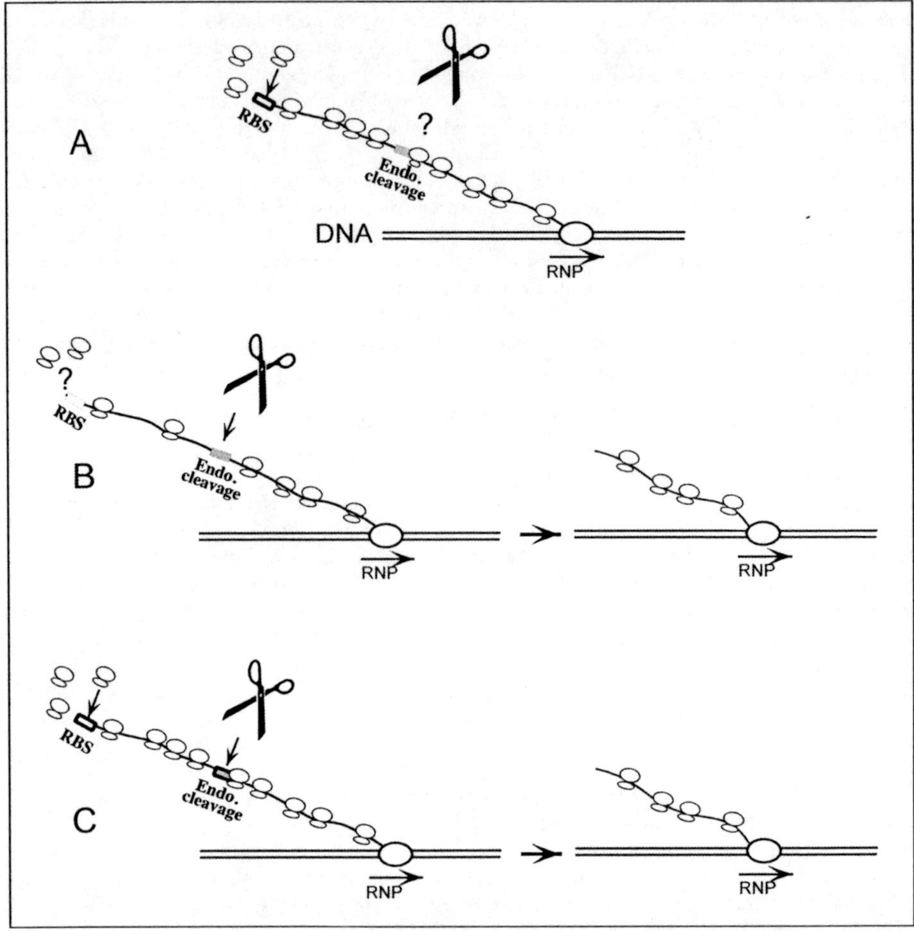

Figure 1. Cartoon showing the effect of a mutation that either decreases the efficiency of translation initiation (B) or increases the susceptibility of the mRNA to an endonuclase (C), on mRNA stability and ribosome packing. In (A) the translation and degradation rates are supposed wild-type. Open and light grey boxes symbolize RBS or endonuclease cleavage sites, respectively; a thicker trait means that the RBS or cleavage site is efficient, a thinner trait that it is less efficient. The endonuclease (scissors) is assumed to cleave the mRNA when the delay between two successive initiations exceeds a certain threshold. Initiation being a stochastic process, this event statistically occurs earlier if translation is inefficient (B) rather than efficient (A). The lifetime of the mRNA will therefore be reduced in (B). However, before cleavage occurs, the successive initiations must have been close enough to prevent cleavage, hence the equal ribosome packing ahead of the cleavage site in (A) and (B). In (C), the cleavage occurs earlier than in (A) because the cleavage site is recognized more efficiently by the nuclease. It is notable that (B) is indistinguishable from (C) from the viewpoint of both ribosome packing and mRNA stability.

due to a competition between ribosome binding and RNase E cleavage.[33] More generally, in vitro studies under conditions where translation-degradation coupling does not take place have been found useful in fixing the above ambiguity. For example, several r-proteins repress the expression of their own operons at a post-transcriptional level, and in vivo the corresponding mRNAs are destabilized under repression conditions.[34,35] However, in vitro studies show unambiguously that repression operates at the translational level, and therefore that destabilization is a consequence of the lack of translation.[36] Using the same in vitro approach, McCormick et al showed that a series of 5'UTR deletions that affected the stability of a hybrid S10'-'lacZ mRNA in vivo, had in fact a direct effect on translation, with stability being affected secondarily.[37] The reverse situation is illustrated by the rnc mRNA, which is destabilized after RNase III processing. Using again an in vitro approach, Matsunaga et al were able to show that the primary effect of this processing is to increase the susceptibility of the mRNA to an endonuclease (presumably RNase E), rather than to depress translation.[38]

Practical Aspects: How to Measure mRNA Stability?

The stability of a mRNA can be measured in a number of ways. Some of them do not require that cell growth be interrupted. For instance, stability can be assessed by comparing the steady-state concentration of the mRNA with the rate of its synthesis; the latter can be measured directly by hybridization after pulse-labeling the RNA[29] or deduced from the expression of a suitable reporter gene.[39] Alternatively, if the synthesis of the mRNA is controlled by an inducible promoter, its stability can be inferred from the kinetics of mRNA appearance or disappearance after transcription is switched on or off, respectively.[31,34,40] However, by far the most commonly used method consists of blocking all cellular transcription with suitable drugs (e.g., rifampicin) and subsequently following the decay of the mRNA of interest. This method, in which steady-state growth is effectively interrupted at the time of measurement, can yield grossly distorted results due, precisely, to the translation-degradation interplay. After rifampicin treatment, certain mRNAs remain apparently stable during extensive periods of time before exponential decay begins.[23,37,40,41] It has been proposed that under these conditions free ribosomes rapidly accumulate due to their release from fast-decaying mRNA; this accumulation, in turn, transiently enhances the translation of the surviving mRNAs and hence their stability.[37,41] In sharp contrast, the rplN mRNA appears to be destabilized by rifampicin addition.[40] In normally growing cells, the translation and stability of this mRNA are down regulated by the r-protein S8. In the presence of rifampicin, the synthesis of 16S rRNA, the primary target of S8, would stop before that of S8 itself; the pool of free S8 protein would then expand, ultimately resulting in translational repression and destabilization of the rplN mRNA[40] (see ref. 34 for a similar argument in the case of the L11 operon). These cases illustrate how treatments that interrupt growth can have unforeseen effects on mRNA stability. A further example is illustrated below in the section 'Translation and degradation: the global relationship'.

The *lacZ* Gene, or the Importance of Comparing Comparable Items

From the viewpoint of the translation-degradation interplay, no procaryotic gene has been more extensively studied than the *E. coli* lacZ gene (see refs. 6, 7, 20, 37, 42 and references therein), often with divergent results. Starting from the genuine chromosomal lac operon, we have introduced a variety of mutations in the lacZ RBS that decrease its efficiency (i.e., by altering the Shine-Dalgarno (SD) sequence or initiation codon, or by creating local secondary structures), and we have recorded the effect of these mutations upon the stability of the mRNA. To this end, we have compared in each case the steady-state level of the mRNA with the rate of its synthesis.[20,43,44] Whatever the mutation used, stability decreases steadily as translation declines, until it becomes invariant for very low translation levels. However, the steady-state mRNA level continues to decrease in this translation range due to transcriptional effects ("polarity"), and we have proposed a plausible interpretation for the shift from stability to polarity changes. Several reports in the literature are consistent with these results; those which are not can possibly be explained by alterations in the 5' region of the mRNA that may on their own affect stability, or by other differences in experimental conditions (see refs. 20 and 37 for discussion). However, one of these reports warrants a special mention given our forthcoming discussion (see below, 'local versus distal protection'). It has been reported that the presence of a ribosome lingering over the first few codons of lacZ can stabilize the whole mRNA.[42] However, only the stability of the 5' mRNA region was probed in this study. Small fragments encompassing this region are known to accumulate in the cell, presumably because they are relatively stable; in case of inefficient translation, they are much more abundant than the full-length message itself.[45-47] Moreover, in the complete absence of translation, no synthesis of the full-length message can be observed unless special precautions are taken to eliminate polarity.[20,37,45,46] When polarity is eliminated, we have observed that the presence of translating ribosomes over the beginning of the coding sequence cannot protect the entire mRNA, although fragments corresponding to the translated region are indeed stabilized (see 'local versus distal protection' below).

Translation and Degradation Relationship: Individual mRNAs

In this section, we describe how the stability of individual mRNAs is affected by their translation, and we tentatively interpret the data in the light of our current understanding of mRNA decay. In this latter respect, we heavily rely on *E. coli* work because information on other bacteria is still quite scarce.

Protective or Killer Ribosomes

Early biochemical work suggested that translating ribosomes, or a subpopulation of them, may play an active role in mRNA degradation.[48] For a while, this idea seemed consistent with the finding that drugs such as chloramphenicol, tetracycline, or fusidic acid, which block the movement of translating ribosomes, also block bulk mRNA degradation.[48,49] However, subsequent work failed to support the notion that most mRNAs are degraded by ribosome-bound nucleases; moreover, other inhibitors of protein synthesis which act by stripping ribosomes off mRNAs, such as kasugamycin or puromycin, were found to destabilize, rather than stabilize, bulk mRNA.[26,49,50] This latter observation led to the opposite view that ribosomes generally exert a protecting effect and that this protection is somehow enhanced when their movement is stopped. We now know that these experiments should have been interpreted with caution, because translation inhibitors exhibit pleiotropic effects: in particular, besides affecting the packing of ribosomes on mRNAs, they also affect the performances of the degradation machinery itself [51] (see below 'Translation and degradation: the global relationship'). Nevertheless, subsequent work confirmed that the coverage of mRNAs by ribosomes generally exerts a protective effect. Indeed, a variety of situations that decrease this coverage have been found to destabilize individual mRNAs, in particular:

1. the introduction of point mutations that depress the efficiency of the RBS;[20,22,23,27]
2. the introduction of premature stop codons;[24,28,52,53]
3. the desynchronization of transcription and translation; in this case, a naked mRNA region is created between the RNA polymerase and the leading ribosome;[7,54]
4. the binding of a translational repressor.[9,34,35]

In some cases, the ribosome-free mRNA is so unstable that it is simply undetectable even though there is evidence that it is synthesized.[7,9] All these examples support the case for a protective effect of ribosomes.

Yet, at least one clear case exists in which translation facilitates mRNA cleavage in *E. coli*.[55] The polycistronic daaA-E mRNA, which encodes the F1845 fimbriae, is processed endonucleolytically within a small open reading frame (daaP) which lies between the penultimate and last genes of the operon. This processing presumably contributes to the differential expression of individual daa genes: the downstream fragment, which encodes the fimbrial subunit, is much more stable than the upstream fragment encoding regulatory or accessory proteins. The presence of translating ribosomes over the daaP sequence is mandatory for cleavage. In particular cleavage requires the synthesis of a tripeptide (GPP) encoded near the end of the daaP gene,[56] and it occurs at a fixed distance upstream of the GPP-coding sequence irrespectively of the local nucleotide sequence. Presumably the newly-synthesized GPP tripeptide somehow «modifies» the translating ribosome, converting it into a form which can either cleave the mRNA itself, or deliver a ribosome-associated nuclease to it. The authors consider the possibility that this «modification» is nothing else but stalling during tripeptide synthesis. Indeed, there are indications that ribosome stalling may cause endonucleolytic cleavage in some cases (see below 'The case of gram-positive organisms'), although this property is certainly not generally valid.[57] To some extent, the daaP processing is reminiscent of situations already documented in higher cells: the β-tubulin mRNA is rapidly degraded in the presence of excess β-tubulin in a process which requires the translation of the first four codons of the gene.[58] Although many points remain unclear—the translation of the beginning of the daaP sequence plays a major but undefined role in cleavage, and the endonuclease(s) involved has not been identified—these results are important because for the first time they document the «killing» capacity of the *E. coli* ribosome. The late discovery of this capacity points out the persisting difficulty of establishing dogmas in the field of mRNA catabolism, even with an organism as thoroughly studied as *E. coli*.

Interference between Translation and RNase E-Mediated Decay: Current Paradigms

Before discussing how RNase E activity is affected by translation, we describe some relevant information on the enzyme itself and its mechanism of cleavage. This section is kept to a minimum as the field is extensively reviewed elsewhere[3, 59] (see Chapter 9 by R. K. Beran et al).

Mechanism of RNase E Cleavage: The 5' Tethering and Internal entry Pathways

As stated above, *E. coli* mRNAs are stabilized in bulk after the inactivation of RNase E, hence the belief that this enzyme controls the decay of many or most individual mRNAs. Since no 5'-3' exonucleases have been identified in bacteria[60] the resulting fragments can either be further cleaved endonucleolytically or degraded by 3'→5' exonucleases, particularly RNase II and polynucleotide phosphorylase (PNPase).[61] Significantly, RNase E and PNPase are associated within a multienzymatic complex, the *E. coli* 'degradosome',[62-64] which may help co-ordinating their activities. For polycistronic mRNAs, the fragments resulting from the initial cleavage are often processed into smaller (eventually monocistronic) mRNAs which are temporarily protected from further endo- or exonucleolytic attack by secondary structures at their 5' or 3' ends (see e.g. refs. 65 and 66). In contrast, for monocistronic mRNAs (whether synthesized as such or processed from polycistronic mRNAs), fragments are usually not detected in strains that are wild type for exonucleases, suggesting that the initial cleavage is rate-limiting in decay. It follows that the effect of translation upon the decay of these mRNAs usually reflects the effect of translation upon the initial cleavage. It should be kept in mind, however, that in many cases a direct implication of RNase E has not been demonstrated and that, when it has, the position of the initial cleavage site is often unknown.

Even though RNase E is an endonuclease, its activity is sensitive to the nature of the 5' end of its substrates. In vitro, mRNA molecules carrying a free 5' monophosphate end (5'p) are cleaved much faster than their counterparts which carry a 5' triphosphate (5'ppp) or a base-paired 5'p end, or which have been circularized prior to assay.[67] In vivo work has revealed further hierarchies: substrates carrying a 5'ppp are themselves cleaved faster than circular molecules.[27] That the presence of a free 5'p end (or, to a lesser extent, of a free 5'ppp) can accelerate RNase E attack suggests that the enzyme can bind these 5' extremities, and that this binding facilitates its subsequent access to internal cleavage sites ("5' tethering" mechanism[3]; Fig. 2A). Due to this preference for 5'p ends, the first RNAse E cleavage on substrates carrying a 5'ppp end is expected to be slower than subsequent ones, explaining in part why this initial cleavage is usually rate-limiting in the decay of monocistronic mRNAs. More generally, the 5' tethering mechanism explains why many mRNAs can be stabilized by appending hairpins at the 5' end.[21,68,69] However, the fact that an mRNA can be stabilized by a 5' hairpin does not necessarily means that it decays via an RNase E-controlled pathway.[22] Interestingly, RNase G, an RNase E homologue which may similarly play a role in mRNA decay, also shows preference for 5'p.[70]

The 5' tethering pathway is obviously not the only route to RNase E cleavage. Circular mRNAs, which lack a 5' extremity, are still cleaved by RNase E at a significant rate in vivo and in vitro.[27,67] In this case, the initial binding of the enzyme must occur internally. The binding site may or may not coincide with the cleavage site, so that this 'internal entry' pathway eventually involves a migration step, like the 5' tethering pathway (Fig. 2B). There is evidence that internal entry occurs not only with circular, but also with certain genuine, linear mRNA molecules. Thus, the RNase E-mediated cleavage of the rne mRNA itself, which takes place within the long 5'UTR region, is largely dependent upon the presence of a conserved hairpin structure ('hp2' in ref. 71) which presumably serves as a docking site for RNase E.[71] More generally, linear mRNAs that cannot be protected from RNase E by appending a 5' hairpin,[47] or that are cleaved by RNase E in spite of the presence of a 5' hairpin (see below 'Local versus distal protection'), are likely candidates for the internal entry pathway.

The endonucleolytic activity is carried by the N-terminal half of RNase E; the C-terminal half contains an arginine-rich RNA-binding site and is also a scaffold for degradosome assembly.[72] Interestingly, those substrates that presumably follow the internal entry pathway appear to require the C-terminal domain for efficient cleavage,[19,73] whereas those following the 5' tethering mechanism (e.g. the 9S rRNA, which carries a 5'p) do not.[19,74] Thus, binding to either the 5' end or internal sites may involve different regions of the RNase E polypeptide.[75]

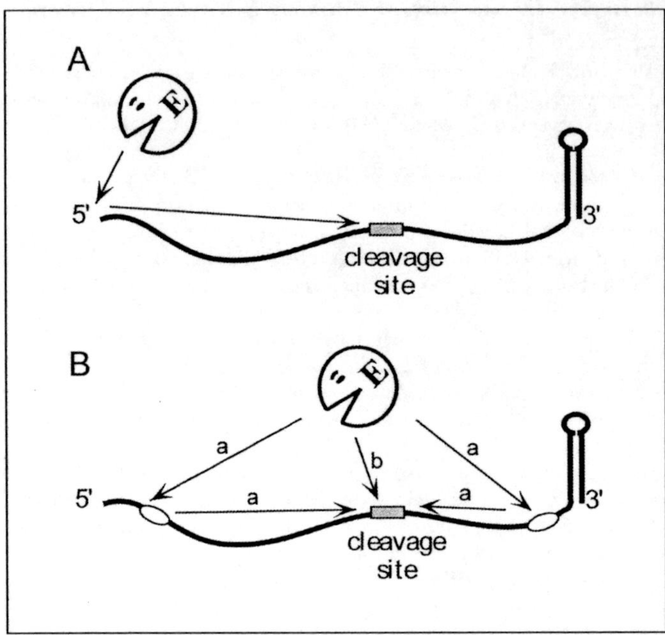

Figure 2. Schematic illustration of the 5' tethering (A) and internal entry (B) pathways. Thin arrows show the hypothetical movement of RNase E («E»). In (B) we have considered the possibility that RNase E either docks on the mRNA at a distance fom the cleavage site (variant 'a'; docking sites are shown as ovals), or in its immediate proximity ('b'). The 5' tethering pathway is comparatively favored by the presence of a free 5'p end, as well as by the presence of translating ribosomes that mask internal docking sites. Conversely, internal entry is favored by the lack of a free 5' end and by the absence of translation (see text).

Local Versus Distal Protection

The facts. Although the presence of ribosomes usually protects *E. coli* mRNAs from RNase E attack, a direct shielding of the cleavage sites is not always required, i.e., ribosomes can eventually provide protection 'at a distance'. For instance, in the bla mRNA, the translation of the first 20% of the coding sequence exerts a protection over the remaining 80%, which otherwise would be labile (the RNase—possibly RNase E—that controls decay in this case has not been characterized).[24] Another example is the rpsT mRNA. This mRNA decays via two nonexclusive pathways which may both be mediated by the degradosome, i.e., PNPase trimming from the 3' end or RNase E cleavage within the coding sequence, at the two-thirds of the mRNA length.[3,76] The presence of ribosomes over the first 15 codons, far away from the targets of RNase E and PNPase, confers some protection to the whole mRNA by inhibiting both pathways.[25] Protection at a distance has also been observed in gram-negative bacteria other than *E. coli*. In the *Rhodobacter capsulatus* puf operon, the stability of the mRNA encoding the promoter-distal pufL and pufM cistrons is insensitive to whether or not these cistrons themselves are translated, but sensitive to the translation of two promoter-proximal cistrons, pufA and pufB.[77] Interestingly, the decay of the puf mRNA is controlled by a *R. capsulatus* enzyme that is closely related to RNase E and also assembles into a degradosome.[78,79] Less precise observations, in which translation is not interrupted prematurely but depressed by mutations in the RBS region, are also suggestive of a protection at a distance. Indeed, in several cases, the presence of only a few translating ribosomes, if not a single one, can significantly protect a whole coding region against RNase E.[22,23] Although the exact location of the RNase E-sensitive sites is not known in these cases, it seems unlikely that protection can be explained by the direct shielding of these sites by the (rare) translating ribosomes: rather, protection at a distance must occur.

In contrast, other cases are known where mRNA stability obviously requires the direct shielding of the cleavage sites by ribosomes. In the rpsO mRNA, the major RNase E cleavage site is located only 10 nt downstream of the stop codon and thus is presumably directly shielded by terminating ribosomes. Increasing the spacing between the stop and the cleavage site to 20 nt is enough to substantially accelerate cleavage, indicating that ribosome shielding is required for protection.[28] Another particularly clear example is provided by recent observations on a variant of the rpsT mRNA (K. Baker and G. Mackie, personal communication). The rpsT mRNA can be markedly (but not completely) stabilized against RNase E by appending an hairpin at its 5' end. However, besides the presence of the hairpin, stabilization also requires translation across the major RNase E cleavage site: the introduction of a stop codon at 16 nt upstream of this site very markedly labilizes the mRNA, whereas its introduction downstream of this site has a far milder effect. The lacZ mRNA is still another example[47] (SJ and MD, unpublished). To study the effect of prematurely interrupting translation upon the stability of this mRNA, we resorted to T7 RNA polymerase (T7 RNAP) for transcribing the corresponding gene, since this enzyme is resistant to polarity.[46] The resulting transcript is prone to RNase E attack even when translation is uninterrupted, because T7 RNAP outpaces ribosomes (see above 'protective or killer ribosomes'). However, the complete absence of translation renders the mRNA even more sensitive to RNase E. This sensitivity is not relieved by the presence of a ribosome stably bound near the 5' end: only the region surrounding the bound ribosome is stabilized.[47] Similarly, partial translation (down to codons 18 or 307) causes protection of fragments encompassing the translated region only, indicating local rather than distal protection. Obviously, in these three cases, ribosomes must be present at the cleavage site to achieve protection.

Possible interpretations. Let us first assume that the initial binding site of RNase E on the mRNA is distinct from the cleavage site (this situation corresponds to the 5' tethering pathway, or to variant 'a ' of the internal entry pathway; Fig. 2). The pathway leading to the initial cleavage can then be operationally divided into three steps, i.e., initial binding, migration to the cleavage region, and cleavage stricto sensu (Fig. 3A). We further assume that each of these three successive steps is sensitive to the presence of ribosomes in the corresponding mRNA region, i.e., in the binding site region, between the binding and the cleavage sites, and near the cleavage site, respectively. Each step will then be facilitated if ribosomes are removed from the corresponding region. However, mRNA destabilization will not necessarily ensue. In a multistep pathway such as the one considered on Figure 3A, only the acceleration of the rate-limiting step (or eventually of steps that precede the rate limiting step; see comment on scheme V in Fig. 3 legend) can yield an increase in the overall reaction rate. In particular, removing the ribosomes from the cleavage region will only destabilize the mRNA if the cleavage itself is rate-limiting in the whole pathway (kinetic scheme III). If it is not, i.e., if cleavage is already a fast step when ribosomes are present, then this removal will not affect the stability further (schemes I and II). In this case, only the presence of ribosomes around the binding site, or between the binding and cleavage sites (depending upon whether step 1 or 2 is rate-limiting) will be important for mRNA stability. In other words, protection 'at a distance' will occur (Fig. 3A).

Alternatively, in the case of the internal entry pathway, RNase E may dock directly at or near the cleavage site (variant 'b' of Fig. 2B). The reaction pathway then reduces to two steps, i.e., binding and cleavage (Fig. 3B). It is noteworthy that, whichever of these step is rate-limiting, the removal of ribosomes from the cleavage region will accelerate the overall reaction rate, destabilizing the mRNA. In this case, no protection at a distance can be observed (Fig. 3B).

According to the above interpretation, mRNAs which show 'protection at a distance'—the bla, rpsT, or puf mRNAs—should decay via the three-step pathway (Fig. 3A) with cleavage being a fast step. In contrast, mRNAs that require direct ribosome shielding for stability—the rpsO mRNA, the rpsT mRNA carrying a 5' hairpin, or the lacZ mRNA synthesized by the T7RNAP-may follow either pathway 10-3A with cleavage being rate-limiting, or pathway 10-3B. We can only speculate why this should be so. Amongst all the above mRNAs, and the rpsT mRNA carrying a 5' hairpin and the T7RNAP synthesized by lacZ mRNA are the best candidates for the internal entry pathway. Indeed, the former cannot be stabilized by appending a 5' hairpin,[47] whereas the latter lacks a free 5' end altogether. It is then tempting to speculate that these mRNA's follow the variant of the internal entry pathway where RNase E docks directly to the cleavage site (variant 'b' in Fig 2B), since this situation explicitly excludes protection at a distance (Fig. 3B). As for the rpsO mRNA, there may be other reasons why it differs from the bla, rpsT, or puf mRNAs. Stop codons are presumably cleared by ribosomes more slowly than other sites within coding sequences.[80] Therefore, compared to other RNase E sites which are cleaved via the three-step pathway (Fig. 3A), the site flanking the

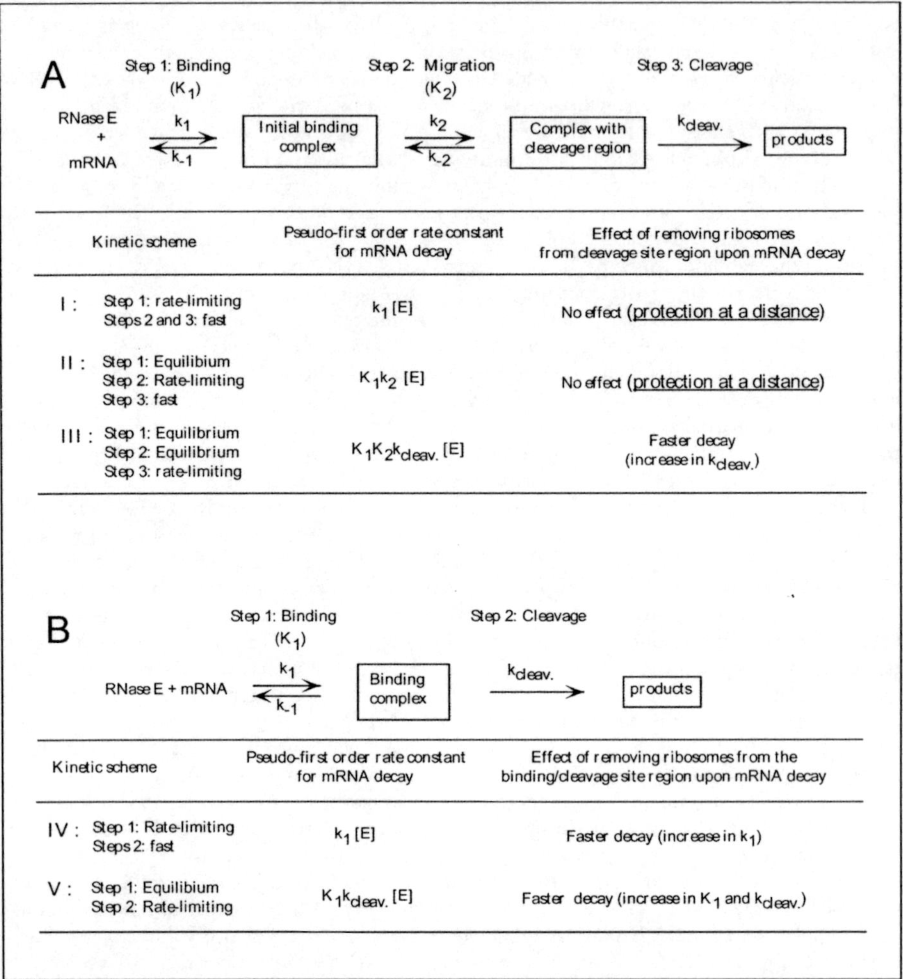

Figure 3. The removal of ribosomes from the region corresponding to the initial RNase E cleavage, may or may not cause mRNA destabilization (the initial cleavage is assumed to control the whole decay process; see Text). In (A) we assume that the pathway to initial cleavage consists of three kinetically distinct steps, i.e., binding, migration to the cleavage region, and cleavage itself (cf. Figs. 2A and 2B, variant 'a'). The first two steps are characterized by equilibrium constants $K_1 = k_1/k_{-1}$ and $K_2 = k_2/k_{-2}$, whereas the last one is irreversible. In (B), we assume that the initial binding and cleavage sites coincide (cf. Fig. 2B, variant 'b'). Pseudo first order rate constants for mRNA decay are derived by assuming that the concentrations of the binding complexes are at a steady state. Further assumptions are: $k_{-1} \ll k_2$, $k_{-2} \ll k_{cleav.}$ (Scheme I), $k_{-1} \gg k_2$, $k_{-2} \ll k_{cleav.}$ (Scheme II), $k_{-1} \gg k_2$, $k_{-2} \gg k_{cleav.}$ (Scheme III), $k_{-1} \ll k_{cleav.}$ (Scheme IV), and $k_1 \gg k_{cleav.}$ (Scheme V). In Scheme V, we have assumed that the removal of ribosomes from the binding site region facilitates binding (i.e., k_1 increases) but has no effect on dissociation once RNase E is bound (k_{-1} invariant), so that K_1 increases.

rpsO stop codon may be inaccessible to RNase E for extensive periods of time, slowing the cleavage step to the point that it becomes rate-limiting.

We end with a word of caution. 'Protection at a distance' is interpreted here in the light of recent work showing that the initial binding site of RNase E can be distinct from its cleavage site, so that the decay rate is influenced by ribosomes located away from the cleavage region. However, before this fact was recognized, 'protection at a distance' had been given a completely different interpretation[23] (see also refs. 22 and 24). According to this view, only certain conformations of an mRNA would be sensitive to RNase E. Newly synthesized mRNA molecules would be resistant but would slowly refold into more sensitive conformations. The crossing of even a single translating ribosome would then 'reset the folding clock' and render the message resistant for another period of time. In this model, the ribosome does not necessarily need to cross the mRNA regions that will ultimately be cleaved by RNase E for providing protection (**Fig. 4**). The observed result will then be 'protection at a distance' but here the ribosome does not interfere directly with any step involving RNase E. At present, this hypothetical mechanism is simply an ad hoc explanation for protection at a distance, but it is certainly too early to dismiss the idea that it plays a major role here.

The Absence of Translation May Favor Internal Entry Over 5' Tethering Mechanism

The models presented in Figure 3 suggest that, whatever the kinetic scheme, the rate of RNase E binding (k_1) will impact the overall kinetics of cleavage (eventually the impact may be indirect, via the binding equilibrium constant K_1; see Fig. 3 legend). Now, this binding rate is expected to be far more sensitive to translation in the internal entry than in the 5' tethering pathway. In the former case, RNase E docking usually occurs within coding sequences, so that binding is directly affected by the presence or absence of ribosomes. In contrast, in the 5' tethering pathway, binding will only be affected if a RBS is located close enough to the 5' end. Because of this difference, mRNAs might shift from the 5' tethering to the internal entry mechanism as translation becomes less efficient. Consistently, whereas a translated lacZ mRNA can be stabilized by appending a 5' hairpin, suggesting that it decays via the 5' tethering pathway,[21] the (very unstable) lacZ mRNA resulting from the removal of a functional RBS cannot be stabilized in that way, indicating internal RNase E entry[47] (see above). This behavior is not always observed, however. The stability of the circular rpsT mRNA, which obviously decays via the internal entry mechanism, is not more sensitive to translation than the stability of the cognate linear mRNA, which presumably decays via the 5' tethering pathway.[27] In this particular case, both modes of RNase E binding are presumably equally facilitated by ribosome removal.

Transcription, Translation, or RNase E Preferences: The Source of Directionality of Decay

It is often stated that mRNAs decay with an overall 5' to 3' direction. Since decay generally consists of endonucleolytic hits followed by exonucleolytic trimming of the upstream fragment, this notion requires clarification. By "decay in the 5'→3' direction," we refer here to the specific situation in which the successive hits occur in an orderly 5' to 3' wave with the first one taking place near the 5' mRNA end. Directionality has been occasionally observed in polycistronic mRNAs,[53,81] but obviously it is not generally valid in this case.[66,82] In contrast, for monocistronic mRNAs (whether synthesized as such or processed from polycistronic mRNAs), directionality is likely to be the rule, if only because of biological efficiency. Indeed, it is the only decay mode which avoids the wasteful synthesis of incomplete polypeptides from truncated mRNAs.[82] Consistently, for several long mRNAs, the 5' end starts decaying before mRNA synthesis is completed.[83-85] Some monocistronic mRNAs appear to break this rule, as exemplified by the rpsO and rpsT mRNAs which are initially attacked by RNase E near their 3' ends,[3,76,86] but these exceptions should be relativized given the very small size of these mRNAs.

What is the molecular basis for directionality? The answer to this question is probably not unique. It has long been thought that directionality simply reflects the vectorial nature of transcription and translation, which themselves proceed in the 5'→3' direction. Thus, the 5' region is synthesized first, and if susceptibility to endonucleolytic cleavage were uniform over the mRNA, it should also statistically disappear first. This argument is particularly relevant for long mRNAs for which the synthesis of the 3' end lags significantly behind that of the 5' end. As concerns the role of translation, it has been proposed that once the RBS has been removed, preventing further ribosome loading, the naked mRNA region lagging behind the last ribosome is rapidly chopped off by endonucleases (e.g., RNase E), resulting in an orderly 5'→3' degradation.[82] While these phenomena—and particularly the first—almost certainly contributes to directionality, it is likely that the mechanism of degradation itself also plays its role. First, whereas the mode of degradation—endonucleolytic hits followed

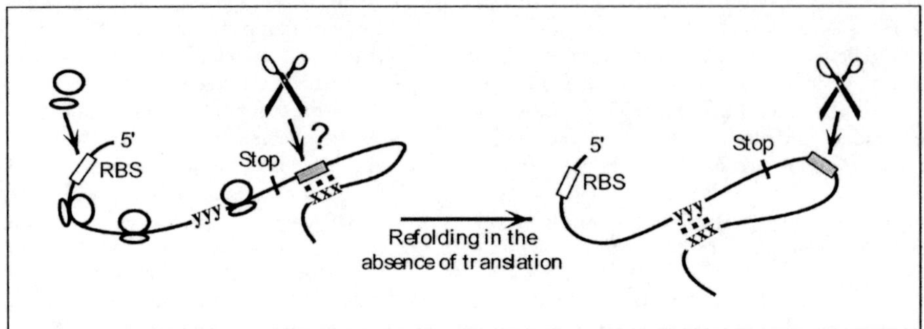

Figure 4. A possible mechanism for protection 'at a distance' by translating ribosomes. When translated (left), the mRNA adopts a conformation in which the RNase E cleavage site is inaccessible, due to base-pairing with sequence xxx. When translation is interupted (right), the mRNA refolds into an alternative conformation in which xxx now pairs with yyy, so that the cleavage site becomes accessible. In this model, ribosomes can provide protection without ever crossing the cleavage site itself (see text). All symbols are as in Fig. 1.

by rapid trimming of upstream fragments—does not in itself guarantee that the first hit occurs near the 5' end, it will in a sense contribute to directionality, insofar as regions located downstream of the hits are more stable than upstream ones. This will be particularly true for the mRNA fragment located downstream of the most distal endonucleolytic hit: this fragment is usually genuinely resistant to exonucleases and therefore should be degraded last (for an illustration see[53]). Second, the 5' region of the mRNA, encompassing the 5'UTR and the RBS, have distinctive features which may favor RNase E cleavage, so that the first hit is likely to occur there. Cleavage within the 5'UTR may be facilitated by the absence of ribosomes. As for the RBS, it is known that efficient ribosome binding usually requires that this region is unstructured[87] and statistical analysis has indeed confirmed this local absence of structure.[88] As such, the RBS is expected to be intrinsically prone to cleavage by single-strand endonucleases such as RNase E. The third and most intriguing possibility is that the affinity of RNase E for 5' ends directly contributes to directionality. Assuming that it follows the 5' tethering pathway, RNase E may have a higher probability of cleaving the first sites it encounters from the 5' end compared to more distant ones. To date, this attractive hypothesis has not yet received experimental support.

The Case of Gram-Positive Organisms

By far the most impressive examples of a 'protection at a distance' have been observed in gram-positive bacteria, notably *Bacilli*. The stalling of ribosomes over the small open reading frames present upstream of the erm genes, ermA and ermC, can markedly stabilize the downstream coding sequences.[15,89] Subsequently, it has been observed that in the cryIIIA gene of *B. thuringiensis* or the SP82 phage of *B. subtilis*, the mere presence of a long SD element without an associated start codon ('STAB-SD') can stabilize kilobases of downstream sequence, presumably by stably binding a small ribosomal subunit.[90,91] A variety of sequences can be stabilized in this way, including long untranslated regions such as the *E. coli* lacZ mRNA lacking a RBS.[91,92] Finally, stalled or bound ribosomes can exert their protecting effect not only when present near the 5' mRNA end but also when located internally,[91,93] although stabilization may be somewhat less efficient in this case.[90,94] Only regions located downstream of the internally-located ribosome are stabilized, whereas upstream regions remain labile (Fig. 5). The 5' end of the protected region has been mapped at only four nt upstream of the STAB-SD,[91] whereas in the case of the ribosome stalled on the ermA mRNA it has been tentatively localized a few nucleotides upstream of the A site codon.[93] Presumably, the nuclease responsible for generating such extremities penetrates deep into the trailing edge of the ribosome (Fig. 5). Moreover, in the case of the cryIIIA STAB-SD, cleavage occurs at a fixed distance from the STAB-SD whatever the local sequence.[91]

These striking results are reminiscent of the 'protection at a distance' observed in *E. coli* (see above). Most similar in this respect is a report by Björnsson and Isaksson; studying an artificial gene

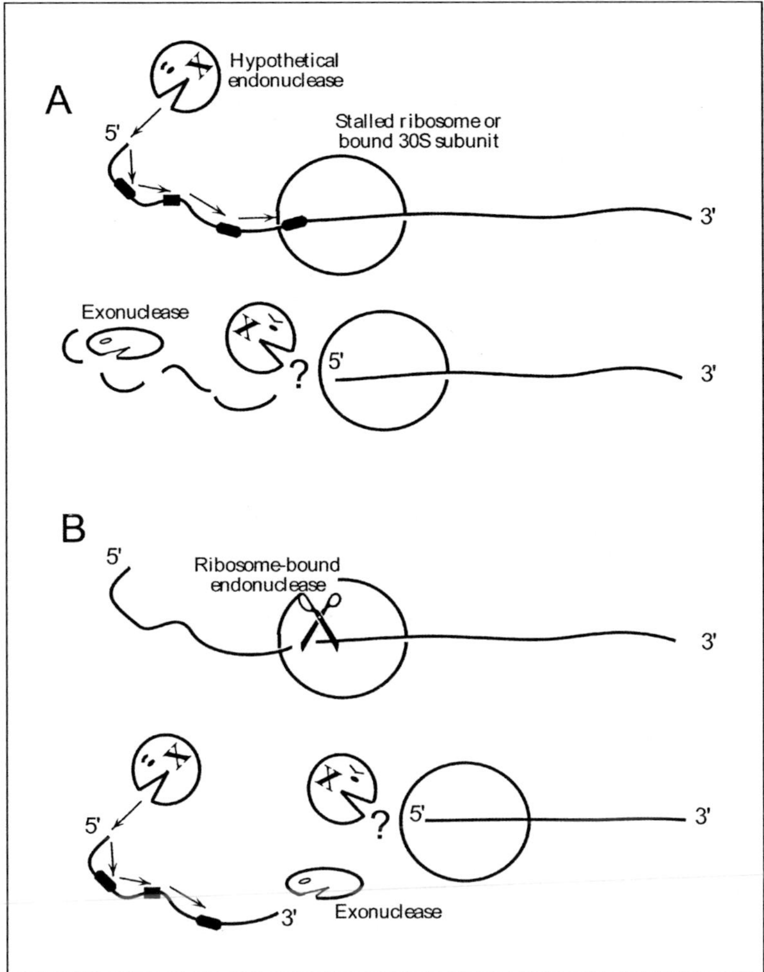

Figure 5. Possible mechanisms for the interplay between translation and mRNA stability in *Bacilli*. In these organisms, a bound 30S subunit or a stalled ribosome (large circle) can protect extensive stretches of downstream sequences whereas the upstream sequence remains labile (see text). We hypothesize that *Bacilli* possess an uncharacterized endonuclease (noted X) that strictly abides by the 5' tethering pathway (cf. Fig. 2A). In (A), X is assumed to be responsible for the decay of the mRNA of interest (X cleavage sites are shown as closed boxes), but to be unable to skip obstacles such as a bound ribosome. In (B), the initial cleavage between the upstream (unstable) and downstream (stable) regions is mediated by the ribosome itself (scissors). X then degrades the upstream region, whereas the downstream region is stable because the bound ribosome prevents the binding of X to the 5' end. In both cases, 3'→5' exonucleases (ovals) presumably participate to the decay of the upstream region.

in *E. coli*, these authors observed that the stalling of a ribosome over an inefficient stop codon resulted in a moderate stabilization of the 261 nt-long downstream sequence.[80] The 5' end of the protected region lay 13 nt upstream of the stop (i.e., A-site) codon, a result reminiscent of the ermA case. However, compared to *B. subtilis*, the magnitude of the stabilizing effects observed in *E. coli* appear modest. In *E. coli*, the mere presence of a ribosome stably bound at the 5' end of the lacZ mRNA is unable to protect the whole message against RNase E[47] (see above 'Local Versus Distal Protection'). Moreover, the STAB-SD stabilizers from SP82 or cryIIIA do not work efficiently in *E.*

coli.[90, 91] Excluding the (unlikely) possibility that *B. subtilis* possesses an unrecognized 5'→3' exonuclease which is at work here and is absent from *E. coli*, then *B. subtilis* must contain an endonuclease (noted 'X' in Fig. 5) that differs from RNase E in at least two respects. First, it must have a much stronger preference for 5' tethering over internal entry pathway. Only in this case can we explain that even long, normally labile regions are quite stable when located downstream of the bound ribosome. In this respect, the hypothetical endonuclease may resemble RNase E lacking the C-terminal region, since this truncated enzyme seems unable to go through the internal entry pathway (see above 'Mechanism of RNase E cleavage'). Second, to explain that a bound ribosome constitutes such a strong barricade to its action, the hypothetical nuclease must scan down the mRNA from the 5' end without any possibility of 'skipping' obstacles (Fig. 5A). This is unlikely to be the case for RNase E, because RNase E substrates carrying an accessible 5'p end, which presumably follow the 5' tethering pathway, are cleaved efficiently even when strong secondary structures are present between the 5' end and the cleavage site (see ref. 3 for discussion). As concerns the 5' ends of the fragments that are protected by stalled or bound ribosome in *B. subtilis*, it is possible that they are generated by the action of this same hypothetical endonuclease (X); however, this enzyme must then have a very low sequence specificity and also be able to penetrate deep into the ribosome-protected region (Fig. 5A). Alternatively (and perhaps more likely) the 5' end of the protected fragment may be generated by an endonuclease activity associated with the stalled ribosome, as suggested for the daaP gene (see above 'Protective or killer ribosomes'). Were this scenario correct, it would explain why the hypothetical RNase E-like enzyme 'X' cannot skip the ribosome: simply, after having mediated the cleavage, the ribosome would remain bound at the 5' end of the downstream fragment, denying access to a nuclease that strictly conforms to the 5' tethering pathway. In contrast, the upstream fragment would be freely accessible to endo- or exonucleolytic cleavage, and would be rapidly degraded (Fig. 5B).

The analysis of the *B. subtilis* genome shows no evidence for an RNase E homologue,[95] but RNase E-like activities obviously exist in this organism.[96] Only the characterization of the mRNA degradation machinery will permit a better understanding of the translation-degradation relationship in *Bacilli*.

Translation and Degradation: The Global Relationship

Translation Inhibitors Stabilize mRNAs whether They Are Translated or Not

As noted above ('Protective or killer ribosomes') drugs such as chloramphenicol, tetracycline, or fusidic acid, which block the movement of translating ribosomes, cause stabilization of *E. coli* mRNAs both individually and in bulk (see ref. 6 for compilation). Interestingly, mutations that slow down ribosome movement also produce similar effects.[97] In contrast, drugs such as puromycin and kasugamycin, which strip ribosomes off mRNAs, destabilize mRNAs.[6,49,50] Traditionally, these effects are interpreted in term of mRNA protection by translating ribosomes (see 'Protective or killer ribosomes' above): stalled ribosome would protect cleavage sites directly or at a distance, whereas stripping mRNAs would make these sites more accessible ('cis' effects). Parallel observations have been made in yeast and higher eucaryotes, with however an important difference: here, mRNAs are stabilized not only by cycloheximide (a ribosome-stalling drug) but also by puromycin.[98] Moreover, even untranslated mRNAs can be stabilized by cycloheximide.[4] On this basis, it has been proposed that drugs stabilize eucaryotic mRNAs not by altering the packing or activity of translating ribosomes, but by somehow inhibiting the mRNA degradation machinery itself ('trans' effect).[98]

By analogy, we reasoned that trans effects might also exist in bacteria, perhaps superimposed with cis effects. Again, trans effects should be revealed most clearly by testing the effect of inhibitors upon the stability of untranslated mRNAs: were cis effects exclusive, this stability should be insensitive to translation inhibitors, whereas if trans effects existed, it would remain sensitive.[51] This led us to investigate the effect of translation inhibitors upon the stability of two untranslated mRNAs, i.e., the lacZ mRNA lacking a RBS (precautions were taken to eliminate polarity; see 'Local versus distal protection' above) and RNAI, a small untranslated RNA which first undergoes RNase E cleavage near its 5' end and is then quickly trimmed by PNPase assisted by poly(A) polymerase.[69, 99] The results were unambiguous: both untranslated RNAs were stabilized by translation inhibitors, whether they cause ribosome stalling or mRNA stripping. Moreover, in the case of RNAI, both the RNase E cleavage and the subsequent PNPase-mediated pathways were inhibited. These results show that blocking translation results in an inhibition of both RNase E and PNPase, i.e., that trans effects exist in *E. coli*. Presumably, with translated mRNAs, both cis and trans effects coexist: in the case of ribosome-stalling drugs, they cooperate in stabilizing mRNAs, whereas for mRNA-stripping drugs they antagonize each other.[51]

A Plausible Model

The model proposed to explain the apparent inhibition of RNase E and PNPase after a translation block rests on two known facts.[51] First, both enzymes autoregulate their own synthesis by degrading their own mRNAs.[85, 100] As a consequence, they are presumably never present in excess; rather, they continuously adjust their synthesis to that of their substrates via the autoregulation loop—except of course when translation is blocked. Second, it is known that rRNA synthesis is boosted after a translational block. Moreover, the newly synthesized rRNA is unstable, presumably because it cannot assemble with r-proteins,[101] and RNase E and PNPase appear to participate in this degradation.[102] We then hypothesized that the apparent inhibition of both enzymes after a translational block, simply reflects the increased synthesis of their substrates under conditions where their pool cannot expand. These enzymes would then become permanently titrated[51] (Fig. 6).

The above model yields a simple prediction. Were it correct, it should be possible to titrate RNase E or PNPase even in the absence of any translational block, simply by inducing the synthesis of a highly expressed substrate in growing cells. Under these conditions, RNAs that are degraded by

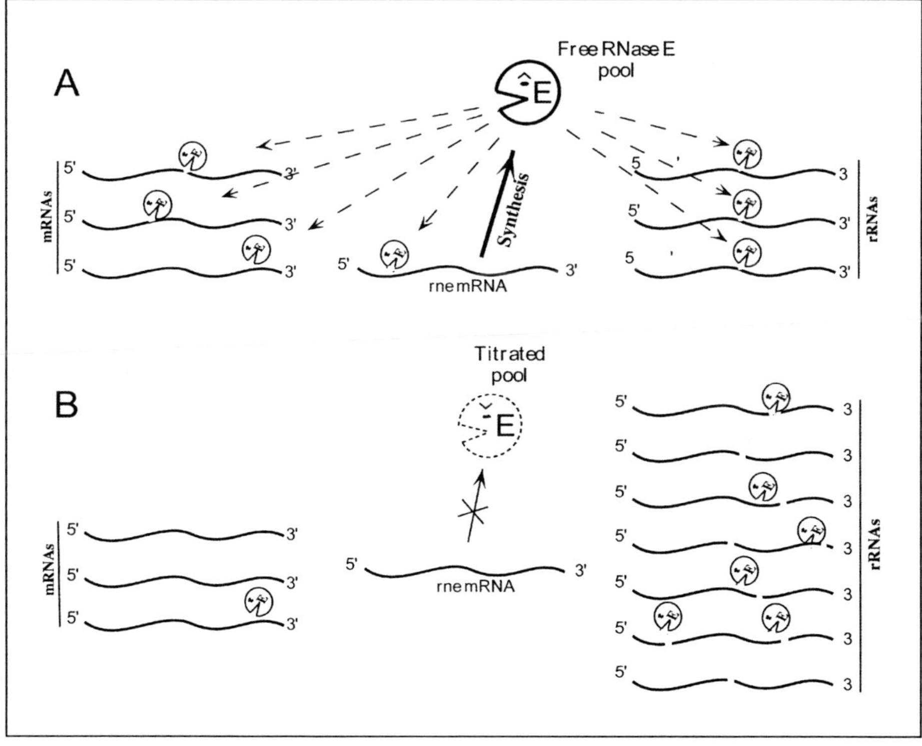

Figure 6. Mechanism of bulk mRNA stabilization after a translational block: an hypothesis. In growing cells (A), RNase E (E) is synthesized from the rne mRNA (thick arrow) and used for both the degradation of mRNAs and the maturation of rRNA (dotted arrows). Its free pool is limited because it degrades its own mRNA, autoregulating its production. After a translational block (B), the synthesis of rRNA is boosted and the newly rRNA is unstable (nicks). In the meantime RNase E synthesis is discontinued (interrupted arrow). We assume that RNase E is then titrated, i.e., its free pool disappears (dotted circle), hence mRNA stabilization.

these enzymes—and in particular their own mRNAs—should be stabilized. However, the titration (and hence mRNA stabilization) should then be transient, because the pools of RNase E or PNPase are now free to expand in response to the stabilization of their own mRNAs. In this respect, RNase E or PNPase would behave like other autoregulated proteins which adjust their synthesis to that of their substrates ('homeostasis'[103]). Recently, the above prediction has been substantiated in our laboratory.[104] When the synthesis of an artificial RNase E substrate was induced to high levels—we estimated that the induction caused an increase of total cellular RNase E substrates by 20 to 50%— the rne mRNA encoding RNase E was markedly stabilized, boosting RNase E synthesis. However, consistent with the above prediction, this stabilization and the increase in RNase E synthesis were only transient, presumably reflecting the delay required for expansion of the RNase E pool. These experiments show that titration of RNase E by excess substrate occurs readily in vivo, and thus constitutes a plausible explanation for the apparent inhibition of this enzyme after a translational block. Whether or not these observations can be extended to PNPase is not known, nor is it known whether similar models can explain the stabilization of mRNAs by translation inhibitors in eucaryotes.

Acknowledgements

Because of space limitation, this essay cannot be comprehensive; we apologize to the many colleagues whose important contributions have not been discussed. We thank Drs. G. Mackie and M. Springer for valuable comments. Work in our Laboratory is funded by CNRS and Ecole Normale Supérieure, as well as by grants from the Association pour la Recherche sur le Cancer (ARC) and the Ministère de la Recherche (programme «PRFMMIP »).

Note added on proofs. Recently, another example of protection downstream sequences by a stalled ribosome in *B. subtilis* (Cf. Fig. 5) has been described (Drider D, DiChiara JM, Wei J et al. Endonuclease cleavage of mRNA in *Bacillus subtilis*. Mol Microbiol 2002; 43:1319-1329).

References

1. Jacob F, Monod J. Genetic regulatory mechanisms in the synthesis of proteins. J Mol Biol 1961; 3:318-356.
2. Wilusz CJ, Wormington M, Peltz SW. The cap-to-tail guide to mRNA turnover. Nature Reviews/Molec Cell Biol 2001; 2:237-246.
3. Coburn GA, Mackie GA. Degradation of mRNA in *Escherichia coli*: an old problem with some new twists. Prog Nucleic Acid Res Mol Biol 1999; 62:55-108.
4. Beelman CA, Parker R. Differential effects of translational inhibition in cis and in trans on the decay of the unstable yeast MFA2 mRNA. J Biol Chem 1994; 269:9687-9692.
5. Grosset C, Chen CY, Xu N et al. A mechanism for translationally coupled mRNA turnover: interaction between the poly(A) tail and a c-fos RNA coding determinant via a protein complex. Cell 2000; 103:29-40.
6. Petersen C. Translation and mRNA stability in bacteria: a complex relationship. In: Belasco JG, Brawerman G, eds. Control of mRNA stability. San Diego CA.: Academic Press, 1993:117-145.
7. Iost I, Dreyfus M. The stability of *Escherichia coli* lacZ mRNA depends upon the simultaneity of its synthesis and translation. EMBO J 1995; 14:3252-3261.
8. Sanson B, Uzan M. Dual role of the sequence-specific bacteriophage T4 endoribonuclease RegB: mRNA inactivation and mRNA destabilization. J Mol Biol 1993; 233:429-446.
9. Nogueira T, de Smit M, Graffe M et al. The relationship between translational control and mRNA degradation for the *Escherichia coli* threonyl-tRNA synthetase gene. J Mol Biol 2001; 310:709-722.
10. Poot RA, Tsareva NV, Boni IV et al. RNA folding kinetics regulates translation of phage MS2 maturation gene. Proc Natl Acad Sci USA 1997; 94:10110-10115.
11. Nierlich DP, Murakawa GJ. The decay of bacterial messenger RNA. Prog Nucl Acids Res Mol Biol 1996; 52:153-216.
12. Schwartz T, Craig E, Kennell D. Inactivation and degradation of messenger ribonucleic acid from the lactose operon of *Escherichia coli*. J Mol Biol 1970; 54:299-311.
13. Puga A, Borras MT, Tessman ES et al. Difference between functional and structural integrity of mRNA. Proc Natl Acad Sci USA 1973; 70:2171-2175.
14. Yamamoto T, Imamoto F. Differential stability of trp messenger RNA synthesized originating at the trp promoter and P_L promoter of lambda trp phage. J Mol Biol 1975; 92:289-304.
15. Sandler P, Weisblum B. Erythromycin-induced stabilization of ermA messenger RNA in *Staphylococcus aureus* and *Bacillus subtilis*. J Mol Biol 1988; 203:905-915.
16. Hue KK, Bechhofer DH. Effect of ermC leader region mutations on induced mRNA stability. J Bacteriol 1991; 173:3732-3740.
17. Mudd EA, Krisch HM, Higgins CF. RNase E, an endoribonuclease, has a general role in the chemical decay of *Escherichia coli* mRNA: evidence that rne and ams are the same genetic locus. Mol Microbiol 1990; 4:2127-2135.

18. Kido M, Yamanaka K, Mitani T et al. RNase E polypeptides lacking a carboxyl-terminal half supress a mukB mutation in *Escherichia coli.* J Bacteriol 1996; 178:3917-3925.

19. Lopez PJ, Marchand I, Joyce SA et al. The C-terminal half of RNase E, which organizes the *Escherichia coli* degradosome, participates in mRNA degradation but not rRNA processing in vivo. Mol Microbiol 1999; 33:188-199.

20. Yarchuk O, Jacques N, Guillerez J et al. Interdependence of translation, transcription and mRNA degradation in the lacZ gene. J Mol Biol 1992; 226:581-596.

21. Hansen MJ, Chen L-H, Fejzo ML et al. The ompA 5' untranlated region impedes a major pathway for mRNA degradation in *Escherichia coli.* Mol Microbiol 1994; 12:707-716.

22. Arnold TE, Yu J, Belasco JG. mRNA stabilization by the ompA 5' untranslated region: two protective elements hinder distinct pathways for mRNA degradation. RNA 1998; 4:319-330.

23. Jain C, Kleckner N. IS10 mRNA stability and steady state levels in *Escherichia coli:* indirect effects of translation and role of rne function. Mol Microbiol 1993; 9:233-247.

24. Nilsson G, Belasco JG, Cohen SN et al. Effect of premature termination of translation on mRNA stability depends on the site of ribosome release. Proc Natl Acad Sci USA 1987; 84:4890-4894.

25. Rapaport LR, Mackie GA. Influence of translational efficiency on the stability of the mRNA for ribosomal protein S20 in *Escherichia coli.* J Bacteriol 1994; 176:992-998.

26. Baumeister R, Flache P, Melefors O et al. Lack of a 5' non-coding region in Tn1721 encoded TetR mRNA is associated with a low efficiency of translation and a short half-life in *Escherichia coli.* Nucl Acids Res 1991; 19:4595-4600.

27. Mackie GA. Stabilization of circular rpsT mRNA demonstrates the 5' end dependence of RNase E action in vivo. J Biol Chem 2000; 275:25069-25072.

28. Braun F, Le Derout J, Régnier P. Ribosomes inhibit an RNase E cleavage which induces the decay of the rpsO mRNA of *Escherichia coli.* EMBO J 1998; 17:4790-4797.

29. Cho KO, Yanofsky C. Sequence changes preceding a Shine-Dalgarno region influence trpE mRNA translation and decay. J Mol Biol 1988; 204:51-60.

30. Petersen C. The functional stability of the lacZ transcript is sensitive towards sequence alterations immediately downstream of the ribosome binding site. Mol Gen Genet 1987; 209:179-187.

31. Petersen C. Multiple determinants of functional mRNA stability: sequence alterations at either end of the lacZ gene affect the rate of mRNA inactivation. J Bacteriol 1991; 173:2167-2172.

32. Lundberg U, Nilsson G, von Gabain A. The differential stability of the *E. coli* ompA and bla mRNA at various growth rates is not correlated to the efficiency of translation. Gene 1988; 72:141-149.

33. Vytvytska O, Moll I, Kaberdin VR et al. Hfq (HF1) stimulates ompA mRNA decay by interfering with ribosome binding. Genes Dev 2000; 14:1109-1118.

34. Cole JR, Nomura M. Changes in the half-life of ribosomal protein messenger RNA caused by translational repression. J Mol Biol 1986; 188:383-392.

35. Singer P, Nomura M. Stability of ribosomal protein mRNA and translational feedback regulation in *Escherichia coli.* Mol Gen Genet 1985; 199:543-546.

36. Yates JL, Nomura M. Feedback regulation of ribosomal protein synthesis in Escherichia coli: localization of the mRNA target sites for repressor action of ribosomal protein L1. Cell 1981; 24:243-249.

37. McCormick JR, Zengel JM, Lindahl L. Correlation of translation efficiency with the decay of lacZ mRNA in *E. coli.* J Mol Biol 1994; 239:608-622.

38. Matsunaga J, Simons EL, Simons RW. *Escherichia coli* RNase III (rnc) autoregulation occurs independently of rnc translation. Mol Microbiol 1997; 26:1125-1135.

39. Lopez PJ, Iost I, Dreyfus M. The use of a tRNA as a transcriptional reporter: the T7 late promoter is extremely efficient in *Escherichia coli* but its transcripts are poorly expressed. Nucleic Acids Res 1994; 22:1186-1193.

40. Liang ST, Ehrenberg M, Dennis P et al. Decay of rplN and lacZ mRNA in *Escherichia coli.* J Mol Biol 1999; 288:521-538.

41. Pease AJ, Wolf REJ. Determination of the growth rate-regulated steps in expression of the Escherichia coli K-12 gnd gene. J Bacteriol 1994; 176:115-122.

42. Wagner LA, Gesteland RF, Dayhuff TJ et al. An efficient Shine-Dalgarno sequence but not translation is necessary for lacZ mRNA stability in Escherichia coli. J Bacteriol 1994; 176:1683-1688.

43. Guillerez J, Gazeau M, Dreyfus M. In the Escherichia coli lacZ gene the spacing between the translating ribosomes is insensitive to the efficiency of translation initiation. Nucl Acids Res 1991; 19:6743-6750.

44. Jacques N, Guillerez J, Dreyfus M. Culture conditions differentially affect the translation of individual *Escherichia coli* mRNAs. J Mol Biol 1992; 226:597-608.

45. Stanssens P, Remaut E, Fiers W. Inefficient translation initiation causes premature transcription termination in the lacZ gene. Cell 1986; 44:711-718.

46. Chevrier-Miller M, Jacques N, Raibaud O et al. Transcription of single-copy hybrid lacZ genes by T7 RNA polymerase in *Escherichia coli:* mRNA synthesis and degradation can be uncoupled from translation. Nucleic Acids Res 1990; 18:5787-5792.

47. Joyce SA, Dreyfus M. In the absence of translation, RNase E can bypass 5' mRNA stabilizers in *Escherichia coli.* J Mol Biol 1998; 282:241-254.

48. Kuwano M, Kwan CN, Apirion D et al. Ribonuclease V of *Escherichia coli:* I. Dependance of ribosomes and translocation. Proc Natl Acad Sci USA 1969; 64:693-700.

49. Pato ML, Bennett PM, von Meyerburg K. mRNA synthesis and degradation in *Escherichia coli* during inhibition of translation. J Bacteriol 1973; 116:710-718.

50. Schneider E, Blundell M, Kennell D. Translation and mRNA decay. Mol Gen Genet 1978; 160:121-129.

51. Lopez PJ, Marchand I, Yarchuk O et al. Translation inhibitors stabilize *Escherichia coli* mRNAs independently of ribosome protection. Proc Natl Acad Sci USA 1998; 95:6067-6072.
52. Morse DE, Yanofsky C. Polarity and the degradation of mRNA. Nature 1969; 224:329-331.
53. Goodrich AF, Steege DA. Roles of polyadenylation and nucleolytic cleavage in the filamentous phage mRNA processing and decay pathways in *Escherichia coli*. RNA 1999; 5:972-985.
54. Makarova OV, Makarov EM, Sousa R et al. Transcribing *Escherichia coli* genes with mutant T7 RNA polymerases: stability of lacZ mRNA inversely correlates with polymerase speed. Proc Natl Acad Sci USA 1995; 92:12250-12254.
55. Loomis WP, Moseley SL. Translational control of mRNA processing in the F1845 fimbrial operon of *Escherichia coli*. Mol Microbiol 1998; 30:843-853.
56. Loomis WP, Koo JK, Cheung TP et al. A tripeptide sequence within the nascent DaaP protein is required for mRNA processing of a fimbrial operon in *Escherichia coli*. Mol Microbiol 2001; 39:693-707.
57. Roche ED, Sauer RT. SsrA-mediated peptide tagging caused by rare codons and tRNA scarity. EMBO J 1999; 18:4579-4589.
58. Theodorakis NG, Cleveland DW. Physical evidence for cotranslational regulation of beta-tubulin mRNA degradation. Mol Cell Biol 1992; 12:791-799.
59. Régnier P, Arraiano CM. Degradation of mRNA in bacteria: emergence of ubiquitous features. Bioessays 2000; 22:235-244.
60. Zuo Y, Deutscher MP. Exoribonuclease superfamilies: structural analysis and phylogenetic distribution. Nucl Acids Res 2001; 29:1017-1026.
61. Donovan WP, Kushner SR. Polynucleotide phosphorylase and ribonuclease II are required for cell viability and mRNA turnover in *Escherichia coli* K12. Proc Natl Acad Sci USA 1986; 83:120-124.
62. Carpousis AJ, Van Houwe G, Ehretsmann C et al. Copurification of *E. coli* RNase E and PNPase: Evidence for a specific association between two enzymes important in RNA processing and degradation. Cell 1994; 76:889-900.
63. Miczak A, Kaberdin VR, Wei CL et al. Proteins associated with RNase E in a multicomponent ribonucleolytic complex. Proc Natl Acad Sci USA 1996; 93:3865-3869.
64. Py B, Higgins CF, Krish HM et al. A DEAD-box RNA helicase in the *Escherichia coli* RNA degradosome. Nature 1996; 381:169-172.
65. Bricker AL, Belasco JG. Importance of a 5' stem-loop for longevity of papA mRNA in *Escherichia coli*. J Bacteriol 1999; 181:3587-3590.
66. Klug G, Cohen SN. Combined actions of multiple hairpin loop structures and sites of rate-limiting endonucleolytic cleavage determine differential degrdation rates of individual segments within polycistronic puf operon mRNA. J Bacteriol 1990; 172:5140-5146.
67. Mackie G. Ribonuclease E is a 5'-end-dependent endonuclease. Nature 1998; 395:720-723.
68. Emory SA, Bouvet P, Belasco JG. A 5'-terminal stem-loop structure can stabilize mRNA in *Escherichia coli*. Genes Dev 1992; 6:135-148.
69. Bouvet P, Belasco JG. Control of RNase E-mediated RNA degradation by 5'-terminal base pairing in *E. coli*. Nature 1992; 360:488-491.
70. Tock MR, Walsh AP, Carroll G et al. The CafA protein required for 5'-maturation of 16S rRNA is a 5'-end-dependent ribonuclease that has context-dependent broad sequence specificity. J Biol Chem 2000; 275:8726-8732.
71. Diwa A, Bricker AL, Jain C et al. An evolutionary conserved RNA stem-loop functions as a sensor that directs feedback regulation of RNase E gene expression. Genes Dev 2000; 14:1249-1260.
72. Vanzo NF, Li YS, Py B et al. Ribonuclease E organizes the protein interactions in the *Escherichia coli* RNA degradosome. Genes Dev 1998; 12:2770-2781.
73. Jiang X, Diwa A, Belasco JG. Regions of RNase E important for 5'-end dependent RNA cleavage and autoregulated synthesis. J Bacteriol 2000; 182:2468-2475.
74. Ow MC, Liu Q, Kushner SR. Analysis of mRNA decay and rRNA processing in *Escherichia coli* in the absence of RNase E-based degradosome assembly. Mol Microbiol 2000; 38:854-866.
75. Marchand I, Nicholson AW, Dreyfus M. Bacteriophage T7 protein kinase phosphorylates RNase E and stabilizes mRNAs synthesized by T7 RNA polymerase. Mol Microbiol 2001; 42:767-776.
76. Mackie GA. Stabilization of the 3' one third of *Escherichia coli* ribosomal protein S20 mRNA in mutants lacking polynucleotide phosphorylase. J Bacteriol 1989; 171:4112-4120.
77. Klug G, Cohen SN. Effects of translation on degradation of mRNA segments transcribed from the polycistronic puf operon of *Rhodobacter capsulatus*. J Bacteriol 1991; 173:1478-1484.
78. Jäger S, Fuhrmann O, Heck C et al. An mRNA degrading complex in *Rhodobacter capsulatus*. Nucl Acids Res 2001; 29:4581-4588.
79. Fritsch J, Rothfuchs R, Rauhut R et al. Indentification of an mRNA element promoting rate-limiting cleavage of the polycistronic puf mRNA in *Rhodobacter capsulatus* by an enzyme similar to RNase E. Mol Microbiol 1995; 15:1017-1029.
80. Björnsson A, Isaksson LA. Accumulation of an mRNA decay intermediate by ribosomal pausing at a stop codon. Nucl Acids Res 1996; 24: 1753-1757.
81. Morse DE, Mosteller R, Baker RF et al. Degradation of tryptophan messenger. Nature 1969; 223:37-43.
82. Kennell DE. The instability of messenger RNA in bacteria. In: Reznikoff W, Gold L, eds. Maximizing gene expression. Butterworth, USA: 1986:101-142.
83. Cannistraro VJ, Kennell D. Evidence that the 5' end of lac mRNA starts to decay as soon as it is synthesised. J Bacteriol 1985; 161:820-822.

84. Yarchuk O, Jacques N, Guillerez J et al. Interdependence of translation, transcription and mRNA degradation in the lacZ gene. J Mol Biol 1992; 226:581-596.
85. Jain C, Belasco JG. RNase E autoregulates its synthesis by controlling the degradation rate of its own mRNA in *E. coli*: unusual sensitivity of the rne transcript to RNase E activity. Genes Dev 1995; 9:84-96.
86. Régnier P, Hajnsdorf E. Decay of mRNA encoding ribosomal protein S15 of *Escherichia coli* is initiated by an RNase E-dependent endonucleolytic cleavage that removes the 3' stabilizing stem and loop structure. J Mol Biol 1991; 217:283-292.
87. de Smit MH, van Duin J. Secondary structure of the ribosome binding site determines translational efficiency: a quantitative analysis. Proc Natl Acad Sci USA 1990; 87:7668-72.
88. Ganoza MC, Kofoid EC, Marlière P et al. Potential secondary structure at translation-initiation sites. Nucl Acids Res 1987; 15:345-360.
89. Bechhofer DH, Dubnau D. Induced mRNA stability in *B. subtilis*. Proc Natl Acad Sci USA 1987; 84:498-502.
90. Hue KK, Cohen SD, Bechhofer DH. A polypurine sequence that acts as a 5' mRNA stabilizer in *Bacillus subtilis*. J Bacteriol 1995; 177:3465-3471.
91. Agaisse H, Lereclus D. STAB-SD: a Shine-Dalgarno sequence in the 5' untranslated region is a determinant of stability. Mol Microbiol 1996; 20:633-643.
92. DiMari JF, Bechhofer DH. Initiation of mRNA decay in *Bacillus subtilis*. Mol Microbiol 1993; 7:705-717.
93. Sandler P, Weisblum B. Erythromycin-induced ribosome stall in the ermA leader: a barricade to a 5'-to-3' nucleolytic cleavage of the ermA transcript. J Bacteriol 1989; 171:6680-6688.
94. Bechhofer DH, Zen KH. Mechanism of erythromycin-induced ermC mRNA stability in *Bacillus subtilis*. J Bacteriol 1989; 171:5803-5811.
95. Kunst F, Ogasawara N, Moszer I et al. The complete genome sequence of the gram-positive bacterium *Bacillus subtilis*. Nature 1997; 390:249-256.
96. Condon C, Putzer H, Luo D et al. Processing of the *Bacillus subtilis* thrS leader mRNA is RNase E-dependent in *Escherichia coli*. J Mol Biol 1997; 268:235-242.
97. Gupta RS, Schlessinger D. Coupling of rates of transcription, translation and mRNA degradation in streptomycin-dependent mutants of *E. coli*. J Bacteriol 1976; 125:84-93.
98. Brawerman G. mRNA degradation in eucaryotic cells: an overview. In: Belasco JG, Brawerman G, eds. Control of mRNA stability. San Diego CA.: Academic Press, 1993:149-159.
99. Xu F, Lin-Chao S, Cohen SN. The Escherichia coli pcnB gene promotes adenylylation of antisense RNAI of ColE1-type plasmids in vivo and degradation of RNAI decay intermediates. Proc Natl Acad Sci USA 1993; 90:6756-6760.
100. Jarrige A-C, Mathy N, Portier C. PNPase autocontrols its expression by degrading a double stranded structure in the pnp mRNA leader. EMBO J 2001; 20:6845-6855.
101. Shen V, Bremer H. Chloramphenicol-induced changes in the synthesis of ribosomal, transfer, and messenger RNA in *Escherichia coli* B/r. J Bacteriol 1977; 130:1098-1108.
102. Bessarab DA, Kaberdin VR, Wei C-L et al. RNA component of *Escherichia coli* degradosome: evidence for rRNA decay. Proc Natl Acad Sci USA 1998; 95:3157-3161.
103. Craig EA, Gross CA. Is hsp70 the cellular thermometer? Trends Biochem Sci 1991; 16:135-140.
104. Sousa S, Marchand I, Dreyfus M. Autoregulation allows *Escherichia coli* RNase E to adjust continuously its synthesis to that of its substrates. Mol Microbiol 2001; 42:867-878.

CHAPTER 11

Polyadenylation and Degradation of RNA in Prokaryotes

Philippe Régnier and Paulo E. Marujo

Summary

Polyadenylation is a postranscriptional modification of RNA found in all cells and in organelles. In bacteria, a small fraction of RNA harbors oligo(A) tails which are mostly shorter than 20 As. Poly(A) polymerase I of *Escherichia coli* can adenylate mRNAs, and small RNA regulators originating from the chromosome, from plasmids and from bacteriophages and also precursors and mature forms of tRNAs and rRNAs which have an accessible 3' extremity. The model of poly(A) metabolism presented in this chapter proposes that the ratio of adenylated to nonadenylated RNAs and the length of oligo(A) tails represent the equilibrium of the antagonistic activities of poly(A) polymerase I, which synthesizes poly(A) at the 3' end of all RNAs, and of exoribonucleases which shorten or completely remove the tails. Poly(A) tails provide toeholds where polynucleotide phosphorylase can initiate exonucleolytic degradation of tightly folded RNAs protected from exoribonucleases by 3' stable secondary structures. Polyadenylation promotes degradation of mRNA fragments and controls the intracellular concentration of regulatory RNAs.

Introduction

Poly(A) polymerase, the enzyme responsible for the synthesis of poly(A) tails of RNA was discovered in bacteria long before the identification of polyadenylated mRNAs in eukaryotic cells.[1] The length of poly(A) tails (60 to 200 As) and the abundance of polydenylated mRNA in eukaryotic cells (nearly all mRNAs are adenylated) compared to the situation in bacteria where only a small fraction of RNA harbors short tails led to the assumption that polyadenylation was specific for eukaryotic cells and that it has only a minor function in the metabolism of bacterial mRNA. Eukaryotic polyadenylation is described by T. Preiss in chapter 12 . More recently, several important discoveries such as the characterization of the *Escherichia coli* poly(A) polymerase (PAP I), its gene (*pcnB*) and its implication in the control of mRNA stability renewed the interest of researchers for polyadenylation of RNAs in bacteria.[2] The literature relating the development of this topic has been recently reviewed.[3,4] This chapter focuses primarily on the metabolism of poly(A) tails and on the role of polyadenylation in the control of mRNA stability in *E. coli*.

Characterization of Poly(A) Tails

The repeated observation that pulse labeled RNAs isolated from *E. coli* and several other eubacteria and archaea could be retained on oligo(dT) cellulose was the first indication that polyadenylation takes place in prokaryotes.[3] However, the fraction of polyadenylated molecules was usually very low (less than 3% of total labeled *E. coli* RNA was retained on oligo(dT) cellulose)[5,6] and attempts to determine the length of tails gave very heterogeneous data. Digestion of polyadenylated RNA by RNase T1 and RNase A cleaving, respectively, the GpN, CpN and UpN phosphodiester bonds generates poly(A) tracts whose length can be determined by electrophoresis or from the adenosine:AMP ratio after alkaline hydrolysis. Poly(A) tails ranging from 1[7,8] to 50 As[5,9] were detected in *E. coli*. It must be pointed out, however, that many of these data, based on the binding of

Translation Mechanisms, edited by Jacques Lapointe and Léa Brakier-Gingras.
©2003 Eurekah.com and Kluwer Academic / Plenum Publishers.

polyadenylated RNA to oligo(dT),[3] probably only give approximate idea of the polyadenylation status of bacterial RNA since they did not take in account RNAs harboring tails of less than 20 As that are not retained on oligo(dT). Investigations of specific transcripts confirmed that only a small fraction of bacterial RNAs harbors poly(A) tails which are mostly less than 20 As in length whereas the majority of molecules are not adenylated. Based on the fraction of RNA retained on oligo(dT), it was estimated that only 0.011% of bacteriophage f1 mRNA is polyadenylated[10] whereas 40% of *lpp* and *trpA* mRNAs harbor poly(A) tails of 10-20 As.[11,12] On the other hand, analysis of mRNAs on Northern blot showed that the majority of *rpsO* primary transcripts are not adenylated with only 10% of these molecules harboring short tails of 1-3 nucleotides (P. Marujo, E. Hajnsdorf and P. Régnier, unpublished).[8] In contrast to eukaryotic cells, polyadenylation is not restricted to mRNA in bacteria (Table 1).[13] Reverse transcription experiments using a primer complementary to an oligonucleotide ligated to the 3' end of RNA molecules demonstrated that a few As were post-transcriptionally added at the 3' end of RNA I, the regulator of Col E1 type plasmid replication.[14] Genomic RNA from bacteriophage MS2,[15] the small *oop* RNA of bacteriophage λ,[16] 5S rRNA[17] and precursors of 16S and 23S rRNAs[18] are also polyadenylated under normal physiological conditions. Moreover, many other RNAs including the mature 23S rRNA,[18] tRNA^{Tyr},[17] the tmRNA,[17] the small *sok* regulatory RNA[19] and the 6S, 4.5S and M1 small RNAs[17] can be polyadenylated in strains overproducing PAP I or lacking exonucleases able to degrade single-stranded 3' ends of RNAs (Table 1) (see below). These RNAs have very short tails (Table 1). Not only are stable, regulatory and messenger RNAs polyadenylated but most of them can be adenylated at many different sites resulting from transcription termination, endonucleolytic processing by RNase E and RNase III and digestion by exonucleases engaged in trimming of stable RNAs or degradation of mRNAs.[4,10,18,20] Finally, some regulatory and messenger RNAs (*copA*, *ompA*, *trxA*) stabilized upon PAP I inactivation are also presumed to be polyadenylated (Table 1) (see below).[9, 21] It therefore appears that PAP I can polyadenylate all classes of bacterial RNA but that only a small fraction of molecules, that may be different from an RNA species to another, is actually polyadenylated.

Enzymes of Poly(A) Metabolism

In *E. coli,* the main enzymes of poly(A) metabolism are PAP I, which accounts for the synthesis of most, if not all, poly(A) tails and exonucleases of 3' to 5' polarity, capable of degrading single-stranded stretches of nucleotides found at the 3' end of RNAs.[22,23] PAP I is encoded by the *pcnB* gene and is dispensable for growth.[24-26] It is a monomer of approximatively 53 kDa[24,27] with characteristic features of the nucleotidyltransferase superfamily including eukaryotic poly(A) polymerases as well as eukaryotic and prokaryotic tRNA nucleotidyltransferases, which generate the CCA 3' extremity of tRNAs.[28,29] PAP I requires a divalent cation, Mg^{2+} or Mn^{2+}, to be active.[1,30] The enzyme uses ATP as substrate to polymerize AMP residues at the 3' end of RNA primers. CTP and UTP can also be used by PAP I but at a much lower rate (at about 5% the rate of ATP incorporation).[30,31] Synthesis of poly(G) tails by PAP I has also been reported[31] but this reaction is probably very slow.[30] These properties may explain why Cs are sometime incorporated in poly(A) tails (E. Hajnsdorf and P. Régnier, unpublished). The Km values for ATP and tRNA primers are 50 μM and 0.2 μM, respectively.[30] The fact that poly(A) tails are synthesized at the 3' ends of many primary transcripts, processed RNAs and intermediary products of exonucleolytic degradation none of which share common 3' terminal features, suggests that PAP I does not specifically recognize structural motifs of particular RNAs.[10,18,20] However, the variability of the length of tails mentioned above suggests that all RNA species are not adenylated with equal efficiency. Moreover, it has been shown that PAP I preferentially adenylates RNAs primers harboring a 5' monophosphorylated extremity or single-stranded stretches of nucleotides at the 3' or at the 5' end.[31,32] The influence of 5' structures on polyadenylation efficiency could explain why truncation at the 5' end of RNA increases sensitivity to poly(A)-dependent degradation.[33] PAP I is a distributive enzyme (it dissociates from RNAs after addition of each or a few nucleotides) which exhibits a preference for poly(A) primers.[34] Its activity is modified by protein Hfq which interacts with A-rich regions of RNAs[35,36] and is involved in replication of the single-stranded genomic RNA of bacteriophage Qβ[37] and in the translation of the *ompA* messenger[38] and of the σ^s subunit of RNA polymerase, specific of stationary phase and osmotic upshift.[39] Hfq stimulates poly(A) synthesis in vivo and in vitro and converts PAP I into a processive enzyme which remains associated to the molecules that it elongates.[40] Other potential

Table 1. Polyadenylated bacterial RNAs

	Genetic Background	Length of Tails	Effect of PAP I on RNA Stability
mRNAs			
rpsO[8]	RE PH	1-3	no
rpsO[47]	PNP RII RE PH	1-20	destabilizes
rpsO(fragments)[33]	PNP RII PH	-	destabilizes
lpp[12]	WT	10-15	-
lpp[9]	PNP RII RE PH	-	destabilizes
trpA[11]	WT	15-20	-
hag[91]	WT	-	-
phage T7[92]	WT	≈50	-
phage T7[7]	WT	1-3	-
ompA[9]	PH	-	no
"	PNP RII RE PH	-	destabilizes
trxA[9]	PH	-	no
"	PNP RII RE PH	-	destabilizes
rpsT (fragments)[56]	PH	-	destabilizes
phage f1 (fragments)[10]	WT	-	destabilizes
pnp[18]	high PAP	-	stabilizes
rne[18]	high PAP	-	stabilizes
Small and stable RNAs			
oop RNA Phage λ[81]	PH	-	destabilized
RNA I (col E1)[14]	WT	1-3	destabilized
CopA[21]	WT	-	destabilized
Sok (R1)[19]	PNP RII RE	<10	destabilized
tRNA^{Tyr}[17]	T PH D BN	3-5	-
rRNA 5S[17]	WT	1	-
"	T PH D BN	1	-
RNA 6S[17]	T PH D BN	4	-
rRNA 23S[18]	high PAP-PH	-	-
tmRNA[17]	T PH D BN	3-4	-
MS2 genomic RNA[15]	WT	1-7	-
Precursors of stable RNAs			
tRNA^{Tyr}[17]	T PH D BN	1-3	-
rRNA 5S[17]	T PH D BN	1-7	-
RNA 6S[17]	T PH D BN	1-4	-
rRNA 23S[17]	T PH D BN	1-5	-
rRNA 23S[18]	PH	-	-
RNA M1[17]	T PH D BN	1-3	-
RNA 4,5S[17]	T PH D BN	3-4	-
rRNA 16S[17]	T PH D BN	4	-
rRNA 16S[18]	PH	-	-

The third column shows the length of poly(A) tails of RNAs isolated from cells whose phenotype is indicated in the second column. WT corresponds to cells where all ribonucleases are active and PNP, RII, RE, T, PH, D and BN to cells deficient for PNPase, RNase II, RNase E, RNase T, RNase PH, RNase D and RNase BN, respectively. High PAP refers to cells overproducing PAP I. The last column shows the effect of PAP I on RNA stability deduced from life times of the molecules in strains lacking or overproducing PAP I.

cofactors interacting with PAP I are RNase E, which is involved in RNA decay, and DEAD box RNA helicases, which may cooperate with PAP I for poly(A) synthesis.[41] The fact that PAP I competes with exonucleases capable of degrading poly(A) tails may explain why the long tails of 500-1000As that can be synthesized by PAP I in vitro are not detected in the cell.[40,42]

E. coli contains eight 3' to 5' exonucleases capable of degrading single-stranded RNAs or oligonucleotides that may all be involved in degradation of poly(A).[43,44] The fact that poly(A) is much more abundant when two of these enzymes, RNase II and polynucleotide phosphorylase (PNPase), are inactive while inactivation of only one of them does not have such an effect suggests that each of these enzymes can degrade poly(A) tails.[8,45-48] RNase II is a processive hydrolase (it remains bound to the molecules that it degrades) which generates ribonucleotides monophosphate.[49] The abundance of this enzyme, which represents 90% of the poly(A) degrading activity measured in *E. coli* cellular extracts, suggests that it plays a major role in poly(A) catabolism.[50] Consistent with this idea, it has been shown that RNase II removes the poly(A) tails synthesized by PAP I at the 3' end of *rpsO* primary transcripts almost completely thus explaining why 90% of these molecules are unadenylated (P. Marujo and P. Régnier, unpublished).[8] Processive removal of nucleotides by RNase II is blocked by stable secondary structures which cause dissociation of the enzyme from the RNA.[51-55] This property explains why RNase II removes single-stranded stretches of nucleotides and poly(A) tails lying downstream of transcription terminators or of REP sequences but fails to degrade RNAs upstream of these hairpins.[8,52,56-59] RNAs devoid of 3' secondary structures are probably completely degraded by RNase II.[33,60]

PNPase is a phosphorylase which attacks single-stranded 3' end of RNAs that it degrades processively into monoribonucleotides diphosphates (NDP).[61] In the reverse reaction, which takes place at low phosphate concentration, PNPase processively incorporates NDP into heterogenous polynucleotides of random sequences.[61] However, its primary role in the cell is the 3' to 5' degradation of RNA molecules[62] and it is one of the components of a multienzymatic complex refered to as the RNA degradosome which participates in mRNA decay (see Chapter 9, by RK Beran et al).[63,64] Although PNPase is blocked by secondary structures, like RNase II, it is capable of degrading highly structured RNA that are resistant to this latter enzyme.[52-54,56,65] RNAs whose 3' terminal nucleotides are sequestered in a secondary structure are poorly attacked by PNPase free or associated with the degradosome.[66-68] In contrast, this enzyme can use single-stranded stretches of nucleotides lying downstream of stable 3' terminal secondary structures (for example transcriptional terminators) as a toehold to begin the exonucleolytic degradation of such RNAs.[66,67,69] A sequence of five As downstream of a hairpin is sufficient to initiate degradation.[66] The mechanism of degradation of folded RNA by PNPase is described below. Beside its primary role in poly(A) and RNA degradation, it has been proposed that PNPase might account for the synthesis of the few poly(A) tracts which have been detected in PAP I deficient cells.[48] It has also been suggested that PNPase synthesizes heterogenous tails (containing As, Us, Cs and Gs) in vivo in the absence of PAP I and incorporates U and C residues in poly(A) tails.[48] Conversion of PNPase activity from phosphorolysis to polymerization and the reverse might be governed by transient modifications of local phosphate concentration at the 3' end of RNAs. The physiological significance of these alternate phases of RNA elongation and shortening remains mysterious.

RNase E plays a major role in mRNA decay in addition to being involved in the maturation of 5S and 16S rRNA.[70,71] It is the scaffold for the association of PNPase, the RhlB RNA helicase and enolase, a glycolytic enzyme, in the RNA degradosome.[63,64,67,72] It is a 5' end dependent endoribonuclease which preferentially cleaves RNAs harboring single-stranded monophosphorylated 5' extremities.[73] This enzyme, which frequently initiates RNA decay, cleaves molecules in single-stranded A-U rich sequences[70,71] and removes tails of polyadenylated molecules in vitro.[74,75]

In addition to PNPase and RNase II, *E. coli* contains five additional 3' to 5' exoribonucleases, RNase T, RNase PH, RNase D, RNase BN and RNase R, capable of removing single-stranded nucleotides from the 3' extremity of RNAs, that have been implicated in the maturation of many stable and small RNAs.[44,76-78] These enzymes, which are also blocked by secondary structures, could contribute to the degradation of oligo(A) tails added at the 3' ends of mRNAs, regulatory RNAs and precursors of stable RNAs. It appears however that they cannot counteract the synthesis of long poly(A) tails by PAP I.[8,46,47]

A Model of Poly(A) Metabolism

The appearance of long poly(A) tails, not detected under normal conditions, in cells either lacking the two 3' to 5' exoribonucleases PNPase and RNase II or overproducing PAP I, led to the conclusion that the length of bacterial poly(A) tails is determined by a dynamic equilibrium between the opposing activities of PAP I and exoribonucleases.[8,17,22,56] Most, if not all, 3' RNA extremities accessible to PAP I can likely be adenylated. Stable RNAs such as tRNAs and rRNAs are nearly never subject to polyadenylation probably because their 3' termini are masked by aminoacylation or burial within the ribonucleoparticles.[17] In contrast, PAP I polyadenylates the 3' extremities of abnormal tRNA precursors (M. Deutscher, personnal communication) that cannot be aminoacylated and of precursors of the 16S and 23S rRNAs that might emerge from the surface of the ribosome.[18] Moreover, tightly folded RNAs with a triphosphorylated 5' extremity and lacking terminal single-stranded stretches of nucleotides may be poorly adenylated.[31,32] The fact that PAP I is a distributive enzyme, which dissociates from the RNA primer after addition of one or a few adenosine residue(s), implies that 3' ends of tails become accessible to exoribonucleases after each (or a few) step(s) of adenylation. If long tails have been synthesized, RNase II or PNPase can carry out rapid processive degradation of poly(A) until it encounters stable stem-loops which cause stalling and dissociation of both ribonucleases (Fig. 1A). The processive degradation probably produces RNAs with single-stranded stretches of about 9 nucleotides downstream of the hairpin.[51] These molecules and other RNA harboring short poly(A) tails are probably degraded distributively, at a slower rate, by RNase II and PNPase to generate short tails of 1-3 As such as those found in the RNA I and *rpsO* transcripts.[8,14] In the case of the *rpsO* mRNA, RNase II can completely remove the tail while PNPase produces RNAs harboring tails of 2-4 As (Fig. 1A).[8] More generally, RNase II can remove single-stranded nucleotides very close to the base of secondary structures that are not accessible to PNPase.[52] 3' terminal hairpins of different structures and stabilities may block 3' to 5' exoribonucleases at different positions and thus determine the number of As left at the 3' end of the RNA.[53] For example, exonucleases can probably remove As very close to hairpins containing breathing A-U base pairs at the base of the stem while these As may not be accessible if the hairpin is closed by G-C base pairs.[8,53] Such exoribonuclease-hairpin interactions could explain why the *lpp* transcripts harbor tails of 10-15 As.[12] Moreover, the other 3' to 5' exonucleases that are involved in 3' trimming of stable RNA precursors may also contribute to the nibbling of messengers and generate tails of different lengths depending on their capability to approach secondary structures.[77] On the other hand, it has been suggested that the poly(A) tails of some RNAs may be specifically degraded by PNPase or RNase II.[23]

The model described above, which postulates that 3' to 5' processive exonucleases can initiate removal of poly(A) tails after addition of each (or a few) A residue, implies that these enzymes counteract poly(A) tail elongation and therefore prevent the synthesis of long tails. It has therefore been proposed that cofactors similar to mammalian CPSF (Cleavage and Polyadenylation Specificity Factor) and PABP II (a poly(A)-binding protein)[79] may cause processive synthesis and ensure the protection of the long poly(A) tails of up to 50 nucleotides in length that have been detected in *E. coli*.[9] One possible candidate was the host factor Hfq which binds A-rich RNAs (see above). The fact that the poly(A) tails of the *rpsO* mRNAs are slightly shorter in Hfq deficient cells and that Hfq stimulates elongation of poly(A) and converts PAP I into a processive enzyme in vitro strongly suggests that it affects poly(A) synthesis in vivo (Fig. 1A).[40] Two other poly(A)-binding proteins may also affect poly(A) metabolism. CspE prevents poly(A) removal in vitro and ribosomal protein S1 can interact physically with two enzymes of poly(A) metabolism, PNPase and RNase E.[80] It has been proposed that an endonucleolytic cleavage by RNase E, close to the RNA-poly(A) junction may remove poly(A) tails protected by poly(A)-binding proteins (Fig. 1B).[75] However there is no evidence that cofactors such as CspE and ribosomal protein S1 affects poly(A) metabolism in vivo. Moreover, the fact that tails longer than 20 As are unfrequently detected in *E. coli* suggests that they may not be many poly(A)-binding proteins which protect tails from exoribonucleases in prokaryotes. More likely, these long tails could result from the dynamic equilibrium between synthesis by PAP I and Hfq and degradation by 3' to 5' exoribonucleases and probably correspond to a limited number of molecules which can undergo many successive steps of elongation without being attacked by exoribonucleases. One could also imagine that only few RNA species that are preferentially polyadenylated can gain long poly(A) tails.

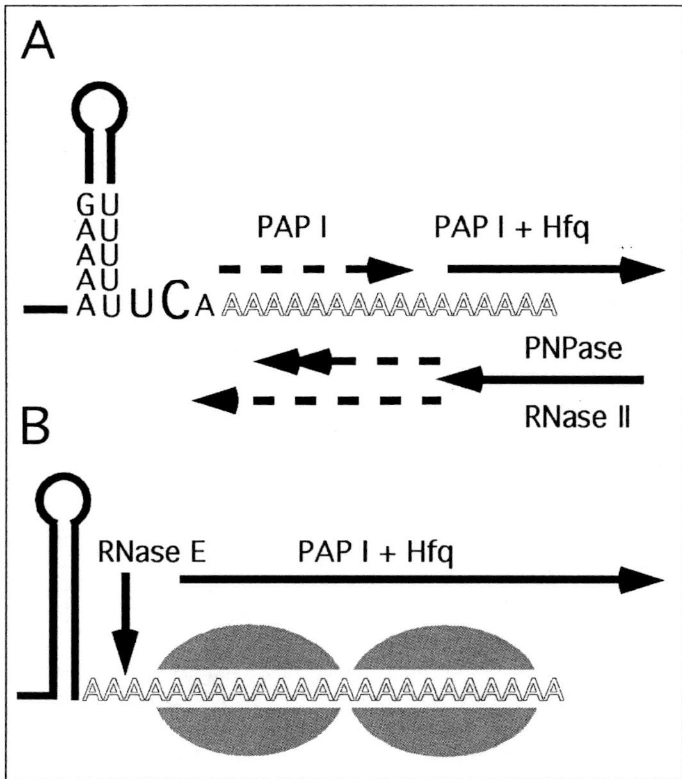

Figure 1. Metabolism of poly(A) in bacteria. A) This model accounts for the synthesis of oligo(A) tails at the 3' end of the *rpsO* mRNA coding for ribosomal protein S15 in vivo. Transcription terminates in the UCA sequence at the bottom of the hairpin of the transcription terminator. Letter size is proportional to the amounts of transcrips terminated at each position. PAP I polymerizes distributively A residues at the 3' end of the primary transcripts and PNPase and RNase II can attack the poly(A) tail after addition of each A. Successive steps of adenylation by PAP I, stimulated by Hfq, could generate long tails that are presumably rapidly degraded processively by PNPase and RNase II into tails of 9-10 As that are then shortened distributively by the two exoribonucleases. Distributive and processive reactions of synthesis and degradation are indicated by dotted and continuous arrows, respectively. PNPase releases RNAs harboring tails of 2-4 As while RNase II completely removes the tails. PNPase can also pass through the hairpin (see Fig. 2). At steady state, 10% of the *rpsO* transcripts harbor tails of 1-3 As and 90% of the molecules are not adenylated. B) Long tails that can contain up to 50 As might be protected by poly(A)-binding proteins (for example CspE) and be removed endonucleolyticaly by RNase E.

Functions of Polyadenylation

The intracellular concentration of several small RNA regulators of different physiological functions is poly(A)-dependent. Originally, a mutation in the *pcnB* gene coding for PAP I reduced the copy number of ColE 1 type plasmids whose replication is negatively controled by RNA I.[25] RNA I is a 108 nucleotide antisense molecule which forms a duplex with the RNA II primer of DNA replication. Inhibition of pBR322 DNA replication in the absence of PAP I was shown to result from the accumulation of an RNase E cleavage product of RNA I, lacking the five 5' nucleotides of the primary transcript, which must be polyadenylated to be degraded.[14] The other small regulatory RNAs whose stability and intracellular concentration also depend on polyadenylation, namely *copA*, *sok* and *oop*, control the replication and maintenance of plasmid R1 and the lysogeny of bacteriophage λ, respectively.[19,21,81] As in the case of RNA I, polyadenylation destabilizes truncated *copA* and *sok* molecules generated by RNase E.

In addition to the small RNA regulators mentioned above, polyadenylation also contributes to the rapid degradation of fragments produced by endonucleolytic digestion of mRNAs.[10,33,56] The mapping of mRNA-poly(A) junctions at multiple locations in several transcripts reinforces the notion that poly(A) polymerase I can polyadenylate and destabilize many different RNA fragments with different 3' extremities thought to be generated by endo and exoribonucleolytic cleavages.[4,10,18,20]

We mentioned above that precursors of stable RNAs, which accumulate in strains deficient for 3' trimming exoribonucleases, are also polyadenylated.[17] It has been proposed that *E. coli* has developed a quality control system that eliminates abnormally folded RNAs which cannot be rapidly processed into active mature molecules. Consistent with this idea, there is experimental evidence that the precursor of a thermosensitive mutant tRNA^{Trp} is polyadenylated and degraded by PNPase (M. Deutscher, personnal communication).

In contrast to eukaryotic cells, where it has been firmy established that poly(A) tails play an important role in translation initiation, there are only very few data supporting the idea that polyadenylation may affect translation in prokaryotes. Based on the observation that ribosomal protein S1 can bind both poly(A) tails and ribosome-binding sites of messengers it has been proposed that S1 could establish a link between the 3' and the 5' ends of the molecule and possibly create a functional interaction between translation and mRNA stability.[80] In this respect, ribosomal protein S1 would be the prokaryotic equivalent of the translation initiation factor/poly(A)-binding protein complex of eukaryotic cells. Moreover, if not fortuitous, the association of PAP I[82] and of polyadenylated RNAs[83] with polysomes could be an indication that poly(A)-dependent degradation affects the stability of mRNAs associated with ribosomes. Polyadenylation, however, does not affect protein synthesis in vitro.[84]

Polyadenylation also allows the rescue of mutant MS2 bacteriophages whose genome is degraded by RNase III. In this case, addition of A residues by PAP I at the 3' end of fragmented RNA genomes (or of the fragmented complementary strands used as template during replication) allows selection of RNase III resistent genomes containing insertions of stretches of As or Us.[15]

Although there is no indication that polyadenylation destabilizes fragments of MS2 RNA, most physiological functions that are poly(A)-dependent are based on the destabilization of small RNAs (Table 1). The stabilization of specific transcripts encoding decay enzymes, RNase E and PNPase, by polyadenylation[18] may be an indirect effect due to the destabilization of another RNA (Table 1). On the other hand, the fact that five As at the 3' end of RNAs are able to promote exonucleolytic degradation[66] suggests that long tails of 50 nucleotides or so may have a different function in mRNA metabolism (see below).

The Role of Polyadenylation in mRNA Decay

The first indication that poly(A) tails destabilize RNA came from the discovery that PAP I controls the stability of the small RNA (RNA I) which regulates the replication of ColE1 plasmids (see above).[14] Further study led to the conclusion that PNPase does not bind RNA I, whose 3' end is sequestered in a secondary structure and that poly(A) tails provide sites where PNPase can bind and initiate the exonucleolytic degradation of RNA I.[68,69] Polyadenylation has since been shown to be involved in the degradation of other RNA species and of RNA in general[9] and it is admitted that the mechanism of degradation of RNA I can be extended to the poly(A)-dependent degradation of any RNA with a 3' secondary structure.[9,10,19,21,47,56] The current idea is that PNPase can carry out the complete processive degradation of RNAs containing weak secondary structures but is blocked when it encounters stable hairpins which cause dissociation of the ribonuclease from its substrate.[22,56,68] It has been proposed that an Exonucleolytic Impeding Factor, referred to as EIF, might provoke PNPase stalling at secondary structures.[85] When blocked at secondary structures, PNPase releases RNAs devoid of a 3' single-stranded stretch of nucleotides that cannot be bound by exoribonucleases (Fig. 2A). The current model of RNA decay postulates that these tail-less RNAs are readenylated thus allowing PNPase to reinitiate exonucleolytic decay. Again, PNPase can generate tail-less RNA or, possibly, continue to degrade the RNA upstream of the tail and remove few nucleotides at the bottom of the hairpin before to dissociate from the RNA (Fig. 2A).[22,56] Localized melting of tightly folded secondary structures probably allows PNPase to invade and nibble the base of the structure.

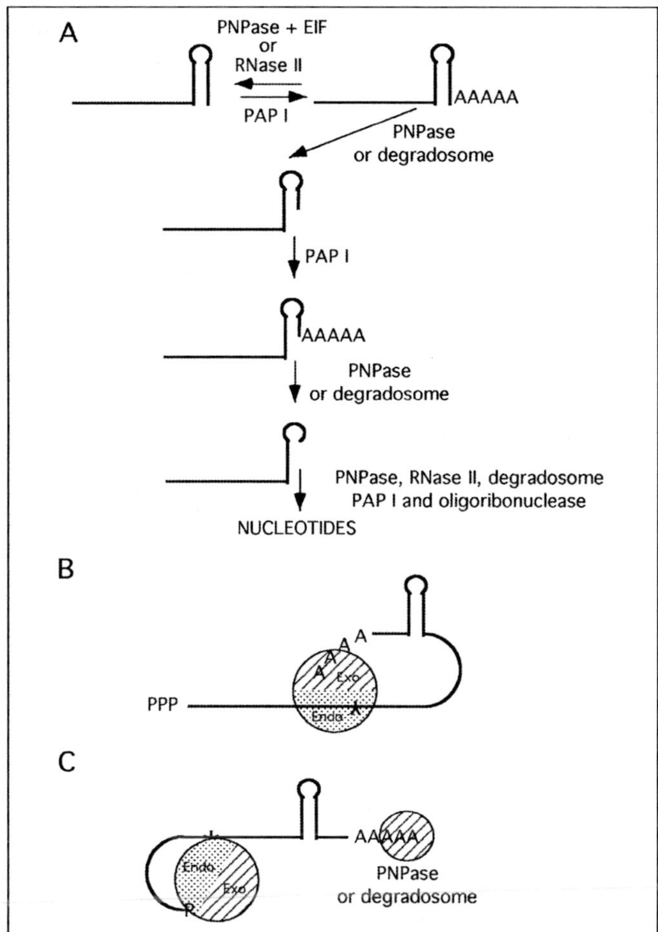

Figure 2. Poly(A)-dependent degradation of RNA harboring 3' stable secondary structures. A) Poly(A) tails synthesized at the 3' ends of RNAs offer a toehold where exonucleases can initiate decay. RNase II removes the poly(A) tails before to be arrested by 3' hairpins. It generates tail-less molecules and therefore protects the RNA from exonucleolytic decay. Similarly, PNPase, alone or in association with other degradosome components including the RhlB RNA helicase, can remove the tail and stop few nucleotides downstream of the secondary structure. A cofactor referred to as EIF (see text) could promote PNPase stalling at secondary structures. PNPase can also slightly nibble the bottom of the hairpin before dissociating from the RNA. Then, the RNA is readenylated by PAP I thus offering to PNPase the opportunity to nibble the descending strand of the hairpin a little further and finally to degrade completely the RNA into nucleotides. Unstructured RNA fragments are degraded by PNPase and RNase II whose activity is facilitated by PAP I. The oligoribonuclease degrades short oligoribonucleotides released by PNPase and RNase II into mononucleotides.[71] B) Degradosome is represented by a circle containing an endonucleolytic shaded domain (RNase E) indicated by "endo" and an exonucleolytic hatched domain (PNPase and RhlB) indicated by "exo". The model postulates that this particule interacts simultaneouly with an internal processing site, shown by a star, through RNase E and with the 3' poly(A) tail through PNPase and RhlB and that this interaction favors degradation mediated by RNase E. C) In contrast, RNA fragments resulting from endonucleolytic cleavages harbor 5' monophosphorylated extremities which promote processing by RNase E or degradosome. The model proposes that degradosome interacting simultaneously with the 5' end of the RNA fragment and an internal processing site looses its affinity for poly(A) tails and that this explains why RNA fragments can be degraded simultaneouly by RNase E and poly(A)-dependent ribonucleases. PNPase, free or associated with degradosome, carrying out the exonucleolytic degradation of the RNA is indicated by a small hatched circle.

Several cycles of exoribonucleolytic degradation and polyadenylation would allow PNPase to progressively shorten the stem and eventually to completely degrade it. Consistent with this hypothesis, it has been shown in vitro that complete degradation of RNAs polyadenylated downstream of a stable hairpin can only be carried out by PNPase if the RNA is continuously adenylated by PAP I.[56] It has been proposed that the most stable of 3' protecting hairpins which cannot be traversed by free PNPase are degraded by PNPase associated with degradosome whose activity is facilitated by the RNA helicase RhlB.[72] This helicase, activated through its association with the Rne polypeptide is thought to break the hydrogen bonds between nucleotides of the hairpin, which are then removed by PNPase in the degradosome.[67] An incremental mechanism of degradation similar to that described above for free PNPase probably accounts for the degradation of stably folded RNAs by degradosome.[56] The clustering of the poly(A) sites in several transcripts observed in vivo is consistent with the idea that successive adenylation events are required for progression of PNPase.[10,18,20] Because polyadenylation also causes destabilization of *rpsO* mRNAs in strains lacking PNPase,[47] it has been proposed that other ribonucleases might be involved in poly(A)-dependent degradation of RNA.[22] One could for example imagine that long tails interfering with internal sequences may induce transconformation of the RNA and expose structural motifs sensitive to endoribonucleases.

The fact that RNase II can completely degrade the poly(A) tails used as a toehold by PNPase to initiate RNA decay implies that this enzyme antagonizes the degradation of polyadenylated RNAs by PNPase (Fig. 2A).[8,22,56] Consistent with this idea, RNase II protects the polyadenylated *rpsT* transcript from PNPase mediated degradation in vitro.[56] This protective role of RNase II explains why the *rpsO* transcript and the RNA OUT, which represses Tn10 transposition, are more rapidly degraded when RNase II is inactive.[8,46,59]

It is striking that the full length primary transcripts which have been examined thus far are not, or are only slightly, dependent on poly(A)-dependent degradation in vivo.[9,47,56] Indeed, PAP I inactivation only causes appreciable stabilization of full length mRNAs in cells that are also deficient for RNase E and exoribonucleases.[9,47] Moreover, poly(A)-dependent ribonucleases cannot participate in the degradation of long polycistronic transcripts which begin to be degraded whereas RNA polymerase has not synthesised yet the 3' end of the molecules.[86] This argues against the idea that polyadenylation could initiate decay of primary transcripts.[9] We propose that primary transcripts harbor preferential cleavage sites which allow RNase E to control the decay of the molecule (Fig. 2B). Degradation initiated by RNase E could mask the destabilizing effect of polyadenylation if the endonucleolytic pathway degrade mRNA more rapidly than the poly(A)-dependent process catalyzed by exoribonucleases.[47] Moreover, elimination of poly(A) tails by RNase II could protect the primary transcript from the exonucleolytic activity of PNPase either free or in the degradosome.[56] On the other hand, it is also possible that a degradosome bound simultaneously at the rate limiting processing site, recognized by RNase E, and at the 3' end of the polyadenylated RNA, through PNPase and RhlB, could impair poly(A)-dependent degradation of full length transcripts (Fig. 2B).[75] Similar multi-site interactions could coordinate the RNase E cleavages which initiate the decay of RNA I and of the *copA* and *sok* RNAs and the subsequent poly(A)-dependent degradation of the processed RNAs.[14,19,87] Indeed, it has been shown that the *copA* RNA must be cleaved by RNase E before becoming a target of poly(A)-dependent degradation[21] and it was proposed that RNase E and PNPase interact functionally during degradation of RNA I.[68] Initial cleavages of primary transcripts at preferential RNase E sites generate RNA fragments harboring 5' monophosphorylated extremities which promote RNase E processing[73] and polyadenylation[32] (Fig. 2C). RNase E probably cleaves these RNA fragments at subsequent sites unmasked by the refolding of the RNA fragment or by the run off of ribosomes. In contrast to full length transcripts, RNA fragments can be simultaneously degraded by RNase E and poly(A)-dependent ribonucleases.[10,33] One can imagine that both pathways of decay are active because RNase E interacting simultaneously with the 5' monophosphorylated extremity and an internal processing site should not prevent the poly(A)-dependent exonucleolytic degradation, which takes place at the other extremity of the molecule (Fig. 2C). Cleavages by endonucleolytic enzymes will eventually generate RNA fragments devoid of RNase E sites that are exclusively degraded by poly(A)-dependent ribonucleases. This is probably the case for small folded RNAs or mRNA fragments that harbor single-stranded segments too short to contain A-U tracks recognized by RNase E.[22,33,56] Although polyadenylation is only required for degradation of tightly folded fragments it seems that it also facilitates decay of unfolded and weakly folded RNAs that can be degraded by RNase II and PNPase even if they are not adenylated.[31,33,60] Cooperation of RNase E or degradosome and poly(A)-dependent ribonucleases probably ensures the rapid degradation of RNA into ribonucleotides available for synthesis of macromolecules.

Conclusions and Perspectives

An important recent improvement of our knowledge of RNA metabolism is the discovery that polyadenylation is a ubiquitous mechanism which takes place not only in nucleus and cytoplasm of eukaryotic cells but also in eubacteria and archaea and in chloroplasts[88] and mitochondria[89] considered as endosymbiotic organelles of prokaryotic origin. Although polyadenylation machineries of eukaryotic cells and bacteria probably derived from a common ancestor,[47] it is striking that both the mechanism and the role of polyadenylation are different in prokaryotes and eukaryotes. In contrast to eukaryotic cells, where the vast majority of mRNAs harbors long poly(A) tails that are covered and protected by poly(A)-binding proteins, most RNAs are not adenylated and harbor short poly(A) tails that are accessible to exoribonucleases in bacteria. An intriguing property of bacterial polyadenylation is its lack of specifity. Indeed, current data suggest that any accessible 3' extremity of any RNA species (tRNA, mRNA, rRNA, regulatory RNAs) can be polyadenylated by PAP I. Since poly(A) tails promote RNA degradation, this implies that poly(A)-dependent decay is a global mechanism which can carry out degradation of all RNAs which exhibit an exposed 3' end. Consistent with this idea, polyadenylation promotes degradation of mRNA fragments, regulatory RNAs and nonfunctional tRNAs. Stable RNAs such as functional tRNAs, rRNAs and full length mRNAs whose stability is controled by RNase E are presumably protected from poly(A)-dependent degradation by aminoacylation, ribosomal proteins, RNA binding proteins and RNA folding. On the other hand, phages and plasmids use this mechanism of degradation of cellular RNAs to control the intracellular concentration of regulatory RNAs. Interestingly, in organelles as in prokaryotes polyadenylation promotes mRNA degradation.[88,89] Poly(A) tails of up to 270 nucleotides in length containing Gs and few Us and Cs have been characterized in chloroplasts.[88] As in bacteria, mRNA, rRNAs and tRNAs are polyadenylated in mitochondria.[90] The function of polyadenylation in bacteria and organelles is therefore very different from that of eukaryotic poly(A) tails which stabilize mRNAs. It is worth pointing out, however, that removal of poly(A) tails is the first step of RNA decay pathways initiated at the 3' end of the molecules in both prokaryotes and eukaryotes.

Bacterial polyadenylation and its function in RNA metabolism have only recently begun to be investigated and despite of the fact that several players in poly(A) metabolism have already been discovered, our current view of this mechanism is probably still very naïve. If one takes eukaryotic polyadenylation as a model, it is likely that bacteria contain factors affecting polyadenylation that have yet to be identified. Cofactors and structural features of RNA could affect either the recognition of RNAs by PAP I or that of poly(A) tails by exoribonucleases and thus affect poly(A) synthesis and nibbling. The interaction between PAP I and degradosome components suggests that RNase E mediated RNA decay and poly(A) synthesis could be functionally related. Such interactions might afford a better understanding of why RNA fragments and full length transcripts harboring similar terminal hairpins are not equally sensitive to poly(A)-dependent decay.[10,33,56] It is also likely that poly(A)-dependent ribonucleases still have to be identified in *E. coli*.[47] Recent improvements of methodologies of RNA structure determination and the availability of mutants affected in genes coding for enzymes of poly(A) metabolism should allow comparison of the fraction of polyadenylated molecules and the lengths of poly(A) tails of different RNA species. Moreover, utilization of cell-free polyadenylation/RNA degradation assays to reconstitute the polyadenylation machinery and its interactions with enzymes involved in RNA degradation will be required to understand completely this important aspect of RNA metabolism.

Acknowledgements

We are grateful to Ciaran Condon for careful critical reading of the manuscript. The authors are supported by grants from CNRS, the Ministère de l'Education Nationale et de la Recherche et de la Technologie and Denis Diderot-Paris 7 University. P.E.M. is recipient of a Ph D grant from Praxis XXI (Portugal).

References

1. August JM, Ortiz PJ, Hurwitz J. Ribonucleotic acid-dependent ribonucleotide incorporation. I. Purification and properties of the enzyme. J Biol Chem 1962; 237:3786-3793.
2. Manley JL. Messenger RNA polyadenylation: A universal modification. Proc Natl Acad Sci USA 1995; 92:1800-1801.
3. Sarkar N. Polyadenylation of mRNA in bacteria. Microbiology 1996; 142:3125-3133.
4. Sarkar N. Polyadenylation of mRNA in prokaryotes. Annu Rev Biochem 1997; 66:173-197.
5. Nakazato H, Venkatesan S, Edmonds M. Polyadenylic acid sequences in *E. coli* messenger RNA. Nature 1975; 256:144-146.
6. Srinivasan PR, Ramanarayanan M, Rabbani E. Presence of polyriboadenylate sequences in pulse-labeled RNA of *Escherichia coli*. Proc Natl Acad Sci U S A 1975; 72:2910-2914.
7. Kramer RA, Rosenberg M, Steitz JA. Nucleotide sequences of the 5' and 3' termini of bacteriophage T7 early messenger RNAs synthesized in vitro: Evidence for sequence specificity in RNA processing. J Mol Biol 1974; 89:767-776.
8. Marujo PE, Hajnsdorf E, Le Derout J et al. RNase II removes the oligo(A) tails which destabilize the *rpsO* mRNA of *Escherichia coli*. RNA 2000; 6:1185-1193.
9. O'Hara EB, Chekanova JA, Ingle CA et al. Polyadenylylation helps regulate mRNA decay in *Escherichia coli*. Proc Natl Acad Sci USA 1995; 92:1807-1811.
10. Goodrich AF, Steege DA. Roles of polyadenylation and nucleolytic cleavage in the filamentous phage mRNA processing and decay pathways in *Escherichia coli*. RNA 1999; 5:972-985.
11. Karnik P, Taljanidisz.J, Sasvari-Szekely M et al. 3'-terminal polyadenylate sequence of *Escherichia coli* tryptophan synthetase a-subunit messenger RNA. J Mol Biol 1987; 196:347-354.
12. Taljanidisz J, Karnik P, Sarkar N. Messenger ribonucleic acid for the lipoprotein of the *Escherichia coli* outer membrane is polyadenylated. J Mol Biol 1987; 193:507-515.
13. Bralley P, Jones HJ. Poly(A) polymerase activity and RNA polyadenylation in *Streptomyces coelicolor*. Mol Microbiol 2001; 40:1155-1164.
14. Xu F, Lin-Chao S, Cohen SN. The *Escherichia coli pcnB* gene promotes adenylation of antisense RNA I of ColE1-type plasmids in vivo and degradation of RNA I decay intermediate. Proc Natl Acad Sci USA 1993; 90:6756-6760.
15. van Meerten D, Zelwer M, Regnier P et al. In vivo oligo(A) insertions in phage MS2: role of *Escherichia coli* poly(A) polymerase. Nucleic Acids Res 1999; 27:3891-3898.
16. Wrobel B, Herman-Antosiewicz A, Szalewska-Palasz S et al. Polyadenylation of *oop* RNA in the regulation of bacteriophage lambda development. Gene 1998; 212:57-65.
17. Li Z, Pandit S, Deutscher MP. Polyadenylation of stable RNA precursors in vivo. Proc Natl Acad Sci U S A 1998; 95:12158-12162.
18. Mohanty BK, Kushner SR. Analysis of the function of *Escherichia coli* poly(A) polymerase I in RNA metabolism. Mol Microbiol 1999; 34:1094-1108.
19. Mikkelsen ND, Gerdes K. Sok antisense RNA from plasmid R1 is functionally inactivated by RNase E and polyadenylated by poly(A) polymerase I. Mol Microbiol 1997; 26:311-320.
20. Haugel-Nielsen J, Hajnsdorf E, Régnier P. The *rpsO* mRNA of *Escherichia coli* is polyadenylated at multiple sites resulting from endonucleolytic processing and exonucleolytic degradation. EMBO J 1996; 15:3144-3152.
21. Söderbom F, Binnie U, Masters M et al. Regulation of plasmid R1 replication: PcnB and RNase E expedite the decay of the antisense RNA, copA. Mol Microbiol 1997; 26:493-504.
22. Hajnsdorf E, Braun F, Haugel-Nielsen J et al. Multiple degradation pathways of the *rpsO* mRNA of *Escherichia coli*. RNase E interacts with the 5' and 3' extremities of the primary transcript. Biochimie 1996; 78:416-424.
23. Mohanty BK, Kushner SR. Polynucleotide phosphorylase, RNase II and RNase E play different roles in the in vivo modulation of polyadenylation in *Escherichia coli*. Mol Microbiol 2000; 36:982-994.
24. Liu J, Parkinson JS. Genetics and sequence analysis of the *pcnB* locus, an *Escherichia coli* gene involved in plasmid copy number control. J Bacteriol 1989; 171:1254-1261.
25. Lopilato J, Bortner S, Beckwith J. Mutations in a new chromosomal gene of *Escherichia coli* K-12, *pcnB*, reduce plasmid copy number of pBR322 and its derivatives. Mol Gen Genet 1986; 205:285-290.
26. Masters M, Colloms MD, Oliver IR et al. The *pcnB* gene of *Escherichia coli*, which is required for ColE1 copy number maintenance is dispensable. J Bacteriol 1993; 175:4405-4413.
27. Cao G-J, Sarkar N. Identification of the gene for an *Escherichia coli* poly(A) polymerase. Proc Natl Acad Sci USA 1992; 89:10380-10384.
28. Raynal LC, Krisch HM, Carpousis AJ. The *Bacillus subtilis* nucleotidyltransferase is a tRNA CCA-adding enzyme. J Bacteriol 1998; 180:6276-6282.
29. Yue D, Maizels N, Weiner AM. CCA-adding enzymes and poly(A) polymerases are all members of the same nucleotidyltransferase superfamily: Characterization of the CCA-adding enzyme from the archaeal hyperthermophile *Sulfolobus shibatae*. RNA 1996; 2:895-908.
30. Sippel AE. Purification and characterization of adenosine triphosphate: ribonucleic acid adenyltransferase from *Escherichia coli*. Eur J Biochem 1973; 37:31-40.
31. Yehudai-Resheff S, Schuster G. Characterization of the *E.coli* poly(A) polymerase: nucleotide specificity, RNA-binding affinities and RNA structure dependence. Nucleic Acids Res 2000; 28:1139-1144.
32. Feng Y, Cohen SN. Unpaired terminal nucleotides and 5' monophosphorylation govern 3' polyadenylation by *Escherichia coli* poly(A) polymerase I. Proc Natl Acad Sci USA 2000; 97:6415-6420.

33. Hajnsdorf E, Régnier P. *E. coli rpsO* mRNA decay: RNase E processing at the beginning of the coding sequence stimulates Poly(A)-dependent degradation of the mRNA. J Mol Biol 1999; 286:1033-1043.
34. Sano H, Feix G. Terminal riboadenylate transferase from *Escherichia coli*. Eur J Biochem 1976; 71:577-583.
35. Senear AW, Argetsinger Steitz J. Site-specific interaction of Qβ host factor and ribosomal protein S1 with Qβ and R17 bacteriophage RNAs. J Biol Chem 1976; 251:1902-1912.
36. Zhang A, Altuvia S, Tiwari A et al. The oxyS regulatory RNA represses *rpoS* translation and binds the Hfq (HF-1) protein. EMBO J 1998; 17:6061-6068.
37. Tsui HC, Leung HC, Winkler ME. Characterization of broadly pleiotropic phenotypes caused by an *hfq* insertion mutation in *Escherichia coli* K-12. Mol Microbiol 1994; 13:35-49.
38. Vytvytska O, Jakobsen JS, Balcunaite G et al. Host factor I, Hfq, binds to *Escherichia coli ompA* mRNA in a growth rate-dependent fashion and regulates its stability. Proc Natl Acad Sci USA 1998; 95:14118-14123.
39. Muffler A, Fischer D, Hengge-Aronis R. The RNA-binding protein HF-1, known as a host factor for Qβ RNA replication, is essential for *rpoS* translation in *Escherichia coli*. Genes & Dev 1996; 10:1143-1151.
40. Hajnsdorf E, Régnier P. Host factor Hfq of *Escherichia coli* stimulates elongation of poly(A) tails by poly(A) polymerase I. Proc Natl Acad Sci USA 2000; 97:1501-1505.
41. Raynal LC, Carpousis AG. Poly(A) polymerase I of *Escherichia coli*: characterization of the catalytic domain, an RNA binding site and regions for the interaction with proteins involved in mRNA degradation. Mol Microbiol 1999; 32:765-775.
42. Raynal LC, Krisch HM, Carpousis AJ. Bacterial poly(A) polymerase: An enzyme that modulates RNA stability. Biochimie 1996; 78:399-398.
43. Nicholson AW. Function, mechanism and regulation of bacterial ribonucleases. FEMS Microbiology Rev 1999; 23:371-390.
44. Deutscher MP, Li Z. Exoribonucleases and their multiple roles in RNA metabolism. Prog Nucleic Acid Res Mol Biol 2000; 66:67-105.
45. Cao G-J, Sarkar N. Poly(A) RNA in *Escherichia coli* : Nucleotide sequence at the junction of the lpp transcript and the polyadenylate moiety. Proc Natl Acad Sci USA 1992; 89:7546-7550.
46. Hajnsdorf E, Steier O, Coscoy L et al. Roles of RNase E, RNase II and PNPase in the degradation of the *rpsO* transcripts of *Escherichia coli*: Stabilizing function of RNase II and evidence for efficient degradation in an *ams pnp rnb* mutant. EMBO J 1994; 13:3368-3377.
47. Hajnsdorf E, Braun F, Haugel-Nielsen J et al. Polyadenylylation destabilizes the *rpsO* mRNA of *Escherichia coli*. Proc Natl Acad Sci USA 1995; 92:3973-3977.
48. Mohanty BK, Kushner SR. Polynucleotide phosphorylase functions both as a 3' right-arrow 5' exonuclease and a poly(A) polymerase in *Escherichia coli*. Proc Natl Acad Sci USA 2000; 97:11966-11971.
49. Spahr PF, Schlessinger D. Breakdown of messenger ribonucleic acid by a potassium-activated phophodiesterase from *Escherichia coli*. J Biol Chem 1964; 238:PC2251-PC2253.
50. Deutscher MP, Reuven NB. Enzymatic basis for hydrolytic versus phosphorolytic mRNA degradation in *Escherichia coli* and *Bacillus subtilis*. Proc Natl Acad Sci USA 1991; 88:3277-3280.
51. Coburn GA, Mackie GA. Overexpression, purification and properties of *Escherichia coli* ribonuclease II. J Biol Chem 1996; 271:1048-1053.
52. McLaren RS, Newbury SF, Dance GSC et al. mRNA degradation by processive 3'-5' exonucleases in vitro and the implication for prokaryotic mRNA decay in vivo. J Mol Biol 1991; 221:81-95.
53. Spickler C, Mackie GA. Action of RNase II and polynucleotide phosphorylase against RNAs containing stem-loops of defined structure. J Bacteriol 2000; 182:2422-2427.
54. Gupta RS, Kasai T, Schlessinger D. Purification and some novel properties of *Escherichia coli* RNase II. J Biol Chem 1977; 252:8945-8949.
55. Braun F, Hajnsdorf E, Régnier P. Polynucleotide phosphorylase is required for the rapid degradation of the RNase E processed *rpsO* mRNA of *Escherichia coli* devoid of its 3' hairpin. Mol Microbiol 1996; 19:997-1005.
56. Coburn GA, Mackie GA. Reconstitution of the degradation of the mRNA for ribosomal protein S20 with purified enzymes. J Mol Biol 1998; 279:1061-1074.
57. Mott JE, Galloway JL, Platt T. Maturation of E. coli tryptophan operon: evidence for 3' exonucleolytic processing after Rho-independent termination. EMBO J 1985; 4:1887-1891.
58. Newbury SF, Smith NH, Robinson EC et al. Stabilization of translationally active mRNA by prokaryotic REP sequences. Cell 1987; 48:297-310.
59. Pepe CM, Maslesa-Galic S, Simons RW. Decay of the IS*10* antisense RNA by 3' exonucleases: evidence that RNase II stabilizes RNA-OUT against PNPase attack. Mol Microbiol 1994; 13:1133-1142.
60. Coburn GA, Mackie GA. Differential sensitivities of portions of the mRNA for ribosomal protein S20 to 3'-exonucleases dependent on oligoadenylation and RNA secondary structure. J Biol Chem 1996; 271:15776-15781.
61. Grunberg-Manago M. Polynucleotide phosphorylase. In: Cohn W, ed. Progress in Nucleic Acids Research. New York: Academic press, 1963:93-133.
62. Donovan WP, Kushner SR. Polynucleotide phosphorylase and ribonuclease II are required for cell viability and mRNA turnover in *Escherichia coli* K-12. Proc Natl Acad Sci USA 1986; 83:120-124.
63. Miczak A, Kaberdin VR, Wei C-L et al. Proteins associated with RNase E in a multicomponent ribonucleolytic complex. Proc Natl Acad Sci USA 1996; 93:3865-3869.
64. Py B, Higgins CF, Krisch HM et al. A DEAD-box RNA helicase in the *Escherichia coli* RNA degradosome. Nature 1996; 381:169-172.

65. Thang MN, Guschlbauer W, Zachau HG et al. Degradation of transfer nucleic acid by polynucleotide phosphorylase. J Mol Biol 1967; 26:403-421.
66. Blum E, Carpousis AJ, Higgins CF. Polyadenylation promotes degradation of 3'-structured RNA by the *Escherichia coli* mRNA degradosome in vitro. J Biol Chem 1999; 274:4009-4016.
67. Coburn GA, Miao X, Briant DJ et al. Reconstitution of a minimal RNA degradosome demonstrates functional coordination between a 3' exonuclease and a DEAD-box RNA helicase. Genes & Dev 1999; 13:2594-2603.
68. Cohen SN. Surprises at the 3' end of prokaryotic RNA. Cell 1995; 80:829-832.
69. Xu F, Cohen SN. RNA degradation in *Escherichia coli* regulated by 3' adenylation and 5' phosphorylation. Nature 1995; 374:180-183.
70. Coburn GA, Mackie GA. Degradation of mRNA in *Escherichia coli*: an old problem with some new twists. Prog Nucleic Acid Res Mol Biol 1999; 62:55-108.
71. Régnier P, Arraiano CM. Degradation of mRNA in bacteria: emergence of ubiquitous features. BioEssays 2000; 22:235-244.
72. Vanzo NF, Li YS, Py B et al. Ribonuclease E organizes the protein interactions in the *Escherichia coli* RNA degradosome. Genes & Dev 1998; 12:2770-2781.
73. Mackie GA. Ribonuclease E is a 5'-end-dependent endonuclease. Nature 1998; 395:720-723.
74. Huang H, Liao J, Cohen SN. Poly(A) and poly(U) specific RNA 3' tail shortening by *E. coli* ribonuclease E. Nature 1998; 391:99-102.
75. Walsh AP, Tock MR, Mallen MH et al. Cleavage of poly(A) tails on the 3'-end of RNA by ribonuclease E of *Escherichia coli*. Nucleic Acids Res 2001; 29:1864-1871.
76. Li Z, Deutscher MP. The tRNA processing enzyme RNase T is essential for maturation of 5S RNA. Proc Natl Acad Sci USA 1995; 92:6883-6886.
77. Li Z, Pandit S, Deutscher MP. 3' exoribonucleolytic trimming is a common feature of the maturation of small, stable RNAs in *Escherichia coli*. Proc Natl Acad Sci USA 1998; 95:2856-2861.
78. Li Z, Pandit S, Deutscher MP. Maturation of 23S ribosomal RNA requires the exoribonuclease RNase T. RNA 1999; 5:139-146.
79. Wahle E, Keller W. The biochemistry of polyadenylation. TIBS 1996; 21:247-250.
80. Feng Y, Huang H, Liao J et al. *Escherichia coli* Poly(A)-binding proteins that interact with components of degradosomes or impede RNA decay mediated by polynucleotide phosphorylase and RNase E. J Biol Chem 2001; 276:31651-31656.
81. Szalewska-Palasz A, Wrobel B, Wegrzyn G. Rapid degradation of polyadenylated *oop* RNA. FEBS Lett 1998; 432:70-72.
82. Ingle C, Kushner SR. Development of an vitro mRNA decay system for *Escherichia coli*: Poly(A) polymerase I is necessary to trigger degradation. Proc Natl Acad Sci USA 1996; 93:12926-12931.
83. Ohta N, Sanders M, Newton A. Characterization of unstable poly (A)-RNA in *Caulobacter crescentus*. Biochim Biophys Acta 1978; 517:65-75.
84. Lee K, Cohen SN. Effects of 3' terminus modifications on mRNA functional decay during in vitro protein synthesis. J Biol Chem 2001; 276:23268-23274.
85. Causton H, Py E, McLaren RS et al. mRNA degradation in *Escherichia coli*: a novel factor which impedes the exoribonucleolytic activity of PNPase at stem-loop structures. Mol Microbiol 1994; 14:731-741.
86. Cannistraro VJ, Kennell D. Evidence that the 5' end of lac mRNA starts to decay as soon as it is synthesized. J Bacteriol 1985; 161:820-822.
87. Söderbom F, Wagner EGH. Degradation pathway of copA, the antisense RNA that controls replication of plasmid R1. Microbiology 1998; 144:1907-1917.
88. Lisitsky I, Klaff P, Schuster G. Addition of destabilizing poly(A)-rich sequences to endonuclease cleavage sites during the degradation of chloroplast mRNA. Proc Natl Acad Sci USA 1996; 93:13398-13403.
89. Gagliardi D, Leaver CL. Polyadenylation accelarates the degradation of the mitochondrial mRNA associated with cytoplasm male sterility in sunflower. EMBO J 1999; 18:3757-3766.
90. Komine Y, Kwong L, Anguera MC et al. Polyadenylation of three classes of chloroplast RNA in *Chlamydomonas reinhadtii*. RNA 2000; 6:598-607.
91. Cao G-J, Sarkar N. Poly(A) RNA in *Bacillus subtilis*. Identification of the polyadenylation site of flagellin mRNA. FEMS Microbiol Letters 1993; 108:281-285.
92. Johnson MD, Popowski J, Cao GJ et al. Bacteriophage T7 mRNA is polyadenylated. Mol Microbiol 1998; 27:23-30.

CHAPTER 12

The End in Sight:
Poly(A), Translation and mRNA Stability in Eukaryotes

Thomas Preiss

Abstract

All nuclear-encoded eukaryotic messenger RNAs possess a 5' cap structure (m^{7}GpppN) and, with a few exceptions, also a 3' poly(A) tail. These modifications are added as part of the mRNA processing pathway during or immediately after transcription in the nucleus. Subsequently, they both influence different aspects of mRNA metabolism including splicing, transport, stability and translation. The cap structure has an important role during the initiation phase of translation as it recruits ribosomes and associated factors to the mRNA. The poly(A) tail can also stimulate translation and cooperates with the cap structure in a synergistic fashion. The eukaryotic initiation factor eIF4G plays a central part as a multifunctional adapter, which brings together various components of the translation apparatus. Through simultaneous interactions with the cap-binding protein eIF4E and the poly(A)-binding protein (PABP), eIF4G is able to bridge the two ends of the mRNA. The resulting pseudo-circular structure of the mRNA is thought to have important functional consequences for the translation process. The importance of the poly(A) tail is further underscored by the fact that the regulated variation of its length on maternal mRNAs is an integral part of gene regulation during oocyte maturation and in early embryonic development. Finally, the majority of cellular mRNAs are degraded by processes that are interconnected with translation and are initiated by poly(A) tail shortening.

Introduction

A common view holds that most control mechanisms to regulate eukaryotic gene expression target the primary step, namely transcription in the nucleus. In contrast to this, it is becoming increasingly apparent that controls acting on the level of translation and mRNA stability are also of critical importance. The translation process is usually divided into three phases: (i) initiation,[1] (ii) elongation,[2] and (iii) termination.[3] The initiation phase represents all processes required for the assembly of a complete (80S) ribosome, consisting of a small (40S) and a large (60S) subunit, at the start codon of the mRNA. During the elongation phase, the actual polypeptide synthesis takes place. When the ribosome reaches the stop codon, this signals termination, including the dissociation of the completed polypeptide and the ribosome from the mRNA. Initiation usually is the rate-limiting step of translation and thus the preferred target of regulatory intervention.[4]

The salient model of cap-dependent translation was originally developed in the mid seventies[5] and has since been continuously refined (for a review, see refs. 1, 6, 7). Initially, the 40S ribosomal subunit is recruited near the 5' end of the mRNA. Next, a lateral movement termed "scanning" along the 5' untranslated region (UTR) leads to the recognition of the usually first AUG triplet as the initiator codon. Then follows the formation of an 80S ribosome and the elongation phase begins (Fig. 1). A large number of eukaryotic initiation factors (eIF) contribute to these processes (for an up-to-date review, see ref. 7 and chapter 18 by F. Poulin and N. Sonenberg). The hetero-trimeric factor eIF4F binds in the cytoplasm to the cap structure at the 5' end of the mRNA. eIF4F consists of the cap-binding protein eIF4E,[8] and the interacting proteins eIF4G and eIF4A (for a review, see

Translation Mechanisms, edited by Jacques Lapointe and Léa Brakier-Gingras.
©2003 Eurekah.com and Kluwer Academic / Plenum Publishers.

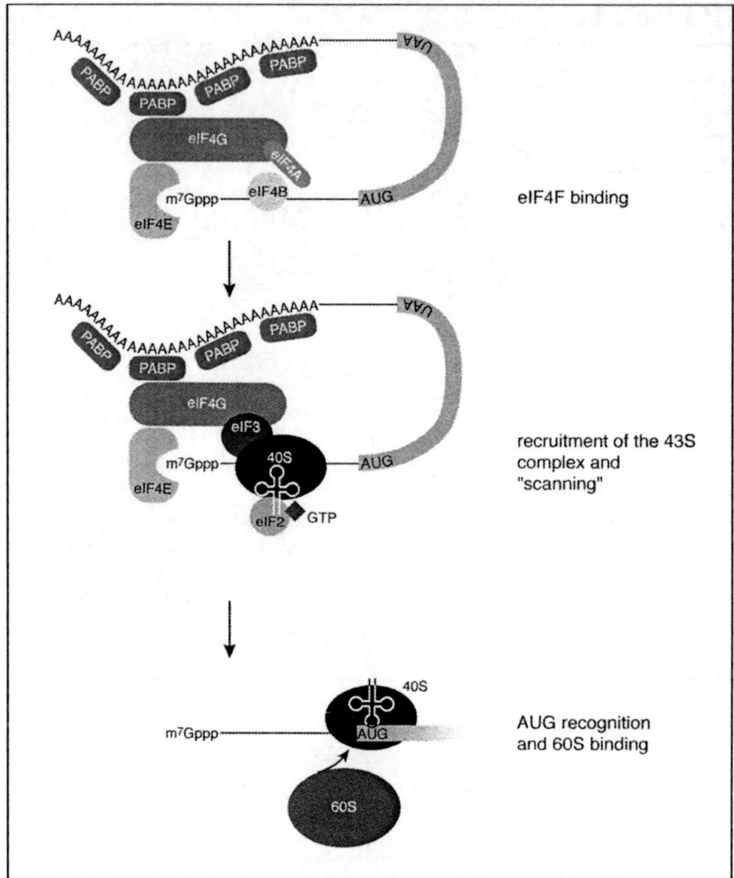

Figure 1. The initiation phase of translation. Shown is a eukaryotic mRNA with the two typical end modifications: the 5' cap structure (m^{7}Gppp) and the 3' poly(A) tail (AAA). The protein-coding open reading frame is marked by a start and stop codon. An early step in initiation is the binding of the eIF4F complex, consisting of eIF4E, eIF4G, and eIF4A, to the cap structure. eIF4A, stimulated by eIF4B, unwinds secondary structure in the 5' UTR. The interaction of the poly(A)-binding protein PABP with eIF4G leads to circularization of the mRNA. In a next step, the small ribosomal subunit (40S) with associated initiation factors eIF2, eIF3, and methionyl-initiator-tRNA$_{Met}$ is recruited to the mRNA. This 43S complex then moves along or "scans" the mRNA in a 3' direction. The codon/anticodon interaction of the initiator-tRNA$_{Met}$ then identifies the AUG start codon which leads to the release of eIF2 and eIF3 and binding of the large ribosomal subunit (60S). The complete (80S) ribosome is now poised for polypeptide synthesis. For clarity, not all factors involved in translation initiation are shown, the emphasis being on factors which group themselves around eIF4G. Adapted with permission from: Preiss T. BIO*spektrum* 2001; 4:315-319. © 2001 Spektrum Akademischer Verlag

ref. 9). eIF4A is an ATP-dependent RNA helicase and is able, upon stimulation by eIF4B, to unwind secondary structure in the cap-proximal region of the mRNA. In addition, eIF4G binds to the hetero-oligomeric factor eIF3, which associates with the 40S ribosomal subunit.[10] The 40S subunit binds, furthermore, the ternary complex comprising initiator-methionyl-tRNA (tRNA$_{Met}$), eIF2 and GTP, which has an important role in start codon recognition (for a review, see ref. 11). Once the 40S subunit and associated factors have arrived at the AUG then 60S subunit joining occurs, aided by eIF5 and eIF5B. eIF5 stimulates GTP hydrolysis of eIF2 at the start codon prior to ejection of all eIFs and joining of the 60S subunit. eIF5B is itself a ribosome-dependent GTPase.[12]

The Mechanistic Role of the Poly(A) Tail during Initiation of Translation

As we now know, the cap structure and the poly(A) tail are of similar importance for the initiation of translation on a typical mRNA (for a review, see refs. 13, 14). Despite the fact that both end modifications had been described at about the same time, it took much longer to develop good models for the function of the poly(A) tail.[14] By 1990, it was known that adding a poly(A) tail to a test mRNA resulted in a modest 2 to 3-fold stimulatory effect on its translation in mammalian cell extracts (for a review, see ref. 15). In the commonly used rabbit reticulocyte lysate system, this was shown to be due to an enhanced 60S joining.[16] Furthermore, a number of studies had documented a tight correlation between cytoplasmic polyadenylation of maternal mRNAs and their translational activation in vertebrate oocytes and developing embryos (for a review, see ref. 17). Collectively, these observations indicated that the poly(A) tail was somehow involved in the translation process. A groundbreaking study then showed in 1991 that exogenous mRNA can be introduced in eukaryotic cells by electroporation and its translation can be monitored *in vivo*.[18] When the mRNA was equipped with a cap structure or a poly(A) tail, this led to a stimulation of translation. Interestingly, the addition of both end modifications led to a much more than additive effect in cells of mammalian, plant and yeast origin. This translational synergy between cap structure and poly(A) tail became one of the focal points of translation research in the following years (for a review, see refs. 9, 13-15). The study of the underlying molecular mechanisms was initially hindered by the fact that all previously known *in vitro* systems were unable to reproduce this phenomenon. This changed with the report of a "synergy-competent" cell-free translation system based on extracts from *Saccharomyces cerevisae*.[19]

Using this system, it was demonstrated that the poly(A) tail, like the cap structure, was able to support the recruitment of the 40S ribosomal subunit (for a review, see ref. 14). This function of the poly(A) tail as well as the functional synergy with the cap structure depends on the poly(A)-binding protein PABP (in yeast: Pab1p).[20] PABP can bind to poly(A) tails with a periodicity of about 25 adenosine residues, although 12 adenosines are sufficient for binding.[21-23] The discovery of an interaction between Pab1p and the N-terminal part of eIF4G in yeast (Fig. 2) then presented a suitable molecular interaction between the 3' end of the mRNA and the process of translation initiation at the 5' end.[24, 25] These observations led to the hypothesis of a pseudo-circular mRNA conformation through simultaneous binding of eIF4E and Pab1p to eIF4G (Fig. 1). This was consistent with earlier observations of circular polyribosomes[26] and was further substantiated through functional data and direct visualization. A set of experiments using the yeast extract system showed that poly(A)-mediated translation by itself displays no preference for the 5' end of the mRNA. Polypeptide synthesis in the isolated poly(A)-dependent mode frequently starts at internal sites of the mRNA. The cap structure exhibits the additional function to tether the ribosome recruitment potential of the poly(A) tail to the 5' end.[27, 28] Adding the minimal components eIF4E, eIF4G, PABP, and a capped and polyadenylated mRNA together in vitro, resulted in the formation of pseudo-circular complexes which could be visualized by atomic force microscopy.[29]

After its discovery in yeast, it did not take long for evidence for a direct interaction between PABP and eIF4G to appear also in plant[30] and mammalian cells. Binding of mammalian PABP to eIF4G was demonstrated in two independent studies.[31, 32] Using the two-hybrid system, it was shown that eIF4G binds to the non-structural rotavirus protein NSP3.[31] NSP3 and PABP compete for the same binding site in a previously overlooked N-terminal region of human eIF4G (Fig. 2). Since NSP3 also interacts at the same time with the (non-adenylated) 3' end of rotavirus mRNA it can provide the virus with a translational advantage in two ways. On the one hand, it selectively blocks PABP-dependent cellular translation and on the other hand, it bridges the two ends of rotavirus mRNAs.[33] In parallel, analyses of cDNAs for both mammalian isoforms of eIF4G also revealed this new N-terminal region and facilitated a direct demonstration of PABP-binding to this region by co-immunoprecipitation.[32] Interestingly, despite the evolutionary conservation of the eIF4G-PABP interaction, there is no apparent sequence homology between the PABP-binding regions of yeast and mammalian eIF4G. Indeed, human PABP cannot bind to yeast eIF4G.[34] Additional evidence for such evolutionary divergence is the existence of two human PABP-interacting proteins, Paip1 and Paip2, with documented roles in translation (Figs. 2 and 3) whereas the yeast genome does not contain any homologous genes. Paip1 was discovered in a screen for PABP-binding proteins and is a protein with similarity to the central third of eIF4G. Paip1 interacts with eIF4A and can co-activate

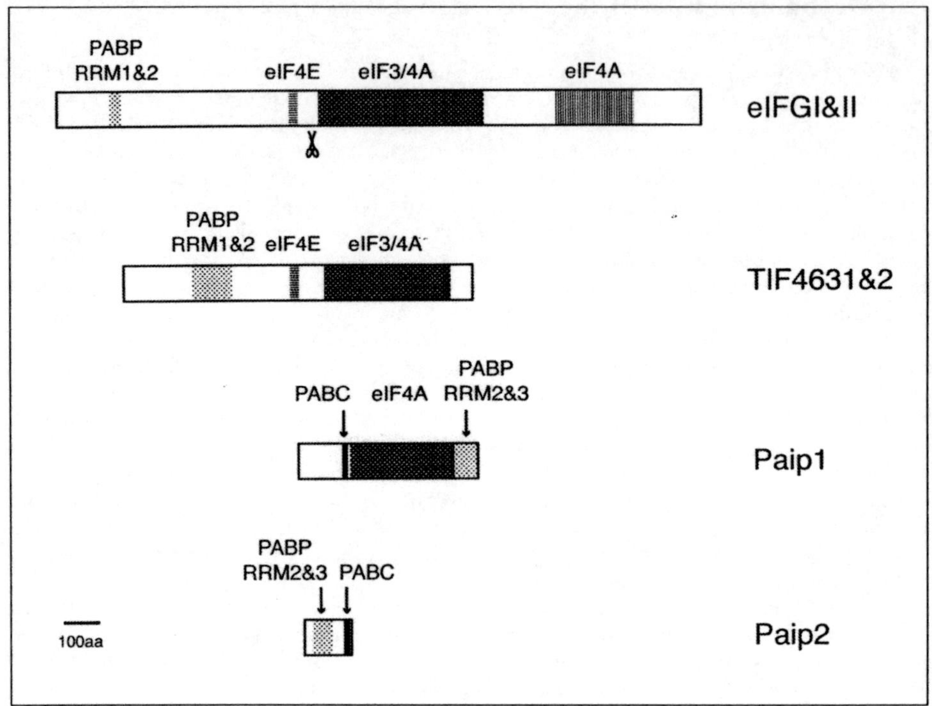

Figure 2. PABP-interacting proteins with a role in translation initiation. Two isoforms of eIF4G are known in human cells (eIF4GI&II), which share the same overall domain structure and interact with other initiation factors in an analogous fashion.[32, 146] eIF4E and PABP bind to the N-terminal third of eIF4G (all binding regions are indicated by different patterning), while the central third interacts with eIF3 and eIF4A. The C-terminal third contains another binding site for eIF4A and for the eIF4E-kinase Mnk-1 (not shown). Many picornaviruses cleave eIF4G between the first and second third using virally encoded proteases (indicated by scissors). The genes TIF4631&2 encode the yeast homologues of eIF4G. The most noticeable difference to mammalian eIF4G is that they lack the C-terminal third. Two other human proteins with translational roles have been discovered through their ability to bind PABP. Paip1 is an activator of translation, exhibits a region of homology with the central third of eIF4G, and binds eIF4A. Deletion analysis has identified a PABP-interacting region at the C-terminus of Paip1. This region has no obvious homology to the PABP-binding domain of eIF4G. Paip2 is a translation repressor and recently, two independent PABP-binding regions were described in Paip2. One is a central region, which binds to RRM2 and RRM3 of PABP, and the other is a conserved peptide motif responsible for interaction with the PABC-domain, the C-terminal domain of PABP. This conserved motif is found in several proteins and also in Paip1 (see the main text for further details). Adapted with permission from: Preiss T. BIO*spektrum* 2001; 4:315-319. © 2001 Spektrum Akademischer Verlag

cap-dependent translation- despite having no eIF4E binding motif.[35] Paip2 is a small acidic protein that acts as a translational repressor, with a preferential effect on translation of polyadenylated mRNAs. It reduces the affinity of PABP for oligo(A) and disrupts the periodicity with which multiple PABP molecules bind to poly(A). Furthermore, Paip1 and Paip2 interact with the same regions on PABP and compete with each other for binding (Fig. 3).[36] Recent years have seen the development of "synergy-competent" cell-free translation systems from higher eukaryotes.[37-39] With these systems in hand, we now have the potential for a biochemical characterization of the translational role of the poly(A) tail in more complex organisms.

The wealth of data on the importance of the eIF4G-PAPB interaction during translation effectively focused much attention on an involvement of the poly(A) tail in 40S subunit recruitment

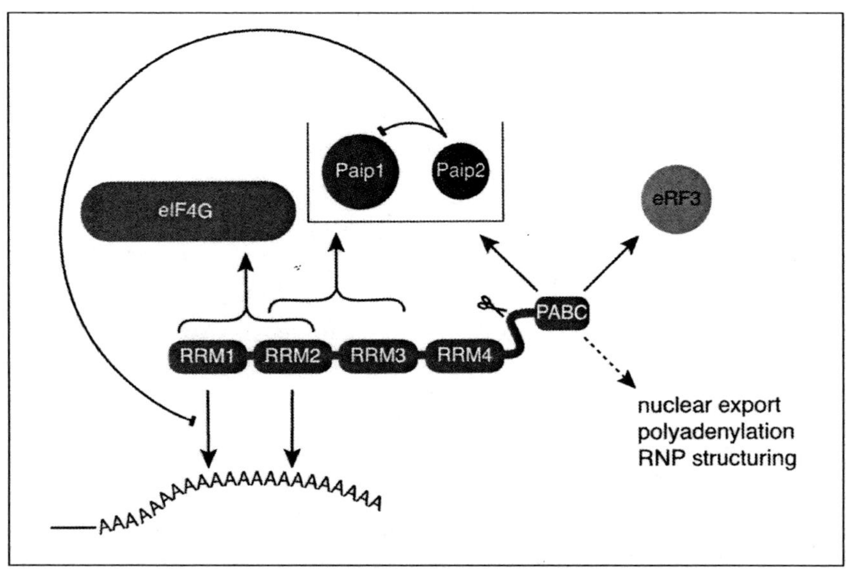

Figure 3. Schematic of the structure of PABP and its interaction partners. The N-terminal part of PABP consists of 4 RRMs, while the PABC domain resides at the C-terminus. The arrows indicate experimentally verified interactions to the binding partners shown while the hatched line symbolizes more generally a connection that has been suggested with other processes in mRNA metabolism. eIF4G and poly(A) can bind simultaneously to RRM1 and -2 while the available evidence suggests that only one partner can bind to the PABC at any given time. Paip2 was shown to interact with RRM2 and RRM3 of PABP as well as with the PABC domain. Paip2 forms complexes of 2:1 stochiometry with PABP in vitro and there are indications that it competes directly with Paip1 for the same binding regions of PABP.[36,74] Paip2 binding to RRM2 and RRM3 of PABP can disrupt the binding of PABP to poly(A) and it interferes with poly(A)-dependent translation. The inhibitory effects of Paip2 are indicated by lines with a vertical bar at the end. Adapted with permission from: Preiss T. BIO*spektrum* 2001; 4:315-319. © 2001 Spektrum Akademischer Verlag

(Fig. 1). There are, however, several indications from experiments in yeast, which suggest that this may not be the whole story. First, early genetic studies had demonstrated that suppressor mutations of a deletion of the essential PAB1 gene in yeast also altered the level of 60S ribosomal subunits,[40,41] consistent with an apparent involvement of the poly(A) tail in the 60S subunit joining step (see above).[16] Second, there are indications for a certain redundancy of the Pab1p-eIF4G interaction. Several mutations in eIF4G or Pab1p that decrease or abolish poly(A)-dependent translation do not have the same deleterious effect on translational synergy between cap structure and poly(A) tail in vitro and do not result in cell inviability.[25] A mutation in Pab1p was isolated that inhibits poly(A)-dependent translation but does not abolish eIF4G-binding.[42] Thus, it is quite possible that Pab1p makes additional contacts to the yeast translation machinery. Third, a number of reports suggest that the poly(A)/Pab1p complex stimulates translation by counteracting an inhibitory complex comprising several yeast super killer (Ski) proteins.[43-47] The main evidence for this is that electroporated, capped, but not adenylated, reporter mRNAs are highly expressed in various ski mutant cells. The SKI genes were identified from mutations allowing the increased production of killer toxin encoded by the M virus, a satellite of the double-stranded RNA L-A virus (for a review, see ref. 48). The L-A virus lacks both 5' cap structure and 3' poly(A) and thus it is not surprising that several SKI genes have roles in mRNA degradation pathways (see below).[49,50] Nevertheless, analysis of mRNA stability substantiated the claim that the Ski proteins modulate the poly(A) tail's effect on translation. Finally, a recent study may link the effects of the SKI genes with the 60S subunit joining step of translation initiation.[51] Using the same mRNA electroporation assay, it was shown that either a deletion of the FUN12 gene or a defect in TIF5 (encoding the yeast homologs of eIF5B and eIF5, respectively) specifically reduced translation of polyadenylated mRNA. Furthermore, deletion of FUN12 in the context of a deletion of SKI2 and the related gene SLH1 reinstated the repression of non-adenylated mRNA. These results suggest a model, which posits that the poly(A)/Pab1p complex inhibits the function of Ski2p and Slh1p, which in turn inhibit Fun12p and Tif5p, which are required for 60S subunit joining. Another recent study, however, describes an additional function of yeast eIF5 in an earlier step of the initiation pathway, namely in binding of mRNA to 40S subunits through a bridging interaction between eIF3 and the C-terminal part of eIF4G.[52] The mechanistic involvement of eIF5 and eIF5B in poly(A)-tail-mediated translation can now be tested in cell-free translation experiments and a future challenge will be to make a more complete model of the role of the poly(A) tail in translation from the above hypotheses.

Structural Information on the Building Blocks of the Bridge between Cap Structure and Poly(A) Tail

The multivalent adapter molecule eIF4G is the centerpiece of a bridge that forms between the ends of the mRNA during translation (for a review, see ref. 53). The binding regions for several interaction partners of eIF4G have been mapped by deletion and mutation analysis (Fig. 2). Apart from the already mentioned factors with a direct role in the translation initiation process, these also include a number of proteins with a modulating influence on translation (i.e. the eIF4E-kinase Mnk-1[54,55]), or the potential to link it to other aspects of mRNA metabolism (i.e. the nuclear cap-binding complex CBC;[56, 57] or the decapping enzyme Dcp1p[58]). Collectively, this forms the picture of a modular structure of eIF4G (Fig. 2). The N-terminal third with binding sites for PABP and eIF4E serves to latch on to the mRNA. The central third with binding sites for eIF4A, eIF3, and RNA primarily functions in ribosome recruitment while the C-terminal third with binding sites for eIF4A and Mnk-1 has a regulatory role.[59, 60] The structure of the phylogenetically conserved middle third of eIF4G[61, 62] was recently solved. A region of 259 amino acids folds into 10 α-helices, which are arranged into 5 HEAT motifs.[63] Proteins with HEAT repeats are commonly involved in the formation of multi-protein complexes.[64]

The amino acid sequences of all known PABPs exhibit the same domain structure.[65] The N-terminal part of the protein consists of 4 RNA-Recognition-Motifs (RRM), joined together by highly conserved linker sequences. The C-terminus forms a further conserved domain for protein-protein contacts (Fig. 3). Systematic biochemical and genetic studies have shown that a fragment consisting of RRM-1 and –2 is responsible for binding to poly(A)[66,67] and also interacts with eIF4G.[42,65,68] X-ray crystallography of a complex between the PABP-RRM-1 and -2 fragment and poly(A) revealed the molecular details of the interaction between single-stranded poly(A) and the RNA-binding motifs of both RRMs. The surface of the PABP-fragment facing away from the RNA forms a phylogenetically conserved region, which was postulated to contact eIF4G.[65] This hypothesis is supported by the results of targeted mutational studies of conserved amino acids in this region.[34] Structural analyses of the eIF4E-complex with m[7]GDP and a short eIF4G-fragment display a certain analogy. eIF4E has a concave side with a small hydrophobic slot for binding the cap structure and a contiguous region for mRNA binding. On the opposite, the convex face of the protein is the contact region with eIF4G.[69,70]

Very recently, structural data was obtained also for the C-terminal part of PABP.[71] This region of the protein contains a conserved sequence of about 60-70 amino acids in length (PABC), the structure of which has been solved by NMR spectroscopy and X-ray crystallography. The NMR studies showed that a region of 74 amino acids folds into a globular domain.[72] The X-ray structural analysis was carried out on an ortholog of this domain from the human hyperplastic disc protein and shows a very similar structure.[73] The PABC domain consists of 5 α-helices that are arranged in relation to each other in the shape of an arrow. PABC binds specifically through a hydrophobic region to an approximately 12 amino acid peptide motif. This motif is found in a number of proteins that interact with the C-terminal part of PABP, including the aforementioned Paip1 and Paip2, and the eukaryotic release factor eRF3 (Fig. 3).[71] In the case of Paip2, the PABC-interacting motif resides near the C-terminus (Fig. 2). In addition, Paip2 exhibits another PABP-binding site in its central region, which interacts specifically with a region encompassing RRM-2 and 3 of PABP (Fig. 3).[74] The latter interaction displays the higher affinity, is sufficient to promote the characteristic disruption of the

PABP-poly(A) complex, and to repress translation in a poly(A)-responsive in vitro system. As binding between the Paip2 C-terminus and PABC fails to show these effects, it is at present unclear what the physiological function of this interaction might be. It could aid in antagonizing Paip1-binding to PABP and/or regulate other processes than translation initiation. Once again, a parallel can be drawn to the case of the cap-binding protein eIF4E. Binding of translational activator proteins (eIF4G to eIF4E, Paip1 to PABP) is antagonized by translational repressors (the eIF4E-binding proteins, 4E-BPs,[9] and Paip2, respectively). The 4E-BPs[75] and the eIF4E-binding region of eIF4G[76] exist just like the PABC-binding peptide[72] in an unfolded state which folds into an ordered structure upon binding to the target protein.

Collectively, this leads to a working model for the function of the PABP/poly(A) complex as follows (Fig. 3):[72] the two N-terminal RRMs of PABP are responsible for binding to poly(A) and for contacting the 5' end of the mRNA through binding of eIF4G. This leaves the PABC region free for further protein-protein contacts, which may also influence translation or other aspects of mRNA metabolism. Paip1 represents a further link to translation initiation while the interaction with eRF3 has the potential to bridge to the termination of translation. Additional proteins with regions of homology to PABC or the PABC-interacting peptide have already been identified and could hint at functional links to other cellular processes.[72, 73]

Molecular Concepts to Explain Translational Synergy

The available data displays a clear connection between the functional cooperativity of the cap structure and the poly(A) tail during translation and a pseudo-circular structure of the mRNA. Potential advantages of this circular structure are easily recognized. The error rate of translation could be reduced since only intact mRNAs act as efficient templates. A well-characterized mRNA decay pathway starts with a deadenylation step, followed by decapping and exonucleolytic degradation of the mRNA body (Fig. 6). The association of both mRNA ends with the translation machinery may therefore stabilize the mRNA (see below).[77] In addition to this protective effect, the spatial proximity could also have a direct positive effect on translation. Ribosomes may not dissociate away from the mRNA after termination but instead initiate a new round of translation at the 5' end of the same mRNA molecule. So far, however, there is no direct evidence for this attractive concept of ribosome recycling. This model further suggests a mechanistic distinction between the first and subsequent rounds of translation (Fig. 4). The idea of a specialized first round of translation is also appealing as it would provide opportunities for an mRNP remodeling process on newly exported mRNA, from a nuclear to a cytoplasmic form, and for monitoring by mRNA surveillance systems (see chapter 13 by L Maquat).[56, 78]

Experiments in the yeast system have shown that translational synergy at least partly originates from a competition for limiting components of the translation machinery.[28, 79] Translation reactions using nuclease-pretreated yeast extracts display robust translation rates when only one of the two end modifications are appended to the test mRNA, but little or no synergy. In untreated extracts, the endogenous cellular mRNAs are still preserved: the competitive conditions here massively favor test mRNAs that exhibit both end modifications.[27] These competition effects could arise at the level of the first or the subsequent rounds of translation. Various demonstrations of cooperative binding along the sequence cap-eIF4E-eIF4G-PABP-poly(A)[30, 80-83] suggest that the observed synergy effects occur at least in part on the level of initial recruitment of the mRNA. Regarding the molecular causes for translational synergy, there are still many unanswered questions. The determination of the structure of a complex comprising eIF4E, eIF4G and PABP could make a significant contribution to answer these questions. Another goal is the development of strategies in cell-free systems to distinguish experimentally between ribosome recruitment and recycling.

Mechanisms of Translational Control Involving the Poly(A) Tail

Viruses employ a variety of unconventional strategies to usurp the cellular translation machinery (for a review, see ref. 84). eIF4G and its interaction partners are preferred targets for viruses. During infection by certain picorna viruses, the N-terminal third of eIF4G that links the protein to cap structure and poly(A) tail of cellular mRNAs is cleaved from the remainder of the protein by proteolytic attack (Fig. 2).[10, 85, 86] Similarly, the PABC domain is removed from PABP.[87, 88] These

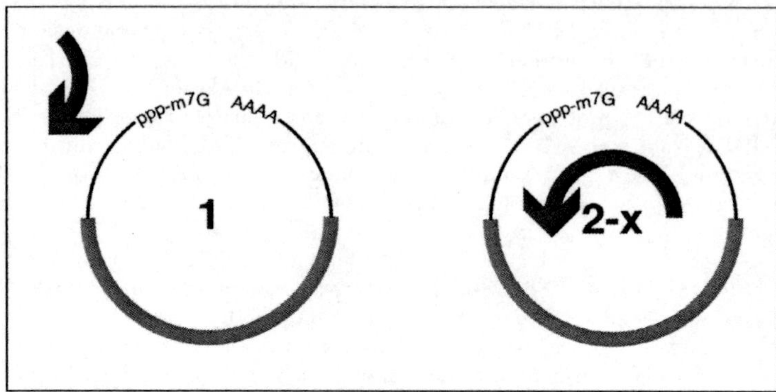

Figure 4. The ribosome-recycling concept. A function of mRNA circularization could be to facilitate a direct recycling of ribosomes or ribosomal subunits, after termination at the stop codon, back to the 5' region of the same mRNA. So far, this remains a speculative model since we have no direct experimental proof for such a recycling. The concept also suggests a distinction between an initial round of translation when an mRNA molecule is recruited to the translation machinery for the first time (1), and the following ones (2-x).

manipulations lead to an inhibition of translation of cellular mRNA in favor of selective translation of the viral mRNA. Another example is the case of the above-mentioned rotaviruses, which employ a translational strategy that directly targets the PABP-eIF4G interaction.[31, 33]

The most recognized area of translational control involving the poly(A) tail, however, is the translational regulation of maternal mRNAs during oocyte maturation and early embryogenesis (for a review, see refs. 13, 89-92). A precise temporal and spatial control of gene expression is particularly important in these early developmental stages. Characteristically, however, there is little or no transcription during this period. Crucial processes during this phase depend therefore on the controlled expression of maternal mRNA molecules, which were stocked in advance in the oocyte cytoplasm. Research into these processes in various genetic model systems of developmental biology is met with an ever-increasing interest. Errors in the post-transcriptional regulation of maternal mRNA expression usually lead to drastic abnormalities in embryonic development.

Typically, these mRNAs are initially stored in the cytoplasm of the developing oocyte in a dormant form with short poly(A) tails (~20-40 adenosines) until translation needs to be activated. This activation usually correlates with an elongation of their poly(A) tail. Cytoplasmic polyadenylation requires two elements in the 3' UTR of these mRNAs (Fig. 5): the "nuclear" polyadenylation motif AAUAAA and the nearby cytoplasmic polyadenylation element (CPE, consensus sequence: UUUUUAU).[93, 94] Investigations of these processes during oocyte maturation in the frog *Xenopus laevis* have yielded interesting insights into the underlying molecular mechanisms. The mos mRNA, for instance, exhibits a CPE and is activated early in the maturation process. A CPE-binding protein (CPEB) is necessary for the concurrent polyadenylation to approximately 150 adenosines. CPEB, a zinc-finger and RRM-containing protein, was first discovered in *Xenopus*[95, 96] but is probably present in all metazoans.[97-99] CPEB is also involved in establishing the preceding inactive state of the mRNAs. This masking activity of CPEB is mediated through the function of a CPEB-interacting protein called maskin.[100] Maskin additionally binds to eIF4E in a manner similar to the aforementioned 4E-BPs.[9] CPEB and maskin form a stable complex whereas the interaction of maskin with eIF4E is strongly reduced during oocyte maturation. It is plausible that this modulates the formation of functional eIF4F complexes on CPE-containing mRNAs and thus their translation (Fig. 5).[90] Furthermore, CPEB is a phosphoprotein and a specific phosphorylation of CPEB at serine 174 by the kinase Eg2 is required for the activation of mos mRNA.[101] This phosphorylation stimulates the interaction of CPEB with the AAUAAA-binding protein CPSF (for Cleavage and Polyadenylation Specificity Factor) and in this way can recruit the cytoplasmic polyadenylation machinery to the mRNA.[102] How does elongation of the poly(A) tail stimulate translation? One possibility relies on the observation that a cap ribose methylation can occur in response to the ongoing process of poly(A) tail elongation. This modification of the 5' end of the mRNA can enhance translation initiation.[103]

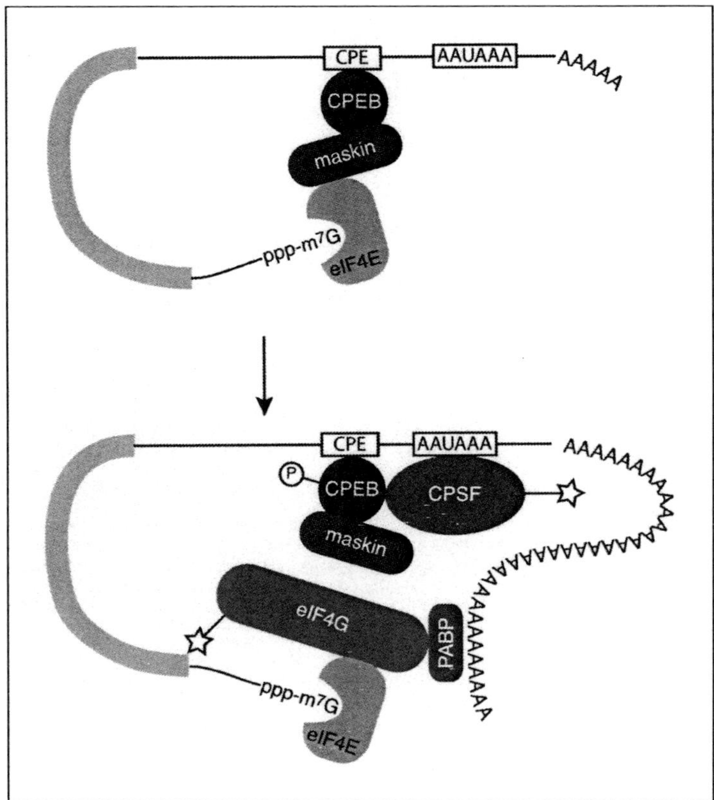

Figure 5. Model of CPEB-dependent masking and activation of mRNA translation in oocytes. Typically, maternal mRNAs initially carry a short poly(A) tail and their CPE motif is bound by CPEB. CPEB forms a complex with maskin, which in turn interacts with eIF4E. This configuration is thought to represent a translationally inactive or "masked" state of the mRNA. The signal for oocyte maturation leads to a phosphorylation of CPEB and a stimulation of binding between CPEB and CPSF. Through further association of CPSF with poly(A)-polymerase, this leads to elongation of the poly(A) tail (indicated by the line with star). The binding between maskin and eIF4E is reduced, possibly as a consequence of polyadenylation. This clears the way for efficient recruitment of eIF4G through binding to eIF4E and PABP and activation of translation (indicated by a further line with star). Adapted with permission from: Preiss T. BIO*spektrum* 2001; 4:315-319. © 2001 Spektrum Akademischer Verlag

However, not all CPE-containing mRNAs are modified in this way.[104] Another possibility is based on the ability of elongated poly(A) tails to recruit multiple PABP molecules to the mRNA and thus enhance the recruitment of eIF4G/eIF4F to the cap structure, probably at the expense of the maskin-eIF4E interaction (Fig. 5). The common form of PABP is only of low abundance in oocytes[105] but its role may be in part carried out by development-specific isoforms.[106] Two studies provide evidence that the interaction of some form of PABP with eIF4G is important for translation in oocytes and for the maturation process. PABP was targeted to test mRNAs in *Xenopus* oocytes using a tethered-function approach and shown to stimulate their translation- apparently through an interaction with eIF4G.[107] Expression of a mutant form of eIF4G, defective in PABP-binding, had a dominant-negative effect on translation of polyadenylated mRNA and drastically inhibited oocyte maturation.[108]

In mammalian oocytes, CPEB seems to serve a similar function as in *Xenopus*, namely to execute the switch between CPE-mediated mRNA repression to polyadenylation-induced activation during maturation.[109-112] The generation of CPEB knockout mice has allowed a number of very interesting observations. Both male and female CPEB null mice are viable and show no obvious growth abnormalities but severe fertility defects.[113] Detailed analyses revealed that both male and female germ cell differentiation is disrupted at a stage of meiosis prior to meiotic maturation- as early as the pachytene stage of meiotic prophase I. This shows that CPEB function is required earlier in germ cell development than was predicted from the known CPEB target mRNAs. Further, CPEB -/- germ cells harbored fragmented chromatin suggesting a defect in homologous chromosome adhesion or synapsis. Indeed, the mRNAs for two proteins involved in synaptonemal complex formation, SCP1 and SCP3, contain functional CPE-motifs and their translational regulation is disrupted by the CPEB deletion.[113] Apart from the reproductive organs, CPEB is also expressed in the hippocampus and is enriched at synapses of neurons in culture.[114, 115] This offers a very interesting extension of this research area, which has not yet been explored with the CPEB knockout mice. Controlled translation enables particularly fast and large changes in the synthesis of proteins, as they are required for synaptic plasticity. Indeed, it was found that the mRNA for calmodulin-dependent protein kinase II (a-CaMKII), an essential factor for activity-dependent synaptic plasticity, contains CPE-motifs in the 3' UTR. During synaptic stimulation, a-CaMKII mRNA is polyadenylated and translationally activated. It is therefore possible that CPEB-mediated local control of translation is involved in synaptic plasticity and memory.

Translation and mRNA Degradation

The cap structure and the poly(A) tail also play important roles in mRNA degradation. It is thus not surprising that there are numerous indications for an intimate connection between translation and mRNA turnover (for a review, see refs. 78, 116, 117). Four distinct pathways of mRNA decay are known in eukaryotes and have been primarily studied in *Saccharomyces cerevisiae* (see chapter 14 by P Mitchell). Two pathways start with deadenylation as the initial step and are thought to occur on many if not all mRNA species. In the predominant pathway (Fig. 6), this is followed by a removal of the cap structure, termed "decapping", and 5' to 3' exonucleolytic degradation of the body of the mRNA. Alternatively, mRNAs can be degraded in 3' to 5' direction following deadenylation. Nonsense-mediated decay (NMD) is a more specific mechanism, which ensures that aberrant mRNA molecules are rapidly decapped and degraded 5' to 3' independently of deadenylation (see chapter 13 by L. Maquat).[118-122] Finally, some mRNAs are known to contain cleavage sites for specific endonucleases that can trigger their degradation.

There are good indications that the predominant mRNA turnover pathway (Fig. 6) is regulated by the status of the translation initiation machinery on the mRNA. Analyses of yeast strains containing mutations in the translation initiation factors eIF4E, eIF4G, eIF4A or a subunit of eIF3 show increased rates of deadenylation and subsequent decapping of both unstable MFA2 and stable PGK1 mRNAs.[77] By contrast, inhibiting the elongation step of translation slows down deadenylation and decapping, perhaps indicating a window of opportunity for mRNA degradation in the translation cycle.[117] An attractive model proposes that deadenylation rates reflect the degree of accessibility of the poly(A) tail for the deadenylase. Likewise, decapping rates may reflect a competition between eIF4E and the decapping enzyme Dcp1p for the cap structure. Several observations provide initial molecular explanations for these models. eIF4E has a much lower intrinsic affinity for cap than Dcp1p.[123] During active translation initiation, this is probably compensated for by an enhancement of eIF4E-binding to the cap structure, through the interaction with eIF4G.[80, 124] Dcp1p also binds directly to eIF4G and Pab1p, either as individual proteins or as part of the eIF4F-Pab1p complex.[58] Pab1p can modulate the interactions of eIF4G and eIF4E with the mRNA[24] while eIF4G can stimulate Dcp1p activity. The latter effect is blocked by eIF4E.[58] It has also been shown that Pab1p is required for the poly(A) tail to function as an inhibitor of decapping.[125, 126] Thus, the emerging picture is one of a dynamic complex that bridges between the mRNA ends and is involved in the switch from translation initiation to mRNA degradation. Loss of Pab1p-binding to the deadenylated 3' end of the mRNA could trigger a rearrangement at the 5' end that allows Dcp1 access to the cap structure.[58]

Components of the mRNA Degradation Machinery

The most complete set of factors involved in mRNA decay has been identified in yeast and many of them have recognized homologs in other eukaryotes (for a review, see refs. 117, 127). The DCP1 gene encodes the decapping enzyme.[128, 129] Dcp2p is another factor required for decapping,

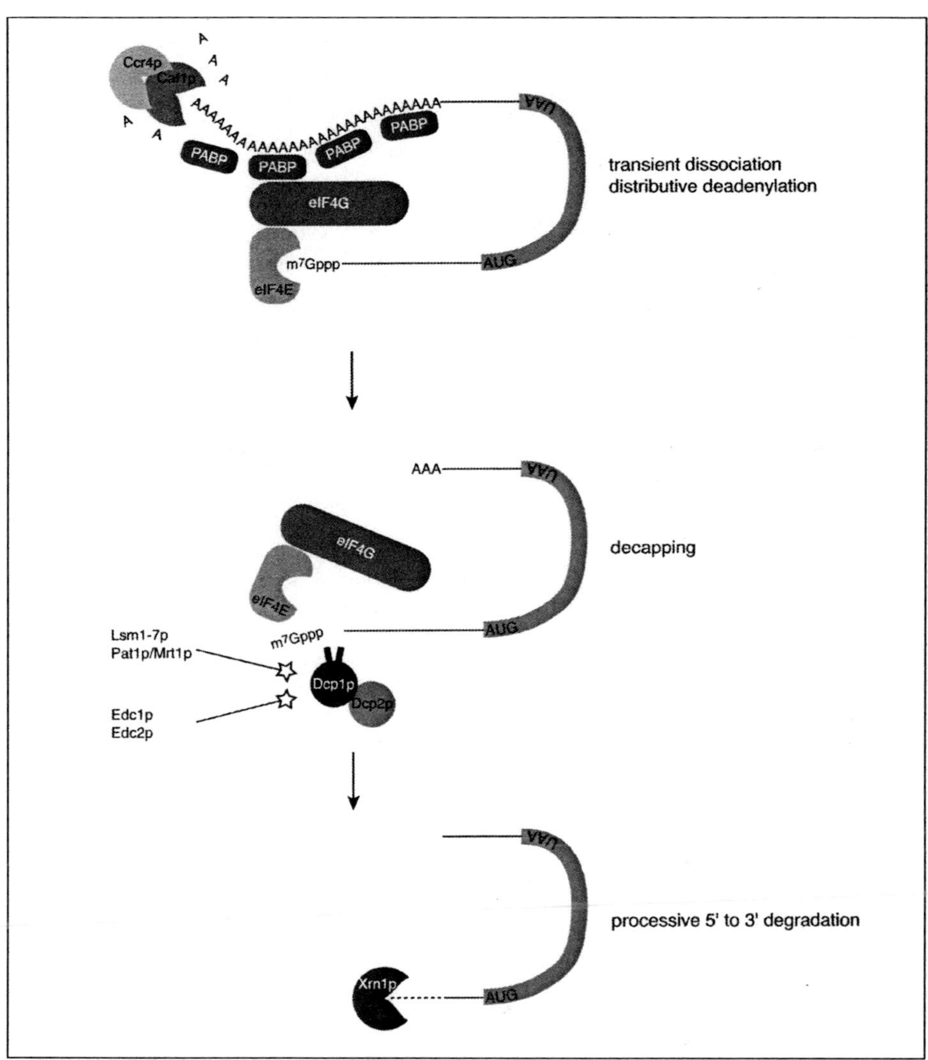

Figure 6. A major mRNA degradation pathway in yeast. The process of translation initiation is stimulated by interactions bridging eIF4E, eIF4G, and Pab1p (see Fig. 1), and this is thought to hinder access of the mRNA degradation machinery to the mRNA. Eventually, however, slow deadenylation mediated by Ccr4p and Caf1p/Pop2p sets in. When the tail is shortened to 10 adenosines or less, this triggers decapping by Dcp1p and Dcp2p followed by 5' to 3' degradation by Xrn1p. Loss of Pab1p binding to the 3' end of the mRNA may induce a rearrangement of factors at the 5' end to allow Dcp1p access to the cap structure. A complex of 7 Lsm proteins and the associated factor Pat1p/Mrt1p, as well as the proteins Edc1p and Edc2p have been shown to activate the decapping step.

probably by regulating Dcp1p activity.[130] The related proteins Edc1p and Edc2p also stimulate decapping.[131] Xrn1p is the exoribonuclease responsible for 5' to 3' degradation of the body of the mRNA.[132, 133] 3' to 5' degradation is carried out by the exosome complex and the accessory proteins Ski2p, 3, 8 (see above and chapter 14 by P. Mitchell).[127, 134, 135] Recently, Ccr4p and Caf1p/Pop2p have been identified as components of the elusive major yeast cytoplasmic deadenylase.[136, 137] Both proteins have nuclease domains, Ccr4p being a member of a magnesium-dependent endonuclease-related family while Caf1p/Pop2p belongs to the RNase D family of 3' to 5' exonucleases. While Caf1p/Pop2p displays significant similarity to the human deadenylase PARN,[138, 139] it does not appear to represent a PARN homologue in yeast. Both Ccr4p and Caf1p/Pop2p are highly conserved in eukaryotic cells.

Additionally, a group of seven Sm-like proteins (Lsm1p–Lsm7p) and the associated factor Pat1p/Mrt1p activate decapping in the major mRNA degradation pathway.[140-143] Sm and Sm-like proteins can form heptameric complexes in the shape of a ring.[144] Consistent with their role in mRNA degradation, several Lsm proteins (Lsm1,2,3,5,6,7) co-purified biochemically with the Xrn1p exonuclease.[142] Lsm1-7 co-immunoprecipitate with Dcp1, Pat1/Mrt1 and mRNA.[143] Two-hybrid screens suggest interactions between various Lsm proteins (Lsm1p–Lsm8p), and components of the mRNA degradation pathway: Pat1p/Mrt1p, Dcp1, Dcp2p, and Xrn1p.[145] While the above results may not all represent bona fide protein-protein interactions, together they make a strong case for a role of the Lsm1-7 complex in mRNA degradation.

Perspectives

Research into the roles of the poly(A) tail in mRNA metabolism has made great progress in the last ten years. A number of interactions between the poly(A) tail and the mRNA translation machinery have been identified and we have good working models for their function. The presence of further PABP-interacting proteins in higher cells are one indication for a more complex network of contacts with the 5' end than suggested by initial experiments in the yeast model system. The biochemical characterization of these processes in poly(A)-dependent cell-free translation systems from higher eukaryotes can make valuable contributions here. It is possible that this may also yield more substantial insights on other functions of mRNA circularization in mRNA metabolism. Studies on the mechanism of translational activation in early developmental stages also promise to further our understanding of mRNA-specific translation control mediated by elements in the 3' UTR and how they functionally interact with the poly(A) tail. Finally, there is a strong case for multiple interrelations between mRNA translation and degradation mechanisms. Much progress in both areas will depend critically on an appreciation of just how the two processes affect each other.

Acknowledgements

The author wishes to thank M. W. Hentze for the longstanding and productive cooperation, and K. Brennan for suggestions on the manuscript. This work was funded by grants from the Deutsche Forschungsgemeinschaft (PR616/1-1&2).

References

1. Hershey JWB, Merrick WC. Pathway and Mechanism of Initiation of Protein synthesis. In: Sonenberg N, Hershey JWB, Mathews MB, eds. Translational Control of Gene Expression. Cold Spring Harbor, New York: Cold Spring Harbor Laboratory Press, 2000:33-88.
2. Merrick WC, J N. The protein biosynthesis elongation cycle. In: Sonenberg N, Mathews MB, Hershey JWB, eds. Translational Control of Gene Expression. Cold Spring Harbor: Cold Spring Harbor Laboratory Press, 2000:89-126.
3. Welch EM, Wang W, Peltz SW. Translation termination: it's not the end of the story. In: Sonenberg N, Hershey JWB, Mathews MB, eds. Translational Control of Gene Expression. Cold Spring Harbor: Cold Spring Harbor Laboratory Press., 2000:467-486.
4. Mathews MB, Sonenberg N, Hershey JBW. Origins and principles of translational control. In: Sonenberg N, Hershey JBW, Mathews MB, eds. Translational Control of Gene Expression. Cold Spring Harbor: Cold Spring Harbor Laboratory Press, 2000:1-31.
5. Shatkin AJ. Capping of eucaryotic mRNAs. Cell 1976;9:645-53.
6. Kozak M. Initiation of translation in prokaryotes and eukaryotes. Gene 1999; 234:187-208.
7. Pestova TV, Kolupaeva VG, Lomakin IB et al. Molecular mechanisms of translation initiation in eukaryotes. Proc Natl Acad Sci USA 2001; 98:7029-36.
8. Sonenberg N, Morgan MA, Merrick WC et al. A polypeptide in eukaryotic initiation factors that crosslinks specifically to the 5'-terminal cap in mRNA. Proc Natl Acad Sci USA 1978; 75:4843-7.
9. Gingras A-C, Raught B, Sonenberg N. eIF4 initiation factors: effectors of mRNA recruitment to ribosomes and regulators of translation. Annu Rev Biochem 1999; 68:913-963.
10. Lamphear BJ, Kirchweger R, Skern T et al. Mapping of functional domains in eukaryotic protein synthesis initiation factor 4G (eIF4G) with picornaviral proteases. J Biol Chem 1995; 270:21975-21983.
11. Hinnebusch AG. Mechanism and regulation of Initiator methionyl-tRNA Binding to Ribosomes. In: Sonenberg N, Hershey JW, Mathews B, eds. Translational Control of Gene Expression. Cold Spring Harbor: Cold Spring Harbor Laboratory Press, 2000:185-245.

12. Pestova TV, Lomakin IB, Lee JH et al. The ribosomal subunit joining reaction in eukaryotes requires eIF5B. Nature 2000; 403:332-335.
13. Preiss T, Hentze MW. From factors to mechanisms: translation and translational control in eukaryotes. Curr Opin Genet Dev 1999; 9:515-21.
14. Sachs AB. Physical and functional interactions between the mRNA cap structure and the poly(A) tail. In: Sonenberg N, Hershey JBW, Mathews MB, eds. Translational Control of Gene Expression. Cold Spring Harbor: Cold Spring Harbor Laboratory Press, 2000:447-465.
15. Jacobson A. Poly(A) metabolism and translation: the closed-loop model. In: Hershey JWB, Mathews MB, Sonenberg N, eds. Translational control. Cold Spring Harbor: Cold Spring Harbor Laboratory Press, 1996:451-480.
16. Munroe D, Jacobson A. mRNA poly(A) tail, a 3' enhancer of translational initiation. Mol Cell Biol 1990; 10:3441-3455.
17. Wickens M. In the beginning is the end: regulation of poly(A) addition and removal during early development. Trends Biochem Sci 1990; 15:320-4.
18. Gallie DR. The cap and poly(A) tail function synergistically to regulate mRNA translational efficiency. Genes Dev 1991; 5:2108-16.
19. Iizuka N, Najita L, Franzusoff A et al. Cap-dependent and cap-independent translation by internal initiation of mRNAs in cell extracts prepared from *Saccharomyces cerevisiae*. Mol Cell Biol 1994; 14:7322-7330.
20. Tarun SZ Jr, Sachs AB. A common function for mRNA 5' and 3' ends in translation initiation in yeast. Genes and Dev 1995; 9:2997-3007.
21. Baer BW, Kornberg RD. Repeating structure of cytoplasmic poly(A)-ribonucleoprotein. Proc Natl Acad Sci USA 1980; 77:1890-2.
22. Baer BW, Kornberg RD. The protein responsible for the repeating structure of cytoplasmic poly(A)-ribonucleoprotein. J Cell Biol 1983; 96:717-21.
23. Sachs AB, Davis RW, Kornberg RD. A single domain of yeast poly(A)-binding protein is necessary and sufficient for RNA binding and cell viability. Mol Cell Biol 1987; 7:3268-3276.
24. Tarun SZ Jr, Sachs AB. Association of the yeast poly(A) tail binding protein with translation initiation factor eIF-4G. EMBO J 1996; 15:7168-7177.
25. Tarun SZ Jr, Wells SE, Deardorff JA et al. Translation initiation factor eIF4G mediates in vitro poly(A) tail-dependent translation. Proc Nat Acad Sci USA 1997; 94:9046-9051.
26. Christensen AK, Kahn LE, Bourne CM. Circular polysomes predominate on the rough endoplasmic reticulum of somatotropes and mammotropes in the rat anterior pituitary. Am J Anat 1987; 178:1-10.
27. Preiss T, Hentze MW. Dual function of the messenger RNA cap structure in poly(A)-tail-promoted translation in yeast. Nature 1998; 392:516-520.
28. Preiss T, Muckenthaler M, Hentze MW. Poly(A)-tail-promoted translation in yeast: implications for translational control. RNA 1998; 4:1321-1331.
29. Wells SE, Hillner PE, Vale RD et al. Circularization of mRNA by eukaryotic translation initiation factors. Mol Cell 1998; 2:135-40.
30. Le H, Tanguay RL, Balasta ML et al. Translation initiation factors eIF-iso4G and eIF-4B interact with the poly(A)-binding protein and increase its RNA binding activity. J Biol Chem 1997; 272:16247-55.
31. Piron M, Vende P, Cohen J et al. Rotavirus RNA-binding protein NSP3 interacts with eIF4GI and evicts the poly(A) binding protein from eIF4F. EMBO J 1998; 17:5811-21.
32. Imataka H, Gradi A, Sonenberg N. A newly identified N-terminal amino acid sequence of human eIF4G binds poly(A)-binding protein and functions in poly(A)-dependent translation. EMBO J 1998; 17:7480-9.
33. Vende P, Piron M, Castagne N et al. Efficient translation of rotavirus mRNA requires simultaneous interaction of NSP3 with the eukaryotic translation initiation factor eIF4G and the mRNA 3' end. J Virol 2000; 74:7064-71.
34. Otero LJ, Ashe MP, Sachs AB. The yeast poly(A)-binding protein Pab1p stimulates in vitro poly(A)-dependent and cap-dependent translation by distinct mechanisms. EMBO J 1999; 18:3153-63.
35. Craig AW, Haghighat A, Yu AT et al. Interaction of polyadenylate-binding protein with the eIF4G homologue PAIP enhances translation. Nature 1998; 392:520-3.
36. Khaleghpour K, Svitkin YV, Craig AW et al. Translational repression by a novel partner of human poly(A) binding protein, Paip2. Mol Cell 2001; 7:205-16.
37. Gebauer F, Corona DF, Preiss T et al. Translational control of dosage compensation in *Drosophila* by Sex-lethal: cooperative silencing via the 5' and 3' UTRs of msl-2 mRNA is independent of the poly(A) tail. EMBO J 1999; 18:6146-54.
38. Michel YM, Poncet D, Piron M et al. Cap-poly(A) synergy in mammalian cell-free extracts: Investigation of the requirements for poly(A)-mediated stimulation of translation initiation. J Biol Chem 2000; 275:32268–32276.
39. Bergamini G, Preiss T, Hentze MW. Picornavirus IRESes and the poly(A) tail jointly promote cap-independent translation in a mammalian cell-free system. RNA 2000; 6:1781-90.
40. Sachs AB, Davis RW. Translation initiation and ribosomal biogenesis: involvement of a putative rRNA helicase and RPL46. Science 1990; 247:1077-9.
41. Sachs AB, Davis RW. The poly(A) binding protein is required for poly(A) shortening and 60S ribosomal subunit-dependent translation initiation. Cell 1989; 58:857-67.
42. Kessler SH, Sachs AB. RNA recognition motif 2 of yeast Pab1p is required for its functional interaction with eukaryotic translation initiation factor 4G. Mol Cell Biol 1998; 18:51-57.
43. Widner WR, Wickner RB. Evidence that the SKI antiviral system of *Saccharomyces cerevisiae* acts by blocking expression of viral mRNA. Mol Cell Biol 1993; 13:4331-41.
44. Masison DC, Blanc A, Ribas JC et al. Decoying the cap- mRNA degradation system by a double-stranded RNA virus and poly(A)- mRNA surveillance by a yeast antiviral system. Mol Cell Biol 1995; 15:2763-71.

45. Benard L, Carroll K, Valle RC et al. Ski6p is a homolog of RNA-processing enzymes that affects translation of non-poly(A) mRNAs and 60S ribosomal subunit biogenesis. Mol Cell Biol 1998; 18:2688-96.
46. Benard L, Carroll K, Valle RC et al. The ski7 antiviral protein is an EF1-alpha homolog that blocks expression of non-Poly(A) mRNA in *Saccharomyces cerevisiae*. J Virol 1999; 73:2893-900.
47. Searfoss AM, Wickner RB. 3' poly(A) is dispensable for translation. Proc Natl Acad Sci USA 2000; 97:9133-7.
48. Wickner RB. Double-stranded RNA viruses of *Saccharomyces cerevisiae*. Microbiol Rev 1996; 60:250-65.
49. Johnson AW, Kolodner RD. Synthetic lethality of sep1 (xrn1) ski2 and sep1 (xrn1) ski3 mutants of *Saccharomyces cerevisiae* is independent of killer virus and suggests a general role for these genes in translation control. Mol Cell Biol 1995; 15:2719-27.
50. Jacobs JS, Anderson AR, Parker RP. The 3' to 5' degradation of yeast mRNAs is a general mechanism for mRNA turnover that requires the SKI2 DEVH box protein and 3' to 5' exonucleases of the exosome complex. EMBO J 1998; 17:1497-506.
51. Searfoss A, Dever TE, Wickner R. Linking the 3' poly(A) tail to the subunit joining step of translation initiation: relations of Pab1p, eukaryotic translation initiation factor 5b (Fun12p), and Ski2p-Slh1p. Mol Cell Biol 2001; 21:4900-8.
52. Asano K, Shalev A, Phan L et al. Multiple roles for the C-terminal domain of eIF5 in translation initiation complex assembly and GTPase activation. EMBO J 2001; 20:2326-37.
53. Hentze MW. eIF4G: a multipurpose ribosome adapter? Science 1997; 275:500-501.
54. Pyronnet S, Imataka H, Gingras AC et al. Human eukaryotic translation initiation factor 4G (eIF4G) recruits Mnk1 to phosphorylate eIF4E. EMBO J 1999; 18:270-9.
55. Waskiewicz AJ, Johnson JC, Penn B et al. Phosphorylation of the cap-binding protein eukaryotic translation initiation factor 4E by protein kinase Mnk1 in vivo. Mol Cell Biol 1999; 19:1871-80.
56. Fortes P, Inada T, Preiss T et al. The yeast nuclear cap-binding complex can interact with translation factor eIF4G and mediate translation initiation. Mol Cell 2000; 6:191-6.
57. McKendrick L, Thompson E, Ferreira J et al. Interaction of eukaryotic translation initiation factor 4G with the nuclear cap-binding complex provides a link between nuclear and cytoplasmic functions of the m(7) guanosine cap. Mol Cell Biol 2001; 21:3632-41.
58. Vilela C, Velasco C, Ptushkina M et al. The eukaryotic mRNA decapping protein Dcp1 interacts physically and functionally with the eIF4F translation initiation complex. Embo J 2000; 19:4372-82.
59. Morino S, Imataka H, Svitkin YV et al. Eukaryotic translation initiation factor 4E (eIF4E) binding site and the middle one-third of eIF4GI constitute the core domain for cap-dependent translation, and the C-terminal one-third functions as a modulatory region. Mol Cell Biol 2000; 20:468-77.
60. De Gregorio E, Preiss T, Hentze MW. Translation driven by an eIF4G core domain in vivo. EMBO J 1999; 18:4865-74.
61. Aravind L, Koonin EV. Eukaryote-specific domains in translation initiation factors: implications for translation regulation and evolution of the translation system. Genome Res 2000; 10:1172-84.
62. Ponting CP. Novel eIF4G domain homologues linking mRNA translation with nonsense- mediated mRNA decay. Trends Biochem Sci 2000; 25:423-6.
63. Marcotrigiano J, Lomakin IB, Sonenberg N et al. A conserved HEAT domain within eIF4G directs assembly of the translation initiation machinery. Mol Cell 2001; 7:193-203.
64. Andrade MA, Bork P. HEAT repeats in the Huntington's disease protein. Nat Genet 1995;11:115-6.
65. Deo RC, Bonanno JB, Sonenberg N et al. Recognition of polyadenylate RNA by the poly(A)-binding protein. Cell 1999; 98:835-45.
66. Burd CG, Matunis EL, Dreyfuss G. The multiple RNA-binding domains of the mRNA poly(A)-binding protein have different RNA-binding activities. Mol Cell Biol 1991; 11:3419-24.
67. Kuhn U, Pieler T. Xenopus poly(A) binding protein: functional domains in RNA binding and protein-protein interaction. J Mol Biol 1996; 256:20-30.
68. Sachs AB, Varani G. Eukaryotic translation initiation: there are (at least) two sides to every story. Nat Struct Biol 2000; 7:356-61.
69. Marcotrigiano J, Gingras AC, Sonenberg N et al. Cocrystal structure of the messenger RNA 5' cap-binding protein (eIF4E) bound to 7-methyl-GDP. Cell 1997; 89:951-61.
70. Matsuo H, Li H, McGuire AM et al. Structure of translation factor eIF4E bound to m7GDP and interaction with 4E-binding protein. Nat Struct Biol 1997; 4:717-24.
71. Varani G. Delivering messages from the 3' end. Proc Natl Acad Sci USA 2001; 98:4288-9.
72. Kozlov G, Trempe JF, Khaleghpour K et al. Structure and function of the C-terminal PABC domain of human poly(A)- binding protein. Proc Natl Acad Sci USA 2001; 98:4409-13.
73. Deo RC, Sonenberg N, Burley SK. X-ray structure of the human hyperplastic discs protein: an ortholog of the C-terminal domain of poly(A)-binding protein. Proc Natl Acad Sci USA 2001; 98:4414-9.
74. Khaleghpour K, Kahvejian A, De Crescenzo G et al. Dual interactions of the translational repressor Paip2 with poly(A) binding protein. Mol Cell Biol 2001; 21:5200-13.
75. Marcotrigiano J, Gingras AC, Sonenberg N et al. Cap-dependent translation initiation in eukaryotes is regulated by a molecular mimic of eIF4G. Mol Cell 1999; 3:707-716.
76. Hershey PE, McWhirter SM, Gross JD et al. The Cap-binding protein eIF4E promotes folding of a functional domain of yeast translation initiation factor eIF4G1. J Biol Chem 1999; 274:21297-304.
77. Schwartz DC, Parker R. Mutations in translation initiation factors lead to increased rates of deadenylation and decapping of mRNAs in *Saccharomyces cerevisiae*. Mol Cell Biol 1999; 19:5247-56.
78. Mitchell P, Tollervey D. mRNA turnover. Curr Opin Cell Biol 2001; 13:320-5.
79. Tarun SZ Jr, Sachs AB. Binding of eukaryotic translation initiation factor 4E (eIF4E) to eIF4G represses translation of uncapped mRNA. Mol Cell Biol 1997; 17:6876-86.
80. Haghighat A, Sonenberg N. eIF4G dramatically enhances the binding of eIF4E to the mRNA 5'-cap structure [published erratum appears in J Biol Chem 1997 Nov 14;272(46):29398]. J Biol Chem 1997; 272:21677-80.

81. Ptushkina M, von der Haar T, Vasilescu S et al. Cooperative modulation by eIF4G of eIF4E-binding to the mRNA 5' cap in yeast involves a site partially shared by p20. EMBO J 1998; 17:4798-808.
82. Ptushkina M, von der Haar T, Karim MM et al. Repressor binding to a dorsal regulatory site traps human eIF4E in a high cap-affinity state. EMBO J 1999; 18:4068-4075.
83. Wei CC, Balasta ML, Ren J et al. Wheat germ poly(A) binding protein enhances the binding affinity of eukaryotic initiation factor 4F and (iso)4F for cap analogues. Biochemistry 1998; 37:1910-6.
84. Pe'ery T, Mathews MB. Viral translational strategies and host defense mechanisms. In: Sonenberg N, Hershey JBW, Mathews MB, eds. Translational Control of Gene Expression. Cold Spring Harbor: Cold Spring Harbor Laboratory Press, 2000:371-424.
85. Imataka H, Sonenberg N. Human eukaryotic translation initiation factor 4G (eIF4G) possesses two separate and independent binding sites for eIF4A. Mol Cell Biol 1997; 17:6940-7.
86. Gradi A, Svitkin YV, Imataka H et al. Proteolysis of human eukaryotic translation initiation factor eIF4GII, but not eIF4GI, coincides with the shutoff of host protein synthesis after poliovirus infection. Proc Natl Acad Sci USA 1998; 95:11089-94.
87. Kerekatte V, Keiper BD, Badorff C et al. Cleavage of Poly(A)-binding protein by coxsackievirus 2A protease in vitro and in vivo: another mechanism for host protein synthesis shutoff? J Virol 1999; 73:709-17.
88. Joachims M, Van Breugel PC, Lloyd RE. Cleavage of poly(A)-binding protein by enterovirus proteases concurrent with inhibition of translation in vitro. J Virol 1999; 73:718-27.
89. Wickens M, Goodwin EB, Kimble J et al. Translational control of developmental decisions. In: Sonenberg N, Hershey JBW, Mathews MB, eds. Translational Control of Gene Expression. Cold Spring Harbor: Cold Spring Harbor Laboratory Press, 2000:295-370.
90. Mendez R, Richter JD. Translational control by CPEB: a means to the end. Nat Rev Mol Cell Biol 2001; 2:521-9.
91. Macdonald P. Diversity in translational regulation. Curr Opin Cell Biol 2001; 13:326-31.
92. Richter J. Influence of Polyadenylation-induced Translation on Metazoan Development and Neuronal Synaptic Function. In: Sonenberg N, Hershey JBW, Mathews B, eds. Translational Control of gene Expression. Cold Spring Harbor: Cold Spring Harbor Laboratory Press, 2000:785-805.
93. McGrew LL, Dworkin-Rastl E, Dworkin MB et al. Poly(A) elongation during *Xenopus* oocyte maturation is required for translational recruitment and is mediated by a short sequence element. Genes Dev 1989; 3:803-15.
94. Fox CA, Sheets MD, Wickens MP. Poly(A) addition during maturation of frog oocytes: distinct nuclear and cytoplasmic activities and regulation by the sequence UUUUUAU. Genes Dev 1989; 3:2151-62.
95. Paris J, Swenson K, Piwnica-Worms H et al. Maturation-specific polyadenylation: in vitro activation by p34cdc2 and phosphorylation of a 58-kD CPE-binding protein. Genes Dev 1991; 5:1697-708.
96. Hake LE, Richter JD. CPEB is a specificity factor that mediates cytoplasmic polyadenylation during Xenopus oocyte maturation. Cell 1994; 79:617-27.
97. Lantz V, Ambrosio L, Schedl P. The *Drosophila orb* gene is predicted to encode sex-specific germline RNA-binding proteins and has localized transcripts in ovaries and early embryos. Development 1992; 115:75-88.
98. Gebauer F, Richter JD. Mouse cytoplasmic polyadenylylation element binding protein: an evolutionarily conserved protein that interacts with the cytoplasmic polyadenylylation elements of c-mos mRNA. Proc Natl Acad Sci USA 1996; 93:14602-7.
99. Luitjens C, Gallegos M, Kraemer B et al. CPEB proteins control two key steps in spermatogenesis in *C. elegans*. Genes Dev 2000; 14:2596-609.
100. Stebbins-Boaz B, Cao Q, de Moor CH et al. Maskin is a CPEB-associated factor that transiently interacts with eIF- 4E [published erratum appears in Mol Cell 2000 Apr;5(4):following 766]. Mol Cell 1999; 4:1017-27.
101. Mendez R, Hake LE, Andresson T et al. Phosphorylation of CPE binding factor by Eg2 regulates translation of c-mos mRNA. Nature 2000; 404:302-7.
102. Mendez R, Murthy KG, Ryan K et al. Phosphorylation of CPEB by Eg2 mediates the recruitment of CPSF into an active cytoplasmic polyadenylation complex. Mol Cell 2000; 6:1253-9.
103. Kuge H, Richter JD. Cytoplasmic 3' poly(A) addition induces 5' cap ribose methylation: implications for translational control of maternal mRNA. EMBO J 1995; 14:6301-10.
104. Gillian-Daniel DL, Gray NK, Astrom J et al. Modifications of the 5' cap of mRNAs during Xenopus oocyte maturation: independence from changes in poly(A) length and impact on translation. Mol Cell Biol 1998; 18:6152-63.
105. Zelus BD, Giebelhaus DH, Eib DW et al. Expression of the poly(A)-binding protein during development of Xenopus laevis. Mol Cell Biol 1989; 9:2756-60.
106. Voeltz GK, Ongkasuwan J, Standart N et al. A novel embryonic poly(A) binding protein, ePAB, regulates mRNA deadenylation in Xenopus egg extracts. Genes Dev 2001; 15:774-88.
107. Gray NK, Coller JM, Dickson KS et al. Multiple portions of poly(A)-binding protein stimulate translation in vivo. EMBO J 2000; 19:4723-33.
108. Wakiyama M, Imataka H, Sonenberg N. Interaction of eIF4G with poly(A)-binding protein stimulates translation and is critical for Xenopus oocyte maturation. Curr Biol 2000; 10:1147-50.
109. Gebauer F, Xu W, Cooper GM et al. Translational control by cytoplasmic polyadenylation of c-mos mRNA is necessary for oocyte maturation in the mouse. EMBO J 1994; 13:5712.
110. Stutz A, Conne B, Huarte J et al. Masking, unmasking, and regulated polyadenylation cooperate in the translational control of a dormant mRNA in mouse oocytes. Genes Dev 1998; 12:2535-48.
111. Oh B, Hwang S, McLaughlin J et al. Timely translation during the mouse oocyte-to-embryo transition. Development 2000; 127:3795-803.

112. Tay J, Hodgman R, Richter JD. The control of cyclin B1 mRNA translation during mouse oocyte maturation. Dev Biol 2000; 221:1-9.
113. Tay J, Richter JD. Germ cell differentiation and synaptonemal complex formation are disrupted in CPEB knockout mice. Dev Cell 2001; 1:201-213.
114. Wu L, Wells D, Tay J et al. CPEB-mediated cytoplasmic polyadenylation and the regulation of experience-dependent translation of alpha-CaMKII mRNA at synapses. Neuron 1998; 21:1129-39.
115. Richter JD. Think globally, translate locally: what mitotic spindles and neuronal synapses have in common. Proc Natl Acad Sci USA 2001; 98:7069-71.
116. Jacobson A, Peltz SW. Interrelationships of the pathways of mRNA decay and translation in eukaryotic cells. Annu Rev Biochem 1996; 65:693-739.
117. Schwartz DC, Parker R. Interaction of mRNA translation and mRNA degradation in *Saccharomyces cerevisiae*. In: Sonenberg N, Hershey JWB, Mathews MB, eds. Translational Control of Gene Expression. Cold Spring Harbor: Cold Spring Harbor Laboratory Press, 2000:807-825.
118. Lykke-Andersen J. mRNA quality control: Marking the message for life or death. Curr Biol 2001; 11:R88-91.
119. Culbertson MR. RNA surveillance. Unforeseen consequences for gene expression, inherited genetic disorders and cancer. Trends Genet 1999; 15:74-80.
120. Hentze MW, Kulozik AE. A perfect message: RNA surveillance and nonsense-mediated decay. Cell 1999; 96:307-10.
121. Jacobson A, Peltz SW. Destabilization of nonsense-containing transcripts in *Saccharomyces cerevisiae*. In: Sonenberg N, Hershey JWB, Mathews MB, eds. Translational Control of Gene Expression. Cold Spring Harbor: Cold Spring Harbor Laboratory Press, 2000:827-847.
122. Maquat LE, Carmichael GG. Quality control of mRNA function. Cell 2001; 104:173-6.
123. Schwartz DC, Parker R. mRNA decapping in yeast requires dissociation of the cap binding protein, eukaryotic translation initiation factor 4E. Mol Cell Biol 2000; 20:7933-42.
124. von Der Haar T, Ball PD, McCarthy JE. Stabilization of eukaryotic initiation factor 4E binding to the mRNA 5'- Cap by domains of eIF4G. J Biol Chem 2000; 275:30551-5.
125. Caponigro G, Parker R. Multiple functions for the poly(A)-binding protein in mRNA decapping and deadenylation in yeast. Genes Dev 1995; 9:2421-32.
126. Coller JM, Gray NK, Wickens MP. mRNA stabilization by poly(A) binding protein is independent of poly(A) and requires translation. Genes Dev 1998; 12:3226-35.
127. Mitchell P, Tollervey D. Musing on the structural organization of the exosome complex. Nat Struct Biol 2000; 7:843-6.
128. Beelman CA, Stevens A, Caponigro G et al. An essential component of the decapping enzyme required for normal rates of mRNA turnover. Nature 1996; 382:642-6.
129. LaGrandeur TE, Parker R. Isolation and characterization of Dcp1p, the yeast mRNA decapping enzyme. Embo J 1998; 17:1487-96.
130. Dunckley T, Parker R. The DCP2 protein is required for mRNA decapping in *Saccharomyces cerevisiae* and contains a functional MutT motif. EMBO J 1999; 18:5411-22.
131. Dunckley T, Tucker M, Parker R. Two related proteins, Edc1p and Edc2p, stimulate mRNA decapping in Saccharomyces cerevisiae. Genetics 2001; 157:27-37.
132. Hsu CL, Stevens A. Yeast cells lacking 5'—>3' exoribonuclease 1 contain mRNA species that are poly(A) deficient and partially lack the 5' cap structure. Mol Cell Biol 1993; 13:4826-35.
133. Muhlrad D, Decker CJ, Parker R. Deadenylation of the unstable mRNA encoded by the yeast MFA2 gene leads to decapping followed by 5'—>3' digestion of the transcript. Genes Dev 1994; 8:855-66.
134. Mitchell P, Petfalski E, Shevchenko A et al. The exosome: a conserved eukaryotic RNA processing complex containing multiple 3'—>5' exoribonucleases. Cell 1997; 91:457-66.
135. Allmang C, Petfalski E, Podtelejnikov A et al. The yeast exosome and human PM-Scl are related complexes of 3' —> 5' exonucleases. Genes Dev 1999; 13:2148-58.
136. Tucker M, Valencia-Sanchez MA, Staples RR et al. The transcription factor associated Ccr4 and Caf1 proteins are components of the major cytoplasmic mRNA deadenylase in *Saccharomyces cerevisiae*. Cell 2001; 104:377-86.
137. Daugeron MC, Mauxion F, Seraphin B. The yeast POP2 gene encodes a nuclease involved in mRNA deadenylation. Nucleic Acids Res 2001; 29:2448-55.
138. Dehlin E, Wormington M, Korner CG et al. Cap-dependent deadenylation of mRNA. EMBO J 2000; 19:1079-86.
139. Korner CG, Wahle E. Poly(A) tail shortening by a mammalian poly(A)-specific 3'- exoribonuclease. J Biol Chem 1997; 272:10448-56.
140. Hatfield L, Beelman CA, Stevens A et al. Mutations in trans-acting factors affecting mRNA decapping in Saccharomyces cerevisiae. Mol Cell Biol 1996; 16:5830-8.
141. Boeck R, Lapeyre B, Brown CE et al. Capped mRNA degradation intermediates accumulate in the yeast spb8-2 mutant. Mol Cell Biol 1998; 18:5062-72.
142. Bouveret E, Rigaut G, Shevchenko A et al. A Sm-like protein complex that participates in mRNA degradation. EMBO J 2000; 19:1661-71.
143. Tharun S, He W, Mayes AE et al. Yeast Sm-like proteins function in mRNA decapping and decay. Nature 2000; 404:515-8.
144. Kambach C, Walke S, Young R et al. Crystal structures of two Sm protein complexes and their implications for the assembly of the spliceosomal snRNPs. Cell 1999; 96:375-87.
145. Fromont-Racine M, Mayes AE, Brunet-Simon A et al. Genome-wide protein interaction screens reveal functional networks involving Sm-like proteins. Yeast 2000; 17:95-110.
146. Gradi A, Imataka H, Svitkin YV et al. A novel functional human eukaryotic translation initiation factor 4G. Mol Cell Biol 1998; 18:334-42.

CHAPTER 13

Nonsense-Mediated mRNA Decay in Mammalian Cells:
From Pre-mRNA Processing to mRNA Translation and Degradation

Lynne E. Maquat

Abstract

Studies of nonsense-mediated messenger RNA decay (NMD) in mammalian cells have lent great insight into splicing-dependent modifications of mRNA and how these modifications recruit factors that are required for NMD. In demonstrating that NMD takes place on cap binding protein (CBP)80-bound mRNA, these studies have also provided the first evidence that CBP80-bound mRNA can be translated, modifying the previously held concept that capped mRNAs require a different cap binding protein, eukaryotic initiation factor (eIF)4E, in order to direct protein synthesis.

Studies of nonsense-mediated messenger RNA decay (NMD) in mammalian cells have lent great insight into splicing-dependent modifications of mRNA and how these modifications recruit factors that are required for NMD. In demonstrating that NMD takes place on cap binding protein (CBP)80-bound mRNA, these studies have also provided the first evidence that CBP80-bound mRNA can be translated, modifying the previously held concept that capped mRNAs require a different cap binding protein, eukaryotic initiation factor (eIF)4E, in order to direct protein synthesis.

Introduction

Nonsense-mediated mRNA decay (NMD), also called mRNA surveillance, is one of several post-transcriptional mechanisms that eukaryotic cells employ to control the quality of mRNA function (reviewed in ref. 1). NMD eliminates abnormal transcripts that prematurely terminate translation and, in so doing, eliminates the production of truncated proteins that could function in dominant-negative or otherwise deleterious ways (reviewed in refs. 2-10). While NMD in mammalian cells was discovered in studies of disease-associated nonsense codons, NMD likely evolved to degrade nonsense-containing transcripts that are a consequence of routine abnormalities in gene expression. For example, the inefficient or inaccurate removal of an intron during the process of pre-mRNA splicing often generates an intron-derived nonsense codon or a shift in the translational reading frame, both of which result in translation termination upstream of the normal termination codon. NMD also targets a number of naturally occurring substrates, including certain alternatively spliced mRNAs[11] and some but not all selenoprotein mRNAs.[12,13] The importance of NMD is illustrated by the finding that mouse embryos inactive in NMD are resorbed shortly after implantation.[14] Furthermore, blastocysts inactive in NMD isolated 3.5 days post-coitum undergo apoptosis in culture after a brief growth period.[14] The inviability of NMD-deficient mammalian cells probably reflects a combined failure to properly regulate natural substrates and eliminate error-generated transcripts. This chapter will focus on the mechanism of NMD in mammalian cells with the goal of presenting current concepts about how pre-mRNA splicing and subsequent metabolic events are critical for generating NMD-susceptible mRNP.

Translation Is Required for NMD

NMD utilizes translation as a means for nonsense codon recognition as evidenced by its sensitivity to inhibitors of translation initiation, elongation or termination. As examples, NMD is inhibited by (i) anisomycin, cycloheximide, emetine, puromycin and pactamycin, all of which bind and inactivate translationally active ribosomes,[15-17] (ii) poliovirus infection, which inactivates eukaryotic initiation factor (eIF) 4G,[17] (iii) a secondary structure in the 5'-untranslated region that either directly or via bound protein blocks access of the initiation codon to 40S ribosomal subunits,[18,19] and (iv) suppressor tRNAs, which direct incorporation of an amino acid at a nonsense codon.[18,20] The finding that NMD is also inhibited by translation reinitiation downstream of and in frame with a termination codon[21] indicates that NMD also requires events subsequent to translation termination that are precluded by reinitiation.

Pre-mRNA Splicing Is Generally Required for NMD

Insight into events required for NMD that occur prior to translation termination was obtained with the finding that NMD requires a splicing-generated exon-exon junction located more than 50-55 nucleotides downstream of a termination codon.[19,22-26] Consistent with this finding, naturally intronless transcripts are immune to NMD.[27] Furthermore, normal termination codons, which generally do not elicit NMD, usually reside within the last exon (i.e., are not followed by an exon-exon junction). Those normal termination codons that do not reside with the last exon generally reside less than 50-55 nucleotides upstream of the 3'-most exon-exon junction.[28] One exception to the >50-55 nucleotide rule is provided by T-cell receptor-β mRNA, which is susceptible to NMD even when the distance between a nonsense codon and a downstream exon-exon junction is less than 50-55 nucleotides.[23] Another exception is provided by uncharacterized *cis*-acting sequences within β-globin mRNA that do not comprise an exon-exon junctions but appear to function comparably to one.[25] Distances >1200 nucleotides between a nonsense codon and a downstream exon-exon junction are capable of supporting NMD,[27,29] suggesting that NMD in mammalian cells is not limited to a distance ≤~200 nucleotides between a termination codon and a downstream destabilizing element as is the case in *Saccharomyces cerevisiae*.[30]

The mechanistic connection between pre-mRNA splicing and NMD is mediated by one or more components of the ~335-kD complex of proteins deposited 20-24 nucleotides upstream of newly synthesized exon-exon junctions as a consequence of splicing.[31,32,33] Immunoprecipitation experiments using HeLa-cell nuclear extract active in splicing revealed that this complex consists of at least five proteins: RNPS1, SRm160, DEK, REF and Y14.[32] Immunoprecipitation experiments using mammalian cells expressing FLAG-tagged constituents of the complex confirmed the presence of all but DEK (see ref. 34, a result corroborated using anti-DEK antibody (Y. Ishigaki, X. Li, and L.E.M., unpublished data; see below)). RNPS1, SRm160 and DEK function in pre-mRNA splicing.[35-37] REF, also called Aly, and Y14 are thought to function in mRNA export, in part because they interact with TAP, which contacts the nuclear pore complex.[33,38-44] Microinjections of pre-mRNAs into *Xenopus laevis* oocyte nuclei demonstrated that the exon-exon junction complex functions in mRNA export by recruiting TAP and its binding partner p15, both of which demonstrably facilitate mRNA export.[33] The complex also recruits Upf2, Upf3 (also called Upf3a) and Upf3X (also called Upf3b) but not Upf1.[33] All four Upf proteins are known to function in NMD.[45-48]

Studies using mammalian cells indicate that RNPS1 and, possibly, Y14 provide the link between the exon-exon junction complex and NMD. First, tethering RNPS1 and, to a lesser extent, Y14 to the 3' untranslated region of β-globin mRNA recapitulated the function of a splicing-generated exon-exon junction in NMD, as does tethering NMD factors Upf1, Upf2, Upf3 or Upf3X.[34,47] Second, FLAG-RNPS1 transiently expressed in HEK293 cells co-immunoprecipitated with Upf1, Upf2, Upf3 and Upf3X.[34] Third, Upf3 and Upf3X associate with Y14, REF and TAP of the exon-exon junction complex.[49] Upf1 is detected in the cytoplasm, Upf2 is perinuclear, and Upf3 and Upf3X are primarily nuclear but shuttle to the cytoplasm.[47,48,50] These data indicate that Upf3 and Upf3X, which to date are functionally indistinguishable, join the exon-exon junction complex in the nucleus, and Upf2 and Upf1 associate with the complex after export to the cytoplasm. While the functions of Upf2, Upf3 and Upf3X are unknown, Upf1 is a group 1 RNA helicase that likely associates with mRNA as a consequence of its role in translation termination.[45,51,52]

NMD Takes Place on CBP80-Bound mRNA

Remarkably, NMD takes place on mRNA bound by cap binding protein CBP80.[53] The heteromeric CBP80/CBP20 complex, which is primarily nuclear but shuttles to the cytoplasm, binds the cap structure of nascent pre-mRNAs shortly after transcription initiation and functions in nuclear processes such as pre-mRNA splicing. (see refs. 54-57; F. Lejeune, Y. Ishigaki, X. Li, and L.E.M., unpublished data). Eukaroytic initiation factor eIF4E, which constitutes the major cytoplasmic cap binding protein but also shuttles to the nucleus, is thought to replace CBP80/CBP20 at an undefined point after mRNA export to the cytoplasm and supports the bulk of cellular translation.[58-60]

Immunopurifications using either anti-CBP80 or anti-eIF4E antibody and RNP from mammalian cells transiently transfected with plasmids producing either nonsense-free or nonsense-containing mRNA revealed that nonsense-containing CBP80-bound mRNA is reduced in abundance to an extent comparable to nonsense-containing eIF4E-bound mRNA.[53] Considering that CBP80-bound mRNA is believed to be a precursor to eIF4E-bound mRNA, these data indicate that NMD takes place on CBP80-bound mRNA. Consistent with this conclusion, both nonsense-free and nonsense-containing CBP80-bound mRNAs, but not their eIF4E-bound counterparts, immunopurified with Upf2 and Upf3 (see ref. 53; Upf3X co-migrated with antibody, precluding its detection). PABP2, which constitutes the primarily nuclear but shuttling poly(A) binding protein, and CPB20 also immunopurified with anti-CBP80 but not anti-eIF4E antibody, providing additional evidence for the specificity of the immunopurifications.[53] Notably, Upf1 was not detected using either antibody,[53] consistent with the finding that anti-Upf1 antibody failed to immunoprecipitate spliced mRNA that derived from pre-mRNA injected into the nuclei of *Xenopus laevis* oocytes.[33] Smg-1, a phosphatidylinositol kinase-related kinase thought to be critical for NMD via its role in phosphorylating Upf1,[52,61,62,63] was also not detected (Y. Ishigaki, X. Li, and L.E.M., unpublished data). Like anti-eIF4E antibody, anti-CBP80 antibody immunopurified eIF4G, consistent with the recent demonstration that CBP80 and eIF4G interact in mammalian cells.[64] Both antibodies also immunopurified ribosomal protein L10, consistent with the requirement of NMD for translation.[53] Confirmation that CBP80-bound mRNA is a template for translation derived from the findings that cycloheximide or suppressor tRNA increased the level of nonsense-containing CBP80-bound mRNA.[53]

Sub-Cellular Location of NMD

Depending on the particular mRNA, NMD in mammalian cells takes place either in association with nuclei (see refs. 17-19, 24, 25) or in the cytoplasm (see refs. 26, 65). Cytoplasmic NMD involves cytoplasmic ribosomes.[65] Whether or not nucleus-associated NMD involves cytoplasmic ribosomes that function during the process of mRNA export or a translation-like mechanism in the nucleoplasm has yet to be resolved. Regardless, both nucleus-associated and cytoplasmic NMD involve CBP80-bound mRNAs.[53]

Additional insight into the mechanism of NMD was obtained in studies that assayed for components of the exon-exon junction complex and Upf proteins in nuclear and cytoplasmic fractions of mammalian cells. Of the five components of the exon-exon junction complex, only DEK did not immunopurify with nucleus-associated CBP80-bound mRNA, and none immunopurified with nucleus-associated eIF4E-bound mRNA (F. Lejeune, Y. Ishigaki, X. Li, and L.E.M., unpublished data). The failure to detect DEK as a component of mammalian-cell RNP also typified immunoprecipitations using mammalian cells expressing FLAG-tagged DEK,[34] suggesting that DEK may not be a stable component of the exon-exon junction complex in mammalian cells. In contrast, anti-DEK antibody did react with mRNP spliced in *Xenopus laevis* oocytes.[33] TAP (F. Lejeune, Y. Ishigaki and L.E.M., unpublished data) Upf2 and Upf3[53] also immunopurified with nucleus-associated CBP80-bound mRNA in mammalian cells, supporting the idea that the exon-exon junction complex serves as a platform for both mRNA transport and NMD factors.[33,34] There is no evidence for SRm160 shuttling using a heterokaryon assay (Y. Ishigaki and L.E.M., unpublished data). Nevertheless, SRm160 and RNPS1, REF, Y14, TAP and Upf3, which do shuttle, all immunopurified with cytoplasmic CBP80-bound mRNA (F. Lejeune, Y. Ishigaki and L.E.M., unpublished data).[34,39,41,42,47,48,66] The finding that SRm160 and Y14 but not RNPS1, REF, TAP or Upf3 were found associated with spliced mRNA in both nuclear and cytoplasmic fractions of *Xenopus laevis* ooctyes[33,66] may reflect differences between mammalian cells and *Xenopus* oocytes. Alternatively, the differences may reflect differences in experimental conditions. Upf2 immunopurified with cytoplasmic CBP80-bound mRNA (F. Lejeune, Y. Ishigaki, X. Li, and L.E.M., unpublished data), consistent with its perinuclear localization, direct interaction with Upf3, and function in NMD.[46-48] Upf2 was also detected on spliced mRNA in the cytoplasm of *Xenopus laevis* oocytes.[33]

mRNP Dynamics Leading Up to NMD: A Model

According to the current model, NMD in mammalian cells takes place during a "pioneer" round of translation, which precedes certain mRNA metabolic events that include (i) the replacement of CBP80 and CBP20 by eIF4E, (ii) the dissociation of Upf2, Upf3 and components of the splicing-dependent complex from upstream of exon-exon junctions, and (iii) the bulk of mRNA translation (Fig. 1). CBP80/CBP20, PABP2, RNPS1, REF, TAP, Y14 and Upf3 associate with mRNA in the nuclear fraction; all but PABP2,which has not been assayed, are associated with mRNA in the cytoplasm. The shuttling of CBP80/CBP20 and Upf3 in association with mRNA is consistent with the findings that the NMD of some mRNAs takes place during translation by cytoplasmic ribosomes.[26,65] Those mRNAs that are subject to nucleus-associated NMD could be degraded in the nucleoplasm, as proposed in a recent study reporting the existence of nuclear translation.[67] Alternatively, nucleus-associated NMD could involve cytoplasmic ribosomes and take place during or immediately after mRNP transport across the nuclear pore complex but prior to release from nuclei into the cytoplasm (reviewed in ref 3).

Considering that there are an estimated 30,000 mammalian genes (reviewed in ref 68), not all of which are expressed at the same time, and an estimated 75,000 transcripts per cell in the process of elongation by the combination of RNA polymerases II and III,[69] the number of CBP80-bound nascent pre-mRNAs at any given time must be on the order of 10^4. Recent estimates indicate that there are an average of four to seven[70] or seven to eight exon-exon junctions per mRNA.[71] Therefore, there are probably no more than ~5 x 10^5 nascent splicing-generated exon-exon junctions per cell, which is on the order of or slightly more than the ~1 x 10^5 molecules of Upf3 and Upf3X.[72] If Upf3 and Upf3X have the potential to bind every exon-exon junction, which is a reasonable hypothesis, then there probably is not enough Upf3 and Upf3X for every junction. This shortfall may at least in part explain why NMD is less than 100% efficient: for example, only 80-85% of nonsense-containing β-globin or triose phosphate isomerase mRNAs is generally subject to NMD.[19,24,25] Nevertheless, the levels of Upf3 and Upf3X appear to be sufficient to support the idea that the majority of newly synthesized exon-exon junctions are bound by either Upf3 or Upf3X.

During the process of mRNA export, exon-exon junction-bound Upf3/3X is thought to recruit cytoplasmic Upf2, possibly near the nuclear periphery, where Upf2 appears to concentrate.[47,48] Considering that Upf2 is approximately 10-fold more abundant than hUpf3/3X,[72] its recruitment is not likely to limit NMD.

A role for Upf1 in translation termination, as suggested by studies of its orthologue in *Saccharomyces cerevisiae*,[73] is consistent with its abundance of ~4 x 10^6 molecules per mammalian cell (i.e., ~0.35 molecules per ribosome). This finding, plus data demonstrating that a dominant-negative Upf1 mutation impairs NMD when Upf2, Upf3 or Upf3X but not Upf1 is tethered downstream of a termination codon, indicate that hUpf1 is the last of the Upf proteins to join the complex required for NMD.[47] Since active translation is required for tethered Upf1 to elicit NMD, and since Upf1 is not detected on CBP80-bound mRNA,[33,53] Upf1 probably functions in NMD as part of a post-termination complex that includes other translation termination factors.

All of these data explain how pre-mRNA splicing contributes to NMD (Fig. 2). In the nucleus, pre-mRNA splicing deposits a ~335-kD complex of proteins consisting of RNPS1, SRm160, REF, Y14 and, possibly, DEK (detected in HeLa-cell extract but not in vivo) ~20-24 nucleotides upstream of the exon-exon junctions of CPB80-bound mRNA (see refs. 32, 34, 53; F Lejeune, Y. Ishigaki, X. Li, and L.E.M., unpublished data). This mRNA is also bound by PABP2.[53] The export factor TAP, probably in association with p15,[33,40,41] and the NMD factor Upf3 or Upf3X (see refs. 34, 53; F. Lejeune, Y. Ishigaki, X. Li, and L.E.M., unpublished data) are then recruited. All factors (except possibly DEK, which if present does not shuttle in heterokaryon assays) are exported with CBP80/CBP20-bound mRNA into the cytoplasm. Whether or not PABP2 is detected on CBP80-bound mRNA in the cytoplasmic has yet to be determined. Next, Upf2 is recruited either during or immediately after export based on its perinuclear location and its presence on newly synthesized mRNP that is competent to serve as a template for the first or "pioneer" round of translation. It is currently unclear if RNPS1, SRm160, REF, Y14, TAP and p15 remain associated with or are removed from mRNA prior to the pioneer round of translation.

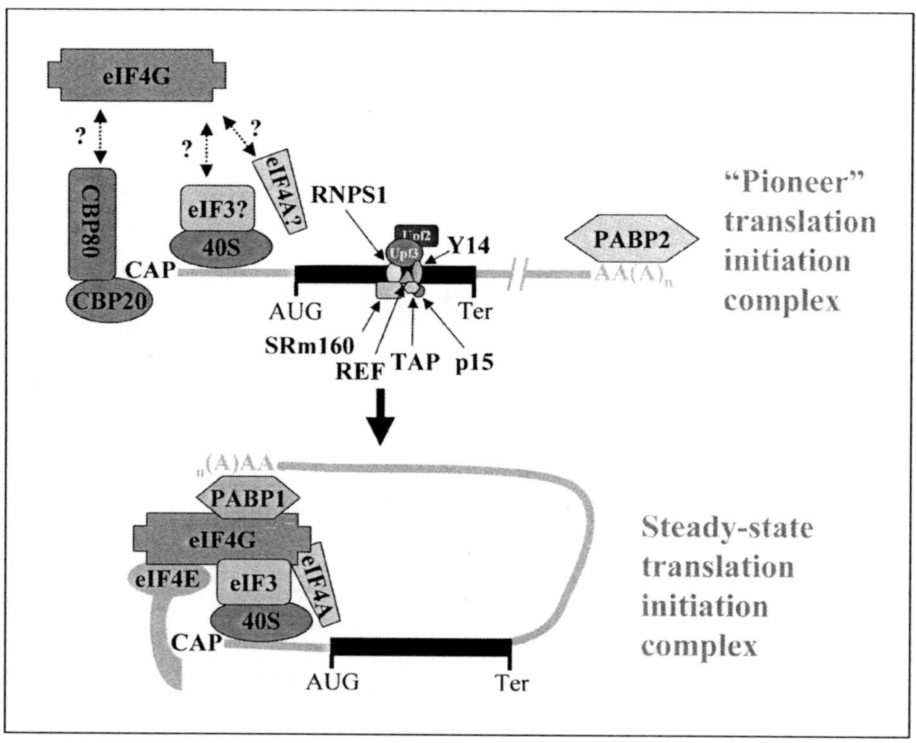

Figure 1. Model for the pioneer translation initiation complex, shown relative to the steady-state translation initiation complex. The pioneer translation initiation complex is the substrate for NMD as well as the first round of translation for both nonsense-free and nonsense-containing mRNAs. It contains: (i) CBP80 and CBP20 bound to the 5' mRNA cap structure; (ii) Upf2 and Upf3 bound ~ 20-24 nucleotides upstream of exon-exon junctions; (iii) RNPS1, SRm160, REF, Y14, TAP and possibly p15 bound ~ 20-24 nucleotides upstream of exon-exon junctions, (iv) PABP2 bound to the poly(A) tail, and (v) eIF4G and, probably, other translation initiation factors that have yet to be identified and that also characterize the steady-state translation initiation complex. At least for mRNAs subject to cytoplasmic NMD, the pioneer translation initiation complex functions in association with cytoplasmic ribosomes. It is uncertain if eIF4G is a multipurpose adaptor as it is in the steady-state translation initiation complex.[59,82,83] Upf1 protein is thought to associate with the pioneer translation initiation complex after translation initiation, probably as a consequence of translation termination (see Fig. 2). The bulk of cellular translation involves the steady-state translation initiation complex, which derives from remodeling of the pioneer translation initiation complex. AUG, initiation codon; Ter, termination codon.

For mRNAs subject to nucleus-associated NMD, the pioneer round may take place either during mRNA export, at a point when mRNA co-purifies with nuclei and could have access to cytoplasmic ribosomes, or in the nucleoplasm prior to export. NMD in the nucleoplasm would necessitate that those NMD factors detected only in cytoplasm also be present in nuclei. It would also require that translation or a related mechanism take place in nuclei, for which there is now some evidence.[67,74] For mRNAs subject to cytoplasmic NMD, the pioneer round of translation takes place in the cytoplasm. Regardless of the cellular site of NMD, if translation terminates more than 50-55 nucleotides upstream of an exon-exon junction that is marked by the splicing-dependent complex, then NMD will occur. In this case, Upf1, which is thought to function in translation termination, either individually or with other components of the translation termination complex,[73] is proposed to elicit NMD by interacting with the exon-exon junction complex containing Upf2 and Upf3. If translation terminates less than 50-55 nucleotides upstream of the 3'-most exon-exon

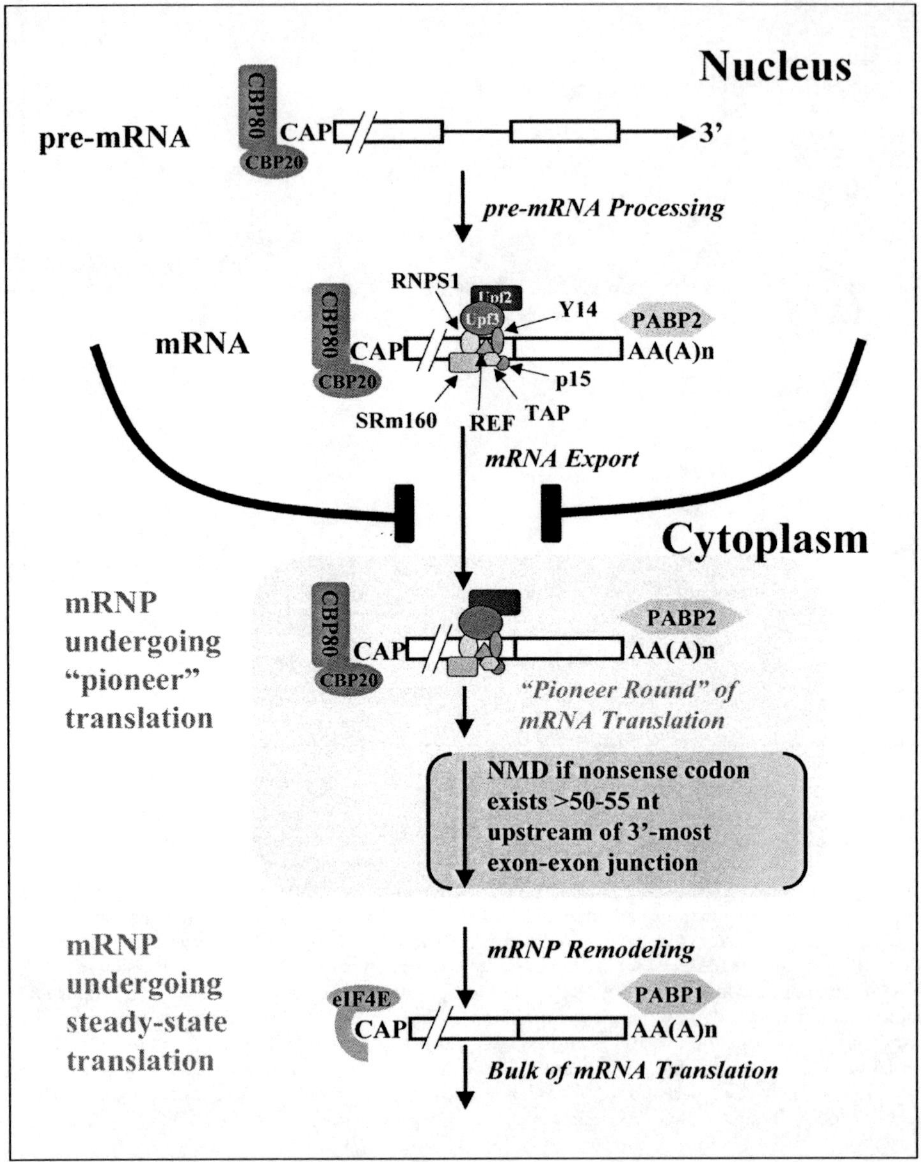

Figure 2. Model for the dynamics of mRNP structure in the nucleus and cytoplasm of mammalian cells. See text for description. Notably, the presence of PABP2 on CBP80-bound mRNA in the cytoplasm has yet to be tested. Proteins labeled with question marks may have been removed from mRNA by the time of the "pioneer" round of translation. If they are, then additional steps of mRNP remodeling occur before the pioneer round of translation. The single exon-exon junction exemplified is likely to typify all splicing-generated exon-exon junctions, regardless of their position relative to the mRNA 3' end.

junction or downstream of the junction, then translating ribosomes are thought to remove all exon-exon junction complexes and render the mRNA immune to NMD. Since NMD is less than 100% efficient (i.e., some nonsense-containing mRNAs escape NMD), it is possible that (i) not all junctions are marked by a functional complex, (ii) not all termination complexes are competent to elicit NMD, or (iii) there are translation-independent mechanisms that eliminate exon-exon junction complexes. Since neither components of the exon-exon junction complex nor factors recruited by the complex are detected on eIF4E-bound mRNA, eIF4E-bound mRNA must be essentially immune to NMD (see ref. 53; F. Lejeune, Y. Ishigaki, X. Li, and L.E.M., unpublished data).

Deadenylation-Independent NMD in Mammalian Cells?

The polarity of NMD has been established only for *Saccharomyces cerevisiae*, where NMD proceeds by deadenylation-independent decapping followed by 5' to-3 'degradation of the mRNA body.[75,76] Considering that constitutive mRNA decay in mammalian cells is now known to involve deadenylation-dependent 3'-to-5' exonucleolytic activity (M. Kiledjian, personal communication), which is surprisingly distinct from the deadenylation-dependent decapping and 5'-to-3' exonucleolytic activity that typifies constitutive mRNA decay in yeast,[75,77] predictions about NMD in mammalian cells can no longer legitimately be made from the NMD pathway in yeast.

To date, the sole data pertaining to the polarity of NMD in mammalian cells derive from studies of the decay intermediates of nonsense-containing human β-globin mRNAs detected in mice transgenic for one of several human β°-thalassemic alleles [78-80] or cultured mouse erythroleukemia cells that stably express a human β°-thalassemic allele (Y. Wang, S. Sekularac, J. Zhang and L.E.M., unpublished data). These intermediates maintain the poly(A) tail of full-length mRNA but are missing more-or-less specific numbers of nucleotides from the mRNA 5' end. Surprisingly, these intermediates also contain a structure at their 5' ends that is indistinguishable from the cap at the 5' end of full-length mRNA.[80] Whether these intermediates are formed as a consequence of decapping followed by 5'-to-3' exonucleolytic activity, as would be predicted from studies of yeast, or endonucleolytic activities has yet to be clarified. Regardless, it is unclear if generalizations about the mechanism of NMD can be made from these intermediates since they are currently the only decay intermediates detected for any nonsense-containing mRNA in mammalian cells.

Conclusions

NMD is a quality control mechanism used by mammalian cells to limit the longevity of mRNAs that prematurely terminate translation and encode potentially harmful proteins. In order to be a substrate for NMD, mRNA must generally be spliced from pre-mRNA so as to acquire a complex of proteins ~20-24 nucleotides upstream of exon-exon junctions that, in turn, recruits NMD factors as well as mRNA transport factors. NMD targets newly synthesized mRNA that is bound by CBP80, offering the first evidence that CBP80-bound mRNA can be translated. To date, the only known NMD factors are: (i) Upf1, which is a group1 RNA helicase that associates with mRNA as a consequence of translation termination, (ii) Upf2, Upf3 and Upf3X, which are recruited to mRNA by one or more components of the splicing-dependent exon-exon junction complex, and (iii) Smg-1, which is the phosphoinositol kinase-related kinase that phosphorylates Upf1. Taking cues from *C. elegans* (see refs. 61, 81), there are likely to be additional *trans*-acting factors required for mammalian-cell NMD.

Future aims include identifying and characterizing all of the *trans*-acting factors critical for NMD and determining how they orchestrate NMD by interfacing with proteins that comprise the splicing-dependent modification of mRNA and proteins that function in mRNA translation. Virtually nothing is known about the polarity of NMD in mammalian cells. Since it has not been possible to engineer mRNAs with built-in barriers to nucleolytic activities, as has been done for yeast,[75,77] insight into the polarity of mammalian-cell NMD may have to await the establishment of NMD in vitro.

Acknowledgements

I thank Melissa Moore and Jens-Lykke Andersen for communicating unpublished results at the time of writing, Linda McKendrick and Fabrice Lejeune for comments on the manuscript, and Yasuhito Ishigaki and Fabrice Lejeune for generating figures. This work was supported by Public Health Service Grants DK 33938 and GM 59614 from the National Institutes of Health.

References

1. Maquat LE, Carmichael GG. Quality control of mRNA function. Cell 2001; 104:173.
2. Maquat LE. When cells stop making sense: effects of nonsense codons on RNA metabolism in vertebrate cells. RNA 1995; 1:453.
3. Maquat LE. Nonsense-mediated RNA decay in mammalian cells: A splicing-dependent means to down-regulate the levels of mRNAs that prematurely terminate translation. In: Sonenberg N, Hershey JWB, Mathews MB, eds. Translational Control of Gene Expression. Cold Spring Harbor: Cold Spring Harbor Press, 2000:849.
4. Jacobson A, Peltz SW. Interrelationships of the pathways of mRNA decay and translation in eukaryotic cells. Annu Rev Biochem 1996; 65: 693.
5. Jacobson A, Peltz SW. Destabilization of nonsense-containing transcripts in *Saccharomyces cerevisiae*. In: Sonenberg N, Hershey JWB, Mathews MB, eds. Translational Control of Gene Expression. Cold Spring Harbor: Cold Spring Harbor Press, 2000:827.
6. Li S, Wilkinson MF. Nonsense surveillance in lymphocytes? Immunity 1998; 8:135.
7. Culbertson, M. RNA surveillance. Unforeseen consequences for gene expression, inherited genetic disorder. Trends Genet 1999; 15(2):74-80.
8. Frischmeyer PA, Dietz HC. Nonsense-mediated mRNA decay in health and disease. Hum Molec Genet 1999; 8:1893.
9. Hentze MW, Kulozik AE. A perfect message: RNA surveillance and nonsense-mediated decay. Cell 1999; 96:307.
10. Hilleren P, Parker R. Mechanisms of mRNA surveillance in eukaryotes. Annu Rev Genet 1999; 33:229-60.
11. Morrison M, Harris KS, Roth M. smg mutants affect the expression of alternatively spliced SR protein mRNAs in *Caenorhabditis elegans*. Proc Natl Acad Sci, USA 1997; 94:9782.
12. Moriarty PM, Reddy CC, Maquat LE. Selenium deficiency reduces the abundance of mRNA for Se-dependent glutathione peroxidase 1 by a UGA-dependent mechanism likely to be nonsense codon-mediated decay of cytoplasmic mRNA. Mol Cell Biol 1998; 18:2932.
13. Sun X, Li X, Moriarty PM et al. Nonsense-mediated decay of mRNA for the selenoprotein phospholipid hydroperoxide glutathione peroxidase is detectable in cultured cells but masked or inhibited in rat tissues. Mol Biol Cell 2001; 12:1009-1017.
14. Medghalchi S, Frischmeyer PA, Mendell JT et al. Rent1, a trans-effector of nonsense- mediated mRNA decay, is essential for mammalian embryonic viability. Hum Mol Genet 2001; 10:99-105.
15. Qian L, Theodor L, Carter M et al. T cell receptor-β mRNA splicing: regulation of unusual splicing intermediates. Mol Cell Biol 1993 13:1686.
16. Menon KP, Neufeld EF. Evidence for degradation of mRNA encoding alpha-L- iduronidase in Hurler fibroblasts with premature termination alleles. Cell Mol Biol 1994; 40:999.
17. Carter MS, Doskow J, Morris P et al. A regulatory mechanism that detects premature nonsense codons in T-cell receptor transcripts in vivo is reversed by protein synthesis inhibitors in vitro. J Biol Chem 1995; 270:28995.
18. Belgrader P, Cheng J, Maquat LE. Evidence to implicate translation by ribosomes in the mechanism by which nonsense codons reduce the nuclear level of human triosephosphate isomerase mRNA. Proc Natl Acad Sci, USA 1993; 90:482.
19. Thermann R, Neu-Yilik G, Deters A et al. Binary specification of nonsense codons by splicing and cytoplasmic translation. EMBO J 1998; 17:3484-3494.
20. Li S, Leonard D, Wilkinson MF. T cell receptor (TCR) mini-gene mRNA expression regulated by nonsense codons: a nuclear-associated translation-like mechanism. J Exp Med 1997; 185:985.
21. Zhang J, Maquat LE. Evidence that translation reinitiation abrogates nonsense-mediated mRNA decay in mammalian cells. EMBO J 1997; 16:826-833.
22. Cheng J, Belgrader P, Zhou X et al. Introns are cis effectors of the nonsense-codon- mediated reduction in nuclear mRNA abundance. Mol Cell Biol 1994; 14:6317.
23. Carter MS, Li S, Wilkinson MF. A splicing-dependent regulatory mechanism that detects translation signals. EMBO J 1996; 15:5965.
24. Zhang J, Sun X, Qian Y et al. At least one intron is required for the nonsense-mediated decay of triosephosphate isomerase mRNA: a possible link between nuclear splicing and cytoplasmic translation. Mol Cell Biol 1998a; 18:5272-5283.
25. Zhang J, Sun X, Qian Y et al. Intron function in the nonsense-mediated decay of β-globin mRNA: indications that pre-mRNA splicing in the nucleus can influence mRNA translation in the cytoplasm. RNA 1998b; 4:801-815.
26. Sun X, Moriarty PM, Maquat LE. Nonsense-mediated decay of glutathione peroxidase 1 mRNA in the cytoplasm depends on intron position. EMBO J 2000; 19:4734-4744.
27. Maquat LE, Li X. Mammalian heat shock p70 and histone H4 transcripts, which derive from naturally intron-less genes, are immune to nonsense-mediated decay. RNA 2001; 7:445.
28. Nagy E, Maquat LE. A rule for termination-codon position within intron-containing genes: when nonsense affects RNA abundance. Trends Biochem Sci 1998; 23:198.
29. Neu-Yilik G, Gehring NH, Thermann R et al. Splicing and 3' end formation in the definition of nonsense-mediated decay-component human β-globin mRNPs. EMBO J 2001; 20:532.
30. Ruiz-Echevarría MJ, González CI, Peltz SW. Identifying the right stop: determining how the surveillance complex recognizes and degrades an aberrant mRNA. EMBO J 1998; 17:575-589.

31. Le Hir H, Moore MJ, Maquat LE. Pre-mRNA splicing alters mRNP composition: evidence for stable association of proteins at exon-exon junctions. Genes Dev 2000a; 14:1098-1108.
32. Le Hir H, Izaurralde E, Maquat LE et al. The spliceosome deposits multiple proteins 20-24 nucleotides upstream of mRNA exon-exon junctions. EMBO J 2000b; 19: 6860-6869.
33. Le Hir H, Gatfield D, Izaurralde E et al. The exon-exon junction complex provides a binding platform for factors involved in mRNA export and NMD. EMBO J 2001; 20:4987-4997.
34. Lykke-Andersen J, Shu MD, Steitz JA. Communication of the Position of Exon-Exon Junctions to the mRNA Surveillance Machinery by the Protein RNPS1. Science 2001; 293:1836-1839.
35. Blencowe BJ, Issner R, Nickerson JA et al. A coactivator of pre-mRNA splicing. Genes & Dev 1998; 12, 996-1009.
36. Mayeda A, Badolato J, Kobayashi R et al. Purification and characterization of human RNPS1: a general activator of pre-mRNA splicing. EMBO J 1999; 18:4560-4570.
37. McGarvey T, Rosonina E, McCracken S et al. The acute myeloid leukemia-associated protein DEK forms a splicing-dependent interaction with exon-exon product complexes. J Cell Biol 2000; 150:309-320.
38. Lou MJ, Reed R. Splicing is required for rapid and efficient mRNA export in metazoans. Proc Natl Acad Sci, USA 1999; 96:14937-14942.
39. Zhou Z, Luo JJ, Sträßer K et al. The protein Aly links pre-messenger-RNA splicing to nuclear export in metazoans. Nature 2000; 407:401-405.
40. Stutz F, Bachi A, Doerks T et al. REF, an evolutionary conserved family of hnRNP-like proteins, interactions with TAP/Mex67p and participates in mRNA nuclear export. RNA 2000; 6:638-650.
41. Rodrigues JP, Rode M, Gatfield D et al. REF proteins mediate the export of spliced and unspliced mRNAs from the nucleus. Proc Natl Acad Sci, USA 2001; 98:1030-1035.
42. Katahira J, Sträßer K, Podtelejnikov A et al. The Mex67p-mediated nuclear mRNA export pathway is conserved from yeast to human. EMBO J 1999; 18:2593-2609.
43. Bachi A, Braun IC, Rodrigues JP et al. The C-terminal domain of TAP interacts with the nuclear complex and promotes export of specific CTE-bearing RNA substrates. RNA 2000; 6:136-158.
44. Kataoka N, Yong J, Kim VN et al. Pre-mRNA splicing imprints mRNA in the nucleus with a novel RNA-binding protein that persists in the cytoplasm. Mol Cell 2000; 6:673-682.
45. Sun X, Perlick HA, Dietz HC et al. A mutated human homologue of yeast Upf1 protein has a dominant-negative effect on the decay of nonsense-containing mRNAs in mammalian cells. Proc Natl Acad Sci, USA 1998; 95:10009-10014.
46. Mendell JT, Medghalchi SM, Lake RG et al. Novel Upf2p orthologues suggest a functional link between translation initiation and nonsense surveillance complexes. Mol Cell Biol 2000; 20:8944-8957.
47. Lykke-Andersen J, Shu MD, Steitz JA. Human Upf proteins target an mRNA for nonsense-mediated decay when bound downstream of a termination codon. Cell 2000; 103:1121-1131.
48. Serin G, Gersappe A, Black JD et al. Identification and characterization of human orthologues to *S. cerevisiae* Upf2 protein and *S. cerevisiae* Upf3 protein (*C. elegans* SMG-4). Mol Cell Biol 2001; 21:209-223.
49. Kim VN, Kataoka N, Dreyfuss G. Role of the Nonsense–Mediated Decay Factor hUpf3 in the Splicing-Dependent Exon-Exon Junction Complex. Science 2001; 293:1832-1836.
50. Applequist SE, Selg M, Raman C et al. Cloning and characterization of HUPF1, a human homolog of the *Saccharomyces cerevisiae* nonsense of mRNA-reducing UPF1 protein. Nucleic Acids Res 1997; 25:814-821.
51. Bhattacharya A, Czaplinski K, Trifillis P et al. Characterization of the biochemical properties of human Upf1 gene product that is involved in nonsense-mediated mRNA decay. RNA 2000; 6:1226-1235.
52. Pal M, Ishigaki Y, Nagy E et al. Evidence that phosphorylation of human Upf1 protein varies with intracellular location and is mediated by a wortmannin-sensitive and rapamycin-sensitive PI 3-kinase-related kinase signaling pathway. RNA 2001; 5:5-15.
53. Ishigaki Y, Li X, Serin G et al. Evidence for a pioneer round of translation: mRNAs subject to nonsense-mediated decay in mammalian cells are bound by CBP80 and CBP20. Cell 2001; 106:607-617.
54. Izaurralde E, Lewis J, McGuigan C et al. A nuclear cap binding protein complex involved in pre-mRNA splicing. Cell 1994; 78:657-668.
55. Lewis JD, Izaurralde E. The role of the cap structure in RNA processing and nuclear export. Eur J Biochem 1997; 247:461-469.
56. Visa N, Izaurralde E, Ferreira J et al. A nuclear cap-binding complex binds Balbiani ring pre-mRNA cotranscriptionally and accompanies the ribonucleoprotein particle during nuclear export. J Cell Biol 1996; 133:5-14.
57. Shen EC, Stage-Zimmermann T, Chui P et al. The yeast mRNA-binding protein Npl3p interacts with the cap-binding complex. J Biol Chem 2000; 275:23718-23724.
58. Lejbkowicz F, Goyer C, Darveau A et al. A fraction of the mRNA 5' cap-binding protein, eukaryotic initiation factor 4E, localizes to the nucleus. Proc Natl Acad Sci, USA 1992; 89:9612-9616.
59. Gingras AC, Raught B, Sonenberg N. eIF4 initiation factors: effectors of mRNA recruitment to ribosomes and regulators of translation. Ann Rev Biochem 1999; 68:913-963.
60. Dostie J, Lejbkowicz F, Sonenberg N. Nuclear eukaroyotic initiation factor 4E (eIF4E) colocalizes with splicing factors in speckles. J Cell Biol 2000; 148:239-245.
61. Page MF, Carr B, Anders KR et al. SMG-2 is a phosphorylated protein required for mRNA surveillance in *Caenorhabditis elegans* and related to Upf1p in yeast. Mol Cell Biol 1999; 19:5943-5951.

62. Denning G, Jamieson L, Maquat LE et al. Cloning of a novel phosphoinositide kinase-related kinase: characterization of the human SMG-1 RNA surveillance protein. J Biol Chem 2001; 276:22709-22714.
63. Yamashita A, Ohnishi T, Kashima I et al. Human SMG-1, a novel phosphatidylinositol 3-kinase-related protein kinase, associates with components of the mRNA surveillance complex and is involved in the regulation of nonsense-mediated mRNA decay. Genes & Dev 2001; 15:2215-2228.
64. McKendrick L, Thompson E, Ferreira J et al. Interaction of eukaryotic translation initiation factor 4G with the nuclear cap-binding complex provides a link between nuclear and cytoplasmic functions of the m(7) guanosine cap. Mol Cell Biol 2001; 21:3632-3641.
65. Rajavel KS, Neufeld EF. Nonsense-mediated decay of HexA mRNA. Mol Cell Biol 2001; 16:5512-5517.
66. Kim VN, Yong J, Kataoka N et al. The Y14 protein communicates to the cytoplasm the position of exon-exon junctions. EMBO J 2001; 20:2062-2068.
67. Iborra F, Jackson DA, Cook RP. Coupled transcription and translation within nuclei of mammalian cells. Science 2001; 293:1139-1142.
68. Claverie JM. What if there are only 30,000 human genes? Science 2001; 291:1257.
69. Jackson DA, Iborra FJ, Manders EMM et al. Numbers and organization of RNA polymerase, nascent transcripts, and transcription units in HeLa nuclei. Mol Biol Cell 1998; 9:1523.
70. Venter JC, Adams MD, Myers EW et al. The sequence of the human genome. Science 2001; 291:1304.
71. Lander ES, Linton LM, Birren B et al. Initial sequencing and analysis of the human genome. Nature 2001; 209:860-921.
72. Maquat LE, Serin G. Nonsense-mediated mRNA decay: Insights into mechanism from the cellular abundance of human Upf1, Upf2, Upf3 and Upf3X proteins. Cold Spring Harbor Symposia on Quantitative Biology: The Ribosome. Cold Spring Harbor: Cold Spring Harbor Press, 2001: 66:313-320.
73. Czaplinski K, Ruiz-Echevarría MJ, Paushkin SV et al. The surveillance complex interacts with the translation release factors to enhance termination and degrade aberrant mRNAs. Genes & Dev 1998; 12:1665-1677.
74. Mühlemann O, Mock-Casagrande CS, Wang J et al. Precursor RNAs harboring nonsense codons accumulate near the site of transcription. Mol Cell 2001; 8,33-44.
75. Muhlrad D, Parker R. Premature translational termination triggers mRNA decapping. Nature 1994; 370:578-581.
76. Hagan KW, Ruiz-Echevarría MJ, Vasudevan S et al. Characterization of *cis*-acting sequences and decay intermediates involved in nonsense-mediated mRNA turnover. Mol Cell Biol 1995; 15:809-823.
77. Decker CJ, Parker R. A turnover pathway for both stable and unstable mRNAs in yeast: evidence for a requirement for deadenylation. Genes & Dev 1993; 7:1632-1643.
78. Lim S, Mullins JJ, Chen CM et al. Novel metabolism of several β°-thalassemic β-globin mRNAs in the erythroid tissues of transgenic mice. EMBO J 1989; 8:2613-2619.
79. Lim SK, Sigmund CD, Gross KW et al. Nonsense codons in human β-globin mRNA result in the production of mRNA degradation products. Mol Cell Biol 1992a; 12:1149-1161.
80. Lim SK, Maquat LE. Human β-globin mRNAs that harbor a nonsense codon are degraded in murine erythroid tissues to intermediates that have a 5' cap-like structure. EMBO J 1992b; 11:3271-3278.
81. Cali B, Kuchma SL, Latham J et al. *Smg-7* is required for mRNA surveillance in *Caenorhabditis elegans*. Genetics 1999; 151;605-616.
82. Hentze MW. eIF4G: A multipurpose ribosome adapter? Science 1997; 275:500-501.
83. Sachs AB, Varani G. Eukaryotic translation initiation: there are (at least) two sides to every story. Nature Struct Biol 2000; 7:356-361.

CHAPTER 14

The Role of the Exosome and Ski Complexes in mRNA Turnover

Philip Mitchell

Abstract

The amount of protein synthesized from an individual mRNA reflects both its translation efficiency and stability. The translation and turnover of eukaryotic mRNAs are closely linked; alterations in translation can influence decay and vice versa, and the cap and poly(A) tail structures of an mRNA impart both increased stability and translation efficiency. Two general mRNA decay pathways have been analyzed in yeast. Initial deadenylation in both pathways is followed either by decapping and 5'->3' exonucleolytic degradation, or by continued exonucleolytic degradation in the 3'->5' direction. The latter pathway is dependent upon a complex of 3'->5' exoribonucleases, known as the exosome. By analogy with the structure of bacterial polynucleotide phosphorylase, the RNase PH homologues of the exosome complex may form a ring that encloses a central channel sufficiently large to accommodate the RNA substrate. Exosome function in mRNA decay requires additional factors, including the Ski complex and the putative GTPase Ski7p. Recent data suggest that Ski7p functionally couples the exosome and Ski complexes.

Introduction

Eukaryotic cells contain a large number of 3'->5' exoribonucleases, many of which are physically associated in a single multienzyme complex known as the exosome.[1,2] The exosome complex was first identified in the budding yeast *Saccharomyces cerevisiae* and this complex is the best characterized to date, although related complexes have been reported in other eukaryotes. The exosome of *S. cerevisiae* consists of 10 proteins (see Table 1), 9 of which either exhibit 3'->5' exoribonuclease activity in vitro or are predicted to belong to this class of enzyme on the basis of sequence homology. Immunofluorescence analyses indicate that the exosome is found in both the nuclear and the cytoplasmic compartments of the cell. An additional 3'->5' exoribonuclease, Rrp6p, is associated with a subfraction of the exosome and is specifically localized to the nucleus.[2]

Many of the functions of the exosome complex characterized thus far have an impact on messenger RNA (mRNA) processing and turnover. The exosome has a direct role in the degradation of cytoplasmic mRNAs.[3] In the nucleus, the exosome functions in the 3' end maturation of small stable RNAs, including small nuclear RNAs (snRNAs).[4] The snRNAs are essential components of the spliceosome machinery that removes intronic sequences from primary mRNA transcripts. Nonspliced mRNA precursors are generally poor substrates for the mRNA export pathway and are rapidly degraded in the nucleus. This pre-mRNA discard pathway is primarily dependent upon the activity of the exosome.[5] The exosome is also required for ribosome biogenesis and can therefore influence translation efficiency, which is itself closely linked to mRNA turnover (see below). The exosome was initially characterized as the exonucleolytic activity required for the 3' end maturation of 5.8S rRNA.[1] Strains mutant for components of the exosome accumulate aberrant 3' extended forms of 5.8S rRNA. At least in the case of the rrp6-1 mutant, these are incorporated into polysomes[6] but it is unclear to what extent the ribosomes are translationally active. Mutations in the exosome complex can also decrease the 60S:40S subunit ratio, causing the accumulation of half-mer polysomes that are stalled during translation initiation.[6]

Translation Mechanisms, edited by Jacques Lapointe and Léa Brakier-Gingras.

Table 1. *Components and cofactors of the yeast exosome complex. [a]Activities are listed that have been ascribed to recombinant proteins; activities predicted on the basis of sequence homology are given in brackets. [b]The percentage identity indicated in brackets for homologous human proteins is for the entire length of the protein.*

Protein	Gene	Activity[a]	Motifs/ Homologues	Deletion Phenotype	Human Homologue[b]	Comments
"Core"Complex						
Rrp4p	YHR069c	hydrolytic, distributive 3->5' exonuclease	S1 RBD	essential	hRrp4p (43%)	hRrp4p complements rrp4-1
Rrp40p	YOL142w	(hydrolytic, distributive 3->5' exonuclease)	S1 RBD	essential	hRrp40p (30%)	homologous to Rrp4p
Csl4p/Ski4p	YNL232w	?	S1 RBD	essential	hCsl4p (48%)	hCsl4p complements csl4-1
Rrp41p/Ski6p	YGR195w	phosphorolytic, processive 3->5' exonuclease	RNase PH	essential	hRrp41p (35%)	hRrp41p complements GAL::rrp41
Rrp42p	YDL111c	(phosphorolytic, processive 3->5' exonuclease)	RNase PH	essential		
Rrp43p	YCR035c	(phosphorolytic, processive 3->5' exonuclease)	RNase PH	essential		
Rrp45p	YDR280w	(phosphorolytic, processive 3->5' exonuclease)	RNase PH	essential	PM-Scl 75 (38%)	Human KIAA0116 and OIP2 also homologous
Rrp46p	YGR095c	(phosphorolytic, processive 3->5' exonuclease)	RNase PH	essential	hRrp46p (26%)	
Mtr3p	YGR158c	(phosphorolytic, processive)	RNase PH	essential		

continued on next page

Table 1. Continued

Protein	Gene	Activity[a]	Motifs/ Homologues	Deletion Phenotype	Human Homologue[b]	Comments
Additional Components						
Rrp44p/Dis3p	YOL021c	hydrolytic, processive 3->5' exonuclease	RNase R (vacB)	essential	hDis3p (45%)	hDis3p complements dis3-81
Rrp6p	YOR001w	hydrolytic 3->5' exonuclease	RNase D	ts-lethal	PM-Scl 100 (32%)	Component only of nuclear complex
Cofactors						
Mtr4p/Dob1p	YJL050W	(ATP-dependant helicase)	DEAD box	essential		
Ski2p	YLR398C	(ATP-dependant helicase)	DEVH box	nonessential	SKIV2L (38%)	
Ski3p	YPR189W		TPR repeat	nonessential	KIAA0372 (20%)	
Ski7p	YOR076c	(GTPase)		nonessential		High similarity to Hbs1p
Ski8p	YGL213C		WD repeat	nonessential		

Given the extremely rapid nature of most exonucleolytic RNA processing and degradation pathways in vivo (intermediates in such processes are rarely detected or are present at very low levels), it is perhaps surprising that immunoaffinity-purified exosome from yeast lysates exhibits only a weak exonucleolytic activity in vitro.[7] It has been proposed that exosome-mediated RNA processing and degradation is facilitated in vivo by transient interactions between the complex and additional co-factors.[1,3] Consistent with this model, all functions of the nuclear exosome characterized thus far require the activity of a putative RNA helicase, Mtr4p/Dob1p,[8,9] which is not detected in purified preparations of the exosome complex. Furthermore, the exosome-mediated 3'->5' cytoplasmic mRNA decay pathway is dependent on another putative RNA helicase, Ski2p, which is related to Mtr4p, together with the proteins Ski3p and Ski8p.[3] The Ski2p, Ski3p and Ski8p proteins are associated together in the "Ski" complex[10] that does not copurify with the exosome. In addition to the Ski complex, 3'->5' mRNA decay requires Ski7p,[11] a putative GTPase. In contrast to Mtr4p and the components of the Ski complex, Ski7p copurifies with the exosome[12] and may function to bridge the exosome and Ski complexes during mRNA decay.

Here I will review what is known about exosome function in mRNA metabolism and discuss data relating to the mechanism of exosome-mediated cytoplasmic mRNA decay.

The Complement of the Yeast Exosome

The core yeast exosome complex contains nine proteins, six of which are highly homologous to the phosphorolytic exoribonuclease RNase PH of *Escherichia coli*: Rrp41p/Ski6p, Rrp42p, Rrp43p, Rrp45p, Rrp46p and Mtr3p[13] (see Table 1). Moreover, recombinant Rrp41p from both yeast and *Arabidopsis thaliana* exhibits phosphorolytic exonuclease activity in vitro.[1,14] The other three proteins of the core structure (Rrp4p, Rrp40p and Csl4p/Ski4p) contain domains related to the RNA binding domain (RBD) of the *E. coli* ribosomal protein S1.[2,15]

Recombinant Rrp4p itself exhibits 3'->5' exoribonuclease activity in vitro and is therefore also a bona fide exoribonuclease.[1] Rrp40p, which is closely related to Rrp4p, is predicted to be a 3'->5' exoribonuclease. On the other hand, Csl4p lacks a number of residues that are conserved between Rrp4p, Rrp40p and their homologues from other eukaryotes and archaea and it is unclear whether it functions as an exonuclease or an RNA binding protein.

Two additional 3'->5' exoribonucleases, Rrp44p/Dis3p and Rrp6p, are associated with the exosome core structure. Rrp44p is homologous to RNase II of *E. coli* and recombinant yeast Rrp44p exhibits a processive exonucleolytic activity in vitro[1], consistent with the characterized in vitro activity of the *E. coli* enzyme. Rrp44p can be released from the rest of the complex upon treatment with moderate salt concentration (~0.5M $MgCl_2$). Rrp6p is tightly associated with the complex but is present at substoichiometric levels and is therefore not required for the core structure.[2] Rrp6p is one of many RNase D related proteins found in eukaryotic cells[16] and recombinant Rrp6p exhibits 3'->5' exoribonuclease activity in vitro.[17]

The Relationship between Rrp6p Function and the Exosome

Two structurally distinct forms of the exosome complex have been reported that differ by the presence or absence of Rrp6p.[2] The exosome components Rrp4p, Rrp43p and Csl4p/Ski4p have been localized to both nuclear and cytoplasmic compartments of the cell, whereas Rrp6p is localized specifically to the nucleus.[2,11,18] Consistent with this, rrp6–Δ strains are defective in all characterized nuclear functions of the exosome but do not exhibit defects in cytoplasmic 3'->5' mRNA decay.[19] The commonly held notion that all defects observed in rrp6 mutants can be attributed to the nuclear exosome complex may, however, be misleading. The observed rrp6–Δ RNA processing phenotypes are related to, but nevertheless clearly distinct from, those of other exosome mutants.[4] In contrast to other exosome components, Rrp6p is not essential for cell viability but rrp6–Δ mutants exhibit a ts-lethal growth phenotype. However, a conditional RNA processing phenotype has not yet been identified for a characterized exosome function in rrp6–Δ strains. Furthermore, Rrp6p is involved in biochemical and genetic interactions with poly(A) polymerase that currently distinguish it from other components of the exosome.[6,20] Taken together, these observations suggest that Rrp6p may perform functions in nuclear mRNA processing that are distinct from the functions of other components of the exosome complex.

Exosome Complexes from Other Organisms

A compelling amount of data suggest that a complex equivalent to the yeast exosome exists in the cells of other eukaryotic organisms. A complex is observed in HeLa cell extracts that contains human homologues of Rrp4p, Rrp6p, Rrp45p, Rrp40p, Rrp41p and Rrp46p and which exhibits exonuclease activity in vitro.[2,21] Furthermore, genetic analyses have revealed that human Rrp4p, Rrp44p/Dis3p, Rrp41p and Csl4p complement growth phenotypes exhibited by mutants of the corresponding genes in *S. cerevisiae*. [1,21-23] In a similar vein, the Rrp41p/Ski6p homologue from *A. thaliana* (AtRrp41p) is found in a comparably sized complex in plant extracts, the recombinant protein exhibits a phosphate-dependent exoribonuclease activity in vitro, and its expression in yeast complements the growth phenotypes of the ts-lethal ski6-100 and carbon source dependent GAL10:rrp41 alleles.[14] Although little is known about the detailed structure of the exosome from any organism, the cross-species complementation data strongly suggest that the structure of the complex has been conserved during eukaryotic evolution.

A number of differences have been pointed out between the complement of exosome homologues in different organisms. For instance, 5 RNase PH homologues have been identified in humans, compared with the 6 present in *S. cerevisiae*.[24] More notably, the genome of *C. elegans* encodes only 4 RNase PH homologues and only one Rrp4p/Rrp40p homologue. It is possible that these proteins are present in more than one copy per complex in order to maintain the overall structure of the exosome throughout evolution. The purified exosome of *Trypanosoma brucei* contains only 5 polypeptides, consisting of homologues to Rrp4p, Rrp45p, Csl4p and Rrp6p, and an additional unidentified protein.[25] Homologues of Rrp40p, Rrp44p and two homologues of Rrp41p have also been identified in *T. brucei*. These proteins do not remain physically associated with the purified complex but all identified homologues are required for 5.8S rRNA maturation. In contrast to *S. cerevisiae*, the *T. brucei* Rrp6p homologue is encoded by an essential gene.

Structural Organization of the Exosome

In the absence of any detailed structural data, proposed models of the organization of the exosome and the mechanisms of its activation have been rather speculative.[13,26] However, the crystallographic structure of polynucleotide phosphorylase (PNPase) from *Streptomyces antibioticus*[27] suggests a plausible structural organization for the exosome core complex. PNPase is a phosphorolytic 3'->5' exoribonuclease and is a component of the degradosome complex[28,29] that plays a central role in mRNA degradation in prokaryotes. The enzyme has a trimeric quaternary structure, the subunits associating around a narrow central channel (~9Å at its narrowest point). PNPase was known to contain a central domain with high sequence homology to RNase PH[15,30] but it became clear upon crystallographic analysis that this structural fold is duplicated in the N-terminal region of the protein. The symmetrical ring structure formed by the PNPase trimer therefore consists of 6 domains homologous to RNase PH. However, only the central domain of each polypeptide is likely to contain a catalytic active site. PNPase also contains KH and S1 RNA binding domains (RBDs) at its C-terminal end that are thought to contribute to the high processivity of the enzyme.[31] The RBDs are organized toward the outer side of the upper surface of the ring structure and the active site is located towards the cavity on the lower side, suggesting that the substrate may pass through the central channel.

A structural organization similar to that of PNPase could be easily envisaged for the core exosome complex (Fig. 1) that contains 6 RNase PH homologues and 3 S1 RBD proteins. The arrangement of the RBD-containing proteins towards the outer rim of the PH ring generates a funnel-shaped structure with a central channel. PNPase requires a minimal 3' single stranded region in order to bind the substrate and can overcome structural elements within the RNA substrate by the associated ATP-dependent RNA helicase activity of RhlB.[28] One possibility is that substrates of the exosome must be similarly unfolded by its putative helicase cofactors Mtr4p or Ski2p in order to allow entry of the substrate into the channel. An alternative model would involve activation of the exosome by structural rearrangements that widen the central channel of the complex. In this case, a "closed" structure would permit binding and degradation of the RNA by the S1 RBD-containing exonucleases Rrp4p and Rrp40p, whereas a more open structure would allow access of the substrate to the phosphorolytic enzymes. This model is consistent with in vitro data showing that purified exosome exhibits a hydrolytic activity similar to recombinant Rrp4p and lacks detectable phosphorolytic

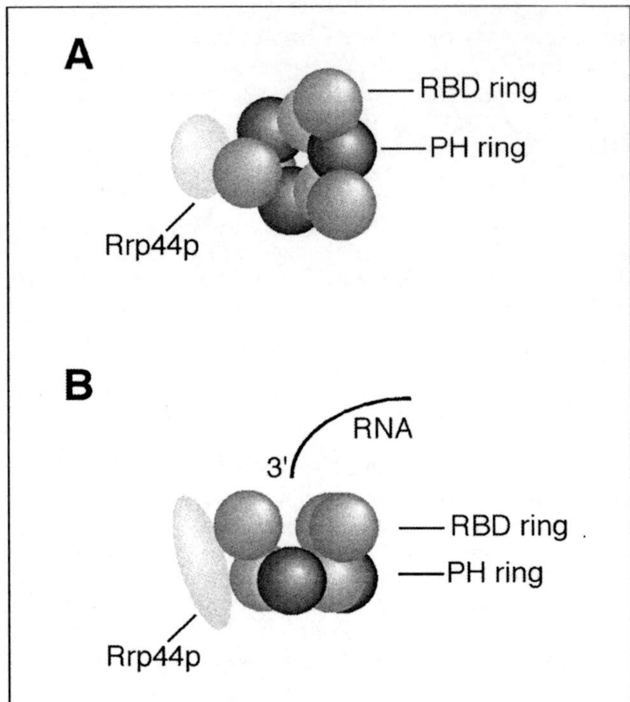

Figure 1. Model of the exosome complex. A possible organisation of the exosome complex viewed (A) from above and (B) from the side. By analogy with the crystal structure of PNPase, the six RNase PH homologues (light tone) are envisaged to form a hexameric ring around a central channel. An S1 RBD protein is associated with every second RNase PH homologue of the ring and together they constitute a second ring (the "RBD ring", mid-gray tone), on the upper surface of the "PH ring". The RBD proteins are located towards the outside of the rim of the PH ring, generating a funnel-shaped complex. The catalytic sites of the RNase PH homologues lie towards the central channel, deep within the ring. The RNA substrate may be fed into the central channel from the upper side. Rrp44p (dark tone) is not required for the stability of the core complex and is shown binding to the side of the complex.

activity.[1] Notably, the activity of the highly processive, hydrolytic exonuclease Rrp44p is also largely, if not completely, repressed in the purified complex. It is difficult to envisage how the active site of the peripherally associated Rrp44p could be buried within the structure of the complex. Rrp44p may be activated by a distinct allosteric mechanism that allows access of the substrate to its active site.

mRNA Decay in Yeast

As noted above, the processes of mRNA translation and mRNA turnover are closely linked in eukaryotic cells. The structural characteristics of the messenger ribonucleoprotein (mRNP) particle that stimulate translation both in vivo and in vitro, the poly(A) tail and the 7-monomethylguanosine cap, also function to protect the transcript from the mRNA turnover machinery.[32,33] The functional lifetime of an mRNP, the time during which it can be translated into protein, therefore largely reflects the rates at which the transcript is deadenylated and the cap structure is removed.

Two general mRNA decay pathways that are responsible for the degradation of a large number of transcripts in *S. cerevisiae* both involve initial shortening of the poly(A) tail.[34,35] Following deadenylation, either the cap structure is removed by Dcp1p,[36,37] generating an entry site for the 5'->3' exoribonuclease Xrn1p,[38] or the deadenylated mRNA is degraded further by a 3'->5' mechanism involving the exosome complex and its Ski cofactors (Fig. 2).[3,11] Analysis of mRNA decay pathways in eukaryotic organisms other than *S. cerevisiae* is limited but the available evidence

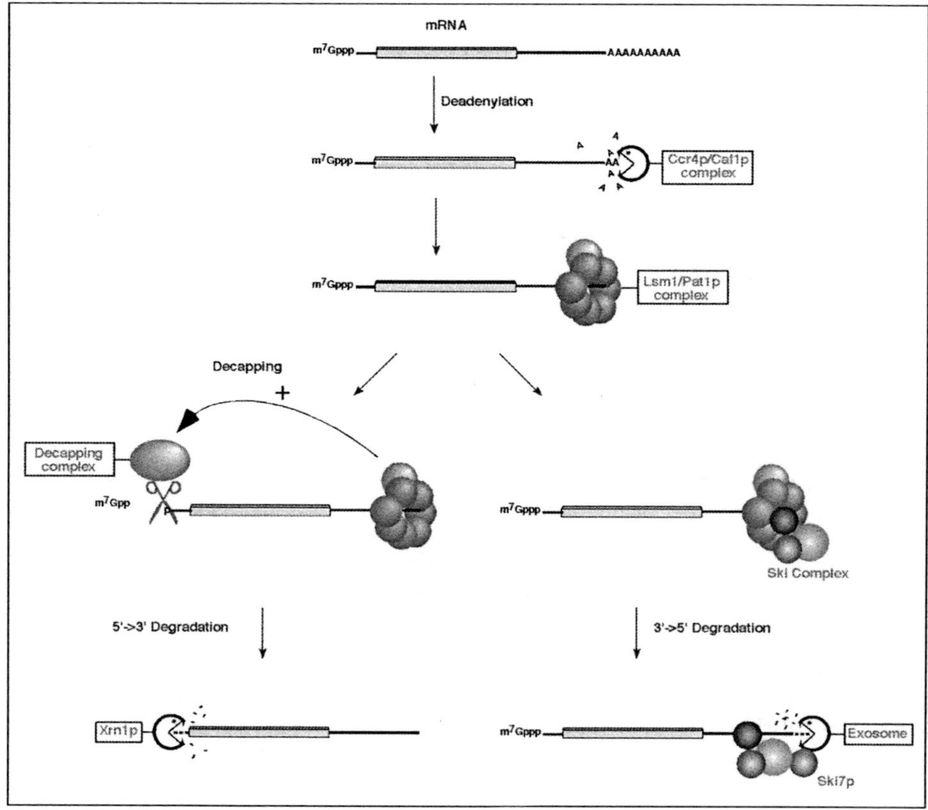

Figure 2. Major mRNA decay pathways in yeast. Schematic representation of the general 5'->3' and 3'->5' decay pathways. Deadenylation by the Ccr4p/Caf1p complex, and to a lesser extent by PAN, generates a capped, deadenylated transcript. Binding of the Lsm1/Pat1p complex, probably to the 3' end of the mRNA, stimulates decapping. Removal of the cap by Dcp1p allows 5'->3' degradation by Xrn1p. The Lsm1/Pat1p complex inhibits nonspecific 3' trimming, promoting normal 3'->5' decay. Ski7p may assemble the Ski and exosome complexes into an activated ternary complex on the mRNP. The exosome then degrades the mRNA in a 3'->5' direction. In yeast the 5'->3' decay pathway is the predominant decay mechanism, whereas in human cells the 3'->5' decay pathway may be more important.

suggests that the two pathways established in yeast are conserved in other eukaryotes. Thus, deadenylation generally precedes degradation of the mRNA body for transcripts analyzed in mammalian cells,[39-41] and functional homologues of Dcp1p, Xrn1p and the exosome complex are found in higher eukaryotes.[2,42,43]

Normal deadenylation in yeast requires Ccr4p and Caf1/Pop2p, two nuclease-related proteins that copurifiy in multimeric complexes with an associated poly(A)-specific exonuclease activity.[44] At present it is unclear which component(s) of this complex is directly responsible for the deadenylase activity. Caf1p is a member of the RNase D family of putative 3'->5' exoribonucleases.[16] Intriguingly, despite having amino acid substitutions at 3 of the 5 residues predicted to be important for catalytic activity on the basis of sequence alignments, recombinant yeast Caf1p exhibits exoribonuclease activity in vitro[45] and is therefore a good candidate for a catalytic subunit of the complex. However, Ccr4p is related to a different family of nucleases and is essential for the Caf1p-associated poly(A)-specific exonuclease activity purified from yeast lysates.[44]

Pan2p is another member of the RNase D-related family of putative exoribonucleases[16] and copurifies with Pan3p, both subunits being required for poly(A)-binding protein (Pab1p)-dependent nuclease (PAN) activity.[46,47] Little or no deadenylation of the MFA2 and PGK1 reporter transcripts is observed in a ccr4–Δ, pan2–Δ mutant and the transcripts are extremely stable,[44] suggesting that the residual deadenylation activity observed in the ccr4–Δ strain is due to PAN.

In wild-type cells, the poly(A) tail is removed almost to completion before decapping and rapid degradation are triggered, a phenomenon described as deadenylation-dependent decapping. These two steps are partially uncoupled in strains mutant for Pab1p or lacking the deadenylase components Caf1p or Ccr4p.[44,48] In these mutants, decapping is triggered before deadenylation is complete. Moreover, decapping is completely uncoupled from deadenylation in the ccr4–Δ, pan2–Δ double mutant, which exhibits essentially no deadenylation activity on reporter constructs. Pab1p is required for PAN activity in vitro and for efficient deadenylation rates in vivo.[48] These results are consistent with models that invoke an inhibition of decapping for translated mRNPs during deadenylation. PARN is an exonuclease related to Caf1p and Pan2p that is required for poly(A) tail shortening during meiotic maturation in Xenopus oocytes and is the major deadenylation activity in mammalian tissue extracts.[49,50] Interestingly, PARN activity is stimulated by the presence of the cap structure and the enzyme interacts directly with the cap in a poly(A)-dependent manner.[51-53] It is therefore plausible that actively engaged deadenylases inhibit mRNA decapping by association with the mRNA cap.

mRNA Decapping and the Lsm Complex

dcp1–Δ mutants are severely inhibited in mRNA decapping and highly purified Dcp1p exhibits decapping activity in vitro, suggesting that Dcp1p is the major cytoplasmic decapping enzyme.[36,37] Decapping in vivo also requires Dcp2p, which copurifies with Dcp1p and is thought to activate Dcp1p.[54] Furthermore, decapping is stimulated by, but is not absolutely dependent upon, a large number of factors that copurify with Dcp1p, including Mrt3p, Vsp16p, Edc1p and Edc2p.[55-57] This suggests that efficient decapping by Dcp1p requires the assembly of a large decapping complex on the mRNP.

An important group of factors that stimulate mRNA decapping form a distinct complex known as the Lsm1/Pat1p complex. Lsm (like-Sm) proteins are a family of small proteins with a conserved domain characteristic of the Sm proteins of snRNPs.[58,59] There are two functionally distinct Lsm complexes characterized in yeast; a nuclear complex comprising Lsm2p-Lsm8p is associated with U6 snRNA and is required for pre-mRNA splicing, whereas a cytoplasmic complex comprising Lsm1p-Lsm7p, together with Pat1p/Mrt1p, stimulates mRNA decapping.[60-63] When visualized by electron microscopy, the RNA-free Lsm complex purified from spliceosomal particles appears as a ring structure. This complex specifically associates with the 3'-end of U6 snRNA[64] and is thought to induce structural alterations in the U6snRNP, thereby facilitating its interaction with the U4 and U5 snRNPs during assembly of the spliceosome.[60,64] A similar reorganization of mRNP structure upon association of the Lsm1/Pat1p complex is envisaged to promote decapping.[63]

Decapping activity is associated with ribosomes and Dcp1p has been reported to interact directly with, and be enzymatically stimulated by, the translation initiation factor eIF4G.[65,66] Interestingly, Dcp1p decapping activity is competitively inhibited by the cap-binding protein eIF4E, which also interacts with eIF4G.[67] Mutations in several translation initiation factors which affect the status of the cap-binding complex cause an increase in the rates of both deadenylation and decapping.[68] Furthermore, the decapping defect observed for the partial loss of function dcp1-1 mutant in vivo was suppressed by the cdc33-42 allele in the cap-binding protein eIF4E. The rate of decapping for a given mRNP may therefore partially reflect a competition between Dcp1p and eIF4E for binding to eIF4G and/or the cap. The cdc33-42 mutation did not, however, suppress the decapping defect seen in lsm1–Δ and pat1–Δ mutants, suggesting that the Lsm1/Pat1p complex stimulates decapping by a mechanism other than competition for the cap.

An optimal context of the translation initiation codon also serves to increase the stability of the mRNA,[69] suggesting that mRNA decapping may also be influenced by events that occur at later stages during translation initiation. In contrast, a decrease in translation elongation in strains mutant for tRNA nucleotidyltransferase activity or by the addition of cycloheximide inhibits decapping and increases mRNA stability.[70,71] This suggests that the mRNA is susceptible to decapping during specific stages during the structurally dynamic process of translation initiation.

The Pat1p component of the Lsm1/Pat1p complex has independently been characterized as a translation initiation factor by a number of criteria: it is found mainly associated with preinitiation complexes upon polysome gradient analyses, it is required for normal rates of translation initiation in vivo and in vitro and is required for mRNP association with the 40S ribosomal subunit particles in vitro.[72] It is presently unclear whether the Lsm1/Pat1p complex functions in translation initiation or if this function reflects an independent activity of Pat1p. Pat1p is required for the association of Lsm1p with polysomes[73], suggesting that Pat1p may recruit the Lsm1 complex to the polysomes at some time before mRNA decapping occurs. Notably, the 5'->3' exoribonuclease Xrn1p is directly associated with the Lsm1/Pat1p complex (62). Hence, activities that function in the sequential events of mRNA decapping and 5'->3' degradation are physically associated. Dcp1p also copurifies with the components of the Lsm1/Pat1p complex but in an RNase-sensitive manner.[63]

Both pat1−Δ and lsm1−Δ mutants are temperature-sensitive for growth whereas dcp1−Δ strains, which are more strongly inhibited in decapping, exhibit a nonconditional slow growth phenotype. This suggests that the Lsm1/Pat1p complex performs a function in addition to stimulating decapping that is crucial for growth at elevated temperatures. In addition to showing defects in decapping, mutants in the Lsm1/Pat1p complex, but not dcp1−Δ, accumulate capped, deadenylated mRNAs which are shortened at the 3' end.[67,74,75] These trimmed mRNAs accumulate upon shift to the nonpermissive temperature and are also apparent in wild-type cells during growth at elevated temperatures. These data suggest that defects in the Lsm1/Pat1p complex affect the status of the 3' end of deadenylated mRNAs, in addition to stimulating decapping. The ts-lethal phenotype of the lsm1−Δ mutant was partially suppressed by the ski2−Δ, ski3−Δ, ski4-1/csl4, ski7−Δ or ski8−Δ mutation that impede the 3'->5' mRNA turnover pathway, whilst the ski4-1/csl4 mutation partially suppressed the ts-lethal growth phenotype of the lsm2−Δ, lsm5−Δ, lsm7−Δ, and pat1−Δ mutants.[75] Furthermore, trimmed mRNAs accumulated to higher levels in the lsm1−Δ, ski double mutant strains, compared with the lsm1−Δ single mutant. These data suggest that the trimmed mRNAs are substrates for the exosome and Ski complexes, and that ski mutants suppress the temperature-sensitive growth phenotype of lsm mutants by inhibiting mRNA turnover.

One possibility is that the Lsm1/Pat1p complex protects the 3' end of the mRNA from premature degradation by the 3'->5' decay machinery. This model invokes a novel, degradative activity that shortens the 3' end of the mRNA in the absence of the Lsm1/Pat1p complex, thereby generating a substrate for the exosome. An alternative interpretation of this data is that the Lsm1/Pat1p complex promotes the onset of normal 3'->5' decay. In this model, the mRNP is directed to a distinct, exosome-dependent discard pathway in the absence of the Lsm1/Pat1p complex. The involvement of the exosome in distinct 3'->5' pathways has been previously suggested for the processing of snoRNAs and snRNAs.[4,76] In the latter case, the competition between a degradative and synthetic pathway is also influenced by association with RNA-binding proteins, including the Lsm complex.

The 3'->5' mRNA Decay Pathway

The 5'->3' degradation pathway is the major cytoplasmic decay mechanism in yeast for all transcripts reported. In fact, the 3'->5' decay pathway can only be analyzed in vivo by using mutants defective in 5'->3' decay or if the progression of the Xrn1p 5'->3' exonuclease is impeded by the insertion of a poly(G) cassette within mRNA reporter construct.[77] Degradation by Xrn1p up to the poly(G) sequence generates a 5' protected poly(G)-3' fragment that remains a substrate for the 3'->5' decay pathway. However, the 3'->5' decay pathway can support mRNA degradation in cells defective for the 5'->3' decay pathway[78] and may be the principal mechanism of mRNA decay in yeast for specific transcripts or under certain growth conditions. Furthermore, the 3'->5' decay pathway appears to be the major turnover mechanism for the decay of transcripts analyzed in mammalian cells.[79,80] Pairs of mutations which inhibit both the 5'->3' and 3'->5' decay pathways are synthetic lethal,[3,11,81] suggesting that mRNA turnover is an essential cellular process.

Ski2p is a DEVH box putative ATP-dependent RNA helicase closely related to the nuclear exosome cofactor Mtr4p/Dob1p.[82,83] Since there is no functional overlap between Mtr4p and Ski2p, these proteins have been suggested to act as substrate-specific adaptors, targeting the exosome to its nuclear and cytoplasmic substrates, respectively.[1, 3, 9] The substrate specificity of RNA helicases is generally dependent upon the RNP context of the substrate and is conferred by proteins that interact with domains flanking the conserved bipartite catalytic domain,[84,85] although some helicases such as DpbA directly recognize structural elements within their RNA substrate.[86]

Ski2p is associated together with Ski3p and Ski8p in a complex with 1:1:1 stoichiometry, based on immunoprecipitation analyses of extracts from [^{35}S]methionine labelled cells.[10] Ski3p can interact with Ski8p in the absence of Ski2p but Ski2p binding requires both Ski3p and Ski8p. Formation of the heterotrimeric Ski complex did not require Ski7p and was not affected by the other ski mutations ski4-1 and ski6-2 which affect components of the exosome complex. Ski3p is a large 164 kDa protein[87] containing 10 copies of a tetratricopeptide repeat (TPR) distributed along the length of the protein and a canonical leucine zipper motif, suggesting that the protein may oligomerize.[10] Ski8p is a 44 kDa protein that contains five WD repeats.[88] TPRs are motifs commonly involved in protein-protein interactions and WD repeats are often found in proteins interacting with TPR-containing proteins. Yeast Ski2p and Ski3p have been localized to the cytoplasmic compartment by subcellular fractionation.[10] However, immunolocalization of Ski3p indicated enrichment in the perinuclear region[10] and Ski3p-LacZ was reported to cofractionate with nuclei on Percoll gradients.[87]

Ski7p is an 85 kDa protein comprising of an N-terminal domain and a central GTPase domain.[89] The N-terminal domain of Ski7p is required and is sufficient for 3'->5' mRNA decay and contains different regions that interact with the Ski complex and the exosome complex.[12] A stable interaction has not been observed between the Ski complex and the exosome[10,13] but Ski7p can be copurified with the exosome in the absence of the Ski complex.[12] One possibility is that 3'->5' decay is mediated by a large complex containing the Ski complex, the exosome and Ski7p that is only transiently formed or is disrupted upon purification. Alternatively, Ski7p may associate with the Ski and exosome complexes in a mutually exclusive manner, reflecting distinct steps during activation of the 3'->5' mRNA turnover pathway.

The function of the putative GTPase domain of Ski7p, which is dispensible for the 3'->5' mRNA decay pathway, is presently unclear.[12] Sequence analyses reveal that Ski7p shows strong homology to the translation termination factor Sup35p/eRF3, and Tef2p, the yeast homologue of elongation factor 1A/EF-Tu.[89] Tef2p brings the aminoacylated tRNA to the ribosomal A site during translation elongation. Sup35p directly interacts with Sup45p/eRF1, the stop codon recognition factor that mimics the structure of tRNA, and has an eRF1- and ribosome-dependent GTPase activity. Although eRF3 proteins typically have an additional N-terminal domain, the C-terminal EF-1A like GTPase domain of Sup35p is necessary and sufficient for translation termination and cell viability.[90,91] Ski7p may therefore interact directly with the ribosome and sense the translational status of the mRNA, allowing docking of the exosome complex onto deadenylated mRNP substrates. The sequence homology between Ski7p and the translation factor GTPases is, however, restricted to the GTPase catalytic domain, whereas eEF1A and eRF3 proteins show extensive homology throughout the C-terminal portion that contributes to the binding of tRNA and eRF1, respectively.[92,93] This suggests that the observed similarity between Ski7p and the translation factors may possibly reflect an evolutionary, rather than functional, relationship.

Two ski alleles, ski4-1 and ski6-2, map to the Csl4p and Rrp41p components of the core exosome complex, respectively.[11,94] Subsequent analyses have shown that the 3'->5' decay pathway is also inhibited in the rrp4-1 mutant but not in mutants of the nuclear proteins Rrp6p or Mtr4p.[3,19] Notably, the ski4-1/csl4 mutant is defective in 3'->5' mRNA decay but does not exhibit a phenotype in the other characterized exosome functions of 5.8S rRNA processing, U4 snRNA synthesis or degradation of the 5' external transcribed spacer (5' ETS) fragment of the precursor ribosomal RNA transcript.[11] The nonconditional ski4-1/csl4 allele results in the mutation of a glycine residue, conserved in Csl4p homologues from plants to humans, located four residues from the C-terminal end of the S1 RNA binding domain. This suggests that RNA binding by Csl4p may be required for 3'->5' mRNA decay but not for nuclear functions of the exosome.

The Ski Mutants: Translation or Turnover?

The ski mutations were initially identified genetically as resulting in overexpression of the killer toxin polypeptide encoded by the nonadenylated, noncapped double-stranded RNA virus M, a satellite of the L-A virus.[95] The ski2, ski3, ski6, ski7 and ski8 mutants were subsequently shown in an electroporation assay to increase the expression of nonadenylated (A⁻) luciferase reporter mRNAs by 8- to 34-fold.[89,94,96] It is assumed that the initial rates of translation observed for electroporated mRNAs reflect their translational efficiency, whereas their functional half-life, the time required to achieve 50% maximal luciferase activity, reflects mRNA stability.[97] The ski mutants have been proposed to increase the translation of A⁻ mRNAs by causing abnormalities in ribosome structure, disturbing a cellular mechanism which acts to translate polyadenylated mRNAs in preference to deadenylated mRNAs.[96] A recent report suggests that the translation factor eIF5B, a putative GTPase required for 60S subunit binding during translation initiation, may function in both polyA-dependent stimulation of translation and in SKI⁺-dependent discrimination against A⁻ mRNAs.[98]

Although the primary effect of the ski1/xrn1 mutant on killer toxin overexpression is generally accepted to be due to increased stability of the viral mRNA, it remains controversial whether the other ski mutants primarily affect mRNA translation or mRNA stability. One possibility is that the observed derepression of nonadenylated luciferase reporter mRNAs is independent of Xrn1p because these transcripts are degraded primarily by the exosome-mediated 3'->5' decay pathway. The ski6-2 mutant has been reported to have a defect in 60S subunit assembly.[94] The polysome profiles of ski2–Δ mutants are, however, identical to wild-type cells[96] and therefore the ski mutants do not cause an apparent common defect in ribosome biogenesis which could explain the observed derepression of nonadenylated transcripts. More to the point, the killer toxin overexpression screen failed to identify any of the myriad of proteins that are known to be required for correct ribosome synthesis and assembly.[99] Instead, a highly specific group of mRNA decay factors were identified that consist of the 5'->3' exoribonuclease Xrn1p/Ski1p, the exosome components Rrp41p/Ski6p and Csl4p/Ski4p, the exosome interacting protein Ski7p and the Ski7p-interacting proteins Ski2p, Ski3p and Ski8p.

Perspectives

It remains to be seen whether the exosome, the Ski complex and Ski7p represent a comprehensive list of factors required for 3'->5' mRNA decay.

Notably, immunoprecipitates of Ski3p are associated with substoichiometric amounts of Ski2p and a number of additional proteins, suggesting that Ski3p may exist in complexes in addition to the Ski complex.[10] Furthermore, it is not clear whether the copurification of Ski7p with the exosome and the Ski complex[12] reflects a direct association or the existence of a larger complex. An important goal in future research will be the detailed characterization of the distinct complexes containing these decay factors. It seems probable that the exosome and Ski complexes are loaded onto polysomes at some point before mRNA deadenylation is complete, although direct evidence for such an association is lacking. Indeed, it is a moot point whether the 3'->5' decay pathway is coupled to translation in a manner similar to deadenylation and decapping,[33] or if the deadenylated mRNA is degraded by the exosome after release from the translational apparatus.

Recent work suggests a pivotal role for Ski7p at the onset of the 3'->5' mRNA decay pathway,[12] insofar as it independently interacts with both the exosome and the Ski complexes. It is unclear at present whether a ternary complex containing the exosome, Ski7p and the Ski complex is formed, or if these two complexes reflect sequential steps in the pathway. One possibility is that Ski7p functions as a coupling factor in the association of the exosome and Ski complexes. An alternative possibility is that Ski7p may somehow regulate the degradative activity of the exosome or the helicase activity of Ski2p. A surprising finding was that the GTPase domain of Ski7p is not essential for 3'->5' decay of MFA2 mRNA and the function of this putative activity therefore remains unclear. GTP hydrolysis by GTP binding proteins normally induces a structural switch within associated proteins that triggers changes in their biomolecular interactions and/or modulates their biological activity. One possibility is that Ski7p GTPase activity has a stimulatory role, rather than a prerequisite function, in 3'->5' decay. Alternatively, a requirement for GTP hydrolysis by Ski7p may simply not be revealed by analyzing the 3' fragment of MFA2pG reporter constructs. The latter point stresses the problems currently associated with analyses of the 3'->5' decay pathway in yeast, which rely on the inhibition of the more active 5'->3' decay pathway. Future research will be greatly aided by the identification of specific transcripts or the establishment of growth conditions that allow the independent analysis of the 3'->5' decay pathway.

Acknowledgements

I would like to thank David Tollervey for discussions and comments on the manuscript. This work was supported by the Wellcome Trust.

References

1. Mitchell P, Petfalski E, Schevchenko A et al. The exosome: a conserved eukaryotic RNA processing complex containing multiple 3'->5' exoribonucleases. Cell 1997; 91:457-466.
2. Allmang C, Petfalski E, Podtelejnikov A et al. The yeast exosome and human PM-Scl are related complexes of 3'->5' exonucleases. Genes Dev 1999; 13:2148-2158.
3. Jacobs Anderson JS, Parker R. The 3' to 5' degradation of yeast mRNAs is a general mechanism for mRNA turnover that requires the SKI2 DEVH box protein and 3' to 5' exonucleases of the exosome complex. EMBO J 1998; 17:1497-1506.
4. Allmang C, Kufel J, Chanfreau G et al. Functions of the exosome in rRNA, snoRNA and snRNA synthesis. EMBO J 1999; 18:5399-5410.
5. Bousquet-Antonelli C, Presutti C et al. Identification of a regulated turnover pathway for nuclear pre-mRNA. Cell 2000; 102:765-775.
6. Briggs MW, Burkard KTD, Butler JS. Rrp6p, the yeast homologue of the human PM-Scl 100 kDa autoantigen, is essential for efficient 5.8S rRNA 3' end formation. J Biol Chem 1998; 273:13255-13263.
7. Mitchell P, Petfalski E, Tollervey D. The 3' end of yeast 5.8S rRNA is generated by an exonuclease processing mechanism. Genes Dev 1996; 10:502-513.
8. Liang S, Hitomi M, Hu Y-H et al. A DEAD-box-family protein is required for nucleocytoplasmic transport of yeast mRNA. Mol Cell Biol 1996; 16:5139-5146.
9. de la Cruz J, Kressler D, Tollervey D et al. Dob1p (Mtr4p) is a putative ATP-dependent RNA helicase required for the 3' end formation of 5.8S rRNA in *Saccharomyces cerevisiae*. EMBO J 1998; 17:1128-1140.
10. Brown JT, Bai X, Johnson AW. The yeast antiviral proteins Ski2p, Ski3p, and Ski8p exist as a complex in vivo. RNA 2000; 6(3):449-457.
11. van Hoof A, Staples RR, Baker RE et al. Function of the Ski4p (Csl4p) and Ski7p proteins in 3'-to-5' degradation of mRNA. Mol Cell Biol 2000; 20:8230-8243.
12. Araki Y, Takahashi S, Kobayashi T et al. Ski7p G protein interacts with the exosome and the Ski complex for 3'-to-5' mRNA decay in yeast. EMBO J 2001; 20:4684-4693.
13. Mitchell P, Tollervey D. Musing on the structural organization of the exosome complex. Nat Struct Biol 2000; 7:843-846.
14. Chekanova JA, Shaw RJ, Wills MA et al. Poly(A) tail-dependent exonuclease AtRrp41p from *Arabidopsis thaliana* rescues 5.8S rRNA processing and mRNA decay defects of the yeast *ski6* mutant and is found in an exosome-sized complex in plant and yeast cells. J Biol Chem 2000; 275:33158-33166.
15. Zuo Y, Deutscher MP. Exoribonuclease superfamilies: structural analysis and phylogenetic distribution. Nucleic Acids Res 2001; 29:1017-1026.
16. Moser MJ, Holley WR, Chatterjee A et al. The proofreading domain of Escherichia coli DNA polymerase I and other DNA and/or RNA exonuclease domains. Nucleic Acids Res 1997; 25:5110-5118.
17. Burkard KTD, Butler JS. A nuclear 3'-5' exonuclease involved in mRNA degradation interacts with poly(A)-polymerase and the hnRNA protein Npl3p. Mol Cell Biol 2000; 20:604-616.
18. Zanchin NIT, Goldfarb DS. Nip7p Interacts with Nop8p, an Essential Nucleolar Protein Required for 60S Ribosome Biogenesis, and the Exosome Subunit Rrp43p. Mol Cell Biol 1999; 19:1518-1525.
19. van Hoof A, Lennertz P, Parker R. Yeast exosome mutants accumulate 3'-extended polyadenylated forms of U4 small nuclear RNA and small nucleolar RNAs. Mol Cell Biol 2000; 20:441-452.
20. Burkard KTD, Butler JS. A nuclear 3'-5' exonuclease involved in mRNA degradation interacts with poly(A) polymerase and the hnRNA protein Npl3p. Mol Cell Biol 2000; 20:604-616.
21. Brouwer R, Allmang C, Raijmakers R et al. Three novel components of the human exosome. J Biol Chem 2001; 276:6177-6184.
22. Baker R, Harris K, Zhang K. Mutations synthetically lethal with *cep1* target *S. cerevisiae* kinetochore components. Genetics 1998; 149:73-85.
23. Shiomi T, Fukushima K, Suzuki N et al. Human dis3p, which binds to either GTP- or GDP-Ran, complements *Saccharomyces cerevisiae* dis3. J Biochem 1998; 123:883-890.
24. Brouwer R, Pruijn GJM, van Venrooij WJ. The human exosome: an autoantigenic complex of exoribonucleases in myositis and scleroderma. Arthritis Research 2000; 3:102-106.
25. Estévez AM, Kempf T, Clayton C. The exosome of *Trypanosoma brucei*. EMBO J 2001; 20:3831-3839.
26. van Hoof A, Parker R. The Exosome: A proteasome for RNA? Cell 1999; 99:347-350.
27. Symmons MF, Jones GH, Luisi BF. A duplicated fold is the structural basis for polynucleotide phosphorylase catalytic activity, processivity, and regulation. Structure 2000; 8:1215-1226.
28. Py B, Higgins CF, Krisch HM et al. A DEAD-box RNA helicase in the *Escherichia coli* RNA degradosome. Nature 1996; 381:169-172.
29. Miczak A, Kaberdin VR, Wei C-L et al. Proteins associated with RNase E in a multicomponent ribonucleolytic complex. Proc Natl Acad Sci USA 1996; 93:3865-3869.
30. Mian S. Comparative sequence analysis of ribonucleases HII, III, II, PH and D. Nucleic Acids Res 1997; 25:3187-3195.
31. Godefry T. Kinetics of polymerization and phosphorolysis reactions of *Escherichia coli* polynucleotide phosphorylase. Eur J Biochem 1970; 14:222-231.
32. Gallie DR. A tale of two termini: a functional interaction between the termini of an mRNA is a prerequisite for efficient translation initiation. Gene 1998; 216:1-11.

33. Tucker M, Parker R. Mechanisms and control of mRNA decapping in *Saccharomyces cerevisiae*. Annu Rev Biochem 2000; 69:571-595.
34. Decker CJ, Parker R. A turnover pathway for both stable and unstable mRNAs in yeast: evidence for a requirement for deadenylation. Genes Dev 1993; 7:1632-1643.
35. Muhlrad D, Decker CJ, Parler R. Turnover mechanisms of the stable yeast *PGK1* mRNA. Mol Cell Biol 1995; 15:2145-2156.
36. Beelman CA, Stevens A, Caponigro G et al. An essential component of the decapping enzyme required for normal rates of mRNA turnover. Nature 1996; 382:642-646.
37. LaGrandeur TE, Parker R. Isolation and characterization of Dcp1p, the yeast mRNA decapping enzyme. EMBO J 1998; 17:1487-1496.
38. Muhlrad D, Decker CJ, Parker R. Deadenylation of the unstable mRNA encoded by the yeast *MFA2* gene leads to decapping followed by 5'->3' digestion of the transcript. Genes Dev 1994; 8:855-866.
39. Wilson T, Triesman R. Removal of poly(A) and consequent degradation of c-fos mRNA faciliated by 3' AU-rich sequences. Nature 1988; 336:396-399.
40. Shyu A-B, Belasco JG, Greenberg ME. Two distinct destabilizing elements in the *c-fos* message trigger deadenylation as a first step in rapid decay. Genes & Dev. 1991; 5:221-231.
41. Couttet P, Fromont-Racine M, Steel D et al. Messenger RNA deadenylation precedes decapping in mammalian cells. Proc Natl Acad Sci USA 1997; 94:5628-5633.
42. Bashkirov VI, Scherthan H, Solinger JA et al. A mouse cytoplasmic exoribonuclease (mXRN1p) with preference for G4 tetraplex substrates. J Cell Biol 1997; 136:761-773.
43. Gao M, Wilusz CJ, Peltz SW et al. A novel mRNA decapping activity in HeLa cytoplasmic extracts is regulated by AU-rich elements. EMBO J 2001; 20: 1134-1143.
44. Tucker M, Valencia-Sanchez MA, Staples RR et al. The transcription factor associated Ccr4p and Caf1p proteins are components of the major cytoplasmic mRNA deadenylase in *Saccharomyces cerevisiae*. Cell 2001; 104:377-386.
45. Daugeron M-C, Mauxion F, Séraphin B. The yeast POP2 gene encodes a nuclease involved in mRNA deadenylation. Nucleic Acids Res 2001; 29:2448-2455.
46. Boeck R, Tarun JS, Rieger M et al. The yeast Pan2 protein is required for poly(A)-binding protein-stimulated poly(A)-nuclease activity. J Biol Chem 1996; 271:432-438.
47. Brown CE, Tarun Jr. SZ, Boeck R et al. *PAN3* encodes a subunit of the Pab1p-dependent poly(A) nuclease in *Saccharomyces cerevisiae*. Mol Cell Biol 1996; 16:5744-5753.
48. Caponigro G, Parker R. Multiple functions for the poly(A)-binding protein in mRNA decapping and deadenylation in yeast. Genes Dev 1995; 9:2421-2432.
49. Körner CG, Wormington M, Muckenthaler M et al. The deadenylating nuclease (DAN) is involved in poly(A) tail removal during the meiotic maturation of *Xenopus* oocytes. EMBO J 1998; 17:5427-5437.
50. Körner CG, Wahle E. Poly(A) tail shortening by a mammalian poly(A)-specific 3'-exoribonuclease. J Biol Chem 1997; 272:10448-10456.
51. Gao M, Fritz DT, Ford LP et al. Interaction between a poly(A)-specific ribonuclease and the 5' cap influences mRNA degradation rates in vitro. Mol Cell 2000; 5:479-488.
52. Dehlin E, Wormington M, Körner CG et al. Cap-dependent deadenylation of mRNA. EMBO J 2000; 19:1079-1086.
53. Martínez J, Ren Y-G, Thuresson A-C et al. A 54-kDa fragment of the poly(A)-specific ribonuclease is an oligomeric, processive, and cap-interacting poly(A)-specific 3' exonuclease. J Biol Chem 2000; 275:24222-24230.
54. Dunckley T, Parker R. The DCP2 protein is required for mRNA decapping in *Saccharomyces cerevisiae* and contains a functional MutT motif. EMBO J 1999; 18:5411-5422.
55. Hatfield L, Beelman CA, Stevens A et al. Mutations in *trans*-acting factors affecting mRNA decapping in *Saccharomyces cerevisiae*. Mol Cell Biol 1996; 16:5830-5838.
56. Zhang S, Williams CJ, Hagan K et al. Mutations in VPS16 and MRT1 stabilize mRNAs by activating an inhibitor of the decapping enzyme. Mol Cell Biol 1999; 19:7568-7576.
57. Dunckley T, Tucker M, Parker R. Two related proteins, Edc1p and Edc2p, stimulate mRNA decapping in *Saccharomyces cerevisiae*. Genetics 2001; 157:27-37.
58. Hermann H, Fabrizio P, Raker V et al. snRNP Sm proteins share two evolutionarily conserved sequence motifs which are involved in Sm protein-protein interactions. EMBO J 1995; 14:2076-2088.
59. Séraphin B. Sm and Sm-like proteins belong to a large family: identification of proteins of the U6 as well as the U1, U2, U4 and U5 snRNPs. EMBO J 1995; 14:2089-2098.
60. Mayes AE, Verdone L, Legrain P et al. Characterization of Sm-like proteins in yeast and their association with U6 snRNA. EMBO J 1999; 18:4321-4331.
61. Salgado-Garrido J, Bragado-Nilsson E, Kandels-Lewis S et al. Sm and Sm-like proteins assemble in two related complexes of deep evolutionary origin. EMBO J 1999; 18:3451-3462.
62. Bouveret E, Rigaut G, Shevchenko A et al. A Sm-like protein complex that participates in mRNA degradation. EMBO J 2000; 19:1661-1671.
63. Tharun S, He W, Mayes AE et al. Yeast Sm-like proteins function in mRNA decapping and decay. Nature 2000; 404:515-518.
64. Achsel T, Brahms H, Kastner B et al. A doughnut-shaped heteromer of human Sm-like proteins binds to the 3'-end of U6 snRNA, thereby facilitating U4/U6 duplex formation in vitro. EMBO J 1999; 20:5789-5802.
65. Stevens A. mRNA decapping enzyme from *Saccharomyces cerevisiae*: purification and unique specificity for long RNA chains. Mol Cell Biol 1988; 8:2005-2010.
66. Vilela C, Velasco C, Ptushkina M et al. The eukaryotic mRNA decapping protein Dcp1 interacts physically and functionally with the eIF4F translation initiation complex. EMBO J 2000; 19:4372-82.

67. Schwartz D, Parker R. mRNA decapping in yeast requires dissociation of the cap binding protein, eukaryotic translation initiation factor 4E. Mol Cell Biol 2000; 20:7933-7942.
68. Schwartz DC, Parker R. Mutations in translation initiation factors lead to increased rates of deadenylation and decapping of mRNAs in *Saccharomyces cerevisiae*. Mol Cell Biol 1999; 19:5247-5256.
69. LaGrandeur T, Parker R. The *cis* acting sequences responsible for the differential decay of the unstable MFA2 and the stable PGK1 transcripts in yeast include the context of the translational start codon. RNA 1999; 5:420-433.
70. Beelman CA, Parker R. Differential effects of translational inhibition in cis and in trans on the decay of the unstable yeast MFA2 mRNA. J Biol Chem 1994; 269:9687-9692.
71. Peltz SW, Donahue JL, Jacobson A. A mutation in the tRNA nucleotidyltransferase gene promotes stabilization of mRNAs in *Saccharomyces cerevisiae*. Mol Cell Biol 1992; 12:5778-5784.
72. Wyers F, Minet M, Dufour ME et al. Deletion of the *PAT1* gene affects translation initiation and suppresses a *PAB1* gene deletion in yeast. Mol Cell Biol 2000; 20:3538-3549.
73. Bonnerot C, Boeck R, Lapeyre B. The two proteins Pat1p (Mrt1p) and Spb8p interact in vivo, are required for mRNA decay, and are functionally linked to Pab1p. Mol Cell Biol 2000; 20:5939-5946.
74. Boeck R, Lapeyre B, Brown Cet al. Capped mRNA degradation intermediates accumulate in the yeast *spb8-2* mutant. Mol Cell Biol 1998; 18:5062-5072.
75. He W, Parker R. The yeast cytoplasmic Lsm1/Pat1p complex protects RNA 3' termini from partial digestion. Genetics 2001; 158:1445-1455.
76. Kufel J, Allmang C, Chanfreau G et al. Precursors to the U3 snoRNA lack snoRNP proteins but are stabilized by La binding. Mol Cell Biol 2000; 20:5415-5424.
77. Caponigro G, Parker R. Mechanisms and control of mRNA turnover in *Saccharomyces cerevisiae*. Microbiol Rev 1996; 60:233-249.
78. Muhlrad D, Decker C, Parker R. Turnover mechanisms of the stable yeast *PGK1* mRNA. Mol Cell Biol 1995; 15:2145-2156.
79. Ross J, Peltz SW, Kobs G et al. Histone mRNA degradation in vivo: the first detectable step occurs at or near the 3' terminus. Mol Cell Biol 1986; 6:4362-4371.
80. Brewer G. Evidence for a 3'-5' decay pathway for *c-myc* mRNA in mammalian cells. J Biol Chem 1999; 274:16174-16179.
81. Johnson AW, Kolodner RD. Synthetic lethality of sep1 (xrn1) ski2 and sep1 (xrn1) ski3 mutants of *Saccharomyces cerevisiae* is independent of killer virus and suggests a general role for these genes in translation control. Mol Cell Biol 1995; 15:2719-2727.
82. Widner WR, Wickner RB. Evidence that the *SKI* antiviral system of *Saccharomyces cerevisiae* acts by blocking expression of viral mRNA. Mol Cell Biol 1993; 13:4331-4341.
83. Laing S, Hitomi M, Hu Y-H et al. A DEAD-box-family protein is required for nucleocytoplasmic transport of yeast mRNA. Mol Cell Biol 1996; 16:5139-5146.
84. Wang Y, Guthrie C. PRP16, a DEAH-box RNA helicase, is recruited to the spliceosome primarily via its nonconserved N-terminal domain. RNA 1998; 4:1216-1229.
85. Tanner NK, Linder P. DExD/H box RNA helicases: from generic motors to specific dissociation functions. Mol Cell 2001; 8:251-262.
86. Tsu CA, Kossen K, Uhlenbeck OC. The *Escherichia coli* DEAD protein DbpA recognizes a small RNA hairpin in 23S rRNA. RNA 2001; 7:702-709.
87. Rhee S, Icho T, Wickner R. Structure and nuclear localization signal of the SKI3 antiviral protein of *Saccharomyces cerevisiae*. Yeast 1989; 5:149-158.
88. Matsumoto Y, Sarkar G, Sommer SS et al. A yeast antiviral protein, SKI8, shares a repeated amino acid sequence pattern with beta-subunits of G proteins and several other proteins. Yeast 1993; 9:43-51.
89. Benard L, Carroll K, Valle RCP et al. The Ski7 antiviral protein is an EF1-a homolog that blocks expression of non-Poly(A) mRNA in *Saccharomyces cerevisiae*. J Virol 1999; 73:2893-2900.
90. Ter-Avanesyan MD, Kushnirov VV, Dagkesamanskaya AR et al. Deletion analysis of the SUP35 gene of the yeast *Saccharomyces cerevisiae* reveals two non-overlapping functional regions in the encoded protein. Mol Microbiol 1993; 7:683-692.
91. Inagaki Y, Doolittle WF. Evolution of the eukaryotic translation termination system: origins of release factors. Mol Biol Evol 2000; 17:882-889.
92. Andersen GR, Pedersen L, Valente L et al. Structural basis for nucleotide exchange and competition with tRNA in the yeast elongation factor complex eEF1A:eEF1Ba. Mol Cell 2000; 6:1261-1266.
93. Merkulova TI, Frolova LY, Lazar M et al. C-terminal domains of human translation termination factors eRF1 and eRF3 mediate their in vivo interaction. FEBS Lett 1999; 443:41-47.
94. Benard L, Carroll K, Valle RCP et al. Ski6p is a homolog of RNA-processing enzymes that affects translation on non-poly(A) mRNAs and 60S ribosomal subunit biogenesis. Mol Cell Biol 1998; 18:2688-2696.
95. Ridley SP, Sommer SS, Wickner RB. Superkiller mutations in *Saccharomyces cerevisiae* suppress exclusion of M2 double-stranded RNA by L-A-HN and confer cold sensitivity in the presence of M and L-A-HN. Mol Cell Biol 1984; 4:761-770.
96. Masison DC, Blanc A, Ribas JC et al. Decoying the cap⁻ mRNA degradation system by a double stranded RNA virus and poly(A)⁻ mRNA surveillance by a yeast antiviral system. Mol Cell Biol 1995; 15:2763-2771.
97. Gallie DR. The cap and poly(A) tail function synergistically to regulate mRNA translational efficiency. Genes Dev 1991; 5:2108-2116.
98. Searfoss A, Dever TE, Wickner R. Linking the 3' Poly(A) tail to the subunit joining step of translation initiation: relations of Pab1p, eukaryotic translation initiation factor 5B (Fun12p), and Ski2p-Slh1p. Mol Cell Biol 2001; 21:4900-4908.
99. Venema J, Tollervey D. Ribosome synthesis in *Saccharomyces cerevisiae*. Ann Rev Genet 1999; 33:261-311

CHAPTER 15

Crystal Structures of the Ribosome and Ribosomal Subunits

Brian T. Wimberly

Abstract

Efforts to understand the structural basis for translational mechanisms have long been hampered by the low resolution of ribosome structures. This impasse suddenly and dramatically changed in 2000 as a result of the determination of crystal structures of entire ribosomes and of ribosomal subunits. Near-atomic resolution structures of the 30S and 50S subunits revealed the architecture of the subunits and have provided detailed models for interactions between the ribosome and its tRNA, mRNA, and antibiotic ligands. The high-resolution subunit structures have also made possible a more detailed interpretation of complementary lower-resolution structures of the entire 70S ribosome bound to tRNAs and mRNA than would have been possible otherwise. The 70S structure in turn validates many of the models based on the isolated subunit structures. All of the structures highlight the importance of the RNA component: the ribosome is a sophisticated RNA-based polymerase with crucial protein additions. These and future structures will be invaluable for directing new functional experiments to determine how ribosomes really work.

Introduction

In all organisms, template-directed protein synthesis is performed by sophisticated polymerases known as ribosomes. Ribosomes catalyze the formation of a peptide bond between two amino acids, each covalently attached to a transfer RNA substrate, according to instructions encoded in a messenger RNA template.[1,2] The substrate tRNAs are: the aminoacyl-tRNA which bears the next amino acid to be added to the peptide chain; the peptidyl-tRNA which bears the partially complete peptide chain, and a third tRNA, deacylated, that has already donated its amino acid to the nascent peptide chain. At the beginning of the translational elongation cycle, the ribosome chooses an incoming aminoacyl-tRNA on the basis of correct base-pairing interactions between the aminoacyl-tRNA anticodon and an mRNA codon. This selection process, known as decoding, is crucial for maintaining translational accuracy. Decoding occurs in several steps and may include a proofreading step, which is thought to occur prior to peptide bond formation catalyzed by the peptidyltransferase. After peptide bond formation, the ribosome moves the substrate tRNAs and associated mRNA by precisely one codon, in order to make room for the next incoming aminoacyl-tRNA. During this movement, known as translocation, the ribosome must prevent slippage of the tRNA substrate relative to the mRNA template. Additional ribosomal functions are necessary for accurate initiation and termination of translation. Normal translation occurs with the help of several non-ribosomal proteins, the initiation, elongation, and termination factors, the most important of which are GTPases activated by the ribosome at appropriate times.

What structural features of the ribosome could one have guessed from these functional requirements alone? In particular, it is difficult to imagine a small ribosome, because the mRNA template and each of the three tRNA substrates are themselves macromolecules. A single tRNA has a mass of some 25 kDa and an extended L-shape, with a distance of over 70 Angstroms between the anticodon and the covalently bound amino acid. The location of the amino acid defines the location of the ribosome's peptidyltransferase site, while the anticodon determines the location of a ribosomal decoding site that decides whether the incoming aminoacyl-tRNA is a correct one. Substrate selection must therefore involve transfer of information over at least 70 Angstroms. In effect, the ribosome must have an active site that spans this huge distance. All other enzymes—even other sophisticated template-directed polymerases, such as the DNA polymerases—use much smaller substrates, whose distinguishing characteristics are much closer to the site of chemical catalysis. Great size is not, however, enough; the ribosome must also orchestrate coordinated movements of three tRNA substrates and the associated mRNA. Thus the ribosome is not only an enzyme; it must also be a macromolecular machine, which implies additional structural complexity. Finally, because the function of the ribosome is to make proteins, it would have been logical to entertain the possibility that the core ribosomal machinery should not consist of protein—otherwise, how could the ribosome have evolved?

The complex functions of ribosomes are indeed performed by a correspondingly complex structure. All ribosomes consist of two loosely associated subunits comprised of both RNA and protein. The simplest ribosomes from bacteria have a mass of some 2.5 megadaltons and contain about two-thirds RNA and one-third protein. The bacterial small and large subunits are named 30S and 50S, respectively (according to their sedimentation coefficients), and the entire ribosome is denoted 70S. The 30S subunit contains one large RNA (16S rRNA) of some 1500 nucleotides, and about 21 proteins, denoted S1-S21. The 50S subunit contains one large RNA (23S rRNA) of some 2900 nucleotides, one small RNA (5S rRNA) of some 120 nucleotides, and about 34 proteins, denoted L1-L34. The number of proteins may vary depending upon the bacterial species.

The size and complexity of the ribosome has been an enormous challenge for high-resolution structure determination. By the early 1990s, over thirty years of genetic, biochemical and biophysical experiments had revealed increasingly detailed information about ribosome function, but only low-resolution ribosome structures.[1-3] The functional importance of the RNA component—and particularly of a few tens of individual RNA residues—was increasingly recognized, but it in the absence of high-resolution structures, it was not possible to exploit fully this hard-won wealth of functional data to direct high-resolution mechanistic experiments. The available low-resolution structures of that era had been determined by electron microscopy, which played a crucial role in the discovery and early characterization of ribosomes.[4] In fact, most of the gross morphological features of the ribosome had been determined by electron microscopy by the mid-1970s (Fig. 1). The 30S subunit is rather thin, with a squat "body", a distinct "head" that tapers to a "beak" on one side, and a less distinct "platform". At the base of the head is a narrow "neck", and near the bottom of the body is a "spur". The 50S subunit is much thicker, indeed nearly hemispherical, with three appendages: the "L1 arm", the "central protuberance", and the L7/L12 "stalk". Biochemical and biophysical studies indicated that the stalk is very mobile, that it plays an important role in stimulating the GTPase activity of the extrinsic factors, and that the peptidyltransferase center is located near its base. The 50S subunit also contains a "tunnel" through which the nascent peptide chain traverses the ribosome. The three tRNA substrates bind between the subunits, such that each subunit has three tRNA binding sites named after the resident tRNA: the A (for aminoacyl), P (peptidyl) and E (exit) sites. The 30S subunit binds mRNA and the anticodon ends of the tRNAs, so that it can directly monitor the base pairing between codon and anticodon during the decoding process, and maintain the reading frame during translocation. The 50S subunit binds the acceptor arm portion of the tRNAs and catalyzes peptide bond formation. Clearly, both subunits must cooperate to effect translocation, and the interface between the subunits constitutes a large and presumably mobile active site that handles the tRNA and mRNA ligands.

During the 1990s, new cryoelectron microscopy and single particle reconstruction methods made it possible to obtain higher-resolution images that directly revealed important structural features, such as the positions of tRNA substrates and elongation factors.[5-10] However, to date, it has not been possible to push these methods to a resolution (3.0 to 3.5 Angstroms) that would allow tracing of RNA or protein chains. During this time, in a complementary "divide and conquer" approach to understanding ribosome structure, crystal structures of individual ribosomal proteins[11,12] and structures of small RNA fragments[13-15] provided high-resolution views of small portions of the ribosome. For the 30S subunit in particular, these high-resolution structures were combined with other data to derive useful models of 30S structure. It was not clear, however, how a high-resolution structure of an entire subunit or of the entire ribosome could result from such approaches. Finally, in

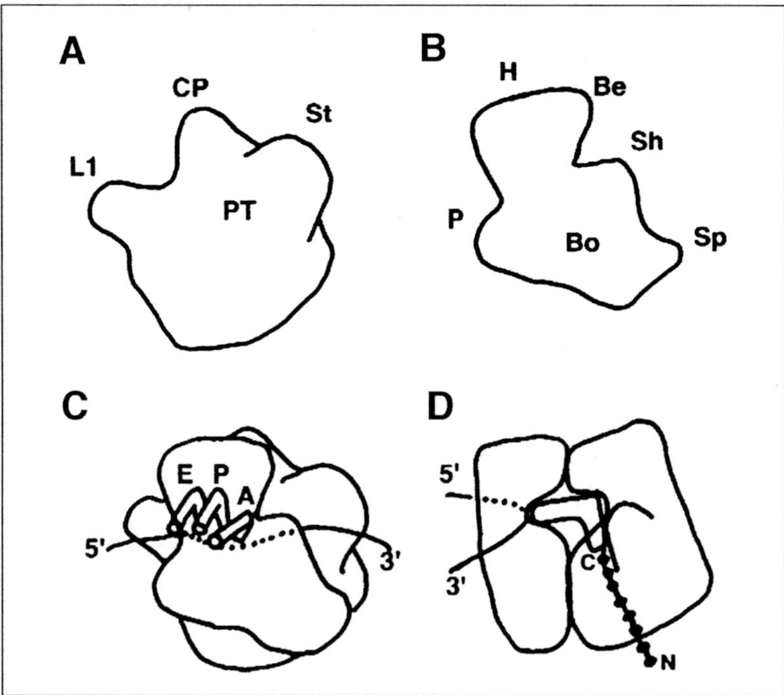

Figure 1. Gross morphology of the ribosome as determined by electron miscroscopy.[3]
   A) The 50S subunit as viewed from the 30S side, with the L1 arm, central protuberance (CP), L7/L12 stalk (ST) and peptidyltransferase site (PT) labeled.
   B) The 30S subunit as viewed from the solvent side, with the head (H), beak (Be), shoulder (Sh), platform (P), body (Bo), and spur (Sp) features labeled. The neck, where tRNA and mRNAs bind, is the narrow region at the junction of the head, platform, and body.
   C) The 70S ribosome viewed from the solvent side of the 30S subunit. tRNA and mRNA ligands pass through the ribosome from right to left. The 5' and 3' ends of the mRNA are labeled, as are the aminoacyl (A), peptidyl (P), and exit (E) tRNAs.
   D) A perpendicular view of the 70S ribosome. The tRNA and mRNA ligands bind between the 50S and 30S subunits, with the tRNA anticodons base-paired with mRNA, and the tRNA aminoacyl ends near the peptidyltransferase site on the 50S subunit. Also shown is a schematic of the nascent peptide chain, which is extruded from the bottom of the 50S subunit through a tunnel.

2000, near-atomic resolution was achieved with the determination of high-resolution crystal structures of both ribosomal subunits. In 2001, a medium-resolution structure of the entire ribosome was also determined (reviewed in refs. 16-18). These structures have revealed the details of ribosome architecture, and they are revolutionizing understanding of ribosome function. Ultimately, however, high-resolution structures of each functional state of the ribosome—and new experiments guided by these structures—will be required to fully understand the structural basis for ribosome function.

It is important to recognize that all of these crystal structures have their roots in pioneering work by Yonath and coworkers, who obtained ribosome crystals in the 1980s.[19] Crucial early contributions were also made by the Puschinow group, who crystallized both the 30S subunit and the 70S ribosome in the late 1980s.[20] However, many more years of work were required to obtain crystals that diffract to a resolution that would allow tracing of RNA and protein chains.[21-24] High-quality crystals are not enough to determine a structure, however; equally important were advances in synchrotron X-ray sources, large and fast X-ray detectors, cryoprotection methods to prolong the lifetime of crystals that decay in intense X-ray beams, the discovery of suitable heavy-atom compounds for crystals of large RNAs, and computing hardware and software.[16]

Structure of the 50S Subunit

After several years of effort,[25-26] the structure of the 50S subunit from the halophilic archaeon *Haloarcula marismortui* was determined by Moore, Steitz and coworkers at a resolution of 2.4 Angstroms.[27] Despite the presence of many RNA insertions and an expanded protein repertoire, this archaeal 50S subunit is sufficiently similar to the *E. coli* subunit to allow exploitation of the vast literature describing the bacterial subunit. These similarities extend to the gross morphology of the *H. marismortui* 50S structure, which is remarkably similar to that of the *E. coli* subunit (Figs. 2C and 1A). As expected, the architecture of the subunit reflects the functional importance of the RNA component. The overall distribution of the proteins is not uniform: most of the globular proteins occupy gaps on the periphery of a rather solid RNA core. The proteins are concentrated on the back and sides of the subunit, with little protein at the functionally important subunit interface region that interacts with 30S and tRNA ligands (Fig. 2C,D). The proteins that do comprise a significant portion of gross morphological features—notably L1 in the L1 arm, and L10, L11 and L7/L12 in the stalk—are partly disordered, and known to be functionally important (RNA protrusions in these and other functionally important features are also disordered.) Many of the proteins contain both a globular component on the periphery of the particle, and one or more narrow extended peptide "fingers" that penetrate deeply into crevices of the RNA core. This architecture suggests that the assembly of the 50S subunit follows an ordered pathway; indeed, proteins are often found to interact with RNA multistem junctions and tight bends in the RNA, which suggests particular functional roles in 50S assembly. Most importantly, there is no protein at all within 18 Angstroms of the peptidyltransferase site, which was localized crystallographically by soaking in a transition-state analog: definitive proof at last that the ribosome is a ribozyme.[28] This structural work and supporting biochemical data[29] were also used to propose a mechanism for peptidyltransferase activity, with a catalytic role for nucleotide A-2486 (A-2451 in the *E. coli* numbering). However, subsequent genetic and biochemical experiments show that A-2451 does not play an inalienable role.[30-33] More work is needed to nail down the details of the peptidyltransferase mechanism.

The architecture of the 23S RNA component is complex. The six so-called "domains" in the 23S rRNA secondary structure are tightly associated in layers to form the nearly hemispherical RNA core of the subunit (Fig. 2A,B). A common strategy for packing RNA helices makes use of patches of adenine residues, whose bases and 2' OH moieties pack favorably against the minor groove of another RNA helix.[34] Four different classes of this "A-minor motif" are defined, some of which had been previously observed in other structures.[35] Another common RNA structural motif is the "kink-turn", a tight bend that is often stabilized, though in various ways, by a bound protein.[36] Analysis of the 50S structure is continuing, and may lead to improvements in the prediction of RNA tertiary structure. Note that the resolution of this *H. marismortui* 50S structure—higher than obtained for the 30S structures, and much higher than that of the 70S structures—allows reliable identification of metal ions and water molecules, which play important roles in stabilizing RNA structure. The relatively high resolution also allows the finer details of the conformation of RNA and protein to be well determined. The *H. marismortui* 50S structure will therefore serve as the "gold standard" for ongoing efforts to mine the recently vastly expanded RNA structure database.

More recently, a structure of the 50S subunit from the eubacterium *Deinococcus radiodurans* has been solved by Yonath, Francheschi and coworkers at lower resolution (3 Angstroms).[37] Most of the structural results are in good agreement with the *H. marismortui* structure. Interestingly, however, there are significant differences in the positioning of the L1 arm and in the conformation of RNA residues surrounding the peptidyltransferase center. The *D. radiodurans* 50S subunit structure has also been solved in the presence of five different antibiotics—four macrolides and chloramphenicol—at reported resolutions of 3.1–3.5 Angstroms.[38] These and other 50S-antibiotic structures will be of great utility in the rational design of improved antibacterial drugs.

Structure of the 30S Subunit

The structure of the 30S subunit from the thermophilic eubacterium *Thermus thermophilus* has been solved independently by groups from the MRC-LMB and from the Max Planck/Weizmann institutes, at resolutions of 3.0 and 3.3 Angstroms, respectively[39,40] Although there are differences between the two 30S structures reported in 2000,[16] more recent structures from the two groups are in good agreement. The final consensus structure includes virtually all of the 16S rRNA and almost

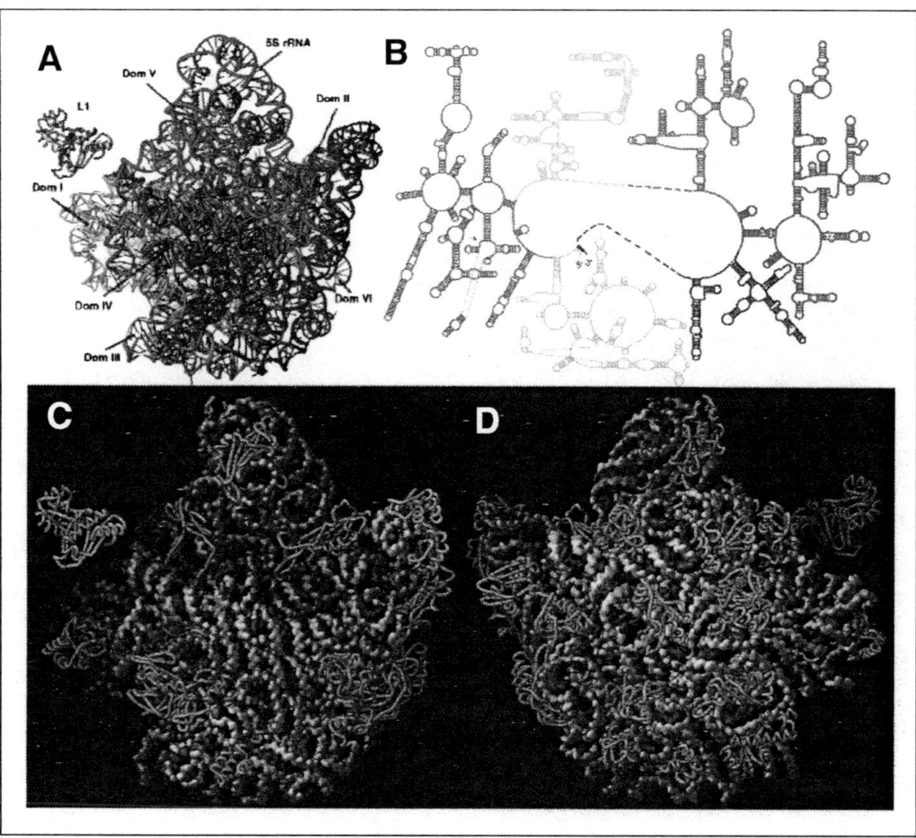

Figure 2. Crystal structure of the 50S subunit from *Haloarcula marismortui* (reprinted from ref. 27, ©2000 American Association for the Advancement of Science, http://www.sciencemag.org/).

A) Tertiary structure of the 23S and 5S rRNAs. Each of the six secondary structure domains of 23S rRNA is colored differently. Although disordered in the crystal structure, protein L1 is also shown in the position determined by cryoelectron microscopy.

B) Secondary structure of the 23S rRNA. Domain colors correspond to those in (A). The secondary structure of the rRNA is adapted from material available from the Comparative RNA Web Site from Dr. Robin Gutell (http://www.rna.icmb.utexas.edu/).

C) The 50S subunit as viewed from the 30S side. RNA residues are represented by a space-filling model, while protein chains are depicted with ribbons. The protein-free region near the middle of the subunit is involved in interactions with the 30s subunit.

D) The 50S subunit as viewed from the solvent side.

A color version of this figure can be viewed at http://www.eurekah.com/abstract.php?chapid=974&bookid=59&catid=54.

all of the small subunit proteins, except for protein S1. As expected, the gross morphology of this eubacterial 30S structure is in good agreement with the *E. coli* 30S features determined by electron microscopy (Figs. 3C and 1B). As observed in the structures of the 50S subunit, the gross morphology is determined by the RNA component, reflecting its functional importance. Once again, the globular protein domains are concentrated on the back and sides of the subunit, away from the most functionally important portions of the 30S subunit, the neck region and the subunit interface area. Again, long thin fingers of extended peptide are seen to penetrate deeply into the RNA core. A few such fingers—highly conserved ones—do reach down into the functionally important tRNA/mRNA-binding neck region. As in the 50S structure, proteins are often found to bind at RNA multistem junctions and at tight bends in the RNA, which suggests a particular role for these proteins in 30S assembly. In the case of the 30S subunit, there are extensive data on mechanisms of 30S assembly, and an analysis of the crystal structure is in good overall agreement with these data. Finally, the detailed strategies used to stabilize RNA-RNA interactions are those seen in the 50S structure: A-minor motifs, adenine patches, kink-turns, and a special role for G-U pairs in mediating some helix-helix packings.[40]

There is a large difference between the large-scale architectures of the 30S and 50S subunits that is probably functionally important. While the secondary structure domains of the 23S rRNA interact closely to form a monolithic, nearly hemispherical mass, the 16S rRNA secondary structure domains are for the most part independent globular entities—i.e., true three-dimensional domains—that make relatively few interactions with each other. One domain of the 16S rRNA constitutes the head, another the platform, and another the bulk of the body (Fig. 3A,B). This architecture is consistent with observations that the individual secondary structure domains of the 16S rRNA, when added to an appropriate mixture of small-subunit proteins, can be assembled into particles that resemble these gross morphological features of the intact 30S subunit. More importantly, this architecture also suggests that independent movement of one or more of the domains—e.g., the head or platform domains, as has been observed in cryo-EM work—may be functionally important. For example, several groups have suggested that rotation of the head may accompany translocation.

These 30S crystal structures reported in 2000 were determined in the absence of any explicitly added tRNA or mRNA ligands. However, the relatively high affinity of the 30S subunit for an RNA stem-loop in its P site resulted in a crystal packing arrangement that provides a high-resolution model for how the P-site tRNA and mRNA interact with the P site of 30S subunit.[41] Helix 6, which constitutes the spur feature, was found to pack into the P site of a neighboring 30S subunit, in a manner that mimics how the P-site tRNA anticodon stem-loop would bind. In addition, the hairpin loop of this helix was found to base-pair with a segment of single-stranded RNA—a mimic of the P-site mRNA. For several reasons, this model is only approximate, but it is nevertheless in good agreement with the lower-resolution view of tRNA/mRNA-30S interactions seen in the 70S crystal structures from Noller and coworkers (see below).

The 30S subunit plays a crucial role in decoding, by monitoring base-pairing interactions between the tRNA anticodon and the mRNA codon in the A site. In 2001, a structure of the *T. thermophilus* 30S subunit was determined with a soaked-in anticodon stem-loop and a single-stranded RNA as models for the A-site tRNA and mRNA, respectively.[42] In the presence of a correct (or "cognate") A-site tRNA anticodon stem-loop, two conformational changes occur in the 16S rRNA: G-530 and A-1492/A-1493 (*E. coli* numbering) swing out and dock into the minor groove of the A-site codon-anticodon minihelix. Together with other conserved elements of the head (helix 34) and body (the 530 loop) previously implicated in decoding, these mobile 16S rRNA residues directly sense the "correctness" of the codon-anticodon interaction (see also Chapter 16 by L. Brakier-Gingras et al). How this information is transmitted to EF-Tu (the elongation factor that delivers the incoming candidate tRNAs to the A site) remains a mystery, however. Interestingly, binding of aminoglycosides that cause miscoding such as paromomycin can induce one of these conformational changes, the swinging out of A1492-3.[41,42] These aminoglycosides were thus proposed to induce miscoding by inducing a conformation of 16S rRNA that has a higher affinity for any A-site tRNA (noncognate as well as cognate). Structures of the *T. thermophilus* 30S subunit bound to many other antibiotics (streptomycin, spectinomycin, tetracycline, hygromycin B, pactamycin, edeine) and two different initiation factors (IF-1 and IF-3) have also been reported.[41-45]

Structure of the 70S Ribosome

Structures of the *T. thermophilus* 70S ribosome with tRNA and mRNA ligands were reported by Noller, Cate, Yusupov and colleagues in 2001, at a resolutions as high as 5.5 Angstroms.[46,47] This apparently rather modest resolution is in fact a phenomenal achievement, given the even greater biochemical and crystallographic difficulties of the 70S system. At this resolution, it is possible to trace RNA (but not protein) chains, albeit with errors in registry, if there are additional data to constrain the chain tracing (e.g., ref. 23). The authors did in fact obtain a correct overall trace of 16S rRNA before the high-resolution 30S structures were reported. Once available, the high-resolution 30S structures were used to correct errors in the registry of the 16S rRNA trace, and as a source for

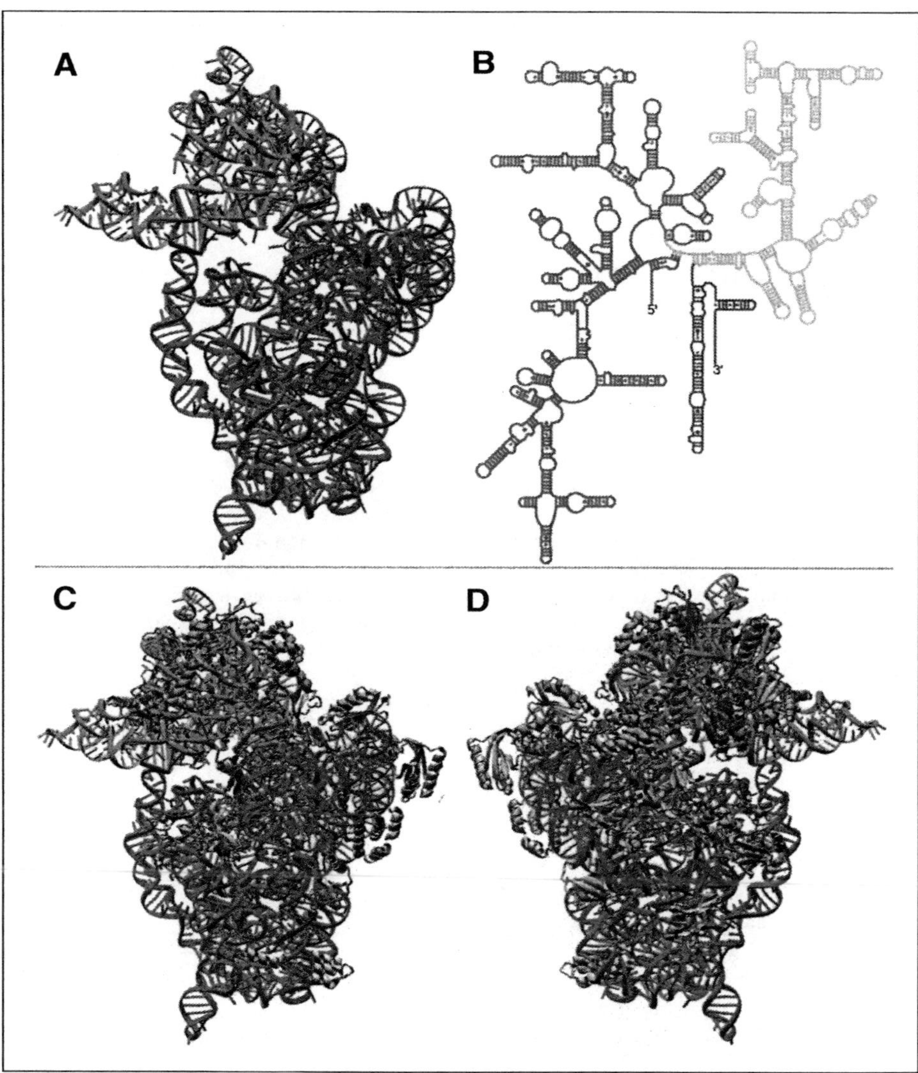

Figure 3. Crystal structure of the 30S subunit from *Thermus thermophilus* (reprinted from ref. 40, with permission from Nature, http://www.nature.com/, ©2000 Macmillan Publishers Ltd.).
  A) Tertiary structure of the 16S rRNA. The secondary structure domains are distinguished by contrasting colors.
  B) Secondary structure of the 16S rRNA, with secondary structure domains colored as in (A). The secondary structure of the rRNA is adapted from material available from the Comparative RNA Web Site from Dr. Robin Gutell (http://www.rna.icmb.utexas.edu/).
  C) The 30S subunit as viewed from the 50S side. RNA residues are depicted as in (A), and protein chains are represented by gray ribbons. The protein-free region near the middle is involved in interactions with the 50S subunit.
  D) The 30S subunit as viewed from the solvent side.
A color version of this figure can be viewed at http://www.eurekah.com/abstract.php?chapid= 974&bookid=59&catid=54.

the structures of the small subunit proteins. Similarly, the high-resolution *H. marismortui* 50S structure was also invaluable for interpretation of the electron density for the large subunit in the 70S maps, although the many differences between the archaeal and eubacterial 50S subunits limited the extent to which the *T. thermophilus* 50S proteins could be built. The 70S crystal structures are largely complementary to the higher-resolution subunit structures described above. The 70S system is clearly more relevant for an understanding of the structure of the entire ribosome; thus, the 70S structures can be used to confirm the relevance of certain aspects of the higher-resolution subunit structures (e.g., the anticodon stem-loop—30S structures). On the other hand, the 70S structures suffer from a lack of resolution. Thus the high-resolution structures are of crucial importance in maximizing the utility of the 70S structures.

The overall morphology of the 70S ribosome crystal structure (Fig. 4) closely resembles the classic EM-derived models of the *E. coli* 70S ribosome (Fig. 1C,D). Comparison of the high-resolution subunit crystal structures with their structures in the context of the 70S particle reveals many small but significant differences, many of which can be ascribed to the effects of crystal packing in the "naked" high-resolution subunit structures. Some of the functionally important protein and RNA features that protrude from the 50S subunit, and that were partly disordered in one or both of the 50S structures, are found to be well-ordered in the 70S structure by virtue of their interactions with the 30S subunit or tRNA ligands. Indeed, the most exciting aspects of the 70S structure are the identification of the 30S-50S contacts, and of ribosome-tRNA contacts. The core of the 30S-50S contact region consists of RNA-RNA contacts, with a more peripheral ring of RNA-protein and protein-protein contacts. As with the overall architecture of each subunit, this RNA-centric manner of subunit association once again emphasizes the overriding functional importance of the RNA component. Analysis of the tRNA-ribosome interactions is equally fascinating. The three tRNA substrates are intimately cradled by intersubunit bridges in the intersubunit space (Fig. 4), in such a way that relative movement of the subunits, as seen in cryo-EM experiments, could effect translocation of the tRNAs.

Clearly, a more complete understanding of translational mechanisms will require more, and higher-resolution, structures of distinct functional states of the ribosome. Both x-ray crystallography and cryoelectron microscopy will be used towards these ends. Equally importantly, the structures now available are being used to design new genetic, biochemical and kinetic experiments. Thus this burst of structural information is by no means the end of the story; it is instead a new beginning.

References

1. Green R, Noller HF. Ribosomes and translation. Annu Rev Biochem 1997; 66:679-716.
2. Garrett RA, Douthwaite SR, Liljas A et al, eds. The ribosome:structure, function, antibiotics, and cellular interactions. Washington, DC: American Society for Microbiology, 2000.
3. Oakes M, Scheinnman A, Atha T et al. In: Hill WE, Dahlberg AE, Garrett RA et al, eds. The ribosome: Structure, function, and evolution. Washington, DC: American Society for Microbiology, Press, 1990:180-193.
4. Tissieres A. Ribosome research: historical background. In: Nomura M, Tissieres A, Lengyel P, eds. Ribosomes. Cold Spring Harbor: Cold Spring Harbor Laboratory Press, 1974:3-12.
5. Agrawal RK, Penczek P, Grassucci RA et al. Direct visualization of A-, P-, and E-site transfer RNAs in the *Escherichia coli* ribosome. Science 1996; 271:1000-1002.
6. Stark H, Orlova EV, Rinke-Appel J et al. Arrangement of tRNAs in pre- and posttranslocational ribosomes revealed by electron cryomicroscopy. Cell 1997; 88:19-28.
7. Stark H, Rodnina MV, Rinke-Appel J et al. Visualization of elongation factor Tu on the *Escherichia coli* ribosome. Nature 1997; 389:403-406.
8. Agrawal RK, Penczek P, Grassucci P et al. Visualization of elongation factor G on the *Escherichia coli* 70S ribosome: the mechanism of translocation. Proc Nat Acad Sci USA 1998; 95:6134-6138.
9. Frank J. Cryo-electron microscopy as an investigative tool: the ribosome as an example. Bioessays 2001; 23:725-732.
10. van Heel M. Unveiling ribosomal structures: the final phases. Curr Opin Struct Biol 2000; 10:259-264.
11. Al-Karadaghi S, Kristensen O, Liljas A. A decade of progress in understanding the structural basis of protein synthesis. Prog Biophys Mol Biol 2000; 73:167-193.
12. Ramakrishnan V, White SW. Ribosomal protein structures: insights into the architecture, machinery and evolution of the ribosome. Trends Biochem Sci 1998; 23:208-212.
13. Moore PB. Structural motifs in RNA. Annu Rev Biochem 1999; 68:287-300.
14. Shen LX, Cai Z, Tinoco I Jr. RNA structure at high resolution. FASEB J 1995; 9:1023-1033.
15. Hermann T, Patel DJ. Stitching together RNA tertiary structures. J Mol Biol 1999; 294:829-849.
16. Ramakrishnan V, Moore PB. Atomic resolution at last: the ribosome in 2000. Curr Opin Struct Biol 2001; 11:144-154.

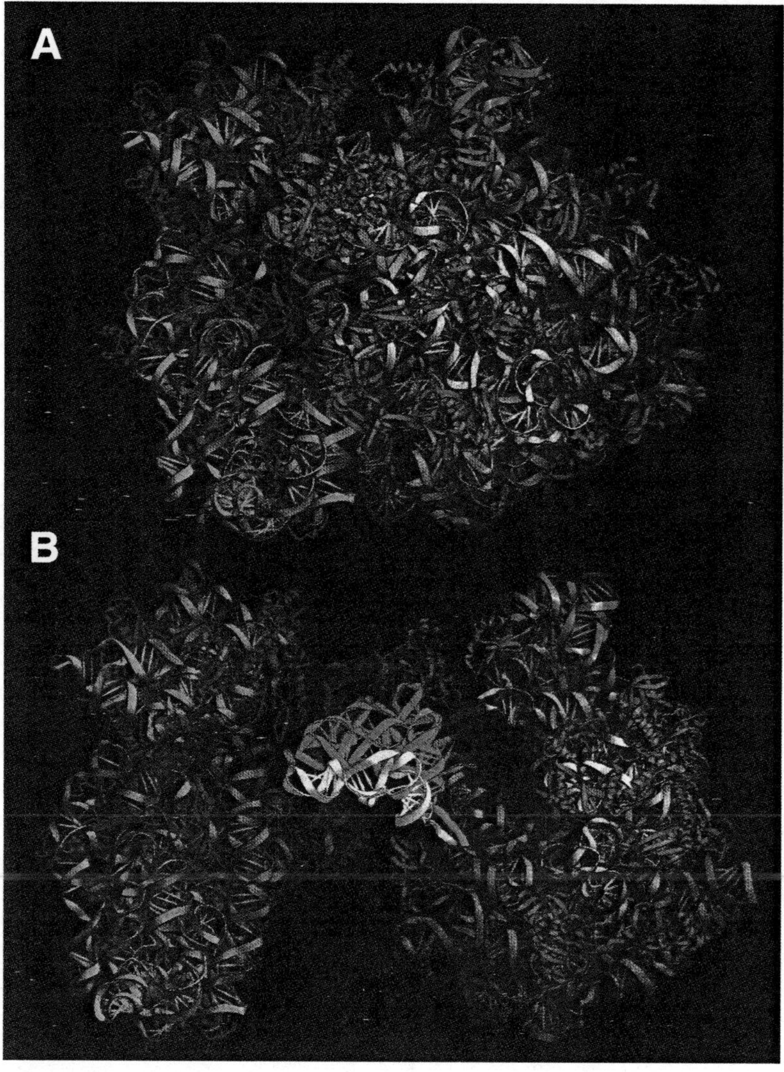

Figure 4. 5.5 Angstrom resolution crystal structure of the *Thermus thermophilus* 70S ribosome, with mRNA and tRNA ligands. This figure is a courtesy from Drs A. Baucom and H. Noller.

A) The 70S ribosome structure, as viewed from the side of the A-site tRNA. The 30S subunit is on the left, with the 16S rRNA in light blue and the small subunit proteins in dark blue. The 50S subunit is on the right, with the 23S rRNA in gray, the 5S rRNA in purple, and the large subunit proteins in magenta. The three tRNA ligands, tightly held between the two subunits, are almost completely hidden from view.

B) The same view of the 70S structure, with both subunits slightly translated and rotated apart from the substrate tRNAs for clarity. The A-site tRNA is yellow, the P-site tRNA is orange, and the E-site tRNA is red.

A color version of this figure can be viewed at http://www.eurekah.com/abstract.php?chapid=974&bookid=59&catid=54.

17. Maguire BA, Zimmermann RA. The ribosome in focus. Cell 2001; 104:813-816.
18. Puglisi JD, Blanchard SC, Green R. Approaching translation at atomic resolution. Nat Struct Biol 2000; 7:855-861.
19. Yonath A, Müssig J, Gewitz HS et al. Crystallization of the large ribosomal subunit from *B. stearothermophilus*. Biochem Internat 1980; 1:428-435.
20. Trakhanov SD. Crystallization of 70S ribosomes and 30S ribosomal subunits from *Thermus thermophilus*. FEBS Lett 1987; 220:319-322.
21. von Böhlen K, Makowski I, Hansen HAS et al. Characterization and preliminary attempts for derivitazation of crystals of large ribosomal subunits from *Haloarcula marismortui* diffracting to 3 Å resolution. J Mol Biol 1991; 222:11-15.
22. Yonath A. Crystallographic studies on the ribosome, a large macromolecular assembly exhibiting severe nonisomorphism, extreme beam sensitivity and no internal symmetry. Acta Crystallogr A 1998; 54:945-955.
23. Clemons WM, May JL, Wimberly BT et al. Structure of a bacterial 30S ribosomal subunit at 5.5 Å resolution. Nature 1999; 400:833-840.
24. Cate JH, Yusupov MM, Yusupova GZ et al. X-ray crystal structures of 70S ribosomal functional complexes. Science 1999; 285:2095-2104.
25. Ban N, Freeborn B, Nissen P et al. A 9-Å resolution x-ray crystallographic map of the large ribosomal subunit. Cell 1998; 93:1105-1115.
26. Ban N, Nissen P, Hansen J et al. Placement of protein and RNA structures into a 5 Å-resolution map of the 50S ribosomal subunit. Nature 1999; 400:841-847.
27. Ban N, Nissen P, Hansen J et al. The complete atomic structure of the large ribosomal subunit. Science 2000; 289:905-920.
28. Nissen P, Hansen J, Ban N et al. The structural basis of ribosomal activity in peptide bond synthesis. Science 2000; 289:920-930.
29. Muth GW, Ortoleva-Donnelly L, Strobel SA. A single adenosine with a neutral pKa in the ribosomal peptidyl transferase center. Science 2000; 289:947-950.
30. Muth GW, Chen L, Kosek AB et al. pH-dependent conformational flexibility within the ribosomal peptidyl transferase center. RNA 2001; 7:1403-1415.
31. Polacek N, Gaynor M, Yassin A et al. Ribosomal peptidyl transferase can withstand mutations at the putative catalytic nucleotide. Nature 2001; 411:498-501.
32. Thompson J, Kim DF, O'Connor M et al. Analysis of mutations at residues A2451 and G2447 of 23S rRNA in the peptidyltransferase active site of the 50S ribosomal subunit. Proc Natl Acad Sci USA 2001; 98 9002-9007.
33. Bayfield MA, Dahlberg AE, Schulmeister U et al. Mechanism of ribosomal peptide bond formation. Proc Nat Acad Sci USA 2001; 98:10096-10101.
34. Nissen P, Ippolito JA, Ban N et al. RNA tertiary interactions in the large ribosomal subunit: the A-minor motif. Proc Natl Acad Sci USA 2001; 98:4899-4903.
35. Doherty EA, Batey RT, Masquida B et al. A universal mode of helix packing in RNA. Nat Struct Biol 2001; 8:339-343.
36. Klein DJ, Schmeing TM, Moore PB et al. The kink-turn: a new RNA secondary structure motif. EMBO J. 2001; 20:4214-4221.
37. Harms J, Scluenzen F, Zarivach R et al. High-resolution structure of the large ribosomal subunit from a mesophilic eubacterium. Cell 2001; 107:679-688.
38. Schluenzen F, Zarivach R, Harms J et al. Structural basis for the interaction of antibiotics with the peptidyl transferase center in eubacteria. Nature 2001; 413:814-821.
39. Schluenzen F, Tocilj A, Zarivach R et al. Structure of a functionally activated small ribosomal subunit at 3.3 Angstroms resolution. Cell 2000; 102:615-623.
40. Wimberly BT, Brodersen DE, Clemons WM et al. Structure of the 30S ribosomal subunit. Nature 2000; 407:327-339.
41. Carter AP, Clemons WM, Brodersen DE et al. Functional insights from the structure of the 30S ribosomal subunit and its interactions with antibiotics. Nature 2000; 407:340-348.
42. Ogle JM, Brodersen DE, Clemons WM et al. Recognition of cognate transfer RNA by the 30S ribosomal subunit. Science 2001; 292:897-902.
43. Brodersen DE, Clemons WM, Carter AP et al. The structural basis for the antibiotics tetracycline, pactamycin, and hygromycin B on the 30S ribosomal subunit. Cell 2000; 103:1143-1154.
44. Carter AP, Clemons WM, Brodersen DE et al. Crystal structure of an initiation factor bound to the 30S ribosomal subunit. Science 2001; 291:498-501.
45. Pioletti M, Schluenzen F, Harms J et al. Crystal structures of the small ribosomal subunit with tetracycline, edeine and IF3. EMBO J 2001; 20:1829-1839.
46. Yusupov MM, Yusupova GZ, Baucom A et al. Crystal structure of the ribosome at 5.5 Å resolution. Science 2001; 292:883-896.
47. Yusupova GZ, Yusupov MM, Cate JHD et al. The path of messenger RNA through the ribosome. Cell 2001; 106:233-241.

CHAPTER 16

Probing the Role of Ribosomal RNA in Protein Synthesis through Mutagenesis

Léa Brakier-Gingras, François Bélanger and Michael O'Connor

Abstract

This review presents a survey of the classic and novel approaches developed for the study of the effects of mutations in ribosomal RNA of *Escherichia coli* on ribosome structure and function. It analyzes representative examples of these mutations at the peptidyl transferase center, at the decoding center and at sites that control translational accuracy, as well as in the intersubunit bridges and the binding site for messenger RNAs. It illustrates the power of the mutagenesis approach to elucidate the role of rRNA in the different steps of protein synthesis.

Introduction

It is now well-established that the ribosome is a ribozyme. The catalytic activity of the ribosome lies in ribosomal RNA and a complete understanding of the mechanisms of translation requires an extensive analysis of the role of the rRNA in protein synthesis.[1,2] Among the variety of approaches that have been used to investigate the role of rRNA, mutagenesis studies have proven to be extremely powerful in identifying and demonstrating the rRNA sites that are important for the different steps of protein synthesis. Initially, the choice of bases for mutagenesis was based on their phylogenetic conservation and location in active sites, together with the positions of crosslinks and protections from chemical modification obtained upon binding of transfer RNAs, factors and other ribosomal ligands. More recently, as detailed in chapter 15 by B. Wimberly, the crystal structures of the 50S subunit from the halophilic archaeon *Haloarcula marismortui*,[3] the 30S subunit from the thermophilic bacterium *Thermus thermophilus*,[4-5] and the 70S ribosome from *T. thermophilus*[6] were solved at atomic resolution, and this has provided a wealth of novel information on the potential importance of specific bases and suggestions about catalytic mechanisms and functions of various regions of rRNA. Such proposals can be tested directly by site-directed mutagenesis of rRNA. Moreover, the striking progress in cryo-electron microscopy has also contributed to the elucidation of the structure of the *Escherichia coli* ribosome in various functional states,[7-8] and this approach has been used to provide information on structural changes induced by mutations in rRNA.[9] In this chapter, we will review classic and novel strategies that have been used to investigate the effects of mutations in rRNA, using as a paradigm the bacterium *E. coli*. We will also examine representative examples illustrating the application of these approaches to important sites of the ribosome, namely the decoding center, the sites controlling translation fidelity, and the peptidyl transferase center. Finally, we will review briefly the effects of mutations in subunit bridges that control the communication between the subunits and mutations in the 16S rRNA that were proposed to influence the binding of messenger RNAs. Given the high degree of conservation of the rRNA, the information obtained with *E. coli* can readily be extrapolated to other organisms.

A comprehensive list of the effects of mutations in 16S and 23S rRNA can be found at the web site of the ribosomal mutation data base developed by K. Triman, Department of Biology, Franklin and Marshall College, Lancaster, PA, USA (http://ribosome.fandm.edu).

Translation Mechanisms, edited by Jacques Lapointe and Léa Brakier-Gingras.
©2003 Eurekah.com and Kluwer Academic / Plenum Publishers.

Strategies for Mutagenesis of Ribosomal RNA

Classic Approaches

The study of rRNA mutations has been limited by the fact that most organisms that are suitable for genetic and biochemical analysis have multiple copies of rRNA operons. *E. coli* has seven such operons in its chromosome, coding for 16S rRNA, 23S rRNA and 5S rRNA as well as several tRNAs. The first classic approach developed in the early 80s to investigate the effects of mutations in rRNA consisted in transforming bacteria with a multicopy plasmid harboring one of the *rrn* operons, with mutations at selected sites (Fig. 1A). The operon was either constitutively expressed, under control of its natural promoters, P_1P_2, or under control of a weaker inducible promoter, such as the thermoinducible P_L promoter from bacteriophage λ (reviewed in Refs 10-12). A further development was the use of the T7 promoter to express rRNA in strains expressing the T7 RNA polymerase from an IPTG-regulated promoter.[13] All these expression systems generated mixed populations of ribosomes, with about 30-70% of the population (depending upon the strength of the promoter) containing plasmid-encoded rRNA and the remainder containing chromosomally-encoded rRNA. With this approach, mutations that affected critical ribosomal functions often resulted in either a dominant lethal or a slow-growth phenotype. In contrast to a commonly held assumption, it must, however, be stressed that a dominant lethal phenotype does not mean that the mutated ribosomes are inactive. On the contrary, dominant lethality requires that the mutant rRNA is expressed and enters the translating pool of ribosomes. The reasons for dominant lethality are not well-understood, with the possible exceptions of the rRNA mutants that increase the level of translational errors and synthesize high levels of aberrant proteins and of 16S rRNAs with alterations in the anti-Shine-Dalgarno sequence, the mRNA binding sequence (see below), that alter radically the level of translation of many mRNAs. Some of the possible explanations that have been considered are effects on ribosome stalling, peptidyl-tRNA drop off, polar effects due to slow elongation and leading to mRNA degradation prior to the completion of translation, and the inability to switch from the initiation to elongation phases of translation. Recessive lethal mutants have also been recovered. Such mutants often have little or no phenotype when expressed in mixed populations of ribosomes. However, under conditions where only the mutant ribosomes are active (see below), such mutants fail to support growth. While one subclass of these mutants were defective in ribosome assembly, many have been found to be well-represented in 70S ribosomes and the actively translating polysome fractions. It is inferred, however, that the activity of these ribosomes is insufficient to support growth. Another class of recessive lethal mutants are true null mutants: while the mutated rRNAs are assembled into subunits, the mutated subunits fail to associate with the other subunit, are excluded from the actively translating pool of ribosomes and consequently have no protein synthesis activity.

To circumvent the problems raised by the heterogeneity of ribosome populations, additional mutations conferring resistance to such antibiotics as spectinomycin, erythromycin or thiostrepton have also been introduced into the plasmid-encoded rRNA. This enables a selective inactivation of the ribosomes containing chromosomally-encoded rRNA in the presence of the antibiotic. A drawback of this procedure is that since antibiotics invariably target critical functional sites in the ribosome, these additional antibiotic resistance mutations may have unanticipated effects on the rRNA function.[14] Structural analysis of mutant rRNA has been facilitated by the development of the allele-specific probing method.[15-16] With this method, silent mutations are introduced into non-conserved regions of the plasmid-encoded rRNA, allowing specific primers to be used in primer extensions to characterize the pattern of RNA modification by chemical reagents and enzymes.

To investigate the consequences of mutations in rRNA without a background of the wild-type rRNA, another strategy termed "synthetic ribosomes" or "in vitro genetics" was developed independently by the groups of Ofengand[17] and Brakier-Gingras[18] for the 16S rRNA (Fig. 1B). With this strategy, mutant rRNA was generated by in vitro transcription of an appropriate plasmid containing the 16S rRNA gene under control of a T7 promoter. This transcript was then assembled into 30S subunits with ribosomal proteins and the activity of the mutated 30S subunits was assessed in vitro in the different steps of protein synthesis. The advantage of the procedure is that the whole population of 30S is mutated, but it is lengthy and laborious compared to the in vivo approach. Also, the transcript lacks the rRNA modifications (base and ribose methylations), and although the 30S

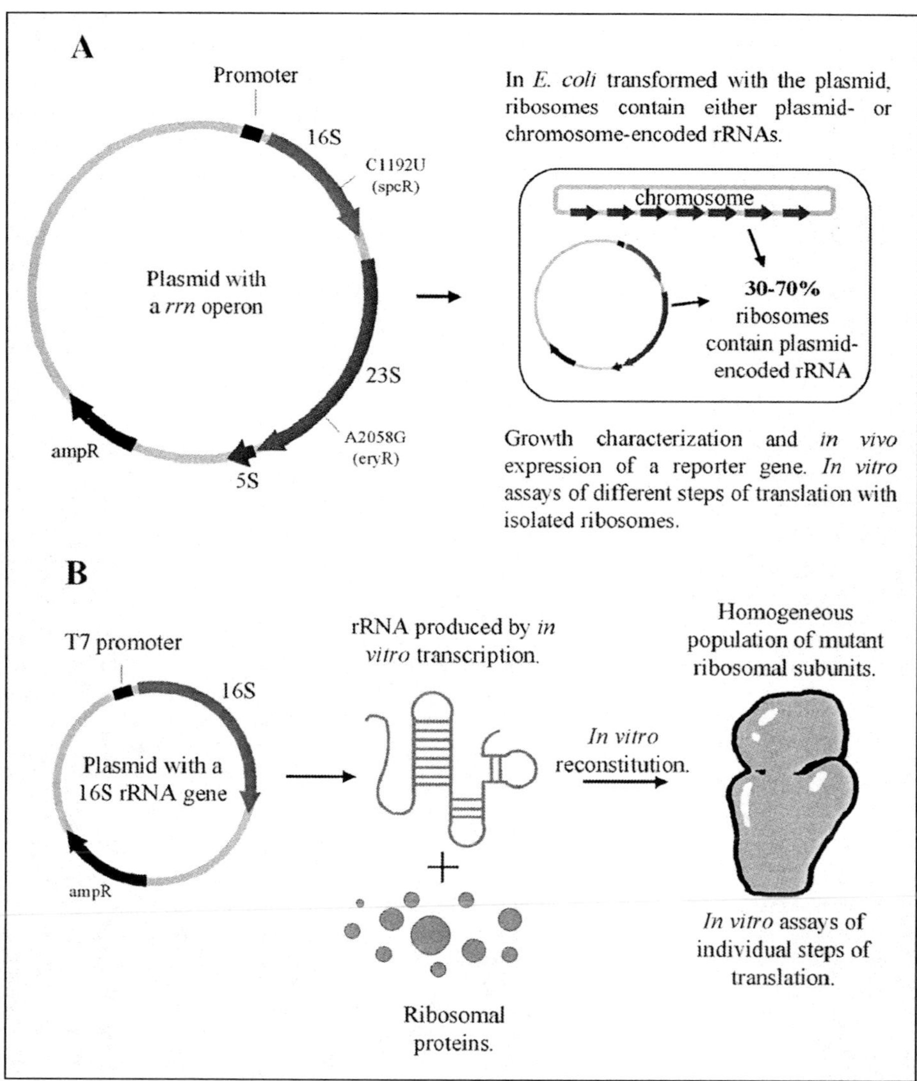

Figure 1. Classic approaches for mutagenesis of rRNA. A) Transformation of bacteria with a multicopy plasmid harboring a *rrn* operon under control of a constitutive or an inducible promoter. Mutations C1192U and A2058G confer respectively resistance to spectinomycin and erythromycin. B) In vitro genetics. Reconstitution of ribosomal subunits with a rRNA generated by in vitro transcription.

subunits reconstituted with the transcript were about 50% as active as the subunits reconstituted with the natural 16S rRNA, the absence of modifications could potentially further compromise the activities of both wild-type and mutant rRNAs.[19] Another potential disadvantage of the synthetic ribosomes is that the slow reaction kinetics typically obtained with in vitro protein synthesis reactions compared to the in vivo situation could mask some important effects of the mutations.

A full-length transcript of *E. coli* 23S rRNA produced by in vitro transcription and assembled with ribosomal proteins generates inactive 50S subunits, which initially prevented the application of this strategy to the study of mutations in 23S rRNA. However, it was observed that a chimeric *E. coli* 23S rRNA transcript where the 3' terminal 600 nucleotides were provided by a RNAse H fragment from natural 23S rRNA, could be assembled to form active 50S subunits.[20] This indicated that some of the modifications were important for forming active 50S subunits, at least in mesophilic bacteria. More recently, it was found that transcripts of 23S rRNA from the thermophilic bacteria *Bacillus stearothermophilus* and *Thermus aquaticus* could be readily assembled into active 50S subunits.[21-22] Thus, mutations in both subunits can now be studied by in vitro genetics.

Novel Approaches

Several recent approaches have improved upon the in vivo mutagenesis schemes described above (see Fig. 2). One approach developed by Squires and her collaborators uses a strain of *E. coli* where all seven chromosomal copies of the rRNA genes have been inactivated (Δ7 prrn).[23] In such a strain, all rRNA is transcribed from a plasmid-encoded *rrn* operon, and this resident plasmid can be displaced by another plasmid encoding either a mutant or wild-type rRNA, using an appropriate selection. This system thus generates a homogeneous population of ribosomes and, in contrast to the classic in vivo approach, avoids the problems encountered with a mixed population. However, mutations that have deleterious effects on ribosome function do not support growth in Δ7 prrn. A second novel strategy involves the use of ribosomes specialized for specific mRNAs. It is well-known that bacterial ribosomes select translational start sites on mRNAs by using the Shine-Dalgarno (SD) interaction, which normally involves base-pairing between a polypurine tract upstream the initiator AUG, the Shine-Dalgarno sequence, and a complementary pyrimidine region at the 3' end of 16S rRNA, the anti-Shine-Dalgarno (ASD) sequence. The specialized ribosome system consists of a plasmid encoding a *rrn* operon, where the ASD sequence on 16S rRNA has been mutated, and a reporter mRNA carrying an altered SD sequence that is now complementary to the mutated ASD sequence of 16S rRNA. Ribosomes containing the rRNA with the altered ASD sequence translate exclusively the reporter mRNA with the complementary SD sequence. This approach was pioneered by de Boer with a reporter gene coding for the human growth hormone[24] and later on for chloramphenicol acetyltransferase (CAT).[25] It was subsequently improved by the group of Cunningham, who showed that several mutated ASD sequences including the ASD previously used by de Boer are lethal, and cause rapid cell lysis, probably because of over-translation of genes that are toxic to the host bacteria. Cunningham and coworkers identified mutated ASD-SD combinations that do not have a toxic effect and enable the translation of the CAT reporter mRNA (carrying the altered SD sequence) exclusively.[26] This has made possible the analysis of the effects of specific rRNA mutations on translation by monitoring the expression of the CAT reporter gene, and since the specialized ribosomes translate only a single mRNA with the altered SD sequence, mutations that would be otherwise extremely deleterious or lethal can be studied with this method. This approach has been used with considerable success to carry out saturation mutagenesis on specific regions of 16S rRNA. The so-called instant evolution procedure consists in randomly mutating all the bases of a region of 16S rRNA under study in the context of an altered ASD sequence and selecting the functional mutants, based on their capacity to translate the CAT reporter mRNA and to support growth on chloramphenicol. Novel functional rRNA mutants carrying multiple base mutations were thus selected.[27] Since multiple rRNA base changes are created simultaneously, the appearance of novel patterns of covariation in the functional mutants (not previously observed in nature) can help uncover complex base-pairing interactions.

Effects of Mutations in Ribosomal RNA

Mutations in the Peptidyl Transferase Center

Peptide bond formation is a crucial step of protein synthesis, which involves tRNA substrates bound at the A and P sites. The central loop in domain V of 23S rRNA as well as two neighboring loops capping helices 80 and 92, called the P and A loop, respectively, have long been linked to the peptidyl transferase (PT) activity (reviewed in Refs 28-31; see Fig. 3). This conclusion was based on the data derived from a variety of biochemical approaches, including crosslinking and chemical protection experiments and also from the study of several mutations within the central loop that confer resistance to inhibitors of the PT activity. The group of Garrett used random PCR mutagenesis in the central loop of domain V to identify nucleotides that are important for the PT activity. Mutant rRNAs were expressed from a multicopy plasmid carrying a *rrn* operon under the control of

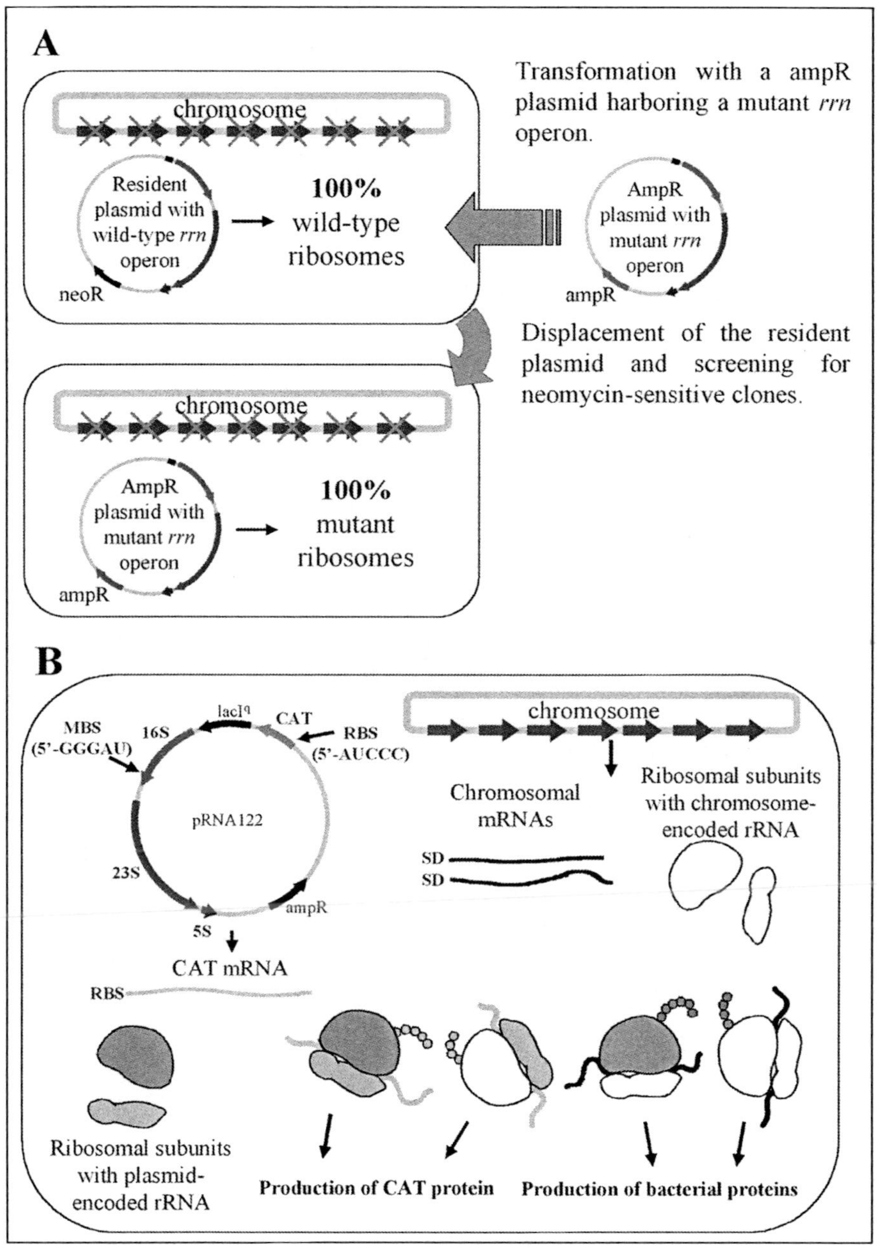

Figure 2. Novel approaches for mutagenesis of rRNA. A) Use of a strain where all the seven chromosomal *rrn* operons have been inactivated and where all rRNAs are expressed from a plasmid-encoded *rrn* operon. NeoR: resistance to neomycin, AmpR: resistance to ampicillin. B) The specialized ribosome system where only ribosomes with plasmid-encoded 16S rRNA translate a reporter mRNA (CAT mRNA coding for chloramphenicol acetyltransferase).

Figure 3. Some important domains in 23S rRNA shown on the following page. A. Portions of the secondary structure of domains IV, V and VI of *E. coli* 23S rRNA are presented. Long arrows indicate bases in the A and P loops which pair to the end of the aminoacyl- and peptidyl-tRNAs, respectively. The boxed bases are G2447 and A2451, that were claimed to directly participate in peptide bond formation. Bases where mutations impair peptidyl transferase (PT) activity are circled. Blue and red dots refer to bases where mutations make, respectively, the ribosomes hyperaccurate and error-prone. Inset: skeleton of the secondary structure of 23S rRNA. B. Overview of the structure of the 23S rRNA elements in domains V and VI presented in A, as seen in the crystal structure of the 50S subunit of *H. marismortui*.[3] Bases 2447 and 2451 (*E. coli* numbering) are orange. White bases correspond to sites where mutations impair peptidyl transferase activity. G2252 (green) and G2553 (yellow) base-pair respectively with the 3'end of the peptidyl- and the aminoacyl-tRNA. The blue and red bases correspond to mutations making, respectively, the ribosomes hyperaccurate and error-prone.

an inducible promoter and containing an additional mutation, A2058G, that confers resistance to erythromycin as well as to clindamycin, an effective PT inhibitor.[32-33] This makes possible a selective inactivation of ribosomes with wild-type chromosomally-encoded rRNA. Peptide bond formation was assessed in vitro with the mixed population of ribosomes in the presence of clindamycin, using the fragment reaction.[34] This reaction uses puromycin, a structural and functional analog of the aminoacylated 3' terminal adenosine of aminoacyl-tRNA, which acts as a nucleophile to attack an N-blocked-aminoacylated fragment of the tRNA in the P-site, usually a fragment of N-formyl- or N-acetyl-methionyl-tRNA[Met] or N-acetyl-phenylalanyl-tRNA[Phe]. The results stressed the importance of several positions in the central loop of domain V for the PT activity, since most changes in these positions perturbed this activity (see Fig. 3 for examples of mutations in 23S rRNA). Moreover, as will be discussed in the next section, domain IV of 23S rRNA interacts with domain V, and random mutagenesis within domain IV also characterized mutants with a decreased PT activity.[35] Nierhaus and coworkers also addressed the problem of the contribution to the PT activity of selected nucleotides in and around the central loop. However, in their case, the mutant 23S rRNAs included the A1067U mutation conferring resistance to the EF-G inhibitor, thiostrepton. PT activity assays were designed to be dependent on the translocation of the donor substrate, N-acetyl-phenylalanyl-tRNA[Phe], from the A to P sites and consequently, the PT activities of mutant ribosomes could be examined in the presence of thiostrepton.[36-38] In vitro genetics where 50S subunits were reconstituted with a mutated RNA transcript of *B. stearothermophilus* or *T. aquaticus* 23S rRNA was also used to examine the contribution of various nucleotides to the PT activity and to the correct positioning of the invariant 3' CCA end of the tRNA in the A and P sites.[21-22] A direct Watson-Crick interaction was demonstrated between C74, the first C at the 3'CCA terminal sequence of P-site bound tRNA, and G2252 of 23S rRNA in the P loop[20], and between C75 of the A-site bound tRNA and G2553 of the A loop.[39-40] The importance of G2252 had been suggested by the dominant lethal phenotype of all mutations of this base. The direct demonstration of its interaction with the 3'end of the tRNA was made by a binding study with a N-blocked oligonucleotide fragment containing the 3'CCA sequence and with mutants of this fragment altered so as to restore a potential base-pairing between position 74 in the tRNA and 2252 in the rRNA. Base-pairing between tRNA and rRNA was then assessed with the mixed population of ribosomes, by investigating the protection conferred by the oligonucleotide fragment on the mutant ribosomes using an allele-specific primer extension assay. This binding assay supported the proposed base-pairing (see Fig. 4). Finally, in vitro genetics with reconstituted 50S subunits containing a chimeric *E. coli* 23S rRNA transcript mutated at position 2252 unambiguously demonstrated the existence of a base-pairing between this position and the tRNA and the requirement of this pairing for PT function. Disruption of the rRNA-tRNA base-pairing by alteration of either component abolished PT activity, which was restored when this pairing was made possible again. The tRNA-rRNA interaction in the A loop at the A site was examined in a similar way. Mutations at position G2553 have a dominant lethal growth defect. Mutant 50S ribosomal subunits were reconstituted in vitro using a 23S rRNA transcript from *B. stearothermophilus* or *T. aquaticus* and assayed in PT assays, with a fragment reaction using as an A site substrate, a puromycin derivative lengthened by one or two nucleotides so as to better mimic the CCA end of an aminoacyl-tRNA. These assays clearly demonstrated a Watson-Crick pairing interaction between position 75 of tRNA at the A-site and position 2553 of 23S rRNA.

A
P-site tRNA
P loop
II
III
IV
V
VI
I
Central loop
Sarcin-ricin
loop
A loop
A-site tRNA
B
Central loop
Sarcin-ricin
loop
P loop
A loop

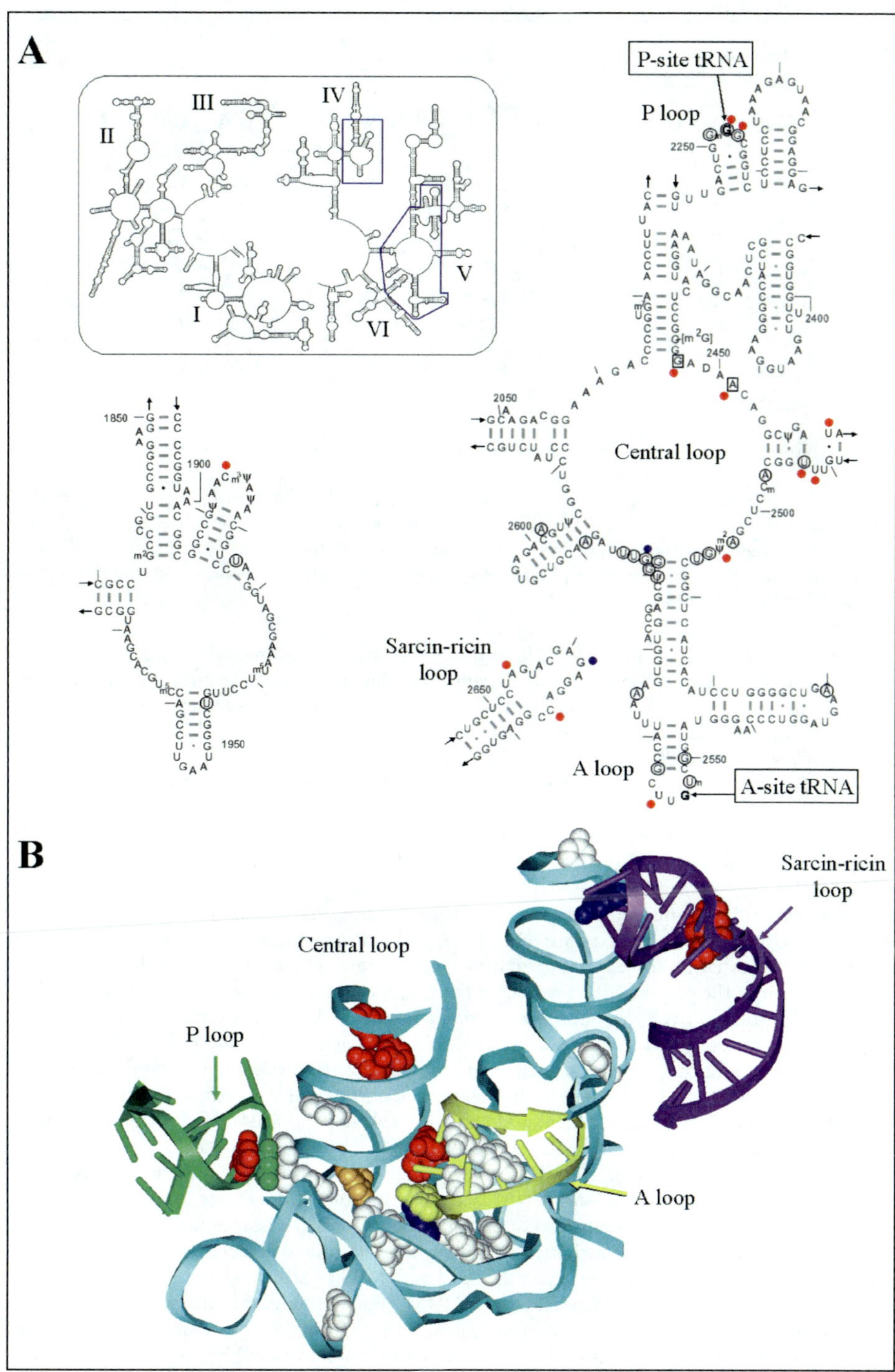

Figure 3.

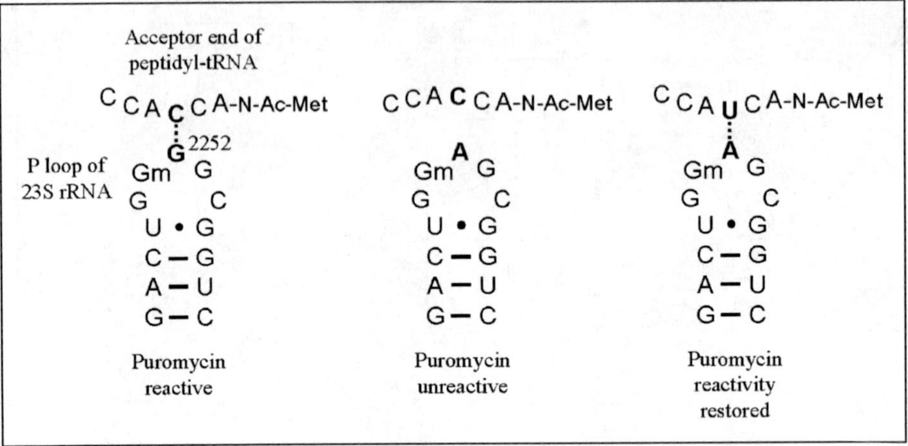

Figure 4. Demonstration of a base-pairing interaction between the P loop in 23S rRNA and an oligonucleotide fragment containing the 3' CCA end of a peptidyl-tRNA. Binding of the oligonucleotide fragment was assessed by measuring protection of 23S rRNA against chemical attack. Puromycin reactivity indicates a functional interaction. The dashed line indicates a base-pairing interaction between positions 74 of the tRNA and 2252 of 23S rRNA.

Site-directed mutagenesis in the A and P loop thus shows that the aminoacyl- and peptidyl-tRNA substrates are positioned in the catalytic PT center via direct base-pairing interactions with 23S rRNA, in full agreement with the recent structures of the 70S ribosome cocrystallized with tRNAs and the 50S subunit cocrystallized with tRNA substrate analogs (see below).[6,41] Curiously, the crystal structure revealed a further tRNA-rRNA base-pairing interaction involving C75 of P-site bound tRNA and G2251, although no such base-pairing interaction was demonstrated by the compensatory mutagenesis approach.[42]

The mechanism of peptide bond formation and the ribosomal components involved in this activity have been studied intensively for many years. Much biochemical and genetic evidence has accumulated to suggest that this activity is RNA based.[1] The high-resolution atomic structure of the *H. marismortui* 50S subunit complexed with substrate analogs have demonstrated unequivocally that the PT active site is composed entirely of RNA and that no protein comes closer than 18Å to the site of the peptide transfer. Crystal structures were obtained where 50S subunits had been crystallized alone, or in the presence of an analog of the tetrahedral carbon intermediate created during peptide bond formation (CCdA-phosphate-puromycin where the CCdA portion binds to the P loop) or in the presence of an N-amino-acylated minihelix that mimics the A site substrate.[41] A careful examination of these crystal structures led to the suggestion that the universally conserved A2451 of 23S rRNA, which is located in the central loop of domain V, proximal to the extremities of both A-site and P-site bound tRNAs, could play a key-role in peptide bond formation through an acid-base catalysis mechanism. Such a mechanism requires that the pK_a of A2451 be raised to the physiological range and this change in pK_a was proposed to be facilitated by a charge relay network involving G2061, G2447, A2450 and A2451. Characterization of the pK_a of A2451, based on its reactivity to dimethylsulfate as a function of pH, showed that it had a near-neutral value, about the same as that reported for the PT activity,[43] which appeared to provide biochemical support for the acid-base catalysis mechanism proposed on the basis of the 50S crystal structures. However, a recent study demonstrated that the pH-dependence of A2451 could be observed only with 50S subunits that had been inactivated by depletion of monovalent cations during their preparation and not with subunits reactivated by adjustment of the ionic balance of the buffers and heating.[44] Furthermore, genetic analyses have now cast considerable doubt on this mechanism. A2451 mutants have a dominant lethal phenotype, stressing the importance of this nucleotide,[43] but this is not unique to A2451. Moreover, despite the universal conservation of A2451, previous work showed that mutations at this position were tolerated and conferred resistance to chloramphenicol in rodent mitochondria.[45] In agreement with this, when ribosomes were obtained from cells expressing rRNA mutated at position 2451 in a plasmid-encoded *rrn* operon under the inducible control of a λP_L promoter and assayed an in vitro translation reaction, translation was resistant to chloramphenicol, showing that the mutated ribosomes retained protein synthesis activity. All three mutations of A2451 had a weak error-promoting activity in vivo, confirming that the mutant ribosomes were active in protein synthesis.[46] Finally, 50S subunits reconstituted with a 23S rRNA transcript mutated at position 2451 were clearly less active than the unmutated control, although they retained a significant PT activity, stressing again that the identity of base 2451 is not absolutely essential for the PT activity.[46-47]

The raised pK_a associated with A2451 had been proposed to be generated through a charge relay system.[41] Critical to this proposed mechanism was the identity of G2447. However, again, mutations at this base had been associated with chloramphenicol resistance in mitochondria and anisomycin resistance in archaea.[48-49] Also, each of the three 2447 mutants had a weak error-promoting activity in vivo when using the inducible expression system of a *rrn* operon. Moreover, G2447A and G2447C mutants were viable in the Δ7 prrn strain where only the mutant rRNA was expressed, and in vitro translation with these mutant ribosomes was completely resistant to the inhibitory effects of chloramphenicol.[46] Lastly, 50S subunits reconstituted with a 23S rRNA transcript mutated at position 2447 retained a significant PT activity.[47] Together, these results demonstrate that the mutant ribosomes were active in vivo and in vitro and that the identity of G2447 was not essential for any ribosome function.

Thus, the proposed acid/base catalysis mechanism can no longer be entertained, although A2451 is clearly important for the PT activity. The crystal structure of the 50S complex indicates that this nucleotide is located at the active site and mutations at A2451 decrease PT activity. However, several other mutations in the active site, as mentioned above, also decrease PT activity. As alternative mechanisms, one should now consider the possibility that the ribosome promotes transpeptidation by either properly positioning the substrates of protein synthesis or via stabilization of the reaction intermediate. The former possibility had already been proposed several years ago.[50]

Mutations in the Decoding Center and in Sites Controlling Translational Accuracy

Another key-function of the ribosome is the selection of the correct aminoacyl-tRNAs. The control of decoding and translational accuracy is a complex phenomenon that was reviewed recently.[51-52] In short, an incoming aminoacyl-tRNA interacts with the ribosome-bound mRNA as a ternary complex with elongation factor EF-Tu and GTP. Correct codon-anticodon interaction in the decoding center in the 30S subunit triggers a succession of events leading to the activation of the GTPase activity of EF-Tu, the release of EF-Tu•GDP, the occupancy of the 50S A site by the acceptor end of the aminoacyl-tRNA, and the activation of the catalytic center for peptide bond formation. Non- and near-cognate aminoacyl-tRNAs can be rejected upon the initial interaction of ternary complexes with the ribosome (initial selection) or, for those tRNAs that escaped the first control, after GTP hydrolysis but prior to peptide bond formation (proofreading). The prevailing scenario is that codon-anticodon interaction induces a conformational change in the decoding center (at the top of helix 44 of 16S rRNA), which is transmitted to other components of the 30S subunit, including helix 27 (see details in chapter 17 by S. J. Lodmell and S.P. Hennelly), the 530 stem/loop and helix 34 of 16S rRNA (see Fig. 5) and subsequently to the 50S subunit.

In vitro genetics demonstrated the importance of the base pairs forming the irregular helical region at the top of helix 44.[53-54] This was further supported by in vivo assays with the inducible expression system of a plasmid-encoded *rrn* operon[55] and with the Δ7 prrn strain where the mutant ribosomes comprised 100% of the ribosomal pool.[56] Perturbations of the structure on top of helix 44 have a very detrimental effect on ribosome activity, which is probably due to an impairment of the communication between subunits.

Many chemical protection studies have implicated bases G530, A1492 and A1493 in the decoding process.[1] Mutations in G530, A1492 and A1493 were found to be dominantly lethal.[57-58] Allele-specific probing of mutant ribosomes showed that the G530 mutants were defective in the EF-Tu mediated binding of the aminoacyl-tRNA,[59] and, more recently, ingenious damage selection experiments with ribosomes bearing biotinylated tRNA, using streptavidin beads, demonstrated that A1492 and A1493 were required for the aminoacyl-tRNA binding at the A site.[58] The current

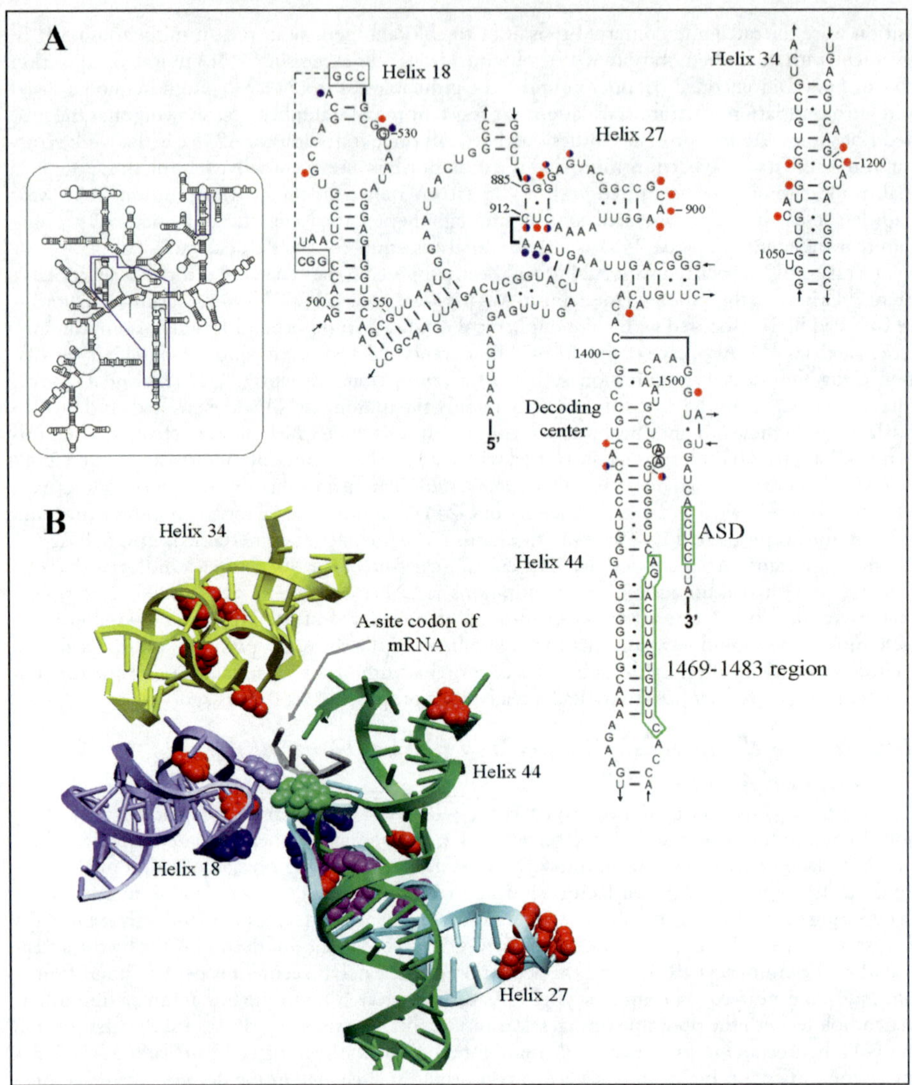

Figure 5. Some important domains in 16S rRNA. A. Portions of subdomains of *E. coli* 16S rRNA that are involved in the control of translational accuracy and that contain sequences demonstrated or proposed to participate in the binding of mRNA are presented. Helix 27 is presented under the 912/885 conformation (see the text). Bases G530, A1492 and A1493 that directly monitor the geometry of the codon-anticodon duplex are circled. The dashed line between the 530 loop and the adjoining bulge in the stem represents the pseudoknot which is suggested to play a critical role for the interaction with EF-Tu. Blue and red dots refer, respectively, to bases where mutations make the ribosomes hyperaccurate and error-prone. Half-blue/half-red dots indicate that ribosomes become hyperaccurate or error-prone, depending upon the nature of the mutation. The regions that are boxed in green correspond to the anti-Shine-Dalgarno (ASD) sequence and the 1469-1483 region. Inset: skeleton of the secondary structure of 16S rRNA. B. Overview of the structure of the 16S rRNA elements involved in the control of translational accuracy as observed in the crystal structure of the 30S subunit of *T. thermophilus*.[5] Blue and red bases correspond to sites where mutations make, respectively, the ribosomes hyperaccurate and error-prone, and purple bases correspond to sites where mutations make ribosomes either hyperaccurate or error-prone, depending upon the nature of the mutation. A1492 and A1493 (*E. coli* numbering) are green and G530 is cyan. The arrow points to the A-site codon of mRNA (in grey).

model, suggested by NMR and X-ray studies, infers that the correct codon-anticodon interaction is monitored by bases A1492, A1493 and G530. Binding of the aminoacyl-tRNA causes A1492 and A1493 to flip out of the internal loop of helix 44 and monitor the geometry of the minor groove of the A site codon-anticodon helix. It causes G530 to convert from a *syn* to an *anti* conformation, thereby also sensing the geometry of the codon-anticodon helix. Base C1054, bulging out of helix 34, is also positioned so as to interact with the anticodon loop of the aminoacyl-tRNA.[6,58,60-61] Not unexpectedly, mutations in the 1400/1490 region of the decoding center, as well as alterations in the 530 stem/loop have all been shown to perturb the fidelity of decoding. Also, mutations have been isolated in both of these regions, that confer resistance to error-promoting aminoglycoside antibiotics.[55,62-69] In the 530 stem/loop (helix 18), a pseudoknot can be formed between the top of the loop and a small adjoining loop bulging out of the stem (Fig. 5). Although the aminoacyl-tRNA itself likely contributes to the transmission of information from the decoding center to the GTP binding domain of EF-Tu,[70] an attractive model that we favor is that the codon-anticodon interaction promotes the formation of the pseudoknot in the 530 stem/loop, which could also contribute to the activation of this GTP binding domain and to the stimulation of the GTPase activity of EF-Tu.[71] This would trigger the release of EF-Tu•GDP and promote the interaction of the 3' end of the aminoacyl-tRNA with the 50S subunit. Cryo-electron microscopy supports this model, indicating some sort of interaction between EF-Tu and the 530 stem/loop region.[72] However, confirmation of this interaction awaits a high-resolution structure of the EF-Tu ternary complex bound to the ribosome. Alternatively, the formation of the pseudoknot could lead to a conformation with a low affinity for EF-Tu•GDP, ensuring that after GTP hydrolysis, EF-Tu•GDP is rapidly released and the aminoacyl-tRNA can occupy the A site on the 50S subunit and participate to peptide bond.[73] Mutations in the 530 stem/loop were found that make the ribosomes hyperaccurate or error-prone. In the light of the models suggesting a role for the pseudoknot, this could be ascribed to the fact that either these mutations interfere with or promote the formation of the pseudoknot, and, consequently, interfere with or promote the occupancy of the 50S A site.[63,68-69]

Helix 34 (the 1050/1200 region) in the 3' major domain of 16S rRNA, which is also part of the decoding center,[74-76] probably interacts with the anticodon loop, thus additionally stabilizing the binding of the aminoacyl-tRNA. Base changes at A1054 and C1200 in helix 34 were isolated genetically as UGA suppressors in *E. coli*.[62,77] To investigate the involvement of this helix in the fidelity of decoding, a range of mutations in helix 34 were constructed by site-directed mutagenesis. Some of these mutations were found to stimulate a wide variety of nonsense and frameshift errors.[66,78] Since frameshift errors can be directly related to errors in the selection of tRNAs[79], this indicates the involvement of bases 1054 and 1200 in the control of translational accuracy and not just in the termination phase of translation. Mutations in this helix probably enhance its interaction with the anticodon loop of the aminoacyl-tRNA, thus stabilizing the binding of this tRNA and making the ribosomes error-prone.

It has been shown that helix 27 switches between two alternate conformations, the 912/888 conformation (where bases 910 to 912 pair with bases 888 to 890), that has a low affinity for aminoacyl-tRNAs and the 912/885 conformation (where bases 910 to 912 pair with bases 885 to 887), that has a high affinity for aminoacyl-tRNAs (see chapter 17 by S. Lodmell and S.P.Hennelly).[80] It is now admitted that the ribosome oscillates between a binding conformation that accept or reject the aminoacyl-tRNAs and a productive conformation that is required for GTP hydrolysis by EF-Tu and occupancy of the 50S A site by the aminoacyl-end of the tRNA.[51-52] The model that we favor, although still speculative, is that an incoming aminoacyl-tRNA encounters ribosomes with the 912/888 conformation of helix 27, that would be the binding conformation (see also Ref. 81). Upon correct codon-anticodon interaction, helix 27 would switch to the 912/885 conformation. However, we cannot exclude the possibility that it is the reverse situation that occurs, and that the incoming aminoacyl-tRNA first encounters the 912/885 conformation that switches then to the 912/888 conformation.[80] Site-directed mutants were constructed in the 910-912 and 885-890 regions, designed to interfere with one or the other conformation. It was found that the mutations favoring the 912/888 conformation make the ribosomes hyperaccurate whereas mutations promoting the 912/885 conformation make the ribosomes error-prone.[80] Using the specialized ribosome system and the instant evolution procedure, it was shown that mutations in the loop capping helix 27 make the ribosomes error-prone, suggesting that they favor the switch to the 912/885 conformation.[82] These mutations were also found to impair subunit association (see below). Mutations in the hinge region adjacent to the switch region of helix 27, the 913-915 region, at the convergence of the major domains of 16S rRNA, make the ribosomes hyperaccurate, suggesting that they promote the switch to the 912/888 conformation of helix 27.[83-84] They also interfere with the formation of the pseudoknot in helix 18[84] as do the mutations in the switch region of helix 27 that make ribosomes hyperaccurate,[80] thus showing a tight interplay between different regions of 16S rRNA involved in decoding .

How conformational changes occurring in the 30S subunit upon binding of an aminoacyl-tRNA transmit a signal to the 50S subunit is still a mystery. As shown by various protection and structural studies, domain IV of 23S rRNA is involved in the formation of several bridges with the 30S subunit. Distinct regions of domain IV interact with the decoding center at the top of helix 44 of 16S rRNA as well as with other portions of this helix, and with the loops capping helices 23, 24 and 27.[6,85-87] It was proposed that the codon-anticodon interaction, by altering the intersubunit bridge contacts, induces conformational changes in domain IV of 23S rRNA. This domain is in a tight communication with domain V, as was clearly shown by X-ray studies.[3] Therefore, changes within domain IV could affect the conformation and catalytic activity of the PT center in domain V, thus controlling the proper position of the peptidyl-tRNA and the incoming aminoacyl-tRNA. Mutations in the conserved 1916 loop of domain IV of 23S rRNA were isolated in a genetic selection for rRNA mutations causing frameshift in a reporter gene and were subsequently shown to affect both nonsense and frameshift errors in *E. coli*.[88] Mutations in the 1916 loop which contacts the decoding center could directly influence decoding by altering the conformation of this center. Alternatively, since mutations in this region have also been shown to have drastic effects on the PT activity,[35] mutations in the 1916 loop could affect the fidelity of translation indirectly by altering the interaction of the 3'end of the tRNAs with the PT center. The same genetic selection used to isolate the domain IV mutations also yielded mutations at position U2555 in the A loop itself, close to the central loop of domain V.[88] Moreover, various mutations in the P loop and at the catalytic center constructed by site-directed mutagenesis were also shown to affect the fidelity of decoding.[65,89-90]

The activity of EF-Tu plays a key role in the delivery of the correct aminoacyl-tRNAs to the ribosome, as was mentioned above. In the 50S subunit, EF-Tu interacts with the sarcin-ricin loop around position 2660 in domain VI.[91-93] Both error-prone and error-restrictive mutations have been isolated in this region. Thus, the G2661C mutation was found to decrease mistranslation in vivo. In vitro characterization of this mutation indicated that this effect was due to an increase in the stringency of the initial selection step although an increase in proofreading efficiency could not be excluded.[94-95] A variety of base changes at positions A2654 and C2666, which are involved in a non-canonical base-pairing, increase the frequency of frameshift and of nonsense errors in vivo.[96] These mutations probably affect tRNA selection by interfering with the correct binding of EF-Tu as well as with its release after GTP hydrolysis and, in this way, alter the relative rates of tRNA rejection and accommodation in the 50S A site.

From the long list of mutations affecting translational accuracy, one can make the general conclusion that these mutations are not limited to the decoding center and that several other sites on both subunits are involved in the process. The fidelity of decoding can also be altered by mutations in either subunit that alter the transmission of signals leading to the occupancy of the A site on the 50S subunit. Mutations that speed up this process can make ribosomes error-prone in contrast to mutations slowing down this signal transmission, which make ribosomes hyperaccurate. The translational fidelity can also be influenced by mutations that affect the placement of the 3'end of tRNAs in the PT center, the structure of this catalytic center and the rate of peptide bond formation. Thus, the tRNA selection and the decoding process require the participation of several rRNA regions, which are tightly interrelated functionally (see Figs. 3 and 5 for a review of mutations affecting translational accuracy).

Mutations in the Intersubunit Bridges and in the Messenger RNA Binding Site

As explained above, the communication between ribosomal subunits is essential for proper ribosome function. Protection data had initially given a general description of the sites in rRNA involved in subunit association,[85-86] to which cryo-electron microscopy[87] and X-ray crystallographic studies subsequently added many essential details.[6] The interface contacts between ribosomal subunits are almost exclusively RNA-RNA bridges, connecting elements of domain IV of 23S rRNA and various regions of 16S rRNA. The upper part of helix 44 plays a major role in the formation of these bridges, which makes mutations within this region highly detrimental (see above Refs 53-56). The specialized ribosome system of Cunningham coupled with the instant evolution method has been used to investigate three regions of 16S rRNA involved in the RNA-RNA bridges, the 790 loop capping helix 24 of 16S rRNA,[27] the 690 loop capping helix 23,[97-98] and the 900 loop capping helix 27.[82] This approach directly establishes the involvement of these loops in subunit association, by demonstrating that the proportion of 30S subunits with plasmid-encoded rRNA decreases in the ribosome and polysome fractions. A covariation analysis of a large number of variants generated with this approach revealed novel interactions that could be confirmed by NMR and X-ray studies.

Lastly, another important function of the 30S subunit is to correctly select translational start sites on mRNAs through the Shine-Dalgarno interaction. Direct demonstration of this interaction was done by nuclease protection experiments[99] and with the use of a specialized ribosome system.[24,26] However, in vitro assays with synthetic ribosomes in which the 16S rRNA transcript lacked the ASD sequence showed that ribosomes could initiate correctly on a mRNA without the Shine-Dalgarno interaction.[100] It is also known that naturally occurring leaderless mRNAs (which begin directly with an initiator AUG triplet) initiate in the absence of any SD-ASD interaction.[101] The existence in several mRNAs of sequences that have been shown by mutagenesis to be important for translation and are partially complementary to regions of rRNA outside of the ASD has given rise to a number of alternative mRNA-rRNA base pairing schemes. A well-known case is the downstream box (DB) sequence, located downstream the initiator codon, that was found in a number of mRNAs and is partially complementary to bases 1469-1483 in helix 44 of 16S rRNA (Fig. 5). This sequence was proposed to promote initiation through base-pairing with the 1469-1483 region. This base-pairing model was supported by mutagenesis experiments in the DB sequence of various mRNA showing that the complementarity of the DB with the 1469-1483 region positively correlated with the efficiency of the gene expression.[102] However, evidence against this model was subsequently provided by mutagenesis experiments with leaderless mRNAs.[103] Furthermore, when the sequence of the 1469-1483 region of helix 44 was altered radically, the mutated rRNAs were viable in the Δ7 prrn strain, where only mutant rRNA was expressed, while the translation of a wide variety of DB-containing mRNAs was unaffected. Thus, although the DB has a clear effect on gene expression, this has no relation with a base-pairing interaction with helix 44 of 16S rRNA.[104-106] Another possible mRNA-rRNA interaction involving a U-rich sequence (termed epsilon) was found in gene 10 of phage T7 and a complementary sequence in the 460 region of 16S rRNA. This interaction, however, has similarly been disproven by showing that radical alterations in the proposed site of mRNA base-pairing in 16S rRNA had no effect on the expression of the epsilon containing mRNAs.[106-107] Thus, although in both these instances, a limited mRNA mutagenesis supported rRNA-mRNA base-pairing, mutagenesis of the corresponding rRNA sequence left mRNA expression unaltered, effectively disproving the proposed mRNA-rRNA interactions. The mechanism underlying the enhancing effect on translation of the epsilon and DB elements remains unknown.

Summary and Perspectives

Mutational analyses have been extremely important in showing how the decoding center signals the formation of the correct codon-anticodon duplex to the other functional centers of the ribosome and in demonstrating the direct involvement of the rRNA in mRNA binding and in the binding of the aminoacyl-tRNA and peptidyl-tRNA. The translocation step, because of its high complexity involving all tRNA binding sites and a relative movement of the two ribosomal subunits, has been less investigated with the mutagenesis approach, but the startling progress in recent years makes this step now ripe for such investigations. However, given the wealth of detailed structural information now available about the ribosome, one may ask whether the mutagenesis approach has any future? Many mechanistic questions remain to be answered and the various strategies that have been developed to investigate mutations in rRNA provide an invaluable tool to address these questions. The elucidation of crystal structures suggests the existence of many elements of tertiary structure whose structural and functional roles can be directly investigated by mutagenesis of rRNA. Several interactions are also suggested between the rRNA and the tRNA or the different factors and the characterization of all these interactions by mutagenesis should contribute to our understanding of the mechanisms that determine the ribosome function.

Acknowledgments

We thank Phil Cunningham, Al Dahlberg, Francis Robert and Sergey Steinberg for critical reading of this review. The work from Léa Brakier-Gingras and coworkers cited herein was supported by a grant from the Canadian Institutes of Health Research (to L. B.-G.). Michael O'Connor was supported by a grant GMS 19756 from the National Institute of Health (to A.E. Dahlberg).

References

1. Green R, Noller HF. Ribosomes and translation. Annu Rev Biochem 1997; 66:679-716.
2. Maguire BA, Zimmermann RA. The ribosome in focus. Cell 2001; 104:813-816.
3. Ban N, Nissen P, Hansen J et al. The complete atomic structure of the large ribosomal subunit at 2.4 Å resolution. Science 2000; 289:905-920.
4. Schluenzen F, Tocilj A, Zarivach R et al. Structure of functionally activated small ribosomal subunits at 3.3 Å resolution. Cell 2000; 102:615-623.
5. Wimberly BT, Brodersen DE, Clemons WM et al. Structure of the 30S ribosomal subunit. Nature 2000; 407:327-339.
6. Yusupov MM, Yusupova GZ, Baucom A et al. Crystal structure of the ribosome at 5.5 Å resolution. Science 2001; 292:883-896.
7. Frank J. Cryo-electron microscopy as an investigative tool: the ribosome as an example. Bioessays 2001; 23:725-732.
8. van Heel M. Unveiling ribosomal structures: the final phases. Curr Opin Struct Biol 2000; 10:259-264
9. Gabashvili IS, Agrawal RK, Grassucci R et al. Major rearrangements in the 70S ribosomal 3D structure caused by a conformational switch in 16S ribosomal RNA. EMBO J 1999; 18:6501-6507.
10. Leclerc D, Brakier-Gingras L. Study of the function of *Escherichia coli* ribosomal RNA through site-directed mutagenesis. Biochem Cell Biol 1990; 68:169-179.
11. Tapprich WE, Göringer HU, De Stasio E et al. Site-directed mutagenesis of *E. coli* rRNA. In: Spedding G, ed. Ribosomes and Protein synthesis. A practical approach. IRL Press at Oxford University, 1990:253-271.
12. Gregory ST, Brunelli CA, Lodmell JS et al. Genetic selection of rRNA mutations. Methods Mol Biol 1998; 77:271-281.
13. Steen R, Dahlberg AE, Lade BN et al. T7 RNA polymerase-directed expression of the *Escherichia coli* rrnB operon. EMBO J 1986; 5:1099-1103.
14. Recht MI, Puglisi JD. Aminoglycoside resistance with homogeneous and heterogeneous populations of antibiotic-resistant ribosomes. Antimicrobial Agents Chemother 2001; 45:2414-2419.
15. Power T, Noller HF. Allele-specific structure probing of plasmid-derived 16S ribosomal RNA from *Escherichia coli*. Gene 1993; 123:75-80.
16. Douthwaite S. Allele-specific priming in the mapping of rRNA. Methods Mol Biology 1998; 77:11-21.
17. Krzyzosiak W, Denman R., Nurse K et al. In vitro synthesis of 16S ribosomal RNA containing single base changes and assembly into a functional 30S ribosome. Biochemistry 1987; 26:2353-2364.
18. Melançon P, Gravel M, Boileau G et al. Reassembly of active 30S ribosomal subunits with unmethylated in vitro transcribed 16S rRNA. Biochem Cell Biol 1987; 65:1022-1030.
19. Cunningham PR, Richard RB, Weitzmann CJ et al. The absence of modified nucleotides affects both in vitro assembly and in vitro function of the 30S ribosomal subunit of *Escherichia coli*. Biochimie 1991; 73:789-796.
20. Samaha RR, Green R, Noller HF. A base-pair between tRNA and 23S rRNA in the peptidyl transferase centre of the ribosome. Nature 1995; 37:309-314.
21. Green R, Noller HF. Reconstitution of functional 50S ribosomes from in vitro transcripts of *Bacillus stearothermophilus* 23S rRNA. Biochemistry 1999; 38:1772-1779.
22. Khaitovich P, Tenson T, Kloss P et al. Reconstitution of functionally active *Thermus aquaticus* large ribosomal subunits with in vitro-transcribed rRNA. Biochemistry 1999; 38:1780-1788.
23. Asai T, Zaporojets D, Squires C et al. An *Escherichia coli* strain with all chromosomal rRNA operons inactivated : complete exchange of rRNA genes between bacteria. Proc Natl Acad Sci USA 1999; 96:1971-1976.
24. Hui AS, de Boer HA. Specialized ribosome system: preferential translation of a single mRNA species by a subpopulation of mutated ribosomes in *Escherichia coli*. Proc Natl Acad Sci USA 1987; 84:4762-4766.
25. Brink MF, Verbeet MP, de Boer HA. Specialized ribosomes: highly specific translation in vivo of a single targetted mRNA species. Gene 1995; 156:215-222.
26. Lee KS, Holland-Staley CA, Cunningham PR. Genetic analysis of the Shine-Dalgarno interaction: selection of alternative functional mRNA-rRNA combinations. RNA 1996; 2:1270-1285.
27. Lee KS, Varma S, SantaLucia Jr J et al. In vivo determination of RNA structure-function relationships: analysis of the 790 loop in ribosomal RNA. J Mol Biol 1997; 269:732-743.
28. Lieberman KR, Dahlberg AE. Ribosome-catalyzed peptide-bond formation. Prog Nucleic Acid Research Mol Biol 1995; 50:1-23.
29. Garrett RA, Rodriguez-Fonseca C. The peptidyl transferase center. In: Zimmermann RA, Dahlberg, AE, eds. Ribosomal RNA. Structure, Evolution, Processing and Function in Protein Biosynthesis. CRC press, Boca Raton, FL, USA, 1996:327-355.

30. Porse BT, Kirillov SV, Garrett RA. Antibiotics and the peptidyl transferase center. In: Garrett RA, Douthwaite SR, Liljas A, Matheson AT, Moore PB, Noller HF, eds. The Ribosome. Structure, Function, Antibiotics and Cellular Interactions. ASM Press, Washington, DC, 2000: 441-449.

31. Vester B, Douthwaite S. Macrolide resistance conferred by base substitutions in 23S rRNA. Antimicrobiol Agents Chemother 2001; 45:1-12.

32. Porse BT, Garrett RA. Mapping important nucleotides in the peptidyl transferase centre of 23S rRNA using a random mutagenesis approach. J Mol Biol 1995; 249:1-10.

33. Porse BT, Thi-Ngoc HP, Garrett RA. The donor substrate site within the peptidyl transferase loop of 23S rRNA and its putative interactions with the CCA-end of N-blocked aminoacyl-tRNA^{Phe}. J Mol Biol 1996; 264:472-483.

34. Monro RE, Marcker KA. Ribosome-catalysed reaction of puromycin with a formylmethionine-containing nucleotide. J Mol Biol 1967; 25:347-350.

35. Leviev I, Levieva S, Garrett RA. Role for the highly conserved region of domain IV of 23S-like rRNA in subunit-subunit interactions at the peptidyl transferase centre. Nucleic Acids Res 1995; 23:1512-1517.

36. Spahn CMT, Remme J, Schäfer MA et al. Mutational analysis of two highly conserved UGG sequences of 23S rRNA from *Escherichia coli*. J. Biol Chem 1996; 271:32849-32856.

37. Spahn CMT, Schäfer MA, Krayevsky AA et al. Conserved nucleotides of 23S rRNA located at the ribosome peptidyltransferase center. J Biol Chem 1996; 271:32857-32862.

38. Saarma U, Spahn CM, Nierhaus KH et al. Mutational analysis of the donor substrate binding site of the ribosomal peptidyltransferase center. RNA 1998; 4:189-194.

39. Kim DF, Green R. Base-pairing between 23S rRNA and tRNA in the ribosomal A site. Molecular Cell 1999; 4:859-864.

40. Khaitovich P, Mankin AS. Reconstitution of the 50S subunit with in vitro transcribed rRNA: a new tool for studying peptidyltransferase. In: Garrett RA, Douthwaite SR, Liljas A, Matheson AT, Moore PB, Noller HF, eds. The Ribosome. Structure, Function, Antibiotics and Cellular Interactions. ASM Press, Washington DC, 2000:229-243.

41. Nissen P, Hansen J, Ban N et al. The structural basis of ribosome activity in peptide bond synthesis. Science 2000; 289:920-930.

42. Green R, Samaha RR, Noller HF. Mutations at nucleotides G2251 and U2585 of 23S rRNA perturb the peptidyl transferase center of the ribosome. J Mol Biol 1997; 266:40-50.

43. Muth GW, Ortoleva-Donnelly L, Strobel SA. A single adenosine with a neutral pK_a in the ribosomal peptidyl transferase center. Science 2000; 289:947-950.

44. Bayfield MA, Dahlberg AE, Schulmeister U et al. A conformational change in the ribosomal peptidyl transferase center upon active/inactive transition. Proc Natl Acad Sci USA 2001; 98:10096-10101.

45. Kearsey SE, Craig IW. Altered ribosomal RNA genes in mitochondria from mammalian cells with chloramphenicol resistance. Nature 1981; 290:607-608.

46. Thompson J, Kim DF, O'Connor M et al. Analysis of mutations at residues A2451 and G2447 of 23S rRNA in the peptidyltransferase active site of the 50S ribosomal subunit. Proc Natl Acad Sci USA 2001; 98:9002-9007.

47. Polacek N, Gaynor M, Yassin M et al. Ribosomal peptidyl transferase can withstand mutations at the putative catalytic nucleotide. Nature 2001; 411:498-501.

48. Dujon B. Sequence of the intron and flanking exons of the mt 21S rRNA gene of yeast strains having different alleles at the omega and rib-1 loci. Cell 1980; 20:185-197.

49. Hummel H, Bock A. 23S ribosomal RNA mutations in halobacteria conferring resistance to the anti-80S ribosome targeted antibiotic anisomycin. Nucleic Acids Res 1987; 15:2431-2443.

50. Nierhaus KH, Schulze H, Cooperman BS. Molecular mechanisms of the ribosomal peptidyl transferase center. Biochem Int 1980; 1:185-192.

51. Pape T, Wintermeyer W, Rodnina MV. Conformational switch in the decoding region of 16S rRNA during aminoacyl-tRNA selection on the ribosome. Nature Struct Biol 2000; 7:104-110.

52. Rodnina MV, Wintermeyer W. Fidelity of aminoacyl-tRNA selection on the ribosome: Kinetics and structural mechanisms. Annu Rev Biochem 2001; 70:415-435.

53. Cunningham PR, Nurse K, Bakin A et al. Interaction between the two conserved single-stranded regions at the decoding site of small subunit ribosomal RNA is essential for ribosome function. Biochemistry 1992; 31:12012-12022

54. Cunningham PR, Nurse K, Weitzmann CJ et al. Functional effects of base changes which further define the decoding center of *Escherichia coli* 16S ribosomal RNA: Mutation of C1404, G1405, C1496, G1497, and U1498. Biochemistry 1993; 32:7172-7180.

55. Recht MI, Douthwaite S, Dahlquist KD et al. Effect of mutations in the A site of 16S rRNA on aminoglycoside antibiotic-ribosome interaction. J Mol Biol 1999; 286:33-43.

56. Vila-Sanjurjo A, Dahlberg AE. Mutational analysis of the conserved bases C1402 and A1500 in the center of the decoding domain of *Escherichia coli* 16S rRNA reveals an important tertiary interaction. J Mol Biol 2000; 308: 457-463.

57. Powers T, Noller HF. Dominant lethal mutations in a conserved loop in 16S rRNA. Proc Natl Acad Sci USA 1990; 87: 1042-1046.

58. Yoshizawa S, Fourmy D, Puglisi JD. Recognition of the codon-anticodon helix by ribosomal RNA. Science 1999; 285:1722-1725.

59. Powers T, Noller HF. Evidence for functional interaction between elongation factor EF-Tu and 16S ribosomal RNA. Proc Natl Acad Sci USA 1993; 90:1364-1368.

60. Carter AP, Clemons Jr WM, Brodersen DE et al. Functional insights from the structure of the 30S ribosomal subunit and its interactions with antibiotics. Nature 2000; 407:340-348.
61. Ogle JM, Brodersen DE, Clemons Jr WM et al. Recognition of cognate transfer RNA by the 30S ribosomal subunit. Science 2001; 292:897-902.
62. Gregory ST, Dahlberg AE. Nonsense suppressor and antisuppressor mutations at the 1409-1491 base-pair in the decoding region of Escherichia coli 16S rRNA. Nucleic Acids Res 1995; 23:4234-4238.
63. Leclerc D, Melançon P, Brakier-Gingras L. The interaction between streptomycin and ribosomal RNA. Biochimie 1991; 73: 1431-1438.
64. O'Connor M, Göringer HU, Dahlberg AE. A ribosomal ambiguity mutation in the 530 loop of E. coli 16S rRNA. Nucleic Acids Res 1992; 20: 4221-4227.
65. O'Connor M, Brunelli C, Firpo MA et al. Genetic probes of ribosomal RNA function. Biochem Cell Biol 1995; 73:859-868.
66. O'Connor M, Thomas CL, Zimmermann RA et al. Decoding fidelity at the ribosomal A and P sites: influence of mutations in three different regions of the decoding domain in 16S rRNA. Nucleic Acids Res 1997; 25:1185-1193.
67. Powers T, Noller HF. A functional pseudoknot in 16S ribosomal RNA. EMBO J 1991; 10:2203-2214.
68. Santer M, Santer U, Nurse K et al. Functional effects of a G to U base change at position 530 in a highly conserved loop of Escherichia coli 16S RNA. Biochemistry 1993; 32:5539-5547.
69. Santer UV, Cekleniak J, Kansil S et al. A mutation at the universally conserved position 529 in Escherichia coli 16S rRNA creates a functional but highly error-prone ribosome. RNA 1995; 1:89-94.
70. Piepenburg O, Pape T, Pleiss JA et al. Intact aminoacyl-tRNA is required to trigger GTP hydrolysis by elongation factor Tu on the ribosome. Biochemistry 2000; 39:1734-1738.
71. Brakier-Gingras L, Pinard R, Dragon F. Pleiotropic effects of mutations at positions 13 and 914 in Escherichia coli 16S ribosomal RNA. Biochem Cell Biol 1995; 75:907-913.
72. Stark H, Rodnina MV, Rinke-Appel J et al. Visualization of elongation factor Tu on the Escherichia coli ribosome. Nature 1997; 389:403-406.
73. Powers T, Noller HF. The 530 loop of 16S rRNA. A signal to EF-Tu. Trends Genet 1994; 10:27-31.
74. Newcomb LF, Noller HF. Directed hydroxyl radical probing of 16S rRNA to the ribosome: spatial proximity of RNA elements in the 3' and 5' domains. RNA 1999; 5:849-855.
75. Sergiev PV, Lavrik IN, Wlas Soff VA et al. The path of mRNA through the bacterial ribosome: a site-directed crosslinking study using new photoreactive derivatives of guanosine and uridine. RNA 1997; 3:464-475.
76. Yusupova GZ, Yusupov MM, Cate JHD et al. The path of messenger RNA through the ribosome. Cell 2001; 106:233-241.
77. Murgola EJ, Pagel FT, Hijazi KA et al. Variety of nonsense suppressor phenotypes associated with mutational changes at conserved sites in Escherichia coli ribosomal RNA. Biochem Cell Biol. 1995; 73:925-931.
78. Moine H, Dahlberg AE. Mutations in helix 34 of Escherichia coli 16S ribosomal RNA have multiple effects on ribosome function and synthesis. J Mol Biol 1994; 243:402-412.
79. Farabaugh PJ, Björk GR. How translational accuracy influences reading frame maintenance. EMBO J 1997; 18:1427-1434.
80. Lodmell JS, Dahlberg AE. A conformational switch in Escherichia coli 16S ribosomal RNA during decoding of messenger RNA. Science 1997; 277:1262-1267.
81. Pape T, Stark H, Matadeen R et al. Vizualization of the translational elongation cycle by cryo-electron microscopy. In: Garrett RA, Douthwaite SR, Liljas A, Matheson AT, Moore PB, Noller HF eds. The Ribosome. Structure, Function, Antibiotics and Cellular Interactions. ASM Press, Washington, DC, 2000:37-44.
82. Bélanger F, Léger M, Saraija AA et al. Functional Studies of the Goo Tetraloop capping helix 27 of 16S rRNA. J Mol Biol. 2002; 320:979-989.
83. Leclerc D, Melançon P, Brakier-Gingras L. Mutations in the 915 region of Escherichia coli 16S ribosomal RNA reduce the binding of streptomycin to the ribosome. Nucleic Acids Res 1991; 14:3973-3977.
84. Pinard R, Côté M, Payant C et al. Positions 13 and 914 in Escherichia coli are involved in the control of translational accuracy. Nucleic Acids Res 1994; 22:619-624.
85. Merryman C, Moazed D, Mc Whirter J et al. Nucleotides in 16S rRNA protected by the association of 30S and 50S ribosomal subunits. J Mol Biol 1999; 285:97-105.
86. Merryman C, Moazed D, Daubresse G et al. Nucleotides in 23S rRNA protected by the association of 30S and 50S ribosomal subunits. J Mol Biol 1999; 285:107-114.
87. Gabashvili IS, Agrawal RK, Spahn CMT et al. Solution structure of the E. coli 70S ribosome at 11.5 Å resolution. Cell 2000; 100:537-549.
88. O'Connor M, Dahlberg AE. The involvement of two distinct regions of 23S ribosomal RNA in tRNA selection. J Mol Biol 1995; 254:838-847.
89. Saarma U, Remme J. Novel mutants of 23S RNA : characterization of functional properties. Nucleic Acids Res 1992; 20:3147-3152.
90. Gregory ST, Lieberman KR, Dahlberg AE. Mutations in the peptidyl transferase region of E. coli 23S rRNA affecting translational accuracy. Nucleic Acids Res 1994: 22, 279-284.
91. Hausner TP, Atmadja J, Nierhaus KH. Evidence that the G2661 region of 23S rRNA is located at the ribosomal binding site of both elongation factors. Biochimie 1997; 69:911-923.

92. Moazed D, Robertson JM, Noller HF. Interaction of elongation factors EF-G and EF-Tu with a conserved loop in 23S rRNA. Nature 1988; 334:362-364.
93. Wool IG, Correll CC, Chan YL. Structure and function of the sarcin-ricin domain. In: Garrett RA, Douthwaite SR, Liljas A, Matheson AT, Moore PB, Noller HF, eds. The Ribosome. Structure, Function, Antibiotics and Cellular Interactions. ASM Press, Washington DC, 2000:461-473.
94. Bilgin N, Ehrenberg M. Mutations in 23S ribosomal RNA perturb transfer RNA selection and can lead to streptomycin dependence. J Mol Biol 1994; 235:813-824.
95. Melançon P, Tapprich WE, Brakier-Gingras L. Single-base mutations at position 2661 of *Escherichia coli* 23S rRNA increase efficiency of translational proofreading. J Bacteriol 1992; 174:7896-7901.
96. O'Connor M, Dahlberg AE. The influence of base identity and base pairing on the function of the α-sarcin loop of 23S rRNA. Nucleic Acids Res 1996; 24:2701-2705.
97. Morosyuk SV, Lee KS, SantaLucia Jr J et al. Structure and function of the conserved 690 hairpin in *Escherichia coli* 16S ribosomal RNA: analysis of the stem nucleotides. J Mol Biol 2000; 300:113-126.
98. Morosyuk SV, SantaLucia Jr J, Cunningham PR. Structure and function of the conserved 690 hairpin in *Escherichia coli* 16S ribosomal RNA. III. Functional analysis of the 690 loop. J Mol Biol 2001; 307:213-228.
99. Steitz JA, Jakes K. How ribosomes select initiator regions in mRNA: base-pair formation between the 3' terminus of 16S rRNA and the mRNA during initiation of protein synthesis in *Escherichia coli*. Proc Natl Acad Sci USA 1975; 72:4734-4738.
100. Melançon P, Leclerc D, Destroismaisons N et al. The anti-Shine-Dalgarno region of *Escherichia coli* 16S ribosomal RNA is not essential for the correct selection of translational starts. Biochemistry 1990; 29:3402-3407.
101. Gualerzi CO, Brandi L, Caserta E et al. (2000) Translation initiation in bacteria. In: Garrett RA, Douthwaite SR, Liljas A, Matheson AT, Moore PB, Noller HF, eds. The Ribosome. Structure, Function, Antibiotics and Cellular Interactions. ASM Press, Washington, DC, 2000:477-494.
102. Sprengart ML, Porter AG. Functional importance of RNA interactions in selection of translational initiation codons. Mol Microbiol 1997; 24:19-28.
103. Moll I, Grill S, Hubber M et al. Evidence against an interaction between the mRNA downstream box and 16S rRNA in translation initiation. J Bacteriol 2001; 183:3499-3505.
104. O'Connor M, Asai T, Squires C et al. Enhancement of translation by the downstream box does not involve base pairing of mRNA with the penultimate stem sequence of 16S rRNA. Proc Natl Acad Sci USA 1999; 96:8973-8978.
105. La Teana A, Brandi, A, O'Connor M et al. Translation during cold adaptation does not involve mRNA-rRNA base pairing through the downstream box. RNA 2000; 6:1393-1402.
106. O'Connor M, Bayfield M, Gregory ST et al. Probing ribosomal structure and function: analyses with rRNA and protein mutants. In: Garrett RA, Douthwaite SR, Liljas A, Matheson AT, Moore PB, Noller HF, eds. The Ribosome. Structure, Function, Antibiotics and Cellular Interactions. ASM Press, Washington DC, 2000:217-227.
107. O'Connor M, Dahlberg AE. Enhancement of translation by the epsilon element is independent of the sequence of the 460 region of 16S rRNA. Nucleic Acids Res 2001; 29:1420-1425.

CHAPTER 17

Conformational Dynamics within the Ribosome

J. Stephen Lodmell and Scott P. Hennelly

Abstract

The ribosome is a dynamic particle that undergoes an iterative series of conformational changes during translation. Individual structural changes in the ribosome in response to tRNA or mRNA binding, initiation or elongation factor binding, buffer conditions, and antibiotic effects have been observed using a variety of techniques over the years. With new high resolution cryoelectron microscopy and x-ray crystallography structure models available, we are approaching an understanding of how these conformational changes mesh together in the context of the whole ribosome. This review emphasizes conformational changes in the rRNA and their likely roles in various steps of translation.

The notion that the ribosome is dynamic seems intuitive to ribosomologists, yet detailed structural information associated with conformational changes of known function has been rather elusive. To demonstrate a switch, one must show experimentally that a structural change occurs (e.g., an alteration in a base pairing scheme within the rRNA, or a specific change in protein-protein or protein rRNA contacts) then show that the structural change has a tangible effect on some ribosomal function, otherwise the change may simply indicate that the observed region of the ribosome is not constrained. In this review, evidence for several conformational changes within the ribosome, with emphasis on those changes that occur in the rRNA during translation, will be discussed from structural and functional viewpoints. Indeed, the ribosome literature is replete with references to presumed or confirmed conformational changes and a comprehensive review is not possible here. I apologize in advance for omissions. Several excellent reviews on various aspects of conformational dynamics of the ribosome have been published; the reader is encouraged to explore these rich resources and references therein.[1-10]

The "Active-Inactive" Interconversion of 30S and 50S Subunits

Perhaps the most extensively studied conformational change in the ribosome occurs during the so-called active-inactive interconversion of 30S subunits.[11] This conformational change is concomitant with a change in the tRNA binding properties of 30S subunits. In the inactive form, 30S subunits are incapable of binding aminoacyl tRNA (aa-tRNA) that is not complexed with either initiation factor IF-2 or elongation factor EF-Tu. In the active form, 30S subunits bind near-stoichiometric amounts of tRNA. The interconversion is brought about by manipulation of mono- and divalent cations and temperature during incubation in vitro. The inactive form is achieved by incubation of subunits in conditions of low magnesium (ca. 0.5 mM) and/or low monovalent cation (potassium or ammonium). Inactive 30S subunits can be transformed into the active (tRNA binding-competent) form by replenishing the cation concentration and heating to 37-40°C for several minutes.

One might argue that depleting the system of magnesium or potassium nonspecifically denatures the ribosome and hence inhibits tRNA binding. However, there is substantial evidence that this transition is not an artifact of the experimental conditions, but actually represents a discrete, physiologically relevant conformational change. For example, inactive 30S subunits become active when presented with aa-tRNA in complex with EF-Tu, suggesting that, in vitro, adjusting temperature and salt concentration mediates a conformational change normally carried out by an elongation

Translation Mechanisms, edited by Jacques Lapointe and Léa Brakier-Gingras.

or initiation factor.[11] Furthermore, no gross perturbations of the ribosome structure or loss of ribosomal proteins could be detected during the conversion from active to inactive that might be expected if the ribosome were simply denaturing in these conditions.[11-13]

The structural aspects of this interconversion have been monitored by several methods. Oligodeoxyribonucleotide probes complementary to regions of 16S rRNA have been used to assess the availability of the rRNA for hybridization in active and inactive subunits.[14,15] Differences in probe binding were found in the 3' minor domain of 16S rRNA, a region known to interact with tRNA and mRNA called the "decoding domain" (reviewed in ref. 16). Oligonucleotides were targeted to this part of the 16S rRNA because it was known to be important in binding mRNA and tRNA, and it turned out that this region was also affected by the active-inactive transition. In most cases, the availability for probe hybridization was decreased in the inactive subunits, arguing against a general unfolding of the 16S rRNA that would likely increase probe binding. In several cases, oligonucleotide binding was increased in the inactive vs. active subunits, suggesting that a discrete, reciprocating conformational change occurs during the interconversion.[15]

Oligonucleotides tethered to photolabile crosslinking agents or to chemical nucleases have also substantiated conformational changes at the decoding site during the active-inactive interconversion. Cooperman's group noted major changes in the profile of ribosomal proteins crosslinked by an oligonucleotide bound to the decoding region before and after activation.[17] Using the chemical nuclease copper:phenanthroline tethered to an oligonucleotide complementary to the decoding region, Hill's laboratory demonstrated a conformational change in the rRNA environment.[18] Whereas nucleotides 923-929 and 1391-1396, which are located in the secondary structure, and nucleotides 1190-1192 which are located far away in helix 34, were cleaved in inactive subunits, these cleavages disappeared upon activation of the subunits, leaving only cleavages in the region of oligonucleotide binding. This difference in cleavage patterns appears to be a result of conformational changes.

Noller and colleagues used chemical probing of the rRNA to monitor this structural conversion.[19] First, this study clearly indicated that there were not gross changes in the secondary structure of the rRNA after cation depletion, arguing against an overall denaturation during inactivation. Second, the nucleotides whose reactivity toward chemical probes were altered were conspicuously clustered in the functionally important decoding region, not randomly distributed throughout the 16S rRNA. Third, they observed that the transition increased the reactivity of some nucleotides and decreased that of others, indicating a concerted change in structure rather than a random loosening or tightening of the subunit. Finally, since these changes in chemical reactivity occurred in regions already known to be instrumental in binding tRNA and mRNA, this study correlated a function of the ribosome (i.e., substrate or message binding) with a specific conformational change. However, although a partial model of the transition was offered, the exact nature of the conformational change was not known except that the reactivity of several nucleotides in the decoding domain were altered during the transition (see Table 1).

More recently, Wollenzien's group used a combination of site-directed mutagenesis and chemical probing to study this interconversion.[20] Mutations in the G926, C1533, and A1394 regions were introduced into synthetic ribosomes based on the results of previous structure probing[19] and psoralen crosslinking[21] studies and were shown to affect the inactive-active transition as measured by chemical probing and tRNA binding. A model for a structural transition within this region was proposed, and included alternative base pairing schemes in the active and inactive forms. Here again, strong evidence for a conformational switch was presented, this time with a detailed description of base pairing rearrangements (see Fig. 1).

The active-inactive interconversion is not limited to the 30S subunit. The 50S subunit, when depleted of monovalent cations, becomes incapable of binding the peptidyltransferase-inhibiting antibiotic chloramphenicol and renders the subunit incompetent for peptide bond formation.[22-24] The structural basis for this effect was recently shown to be associated with a conformational change at the peptidyltransferase center in the vicinity of nucleotide A2451.[25] In particular, this study showed that the reactivity of several nucleotides toward dimethyl sulfate (DMS) changes upon the active/inactive interconversion. Nucleotide A2451 was shielded in the active and susceptible to attack in the inactive conformation of *E. coli* ribosomes. The data suggest that the increased reactivity of A2451 with DMS in the inactive conformation is a consequence of the loss of a specifically bound potassium ion. This is especially interesting in light of the recent high resolution crystal structures of

Table 1. *Summary of chemical probing results for the active-inactive interconversion of the 305 subunit*[19]

Reactivity in the Inactive Conformation	Higher	Lower
Watson-Crick positions (N-1, N-2, N-3)	U723, G791, A908, A923, C924, G925, G927, U1391, G1392, U1393, A1394, C1395, A 1396, A1398, C1399, G1401, U1414, G1415, G1416, G1417, A1433 G1494, U1495, G1497, A1499, A1500, A1502, G1505	G926, A1503, C1533, C1535, C1536, C1538
N-7 position	G926, G1392, G1494, G1497, A1499, A1500, A1502, A1503, G1517, A1531, A1534	A1398, G1401

the large subunit that demonstrate that nucleotide 2451 is very close to the catalytic active site, and may be a direct participant in peptidyltransfer,[26,27] but the mechanism of peptidyltransfer is still a matter of active debate.[28,29] Furthermore, this study suggests that the previously reported pH-dependence of the A2451 reactivity with DMS was due to a local conformational change rather than to modulation of the pK_a of this nucleotide mediated by interactions with nearby nucleotides.[30]

Structural Changes Associated with mRNA Binding and Initiation Events

Among the ribosome's first tasks in translation is the binding of an appropriate mRNA. This interaction is promoted in prokaryotes by the mRNA-rRNA interaction known as the Shine-Dalgarno (SD) interaction.[31] Highly expressed prokaryotic mRNAs usually have a purine-rich sequence upstream of the AUG initiation codon with complementarity to a conserved pyrimidine-rich region near the 3' end of the rRNA (in the vicinity of nucleotides 1435-1440 of *E. coli* 16S rRNA) called the anti-Shine-Dalgarno (ASD) region. This interaction is a powerful determinant of the efficiency of translation of a particular message; mRNAs with a weak SD sequence (less than perfect complementarity with the rRNA sequence) are translated with low efficiency, whereas highly translated messages almost invariably have strong SD sequences (reviewed in ref. 32).

After the SD-ASD interaction aids in the recruitment of an mRNA for the ribosome, it may be advantageous to sequester the ASD sequence for two reasons. First, the ribosome must be able to release the 5' end of the message so that it can move toward the 3' end of the mRNA without dragging along the 5' end. Second, when a ribosome is actively engaged in translation of a given mRNA, additional Shine-Dalgarno mediated initiation events with a second mRNA should be prevented. Kössel and colleagues proposed that the ASD sequence might hybridize to the 5' region of the 16S rRNA when the ribosome is actively translating.[33] The authors presented sequence comparisons of the 5' and 3' ends of 16S rRNAs from several organisms which were consistent with the idea that these regions could form base pairs, although direct experimental evidence was not presented. More recent compilations of rRNA sequences have not fully supported this interaction inasmuch as canonical base pairing has not been conserved in all species.[34] Another helical switch model for the 5' end region was proposed by the Brakier-Gingras group,[35] based on a combination of phylogenetic and biochemical data including translational fidelity and streptomycin binding with wild-type and mutant ribosomes.[36] In this model, the central pseudoknot, comprised of nucleotides 17-19:918-916, transiently opens and alternate base pairs are formed (nucleotides 12-16 base-pair with 911-915). However, subsequent studies using site-directed mutagenesis of nucleotides

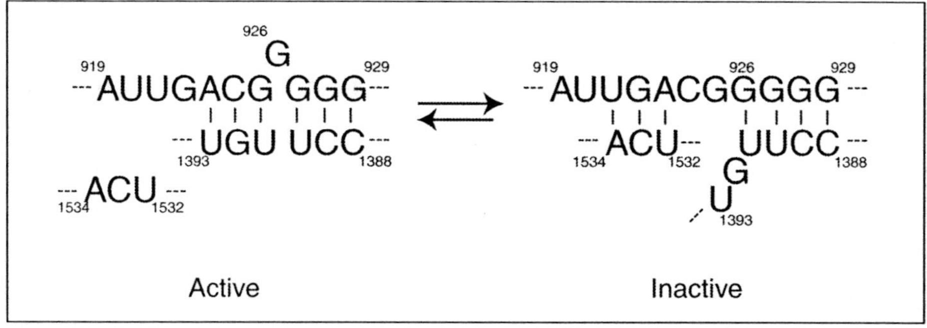

Figure 1. A conformational switch proposed for the active-inactive interconversion based upon chemical probing and mutagenesis.[20] In this model for the conversion of the active to inactive conformation, the base pairing arrangement changes for several nucleotides, notably G926 and nucleotides 1532-1534.

implicated in the proposed switch, while not ruling out a conformational change in the region, did not directly support it.[37]

Another study on the central pseudoknot was performed using the specialized ribosome system.[38] Here it was shown that, at least among ribosomes containing an artificial anti-Shine-Dalgarno sequence, any complementarity between nucleotides 17-19 and 918-916 was necessary and sufficient for normal initiation on the ribosome. If these sequences were engaged in transient alternate base pairing schemes (i.e., a switch), one might expect to find some additional defects that arise in these mutants due to lost complementarity in the other conformation of the switch. However, because the specialized ribosome system utilizes ribosomes containing mutated anti-Shine-Dalgarno sequences, there is potential for interference with the analysis of helical switching involving sequences also in this region. The central pseudoknot is clearly visible in recent crystal structures,[39,40] but alternate conformations may be forthcoming in future crystal structures.

Results from the Gualerzi and Brimacombe laboratories suggest that the structure of the ASD region as well as the path of the mRNA through the ribosome changes upon binding of initiation factors (IFs).[41-43] First it was shown that an oligoribonucleotide mimicking the SD region of a mRNA could effectively compete with a full length mRNA for 30S subunit binding until initiation factors were added. At that point, the SD oligo could no longer compete for binding, suggesting that the SD region had been altered. In the second study, synthetic mRNAs were transcribed using the photolabile nucleotide analog 4-thiouridine. The position of the mRNA on the ribosome could then be monitored by binding the mRNA to the ribosome in the presence or absence of IFs, then irradiating the complexes with UV light and examining the ribosomal components for sites of crosslinking with the mRNA analog. Indeed, two distinct sets of nucleotides in the 16S were crosslinked, depending upon whether or not IFs were present, providing compelling evidence for a positional change of the mRNA relative to the 30S subunit under the influence of initiation factors. The nature of the conformational change was schematized with respect to the proximity of the crosslinked nucleotides and ribosomal proteins.

Another model for a helical switch has been proposed by the Ehresmann group for the ASD region of 16S rRNA, influenced by IF-3.[44] According to this model, the 3' end of 16S rRNA pairs with the 830 region of 16S rRNA unless IF-3 is present. IF-3 competes with the ASD for binding near position 830, allowing the ASD to hybridize with mRNA. This model was proposed based on crosslinking data of the IF-3 to the rRNA and on the apparent complementarity of the ASD and the 830 region. In fact, recent cryoelectron microscopy, crystallographic, and crosslinking data confirmed the ASD-830 region proximity and demonstrated a conformational change of the 30S subunit upon IF-3 binding.[45,46,46a] The cryo-EM study demonstrates a binding site at the subunit interface, with conformational changes seen in the head and platform, while in the crystal structure, the C-terminal domain of IF-3 was localized to the solvent side of the subunit, not at the subunit interface.[46] It is obvious that these two localizations of IF3 do not represent the same physiological state of the ribosome, even if both represent physiologically relevant IF3 binding states. However, in both cases IF3 binding to the 30S subunit causes conformational changes that can at least partially explain IF3's functions in translation.

A recent crystal structure of the 30S subunit with IF-1 from Ramakrishnan's group has helped to shed some light on the role of this initiation factor whose supporting role in initiation is known to be essential, but whose mechanism of function has been elusive.[47,48] Compared to its structure in an empty 30S subunit, helix 44 of the IF-1 bound subunit adopts a markedly different conformation. IF-1 and protein S12 form separate pockets for nucleotides A1492 and A1493, locking them into a splayed-out conformation, in contrast to their stacked, flipped out conformation that is the hallmark of the paromomycin-bound (or cognate codon-anticodon recognition) state.[49,50] The presence of IF-1 at the top of helix 44 not only occludes the A site from tRNA binding, but its effects are propagated down the helix, interfering with normal helical base pairing over a large distance. Helix 44 is observed in the crystal structure to undergo a lateral shift of several angstroms, a phenomenon that was observed previously under various experimental conditions using cryoelectron microscopy (see below).[51,52] IF-1 binding also causes shifts in relative subunit domain positioning. The platform and shoulder rotate toward the A site upon factor binding. Intriguingly, binding of several antibiotics also cause a similar rotation of the shoulder and platform, with the additional effect of the head tilting back and away.[49] These observations lend credence to the idea that some antibiotics target and block normal concerted movements in the ribosome.

Initiation factor 2 (IF-2) is responsible for installing the initiator tRNA^{fMet} into the P site of the initiating 30S subunit. Although a high resolution crystal structure of this interaction has not yet been published, biochemical data indicate a conformational change in the 30S subunit upon IF-2 binding.[53,54] Some of the observed changes in reactivity to chemical probes and crosslinking reagents occurred in the classical decoding and P site domains (notably nucleotides A1493 and G1401), while others were observed in more distal regions of the 16S rRNA, suggesting an alteration of the overall topology.

A different helical switch involving mRNA-rRNA and rRNA-rRNA base pairing, called the "downstream box" interaction, has been debated in the literature[55-59] and is described in detail in Chapter 16 by Brakier-Gingras et al. Although mutations in this region of rRNA did have effects on cell growth, the ensemble of experiments with mutations in rRNA and mRNA have not supported the idea of a simple helical switch involving rRNA-mRNA base pairing interactions.

Conformational Changes during Ribosome-tRNA Interactions

The possibilities for conformational changes are numerous during the tRNA's transit through the ribosome: initial recognition of ternary complex, selection at the A site, peptide transfer from one tRNA to the next, translocation, and ejection of the spent deacylated tRNA from the E site. Conformational changes within the ribosome have been proposed for most of these steps, supported with various types of experimental evidence.

Extensive analysis of tRNA-rRNA interactions from Noller and colleagues using chemical probing of rRNA structure in the presence and absence of ribosome binding has yielded an extensive list of potential contact points (reviewed in ref. 60). They identified a set of highly conserved nucleotides in 16S and 23S rRNA whose reactivities toward chemical probes changed when the ribosome was bound by tRNAs and antibiotics known to inhibit various steps in translation. By repeating this approach with different types of tRNA under different experimental conditions, a path of changes in nucleotide reactivities caused by tRNA binding could be followed through the ribosome.[61] Because some of the changes in chemical reactivity were difficult to explain as a result of direct shielding of nucleotides by tRNA, conformational changes were strongly implicated. For example, it was determined that small acylated oligoribonucleotide tRNA analogs caused almost the same footprints on the rRNA as the intact tRNA.[62] Because it is unlikely that the small tRNA analog physically contacted all of the same sites that an intact tRNA would have, it appeared that the changes in reactivity were part of a coordinated conformational change in response to the binding of tRNA or of a tRNA analogue.

Ribosome : aa-tRNA Interactions and Translational Fidelity

Aminoacyl-tRNA selection probably involves at least two major steps, initial selection and proofreading, and each step has potential for conformational alterations. First, as the ternary complex binds the ribosome, a structural change may be necessary to accommodate it. Then, the ribosome-dependent GTP hydrolysis on the EF-Tu may be preceded by, accompanied by, or followed by structural changes in the ribosome, the EF-Tu, the tRNA, or a combination thereof. The EF-Tu-GDP complex then dissociates from the ribosome, leaving the aminoacyl-tRNA at the A site (or at a proofreading site) where it will either be used in the peptidyl transfer reaction or will be rejected prior to the peptidyl transfer reaction; each of these increments could involve conformational changes.

During tRNA selection, the ribosome must bind incoming tRNAs with minimal nonspecific interactions in order to emphasize the codon-anticodon interaction relative to the overall binding energy. A likely mechanism for this discrimination relies upon differences in binding geometry of the fully complementary versus the noncognate tRNAs. One can envision that a fully complementary codon-anticodon interaction would allow the tRNA to settle completely into a binding pocket and make contacts with the mRNA, rRNA and/or ribosomal proteins not available to noncognate tRNAs. These additional contacts could enhance stability of the tRNA-ribosome complex or may trigger the ribosome to proceed to the next step of translation.

Several models of aminoacyl-tRNA selection invoke two or more functional states of the A site and several lines of evidence suggest that structural changes occur when the tRNA-GTP-EF-Tu complex encounters the ribosome. Recent cryoelectron microscopy data have confirmed our suspicions that there are conformational adjustments on the ribosome upon EF-Tu binding. In particular, significant rearrangements in the 50S subunit were detected in the vicinity of proteins L7/L12 and the A site tRNA binding region upon binding of a kirromycin-stalled ternary complex.[63] At the biochemical level, Moazed and Noller demonstrated that the ternary complex interacts with a different set of nucleotides in the rRNA compared to after GTP hydrolysis and factor release.[61] The footprint for A site aa-tRNA binding did not appear on the 23S rRNA until after hydrolysis of GTP and release of EF-Tu, suggesting that binding of charged tRNA to the A site is at least a two-step process with an associated conformational change. A similar two-step scenario with intermediate hybrid states of the ribosome was proposed for the translocation event, described below.

The 530 loop's involvement in translational fidelity has been extensively studied, and several lines of evidence suggest that an "open/closed" structural change takes place here during decoding. Neomycin binding causes increased reactivity at C525,[64] while streptomycin aids tRNA protection of G530. Based upon chemical probing of ribosome-tRNA complexes in the presence and absence of antibiotics and hyperaccurate S12 alleles, it was proposed that the loop adopts an open conformation during proofreading that correlated with a diminished off rate for EF-Tu GDP[65,66] In this state, the tRNA is able to diffuse from the ribosome prior to peptidyl transferase. However, in contrast to this model, one study with chemical and enzymatic probing suggested that an open form of the 530 loop promotes translational errors.[67] Conformational adjustment of this loop, especially a *syn* to *anti* rearrangement of G530 upon codon recognition seen in recent crystal structures, attests its importance in translational fidelity.[50]

Kinetic analysis of tRNA and ternary complex interactions with the ribosome by the group of Rodnina and Wintermeyer has made possible the dissection of aa-tRNA selection into several discrete steps with intriguing implications for an induced fit mechanism for tRNA selection (Fig. 2). First, complete kinetic descriptions of aminoacyl-tRNA binding, accommodation, and peptidyl transfer were established using stopped-flow fluorometric techniques to examine tRNA interactions with EF-Tu and with the ribosome, and quench-flow measurements using radioactive detection to examine rates of GTP hydrolysis and peptide bond formation (see ref. 68 and references therein). This analysis of each kinetically identifiable step in tRNA selection formed the foundation for a new model that has clarified our view of how a two-step selection mechanism might function to optimize tRNA selection with respect to speed and accuracy.

Particularly interesting in the kinetic data is the observation that the rates of GTPase activation prior to EF-Tu departure, and accommodation of the aa-tRNA into the peptidyltransferase-active A site when it has been released by EF-Tu, are vastly different between cognate and near-cognate tRNAs.[68,69] This demonstrates that not only is a correct codon/anticodon fit necessary for stable initial binding of the ternary complex to the ribosome, but that correct geometry of the fit actually accelerates the next step in the sequence. Such an acceleration of catalytic activity most likely results from a change in the conformation of the ribosome, tRNA, or EF-Tu, or

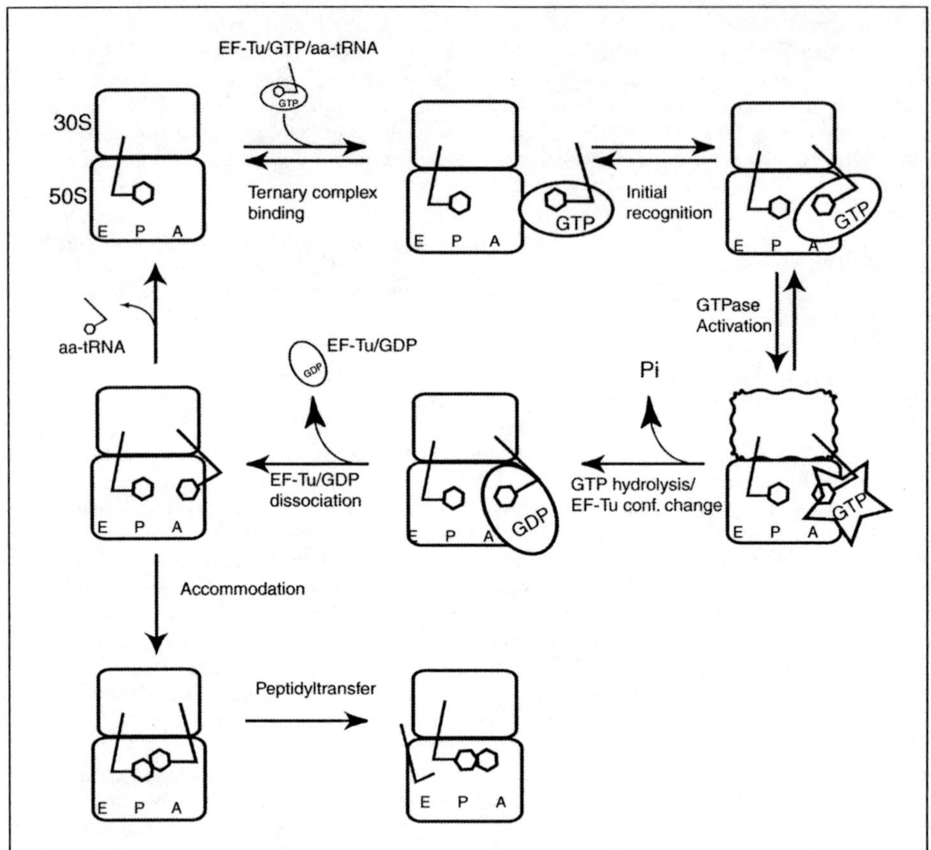

Figure 2. A model for aa-tRNA recognition and proofreading based upon analysis of each kinetically identifiable step by Pape et al.[69] Particularly noteworthy is the conformational change in the ribosome (designated by wavy lines) that occurs upon cognate codon recognition or paromomycin binding, but less readily upon near- or noncognate aa-tRNA binding. This conformational change in turn enhances EF-Tu GTPase activity. The star-shaped EF-Tu indicates the GTPase-activated form. Cognate aa-tRNA was also shown to accelerate the accommodation step relative to near-cognate aa-tRNA during proofreading. The conformational changes that lead to enhanced rates of GTPase and accomodation only with cognate aa-tRNA are representative of an induced fit mechanism.

a combination thereof in response to a certain favorable geometry of tRNA-mRNA binding. These data are best explained by an induced fit mechanism that was proposed by Pape et al.[69] According to this model, a correct codon-anticodon interaction promotes 16S rRNA to adopt a conformation that then accelerates GTPase activation and accommodation of the aminoacyl-tRNA –CCA end into the peptidyl transferase active site. In fact, this activated 16S conformation was also shown by kinetic methods to be induced by the translational error-inducing antibiotic paromomycin,[70] with clear mechanistic implications for this class of antibiotics. These authors suggested that the structural changes are likely to be related to conformational changes observed at nucleotides 1492, 1493 and 530 (*E. coli* numbering) of the decoding region.[9,49,71,72] This argument has been strongly supported by recent crystallographic evidence of an induced fit by cognate tRNA and potentiated by paromomycin.[50]

The Nierhaus group's allosteric three site model has been useful in helping to rationalize results of diverse tRNA binding experiments and has helped to emphasize the importance of the E site in translation.[73] According to this model, the presence of an aminoacyl-tRNA in the A site causes a decrease in the affinity for deacylated tRNA in the E site and vice versa. Such an interconversion of affinities would suggest a simultaneous structural change in each of these sites. Modulation of the A site affinity was proposed to aid in correct aa-tRNA selection. In fact, the allosteric three site model has undergone some revision in recent years. However, a new model, called the α/ε model, that is largely compatible with the older allosteric model, also predicts conformational changes resulting in high and low affinity tRNA binding to the A and E sites, with obvious implications for aa-tRNA selection.[4,74,75]

Evidence for the α/ε conformational change was derived from several types of experiments. Originally, different species of radiolabelled tRNAs were used in binding experiments to monitor the occupancy of ribosomes under different conditions. Then biophysical and biochemical structural studies, including iodine cleavage of phosphorothioate containing tRNAs, neutron scattering, and cryoelectron microscopy, were applied to monitor the local environment of the tRNAs as they passed through the ribosome.[75-77] These studies showed an apparent lack of change of the local environment surrounding the tRNAs in steps prior to or after translocation. Because the structure of the tRNAs themselves do not change during translocation, the authors propose a movable domain wherein the A and P site bound tRNAs are transported together by a conveyor within the ribosome to the P and E sites. In this model, the A site and E site alternate between high affinity and low affinity states in a reciprocal fashion, although the interaction is not allosteric. Rather, the affinity state is determined by the relative positions of the decoding center and the movable domain (see ref. 4 for a detailed description of the α/ε model). While the particular nucleotides or proteins responsible for the modulation of A site affinity have not been identified, cryoelectron microscopy has suggested a narrowing of the A site entrance and the appearance of unidentified density that could affect ternary complex binding and aa-tRNA selection.[78,79]

Evidence for reciprocal, competing conformers related to tRNA selection was demonstrated with chemical probing techniques by Allen and Noller.[80] They showed that nucleotides A908 and A909 in helix 27 of 16S rRNA were more reactive to dimethylsulfate in ribosomes isolated from a mutant strains of bacteria that have enhanced rates of translational errors, and were less reactive in strains that have lower levels of translational errors. The mutations responsible for these translational fidelity phenotypes are in ribosomal proteins S4 and S12. Mutations in S12 confer resistance to or dependence upon the translational error inducing drug streptomycin.[81] Some of these "restrictive" mutations, so-called because they restrict the suppression of nonsense codons, can also cause ribosomes to translate slowly and with increased translational accuracy.[82] Another class of mutations was subsequently discovered that could counteract the hyperaccurate, streptomycin dependent phenotype of the S12 mutations. These second-site mutations were mapped to ribosomal proteins S4 and S5. When segregated from the S12 restrictive mutations, these S4 and S5 mutations were found to increase the level of translational errors and were designated *ram* mutations, for *r*ibosomal *am*biguity.

Allen and Noller proposed that the 908-909 region was involved in an equilibrium between two conformational states in 16S rRNA. The balance of the equilibrium was controlled at least in part by ribosomal proteins S4 and S12 (they did not test the influence of S5), hence mutations in either of these proteins could perturb the equilibrium. Interestingly, streptomycin could also influence the equilibrium similarly to the S4 *ram* mutations. They proposed that the reactivity of A908-A909 was an indicator of the position of the equilibrium, although with the limited amount of structural data available, they did not propose the specific structures involved. Recent high resolution glimpses at the 30S subunit clearly indicate that S4, S5, and S12 are close to the 900 region and close to the decoding region, making the proposition that nucleotides 908-909 have a direct or indirect role in tRNA selection appear very reasonable.[39,40,45,49,83]

Another conformational change associated with translational fidelity has been proposed for nucleotides 910-912, also in helix 27.[84] Through mutagenesis, chemical structure probing, and complementation with different mutant ribosomal protein alleles, it was proposed that nucleotides 910-912 base pair not only with nucleotides 885-887, as is commonly depicted on the secondary structure maps, but also transiently with nucleotides 888-890 (Fig. 3). It is notable that mutations that favored strictly one or the other of the helix 27 conformations were deleterious or lethal, while mutations that permitted both base-pairing arrangements were tolerated. This suggests that both arrangements are utilized during translation. Mutations in the rRNA that favored the 912:885 conformation

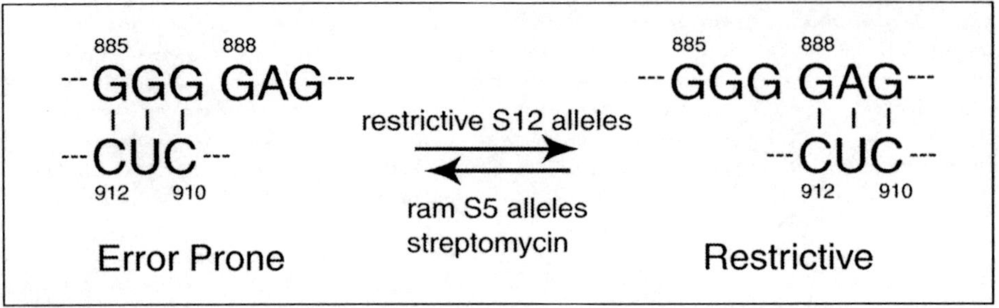

Figure 3. Alternate base pairing schemes for nucleotides 912-910 and the 885-890 region of helix 27 in *E. coli*.[84] Mutations that favor the 912-885 conformation on the left have an error-prone phenotype and are complemented by restrictive S12 mutations. Mutations that favor the 912-888 conformation on the right result in a hyperaccurate phenotype that is complemented by S5 *ram* mutations. The equilibrium between these states may be perturbed by S12 restrictive mutations and S5 *ram* mutations, which could favor the 912-888 or 912-885 structures, respectively. See text for details.

exhibited an error-prone phenotype. Furthermore, these rRNA mutations were very deleterious when combined with *ram* S5 alleles, but were completely compatible with restrictive mutations in S12. On the other hand, the growth of cells harboring rRNA mutations that favored the 912-888 conformation were incompatible with restrictive S12 mutations, but were complemented with *ram* S5 mutations. Since the mutations affected translational fidelity, these mutations presumably affect A-site tRNA binding or proofreading, although a connection of this three nucleotide switch with translocation cannot be ruled out. Recent data from Ganoza's group, based upon chemical reactivity changes in helix 27 and the decoding region upon binding of the ribosomal ATPase RbbA and the translocation-inhibiting and error-inducing antibiotic hygromycin B, raise the possibility that RbbA may have a role in the dynamics of this helix.[85,86]

Chemical structure probing of ribosomes containing helix 27 mutations showed differences in reactivities in several regions of the ribosome implicated in decoding, including the decoding center, helix 34, and the 530 loop.[84] Mutations in the rRNA that favored the 912-885 conformation affected the chemical reactivities oppositely to those that favored the 912-888 conformation. Cryoelectron microscopy studies on ribosomes carrying these two types of mutations in helix 27 also revealed discrete structural changes in several regions of the 30S subunit, and indeed some changes in the 50S subunit of a 70S ribosome.[51] This suggests that a change in the state of helix 27 could be sensed at distal regions of the ribosome, and therefore could be part of a signal relay system.

Consistent with its putative role in aa-tRNA selection, crystal structure models of the 30S subunit place helix 27 at the floor of the A-site.[39,40,45,49,83] In these high resolution structure models, helix 27 has been observed only in its 912-885 conformation so far. Crystallization of ribosomes in the 912-888 restrictive form would be most interesting with respect to rigorously defining the proposed signal relay system. These efforts are hampered somewhat by the deleterious nature of mutant ribosomes harboring mutations that favor only the 912-888 conformation.

Liebman and coworkers have constructed a series of site-directed and selected helix 27 mutations in yeast.[87] While these mutations affected translational fidelity and sensitivity to error-inducing antibiotics, the pattern of phenotypes did not match the pattern observed in *E. coli*. Therefore, while this study lends further support to the notion that helix 27 is important in the decoding process, it does not support the *E. coli*-style helix 27 switching mechanism in yeast. Further studies will be required to determine whether the proposed switch is a conserved feature among different species.

Further support for alternating conformations and affinities during tRNA selection comes from Ehrenberg's group. They performed a series of studies of the kinetics and thermodynamics of tRNA binding to wild-type and mutant ribosomes in the presence or absence of error-inducing antibiotics. First, it was demonstrated that error-prone mutant ribosomes harboring mutations in ribosomal proteins S4 or S5, or ribosomes treated with error-inducing antibiotics bound their aa-tRNAs more tightly in the A-site, and the proofreading step was diminished.[88] Furthermore, they showed that the same mutations or antibiotics that caused increased binding to the A site decreased the stability of binding to the P site, as measured by the propensity for the peptidyl-tRNA to drop off the ribosome.[89] In fact, the reverse was true when mutant ribosomes with mutations in ribosomal protein S12 conferring a hyperaccurate phenotype were tested. The authors suggested that such reciprocity reflects a movement in the 16S that is opposite for the *ram* and restrictive phenotypes.

Conformational Changes during Translocation

During translocation, the ribosome moves along the mRNA and the tRNAs move from one functional site to another within the ribosome and the implications for conformational change are obvious. Spirin's group demonstrated using small angle neutron scattering that the radius of gyration differed between pre- and post-translocational ribosomes.[90] Noller's hybrid-states model provided an alternative to the classic A-site then P-site view of translocation.[61] As determined by the footprints of tRNA, translocation occurs at different times in the 30S and 50S subunits. According to this model, translocation occurs spontaneously in the 50S subunit upon peptidyl transfer (i.e., A site tRNA moves to P site and P site tRNA moves to E site). In the 30S subunit, however, translocation does not occur until or during the hydrolysis of GTP upon EF-G, after peptidyl transfer. In the interim between these two translocation events, one tRNA is in the A site of the 30S subunit and the P site of the 50S subunit (the A/P site), while the other tRNA is in the P site of the 30S subunit and the E site of the large subunit (P/E site).

These distinctive changes in the footprints indicated that a structural shift occurs. However, because of the limited resolution of the technique, the exact nature of the conformational change was not obvious. The crystal structures of the 70S ribosome with tRNAs bound have so far demonstrated only classic A/A, P/P, and E site positioning of the tRNAs, [83,91] although this snapshot cannot be considered the only physiological state of the ribosome. In fact, cryoelectron microscopy has shown a P/E hybrid state for deacylated tRNA under some buffer conditions,[92] but did not provide clear evidence of hybrid states during the intersubunit ratcheting motion observed in the pre- and post-translocational states.[52] Furthermore, recent crosslinking results[93] and cryoelectron microscopy studies[79] suggest the existence of an additional exit site (called the F or E2 site) that could add some complexity to existing models of tRNA transit through the ribosome.

Wintermeyer's group proposed that binding of EF-G to the ribosome causes a conformational change in the E site that enhances its affinity for deacylated tRNA, providing a thermodynamic "escape route" for P site bound tRNA.[94] Furthermore, the binding of the 3' end of deacylated tRNA to 23S rRNA was proposed to stimulate GTP hydrolysis and translocation. The nature of the linkage between the EF-G binding site on the ribosome and the E site was not known, but the distance between the two regions of the ribosome based on models of the 50S subunit suggest that it is a long-range interaction. In a different study, it was shown that the accessibility of the E site for binding of an oligonucleotide increased when deacylated tRNA was bound to the P site relative to vacant or aa-tRNA bound ribosomes.[95] These data suggested there was a structural change in the ribosome that made the E site more available for binding of an oligonucleotide, and by extension, a tRNA, but only in response to the presence of a deacylated tRNA in the P site, a situation ripe for translocation. Such a structural change may also be correlated with improving the energetics of translocation.

In addition to features relevant to aminoacyl-tRNA selection, Nierhaus's α/ε model proposes a novel way of looking at translocation that contrasts with the hybrid states model. Data supporting this model include phosphorothioate footprinting, neutron scattering and cryoelectron microscopy data, and have been described above in the A site-tRNA interactions section.[75-77,79] Of particular relevance to translocation are the cryoelectron microscopy data where the A site density corresponding to the 1492-1493 region in helix 44 is seen to move the distance of one codon toward the P site when one compares the images of ribosomes complexed with EF-G plus a nonhydrolyzable GTP analogue and EF-G-GDP frozen with fusidic acid.[77] This could be interpreted as movement of the conveyor from the A to P site, although higher resolution structural analysis of various conformational states of ribosome-tRNA complexes will be necessary to fully describe translocation.

Other tRNA-rRNA Interactions

Crosslinking experiments have provided a wealth of information about the positions of tRNAs within the ribosome, and some of the patterns of crosslinks are highly suggestive of conformational changes in the ribosome (for recent reviews, see refs. 93,96-98). For example, the anticodon of tRNA[Arg] (derivatized at position 32 with a photoaffinity crosslinking moiety) was shown to form crosslinks to position 1378 of 16S rRNA from both the A and E sites, but not from the P site, suggesting that the position of this region of rRNA can be altered rather dramatically under different tRNA binding conditions.[99] These data have been supported by other crosslinking experiments that gave compatible contact patterns, arguing against the possibility that the position of the tRNA in the ribosome was not properly assigned (e.g., ref. 100). Likewise, crosslinking to the peptidyltransferase region of 23S rRNA from two widely separated residues of P site-bound tRNA (position 47, near the "elbow" of tRNA, and a derivatized aminoacyl moiety at the -CCA end of tRNA) is suggestive of another rRNA conformational change in another functionally important part of the rRNA,[96] although assignment of an exact position for a given tRNA on the ribosome is a delicate task (see ref. 92 for example).

Conformational Changes Related to Antibiotic Binding

Ribosome-binding antibiotics block specific steps of translation and are therefore useful for both their clinical applications and as probes of ribosomal function. Numerous accounts exist in the literature of a change in conformation of the ribosome upon antibiotic binding. With new high resolution structural information on the ribosome and ribosome-antibiotic complexes, many more conformational changes have been detected. For example, streptomycin was shown in the 1970's to promote the conversion of inactive to active 30S ribosomal subunits,[101] and was shown shortly thereafter by biophysical methods to alter the structure of the 30S subunit.[102] More recently, Jerinic and Joseph demonstrated that the toeprint of ribosomes on a mRNA was altered when streptomycin or other aminoglycoside antibiotics was added.[103] This effect could be modulated by the presence of restrictive or error-prone S12 or S4 alleles. The crystal structure model of the 30S-streptomycin complex did not reveal any major structural changes compared to the vacant subunit, although this does not preclude the existence of other conformers outside these crystal conditions.[49] The localization of streptomycin to the helix 27 region in the crystal structure together with the biochemical and genetic data discussed above suggest that streptomycin affects an equilibrium of competing conformers in this region.

Paromomycin, another aminoglycoside antibiotic, has also been shown to alter 30S subunit and rRNA structure. At the fine structure level, Puglisi's group has shown by NMR analysis that paromomycin alters the structure of a model RNA mimicking the decoding region of rRNA in a way reminiscent of a cognate codon recognition event.[71,72] Specifically, nucleotides A1492 and A1493 flip out from the helix in response to paromomycin binding. In the 30S crystal structures without or with paromomycin, these nucleotides are disordered or flipped out, respectively.[39,49] Again, it is likely that more conformers of ribosomes will be crystallized in the future and further conformational changes will be seen.

Cocrystal structures with the antibiotics tetracycline and spectinomycin strongly suggest that a mechanism of action of these antibiotics is to interfere with normal ribosomal movements.[45,49] Spectinomycin is proposed to prevent or alter a normal movement of the 30S head relative to the body, possibly during translocation.[49] The spectinomycin binding site in helix 34 is strategically located to prevent such movement. On the other hand, tetracycline's major binding site on the 30S subunit interferes with tRNA entrance to the A-site, while the several alternative sites appear to be able to block other parts of a hypothesized signal relay system.[45,104]

Pactamycin, a potent antibiotic that exerts its effects on prokaryotes, archae, and eukaryotes, was proposed by Mankin, based upon chemical probing and isolation of pactamycin resistant mutants, to inhibit normal conformational changes in the small subunit that may be associated with initiation of translation.[105] Crystallographic data on pactamycin binding did not demonstrate a major conformational change, but indicated that helices 23a and 24a may be frozen together by pactamycin, and that pactamycin may interfere with the Shine-Dalgarno/ anti-Shine-Dalgarno interaction or the normal path of mRNA at the E site.[104] A separate crystallographic study by Franceschi, Yonath and coworkers of the edeine-30S complex demonstrated a remarkable agreement on the mechanism of action of these two antibiotics.[45] Edeine, a universal antibiotic related to pactamycin, was shown to bind the 30S subunit in the same strategic location as pactamycin, effectively tying together regions involved in mRNA, tRNA, and IF-3 binding (helices 23, 24, 28, 44, 45). In fact, by bringing helices 23 and 24 closer together, edeine promoted the formation of a new base pair. Because cryoelectron microscopy studies have implicated movements of these helices relative to each other during translation,[51,77] edeine and pactamycin very likely exert their effects not solely by blocking binding of tRNA, mRNA, or IF-3, but by blocking necessary conformational changes of the ribosome.

Finally, hygromycin B, an aminoglycoside antibiotic that inhibits translocation and causes translational errors, has been localized in co-complexes with 30S subunits near the top of helix 44.[104] Hygromycin B did not cause a large conformational change in the 30S subunit, but because helix 44 has been shown in cryoelectron microscopy studies to oscillate,[52] and because this region constitutes part of the decoding center, hygromycin B likely acts by inhibiting normal movement of helix 44 during decoding and/or translocation. Whereas the crystal structures of most of these antibiotic/ subunit complexes did not reveal large perturbations in the subunit structure, crosslinking and other biochemical experiments suggest that these antibiotics can and do cause conformational changes in ribosomes in solution (see ref. 106 and references therein).

Evidence for Other Conformational Switches

Several switches were proposed in the mid-1980's by examining the double-helical products of partially nuclease-digested ribosomes. This methodology was primarily used to obtain biochemical support for helices proposed in early secondary structure models of the rRNA. Adapted to finding switches, the idea was that if a piece of rRNA normally formed base pairs with two (or more) other sequences, then after digestion, electrophoretic separation, and sequencing of digestion products, it should be possible to identify both helical arrangements. In the digestion products of several studies, some pieces of rRNA were consistently found base paired to two different sequences of rRNA (e.g., see refs. 107-109). Using the rather limited rRNA sequence data available at the time, attempts were made to ensure that only phylogenetically conserved switches were considered. However, these switches have not been tested experimentally.

Several chemical probing studies have been used to examine which nucleotides in rRNA are transiently exposed or protected during translation. Laughrea used in vivo chemical footprinting to examine reactivity of rRNA residues to dimethyl sulfate during translation in actively growing cells.[110] These data showed several transient exposures of nucleotides that were abolished when translation was interrupted with chloramphenicol. Specifically, nucleotides in helices 26, 33, and 44 were shown to be transiently exposed during translation. In an in vitro study, Barta and colleagues[111] examined susceptibility of rRNA to cleavage by lead(II) ions in pre and post-translocational state. Strikingly, they found that despite all of the evidence for conformational changes in the small subunit, they detected very little change in the cleavage pattern for 16S rRNA before and after translocation. Equally remarkable was the observation that significant conformational changes did occur in 23S rRNA in domains V and VI, near the peptidyltransferase center and the sarcin-ricin factor binding site. This was surprising because, compared to the rather fluid and changeable 30S subunit, there historically has been some temptation to think of the 50S as a relatively static entity.

Another kind of conformational multiplicity has been observed in cryoelectron microscopy images of free 30S subunits compared to those complexed with 50S subunits.[112] The major site of structural heterogeneity was at the head/neck region; some of this heterogeneity was lost upon association of the small subunit to the large subunit. It would be interesting in this context to see if antibiotics that affect relative domain positions in the 30S subunit[45,49] might limit some of these conformers.

Gating of both the mRNA through the 30S subunit[52,112] and of the nascent peptide through the exit tunnel(s) in the 50S subunit[113] have been observed by cryoelectron microscopy. In the case of the mRNA channel, a pore is seen to alternately open and close around the mRNA, while for the nascent peptide exit tunnel, the gate may be used to direct a peptide through alternate pathways, or to alternately stop and release the growing peptide chain.

Ribosome Assembly

Whereas the assembly of ribosomes is not per se part of translation, an intriguing helical switching mechanism during 30S subunit assembly has been proposed that merits discussion here.[114] A C to U mutation at position 23 ("C23U") in 16S rRNA was shown to confer cold-sensitivity. Growth of this mutant was inhibited below 26°C but not at 37-42°C. When shifted to the nonpermissive temperature, these cells accumulated malformed 30S ribosomal subunits. The mutation was mapped to a helical region in the 5' domain of 16S rRNA, thus it was surprising that it caused cold sensitivity since its effect would apparently be to destabilize a helix that might be further destabilized by heat. Their explanation, supported by genetic and biochemical data, was that nucleotide G11, while depicted as base paired to position C23 in the secondary structure model of 16S rRNA, must also form a base pair with a nucleotide (U-5) in the upstream precursor sequence of immature 16S rRNA during ribosome assembly. Thus, a mutation at position 23 would perturb an equilibrium between two competing helical conformations and interfere with subunit assembly. In addition to this specific helical switching event, a complex and intriguing series of conformational changes in 16S rRNA structure during sequential assembly events has been documented by the Noller group (e.g., see ref. 115). This large body of work is beyond the scope of the present review.

Concluding Remarks

The idea that the ribosome, and in particular, the ribosomal RNA changes conformation during translation has apparently gained acceptance over the last fifteen years. In a 1986 article, Richard Brimacombe wrote: "...many ribosomologists, while tacitly accepting the concept that the secondary structure of mRNA has to be entirely unfolded loop by loop during the translation process, show a surprising resistance to the idea that secondary structural changes can occur in the ribosomal RNA during the same process."[116] Whereas structural evidence for any conformational change in the ribosome may have stirred much excitement in the early 1980's, there is now ample evidence that the ribosome is dynamic. The question of a moveable ribosome has now shifted from 'whether' to 'how, when, and why?' This field is bristling with excitement as these questions are answered with a balance of experimental techniques. Comparisons of high resolution structure determinations of ribosomes in different states will give us the fine details as no other technique can, while solution biochemical and biophysical approaches combined with genetics, comparative and computational analyses will continue to offer insights not readily accessible in the confines of a crystal lattice.

Acknowledgments

We would like to thank Albert Dahlberg for helpful discussions during the preparation of this manuscript and to acknowledge support for research in our laboratory by the NIH (grant number GM35717 to W.E.Hill).

References

1. Serdyuk IN, Spirin AS. Structural Dynamics of the translating ribosome. In: Hardesty B, Kramer G, eds. Structure, Function, and Genetics of Ribosomes. New York: Springer-Verlag, 1986:425-437.
2. Burma DP, Srivastava S, Srivastava AK et al. Conformational change of 50S ribosomes during protein synthesis. In: Hardesty B, Kramer G, eds. Structure, Function, and Genetics of Ribosomes. New York: Springer-Verlag, 1986:438-453.
3. Laughrea M. Structural dynamics of translating ribosomes: 16S ribosomal RNA bases that may move twice during translocation. Mol Microbiol 1994; 11:999-1007.
4. Spahn CM, Nierhaus KH. Models of the elongation cycle: an evaluation. Biol Chem 1998; 379:753-72.
5. Wilson KS, Noller HF. Molecular movement inside the translational engine. Cell 1998; 92:337-49.
6. Rodnina MV, Savelsbergh A, Wintermeyer W. Dynamics of translation on the ribosome: molecular mechanics of translocation. FEMS Microbiol Rev 1999; 23:317-33.
7. Agrawal RK, Frank J. Structural studies of the translational apparatus. Curr Opin Struct Biol 1999; 9:215-21.
8. van Heel M. Unveiling ribosomal structures: the final phases. Curr Opin Struct Biol 2000; 10:259-264.
9. Rodnina MV, Wintermeyer W. Ribosome fidelity: tRNA discrimination, proofreading and induced fit. Trends Biochem Sci 2001; 26:124-30.
10. Rodnina MV, Wintermeyer W. Fidelity of aminoacyl-tRNA selection on the ribosome: Kinetic and Structural Mechanisms. Annu Rev Biochem 2001; 70:415-435.
11. Zamir A, Miskin R, Elson D. Inactivation and reactivation of ribosomal subunits: amino acyl- transfer RNA binding activity of the 30 s subunit of *Escherichia coli*. J Mol Biol 1971; 60:347-64.
12. Kearney KR, Moore PB. X-ray solution-scattering studies of active and inactive *Escherichia coli* ribosomal subunits. J Mol Biol 1983; 170:381-402.
13. Ginzburg I, Miskin R, Zamir A. N-ethyl maleimide as a probe for the study of functional sites and conformations of 30 S ribosomal subunits. J Mol Biol 1973; 79:481-94.
14. Backendorf C, Ravensbergen CJ, Van der Plas J et al. Basepairing potential of the 3' terminus of 16S RNA: dependence on the functional state of the 30S subunit and the presence of protein S21. Nucleic Acids Res 1981; 9:1425-44.
15. Weller JW, Hill WE. Probing dynamic changes in rRNA conformation in the 30S subunit of the *Escherichia coli* ribosome. Biochemistry 1992; 31:2748-57.
16. Zimmermann RA. The decoding domain. In: Zimmermann RA, Dahlberg AE, eds. Ribosomal RNA: Structure, Evolution, Processing, and Function in Protein Biosynthesis. Boca Raton: CRC Press, 1996.

17. Muralikrishna P, Cooperman BS. A photolabile oligodeoxyribonucleotide probe of the decoding site in the small subunit of the *Escherichia coli* ribosome: identification of neighboring ribosomal components. Biochemistry 1994; 33:1392-8.
18. Muth GW, Hennelly SP, Hill WE. Using a targeted chemical nuclease to elucidate conformational changes in the *E. coli* 30S ribosomal subunit. Biochemistry 2000; 39:4068-74.
19. Moazed D, Van Stolk BJ, Douthwaite S et al. Interconversion of active and inactive 30 S ribosomal subunits is accompanied by a conformational change in the decoding region of 16 S rRNA. J Mol Biol 1986; 191:483-93.
20. Ericson G, Minchew P, Wollenzien P. Structural changes in base-paired region 28 in 16 S rRNA close to the decoding region of the 30 S ribosomal subunit are correlated to changes in tRNA binding. J Mol Biol 1995; 250:407-19.
21. Ericson G, Wollenzien P. An RNA secondary structure switch between the inactive and active conformations of the *Escherichia coli* 30 S ribosomal subunit. J Biol Chem 1989; 264:540-5.
22. Miskin R, Zamir A, Elson D. The inactivation and reactivation of ribosomal-peptidyl transferase of *E. coli*. Biochem Biophys Res Commun 1968; 33:551-7.
23. Vogel Z, Vogel T, Zamir A et al. Correlation between the peptidyl transferase activity of the 50 S ribosomal subunit and the ability of the subunit to interact with antibiotics. J Mol Biol 1971; 60:339-46.
24. Zamir A, Miskin R, Vogel Z et al.. The inactivation and reactivation of *Escherichia coli* ribosomes. Methods Enzymol 1974; 30:406-26.
25. Bayfield MA, Dahlberg AE, Schulmeister U et al. A conformational change in the ribosomal peptidyl transferase center upon active/inactive transition. Proc Natl Acad Sci USA 2001; 98:10096-101.
26. Nissen P, Hansen J, Ban N et al. The structural basis of ribosome activity in peptide bond synthesis. Science 2000; 289:920-30.
27. Ban N, Nissen P, Hansen J et al. The complete atomic structure of the large ribosomal subunit at 2.4 A resolution. Science 2000; 289:905-20.
28. Polacek N, Gaynor M, Yassin A et al. Ribosomal peptidyl transferase can withstand mutations at the putative catalytic nucleotide. Nature 2001; 411:498-501.
29. Thompson J, Kim DF, O'Connor M et al. Analysis of mutations at residues A2451 and G2447 of 23S rRNA in the peptidyltransferase active site of the 50S ribosomal subunit. Proc Natl Acad Sci USA 2001; 98:9002-7.
30. Muth GW, Ortoleva-Donnelly L, Strobel SA. A single adenosine with a neutral pKa in the ribosomal peptidyl transferase center. Science 2000; 289:947-50.
31. Shine J, Dalgarno L. The 3'-terminal sequence of *Escherichia coli* 16S ribosomal RNA: complementarity to nonsense triplets and ribosome binding sites. Proc Natl Acad Sci USA 1974; 71:1342-6.
32. Dreyfus M. What constitutes the signal for the initiation of protein synthesis on *Escherichia coli* mRNAs? J Mol Biol 1988; 204:79-94.
33. Kossel H, Hoch B, Zeltz P. Alternative base pairing between 5'- and 3'-terminal sequences of small subunit RNA may provide the basis of a conformational switch of the small ribosomal subunit. Nucleic Acids Res 1990; 18:4083-8.
34. Maidak BL, Cole JR, Parker CT Jr et al. A new version of the RDP (Ribosomal Database Project). Nucleic Acids Res 1999; 27:171-3.
35. Leclerc D, Brakier-Gingras L. A conformational switch involving the 915 region of *Escherichia coli* 16 S ribosomal RNA. FEBS Lett 1991; 279:171-4.
36. Brakier-Gingras L, Pinard R, Dragon F. Pleiotropic effects of mutations at positions 13 and 914 in *Escherichia coli* 16S ribosomal RNA. Biochem Cell Biol 1995; 73:907-13.
37. Pinard R, Payant C, Melancon P et al. The 5' proximal helix of 16S rRNA is involved in the binding of streptomycin to the ribosome. FASEB J 1993; 7:173-6.
38. Brink MF, Verbeet MP, de Boer HA. Formation of the central pseudoknot in 16S rRNA is essential for initiation of translation. EMBO J 1993; 12:3987-96.
39. Wimberly BT, Brodersen DE, Clemons WM Jr et al. Structure of the 30S ribosomal subunit. Nature 2000; 407:327-39.
40. Schluenzen F, Tocilj A, Zarivach R et al. Structure of functionally activated small ribosomal subunit at 3.3 angstroms resolution. Cell 2000; 102:615-23.
41. Canonaco MA, Gualerzi CO, Pon CL. Alternative occupancy of a dual ribosomal binding site by mRNA affected by translation initiation factors. Eur J Biochem 1989; 182:501-6.
42. La Teana A, Gualerzi CO, Brimacombe R. From stand-by to decoding site. Adjustment of the mRNA on the 30S ribosomal subunit under the influence of the initiation factors. RNA 1995; 1:772-82.
43. Petrelli D, LaTeana A, Garofalo C et al. Translation initiation factor IF3: two domains, five functions, one mechanism? EMBO J 2001; 20:4560-9.
44. Ehresmann C, Moine H, Mougel M et al. Cross-linking of initiation factor IF3 to *Escherichia coli* 30S ribosomal subunit by trans-diamminedichloroplatinum(II): characterization of two cross-linking sites in 16S rRNA; a possible way of functioning for IF3. Nucleic Acids Res 1986; 14:4803-21.
45. McCutcheon JP, Agrawal RK, Philips et al. Location of translational initiation factor IF3 on the small ribosomal subunit. Proc Natl Acad Sci USA 1999; 96:4301-4306.
46. Pioletti M, Schlunzen F, Harms J et al. Crystal structures of complexes of the small ribosomal subunit with tetracycline, edeine and IF3. EMBO J 2001; 20:1829-39.
46a. Shapkina TG, Dolan MA, Babin P et al. Initiation factor 3-induced structural changes in the 30 S ribosomal subunit and in complexes containing tRNA(f)(Met) and mRNA. J Mol Biol 2000; 299:615-28.
47. Carter AP, Clemons WM, Jr., Brodersen DE et al. Crystal structure of an initiation factor bound to the 30S ribosomal subunit. Science 2001; 291:498-501.
48. Gualerzi CO, Brandi L, Caserta E et al. Translation initiation in bacteria. In: Garrett RA, Douthwaite S, Liljas A, eds. The Ribosome: Structure, Function, Antibiotics, and Cellular Interactions. Washington, DC: ASM Press, 2000:477-494.

49. Carter AP, Clemons WM, Brodersen DE et al. Functional insights from the structure of the 30S ribosomal subunit and its interactions with antibiotics. Nature 2000; 407:340-8.
50. Ogle JM, Brodersen DE, Clemons WM Jr et al. Recognition of cognate transfer RNA by the 30S ribosomal subunit. Science 2001; 292:897-902.
51. Gabashvili IS, Agrawal RK, Grassucci R et al. Major rearrangements in the 70S ribosomal 3D structure caused by a conformational switch in 16S ribosomal RNA. EMBO J 1999; 18:6501-7.
52. Frank J, Agrawal RK. A ratchet-like inter-subunit reorganization of the ribosome during translocation. Nature 2000; 406:18-22.
53. Wakao H, Romby P, Ebel JP et al. Topography of the *Escherichia coli* ribosomal 30S subunit-initiation factor 2 complex. Biochimie 1991; 73(7-8):991-1000.
54. Wakao H, Romby P, Laalami S et al. Binding of initiation factor 2 and initiator tRNA to the *Escherichia coli* 30S ribosomal subunit induces allosteric transitions in 16S rRNA. Biochemistry 1990; 29:8144-51.
55. Sprengart ML, Fatscher HP, Fuchs E. The initiation of translation in *E. coli*: apparent base pairing between the 16srRNA and downstream sequences of the mRNA. Nucleic Acids Res 1990; 18:1719-23.
56. Sprengart ML, Fuchs E, Porter AG. The downstream box: an efficient and independent translation initiation signal in *Escherichia coli*. EMBO J 1996; 15:665-74.
57. O'Connor M, Asai T, Squires CL et al. Enhancement of translation by the downstream box does not involve base pairing of mRNA with the penultimate stem sequence of 16S rRNA. Proc Natl Acad Sci USA 1999; 96:8973-8.
58. La Teana A, Brandi A, O'Connor M et al. Translation during cold adaptation does not involve mRNA-rRNA base pairing through the downstream box. RNA 2000; 6:1393-402.
59. Moll I, Huber M, Grill S, Sairafi P et al. Evidence against an Interaction between the mRNA downstream box and 16S rRNA in translation initiation. J Bacteriol 2001; 183:3499-505.
60. Noller HF, Moazed D, Stern S et al. Structure of rRNA and its functional interactions in translation. In: Hill WE, Dahlberg AE, Garrett RA, eds. The Ribosome: Structure, Function, & Evolution. Washington, D.C.: American Society for Microbiology, 1990:73-92.
61. Moazed D, Noller HF. Intermediate states in the movement of transfer RNA in the ribosome. Nature 1989; 342:142-8.
62. Moazed D, Noller HF. Sites of interaction of the CCA end of peptidyl-tRNA with 23S rRNA. Proc Natl Acad Sci USA 1991; 88:3725-8.
63. Stark H, Rodnina MV, Rinke-Appel J et al. Visualization of elongation factor Tu on the *Escherichia coli* ribosome. Nature 1997; 389:403-6.
64. Moazed D, Noller HF. Interaction of antibiotics with functional sites in 16S ribosomal RNA. Nature 1987; 327:389-94.
65. Powers T, Noller HF. Evidence for functional interaction between elongation factor Tu and 16S ribosomal RNA. Proc Natl Acad Sci USA 1993; 90:1364-8.
66. Powers T, Noller HF. Selective perturbation of G530 of 16 S rRNA by translational miscoding agents and a streptomycin-dependence mutation in protein S12. J Mol Biol 1994; 235:156-72.
67. Van Ryk DI, Dahlberg AE. Structural changes in the 530 loop of *Escherichia coli* 16S rRNA in mutants with impaired translational fidelity. Nucleic Acids Res 1995; 23:3563-70.
68. Pape T, Wintermeyer W, Rodnina MV. Complete kinetic mechanism of elongation factor Tu-dependent binding of aminoacyl-tRNA to the A site of the *E. coli* ribosome. EMBO J 1998; 17:7490-7.
69. Pape T, Wintermeyer W, Rodnina M. Induced fit in initial selection and proofreading of aminoacyl-tRNA on the ribosome. EMBO J 1999; 18:3800-7.
70. Pape T, Wintermeyer W, Rodnina MV. Conformational switch in the decoding region of 16S rRNA during aminoacyl-tRNA selection on the ribosome. Nat Struct Biol 2000; 7:104-7.
71. Yoshizawa S, Fourmy D, Puglisi JD. Recognition of the codon-anticodon helix by ribosomal RNA. Science 1999; 285:1722-5.
72. Fourmy D, Yoshizawa S, Puglisi JD. Paromomycin binding induces a local conformational change in the A-site of 16 S rRNA. J Mol Biol 1998; 277:333-45.
73. Rheinberger HJ, Nierhaus KH. Testing an alternative model for the ribosomal peptide elongation cycle. Proc Natl Acad Sci USA 1983; 80:4213-7.
74. Nierhaus KH, Beyer D, Dabrowski M et al. The elongating ribosome: structural and functional aspects. Biochem Cell Biol 1995; 73:1011-21.
75. Dabrowski M, Spahn CM, Schafer MA et al. Protection patterns of tRNAs do not change during ribosomal translocation. J Biol Chem 1998; 273:32793-800.
76. Nierhaus KH, Wadzack J, Burkhardt N et al. Structure of the elongating ribosome: arrangement of the two tRNAs before and after translocation. Proc Natl Acad Sci USA 1998; 95:945-50.
77. VanLoock MS, Agrawal RK, Gabashvili IS et al. Movement of the decoding region of the 16 S ribosomal RNA accompanies tRNA translocation. J Mol Biol 2000; 304:507-15.
78. Agrawal RK, Lata RK, Frank J. Conformational variability in *Escherichia coli* 70S ribosome as revealed by 3D cryo-electron microscopy. Int J Biochem Cell Biol 1999; 31:243-54.
79. Agrawal RK, Spahn CM, Penczek P et al. Visualization of tRNA movements on the *Escherichia coli* 70S ribosome during the elongation cycle. J Cell Biol 2000; 150:447-60.
80. Allen PN, Noller HF. Mutations in ribosomal proteins S4 and S12 influence the higher order structure of 16 S ribosomal RNA. J Mol Biol 1989; 208:457-68.
81. Gorini L. Streptomycin and misreading of the genetic code. In: Nomura M, Tissieres A, Lengyel P, eds. Ribosomes. Cold Spring Harbor: Cold Spring Harbor Laboratory Press, 1974:791-803.
82. Bohman K, Ruusala T, Jelenc PC et al. Kinetic impairment of restrictive streptomycin-resistant ribosomes. Mol Gen Genet 1984; 198:90-9.
83. Yusupov MM, Yusupova GZ, Baucom A et al. Crystal structure of the ribosome at 5.5 A resolution. Science 2001; 292:883-96.

84. Lodmell JS, Dahlberg AE. A conformational switch in *Escherichia coli* 16S ribosomal RNA during decoding of messenger RNA. Science 1997; 277:1262-7.
85. Kiel MC, Ganoza MC. Functional interactions of an *Escherichia coli* ribosomal ATPase. Eur J Biochem 2001; 268:278-86.
86. Ganoza MC, Kiel MC. A Ribosomal ATPase Is a Target for Hygromycin B Inhibition on *Escherichia coli* Ribosomes. Antimicrob Agents Chemother 2001; 45:2813-9.
87. Velichutina IV, Dresios J, Hong JY et al. Mutations in helix 27 of the yeast *Saccharomyces cerevisiae* 18S rRNA affect the function of the decoding center of the ribosome. RNA 2000; 6:1174-84.
88. Karimi R, Ehrenberg M. Dissociation rate of cognate peptidyl-tRNA from the A-site of hyper- accurate and error-prone ribosomes. Eur J Biochem 1994; 226:355-60.
89. Karimi R, Ehrenberg M. Dissociation rates of peptidyl-tRNA from the P-site of *E.coli* ribosomes. EMBO J 1996; 15:1149-54.
90. Serdyuk I, Baranov V, Tsalkova T et al. Structural dynamics of translating ribosomes. Biochimie 1992; 74:299-306.
91. Cate JH, Yusupov MM, Yusupova GZ et al. X-ray crystal structures of 70S ribosome functional complexes. Science 1999; 285:2095-104.
92. Agrawal RK, Penczek P, Grassucci RA et al. Effect of buffer conditions on the position of tRNA on the 70 S ribosome as visualized by cryoelectron microscopy. J Biol Chem 1999; 274:8723-9.
93. Wower J, Kirillov SV, Wower IK et al. Transit of tRNA through the *Escherichia coli* ribosome. Cross-linking of the 3' end of tRNA to specific nucleotides of the 23 S ribosomal RNA at the A, P, and E sites. J Biol Chem 2000; 275:37887-94.
94. Lill R, Robertson JM, Wintermeyer W. Binding of the 3' terminus of tRNA to 23S rRNA in the ribosomal exit site actively promotes translocation. EMBO J 1989; 8:3933-8.
95. Lodmell JS, Tapprich WE, Hill WE. Evidence for a conformational change in the exit site of the *Escherichia coli* ribosome upon tRNA binding. Biochemistry 1993; 32:4067-72.
96. Mueller F, Doring T, Erdemir T et al. Getting closer to an understanding of the three-dimensional structure of ribosomal RNA. Biochem Cell Biol 1995; 73:767-73.
97. Wower J, Wower IK, Kirillov SV et al. Peptidyl transferase and beyond. Biochem Cell Biol 1995; 73:1041-7.
98. Nagano K, Nagano N. Transfer RNA docking pair model in the ribosomal pre- and post- translocational states. Nucleic Acids Res 1997; 25:1254-64.
99. Doring T, Mitchell P, Osswald M et al. The decoding region of 16S RNA; a cross-linking study of the ribosomal A, P and E sites using tRNA derivatized at position 32 in the anticodon loop. EMBO J 1994; 13:2677-85.
100. Wower J, Scheffer P, Sylvers LA et al. Topography of the E site on the *Escherichia coli* ribosome. EMBO J 1993; 12:617-23.
101. Miskin R, Zamir A. Effect of streptomycin on ribosome interconversion, a possible basis for the action of the antibiotic. Nat New Biol 1972; 238:78-80.
102. Noreau J, Grise-Miron L, Brakier-Gingras L. Comparison of the action of streptomycin and neomycin on the structure of the bacterial ribosome. Biochim Biophys Acta 1980; 608:72-81.
103. Jerinic O, Joseph S. Conformational changes in the ribosome induced by translational miscoding agents. J Mol Biol 2000; 304:707-13.
104. Brodersen DE, Clemons WM Jr., Carter AP et al. The structural basis for the action of the antibiotics tetracycline, pactamycin, and hygromycin B on the 30S ribosomal subunit. Cell 2000; 103:1143-54.
105. Mankin AS. Pactamycin resistance mutations in functional sites of 16 S rRNA. J Mol Biol 1997; 274:8-15.
106. Noah JW, Dolan MA, Babin P et al. Effects of tetracycline and spectinomycin on the tertiary structure of ribosomal RNA in the *Escherichia coli* 30 S ribosomal subunit. J Biol Chem 1999; 274:16576-81.
107. Atmadja J, Brimacombe R, Maden BE. *Xenopus laevis* 18S ribosomal RNA: experimental determination of secondary structural elements, and locations of methyl groups in the secondary structure model. Nucleic Acids Res 1984; 12:2649-67.
108. Glotz C, Brimacombe R. An experimentally-derived model for the secondary structure of the 16S ribosomal RNA from *Escherichia coli*. Nucleic Acids Res 1980; 8:2377-95.
109. Spitnik-Elson P, Elson D, Avital S et al. Long range RNA-RNA interactions in the 30 S ribosomal subunit of *E. coli*. Nucleic Acids Res 1985; 13:4719-38.
110. Laughrea M, Tam J. In vivo chemical footprinting of the *Escherichia coli* ribosome. Biochemistry 1992; 31:12035-41.
111. Polacek N, Patzke S, Nierhaus KH et al. Periodic conformational changes in rRNA: monitoring the dynamics of translating ribosomes. Mol Cell 2000; 6:159-71.
112. Gabashvili IS, Agrawal RK, Grassucci R et al. Structure and structural variations of the *Escherichia coli* 30 S ribosomal subunit as revealed by three-dimensional cryo-electron microscopy. J Mol Biol 1999; 286:1285-91.
113. Gabashvili IS, Gregory ST, Valle M et al. The polypeptide tunnel system in the ribosome and its gating in erythromycin resistance mutants of L4 and L22. Mol Cell 2001; 8:181-8.
114. Dammel CS, Noller HF. A cold-sensitive mutation in 16S rRNA provides evidence for helical switching in ribosome assembly. Genes Dev 1993; 7:660-70.
115. Powers T, Daubresse G, Noller HF. Dynamics of in vitro assembly of 16 S rRNA into 30 S ribosomal subunits. J Mol Biol 1993; 232:362-74.
116. Brimacombe R. The three dimensional organization of *Escherichia coli* ribosomal RNA. In: Van Knippenberg PH, Hilbers CW, eds. Structure and Dynamics of RNA. New York: Plenum Press, 1986:239-252.

CHAPTER 18

Mechanism of Translation Initiation in Eukaryotes

Francis Poulin and Nahum Sonenberg

Abstract

The initiation of protein synthesis consists in the recruitment of a ribosome•initiator tRNA complex to the initiation codon of a messenger RNA. In prokaryotes, this process involves the direct interaction of the ribosomal RNA with the mRNA. In contrast, eukaryotes have evolved a sophisticated mechanism that relies mostly on protein-RNA and protein-protein interactions. Eukaryotes have taken advantage of the evolution of novel mRNA structures, such as the 5' cap and the poly (A) tail, to develop new mechanisms for the recruitment of the ribosome to the mRNA. As a result, the eukaryotic translation initiation apparatus is now a complex machinery comprising at least eleven factors. This complexity provides a fertile ground for enhanced regulation, and many new mechanisms have been adopted by eukaryotes to control proteins synthesis. Indeed, many translation factors are phosphoproteins whose function can be regulated by extracellular signals. We will describe here the mechanism of translation initiation in eukaryotes, with a particular emphasis on translation factors and their function.

Introduction

With the structural confirmation that the ribosome is a ribozyme (see Chapter 15 by B.T. Wimberly), translation emerges as a process driven for the most part by RNA. As such, the initiation of protein synthesis appears to be devoted to the assembly of the catalytic rRNA and an initiator tRNA at the correct AUG codon of a template mRNA. These RNAs may now be widely acknowledged as the stars but they nevertheless cannot perform alone. Indeed the translation machinery, especially in eukaryotes, has evolved to require a host of protein factors.[1] The focus of this review is on the essential parts played by the eukaryotic initiation factors (eIFs) in bringing together the initiator tRNA, ribosome and mRNA.

In eukaryotes, at least eleven different initiation factors are required to properly initiate translation.[1] Collectively, they ensure that the methionyl-initiator tRNA (Met-tRNA$_i^{Met}$) is brought in the P site of the ribosome to the initiator AUG of an mRNA. Conceptually, this process can be divided in four steps:

1. formation of the 43S pre-initiation complex, when the Met-tRNA$_i^{Met}$ is delivered by eIF2 to the P site of the 40S ribosomal subunit;
2. recruitment of the 43S complex to the 5' end of the mRNA by eIF3 and the eIF4 factors;
3. scanning of the 5' untranslated region (UTR) and recognition of the AUG codon, and
4. assembly of the 80S ribosome (Fig. 1).

Only the standard "scanning" mechanism utilized by most cytoplasmic mRNAs will be considered here. The reader is directed to recent reviews by Jackson[2] and Hellen & Sarnow[3] for a detailed discussion of alternative initiation pathways.

The sequence of several archaeal genomes revealed that translation in archaea is a hybrid between the bacterial and eukaryotic systems.[4,5] Archaeal mRNAs are uncapped, polycistronic, lack long poly (A) tails, and their coding sequences are most often preceded by a Shine-Dalgarno element. In

Translation Mechanisms, edited by Jacques Lapointe and Léa Brakier-Gingras.

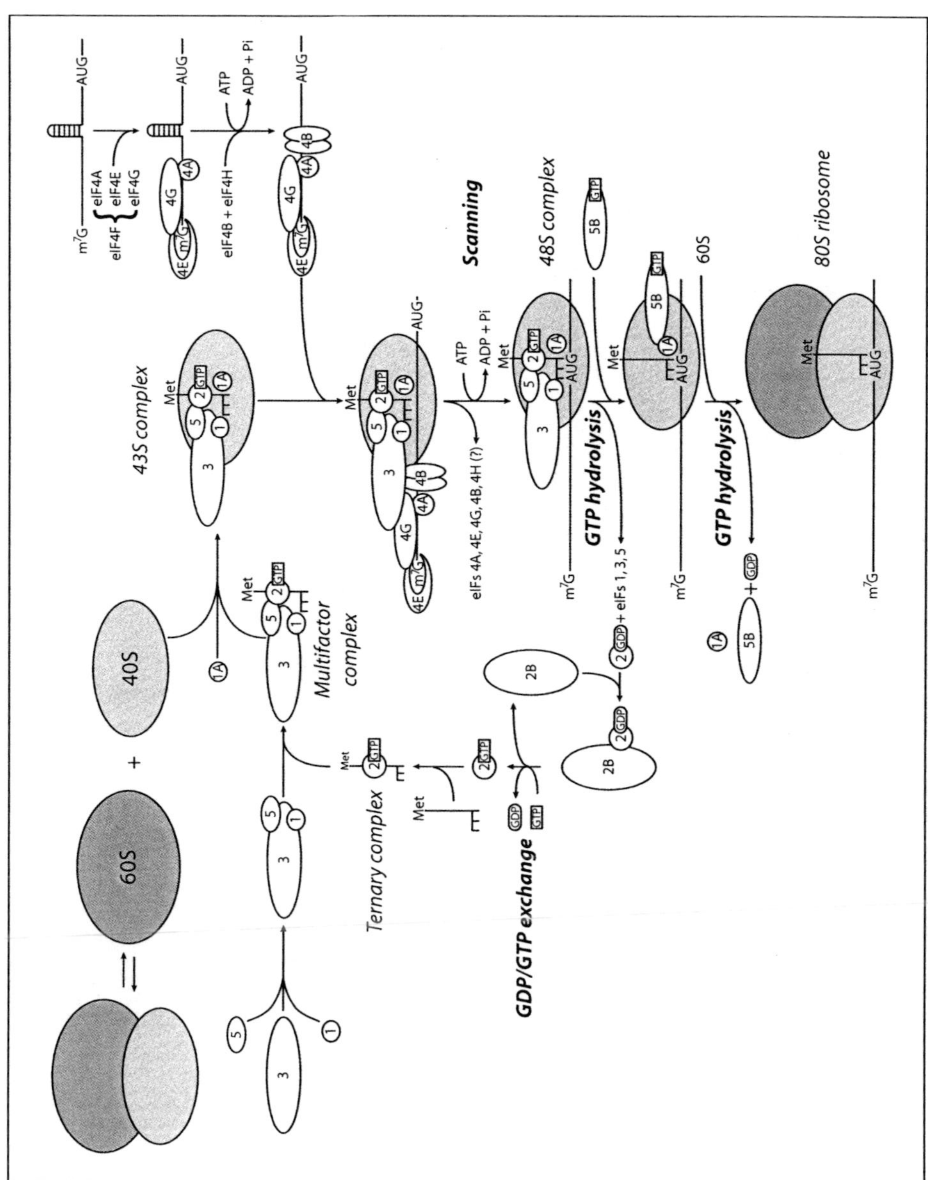

Figure 1. Scanning model of eukaryotic translation initiation. The Met-tRNA$_i^{Met}$ interacts with eIF2•GTP to form the ternary complex. The multifactor complex is an intermediate facilitating Met-tRNA$_i^{Met}$ recruitment to the 40S ribosomal subunit, which generates the 43S complex. The eIF4 factors promote the recruitment of the 43S complex to the mRNA 5' end. The 43S complex then scans the mRNA 5' untranslated region to localize the initiator AUG (48S complex). It is not known whether the eIF4 factors participate in the scanning process. Base pairing between the Met-tRNA$_i^{Met}$ anticodon and the AUG codon activates eIF2 GTPase, which causes the release of the bound factors. eIF1A and eIF5B interact to promote ribosomal subunits joining. 60S joining activates the GTPase activity of eIF5B, leading to its release from the 80S ribosome. The ribosome is then ready to accept the first elongating aminoacyl-tRNA in the A site. Initiation factors are labeled with their respective number, and ribosomal subunits are depicted as shaded ovals.

contrast to the bacterial mechanism for initiation codon selection, archaeal genomes contain homologues of many initiation factors that were thought to be exclusively eukaryotic.[6-8] We will not discuss translation in archaea in more detail, except to point out that archaeal factors have been utilized in the last years to derive important structural and functional insights into the mechanisms of eukaryotic translation initiation (see below).

Formation of the 43S Preinitiation Complex

Initiator tRNA

Translation is initiated by a special Met-tRNA in every organism.[9] Because of its unique function, the initiator Met-tRNA (Met-tRNA$_i^{Met}$) presents many functional and structural features that distinguish it from the elongator Met-tRNA. In eukaryotes, the Met-tRNA$_i^{Met}$ interacts specifically with an eIF2•GTP complex (see below), which delivers it exclusively to the ribosomal P site.[10] This particular interaction can be explained by sequence divergence between the initiator and elongator form of Met-tRNA (Fig. 2). Some sequence specificities promote the activity of the Met-tRNA$_i^{Met}$ in initiation (Fig. 2). These include the A1:U72 base pair which is critical for initiator function and cell viability in *S. cerevisiae*[11] and, when mutated to the G1:C72 present in elongator Met-tRNA, weakens the affinity of the human Met-tRNA$_i^{Met}$ for eIF2 by a factor of 10.[12] The critical function of the A1:U72 base pair in initiation is underscored by its almost universal conservation amongst cytoplasmic eukaryotic initiator tRNAs, and its exclusion from all eukaryotic elongator tRNAs.[13] In addition, the Met-tRNA$_i^{Met}$ loop IV contains A60 in place of pyrimidine 60, and A54 instead of the T54 present in the TψC sequence found in elongator tRNAs (Fig. 2). The presence of an A at positions 54 and 60 is not absolutely required, as only the A54U substitution is lethal in yeast.[11] In comparison, the double substitution A54, A60 to T54, U60 in the human tRNA has minimal effect on an in vitro initiation assay.[14] A yeast Met-tRNA$_i^{Met}$ with both G1:C72 and T54 mutations rescues an elongator tRNA^{Met}-depleted strain, suggesting a requirement of these residues for initiator functions and discrimination against elongator tRNA^{Met}.[15] Finally, the Met-tRNA$_i^{Met}$ contains three consecutive G:C base pairs in the anticodon stem (Fig. 2). Substituting these nucleotides with those found in the elongator tRNA reduces the initiator activity in vitro.[14] Furthermore, introduction of G29:C41 and G31:C39 in a yeast elongator tRNA with the A1:U72 and A54 mutations increases its activity in initiation.[15]

Other sequence specificities are important for the discrimination of initiator and elongator tRNA^{Met} (Fig. 2). Position 64 in plants and fungi has a unique 2'-O-phosphoribosyl group, which disrupts the structure of the TψC stem.[9,16] This modification prevents the elongator function of Met-tRNA$_i^{Met}$,[17,18] probably by inhibiting its interaction with the eEF1A•GTP complex.[19,20] In vertebrates, it has been proposed that the TψC stem perturbation is due to the nature of the base pairs at the base of the stem.[21] Indeed, mutation in base pairs A50:U64 and U51:A63 yield an initiator tRNA that participates in elongation and, when combined with the A1:U72 -> G1:C72 mutation, results in almost wild-type elongator function.[21]

eIF2

The initiator tRNA is delivered directly to the P site of the 40S ribosomal subunit as part of a ternary complex that also contains the heterotrimeric G protein eIF2 and GTP (Fig. 1). Subsequent pairing between the AUG initiation codon and the Met-tRNA$_i^{Met}$ anticodon elicits hydrolysis of the GTP by eIF2, which requires the GTPase activating protein (GAP) eIF5 (Fig. 1; see below). The eIF2•GDP complex does not bind to the Met-tRNA$_i^{Met}$, and recycling of the GTP necessitates the guanine nucleotide exchange factor (GEF) eIF2B (Fig. 1; see below). An eIF2 homologue is present in archaea but not in bacteria, where IF2 delivers the fMet-tRNA$_f^{Met}$ to the 30S subunit.[7]

Essential genes encode the α, β and γ subunits of eIF2 in *S. cerevisiae*. Mutations in all three subunits affect the accuracy of AUG initiation codon selection.[22] Biochemical and genetic analyses indicate that eIF2γ likely binds to GTP and Met-tRNA$_i^{Met}$.[10,22] Interaction between eIF2γ and the initiator tRNA is reduced by the Y142H and N135K mutations in yeast.[23,24] The N135K mutant also increases GTP hydrolysis,[24] whereas the K250R mutation increases the off-rate for GDP and GTP.[23] These observations have recently been corroborated by the crystal structure of the archaeal eIF2γ (a-eIF2γ) from *Pyrococcus abyssi*.[25] Sequence similarity[7] suggests that a-eIF2γ is a close structural

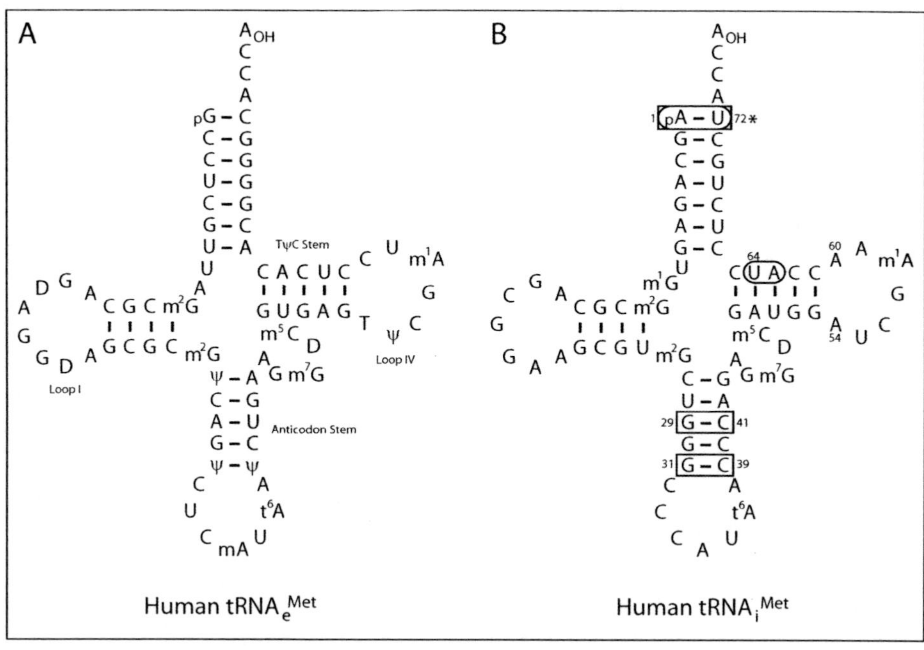

Figure 2. Specific elements in human initiator tRNA. Structure of the human elongator (A) and initiator (B) Met-tRNA. Bases important for promoting initiator function are boxed, whereas those discriminating against elongator function are circled. An asterisk denotes the binding site of eIF2. Position 64 in fungi and plants Met-tRNA$_i^{Met}$ has a phosphoribosyl group attached to the ribose 2'-OH. Position 54 is critical for the initiator function of yeast Met-tRNA$_i$Met. (Adapted from ref. 14)

homologue of the elongation factors eEF1A and the bacterial EF1A (see Chapter 20 by P. Nissen et al and ref. 25). This is most apparent in the nucleotide-free structure, which shows that the three domains of a-eIF2γ are very similar to their equivalents in eEF1A and EF1A, although their relative orientation is somewhat different.[25] Accordingly, the GDP binding pocket of a-eIF2γ is identical to that of EF1A. Superimposition of domains 2 and 3 from the EF1A•GTP•Phe-tRNA^{Phe} structure on the corresponding domains in a-eIF2γ indicates that the Phe-tRNA^{Phe} can readily be accommodated by the structure, except for the Phe side chain.[25] This in turn suggests a binding pocket for the Met side chain of Met-tRNA$_i^{Met}$, which might explain why the methionyl group enhances initiation.[26] The structure also shows that a-eIF2γ lacks features allowing EF1A to bind aminoacyl-tRNA nonspecifically, helping to explain the exclusion of elongator tRNA from eIF2.[25] Unfortunately, no difference could be observed when comparing the GDP-bound structure of a-eIF2γ to the GMP-PNP bound form, which might indicate that eIF2β has a role in tRNA$_i^{Met}$ binding to eIF2. Therefore, the structural explanation of how the hydrolysis of GTP causes the release of the Met-tRNA$_i^{Met}$ will probably require a co-crystal of a-eIF2γ with a-eIF2β, a-eIF2α or both, along with the initiator tRNA. An interesting aspect of the a-eIF2γ structure is the occurrence of a conserved four cysteine cluster that was found to coordinate a zinc atom.[25] The specific role of this motif is not known, but its significance has been highlighted in yeast where mutating the second cysteine (C155) to a serine causes a slow growth phenotype.[27]

While eIF2γ forms the structural core of eIF2, essential functions are also performed by eIF2β.[10] Genetic screening in yeast identified eIF2β mutations in an essential Cys$_2$-Cys$_2$ zinc-finger motif found in the carboxy-terminal domain of eIF2β.[22] Two of these mutants (S264Y and L254P) were biochemically characterized and confer an intrinsic GTPase activity to eIF2, implicating eIF2β in promoting eIF2γ GTPase activity.[24] Independent of GTP hydrolysis, the S264Y mutant also increases initiator tRNA dissociation from eIF2.[24] In addition, the zinc-finger motif is part of a larger mRNA binding region, and could therefore be important for facilitating codon-anticodon interaction in eIF2.[28] mRNA binding properties were also ascribed to the three lysine repeats found at the amino-terminus of eIF2β.[28] Moreover, these repeats mediate the mutually exclusive interaction of eIF5 and eIF2Bε with eIF2β.[29,30] In agreement with the absence of eIF2B and eIF5 in archaea, a-eIF2β does not possess the region containing the lysine repeats. The solution structure of a-eIF2β from *Methanococcus jannaschii* was recently solved.[31] a-eIF2β consists of two independent structural domains connected by a conserved helical linker.[31] The N-terminal domain of a-eIF2β corresponds to the central domain of yeast eIF2β, which is responsible for the interaction with eIF2γ.[32] The structure of the C-terminal domain reveals a critical role for the zinc atom in maintaining the structural integrity of the domain.[31] Yeast eIF2β mutations implicated in initiation codon selection[22] affect residues that are conserved in a-eIF2β. Four of these residues (Lys122, Arg125, Val 126 and Ile140) are located in close proximity in the structure of the C-terminal domain,[31] and might serve as a binding site for other translational components implicated in initiation codon recognition.[31]

Phosphorylation of the conserved serine residue (Ser51) in eIF2α converts eIF2•GDP from a substrate to a competitive inhibitor of eIF2B.[10] Whereas eIF2α is required for yeast survival under normal conditions, it is not essential as long as the β and γ subunits as well as the initiator tRNA are overexpressed.[33] Removal of the α subunit of eIF2 causes an 18-fold increase in the K_m of eIF2B catalyzed nucleotide exchange, indicating that eIF2α is required for interactions between eIF2 and eIF2B that promote wild-type rates of nucleotide exchange.[34] Consistent with this, phosphorylation of eIF2α on Ser51 increases its affinity for the regulatory subcomplex of eIF2B (α, β, δ), in a way that hampers the function of the eIF2B catalytic subcomplex (γ, ε).[35] Taken together, these data clearly indicate that the α subunit of eIF2 has evolved to become first and foremost a regulator of translation initiation. The X-ray structure of the N-terminal two-thirds of human eIF2α has been solved.[36] This structure contains two domains: an N-terminal oligonucleotide binding (OB) domain similar to that in IF1[37] and eIF1A,[38] and a C-terminal helical domain.[36] The Ser51 residue is located in an unstructured loop connecting β3 and β4 of the OB domain. The putative RNA binding site of the OB domain lacks the cluster of positive charges that is observed in other members of the OB family, which is consistent with eIF2α not interacting with RNA.[36] Protein-protein interactions involving eIF2α could be mediated by a highly conserved groove that is observed between the OB and helical domains.[36]

eIF2B

eIF2B is a heteropentameric GEF that is required to recycle the inactive eIF2•GDP complex into eIF2•GTP (Fig. 1). eIF2B is present in cells in lower amounts than eIF2.[39] Phosphorylation of eIF2 on Ser51 of its α subunit sequesters eIF2B, thus repressing translation initiation.[10] Genetic and biochemical methods indicate that eIF2B is divided into a catalytic subcomplex comprising eIF2Bγ and eIF2Bε, and a regulatory subcomplex consisting of eIF2Bα, eIF2Bβ and eIF2Bδ.[40] The catalytic subcomplex can perform the guanylate exchange reaction even when eIF2α is phosphorylated, unless the regulatory subcomplex is present to modulate its activity.[40]

eIF2Bε and eIF2γ display sequence homology throughout their length, and both subunits are essential in yeast.[10] eIF2Bγ does not exhibit catalytic activity, but it enhances the weak intrinsic guanylate exchange activity of eIF2Bε.[40,41] Indeed, the catalytic subcomplex exhibits a higher activity than the holoenzyme, probably owing to the absence of regulation from the other subunits.[40] The catalytic site is located in the C-terminal region of eIF2Bε, and is independent from the bipartite motif that is required for binding to eIF2β.[30,42] The N-terminus of eIF2Bε contains elements that are necessary for the assembly of eIF2Bε and eIF2Bγ into the eIF2B complex.[42]

eIF2Bα, eIF2Bβ and eIF2Bδ also display sequence homology to each other. These three subunits are required for the regulatory subcomplex to bind to eIF2α and differentiate between its phosphorylated and nonphosphorylated forms.[40] eIF2Bβ and eIF2Bδ are essential in yeast, but eIF2Bα is dispensable.[10] However, the absence of eIF2Bα renders eIF2B insensitive to the phosphorylation status of eIF2α, both with yeast and rat factors.[40,43] In summary, eIF2B appears to have evolved to function in two roles: a GEF to recycle eIF2•GDP, and a regulator to respond to eIF2α phosphorylation.

Interestingly, mutations in the five subunits of human eIF2B are associated with leukoencephalopathy with vanishing white matter (VWM), a rare brain disease usually affecting children.[44,45]

This is the first documented disease to be associated with mutations in a translation factor. Many missense mutations were observed in the different eIF2B subunits, but nonsense mutations were always found to be heterozygous.[44,45] These data suggest that mutations completely inactivating eIF2B are lethal when homozygous, in agreement with the essential role of eIF2B.[10] The central nervous system specificity of the disease is surprising. Translational regulation by eIF2B involves the phosphorylation of eIF2 on the α subunit in response to various stress signals (see below). Thus, it is possible that the brain is more sensitive to an altered stress response brought about by mutated eIF2B. Indeed, neurological deterioration is observed during or after episodes of fever in VWM patients.[44,45]

eIF3

The largest and most complex initiation factor, eIF3, is composed of up to 11 nonidentical subunits in mammals, 6 of which are conserved in yeast (Table 1).[1,10] In agreement with its elaborate constitution, eIF3 has multiple functions in translation initiation:

1. it is thought to promote the dissociation of 40S and 60S ribosomal subunits,
2. it participates in the recruitment of the ternary complex to the 40S subunit, and
3. it is required for mRNA recruitment to the preinitiation complex.[1,10]

eIF3 and eIF1A are thought to promote ribosome dissociation, but the mechanism is still poorly characterized.[1] Recent work suggests that eIF3 prevents 60S subunits from disrupting the 43S preinitiation complex.[46] This is consistent with earlier observations of eIF3 being associated with native 40S subunits,[47] which implies that the ternary complex binds to a preformed 40S•eIF3 complex. Indeed, a ternary complex bound to 40S subunits in the absence of eIF3 was readily dissociated by 60S subunits.[46] Thus, eIF3 appears to be required for the stabilization of ternary complex binding to the small ribosomal subunit.

The study of eIF3 in yeast has recently clarified its role in the recruitment of the ternary complex to 40S subunits. eIF3 from *S. cerevisiae* is composed of a core of five essential subunits that are conserved in mammals (Table 1): eIF3a, eIF3b, eIF3c, eIF3g and eIF3i.[10,48,49] The core complex co-purifies with eIF5,[48] an interaction that has also been observed in mammalian cells.[50] This interaction is mediated by the amino-terminus of eIF3c,[30,51] and is important for bridging eIF3 to eIF2, since both can concurrently bind to the C-terminal bipartite motif of eIF5 (Fig. 3).[51] eIF1 can also interact with eIF3c,[49,52] and a large multifactor complex consisting of eIF1/eIF2/eIF3/eIF5/Met-tRNA$_i^{Met}$ is observed in vivo in the absence of 40S subunits (Fig. 1).[51] Since eIF1, eIF5 and the three eIF2 subunits promote stringent AUG selection,[10,22,24] eIF3 could potentially have a structural function in AUG recognition. Thus, the multiple protein-protein interactions mediated by eIF3 appear to be critical in the recruitment or stable binding of the ternary complex to 40S subunits (Fig. 3).

Table 1. eIF3 subunit composition

| | Former Designation | |
Subunit Name	Human	*S. cerevisiae*
eIF3a	p170	Tif32p/Rpg1p
eIF3b	p116	Prt1p
eIF3c	p110	Nip1p
eIF3d	p66	None
eIF3e	p48	None
eIF3f	p47	None
eIF3g	p44	Tif35p
eIF3h	p40	None
eIF3i	p36	Tif34p
eIF3j	p35	Hcr1p
eIF3k	p28	None

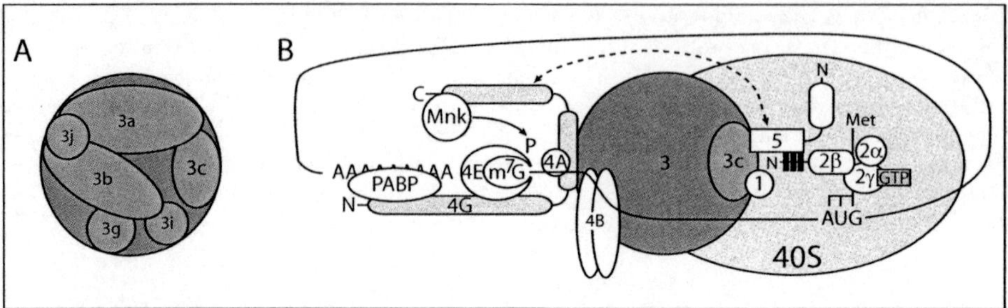

Figure 3. eIF3 and eIF4G coordinate protein-protein interactions among initiation factors. (A) Summary of the protein-protein interactions within yeast eIF3.[10,49,51,53] (B) The assembly of the mammalian translation initiation complex is coordinated by eIF3 and eIF4G. eIF4G functions as a molecular scaffold that binds to PABP, eIF4E, eIF4A, eIF3 and Mnk1 (see text for details). eIF4G participates in ribosomal recruitment, and brings together the cap and poly (A) tail of the mRNA. eIF3 interacts with eIF1, eIF5, eIF4B, eIF4G and 40S ribosome. eIF3 mediates the binding of the ternary complex to the 40S subunit, probably via eIF5. Moreover, the interaction of eIF3 to eIF4G is thought to act as a bridge between the 43S complex and the 5' cap. The subunit of eIF3 that interacts with eIF4G is not known. Interaction between eIF3 and eIF4G has not been observed in yeast. Instead, the interaction of eIF5 with the C-terminus of eIF4G (dashed arrow) could promote 43S complex recruitment to the mRNA 5' end. The bipartite motif at the C-terminus of eIF5 interacts with the three lysine repeats (black boxes) at the N-terminus of eIF2β. (Adapted from ref. 134)

A yeast homologue of the mammalian eIF3j is not associated with the purified eIF3 core complex.[48] However, it appears that at least a fraction of eIF3j is incorporated into eIF3.[53] Like its mammalian counterpart,[54] yeast eIF3j interacts with eIF3a.[53] Interestingly, the amino acid sequence of eIF3j is similar to the middle region of eIF3a, and forms a motif that is sufficient for binding to the RNA recognition motif (RRM) of eIF3b. Deletion of eIF3j is not lethal but causes a destabilization of eIF3 and its associated factors, suggesting that it is important in stabilizing the interaction of eIF3a with eIF3b.[53] The latter interaction is critical in yeast, since deletion of the eIF3b RRM releases eIF3a and induces the breakdown of the multifactor complex, with the remaining eIF3b-eIF3g-eIF3i being unable to bind 40S subunits.[53] In support of these observations, a purified subcomplex of eIF3 containing eIF3a, eIF3b and eIF3c is found to integrate eIF3j, eIF1 and eIF5, and can rescue translation in vitro almost as efficiently as purified eIF3 core complex.[55]

Finally, eIF3 is required for the recruitment of mRNA to the 43S preinitiation complex. In mammals, this is thought to be due to the interaction of eIF3 with eIF4F, which acts as a bridge to the mRNA through the interaction of eIF4E with the cap structure.[56,57] The subunit of eIF3 responsible for this interaction has not been identified yet, and there is no report of a similar interaction in yeast. However, it was recently reported that yeast eIF5 can interact with the carboxy-terminal half of eIF4G, and could act as a link between eIF3 and the mRNA (Fig. 3).[58] Another factor that could be involved in recruiting the 43S preinitiation complex to the mRNA is eIF4B (see below), which interacts with eIF3a in mammals,[59] and with eIF3g in *S. cerevisiae*.[60]

Recruitment of the 43S Complex to the mRNA

Cap Structure

Eubacterial and archaeal mRNAs possess a Shine-Dalgarno sequence that facilitates ribosome recruitment by interacting directly with the complementary sequence at the 3' end of the 16S ribosomal RNA. Despite having a very similar 3' end, the eukaryotic 18S rRNA lacks such an element.[2] To identify the 5' end of mRNAs, eukaryotic cells have instead opted for a specific structure called a cap. This m[7]GpppX sequence (where X is any nucleotide) is present at the 5' end of all nuclear transcribed mRNAs from eukaryotes,[61] and has possibly evolved to stabilize mRNAs. The cap facilitates translation initiation, but also participate in other processes such as mRNA splicing and nucleocytoplasmic transport.[62] The recruitment of the 43S complex to the mRNA is mediated by the eIF4 group of factors. The function of these factors has been thoroughly reviewed elsewhere,[63] and thus only a summary of our current understanding will be presented here.

eIF4E

eIF4E interacts with the cap structure and, together with eIF4G and eIF4A, forms the eIF4F complex (Fig. 1). eIF4E is an essential factor that is conserved across all eukaryotic kingdoms. The structures of yeast and mouse eIF4E bound to m^7GDP were solved,[64,65] and demonstrate the conservation of the cap-binding mode. Indeed, all the amino acids involved in cap recognition are conserved from yeast to humans, which strongly suggests a common ancestral origin. In light of the fact that an eIF4E homologue is absent from archaea, it will be very interesting to address the question of the origin of this factor.[66] Why did the eukaryotes select to use the cap structure to initiate translation? What selective advantage is achieved by adopting such a radically different ribosome recruitment mode?

In mouse and human, eIF4E is phosphorylated on one major site, serine 209.[67] The function of this phosphorylation in not completely understood. Phosphorylation of eIF4E is correlated with increased translation rates, but it is still disputed whether phosphorylation increases the cap binding affinity of eIF4E.[68-70] The structure suggests that phosphorylated Ser209 could form a salt bridge with Lys159, which would act as a clamp above the proposed trajectory of the mRNA path.[64] However, a more recent structural analysis of the full-length eIF4E complexed with m^7GpppA indicates that the distance between Ser209 and Lys159 would be too long to form a salt bridge.[68] In fact, the C-terminal loop region of eIF4E, which encompasses Ser209, interacts with the second nucleotide of the mRNA. Hence, phosphorylation of Ser209 could still have an effect on the affinity of eIF4E for capped mRNA.[68]

Notwithstanding its structural basis, the biological significance of eIF4E phosphorylation is not immediately clear. Studies on eIF4E phosphorylation mutants have demonstrated that they can restore translation as well as the wild-type protein in an in vitro assay.[71] These eIF4E variants can also rescue the lethal phenotype of eIF4E deletion in yeast.[71] In contrast, transgenic *Drosophila melanogaster* expressing an eIF4EI phosphorylation mutant (S251A, analogous to S209A) displays reduced viability (65% survival) in an eIF4EI mutant background, with the surviving flies having growth retardation defects.[72] These defects were not observed when a S251D mutant was used to mimic a constitutively phosphorylated residue.

eIF4E is phosphorylated in vivo by the kinase Mnk1.[73,74] In order to phosphorylate eIF4E, Mnk1 needs to be recruited to the C-terminal part of eIF4G (Fig. 3).[75] It appears highly unlikely that such a sophisticated mechanism would have evolved if it did not have a purpose. Therefore we conclude that eIF4E phosphorylation must play an important regulatory role in vivo, at least in higher eukaryotes. This will probably be more fully appreciated by performing in vivo studies of a mouse knock-in bearing an eIF4E Ser209Ala mutation.

eIF4G

Two related proteins, eIF4GI and eIF4GII, function as modular adaptor proteins that mediate the series of protein-protein interactions culminating in the recruitment of the 43S complex to the mRNA 5' end (Fig. 3).[63,76,77] eIF4G has evolved around a phylogenetically conserved central core (Fig. 4).[66] This core is contained within the middle fragment of the protein and possesses two interaction domains: the first interacts with the eIF4A RNA helicase,[78,79] and the second is an RNA binding domain.[80,81] In addition, this fragment of mammalian eIF4G also contains an eIF3 interaction domain.[56,78] The X-ray structure of the middle portion of human eIF4GII has been solved, and exhibits a crescent-shaped domain consisting of ten alpha helices organized into five HEAT repeats.[82] Proteins containing HEAT repeats are involved in the assembly of large multiprotein complexes, which is probably facilitated by the large and accessible surface area of the domain.[83]

Extension of the central core of eIF4G to include the eIF4E-binding site generates an eIF4G domain that is necessary and sufficient for cap-dependent translation initiation.[84] The 4E-binding site is apparently only required for targeting eIF4G to the mRNA. Indeed cap-independent translation of an mRNA containing an iron response element (IRE) can be directed by the central core of eIF4G, if this domain is fused to the IRE-binding protein IRP-1.[85,86]

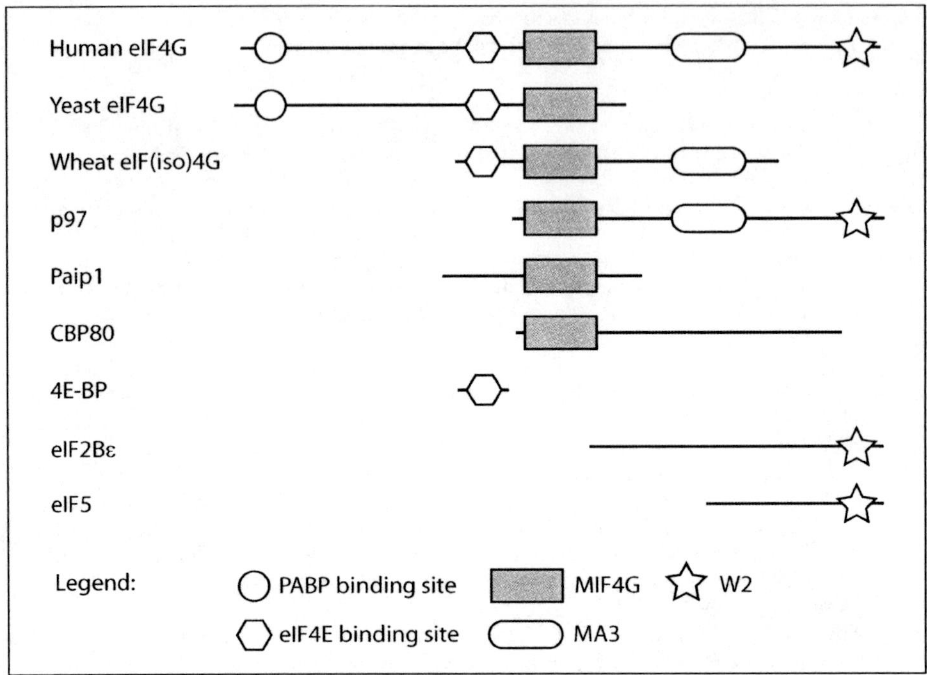

Figure 4. eIF4G evolved around a conserved core domain. Mammalian eIF4G contains five distinct structural elements. Its amino-terminal third contains a motif for PABP binding, and an eIF4E binding site. The middle fragment of eIF4G (MIF4G) is a HEAT domain and constitutes the conserved core of the protein. The carboxy-terminal third of mammalian eIF4G consists of an MA3 domain (named after the Mouse Apoptosis gene-3, which is induced upon cell death[149]), and a W2 domain (named after two conserved tryptophans). The MIF4G is common to many proteins involved in translation and nonsense-mediated mRNA decay. Amongst others, an MIF4G homologue has been observed in the translational repressor p97,[150] the translational activator Paip1,[151] the nuclear cap binding protein CBP80,[152] and the nonsense-mediated mRNA decay 2 protein (NMD2/Upf2; see Chapter 13).[66,153] The MA3 and W2 domains appear to have been added later in evolution, and are mostly involved in regulatory interactions.[84] In mammalian eIF4G, the MA3 domain interacts with eIF4A,[78] and the W2 domain with the eIF4E kinase Mnk.[75] eIF5 and eIF2Bε also display a W2 domain,[93] but in this case it is important for the interaction with eIF2β.

The amino-terminal portion of eIF4G also contains an interaction domain for the poly(A) binding protein (PABP).[87,88] PABP is a cytoplasmic protein that binds to the poly(A) tail present at the 3'-end of messenger RNAs. Its interaction with eIF4G provides a physical link between the cap and the poly(A) tail (Fig. 3). It also explains the observation that the poly(A) tail can operate as an enhancer of translation initiation.[89,90] The significance of the cap-poly(A) tail interaction is discussed in more detail in Chapter 12.

The C-terminus of eIF4G has been considerably extended during the expansion of eukaryotes.[66] A second eIF4A interaction domain is present in human eIF4G.[78] This domain is absent from yeast eIF4G, but can be found in *Arabidopsis* and *Drosophila* eIF4G homologues. The related protein p97/NAT1/DAP-5 has a similar domain,[66] which on the other hand does not bind to eIF4A.[78] In eIF4G, the C-terminal eIF4A binding domain is not essential for translation, but instead plays a role in modulating the function of the core domain.[84] There is no agreement as to whether eIF4G interacts with one or two eIF4A molecules.[84,91,92] The extreme C-terminus of eIF4G from metazoans displays another conserved motif which, in mammals, is necessary for the interaction with the eIF4E kinase Mnk1.[75] It is intriguing that this motif is conserved in the C-termini of eIF5 and eIF2Bε,[93] and participates in their interaction with eIF2β.[30] This may represent a conserved module important for protein-protein interactions regulating the assembly of the translational apparatus.[66]

eIF4A

eIF4A is a member of the large family of RNA helicases termed DEAD-box, after one of their 8 conserved motifs.[94] In yeast, eIF4A is encoded by two genes whose simultaneous inactivation is lethal.[95] There are three eIF4A isoforms in mammals, but only eIF4AI and eIF4AII appear to be involved in translation initiation.[96] Interestingly, mammalian eIF4A can interact with yeast eIF4G1 in vitro,[79] even if it cannot complement yeast mutants defective in eIF4A.[97] In fact mammalian eIF4A appears to act as a translational inhibitor in yeast,[79] much like eIF4AIII inhibits translation in mammals.[96] eIF4A is the only eIF4 factor to have a homologue in archaea.[6]

eIF4A is an RNA dependent ATPase and can unwind RNA duplexes by itself or, much more efficiently, as part of the eIF4F complex.[63] The intrinsic helicase activity of eIF4A is relatively weak, but it is strongly stimulated by eIF4B, although the two proteins do not appear to interact directly.[98] Recent studies suggest that eIF4A is a nonprocessive helicase that is limited to a single round of unwinding.[99,100] The presence of eIF4B (or eIF4H) may limit RNA re-association and allow the unwinding of longer duplexes.[99,100] It is noteworthy that eIF4A functions primarily as part of the eIF4F complex, but that it apparently needs to be exchanged with free eIF4A to effect unwinding.[101] It is possible that this recycling process is what conveys processivity to the complex. An unanswered question is how does eIF4A facilitate ribosome binding? eIF4F is generally believed to target eIF4A to the mRNA, which then unwinds secondary structures to allow ribosome recruitment. The role of eIF4A could be to utilize the energy from ATP hydrolysis to rearrange RNA structures, or maybe RNA-protein complexes, thus allowing 40S subunit binding to the mRNA.[102,103]

eIF4B

The function of eIF4B in translation initiation is not completely clear. In vitro, eIF4B enhances the helicase activity of eIF4A and eIF4F,[98] and promotes 48S complex assembly.[84,104] However, eIF4B is not required in ribosome binding assays,[84,104] and yeast strains with a disrupted eIF4B gene display translational defects and a slow-growth phenotype, but are viable.[105,106]

eIF4B homodimerizes and interacts with RNA through a N-terminal RRM and a C-terminal arginine-rich motif (ARM).[59,107,108] Since eIF4B also demonstrates an RNA-annealing activity, it was proposed that it could play a role in mediating interactions between the rRNA and mRNA.[109] Consistent with this hypothesis, eIF4B can interact with the 18S rRNA in vitro, and can simultaneously bind to two RNA molecules.[110]

eIF4H

eIF4H is a recently identified factor that shares functional and sequence homology with eIF4B, mostly with the RRM domain.[111] It weakly binds RNA and stimulates protein synthesis in vitro, as well as the ATPase and helicase activity of eIF4A and eIF4F.[99,100,111,112] A recent report shows that the virion host shutoff protein (Vhs) of herpes simplex virus interacts with eIF4H.[113] This protein accelerates mRNA degradation, which modulates the levels of viral and cellular gene expression. It appears that Vhs is targeted to the mRNA 5' end through its interaction with eIF4H.[113]

Scanning and Localization of the Initiator AUG

mRNA 5' UTR

After binding to the 5' end of the mRNA, the preinitiation complex must locate the AUG codon. It is believed that the ribosome and associated factors migrate through the 5' UTR until the Met-tRNA$_i^{Met}$ can form a productive base pairing with an initiation codon. This has been defined as the scanning model,[114,115] and is consistent with the experimental data. For example, the insertion of an upstream AUG usually creates a new translation start site. In addition, insertion of a stable secondary structure in the 5' UTR dramatically reduces translational efficiency, probably by hampering ribosome movement.[116] The minimal set of factors required for 48S complex assembly at an initiation codon has been better defined in recent years (see below).[117,118] Nevertheless, scanning has never been biochemically assayed and little is known about its mechanism.

eIF1

Sequence homologues of eIF1 have been identified in archaea and bacteria, indicating an ancient origin for this factor.[8] In yeast, eIF1 is essential for viability,[119] and genetic screenings have identified a number of mutations affecting the fidelity of initiation codon selection.[22] These mutations affect residues that are conserved between eukaryotes, archaea and bacteria,[8] and the NMR structure of eIF1 suggest that they represent an interaction surface for other molecules.[52] A fraction of eIF1 from *S. cerevisiae* copurifies with eIF3, through an interaction with eIF3c.[48,49,120] This might explain how eIF5 mutants can suppress the effect of eIF1 mutations in yeast,[22] because eIF3c can interact simultaneously with eIF1 and eIF5 (Fig. 3).[51] Thus, eIF1 is in close proximity to eIF5 and could affect its GTPase activating function. Toe-printing analysis of pre-initiation complexes bound to mRNA have provided evidence that eIF1 is required for the formation of 48S complexes.[117] In this assay, which also requires eIF2, eIF3, eIF4A, eIF4B, eIF4F and Met-tRNA$_i^{Met}$, eIF1 can weakly promote the correct positioning of 43S complex at the initiation codon. This activity is strongly enhanced by eIF1A (see below).[117] The molecular mechanism remains however unclear, and thus the exact function of eIF1 could be to participate in scanning, destabilize incorrectly positioned complexes, or stabilize correctly positioned ones.

eIF1A

eIF1A is the eukaryotic homologue of the bacterial IF1 and the archaeal a-eIF1A.[8] The 3D structures of eIF1A shows that it contains an OB domain similar to that of IF1.[37,38] eIF1A is essential in *S. cerevisiae* and appears to have multiple roles in translation initiation.[1] Pestova et al reported that eIF1A acts in synergy with eIF1 to promote the assembly of 48S complex at the initiation codon.[117] However, eIF1A without eIF1 can only promote the formation of an aberrant complex located close to the cap. This cap-proximal complex is not a translational intermediate, and it is unable to reach the AUG codon.[117] eIF1A also promotes the binding of the ternary complex to isolated 40S subunits.[46,121] In addition, eIF1A interacts with eIF5B.[122] Along with structural studies of a 30S•IF1 complex,[123] this suggests that eIF1A occupies the A site of the small ribosomal subunit and, in conjunction with eIF5B, directs the initiator tRNA to the P site.[124]

60S Ribosomal Subunit Joining

Once the preinitiation complex has reached an initiator codon, base pairing between the AUG and the Met-tRNA$_i^{Met}$ anticodon elicits a series of events that culminate in the joining of the 60S ribosomal subunit to form an active ribosome that is competent for elongation. This involves the release of the initiation factors bound to the 40S subunits, which requires GTP hydrolysis. Recruitment of the 60S subunit is not spontaneous after the factors release, and it necessitates additional initiation factors.

eIF5

Once the 48S complex is correctly assembled at the initiation codon, the bound initiation factors must be displaced to allow the joining of the 60S subunit. This step requires the hydrolysis of the GTP molecule bound to eIF2 (Fig. 1), since its substitution with the nonhydrolyzable GMP-PNP arrests initiation at the 48S stage.[125] Activation of the GTPase function of eIF2 requires its association with eIF5 and the 40S subunit.[10] eIF5 is essential in yeast, and an activated eIF5 mutant (G31R) can initiate translation at non-AUG codons.[22] This mutant demonstrates a two-fold stimulation of GTP hydrolysis. Combined with the studies on eIF2 mutants showing intrinsic GTPase activity, it appears that eIF2 and eIF5 act in concert to maintain the accuracy of initiation codon recognition in eukaryotes. Intriguingly, eIF5 has not been found to interact with the GTP-bound eIF2γ, but rather with eIF2β, at the same site which was demonstrated to interact with eIF2Bε.[29,30] This argues that GTP hydrolysis must be triggered by a conformational change in eIF2, when the initiator tRNA is correctly paired with the AUG codon. The amino-terminus of eIF5 is similar to the carboxy-terminus of eIF2β, including the Cys$_2$-Cys$_2$ zinc-finger motif, and is expected to adopt a similar fold.[31]

Recent observations suggest that eIF5 acts as a bridge between eIF3 and eIF2, and would therefore participate in the recruitment of the ternary complex to the 40S subunit (see above).[51] It has also been demonstrated that yeast eIF5 can interact simultaneously with eIF3 and eIF4G, and could thus take part in 43S complex recruitment to the mRNA.[58] These interactions are all mediated by the C-terminal domain of eIF5, while the GTPase activating region apparently resides in the N-terminus.[58] Interestingly, the amino-terminus of eIF5 shares sequence homology with the carboxy-terminal fragment of eIF2β.[29,31]

eIF5B

eIF5B was first identified as a translation factor in yeast, and termed yeast IF2 (yIF2) for its homology to the prokaryotic IF2.[126] Human,[127,128] *Drosophila*,[129] and archaeal[7] homologues were also identified, making IF2/eIF5B a universally conserved translation factor. Disruption of the eIF5B gene (*FUN12*) in yeast causes a severe slow-growth phenotype, associated with a defect in translation initiation.[126] It was later demonstrated that eIF5B has a function analogous to prokaryotic IF2 in mediating the joining of the 60S ribosomal subunit.[130] The requirement for both eIF2 and eIF5B strongly suggest that translation initiation in eukaryotes requires the hydrolysis of two GTP molecules (Fig. 1).[125] This is consistent with the kinetic analysis of translation initiation in vitro,[131] and offers additional possibilities for regulation.

The structure of eIF5B has revealed an unusual architecture.[132] It consists of three N-terminal domains (I, II, III) connected by a long α helix to domain IV. Domain I is a G domain, domains II and IV are β-barrels and domain III has a novel α/β/α sandwich fold.[132] The G domain and the β-barrel domain II display a similar structure and arrangement to the homologous domains in EF1A, eEF1A and a-eIF2γ.[25,132] This suggests that they form a core structure present in all GTPases involved in translation.[124] GTP-bound eIF5B facilitates 60S subunit joining, but GTP hydrolysis occurs after 80S formation and is required for the release of eIF5B.[130] The comparison of eIF5B•GTP and eIF5B•GDP demonstrates that, like other GTPases, GTP hydrolysis induces a modest conformational switch in the G domain. This small modification is however amplified via a coordinated rearrangement of domains II-IV, resulting in the movement of domain IV.[132] Domain IV is essential for the in vivo function of eIF5B and interacts with eIF1A, suggesting that the release of eIF1A and eIF5B from the ribosome could be coupled.[122]

Regulation of Translation Initiation

As with any multi-step process, eukaryotic translation can be regulated at various levels. However, it is generally more efficient to regulate complex pathways at their initiation step, and this is where translational control most often occurs.[133] Two factors are known to play a critical role in the regulation of translation initiation: eIF2 and eIF4E. The activity of eIF2 is controlled by the rate of GTP recycling, and thus by the availability of eIF2B.[10] The function of eIF4E is regulated by its incorporation into eIF4F, which is inhibited by the eIF4E-Binding Proteins (4E-BPs).[63] Both of these mechanisms are subjected to regulatory phosphorylation and can thus provide rapid and reversible means to maintain cellular homeostasis in response to environmental signals. We will give a brief outline of how the activity of eIF2 and eIF4E can be regulated. The reader is directed to a recent review by Dever[134] for a more detailed discussion of how gene-specific regulation can be achieved by modulating the activity of general translation factors. An in depth review of the multiple translational control mechanisms can be found in reference 135.

eIF2α Kinases

As mentioned above, the phosphorylation of eIF2α on Ser51 results in the conversion of eIF2 from a substrate to a competitive inhibitor of eIF2B. The critical importance and versatility of this regulatory event is emphasized by the successive appearance of the four known eIF2· kinases through the eukaryotic lineages: mammals acquired the double-stranded RNA-activated protein kinase (PKR); vertebrates, the hemin-regulated inhibitor kinase (HRI); metazoans, the PKR-like ER kinase (PERK); while all eukaryotes possess the kinase GCN2.[10,136-138] These four kinases are activated by environmental stress.[139] PKR is part of the anti-viral response, since its expression is induced by interferon and its kinase function is activated by double-stranded RNA.[138] HRI regulates protein synthesis in erythroid cells in response to heme deprivation, heat shock and oxidative and osmotic stress.[137,140] PERK is activated as part of the unfolded protein response, which is caused by stress to the endoplasmic reticulum.[136] Finally the activity of GCN2 can be stimulated by a variety of conditions, the best characterized being amino acid starvation.[10,141]

eIF4E-Binding Proteins

The complex m[7]GpppX•eIF4E•eIF4G directs the 43S preinitiation complex to the mRNA 5' end. The 4E-BPs specifically inhibit cap-dependent translation initiation by preventing the interaction of eIF4E with eIF4G.[63] Three 4E-BPs were identified and found to have similar functions,[142,143] but 4E-BP1 has been more thoroughly characterized. The binding of 4E-BP1 to eIF4E is reversible, and is regulated by the phosphorylation state of 4E-BP1. Upon stimulation of cells with serum, growth factors or hormones, 4E-BP1 becomes phosphorylated on specific serine/threonine residues and dissociates from eIF4E to relieve translational inhibition.[63,144] Conversely, cellular stresses such as nutrient deprivation or viral infections will cause a decrease in 4E-BP1 phosphorylation and increase its affinity for eIF4E.[63,67] The phosphorylation events leading to the dissociation of 4E-BP1 from eIF4E are mediated by the FKBP12-rapamycin associated protein/mammalian target of rapamycin (FRAP/mTOR), and probably other kinase(s) that have yet to be identified.[144-146] The activity of FRAP/mTOR is modulated by the availability of nutrients such as amino acids, and by the intracellular concentration of ATP.[146-148] Given that FRAP/mTOR signals not only to 4E-BP1 but to other translational targets (eIF4B, eIF4GI, eEF2 and S6K1), it emerges as a molecular sensor capable of integrating diverse inputs to modulate protein synthesis in response to a cell's nutritional and energetic status.[146,148]

Conclusion

The structural and functional analysis of eukaryotic translation initiation factors reveals that they have most likely evolved to take advantage of the many RNA structural elements that are specific to eukaryotes. Archaeal translation illustrates the evolution of the protein synthesis machinery from eubacteria to eukaryotes, exhibiting characteristics from both systems. For example, archaeal mRNAs have a Shine-Dalgarno sequence but utilize a factor homologous to eIF2 for Met-tRNA$_i^{Met}$ delivery to the ribosome. Many of the eukaryotic translation initiation factors do not have counterparts in prokaryotes, and must have evolved together with the unique features of eukaryotic mRNA (cap structure, poly(A) tail). Archaea lack a 5' cap or long poly(A) tail and, accordingly, are devoid of factors related to eIF3, eIF4G or eIF4E. It can therefore be envisaged that eIF4G has evolved through the accretion of various interaction domains around a conserved central core. An eIF4E-binding site could thus have been added after the 5' cap structure has evolved, and a PABP interaction domain after the poly(A) tail was developed. Thus, RNA structural elements appear to have played a critical role in directing the evolution of the eukaryotic translational machinery. The increase in complexity that has resulted served to develop new regulatory mechanisms to control gene expression, such as the sequestration of eIF4E by the 4E-BPs or the phosphorylation of eIF2α. Once evolved, regulatory mechanisms could be subjected to further refinement. For instance, mammals have evolved four different eIF2α kinases. As a consequence of the increased regulatory capabilities, translation initiation has become a very dynamic process that can be delicately and rapidly modulated in response to environmental changes. Thus, the ability to better control translation is likely to have provided eukaryotes with increased adaptive flexibility.

Acknowledgements

We would like to thank Thomas Dever, Emmanuel Petroulakis, Mathieu Miron and Park Cho-Park for critical reading of the manuscript. F.P. is supported by a Doctoral Award from the Canadian Institutes of Health Research (CIHR). N.S. is a CIHR Distinguished Investigator and a Howard Hughes Medical Institute International Scholar.

References

1. Hershey JWB, Merrick WC. The pathway and mechanisms of initiation of protein synthesis. In: Sonenberg N, Hershey JWB, Mathews MB, eds. Translational Control of Gene Expression. Cold Spring Harbor: Cold Spring Harbor Laboratory Press, 2000:33-88.
2. Jackson RJ. A comparative view of initiation site selection mechanisms. In: Sonenberg N, Hershey JWB, Mathews MB, eds. Translational control of gene expression. Cold Spring Harbor: Cold Spring Harbor Laboratory Press, 2000:127-183.
3. Hellen CU, Sarnow P. Internal ribosome entry sites in eukaryotic mRNA molecules. Genes Dev 2001; 15:1593-1612.
4. Bult CJ, White O, Olsen GJ et al. Complete genome sequence of the methanogenic archaeon, *Methanococcus jannaschii*. Science 1996; 273:1058-1073.

5. Klenk HP, Clayton RA, Tomb JF et al. The complete genome sequence of the hyperthermophilic, sulphate-reducing archaeon *Archaeoglobus fulgidus*. Nature 1997; 390:364-370.
6. Dennis PP. Ancient ciphers: translation in Archaea. Cell 1997; 89:1007-1010.
7. Kyrpides NC, Woese CR. Archaeal translation initiation revisited: the initiation factor 2 and eukaryotic initiation factor 2B alpha-beta-delta subunit families. Proc Natl Acad Sci USA 1998; 95:3726-3730.
8. Kyrpides NC, Woese CR. Universally conserved translation initiation factors. Proc Natl Acad Sci USA 1998; 95:224-228.
9. RajBhandary UL, Chow CM. Initiator tRNAs and initiation of protein synthesis. In: Söll D, RajBhandary UL, eds. tRNA: structure, biosynthesis, and function. Washington, DC: American Society for Microbiology; 1995:511-528.
10. Hinnebusch AG. Mechanism and regulation of initiator Methionyl-tRNA binding to ribosomes. In: Sonenberg N, Hershey JWB, Mathews MB, eds. Translational control of gene expression. Cold Spring Harbor: Cold Spring Harbor Laboratory Press, 2000:185-243.
11. von Pawel-Rammingen U, Astrom S, Bystrom AS. Mutational analysis of conserved positions potentially important for initiator tRNA function in *Saccharomyces cerevisiae*. Mol Cell Biol 1992; 12:1432-1442.
12. Farruggio D, Chaudhuri J, Maitra U et al. The A1 x U72 base pair conserved in eukaryotic initiator tRNAs is important specifically for binding to the eukaryotic translation initiation factor eIF2. Mol Cell Biol 1996; 16:4248-4256.
13. Sprinzl M, Horn C, Brown M et al. Compilation of tRNA sequences and sequences of tRNA genes. Nucleic Acids Res 1998; 26:148-153.
14. Drabkin HJ, Helk B, RajBhandary UL. The role of nucleotides conserved in eukaryotic initiator methionine tRNAs in initiation of protein synthesis. J Biol Chem 1993; 268:25221-25228.
15. Astrom SU, von Pawel-Rammingen U, Bystrom AS. The yeast initiator tRNAMet can act as an elongator tRNA(Met) in vivo. J Mol Biol 1993; 233:43-58.
16. Basavappa R, Sigler PB. The 3 Å crystal structure of yeast initiator tRNA: functional implications in initiator/elongator discrimination. EMBO J 1991; 10:3105-3111.
17. Astrom SU, Bystrom AS. Rit1, a tRNA backbone-modifying enzyme that mediates initiator and elongator tRNA discrimination. Cell 1994; 79:535-546.
18. Kiesewetter S, Ott G, Sprinzl M. The role of modified purine 64 in initiator/elongator discrimination of tRNA(iMet) from yeast and wheat germ. Nucleic Acids Res 1990; 18:4677-4682.
19. Forster C, Chakraburtty K, Sprinzl M. Discrimination between initiation and elongation of protein biosynthesis in yeast: identity assured by a nucleotide modification in the initiator tRNA. Nucleic Acids Res 1993; 21:5679-5683.
20. Astrom SU, Nordlund ME, Erickson FL et al. Genetic interactions between a null allele of the RIT1 gene encoding an initiator tRNA-specific modification enzyme and genes encoding translation factors in *Saccharomyces cerevisiae*. Mol Gen Genet 1999; 261:967-976.
21. Drabkin HJ, Estrella M, Rajbhandary UL. Initiator-elongator discrimination in vertebrate tRNAs for protein synthesis. Mol Cell Biol 1998; 18:1459-1466.
22. Donahue TF. Genetic approaches to translation initiation in *Saccharomyces cerevisiae*. In: Sonenberg N, Hershey JWB, Mathews MB, eds. Translational control of gene expression. Cold Spring Harbor: Cold Spring Harbor Laboratory Press, 2000:487-502.
23. Erickson FL, Hannig EM. Ligand interactions with eukaryotic translation initiation factor 2: role of the gamma-subunit. EMBO J 1996; 15:6311-6320.
24. Huang HK, Yoon H, Hannig EM et al. GTP hydrolysis controls stringent selection of the AUG start codon during translation initiation in *Saccharomyces cerevisiae*. Genes Dev 1997; 11:2396-2413.
25. Schmitt E, Blanquet S, Mechulam Y. The large subunit of initiation factor aIF2 is a close structural homologue of elongation factors. EMBO J 2002; 21:1821-1832.
26. Drabkin HJ, RajBhandary UL. Initiation of protein synthesis in mammalian cells with codons other than AUG and amino acids other than methionine. Mol Cell Biol 1998; 18:5140-5147.
27. Erickson FL, Harding LD, Dorris DR et al. Functional analysis of homologs of translation initiation factor 2gamma in yeast. Mol Gen Genet 1997; 253:711-719.
28. Laurino JP, Thompson GM, Pacheco E et al. The beta subunit of eukaryotic translation initiation factor 2 binds mRNA through the lysine repeats and a region comprising the C2-C2 motif. Mol Cell Biol 1999; 19:173-181.
29. Das S, Maiti T, Das K et al. Specific interaction of eukaryotic translation initiation factor 5 (eIF5) with the beta -subunit of eIF2. J Biol Chem 1997; 272:31712-31718.
30. Asano K, Krishnamoorthy T, Phan L et al. Conserved bipartite motifs in yeast eIF5 and eIF2Bepsilon, GTPase-activating and GDP-GTP exchange factors in translation initiation, mediate binding to their common substrate eIF2. EMBO J 1999; 18:1673-1688.
31. Cho S, Hoffman DW. Structure of the beta Subunit of Translation Initiation factor 2 from the Archaeon *Methanococcus jannaschii*: A Representative of the eIF2beta/eIF5 Family of Proteins. Biochemistry 2002; 41:5730-5742.
32. Thompson GM, Pacheco E, Melo EO et al. Conserved sequences in the beta subunit of archaeal and eukaryal translation initiation factor 2 (eIF2), absent from eIF5, mediate interaction with eIF2gamma. Biochem J 2000; 347 Pt 3:703-709.
33. Erickson FL, Nika J, Rippel S et al. Minimum requirements for the function of eukaryotic translation initiation factor 2. Genetics 2001; 158:123-132.
34. Nika J, Rippel S, Hannig EM. Biochemical analysis of the eIF2beta gamma complex reveals a structural function for eIF2alpha in catalyzed nucleotide exchange. J Biol Chem 2001; 276:1051-1056.

35. Krishnamoorthy T, Pavitt GD, Zhang F et al. Tight binding of the phosphorylated alpha subunit of initiation factor 2 (eIF2alpha) to the regulatory subunits of guanine nucleotide exchange factor eIF2B is required for inhibition of translation initiation. Mol Cell Biol 2001; 21:5018-5030.
36. Nonato MC, Widom J, Clardy J. Crystal structure of the N-terminal segment of human eukaryotic translation initiation factor 2alpha. J Biol Chem 2002; 277:17057-17061.
37. Sette M, van Tilborg P, Spurio R et al. The structure of the translational initiation factor IF1 from E.coli contains an oligomer-binding motif. EMBO J 1997; 16:1436-1443.
38. Battiste JL, Pestova TV, Hellen CUT et al. The eIF1A solution structure reveals a large RNA-binding surface important for scanning function. Mol Cell 2000; 5:109–119.
39. Trachsel H. Binding of initiator methionyl-tRNA to ribosomes. In: Hershey JWB, Mathews MB, Sonenberg N, eds. Translational Control. Cold Spring Harbor: Cold Spring Harbor Laboratory Press, 1996:113-138.
40. Pavitt GD, Ramaiah KV, Kimball SR et al. eIF2 independently binds two distinct eIF2B subcomplexes that catalyze and regulate guanine-nucleotide exchange. Genes Dev 1998; 12:514-526.
41. Fabian JR, Kimball SR, Heinzinger NK et al. Subunit assembly and guanine nucleotide exchange activity of eukaryotic initiation factor-2B expressed in Sf9 cells. J Biol Chem 1997; 272:12359-12365.
42. Gomez E, Pavitt GD. Identification of domains and residues within the epsilon subunit of eukaryotic translation initiation factor 2B (eIF2Bepsilon) required for guanine nucleotide exchange reveals a novel activation function promoted by eIF2B complex formation. Mol Cell Biol 2000; 20:3965-3976.
43. Kimball SR, Fabian JR, Pavitt GD et al. Regulation of guanine nucleotide exchange through phosphorylation of eukaryotic initiation factor eIF2alpha. Role of the alpha- and delta-subunits of eIF2B. J Biol Chem 1998; 273:12841-12845.
44. Leegwater PA, Vermeulen G, Konst AA et al. Subunits of the translation initiation factor eIF2B are mutant in leukoencephalopathy with vanishing white matter. Nat Genet 2001; 29:383-388.
45. van der Knaap MS, Leegwater PA, Konst AA et al. Mutations in each of the five subunits of translation initiation factor eIF2B can cause leukoencephalopathy with vanishing white matter. Ann Neurol 2002; 51:264-270.
46. Chaudhuri J, Chowdhury D, Maitra U. Distinct functions of eukaryotic translation initiation factors eIF1A and eIF3 in the formation of the 40 S ribosomal preinitiation complex. J Biol Chem 1999; 274:17975-17980.
47. Thompson HA, Sadnik I, Scheinbuks J et al. Studies on native ribosomal subunits from rat liver. Purification and characterization of a ribosome dissociation factor. Biochemistry 1977; 16:2221-2230.
48. Phan L, Zhang X, Asano K et al. Identification of a translation initiation factor 3 (eIF3) core complex, conserved in yeast and mammals, that interacts with eIF5. Mol Cell Biol 1998; 18:4935-4946.
49. Asano K, Phan L, Anderson J et al. Complex formation by all five homologues of mammalian translation initiation factor 3 subunits from yeast *Saccharomyces cerevisiae*. J Biol Chem 1998; 273:18573-18585.
50. Bandyopadhyay A, Maitra U. Cloning and characterization of the p42 subunit of mammalian translation initiation factor 3 (eIF3): demonstration that eIF3 interacts with eIF5 in mammalian cells. Nucleic Acids Res 1999; 27:1331-1337.
51. Asano K, Clayton J, Shalev A et al. A multifactor complex of eukaryotic initiation factors, eIF1, eIF2, eIF3, eIF5, and initiator tRNA(Met) is an important translation initiation intermediate in vivo. Genes Dev 2000; 14:2534-2546.
52. Fletcher CM, Pestova TV, Hellen CU et al. Structure and interactions of the translation initiation factor eIF1. EMBO J 1999; 18:2631-2637.
53. Valasek L, Phan L, Schoenfeld LW et al. Related eIF3 subunits TIF32 and HCR1 interact with an RNA recognition motif in PRT1 required for eIF3 integrity and ribosome binding. EMBO J 2001; 20:891-904.
54. Block KL, Vornlocher HP, Hershey JW. Characterization of cDNAs encoding the p44 and p35 subunits of human translation initiation factor eIF3. J Biol Chem 1998; 273:31901-31908.
55. Phan L, Schoenfeld LW, Valasek L et al. A subcomplex of three eIF3 subunits binds eIF1 and eIF5 and stimulates ribosome binding of mRNA and tRNA(i)Met. EMBO J 2001; 20:2954-2965.
56. Lamphear BJ, Kirchweger R, Skern T et al. Mapping of functional domains in eukaryotic protein synthesis initiation factor 4G (eIF4G) with picornaviral proteases. Implications for cap-dependent and cap-independent translational initiation. J Biol Chem 1995; 270:21975-21983.
57. Gradi A, Imataka H, Svitkin YV et al. A novel functional human eukaryotic translation initiation factor 4G. Mol Cell Biol 1998; 18:334-342.
58. Asano K, Shalev A, Phan L et al. Multiple roles for the C-terminal domain of eIF5 in translation initiation complex assembly and GTPase activation. EMBO J 2001; 20:2326-2337.
59. Methot N, Song MS, Sonenberg N. A region rich in aspartic acid, arginine, tyrosine, and glycine (DRYG) mediates eukaryotic initiation factor 4B (eIF4B) self-association and interaction with eIF3. Mol Cell Biol 1996; 16:5328-5334.
60. Vornlocher HP, Hanachi P, Ribeiro S et al. A 110-kilodalton subunit of translation initiation factor eIF3 and an associated 135-kilodalton protein are encoded by the *Saccharomyces cerevisiae* TIF32 and TIF31 genes. J Biol Chem 1999; 274:16802-16812.
61. Shatkin AJ. Capping of eucaryotic mRNAs. Cell 1976; 9:645-653.
62. Varani G. A cap for all occasions. Structure 1997; 5:855-858.
63. Gingras A-C, Raught B, Sonenberg N. eIF4 initiation factors: effectors of mRNA recruitment to ribosomes and regulators of translation. Annu Rev Biochem 1999; 68:913-963.
64. Marcotrigiano J, Gingras A-C, Sonenberg N et al. Cocrystal structure of the messenger RNA 5' cap-binding protein (eIF4E) bound to 7-methyl-GDP. Cell 1997; 89:951-961.

65. Matsuo H, Li H, McGuire AM et al. Structure of translation factor eIF4E bound to m7GDP and interaction with 4E-binding protein. Nat Struct Biol 1997; 4:717-724.
66. Aravind L, Koonin EV. Eukaryote-specific domains in translation initiation factors: implications for translation regulation and evolution of the translation system. Genome Res 2000; 10:1172-1184.
67. Raught B, Gingras A-C, Sonenberg N. Regulation of ribosomal recruitment in eukaryotes. In: Sonenberg N, Hershey JWB, Mathews MB, eds. Translational control of gene expression. Cold Spring Harbor, N.Y.: Cold Spring Harbor Laboratory Press; 2000:245-293.
68. Tomoo K, Shen X, Okabe K et al. Crystal structures of 7-methylguanosine 5prime prime or minute-triphosphate (m7GTP)- and P1-7-methylguanosine-P3-adenosine-5prime prime or minute,5prime prime or minute-triphosphate (m7GpppA)-bound human full-length eukaryotic initiation factor 4E: biological importance of the C-terminal flexible region. Biochem J 2002; 362:539-544.
69. Minich WB, Balasta ML, Goss DJ et al. Chromatographic resolution of in vivo phosphorylated and nonphosphorylated eukaryotic translation initiation factor eIF-4E: increased cap affinity of the phosphorylated form. Proc Natl Acad Sci USA 1994; 91:7668-7672.
70. Scheper GC, van Kollenburg B, Hu J et al. Phosphorylation of eukaryotic initiation factor 4E markedly reduces its affinity for capped mRNA. J Biol Chem 2002; 277:3303-3309.
71. McKendrick L, Morley SJ, Pain VM et al. Phosphorylation of eukaryotic initiation factor 4E (eIF4E) at Ser209 is not required for protein synthesis in vitro and in vivo. Eur J Biochem 2001; 268:5375-5385.
72. Lachance PE, Miron M, Raught B et al. Phosphorylation of eukaryotic translation initiation factor 4E is critical for growth. Mol Cell Biol 2002; 22:1656-1663.
73. Fukunaga R, Hunter T. MNK1, a new MAP kinase-activated protein kinase, isolated by a novel expression screening method for identifying protein kinase substrates. EMBO J 1997; 16:1921-1933.
74. Waskiewicz AJ, Flynn A, Proud CG et al. Mitogen-activated protein kinases activate the serine/threonine kinases Mnk1 and Mnk2. EMBO J 1997; 16:1909-1920.
75. Pyronnet S, Imataka H, Gingras AC et al. Human eukaryotic translation initiation factor 4G (eIF4G) recruits Mnk1 to phosphorylate eIF4E. EMBO J 1999; 18:270-279.
76. Hentze MW. eIF4G: a multipurpose ribosome adapter? Science 1997; 275:500-501.
77. Dever TE. Translation initiation: adept at adapting. Trends Biochem Sci 1999; 24:398-403.
78. Imataka H, Sonenberg N. Human eukaryotic translation initiation factor 4G (eIF4G) possesses two separate and independent binding sites for eIF4A. Mol Cell Biol 1997; 17:6940-6947.
79. Dominguez D, Kislig E, Altmann M et al. Structural and functional similarities between the central eukaryotic initiation factor (eIF)4A-binding domain of mammalian eIF4G and the eIF4A-binding domain of yeast eIF4G. Biochem J 2001; 355:223-230.
80. Pestova TV, Shatsky IN, Hellen CU. Functional dissection of eukaryotic initiation factor 4F: the 4A subunit and the central domain of the 4G subunit are sufficient to mediate internal entry of 43S preinitiation complexes. Mol Cell Biol 1996; 16:6870-6878.
81. Lomakin IB, Hellen CU, Pestova TV. Physical association of eukaryotic initiation factor 4G (eIF4G) with eIF4A strongly enhances binding of eIF4G to the internal ribosomal entry site of encephalomyocarditis virus and is required for internal initiation of translation. Mol Cell Biol 2000; 20:6019-6029.
82. Marcotrigiano J, Lomakin IB, Sonenberg N et al. A conserved HEAT domain within eIF4G directs assembly of the translation initiation machinery. Mol Cell 2001; 7:193-203.
83. Andrade MA, Bork P. HEAT repeats in the Huntington's disease protein. Nat Genet 1995; 11:115-116.
84. Morino S, Imataka H, Svitkin YV et al. Eukaryotic translation initiation factor 4E (eIF4E) binding site and the middle one-third of eIF4GI constitute the core domain for cap-dependent translation, and the C-terminal one-third functions as a modulatory region. Mol Cell Biol 2000; 20:468-477.
85. De Gregorio E, Preiss T, Hentze MW. Translational activation of uncapped mRNAs by the central part of human eIF4G is 5' end-dependent. RNA 1998; 4:828-836.
86. De Gregorio E, Preiss T, Hentze MW. Translation driven by an eIF4G core domain in vivo. EMBO J 1999; 18:4865-4874.
87. Tarun SZ Jr, Sachs AB. Association of the yeast poly(A) tail binding protein with translation initiation factor eIF-4G. EMBO J 1996; 15:7168-7177.
88. Imataka H, Gradi A, Sonenberg N. A newly identified N-terminal amino acid sequence of human eIF4G binds poly(A)-binding protein and functions in poly(A)-dependent translation. EMBO J 1998; 17:7480-7489.
89. Jacobson A. Poly(A) metabolism and translation: the closed-loop model. In: Hershey JWB, Mathews MB, Sonenberg N, eds. Translational Control. Cold Spring Harbor: Cold Spring Harbor Laboratory Press, 1996:451-480.
90. Sachs AB. Physical and functional interactions between the mRNA cap structure and the poly(A) tail. In: Sonenberg N, Hershey JWB, Mathews MB, eds. Translational control of gene expression. Cold Spring Harbor: Cold Spring Harbor Laboratory Press, 2000:447-465.
91. Korneeva NL, Lamphear BJ, Hennigan FL et al. Characterization of the two eIF4A-binding sites on human eIF4G-1. J Biol Chem 2001; 276:2872-2879.
92. Li W, Belsham GJ, Proud CG. Eukaryotic initiation factors 4A (eIF4A) and 4G (eIF4G) mutually interact in a 1:1 ratio in vivo. J Biol Chem 2001; 276:29111-29115.
93. Koonin EV. Multidomain organization of eukaryotic guanine nucleotide exchange translation initiation factor eIF-2B subunits revealed by analysis of conserved sequence motifs. Protein Sci 1995; 4:1608-1617.
94. de la Cruz J, Kressler D, Linder P. Unwinding RNA in *Saccharomyces cerevisiae*: DEAD-box proteins and related families. Trends Biochem Sci 1999; 24:192-198.

95. Linder P, Slonimski PP. An essential yeast protein, encoded by duplicated genes TIF1 and TIF2 and homologous to the mammalian translation initiation factor eIF-4A, can suppress a mitochondrial missense mutation. Proc Natl Acad Sci USA 1989; 86:2286-2290.

96. Li Q, Imataka H, Morino S et al. Eukaryotic translation initiation factor 4AIII (eIF4AIII) is functionally distinct from eIF4AI and eIF4AII. Mol Cell Biol 1999; 19:7336-7346.

97. Prat A, Schmid SR, Buser P et al. Expression of translation initiation factor 4A from yeast and mouse in *Saccharomyces cerevisiae.* Biochim Biophys Acta 1990; 1050:140-145.

98. Rozen F, Edery I, Meerovitch K et al. Bidirectional RNA helicase activity of eucaryotic translation initiation factors 4A and 4F. Mol Cell Biol 1990; 10:1134-1144.

99. Rogers GW, Jr., Richter NJ, Merrick WC. Biochemical and kinetic characterization of the RNA helicase activity of eukaryotic initiation factor 4A. J Biol Chem 1999; 274:12236-12244.

100. Rogers GW, Jr., Richter NJ, Lima WF et al. Modulation of the helicase activity of eIF4A by eIF4B, eIF4H, and eIF4F. J Biol Chem 2001; 276:30914-30922.

101. Pause A, Methot N, Svitkin Y et al. Dominant negative mutants of mammalian translation initiation factor eIF-4A define a critical role for eIF-4F in cap-dependent and cap-independent initiation of translation. EMBO J 1994; 13:1205-1215.

102. Lorsch JR, Herschlag D. The DEAD box protein eIF4A. 2. A cycle of nucleotide and RNA-dependent conformational changes. Biochemistry 1998; 37:2194-2206.

103. Lorsch JR, Herschlag D. The DEAD box protein eIF4A. 1. A minimal kinetic and thermodynamic framework reveals coupled binding of RNA and nucleotide. Biochemistry 1998; 37:2180-2193.

104. Pestova TV, Hellen CU, Shatsky IN. Canonical eukaryotic initiation factors determine initiation of translation by internal ribosomal entry. Mol Cell Biol 1996; 16:6859-6869.

105. Coppolecchia R, Buser P, Stotz A et al. A new yeast translation initiation factor suppresses a mutation in the eIF-4A RNA helicase. EMBO J 1993; 12:4005-4011.

106. Altmann M, Muller PP, Wittmer B et al. A *Saccharomyces cerevisiae* homologue of mammalian translation initiation factor 4B contributes to RNA helicase activity. EMBO J 1993; 12:3997-4003.

107. Methot N, Pause A, Hershey JW et al. The translation initiation factor eIF-4B contains an RNA-binding region that is distinct and independent from its ribonucleoprotein consensus sequence. Mol Cell Biol 1994; 14:2307-2316.

108. Naranda T, Strong WB, Menaya J et al. Two structural domains of initiation factor eIF-4B are involved in binding to RNA. J Biol Chem 1994; 269:14465-14472.

109. Altmann M, Wittmer B, Methot N et al. The *Saccharomyces cerevisiae* translation initiation factor Tif3 and its mammalian homologue, eIF-4B, have RNA annealing activity. EMBO J 1995; 14:3820-3827.

110. Methot N, Pickett G, Keene JD et al. In vitro RNA selection identifies RNA ligands that specifically bind to eukaryotic translation initiation factor 4B: the role of the RNA recognition motif. RNA 1996; 2:38-50.

111. Richter-Cook NJ, Dever TE, Hensold JO et al. Purification and characterization of a new eukaryotic protein translation factor. Eukaryotic initiation factor 4H. J Biol Chem 1998; 273:7579-7587.

112. Richter NJ, Rogers GW Jr, Hensold JO et al. Further biochemical and kinetic characterization of human eukaryotic initiation factor 4H. J Biol Chem 1999; 274:35415-35424.

113. Feng P, Everly DN Jr, Read GS. mRNA decay during herpesvirus infections: interaction between a putative viral nuclease and a cellular translation factor. J Virol 2001; 75:10272-10280.

114. Kozak M. The scanning model for translation: an update. J Cell Biol 1989; 108:229-241.

115. Kozak M. Initiation of translation in prokaryotes and eukaryotes. Gene 1999; 234:187-208.

116. Pelletier J, Sonenberg N. Insertion mutagenesis to increase secondary structure within the 5' noncoding region of a eukaryotic mRNA reduces translational efficiency. Cell 1985; 40:515-526.

117. Pestova TV, Borukhov SI, Hellen CU. Eukaryotic ribosomes require initiation factors 1 and 1A to locate initiation codons. Nature 1998; 394:854-859.

118. Pestova TV, Hellen CU. Ribosome recruitment and scanning: what's new? Trends Biochem Sci 1999; 24:85-87.

119. Yoon HJ, Donahue TF. The sui1 suppressor locus in Saccharomyces cerevisiae encodes a translation factor that functions during tRNA(iMet) recognition of the start codon. Mol Cell Biol 1992; 12:248-260.

120. Naranda T, MacMillan SE, Donahue TF et al. SUI1/p16 is required for the activity of eukaryotic translation initiation factor 3 in *Saccharomyces cerevisiae.* Mol Cell Biol 1996; 16:2307-2313.

121. Chaudhuri J, Si K, Maitra U. Function of eukaryotic translation initiation factor 1A (eIF1A) (formerly called eIF-4C) in initiation of protein synthesis. J Biol Chem 1997; 272:7883-7891.

122. Choi SK, Olsen DS, Roll-Mecak A et al. Physical and functional interaction between the eukaryotic orthologs of prokaryotic translation initiation factors IF1 and IF2. Mol Cell Biol 2000; 20:7183-7191.

123. Carter AP, Clemons WM Jr, Brodersen DE et al. Crystal structure of an initiation factor bound to the 30S ribosomal subunit. Science 2001; 291:498-501.

124. Roll-Mecak A, Shin BS, Dever TE et al. Engaging the ribosome: universal IFs of translation. Trends Biochem Sci 2001; 26:705-709.

125. Pestova TV, Dever TE, Hellen CU. Ribosomal subunit joining. In: Sonenberg N, Hershey JWB, Mathews MB, eds. Translational control of gene expression. Cold Spring Harbor, N.Y.: Cold Spring Harbor Laboratory Press; 2000:425-445.

126. Choi SK, Lee JH, Zoll WL et al. Promotion of Met-tRNA$_i^{Met}$ binding to ribosomes by yIF2, a bacterial IF2 homolog in yeast. Science 1998; 280:1757-1760.

127. Lee JH, Choi SK, Roll-Mecak A et al. Universal conservation in translation initiation revealed by human and archaeal homologs of bacterial translation initiation factor IF2. PNAS 1999; 96:4342-4347.

128. Wilson SA, Sieiro-Vazquez C, Edwards NJ et al. Cloning and characterization of hIF2, a human homologue of bacterial translation initiation factor 2, and its interaction with HIV-1 matrix. Biochem J 1999; 342 (Pt 1):97-103.
129. Carrera P, Johnstone O, Nakamura A et al. VASA mediates translation through interaction with a *Drosophila* yIF2 homolog. Mol Cell 2000; 5:181-187.
130. Pestova TV, Lomakin IB, Lee JH et al. The joining of ribosomal subunits in eukaryotes requires eIF5B. Nature 2000; 403:332—335.
131. Lorsch JR, Herschlag D. Kinetic dissection of fundamental processes of eukaryotic translation initiation in vitro. EMBO J 1999; 18:6705-6717.
132. Roll-Mecak A, Cao C, Dever TE et al. X-Ray structures of the universal translation initiation factor IF2/eIF5B: conformational changes on GDP and GTP binding. Cell 2000; 103:781-792.
133. Mathews MB, Sonenberg N, Hershey JWB. Origins and principles of translational control. In: Sonenberg N, Hershey JWB, Mathews MB, eds. Translational control of gene expression. Cold Spring Harbor: Cold Spring Harbor Laboratory Press, 2000:1-31.
134. Dever TE. Gene-specific regulation by general translation factors. Cell 2002; 108:545-556.
135. Sonenberg N, Hershey JWB, Mathews MB, eds. Translational control of gene expression. Cold Spring Harbor: Cold Spring Harbor Laboratory Press, 2000.
136. Ron D, Harding HP. PERK and translational control by stress in the endoplasmic reticulum. In: Sonenberg N, Hershey JWB, Mathews MB, eds. Translational control of gene expression. Cold Spring Harbor: Cold Spring Harbor Laboratory Press, 2000:547-560.
137. Chen J-J. Heme-regulated eIF2alpha kinase. In: Sonenberg N, Hershey JWB, Mathews MB, eds. Translational control of gene expression. Cold Spring Harbor: Cold Spring Harbor Laboratory Press, 2000:529-546.
138. Kaufman RJ. The double-stranded RNA-activated protein kinase PKR. In: Sonenberg N, Hershey JWB, Mathews MB, eds. Translational control of gene expression. Cold Spring Harbor: Cold Spring Harbor Laboratory Press, 2000:503-527.
139. Clemens MJ. Initiation factor eIF2alpha phosphorylation in stress responses and apoptosis. Prog Mol Subcell Biol 2001; 27:57-89.
140. Lu L, Han AP, Chen JJ. Translation initiation control by heme-regulated eukaryotic initiation factor 2alpha kinase in erythroid cells under cytoplasmic stresses. Mol Cell Biol 2001; 21:7971-7980.
141. Harding HP, Novoa II, Zhang Y et al. Regulated translation initiation controls stress-induced gene expression in mammalian cells. Mol Cell 2000; 6:1099-1108.
142. Poulin F, Gingras A-C, Olsen H et al. 4E-BP3, a new member of the eukaryotic initiation factor 4E-binding protein family. J Biol Chem 1998; 273:14002-14007.
143. Pause A, Belsham GJ, Gingras A-C et al. Insulin-dependent stimulation of protein synthesis by phosphorylation of a regulator of 5'-cap function. Nature 1994; 371:762-767.
144. Gingras A-C, Raught B, Gygi SP et al. Hierarchical phosphorylation of the translation inhibitor 4E-BP1. Genes Dev 2001; 15:2852-2864.
145. Gingras A-C, Gygi SP, Raught B et al. Regulation of 4E-BP1 phosphorylation: a novel two-step mechanism. Genes Dev 1999; 13:1422-1437.
146. Gingras AC, Raught B, Sonenberg N. Regulation of translation initiation by FRAP/mTOR. Genes Dev 2001; 15:807-826.
147. Dennis PB, Fumagalli S, Thomas G. Target of rapamycin (TOR): balancing the opposing forces of protein synthesis and degradation. Current Opinion In Genetics And Development 1999; 9:49-54.
148. Dennis PB, Jaeschke A, Saitoh M et al. Mammalian TOR: a homeostatic ATP sensor. Science 2001; 294:1102-1105.
149. Shibahara K, Asano M, Ishida Y et al. Isolation of a novel mouse gene MA-3 that is induced upon programmed cell death. Gene 1995; 166:297-301.
150. Imataka H, Olsen HS, Sonenberg N. A new translational regulator with homology to eukaryotic translation initiation factor 4G. EMBO J 1997; 16:817-825.
151. Craig AW, Haghighat A, Yu AT et al. Interaction of polyadenylate-binding protein with the eIF4G homologue PAIP enhances translation. Nature 1998; 392:520-523.
152. Ishigaki Y, Li X, Serin G et al. Evidence for a pioneer round of mRNA translation: mRNAs subject to nonsense-mediated decay in mammalian cells are bound by CBP80 and CBP20. Cell 2001; 106:607-617.
153. Ponting CP. Novel eIF4G domain homologues linking mRNA translation with nonsense-mediated mRNA decay. Trends Biochem Sci 2000; 25:423-426.

CHAPTER 19

Ribosomes on Standby:
A Prelude to Translational (Re)Initiation

Maarten H. de Smit and Jan van Duin

Abstract

Prokaryotic ribosomes, in landing at internal ribosome binding sites, have to deal with the secondary structure that is present in every RNA molecule. In previous work, we have described how the binding of ribosomes competes with the spontaneous folding of the mRNA. The strength of the mRNA structure thus determines the accessibility of the ribosome binding site and consequently the level of expression that is obtained. In this chapter, we consider a problem that was not addressed previously, namely that spontaneous unfolding of the RNA exposes the ribosome binding site only very briefly. This exposure time will often be far too short to recruit a ribosome from the cytoplasm. The fact that high expression levels are nonetheless reached suggests that a ribosome is in fact already present at or near the translational start site when the structure opens. We provide evidence from experimental data on initiation and reinitiation, from structure analysis of bacteriophage RNAs and from theoretical kinetic analysis of the initiation process, that such ribosome standby sites indeed exist and function to overcome the fast folding kinetics of mRNA.

Introduction

This is not a review in the classical sense, in that we do not claim to present a comprehensive overview of the available literature on translation (re)initiation in bacteria. Instead, we will address a paradox that arises from analysis of initiation kinetics. By examining the events taking place in translational reinitiation and extending their properties to de novo initiation, we arrive at a possible solution to this problem.

From Reinitiation to Initiation

Basic Features of de novo Initiation

Proper and efficient translational start site selection in *Escherichia coli* relies on at least three interactions. First, ribosomal protein S1 anchors the 30S subunit in the vicinity of the start site by its nonsequence-specific interaction with regions of single-stranded nucleotides (usually, but not necessarily, pyrimidines). Secondly, the well-known Shine-Dalgarno (SD) base pairing positions the subunit in a more precise manner close to the start codon. Finally, codon-anticodon interaction of tRNA$_f^{Met}$ locks the ribosome in its starting position and sets the reading frame (for recent reviews on translational initiation, see refs. 1-4). This selection process, which does not consume energy, can take place in the absence of initiation factors and the initiator tRNA does not have to be charged with formyl methionine. Moreover, analysis of the 30S initiation complex by the toeprinting technique (inhibition of primer extension by mRNA-bound 30S ribosomes) showed that the bound subunit can be shifted to nearby codons by replacing tRNA$_f^{Met}$ with their cognate tRNAs.[5] This finding makes it clear that S1 and SD interactions still allow some lateral movement of the mRNA over the ribosomal surface. In vivo noninitiator tRNAs are prevented from binding by IF3 proofreading.[6,7]

Translation Mechanisms, edited by Jacques Lapointe and Léa Brakier-Gingras.
©2003 Eurekah.com and Kluwer Academic / Plenum Publishers.

Control of Translation by Reinitiation

Usually not all cistrons in a polycistronic mRNA are directly accessible to ribosomes from solution. Instead, translation of a downstream reading frame may depend on reading of the neighboring upstream cistron.[8-11] In such case the terminated ribosome somehow restarts at the downstream reading frame. In most examples, stop and restart codons are only a few nucleotides apart. In Chapter 21 by R. Buckingham and M.Ehrenberg, it is described how a terminated ribosome is disassembled and we may deduce the sequence of events that lead to reinitiation. The critical step in the disassembly is probably the release of the last tRNA, a reaction catalyzed by IF3. Once this uncharged tRNA is removed, the remaining 30S ribosome is free to align with the nearby SD box and to capture an initiator tRNA. After codon-anticodon pairing, translation of the downstream cistron can begin. The essence of this coupling is that translation brings the ribosome to a region on the mRNA which it cannot reach on its own account from solution because of inhibitory RNA structure or other features disfavoring binding.

We do not understand all the details of reinitiation; in particular, it is not clear what determines its efficiency. Frequently, only a fraction of the terminated ribosomes succeed in a restart. Those that do not make it detach spontaneously from the message and before they can rebind, the mRNA refolds to its inaccessible conformation. This is of course also the path followed at stop codons not serving as a restart site. Ribosomes that do succeed in a restart probably do so by extending their weak interaction with the messenger by pairing with the SD box of the downstream reading frame[12] or by interactions through protein S1. These restarts classify as kinetic escapes; at equilibrium the 30S·mRNA termination intermediate would dissociate into its components and the reverse reaction would not occur because de novo initiation from this site is by definition impossible. In other words, the terminated ribosome is rescued from release by the restart pathway.

Reinitiation at Distant Restart Signals

Not always are stop and restart sites in close proximity. One may find the restart codon rather far down or even upstream from the termination site. Reinitiation in this case is not mechanistically different from the previous one, but we treat it separately because it demonstrates an important property of the initiation process not immediately evident from the previous section.

The most illuminating example is translational coupling between the coat and lysis genes of RNA phage MS2. The genetic map of the RNA phage is shown in Figure 1A. As can be seen, the restart codon of the lysis gene is located 46 nucleotides (nt) upstream of the coat termination site. When the coat gene is not translated, there is no translation of the lysis gene either, because independent ribosomal access to this gene is inhibited by the lysis hairpin (Fig. 1B).[13] Activation is triggered by ribosomes terminating at the coat gene stop codon. In mutants where termination occurred downstream of the normal coat stop site, lysis expression was abolished but when termination codons were engineered more upstream, closer and closer to the lysis start, its translation became progressively enhanced.[9] Activation of the lysis gene is thus clearly coupled to the nearby termination event. It was, in fact, proposed to result from the terminated but not yet released 30S ribosomes. These are thought to start a random walk along the RNA, leading some of them (~ 5%) to the lysis start 46 nt upstream, where they reinitiate. The other 95% are released and return to the free pool. The main argument that ribosomes actually slide along the RNA is that introduction of a new start codon between termination and restart site forms a barrier for this random walk to arrive at the authentic start, causing reinitiation to occur at this new start codon instead.[14] There is a close analogy here with eukaryotic scanning ribosomes that are intercepted at AUG codons introduced between the cap and the true start.[15] The capacity of *E. coli* ribosomes to reach distant start sites after termination is confirmed by in vitro translation of very short messengers. Here the ribosome shuttles back and forth between stop and start without ever leaving the messenger.[16] When the lysis hairpin is stabilized by mutations that create stronger base pairs, the scanning ribosomes are no longer able to reach the lysis start, suggesting that their "melting power" is limited.

Principles of Ribosome Sliding

The important lesson to be learned from the prokaryotic scanning model is that ribosomes bound to mRNA but not yet committed to a reading frame by codon-anticodon interaction can slide in either direction over fairly long distances. We do not know which components participate in

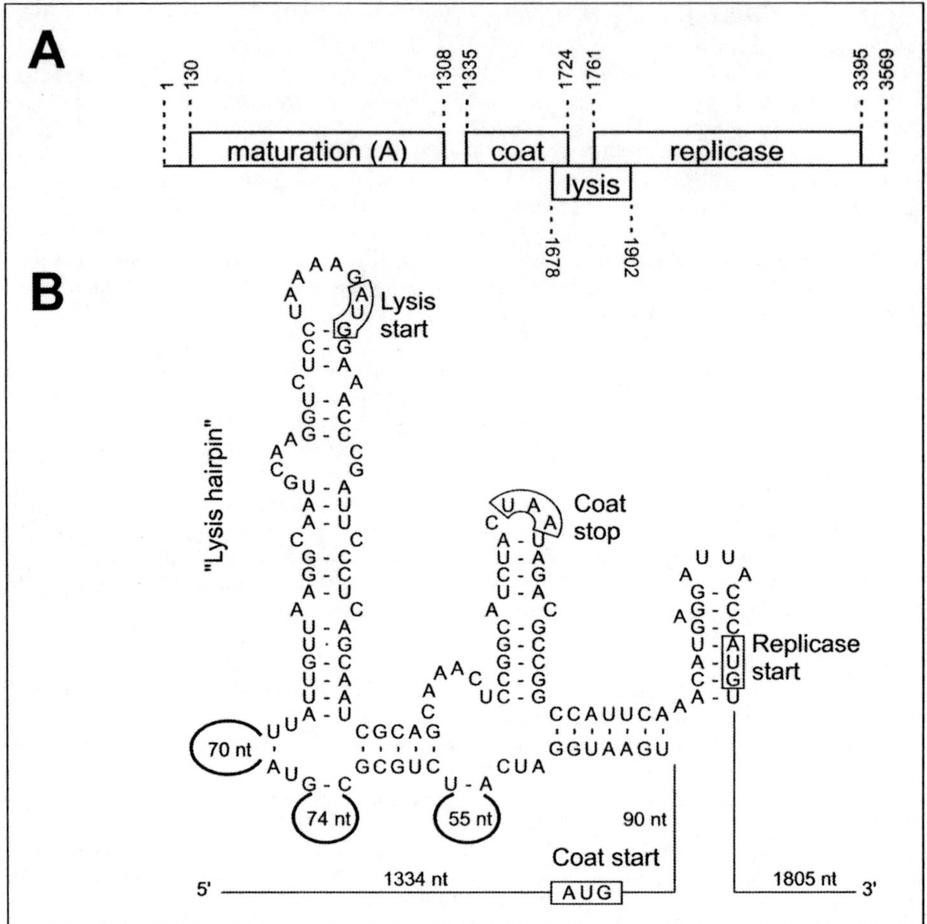

Figure 1. Translational coupling between the coat and lysis genes of RNA bacteriophage MS2. (A) Genetic map of MS2. (B) Structure model of the phage RNA around the coat stop and the lysis start.

this interaction. As continuously different sequences are encountered on the messenger, it is unlikely that the bases are involved. Moreover, these should be available to establish specific contacts such as the SD and codon-anticodon interaction. Thus, it could be either the phosphates or the riboses. On the ribosome side, an obvious stabilizing component is S1, a protein interacting with unstructured RNA,[17] but also the unidentified component that allows ribosomes to bind to leaderless mRNA may be considered.[18] As the interaction between the 30S subunit and the RNA probably does not involve the RNA bases, each new position has the same binding energy. Movement can be very rapid as evidenced by the lysis gene case where the distance between stop and restart site must be traversed before an elongating, coat-synthesizing ribosome blocks passage.

It is of interest to compare the speed of initiation complex formation via reinitiation and via the de novo pathway. To do so, we prepared a mutant containing a destabilized lysis hairpin. As expected, this mutant produced lysis protein even when coat translation was fully blocked. When coat translation was restored, the production of lysis protein remained unchanged. However, now there are two potential sources to feed the lysis start. There can be de novo initiation owing to the destabilized lysis hairpin and there can be restarts resulting from coat gene termination. To distinguish between the two possibilities, we transplanted the coat stop codon in the 3' direction to a position

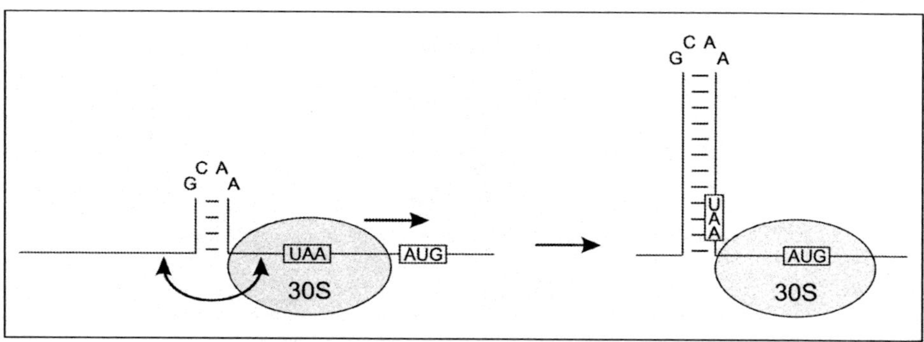

Figure 2. A possible way to confer directionality to the random walk of a terminated 30S ribosomal subunit. The subunit is not drawn to scale.

too far away to allow reinitiation. Now, the lysis gene was not translated at all. Clearly, the rapid flow of coat-translating ribosomes interfered with de novo initiation at the opened-up lysis start.[19] These experiments show that lysis can only be produced by reinitiation and that reinitiation is much faster than de novo initiation, even if it includes scanning. It appears therefore that the initial binding step of free ribosomes is relatively slow.

Although the movement of 30S subunits over the mRNA is basically a heat-driven random walk, nature could conceivably introduce directionality into such movement by evolving specific RNA secondary structures that push the ribosome in one direction (Fig. 2). To our knowledge, the potential influence of RNA secondary structure in termination regions on restart events has not been studied so far.

A Standby Intermediate May Solve a Kinetic Paradox in de Novo Initiation

There is no obvious difference between the 30S·mRNA (IF3) termination intermediate and a 30S·mRNA (IF3) initiation intermediate. This implies that de novo initiation does not necessarily start with binding exactly at the ribosome binding site (RBS; defined as the footprint of the 30S subunit on the mRNA in a 30S initiation complex). The 30S subunits will bind with significant affinity to any piece of nonstructured RNA, like the classic example of poly(U). It is thus quite possible that the 30S subunit has the potential to land at a single-stranded region outside the RBS and then walk randomly to the true start site where it becomes fixed by additional interactions (S1, SD, tRNA$_f^{Met}$). This resembles eukaryotic initiation where lateral movement takes place after the initial landing, be it on the cap or on an internal ribosome entry site (IRES).[20,21]

In *E. coli*, the existence of such a landing pad or "standby site" could resolve a serious problem that exists in the kinetics of initiation. Most RBSs (like all RNAs) adopt a secondary structure of some kind. Productive ribosome binding requires the RBS to be unfolded, but this state lasts too short to allow diffusion-controlled ribosome binding (see below). Having a ribosome on standby close to its target would be a solution to this problem.

Do We Need Standby Binding?

The possibility of standby binding of 30S subunits to mRNA as a mechanism for enhancing their binding rate to the RBS has been proposed earlier by Draper,[22] but the kinetic data available at the time gave no indication as to the actual need for rate enhancement. Estimated ribosome concentrations and association rates (see below) predict that 30S binding is far more rapid than the initiation process and nothing would therefore prevent ribosomes from following each other closely on a well-translated messenger. So, if RBSs were continuously accessible, little could be gained by placing 30S subunits on standby.

While RBSs may in general be less structured than other mRNA parts,[23] they are rarely structure-free. Most of the affinity of the ribosome for the messenger comes however from interactions that require the RNA to be single-stranded. The ribosome must therefore compete with the structure and capture the RBS when it happens to be in the unfolded state.[1,24,25] In the following sections, we will argue that this state exists only during very brief intervals and that therefore 30S binding rates are required that are much faster than can be achieved by simple diffusion. We will show that this paradox may be solved by invoking standby binding.

Flash Exposure of a Ribosome Binding Site

We have previously used the RBS of the coat protein gene from bacteriophage MS2 to examine how simple mRNA structures control expression by limiting access to ribosomes.[25,26] The MS2 coat RBS folds into a hairpin structure that encompasses both the SD region and the AUG initiation codon (Fig. 3A). Extensive mutagenesis revealed that the level of expression from this RBS is dictated directly by the stability of the hairpin structure (Fig. 3B and see below). In thermodynamic terms, the binding of free 30S subunits competes against formation of the structure at the RBS. Even though the wild-type structure is stable enough to exist more than 99% of the time, the high affinity of the 30S subunits for the unfolded RBS allows them to outcompete the structure. Expression of the wild-type gene thus reaches a maximal level that is not limited by the structure, but only by the kinetics of the initiation process and the clearance of the RBS by the elongating ribosome.

Since we proposed this quantitative thermodynamic model, data on the kinetics of spontaneous formation and disruption of hairpin structures have emerged. Not unexpectedly, simple hairpins turn out to fold extremely rapidly, with half-lives of the unfolded state in the order of microseconds. For a rather stable hairpin like the one at the RBS of the MS2 coat gene, this has important implications. Not only is it closed most of the time, but when it happens to open spontaneously, it is only for a very brief time. The question arises if a 30S subunit from the cytoplasmic pool can reach the RBS during this "flash exposure".

The rate at which the coat hairpin refolds can be estimated in two independent ways. First, a rule of thumb has been proposed to calculate the rate of folding of a hairpin from the stability of its loop, including the first base pair.[27,28] This builds on the assumption that the formation of a hairpin is most likely to commence with the formation of the topmost base pair, after which the remainder of the stem closes rapidly in a zipper-like fashion. Using the Turner parameter set 2.3 for calculating ΔG-values (see below), the rule predicts a folding rate k_F at 37°C of about 10^5 s^{-1}. This corresponds to an average lifetime of the open state of $1/k_F = 10$ μs.

An even shorter time is found when we examine the kinetics of translation of the coat gene. Analysis of polysome profiles from *E. coli* cells that were infected with MS2 or R17 (virtually identical bacteriophages) has shown that translated phage RNA is found mostly in dimeric and trimeric polysomes.[29-31] These polysomes virtually disappear when the coat protein gene contains an amber mutation, supporting the notion that they mainly represent translation of the coat protein gene.[32,33] Given an elongation rate of about 16 codons per second,[34] a length of the coat protein gene of 130 codons and the fact that one of the ribosomes will always occupy the RBS, two to three ribosomes per RNA would roughly correspond to one initiation event per six seconds. This is not far off from the value of one initiation per three seconds found by a different approach for another well-translated gene, *lacZ*.[35,36]

An average time span of six seconds between initiation events leaves a window of about four seconds for a new ribosome to attach to the messenger (assuming two seconds for the initiation process itself and the subsequent clearance of the RBS).[37] Since the hairpin folds up immediately upon being liberated by the elongating ribosome, the observed initiation frequency prescribes that the hairpin cannot remain closed for more than four seconds.

The maximal duration of the "flash exposure" of the RBS can now be calculated from the equilibrium constant of the helix-coil transition (K_F = [F]/[U]), which equals the ratio between the "open" and "closed" lifetimes. We can obtain this equilibrium constant from the free energy of formation (ΔG^0_f) of the hairpin ($K_F = e^{-\Delta G/RT}$). Again using parameter set 2.3 of Turner at 37°C, one gets a value of $K_F = 6 \cdot 10^7$ for the MS2 coat hairpin. A "closed" time of less than four seconds thus implies an "open" time of less than $4\,s\,/(6 \cdot 10^7) = 0.7 \cdot 10^{-7}$ s, i.e., the open RBS refolds in less than 0.1 μs. To account for the observed initiation frequency at the coat protein gene, a new ribosome must always manage to initiate within four seconds after the previous one has left the RBS. In other words, there must be a 100% probability that within 0.1 μs a 30S subunit captures the open RBS in

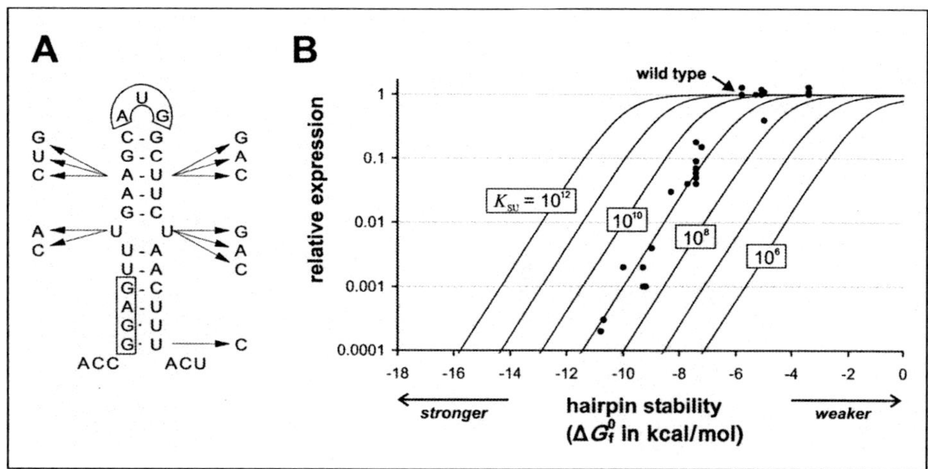

Figure 3. (A) Hairpin structure at the RBS of the MS2 coat gene. The SD region and the initiation codon are boxed. Mutations introduced in various combinations are indicated with arrows. (B) Expression of the coat protein gene as a function of the stability of the hairpin structure at the RBS. Expression was measured by quantitative Western blotting. ΔG^0_f-values were calculated as in ref. 25. Curves show relative expression levels predicted by equilibrium thermodynamics (Fig. 8A) for arbitrary values of K_{SU}, the 30S-mRNA binding constant, as indicated.

a productive manner, resulting in the formation of a functional initiation complex. Note that this time span of 0.1 µs may represent a single exposure, or may be divided over multiple, even shorter exposures within the four-second window.

How Fast Can Free-Floating Ribosomes Be on Site?

To estimate the on-rate for 30S-mRNA association, we need to know the concentration of free 30S subunits in the cell. We have previously used an estimate of 8.5 µM, noting that this is an upper limit because not all free 30S particles may be functional.[25] We assume these subunits to be saturated with initiation factors[38,39] and fMet-tRNA.[40] The binding rate constant k_{SU} has been measured in vitro using a quench-flow technique (K. Andersson and M. Ehrenberg, personal communication). Its observed value of about $3 \cdot 10^7$ M^{-1}s^{-1} is in good agreement with the expected diffusion rate[*,22,41,42] and yields an on-rate $k_{SU} \cdot$ [30S] of less than 250 s^{-1}. The RBS would thus on average take 4 ms to collide with a free 30S subunit, much longer than the available interval of less than 0.1 µs. Even if one takes into account all the uncertainties in the numbers used to arrive at this conclusion, a discrepancy of more than four orders of magnitude is difficult to ignore. It appears that ribosomes dispersed in the cytoplasm are simply too slow to jump onto the RBS within the brief instant that it is accessible.

Where then do the ribosomes that initiate at the coat start come from, if not from solution? Since expression of the coat gene is not coupled to any of the other genes present on the MS2 RNA, we can only envisage that they are sitting somewhere on the RNA, waiting for the RBS to become available. This is where we see a possible role for standby binding as proposed above.

We will now examine what standby binding sites may look like and what evidence we can find for their existence. In the final section, we will show that, given what is known about the physico-chemical aspects of translation, standby binding may indeed provide a quantitatively satisfactory solution to the aforementioned kinetic paradox.

---

[*] In view of the crowded nature of the cytoplasm, in vivo binding may well be even slower. Note that crowding can also increase reaction rates, but only for reactions that are not diffusion limited.[41,42]

Nature and Biology of 30S Standby Binding

Depending on the place where the 30S subunit sits on the mRNA until it can capture an unfolded RBS, we may distinguish three fundamentally different models for standby binding. These are shown schematically in Figure 4 and will be discussed in the following subsections.

On-Site Binding: the Hammock Model

The Ribosome Can Accommodate Certain mRNA Structures

The SD interaction, the codon-anticodon interaction and the binding of S1 all require the mRNA to be locally single-stranded. Until recently, we believed that this had to be true for the entire RBS. New data have however forced us to reconsider this assumption. It now appears that the 30S subunit has the space and ability to accommodate various hairpin structures in the mRNA, as long as the interactions mentioned above are not compromised. Structures that are ignored by bound ribosomes have been found directly upstream of the SD region, between the SD region and the initiation codon and directly downstream of the initiation codon.

Noninhibitory structures upstream of the SD region are typically short, regular hairpins.[1,43] The largest structure of this kind described so far is a 13 base-pair hairpin in the RBS of the *E. coli thrS* gene.[44] It lies 5' adjacent to the SD region and forms part of the binding site for the autorepressor protein ThrRS, but by itself does not affect ribosome binding and initiation. In fact, deletion and mutation studies showed that the single-stranded RBS continues on the 5' side of the hairpin and that this upstream stretch of unpaired nucleotides is crucial for *thrS* translation. Moreover, nuclease and chemical protection analyses clearly showed the hairpin protruding from the bound 30S subunit. The single-stranded stretches on either side of the hairpin may thus be considered as a quasi-continuous RBS.

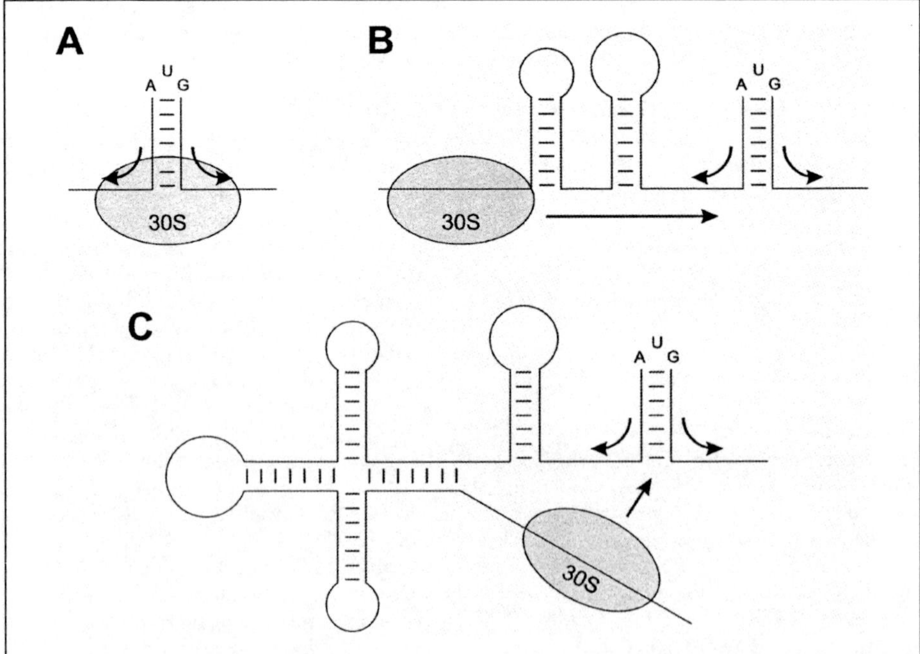

Figure 4. Three possible models for standby binding of a 30S subunit to structured mRNA. (A) "On-site" standby: the subunit can fix the unfolded RBS as soon as it opens. (B) "Off-site" landing pad: the subunit reaches the RBS by linear diffusion. (C) "Long-range enhancer": the subunit reaches the RBS by "handing-over" or by local diffusion in 3D. Subunits are not drawn to scale.

Structures upstream from the SD region that do function as inhibitors of translation are generally large and composite, no doubt precluding the formation of such a quasi-continuous RBS.[1,24] Why very long hairpins (> 20 bp) can also inhibit ribosome binding is as yet unclear.[45,46] Another mystery is the existence of long-range interactions just upstream from the SD that stimulate translation while disrupting local base pairing at the RBS.[47-49]

Stable structures have also been discovered between the SD region and start codon of two genes from bacteriophage T4, where they are essential for fixing the correct distance between these two elements.[50,51] Whether this unusual arrangement serves some regulatory purpose is not clear. The same phenomenon was observed by toeprinting on an artificial mRNA where a structure from phage T4 gene *60* was placed between an SD region and an AUG codon. In the same experiment, a long hairpin was accepted by the ribosome when placed just downstream of the AUG.[52]

Together, these data indicate that the 30S subunit is capable of attaching to RNAs that contain substantial base pairing in the form of stem-loop structures, as long as a quasi-continuous single-stranded stretch of sufficient length is available. Small structures are noninhibitory if they can be oriented so as to protrude from the ribosome. Indeed, various lines of evidence indicate that much of the mRNA that is protected against nuclease and chemical attack is not actually hidden inside the 30S particle, but is attached to its surface on what is known as the platform and the shoulder.[52-59]

The Main Entry Site of RNA-Phage RNAs

Based on the above observations, we now envisage the following scenario. A free-floating 30S subunit may attach to the single-stranded stretches flanking a structured RBS and wait there for the structure to open (Fig. 4A). Once this happens, the 30S snaps into place, preventing the structure from closing again.

Suggestive support for the existence of this type of "on-site" standby binding comes from the secondary structure models for the RNAs of various RNA bacteriophages that were developed by phylogeny and probing.[60-63] These RNAs, which function both as genome and as messenger, have an intricate structure composed of large structural domains that hardly leave any single-stranded stretches longer than a few nucleotides. A remarkable feature of all these RNAs, however much they differ in sequence, is the presence of one region devoid of large-scale structures (Fig. 5A). This region invariably comprises about 120 nt and includes the RBS of the coat protein gene.

The coat gene RBS functions in these phages as the central control point for translational regulation and it is the primary entry site for ribosomes. In all cases the SD region and initiation codon are taken up in a small but fairly stable hairpin structure. The gene is nonetheless translated at a very high level.

The extent of the poorly structured region of 120 nt appears quite meaningless, until one counts only the unpaired nucleotides, ignoring the paired ones and the various loops. Figure 5B shows that in all cases these quasi-continuous single-stranded stretches add up to between 44 and 49 nt. This is only marginally larger than the 30S footprint on the mRNA backbone in the absence of initiator tRNA.[64] We therefore like to picture this region as a kind of hammock, bounded on either side by the multi-branched structural domains of the remainder of the phage RNA, in which a 30S subunit can rest and wait for the structure to open.

Among the wealth of experimental data on in vitro ribosome binding to the various phage RNAs, there is one observation that may relate directly to the standby complex we propose. The 30S subunits can be bound to MS2 or Qβ RNA in the absence of initiator tRNA with an association constant of about 10^6 M^{-1}.[65] When prepared at 37°C, the resulting complex can be chased into translation, even in the presence of inhibitors of de novo initiation such as edeine, aurintricarboxylic acid or anti-S1 antibodies. In striking contrast, this is not possible with a similar complex formed at 0°C.[66,67]

A new interpretation of these observations could be as follows. At 37°C, the hairpin at the coat protein gene RBS is weak enough to allow the 30S subunit to shift rapidly from its standby position to form the true pre-initiation complex, which needs no further reorganization to enter into translation. At 0°C, on the other hand, the hairpin is stabilized to such a degree that it does not open within the time span of the experiment, leaving the 30S subunit sitting on the structured RBS as an on-site standby complex. To allow this complex to enter translation at 37°C, the 30S subunit has to

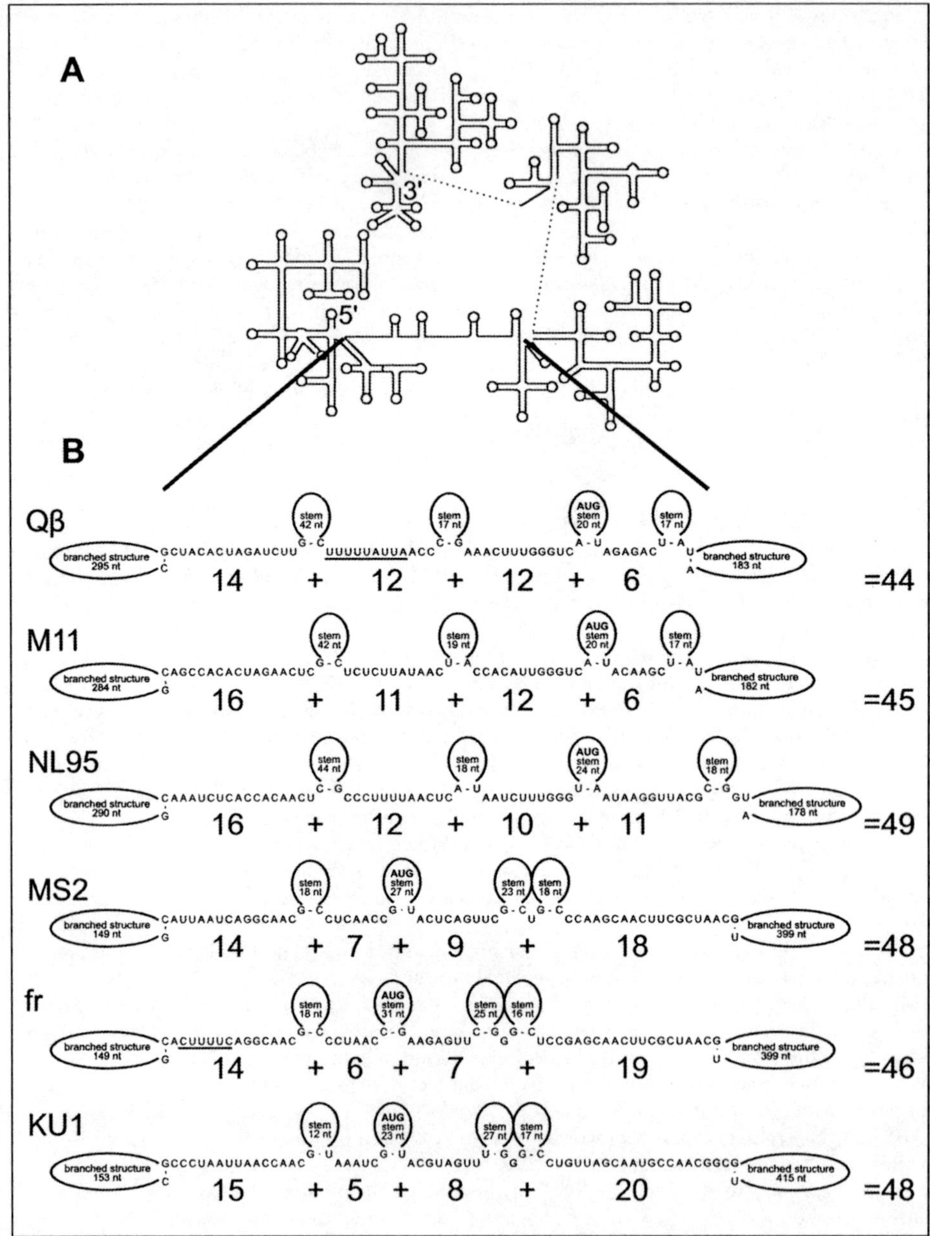

Figure 5. Proposed "on-site" standby sites for 30S binding at the coat gene starts of the single-stranded RNA bacteriophages. (A) Structure model of the entire genomic RNA of phage Qβ (modified after ref. 61). (B) Quasi-continuous single-stranded regions for representatives from all four species of ss-RNA bacteriophage. Coat gene start codons are indicated; underlined sequences in Qβ and fr UV-cross-link to S1 in 30S-mRNA pre-initiation complexes.[105] Structures are based on refs. 60-63.

resettle on the RNA in a somewhat shifted position as soon as the hairpin has opened. During this resettling, the 30S subunit is apparently sensitive to the aforementioned inhibitors. Note that S1 is involved in both the 0°C and the 37°C complex.[68] In agreement with this explanation, the complex formed at 37°C remains insensitive when cooled down to 0°C, indicating that the unfolded RBS cannot refold once the true pre-initiation complex has formed.

A similar interaction of the 30S subunit with single-stranded stretches between hairpins (or with their loops) was recently proposed as an important intermediate in translational autocontrol of the *E. coli rpsA* gene, the gene that encodes ribosomal protein S1.[69] Details of this system await further elucidation. Evidence for a standby complex with the SD interaction already in place was obtained by probing the structure of the *cro* mRNA in 30S·mRNA complexes with or without initiator tRNA. A hairpin structure downstream from the AUG start codon was shown to exist in the binary 30S·mRNA complex, but disappeared upon the addition of initiator tRNA.[70] This confirms that the 30S subunit can indeed bind to a partially structured RBS and capture the RNA in its single-stranded form when it opens.

A Standby Site Activated by Translating Ribosomes?

The RBS of the replicase genes of the single-stranded RNA bacteriophages is by itself inactive, because it lies hidden in a heavily structured region composed of local hairpins and long-range stems (Fig. 6A). It is only activated when ribosomes translating the upstream coat-protein gene pass through the long-range components of these structures, known as the MJ and VD interactions.[71-74] This is believed to open the replicase start region for long enough to allow one round of initiation, yielding the expression ratio of about 1:1 as observed in expression studies with MS2 cDNA plasmids.[75]

Like the coat gene start, the start of the replicase gene forms a hairpin structure comprising both the SD region and the AUG initiation codon (Fig. 6A), but by itself this structure is weak enough to allow high-level translation.[76] However, just like in the case of the coat gene, we are faced with the problem of the relative kinetics of 30S binding and hairpin folding. The replicase RBS is predicted to exist in its unfolded state about 0.02% of the time. This implies that a random collision with a free 30S subunit is five thousand times less likely to be productive than if the RBS would be single-stranded permanently.

If elongation takes place at a rate of 16 codons per second,[34] a ribosome translating the coat gene keeps the long-range interactions open for about one second. To obtain an expression ratio of 1:1, this one second must thus be enough to achieve a successful initiation event at the replicase start. The aforementioned binding rate constant ($3 \cdot 10^7$ M^{-1}s^{-1}) and 30S concentration (8.5 μM) predict that 30S subunits encounter the RBS at a frequency of $3 \cdot 10^7$ x $8.5 \cdot 10^{-6}$ = 250 times per second. At a fraction of open RBSs of 0.02%, we can therefore expect no more than 0.05 productive collisions within the available second, yielding a coat : replicase ratio of 20 : 1. This is not consistent with the experimental observations.

A closer look at the structures proposed around the replicase RBS reveals that the passage of a coat-translating ribosome may in fact create a quasi-continuous single-stranded region, much like the one present at the coat gene start. In Figure 6B, the two long-range interactions have been opened, as they would be by a translating ribosome. As a result, a series of single-stranded stretches has become available, adding up to 49 nt. The possibility to create a region of this size is conserved within the single-stranded RNA phages (Fig. 6B),[60,61] but we need to postulate that the bottom part of the hairpin following the replicase start codon is also open. The unusually asymmetric internal loop that terminates this helix in all the different RNA phages (see Fig. 6A) may function as a hinge to facilitate the opening of the standby site.[77,78]

Until now, it was assumed that opening of the MJ interaction activates the replicase start because it forms part of the RBS, albeit upstream from the SD region. The VD interaction was proposed to contribute to regulation because it might stabilize the MJ interaction by coaxial stacking.[74] The possibility of standby binding to the surroundings of the replicase start while the local hairpins are still intact suggests a different interpretation. Movement of ribosomes through both VD and MJ is now essential to create an "on-site" standby site that is large enough to accommodate a 30S subunit. There, the subunit can wait for as long as its off-rate allows, until the hairpin at the replicase RBS opens and a true initiation complex can be formed.

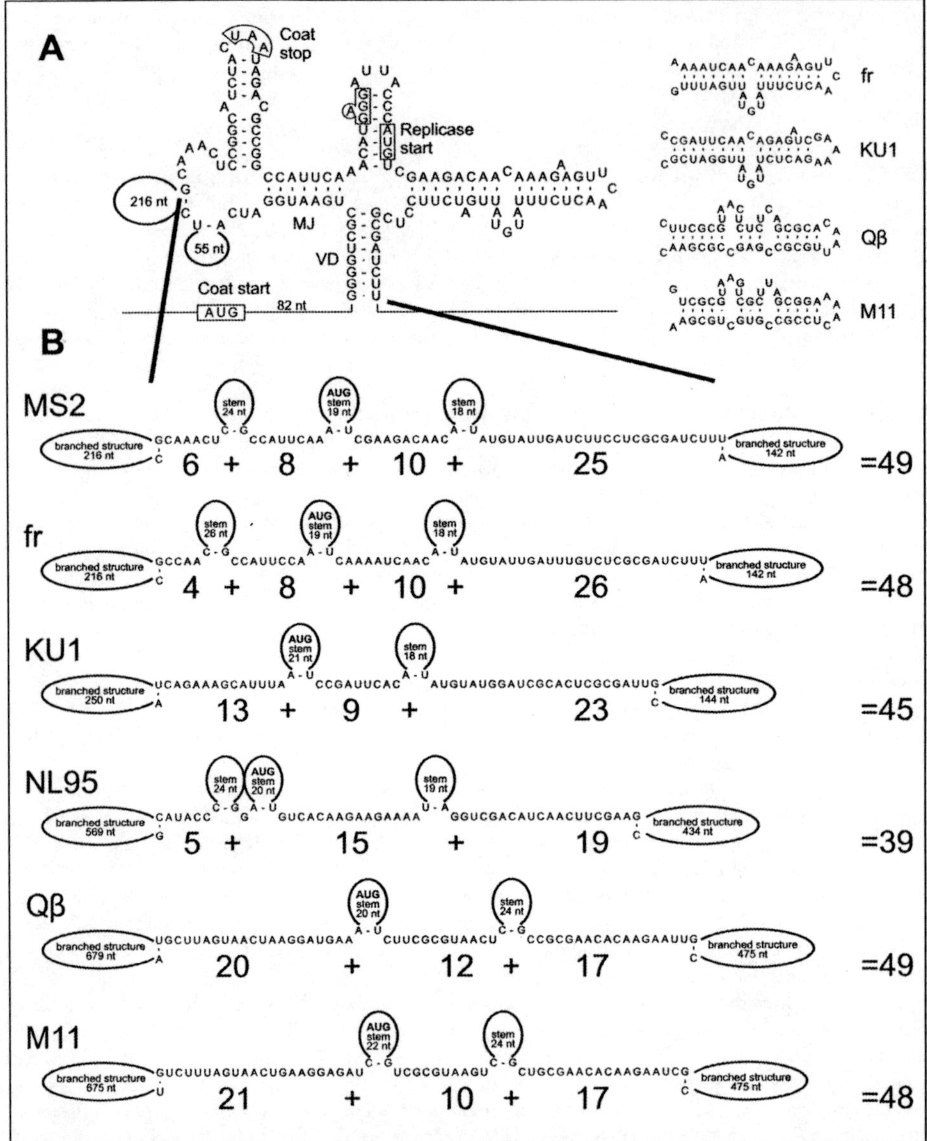

Figure 6. Proposed "on-site" standby sites for 30S binding at the replicase starts of the single-stranded RNA bacteriophages, as exposed upon translation of the coat-protein gene. (A) Structure conferring translational coupling in MS2. MJ = Min Jou interaction;[71] VD = van Duin interaction.[74] Conservation of an asymmetric loop in the rightward hairpin is shown on the right (NL95 lacks the bottom half of this hairpin). (B) Quasi-continuous single-stranded regions for representatives from all four species of ss-RNA bacteriophage. Replicase start codons are indicated. Based on refs. 60-63.

The only reason why we think that an inducible standby binding site is necessary here, is the presence of a small but stable hairpin at the replicase RBS, which limits its accessibility to intervals of only a few microseconds. This hairpin however is crucial to the phage, as it is the main binding site for the MS2 coat protein. Late in the life cycle of the phage, the binding of coat protein at this hairpin blocks replicase translation and initiates packaging of the RNA into capsids. Thus, the replicase RBS has evolved under opposite selection pressures from the coat protein and the ribosome. We suggest that the creation of an inducible standby binding site has allowed the union of three otherwise incompatible functions in one stretch of RNA: a locally structured binding site for the coat protein; a long-range structure permitting translational coupling; and a single-stranded region for the 30S subunit to enter. With every new detail we learn about the RNA phages, it becomes more striking how intricately the many control mechanisms interrelate in these tiny survival machines.

Close-by Binding: Scanning, Shunting and the Landing-Pad Model

Linear diffusion is a common principle in "target location" by DNA-binding proteins.[79] The relative uniformity of the DNA double helix allows these proteins to bind nonspecifically anywhere on the DNA and shift to their final position by random lateral movement. In contrast, the applicability of linear diffusion to ribosome binding has been considered limited because at first sight it would seem to require extensive stretches of structure-free RNA.[22] There is however good evidence that certain secondary structures can be skipped during prokaryotic scanning, obviating the need for the region between a potential landing site and the target site (RBS) to be fully unfolded.

Analysis of reinitiation patterns in nonsense mutants of the *lacI* gene has indicated that the prokaryotic scanning ribosome does not necessarily stop or dissociate when it encounters structures in the mRNA. Instead, it can sometimes skip over them and continue scanning on the other side.[80] Thus, large distances between termination sites and reinitiation sites can be bridged. Here again the presence of a quasi-continuous single-stranded stretch of sufficient length seems pivotal.

The same principle could apply to the binding of free subunits, if they can attach to a single-stranded region at some distance from an RBS that itself is inaccessible (Fig. 4B). By random movement over the single-stranded stretches, they could hit upon the initiation site during its "flash exposure", in a prokaryotic equivalent of ribosome shunting.[21] Prokaryotic scanning must be relatively fast, since in the coupling between the MS2 coat and lysis genes terminated ribosomes manage to reach the upstream lysis start against the current of coat-gene translating ribosomes (see above).

A stretch of accessible RNA could thus function as a ribosomal landing pad for a nearby RBS. It should be noted however that we have no mechanism to give a 30S subunit on standby a push in the right direction at the right moment, *viz.* when the structure at the RBS opens. In contrast to the "on-site" binding discussed in the previous section, the subunit is now not necessarily in place when the RBS is exposed. Such "off-site" standby binding would therefore only function by increasing the probability of a productive encounter and we consider its potential therefore limited compared to the "on-site" model.

Intersegment Transfer and Dissociation/Association

The fast rate at which the LacI transcriptional repressor protein reaches its target on double-stranded DNA is believed to depend in part on a process termed direct intersegment transfer.[81,82] It implies that the protein binds initially to a nonspecific site on the nucleic acid and from there is handed over to its site of action. A similar scenario might be imagined for the binding of ribosomes to mRNA (Fig. 4C). The LacI repressor functions as a tetramer and therefore has multiple identical DNA binding sites, which allows the existence of an intermediary situation where secondary and primary targets are bound at the same time. Although 30S subunits do not form multimers, it is not inconceivable that they may attach to a single-stranded section of the RNA through ribosomal protein S1 and then capture a distant RBS with, for example, the anti-SD region. Subsequently, S1 may follow suit and jump over to the RBS as well. As an aside, we note that the "mRNA-standby site" on the 30S subunit proposed by Gualerzi and coworkers could not be used for this kind of handing over, since it occupies roughly the same space on the subunit as the final binding site.[4,65,83]

Alternatively, even simple dissociation from an initial nonspecific binding site and re-association at the RBS may raise the probability of functional complex formation. Limiting the distance over which the 30S subunits diffuse (from other parts of the same RNA, rather than from anywhere in the cytoplasm) would increase the frequency with which 30S subunits collide with the RBS. In other words, it increases the "local concentration" of subunits in the vicinity of the RBS. This would parallel the way in which the movement of, among others, RNA polymerase is confined to a local DNA domain until it finds its specific binding site.[84,85]

If one applies the concept of local concentration to the MS2 coat gene RBS, then the apparent 10,000-fold increase in binding rate over simple diffusion (see above) would require a local 30S concentration of 10,000 x 8.5 μM. This would imply that the ribosomes be confined to a space smaller than their own dimensions. Confinement of the ribosomes to a limited volume by dissociation/association, involving secondary binding sites elsewhere on the mRNA, is apparently not sufficient to explain this particular case. A more satisfactory explanation would be that there is permanently a 30S subunit in place at the RBS. Its mRNA binding track, already oriented towards the RNA, then indeed covers only a fraction of the ribosome volume. This situation coincides with the on-site binding model proposed earlier (Fig. 4A).

Dissociation/association or intersegment transfer may nonetheless play a supportive role in the MS2 coat gene system. We have previously reported that the upstream 1300 nucleotides of native MS2 sequence strongly enhance the capability of the ribosome to compete with the hairpin structure at the RBS. Under conditions of maximal sensitivity (i.e., strong competition by the RBS structure), expression was 25-fold higher in the presence of the upstream RNA, presumably through an increase in the 30S binding rate.[86] This enhancement could not be localized to a specific segment of the upstream sequence. The native structure of this highly organized piece of RNA appears to guide the 30S subunit to the coat gene RBS. By itself, such guidance would be insufficient to put the subunit in place within the 0.1 μs that the hairpin is unfolded (see above). But it is conceivable that it is needed to put and keep the 30S in its standby position on the quasi-continuous single-stranded region shown in Figure 5. Many DNA-binding proteins also appear to reach their targets by a combination of long-range three-dimensional movement and local linear sliding.[87,88]

Ribosome Recycling in cis

An intriguing way in which intersegment transfer could be employed for translational control, is the recycling of terminated ribosomes back to the original initiation site, where they may reinitiate for another round of translation. Recycling of ribosomes *in cis* has been proposed in eukaryotes, where circularized messengers were directly observed by atomic force microscopy.[89-91] In prokaryotes, recycling is reportedly happening on tripeptide-encoding minigene messengers, but there the ribosomes are simply slipping back over a very short distance.[16] No direct evidence exists, to our knowledge, for the long-range transfer that would be needed in the case of genes of normal length. However, a very suggestive arrangement of termination and initiation codons is found in the *hok*-like operons for plasmid stabilization.

All operons of the *hok* family consist of two open reading frames that overlap by almost their entire lengths (Fig. 7A).[92,93] While the downstream frame encodes a killer protein that has a central function in plasmid stabilization, the polypeptide encoded by the upstream frame has no function. This frame (called *mok* in the prototype *hok* system) only exists to activate the killer gene (*hok*) through translational coupling. When a cell loses the plasmid during cell division, the *mok* start on the remaining mRNA is activated through the degradation of an inhibitory antisense RNA. This leads to expression of *hok* and the subsequent death of the plasmid-free cell.

The structure of the *hok* mRNA has been solved and is shown in Figure 7B.[94,95] Both RBSs are located in stable structures, but the *mok* hairpin is weak enough to allow the low level of *mok* translation that is required to activate the *hok* start.[95] One would expect translational coupling to function perfectly well if the *mok* termination codon were located next to the *hok* start. Nonetheless, the unusual arrangement with the *mok* stop at the end of the *hok* gene (Fig. 7A) is conserved throughout the highly divergent *hok* family. It is intriguing to see that the secondary structure puts the two termination codons opposite the RBS of *hok*. This strongly suggests that termination, rather than elongation, is the crucial event activating the *hok* start. While that may be related to the fact that termination consists of a rather time-consuming sequence of steps, we cannot ignore the possibility that ribosomes actually recycle *in cis* on this particular class of messengers, by being handed over from the *mok* stop codon to the *hok* start (compare Fig. 4C). Moreover, since *hok*-translating ribosomes also terminate opposite the *hok* RBS, activation of *mok* would lead to a self-sustaining avalanche of *hok* translations, ultimately resulting in the killing of the cell. We imagine that such a

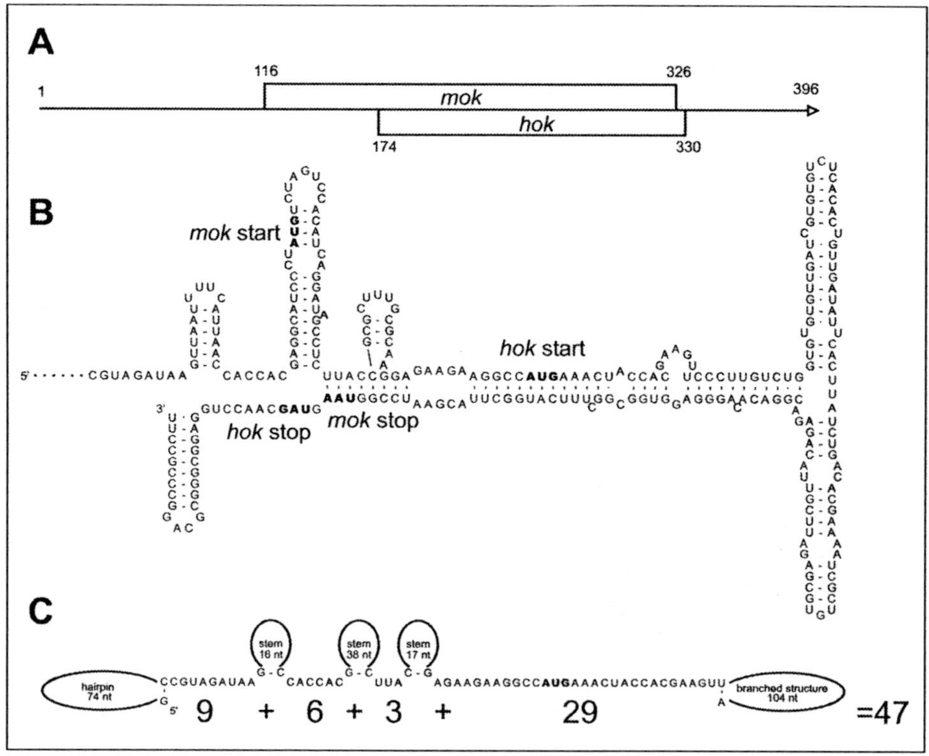

Figure 7. Structure of the *hok* operon from plasmid R1. (A) Genetic map. (B) Relevant part of the active structure. Based on refs. 94-95. (C) Proposed "on-site" standby site for 30S binding at the *hok* start, as exposed upon translation termination of either *mok* or *hok*. Extension of the third stem to 17 nt (5 bp) is based on thermodynamic calculations.

highly unusual positive feedback loop would only be allowed in a situation where rapid cell death is the final aim anyway.

Whether recycling of ribosomes on the *hok* messenger occurs or not, we note that the terminating ribosomes may once again create an "on-site" standby site (Fig. 7C). When we assume that the asymmetric internal loop downstream of the *hok* start functions as a hinge, as proposed for the replicase start (Fig. 6), the stretches of unpaired nucleotides around the *hok* start add up to 47 nt. Standby binding would be necessary here because once the long-range interaction is disrupted, the tiny hairpin upstream from the *hok* start will extend to include the *hok* SD-region (Fig. 7C).[94,95]

Standby Binding and Translational Repression

Draper was the first to point out that the intuitive mechanism for the functioning of translational repressors may not be valid for thermodynamic reasons.[96] Simple competition between a repressor protein and the 30S subunit would require high concentrations of repressor or high affinities for the RNA, neither of which was supported by available quantitative data. A variety of mechanisms have since been proposed that would help the repressor to do its job by taking advantage of mRNA structure, binding kinetics, mRNA degradation and even of the ribosome itself.[22,96-99]

As an additional possibility, we propose that the existence of an obligatory standby 30S·mRNA intermediate may allow translational regulation by direct competition without the need for extreme concentrations or affinities. As argued in the following section, standby binding does not need to be very strong to be functional. But if a repressor would prevent the 30S subunit from occupying its standby site by competing with this low-affinity binding, it would effectively block initiation at the structured RBS.

Once more, we can turn to the RNA bacteriophages for a possible example. During the life cycle of the phage there comes a point when its RNA has to be replicated rather than translated. This functional switch begins with the binding of the phage-encoded replicase just upstream from the coat gene RBS (Fig. 5).[100,101] This apparently suffices to block translation of the coat gene and consequently, through translational coupling, of the lysis and replicase genes. The remaining maturation gene has already switched itself off by that time.[102,103] Remarkably, the replicase uses a recruited copy of ribosomal protein S1 to bind at the coat gene RBS.[104,105] While the replicase may be unable to compete with formation of a 30S initiation complex on a single-stranded RNA, it may well outcompete the proposed weak standby complex, especially if its binding would negate the enhancement of 30S binding by the upstream RNA (see above). A picture now emerges in which the hairpin, the quasi-continuous single-stranded stretch and the stimulatory upstream structure have co-evolved to allow efficient regulation of the translation-replication switch that is central to the life cycle of these phages.

Physical Chemistry of 30S Standby Binding

Ribosomal Affinity and the Sensitivity to RBS Structure

We have previously approached the mechanism by which local mRNA structures control translational initiation as an equilibrium competition scheme (Fig. 8A). For the coat-RBS of phage MS2 and for several other RBSs, we showed that the expression decreases tenfold with every -1.4 kcal/mol added to the stability of the structure (Fig. 3B).[25,106] Theoretically, every -1.4 kcal/mol reduces the fraction of RNA molecules in which the hairpin is open at a given point in time by the same factor of ten. The expression data thus indicated that hairpin formation and 30S binding compete directly for the unfolded RBS. In addition, the affinity of the 30S subunit for the RBS was predicted, and subsequently shown, to determine the sensitivity to inhibition by the structure.[26,107] In other words: the stronger the binding of the 30S to the unfolded mRNA is, the stronger the structure at the RBS has to be to inhibit translation. This is indicated in Figure 3B by theoretical expression curves shifting to the left as the 30S-mRNA affinity (K_{SU}) increases. Relative expression is defined here as the fraction of the RBSs being in complex with a 30S subunit and it thus approaches the value of one as the messengers become saturated with ribosomes.

Standby Binding Relies on Kinetics

Standby binding of ribosomal subunits to the messenger can be introduced into this scheme as shown in light gray in Figure 8A. However, the equilibrium approach now turns out to be too limited. In fact, standby binding would be predicted to decrease translational efficiency, as the complex between the folded RNA and the 30S subunit (30S·F) would form at the expense of the pool that drives translation (30S·U). Moreover, we predicted standby binding only to be effective with fast RNA folding kinetics, an aspect that is completely ignored in thermodynamics (note that equilibrium thermodynamics, in only dealing with equilibrium constants, by definition ignores the time factor).

Figure 8B shows how we have converted the equilibrium scheme into a kinetic model, mainly by splitting the equilibrium constants into their component rate constants and by adding a route for the RBSs and 30S subunits to return to the free pool after taking part in translation. We have also split the initiation process in two steps, so that we can examine the consequences of varying the rate of formation of the true 30S initiation complex (k_i) without affecting the clearance of the RBS (k_c; see ref. 2). Apart from ignoring the time factor, the original equilibrium scheme represents nothing but a special case of the kinetic scheme, namely when k_i is negligible compared to k_{-SU} and k_{-SFU}.

The binding of 30S subunits to the mRNA is now described by no less than eight rate constants, but these are related through the first law of thermodynamics. If k_i is made zero (i.e., formation of the 30S initiation complex is blocked), the circular chain of reactions should be in equilibrium, meaning that the multiplied rate constants clockwise must equal those counterclockwise. With this restriction, the model allows us to freely enter established values for the different parameters, or to vary them one by one and examine how each parameter affects the sensitivity of expression to the RBS structure. We did this by solving the corresponding steady-state equations, entering these into

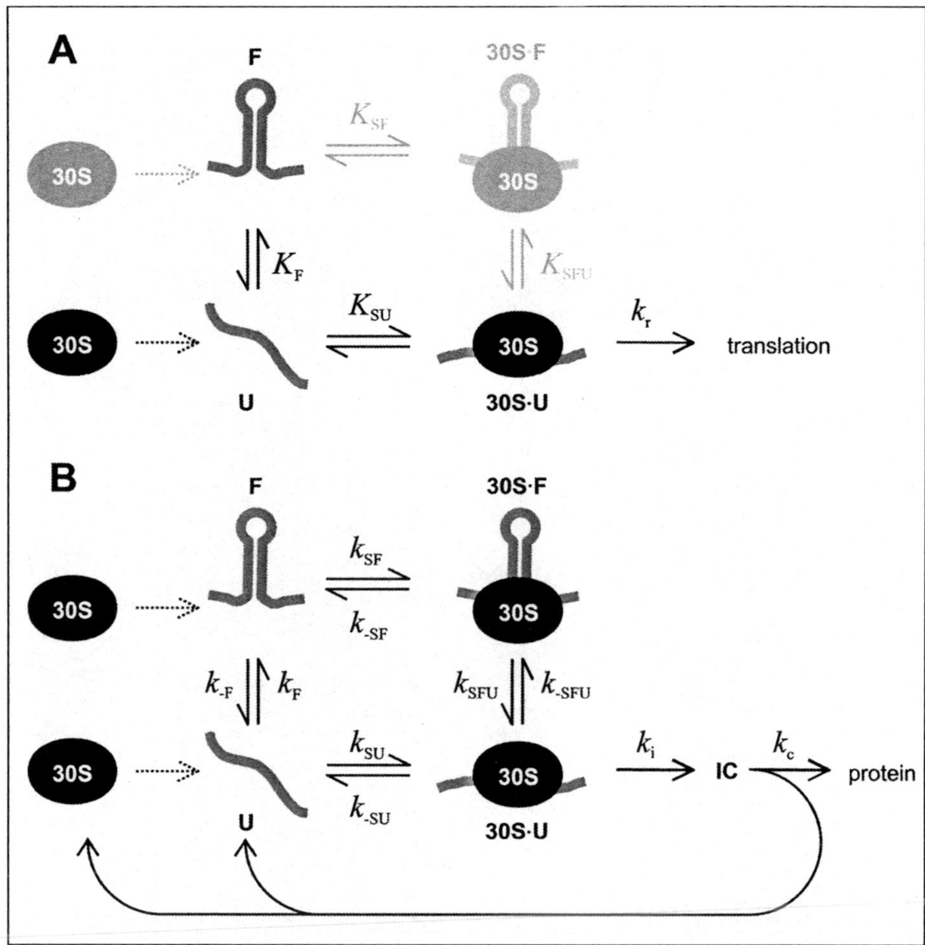

Figure 8. (A) Reaction scheme for ribosome binding to a structured RBS following the thermodynamic approach. F = folded RBS; U = unfolded RBS; 30S = 30S ribosomal subunit. Standby binding is indicated in light gray. (B) Reaction scheme for examining the steady-state kinetics of standby binding. IC = 30S initiation complex.

a spreadsheet and plotting the predicted outcome in the same format as Figure 3B. Table 1 summarizes the justifications for the various parameter values that we used as default.

Figure 9 shows the predictions using the values of Table 1. It is clear that without involvement of standby binding, the predicted expression for the wild-type coat RBS (ΔG^0_f = -9.9 kcal/mol at 42°C; see below) would be practically zero (dashed curve). Standby binding, even with a binding constant as low as 10^5 M^{-1}, is predicted to yield a dramatic increase in the ability of ribosomes to initiate at the coat start (dotted curve). In fact, the model now predicts maximal expression with the wild-type hairpin, as we observed in our experiments.

Standby Binding May Explain Extreme Apparent 30S-mRNA Affinities in Vivo

When we first reported on our mutational analysis of the MS2 coat gene start in 1990, we noted that the obtained expression curve suggested a 30S-mRNA affinity of about -13.5 kcal/mol, i.e., a binding constant of 3·10^9 M^{-1} (Fig. 3B).[25] The fact that this was about two orders of magnitude

Table 1. Parameter values entered into the model of Figure 8B to obtain Figures 9 to 11

Parameter	Value	Explanation	References
k_F	$1.5 \cdot 10^7$ s^{-1}	folding rate based on the kinetics of coat-gene translation; using the folding rate based on the stability of the first loop to be formed ($8 \cdot 10^6$ s^{-1}) shifts the curves marginally to the right	27, 28, 114, see text
k_{-F}	variable	variable on the horizontal axis; k_F / k_{-F} represents the stability of the RBS structure	
k_{SU}	10^7 M^{-1}s^{-1}	on-rate constant 30S-mRNA; measured in vitro	22, K. Andersson and M. Ehrenberg (personal communication)
k_{-SU}	1 s^{-1}	derived as $k_{-SU} = k_{SU}$ / K_{SU}; K_{SU} has been measured in vitro for many different nucleic acids and is usually in the range of 10^7 M^{-1}	96,108-113
k_{SF}	10^7 M^{-1}s^{-1}	taken as identical to k_{SU}, as the on-rate is mainly diffusion controlled	
k_{-SF}	10^2 s^{-1}	we have tried different values and find that standby binding would still work fine with $k_{-SF} = 100 \cdot k_{-SU}$, i.e., hundred-fold weaker binding at the standby site than at the unfolded RBS	
k_{SFU}	$= k_{-F}$	assuming that the 30S immediately snaps onto the unfolded RBS when is becomes available, the rate is determined by spontaneous unfolding	
k_{-SFU}	variable	fixed by the other rate constants in the circular part of the scheme, through the conservation of energy	see text
k_i	10^5 s^{-1}	determines the effectiveness of standby binding, by fixing the 30S·U complex	see text
k_c	0.5 s^{-1}	rate of initiation complex formation, initiation, and clearance of the RBS	see text

higher than the values usually measured in vitro[96,108-113] was blamed, at least in part, on inaccuracies in the parameters used for calculating the hairpin stability.

Since 1990, the parameters for ΔG-calculation have been updated several times, partly on the basis of new experimental data and theoretical considerations, but also by optimizing them for more accurate structure prediction.[114] Presently, the Turner group proposes two alternative sets of parameters designated versions 2.3 and 3.0, where only version 2.3 can be adapted for different temperatures (see http://bioinfo.math.rpi.edu/~zukerm/rna/energy/ for parameters and references). We have a preference for version 2.3, because it is more directly based on experimental data. Moreover, our experimental expression curves show considerably less scattering with set 2.3 than with set 3.0, suggesting that the former better approximates free-energy values in vivo. Recalculated with parameter set 2.3, the ΔG^0_f of the wild-type coat hairpin at 42°C is now -9.9 kcal/mol rather than the earlier −5.6 kcal/mol and the values for the mutants decrease by similar amounts. Shockingly, the updated expression curve would now suggest a 30S-mRNA affinity of -17.3 kcal/mol or $K_{SU} = 10^{12}$ M^{-1} (compare Figs. 3B and 9), which is five orders of magnitude beyond binding constants measured in vitro.

The kinetic scheme of Figure 8B and its predictions in Figure 9 may provide an adequate explanation for this discrepancy. The gain achieved by always having a 30S subunit on standby graphically mimics a large addition to the affinity between the 30S subunit and the RBS, yielding an apparent K_{SU} of almost 10^{12} M^{-1}. Even though this was not a priori introduced with our choice of

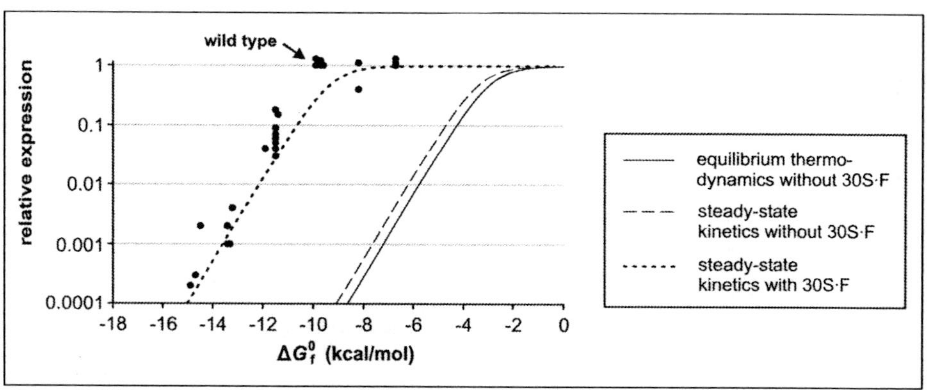

Figure 9. Comparison of expression levels predicted by the various schemes of Figure 8. Experimental data are now plotted against updated values for ΔG^0_f (see text). 30S·F = standby complex of 30S subunits and mRNA with folded RBS. All curves were calculated for $K_{SU} = 10^7$ M^{-1}.

rate-constant values, we thus obtain an expression curve that fits well with the experimental data without requiring ribosome-mRNA affinities that are unreasonably high (Fig. 9; dotted curve).

Standby binding thus solves two problems that are physically related but were encountered independently. Firstly, it answers the theoretical question of how ribosomes can bind to an RBS that is only exposed for a very brief instant. Secondly, it explains the extreme apparent 30S-mRNA affinities observed in in vivo translation from a structured RBS as compared to in vitro ribosome binding.

Further experimentation with the parameters in the standby model indicates that the leftward shift of the predicted curve as shown in Figure 9 is close to the maximum that can be obtained with reasonable parameter values. In Figure 10, we summarize the effects of moderate changes in the various parameter values and thus show that the predictions are not highly dependent on particular estimates (see the legend for more detailed information).

Comparing the experimental data in Figure 9 with the various theoretical curves on Figure 10, it seems that the MS2 coat gene system has evolved to achieve the maximum profit that can physically be gained from standby binding. Only its full exploitation would have allowed the hairpin to evolve to its present high stability without a concomitant reduction in translation. This is not so surprising when one considers the tendency of these phages to develop stable RNA structures on the one hand and their need for high-level synthesis of coat protein on the other. In addition, a hairpin that is only just noninhibitory may sensitize the coat gene RBS to translational repression by the phage replicase (see below). Evolutionary experiments in our laboratory have provided ample evidence for extremely rapid and precise optimization of translational control mechanisms in these organisms.[107,115-122]

Kinetics of Initiation in Vivo and in Vitro

An important difference between the old thermodynamic model and the present kinetic one is the necessary introduction of the fast step indicated with k_i in Figure 8B. When we approached ribosome binding as an equilibrium reaction, we had to assume that the drain of 30S·U complexes into translation was slow compared to the preceding reactions. This assumption was supported by in vitro translation data indicating a slow conversion of the 30S·U complex into a true 30S initiation complex.[108,111] Gualerzi and coworkers have suggested that this slow step may reflect a conformational change in the ribosome, necessary for codon-anticodon base pairing at the initiation codon.[108,123]

Now that the existence of a fast but essentially irreversible step turns out to be necessary to make standby binding work, we need to re-examine these assumptions. In fact, one may wonder about the exact significance of 30S-mRNA affinities as measured in vitro. If a fast and essentially irreversible reaction would follow the binding of 30S subunits in such experiments, this would produce an infinitely high binding constant, but instead values of about 10^7 M^{-1} are usually found. We must therefore assume that the hypothetical fast reaction does not occur in simple binding experiments with purified compounds.

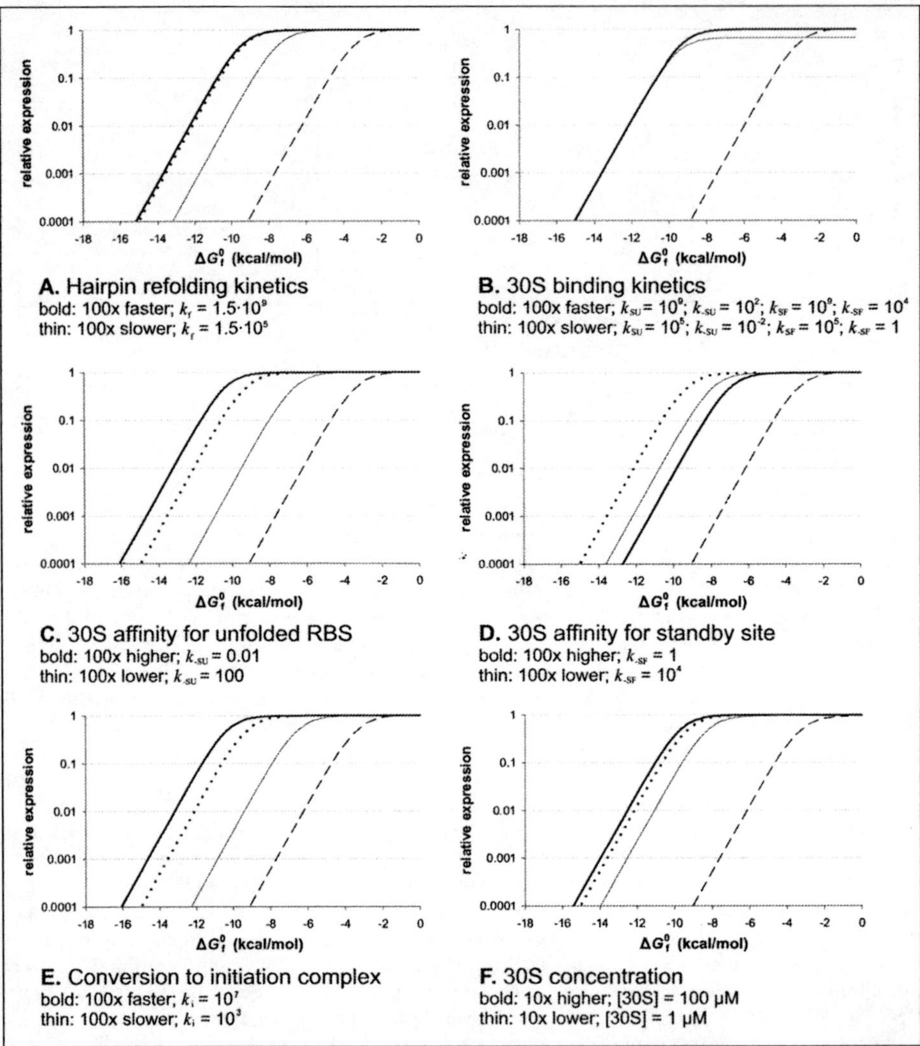

Figure 10. Effects of moderate changes in parameter values on predictions by the kinetic standby model of Figure 8B. The dotted and dashed curves correspond to the standard values in Table 1 with and without standby binding, respectively (compare Fig. 9). (A) Standby binding stimulates expression most when the RBS structure folds fast. (B) Binding kinetics of the 30S subunit are irrelevant because there is always a 30S subunit on standby. (C) Simulated effect of Shine-Dalgarno mutations: increasing the binding strength beyond a certain point no longer stimulates translation. (D) Standby binding should be neither too weak nor too strong to have an optimal effect. (E) Fast and irreversible conversion of the 30S·U complex to the 30S initiation complex is important. (F) Predictions remain valid with different concentrations of free 30S subunits.

We know of only one report where ribosome binding parameters were derived from a complete in vitro translation reaction.[111] The authors estimated the Michaelis constant for the binding of 30S subunits to the R17 coat-gene start (identical to MS2) as 0.07 µM. This corresponds to an apparent binding constant of $1.4 \cdot 10^7$ M^{-1} if one assumes, as they did, that the step following the association is negligibly slow. While at first sight this value seems in good agreement with the in vitro binding

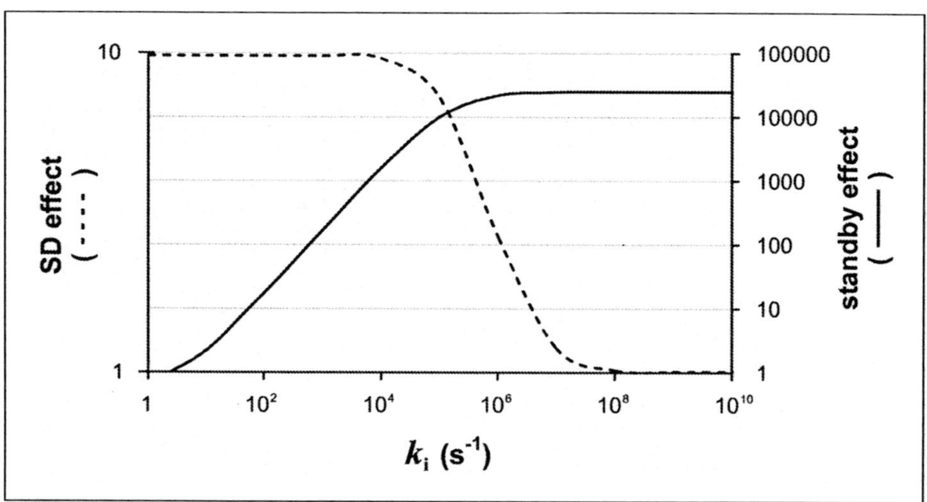

Figure 11. Predicted stimulation of expression by increased SD complementarity and by standby binding, as a function of the rate of formation of the initiation complex (k_i). SD effect = {expression with strong SD (k_{-SU} = 1 s^{-1})} / {expression with weaker SD (k_{-SU} = 10 s^{-1})}; standby effect = {expression with standby} / {expression without standby}. Other parameters used are as in Table 1.

data mentioned above, it was in fact not corrected for the presence of the hairpin at the RBS, implying a serious underestimation of the affinity for the unfolded RBS. It can be calculated that the binding constant to the unfolded RBS (K_{SU}) relates to the apparent binding constant to the structured one ($K_{SU,app}$) and the equilibrium constant of hairpin formation (K_F) as $K_{SU} = K_{SU,app} \cdot (K_F + 1)$. Using the calculated stability of the coat hairpin, this gives a value for K_{SU} of $8 \cdot 10^{14}$ M^{-1}, far higher than the value of 10^7 M^{-1} found in in vitro binding experiments with poly(U) and even higher than the value of 10^{12} M^{-1} derived from our in vivo data. We think that this extreme value may reflect on the one hand the existence of standby binding and on the other hand an overestimation due to a nonnegligible forward rate (k_i), in agreement with our kinetic model. Although inadequate as true binding constants, these high numbers vividly illustrate the efficiency with which ribosomes are recruited for translation.

We have previously found that the strength of the SD-interaction is one of the factors determining the capability of the ribosome to compete against mRNA structure.[26] The fact that expression can thus be modulated by varying the strength of the SD interaction must result from altering the dissociation rate of the 30S subunit, implying that this rate ($k_{-SU} \cdot [30S \cdot U]$) is significant compared to the irreversible entry of the ribosome into translation ($k_i \cdot [30S \cdot U]$). So while the forward rate must be fast for standby binding to be effective, it must be relatively slow for the SD interaction to modulate expression. While no experimental data are available to assess the true value of this parameter, Figure 11 provides a suggestion for how this paradox may be solved. Intrinsic to the reaction scheme of Figure 8B, there exists a value for k_i where expression is almost maximally sensitive to both standby binding and the strength of the SD-interaction. In our numerical example, based upon the coat gene RBS of MS2, this point is at $k_i = 10^5$ s^{-1}.

Conclusion: Target Location in Translation

The problem of target location in molecular biology has fascinated biologists, chemists and physicists alike for decades. Theories on the various routes that a molecule or complex can take to reach its target in a fast and efficient manner, despite the overcrowded nature of the cytoplasm, are manifold.[79,88] Applied to transcription regulators and restriction enzymes in particular, they have helped in understanding quantitative aspects of their binding kinetics that would otherwise have been incomprehensible. Remarkably little of these developments has permeated into the world of translation. Presumably the complexity of the initiation process, especially in eukaryotes, has kept most people from asking this type of quantitative questions on how ribosomes reach their target in and among the tangled clews of mRNA in the cell. Moreover, it has been argued that due to precisely this crowded nature of the cytoplasm standard thermodynamic and kinetic theory, based on ideal solutions, may break down in vivo.[42,124]

While the values for kinetic and thermodynamic constants and concentrations in vivo that we have used may be imprecise, we believe that this "naive" modeling of the reactions involved is nonetheless a useful exercise. Ten thousand-fold discrepancies between theoretical predictions and observed reality are a strong stimulus to further thinking. In the end, this will hopefully lead to the design of targeted experiments to test possible mechanisms, preferably in vivo. Let us not forget that in a world of genomes and proteomes, it is the dynamics of their interactions that defines life.

Acknowledgements

We thank Dr Måns Ehrenberg and his coworkers for stimulating discussions and Dr Kerstin Andersson for her unpublished results. M.S. was paid by the Netherlands Organization for Scientific Research, NWO.

References

1. de Smit MH. Translational control by mRNA structure in eubacteria: molecular biology and physical chemistry. In: Simons RW, Grunberg-Manago M, eds. RNA structure and Function. Cold Spring Harbor: Cold Spring Harbor Laboratory Press, 1998:495-540.
2. Draper DE, Gluick TC, Schlax PJ. Pseudoknots, RNA folding, and translational regulation. In: Simons RW, Grunberg-Manago M, eds. RNA structure and Function. Cold Spring Harbor: Cold Spring Harbor Laboratory Press, 1998:415-436.
3. Kozak M. Initiation of translation in prokaryotes and eukaryotes. Gene 1999; 234:187-208.
4. Gualerzi CO, Brandi L, Caserta E et al. Translation initiation in bacteria. In: Garrett R, Douthwaite SR, Liljas A et al, eds. The Ribosome: Structure, Function, Antibiotics, and Cellular Interactions. Washington: ASM Press, 2000:477-494.
5. Hartz D, McPheeters DS, Traut R et al. Extension inhibition analysis of translation initiation complexes. Methods Enzymol 1988; 164:419-425.
6. Risuleo G, Gualerzi C, Pon C. Specificity and properties of the destabilization, induced by initiation factor IF-3, of ternary complexes of the 30-S ribosomal subunit, aminoacyl-tRNA and polynucleotides. Eur J Biochem 1976; 67:603-613.
7. Hartz D, Binkley J, Hollingsworth T et al. Domains of initiator tRNA and initiation codon crucial for initiator tRNA selection by *Escherichia coli* IF3. Genes Dev 1990; 4:1790-1800.
8. Oppenheim DS, Yanofsky C. Translational coupling during expression of the tryptophan operon of *Escherichia coli*. Genetics 1980; 95:785-795.
9. Berkhout B, Schmidt BF, van Strien A et al. Lysis gene of bacteriophage MS2 is activated by translation termination at the overlapping coat gene. J Mol Biol 1987; 195:517-524.
10. Schümperli D, McKenney K, Sobieski DA et al. Translational coupling at an intercistronic boundary of the *Escherichia coli* galactose operon. Cell 1982; 30:865-871.
11. Ivey-Hoyle M, Steege DA. Mutational analysis of an inherently defective translation initiation site. J Mol Biol 1992; 224:1039-1054.
12. Spanjaard RA, van Duin J. Translational reinitiation in the presence and absence of a Shine and Dalgarno sequence. Nucleic Acids Res 1989; 17:5501-5507.
13. Schmidt BF, Berkhout B, Overbeek G et al. Determination of the RNA secondary structure that regulates lysis gene expression in bacteriophage MS2. J Mol Biol 1987; 195:505-516.
14. Adhin MR, van Duin J. Scanning model for translational reinitiation in eubacteria. J Mol Biol 1990; 213:811-818.
15. Kozak M. Selection of initiation sites by eucaryotic ribosomes: effect of inserting AUG triplets upstream from the coding sequence for preproinsulin. Nucleic Acids Res 1984; 12:3873-3893.
16. Pavlov MY, Freistroffer DV, MacDougall J et al. Fast recycling of *Escherichia coli* ribosomes requires both ribosome recycling factor (RRF) and release factor RF3. EMBO J 1997; 16:4134-4141.
17. Subramanian AR. Structure and functions of ribosomal protein S1. Prog Nucleic Acid Res Mol Biol 1983; 28:101-142.
18. Moll I, Resch A, Bläsi U. Discrimination of 5'-terminal start codons by translation initiation factor 3 is mediated by ribosomal protein S1. FEBS Lett 1998; 436:213-217.
19. Berkhout B, Kastelein RA, van Duin J. Translational interference at overlapping reading frames in prokaryotic messenger RNA. Gene 1985; 37:171-179.
20. Kozak M. Structural features in eukaryotic mRNAs that modulate the initiation of translation. J Biol Chem 1991; 266:19867-19870.
21. Jackson, RJ. A comparative view of initiation site selection mechanisms. In: Hershey JWB, Mathews MB, Sonenberg N, eds. Translational control. Cold Spring Harbor: Cold Spring Harbor Laboratory Press, 1996:71-112.
22. Draper D. Mechanisms of translational initiation and repression in prokaryotes. In: Nierhaus KH, ed. The Translational Apparatus. New York: Plenum Press, 1993:197-207.

23. Ganoza MC, Kofoid EC, Marlière P et al. Potential secondary structure at translation-initiation sites. Nucleic Acids Res 1987; 15:345-360.
24. de Smit MH, van Duin J. Control of prokaryotic translational initiation by mRNA secondary structure. Prog Nucleic Acid Res Mol Biol 1990; 38:1-35.
25. de Smit MH, van Duin J. Secondary structure of the ribosome binding site determines translational efficiency: a quantitative analysis. Proc Natl Acad Sci USA 1990; 87:7668-7672.
26. de Smit MH, van Duin J. Translational initiation on structured messengers. Another role for the Shine-Dalgarno interaction. J Mol Biol 1994; 235:173-184.
27. Tacker M, Fontana W, Stadler PF et al. Statistics of RNA melting kinetics. Eur Biophys J 1994; 23:29-38.
28. Gultyaev AP, van Batenburg FHD, Pleij CWA. The influence of a metastable structure in plasmid primer RNA on antisense RNA binding kinetics. Nucleic Acids Res 1995; 23:3718-3725.
29. Hotham-Iglewski B, Franklin RM. Replication of bacteriophage ribonucleic acid: Alterations in polyribosome patterns and association of double-stranded RNA with polyribosomes in *Escherichia coli* infected with bacteriophage R17. Proc Natl Acad Sci USA 1967; 56:743-749.
30. Koontz SW, Jakubowski H, Goldman E. Control of RNA and protein synthesis by the concentration of Trp-tRNA^{Trp} in *Escherichia coli* infected with bacteriophage MS2. J Mol Biol 1983; 168:747-763.
31. Truden JL, Franklin RM. Polysomal localization of R17 bacteriophage-specific protein synthesis. J Virol 1972; 9:75-84.
32. Phillips LA, Truden JL, Iglewski WJ et al. Replication of bacteriophage ribonucleic acid: Alterations in polyribosome patterns in *Escherichia coli* infected with amber mutants of bacteriophage R17. Virology 1969; 39:781-790.
33. Beremand MN, Blumenthal T. Overlapping genes in RNA phage: a new protein implicated in lysis. Cell 1979; 18:257-266.
34. Bremer H, Dennis PP. Modulation of chemical composition and other parameters of the cell by growth rate. In: Neidhardt FC, ed. Cellular and Molecular Biology of *Escherichia coli* and *Salmonella typhimurium*. Washington DC: Am. Soc. Microbiol., 1987:1527-1542.
35. Kennell D, Riezman H. Transcription and translation initiation frequencies of the *Escherichia coli lac* operon. J Mol Biol 1977; 114:1-21.
36. Petersen C. Translation and mRNA stability in bacteria: a complex relationship. In: Belasco J, Brawerman G, eds. Control of Messenger RNA Stability. San Diego: Academic Press, 1993:117-145.
37. Tomsic J, Vitali LA, Daviter T et al. Late events of translation initiation in bacteria: a kinetic analysis. EMBO J 2000; 19:2127-2136.
38. Grunberg-Manago M. Initiation of protein synthesis as seen in 1979. In: Chambliss G, ed. Ribosomes: Structure, Function and Genetics. Baltimore: University Park Press, 1979:445-477.
39. Gualerzi CO, Pon CL. Initiation of mRNA translation in prokaryotes. Biochemistry 1990; 29:5881-5889.
40. de Smit MH. Regulation of translation by mRNA structure. PhD Thesis. Leiden: Leiden University 1994.
41. Goodsell DS. Inside a living cell. Trends Biochem Sci 1991; 16:203-206.
42. Zimmerman SB, Minton, AP. Macromolecular crowding: Biochemical, biophysical, and physiological consequences. Annu Rev Biophys Biomol Struct 1993; 22:27-65.
43. Shinedling S, Gayle M, Pribnow D et al. Mutations affecting translation of the bacteriophage T4 *rIIB* gene cloned in *Escherichia coli*. Mol Gen Genet 1987; 207:224-232.
44. Sacerdot C, Caillet J, Graffe M et al. The *Escherichia coli* threonyl-tRNA synthetase gene contains a split ribosomal binding site interrupted by a hairpin structure that is essential for autoregulation. Mol Microbiol 1998; 29:1077-1090.
45. Wilson IW, Praszkier J, Pittard AJ. Molecular analysis of RNAI control of *repB* translation in IncB plasmids. J Bacteriol 1994; 176:6497-6508.
46. Malmgren C, Engdahl HM, Romby P et al. An antisense/target RNA duplex or a strong intramolecular RNA structure 5' of a translation initiation signal blocks ribosome binding: The case of plasmid R1. RNA 1996; 2:1022-1032.
47. Wilson IW, Praszkier J, Pittard AJ. Mutations affecting pseudoknot control of the replication of B group plasmids. J Bacteriol 1993; 175:6476-6483.
48. Asano K, Mizobuchi K. An RNA pseudoknot as the molecular switch for translation of the *repZ* gene encoding the replication initiator of Inclα plasmid ColIb-P9. J Biol Chem 1998; 273:11815-11825.
49. Ravnum S, Andersson DI. An adenosyl-cobalamin (coenzyme-B12)-repressed translational enhancer in the *cob* mRNA of Salmonella typhimurium. Mol Microbiol 2001; 39:1585-1594.
50. Hartz D, McPheeters DS, Gold L. Influence of mRNA determinants on translation initiation in *Escherichia coli*. J Mol Biol 1991; 218:83-97.
51. Nivinskas R, Malys N, Klausa V et al. Post-transcriptional control of bacteriophage T4 gene 25 expression: mRNA secondary structure that enhances translational initiation. J Mol Biol 1999; 288:291-304.
52. Ringquist S, MacDonald M, Gibson T et al. Nature of the ribosomal mRNA track: analysis of ribosome-binding sites containing different sequences and secondary structures. Biochemistry 1993; 32:10254-10262.
53. Murakawa GJ, Nierlich DP. Mapping the *lacZ* ribosome binding site by RNA footprinting. Biochemistry 1989; 28:8067-8072.
54. Montessano-Roditis L, Glitz DG. Tracing the path of messenger RNA on the *Escherichia coli* small ribosomal subunit. J Biol Chem 1994; 269:6458-6470.
55. Bhangu R, Wollenzien P. The mRNA track in the *Escherichia coli* ribosome for mRNAs of different sequences. Biochemistry 1992; 31:5937-5944.
56. Gabashvili IS, Agrawal RK, Grassucci R et al. Structure and structural variations of the *Escherichia coli* 30 S ribosomal subunit as revealed by three-dimensional cryo-electron microscopy. J Mol Biol 1999; 286:1285-1291.

57. Schluenzen F, Tocilj A, Zarivach R et al. Structure of functionally activated small ribosomal subunit at 3.3 Å resolution. Cell 2000; 102:615–623.
58. Ogle JM, Brodersen DE, Clemons WM et al. Recognition of cognate transfer RNA by the 30S ribosomal subunit. Science 2001; 292:897-902.
59. Yusupova GZ, Yusupov MM, Cate JDH et al. The path of messenger RNA through the ribosome. Cell 2001; 106:233-241.
60. Groeneveld H. Secondary Structure of Bacteriophage MS2 RNA: Translational Control by Kinetics of RNA Folding. PhD thesis. Leiden: Leiden University, 1997.
61. Beekwilder J. Secondary structure of the RNA genome of bacteriophage Qβ. PhD thesis. Leiden: Leiden University, 1996.
62. Olsthoorn RCL. Structure and evolution of RNA phages. PhD thesis, Leiden: Leiden University, 1996.
63. Skripkin EA, Jacobson AB. A two-dimensional model at the nucleotide level for the central hairpin of coliphage Qβ RNA. J Mol Biol 1993; 233:245-260.
64. Hüttenhofer A, Noller HF. Footprinting mRNA-ribosome complexes with chemical probes. EMBO J 1994; 13:3892-3901.
65. Canonaco MA, Gualerzi CO, Pon CL. Alternative occupancy of a dual ribosome binding site by mRNA affected by translation initiation factors. Eur J Biochem 1989; 182:501-506.
66. van Dieijen G, Zipori P, van Prooijen W et al. Involvement of ribosomal protein S1 in the assembly of the initiation complex. Eur J Biochem 1978; 90:571-580.
67. van Duin J, Overbeek GP, Backendorf C. Functional recognition of phage RNA by 30-S ribosomal subunits in the absence of initiator tRNA. Eur J Biochem 1980; 110:593-597.
68. Szer W, Leffler S. Interaction of *Escherichia coli* 30S ribosomal subunits with MS2 phage RNA in the absence of initiation factors. Proc Natl Acad Sci USA 1974; 71:3611-3615.
69. Boni IV, Artamonova VS, Tzareva NV et al. Noncanonical mechanism for translational control in bacteria: synthesis of ribosomal protein S1. EMBO J 2001; 20:4222-4232.
70. Balakin A, Skripkin E, Shatsky I et al. Transition of the mRNA sequence downstream from the initiation codon into a single-stranded conformation is strongly promoted by binding of the initiator tRNA. Biochim Biophys Acta 1990; 1050:119-123.
71. Min Jou W, Haegeman G, Ysebaert M et al. Nucleotide sequence of the gene coding for the bacteriophage MS2 coat protein. Nature 1972; 237:82-88.
72. Berkhout B, van Duin J. Mechanism of translational coupling between coat protein and replicase genes of RNA bacteriophage MS2. Nucleic Acids Res 1985; 13:6955-6967.
73. van Himbergen J, van Geffen B, van Duin J. Translational control by a long range RNA-RNA interaction; a basepair substitution analysis. Nucleic Acids Res 1993; 21:1713-1717.
74. Licis N, van Duin J, Balklava Z et al. Long-range translational coupling in single-stranded RNA bacteriophages: an evolutionary analysis. Nucleic Acids Res 1998; 26:3242-3246.
75. Berkhout B, van Duin J. Mechanism of translational coupling between coat protein and replicase genes of RNA bacteriophage MS2. Nucleic Acids Res 1985; 13:6955-6967.
76. Berkhout B. Translational Control Mechanisms. PhD thesis. Leiden: Leiden University, 1986.
77. Papanicolaou C, Gouy M, Ninio J. An energy model that predicts the correct folding of both the tRNA and the 5S RNA molecules. Nucleic Acids Res 1984; 12:31-44.
78. Peritz AE, Kierzek R, Sugimoto N et al. Thermodynamic study of internal loops in oligoribonucleotides: symmetric loops are more stable than asymmetric loops. Biochemistry 1991; 30:6428-6436.
79. Shimamoto N. One-dimensional diffusion of proteins along DNA. Its biological and chemical significance revealed by single-molecule measurements. J Biol Chem 1999; 274:15293-15296.
80. Matteson RJ, Biswas SJ, Steege DA. Distinctive patterns of translational reinitiation in the lac repressor mRNA: bridging of long distances by out-of-frame translation and RNA secondary structure, effects of primary sequence. Nucleic Acids Res 1991; 19:3499-3506.
81. Fickert R, Müller-Hill B. How Lac repressor finds *lac* operator in vivo. J Mol Biol 1992; 226:59-68.
82. Ruusala T, Crothers DM. Sliding and intermolecular transfer of the lac repressor: kinetic perturbation of a reaction intermediate by a distant DNA sequence. Proc Natl Acad Sci USA 1992; 89:4903-4907.
83. La Teana A, Gualerzi CO, Brimacombe R. From stand-by to decoding site. Adjustment of the mRNA on the 30S ribosomal subunit under the influence of the initiation factors. RNA 1995; 1:772-782.
84. Park CS, Hillel Z, Wu CW. Molecular mechanism of promoter selection in gene transcription. I. Development of a rapid mixing-photocrosslinking technique to study the kinetics of *Escherichia coli* RNA polymerase binding to T7 DNA. J Biol Chem 1982; 257:6944-6949.
85. Singer P, Wu CW. Promoter search by *Escherichia coli* RNA polymerase on a circular DNA template. J Biol Chem 1987; 262:14178-14189.
86. de Smit MH, van Duin J. Translational initiation at the coat-protein gene of phage MS2: native upstream RNA relieves inhibition by local secondary structure. Mol Microbiol 1993; 9:1079-1088.
87. Ruusala T, Crothers DM. Sliding and intermolecular transfer of the lac repressor: kinetic perturbation of a reaction intermediate by a distant DNA sequence. Proc Natl Acad Sci USA 1992; 89:4903-4907.
88. Stanford NP, Szczelkun MD, Marko JF, Halford SE. One- and three-dimensional pathways for proteins to reach specific DNA sites. EMBO J 2000; 19:6546-6557.
89. Novoa I, Carrasco L. Cleavage of eukaryotic translation initiation factor 4G by exogenously added hybrid proteins containing poliovirus 2Apro in HeLa cells: effects on gene expression. Mol Cell Biol 1999; 19:2445-2454.
90. Wells SE, Hillner PE, Vale RD et al. Circularization of mRNA by eukaryotic translation initiation factors. Mol Cell 1998; 2:135-140.
91. Preiss T, Hentze MW. From factors to mechanisms: translation and translational control in eukaryotes. Curr Opin Genet Dev 1999; 9:515-521.

92. Gerdes K, Poulsen LK, Thisted T et al. The *hok* killer gene family in gram-negative bacteria. New Biol 1990; 2:946-956.

93. Gerdes K, Gultyaev AP, Franch T et al. Antisense RNA-regulated programmed cell death. Annu Rev Genet 1997; 31:1-31.

94. Gultyaev AP, Franch T, Gerdes K. Programmed cell death by *hok/sok* of plasmid R1: coupled nucleotide covariations reveal a phylogenetically conserved folding pathway in the *hok* family of mRNAs. J Mol Biol 1997; 273:26-37.

95. Franch T, Gultyaev AP, Gerdes K. Programmed cell death by *hok/sok* of plasmid R1: processing at the *hok* mRNA 3'-end triggers structural rearrangements that allow translation and antisense RNA binding. J Mol Biol 1997; 273:38-51.

96. Draper D. Translational regulation of ribosomal proteins in *Escherichia coli*. In: Ilan J, ed. Translational Regulation of Gene Expression. New York: Plenum, 1987:1-26.

97. Brunel C, Romby P, Sacerdot C et al. Stabilised secondary structure at a ribosomal binding site enhances translational repression in *E. coli*. J Mol Biol 1995; 253:277-290.

98. Ehresmann C, Philippe C, Westhof E et al. A pseudoknot is required for efficient translational initiation and regulation of the *Escherichia coli rpsO* gene coding for ribosomal protein S15. Biochem Cell Biol 1995; 73:1131-1140.

99. Nogueira T, de Smit M, Graffe M et al. The relationship between translational control and mRNA degradation for the *Escherichia coli* threonyl-tRNA synthetase gene. J Mol Biol 2001; 310:709-722.

100. Kolakofsky D, Weissmann C. Qβ replicase as repressor of Qβ RNA-directed protein synthesis. Biochim Biophys Acta 1971; 246:596-599.

101. Kolakofsky D, Weissmann C. Possible mechanism for transition of viral RNA from polysome to replication complex. Nat New Biol 1971; 231:42-46.

102. Poot RA, Tsareva NV, Boni IV et al. RNA folding kinetics regulates translation of phage MS2 maturation gene. Proc Natl Acad Sci USA 1997; 94:10110-10115.

103. van Meerten D, Girard G, van Duin J. Translational control by delayed RNA folding: identification of the kinetic trap. RNA 2001; 7:483-494.

104. Goelz S, Steitz JA. *Escherichia coli* ribosomal protein S1 recognizes two sites in bacteriophage Qbeta RNA. J Biol Chem 1977; 252:5177-5179.

105. Boni IV, Isaeva DM, Musychenko ML et al. Ribosome-messenger recognition: mRNA target sites for ribosomal protein S1. Nucleic Acids Res 1991; 19:155-162.

106. de Smit MH, van Duin J. Control of translation by mRNA secondary structure in *Escherichia coli*. A quantitative analysis of literature data. J Mol Biol 1994; 244:144-150.

107. Olsthoorn RC, Zoog S, van Duin J. Coevolution of RNA helix stability and Shine-Dalgarno complementarity in a translational start region. Mol Microbiol 1995; 15:333-339.

108. Gualerzi C, Risuleo G, Pon CL. Initial rate kinetic analysis of the mechanism of initiation complex formation and the role of initiation factor IF-3. Biochemistry 1977; 16:1684-1689.

109. Katunin VI, Semenkov YP, Makhno VI et al. Comparative study of the interaction of polyuridylic acid with 30S subunits and 70S ribosomes of *Escherichia coli*. Nucleic Acids Res 1980; 8:403-421.

110. Pon CL, Gualerzi CO. Mechanism of protein biosynthesis in prokaryotic cells: Effect of initiation factor IF1 on the initial rate of 30S initiation complex formation. FEBS Lett 1984; 175:203-207.

111. Ellis S, Conway TW. Initial velocity kinetic analysis of 30S initiation complex formation in an in vitro translation system derived from *Escherichia coli*. J Biol Chem 1984; 259:7607-7614.

112. Calogero RA, Pon CL, Canonaco MA et al. Selection of the mRNA translation initiation region by *Escherichia coli* ribosomes. Proc Natl Acad Sci USA 1988; 85:6427-6431.

113. Czworkowski J, Odom OW, Hardesty B. Fluorescence study of the topology of messenger RNA bound to the 30S ribosomal subunit of *Escherichia coli*. Biochemistry 1991; 30:4821-4830.

114. Mathews DH, Sabina J, Zuker M et al. Expanded Sequence Dependence of Thermodynamic Parameters Improves Prediction of RNA Secondary Structure. J Mol Biol 1999; 288:911-940.

115. Olsthoorn RC, Licis N, van Duin J. Leeway and constraints in the forced evolution of a regulatory RNA helix. EMBO J 1994; 13:2660-2668.

116. Olsthoorn RC, van Duin J. Random removal of inserts from an RNA genome: selection against single-stranded RNA. J Virol 1996; 70:729-736.

117. Olsthoorn RC, van Duin J. Evolutionary reconstruction of a hairpin deleted from the genome of an RNA virus. Proc Natl Acad Sci USA 1996; 93:12256-12261.

118. Klovins J, van Duin J, Olsthoorn RC. Rescue of the RNA phage genome from RNase III cleavage. Nucleic Acids Res 1997; 25:4201-4208.

119. Klovins J, Tsareva NA, de Smit MH et al. Rapid evolution of translational control mechanisms in RNA genomes. J Mol Biol 1997; 265:372-384.

120. Licis N, van Duin J, Balklava Z et al. Long-range translational coupling in single-stranded RNA bacteriophages: an evolutionary analysis. Nucleic Acids Res 1998; 26:3242-3246.

121. Licis N, Balklava Z, Van Duin J. Forced retroevolution of an RNA bacteriophage. Virology 2000; 271:298-306.

122. van Meerten D, Olsthoorn RC, van Duin J et al. Peptide display on live MS2 phage: restrictions at the RNA genome level. J Gen Virol 2001; 82:1797-1805.

123. Wintermeyer W, Gualerzi C. Effect of *Escherichia coli* initiation factors on the kinetics of *N*-AcPhe-tRNA^{Phe} binding to 30S ribosomal subunits. A fluorescence stopped-flow study. Biochemistry 1983; 22:690-694.

124. Kuthan H. Self-organisation and orderly processes by individual protein complexes in the bacterial cell. Prog Biophys Mol Biol 2001; 75:1-17.

CHAPTER 20

Translational Elongation

Poul Nissen, Jens Nyborg and Brian F.C. Clark

Abstract

The elongation phase of protein biosynthesis adds one amino acid at a time to the growing polypeptide according to the sequence information contained in the mRNA. The process is catalyzed by elongation factors of which two are GTP/GDP binding proteins (G-proteins). Several crystal structures illustrating various functional states of these elongation factors have been obtained during the last decade. The growing knowledge of the elongation phase based on these structures is thus becoming increasingly detailed. Some of these details are discussed and possible mechanisms on how the large conformational change of elongation factor Tu is caused by switching between binding of GDP and GTP are postulated. One unexpected observation has been that some translation protein factors, which interact with the ribosomal A site, mimic the structure of tRNA ("macromolecular mimicry").

Introduction

Protein biosynthesis is an essential and central biological process in a living cell. Indeed, it is the final step of the transmission of genetic information in DNA into the production of active proteins useful to the cell. Also, it represents the most complex step of this informational flow as it links nucleic acid chemistry with polypeptide chemistry by an extremely sophisticated mechanism of RNA template-based poly-peptide synthesis bond formation. Thus, protein biosynthesis takes place on ribosome particles where mRNAs are literally translated into polypeptides aided by the concerted action of numerous external factors. By its own nature, the translation apparatus represents the linking role between nucleic acids and proteins. The ribosome is a large protein-RNA enzyme and the functionally associated adapters and factors include tRNAs and a range of protein translation factors which as it turns out may utilize a common L-shape in their interaction with the ribosome ("macromolecular mimicry"). Equally worth noting, the translation apparatus uses a GTPase coupled control mechanism which is quite likely the ancestor of the more general G-protein coupled signalling pathways. The ribosome itself is now being studied in molecular and structural terms at the atomic level. It is established that it is a ribozyme, and details of codon-anticodon recognition are now well understood. This represents the end-point of decades of biochemical studies of the ribosome with the purpose of understanding its structure, while at the same time opening up a new era of detailed functional and structural studies of various functional states of the translation apparatus. Apart from that, the elongation phase of protein synthesis of the ribosome is probably one of the most illuminated biological functions thanks to progress over the last decade involving structural determination of various components in the pathway and the clever design of biochemical experiments and interpretation of resulting biochemical data, partly by using these structures. The other stages of the protein biosynthetic pathway, initiation and termination, are not similarly advanced in our understanding, one reason being that three-dimensional structures of some of the critical factors have only recently been determined. Further, although initiation is rather simple in prokaryotes, it is a far more complex process in eukaryotes, being the major target for translational control.[1] The interaction of the ribosomes with the release factors involved in termination and the function of these factors have only recently been properly addressed.[2]

Translation Mechanisms, edited by Jacques Lapointe and Léa Brakier-Gingras.
©2003 Eurekah.com and Kluwer Academic / Plenum Publishers.

During the elongation phase, aminoacyl-tRNA (aa-tRNA) is brought to the ribosomal recognition site (or preA site) for decoding mRNA as a ternary complex (TC) composed of any elongator aa-tRNA, elongation factor Tu (EF-Tu) and guanosine triphosphate (GTP).[3] In eukaryotes and archaebacteria, the highly conserved equivalent of EF-Tu (systematically, now called EF1A) is called EF-1α (or now eEF1A and aEF1A). The functional cycle of EF-Tu is shown in Figure 1. The primary recognition on the ribosome is governed by base pairing between the exposed anticodon of tRNA in the ternary complex and the exposed mRNA codon at the A site of the small ribosomal subunit. Correct base pairing (as formulated by the genetic code) leads to the elongation cycle, whereas incorrect pairing (non-cognate pairing) leads to silent dissociation of the ternary complex from the ribosome.[4] EF-Tu:GTP binding to aa-tRNA protects the aminoacyl bond of the activated amino acid. Upon binding of a correct ternary complex to the preA site, GTP is hydrolyzed to GDP (guanosine diphosphate).[4] GTP hydrolysis leads to dissociation of the EF-Tu:GDP binary complex from the ribosome, probably because of the large conformational change of EF-Tu going from the GTP to the GDP form (see later section). Thereby aa-tRNA is released from its complex with EF-Tu and docks into the catalytic A site of the large ribosomal subunit. Further proof-reading occurs in the sense that incorrect aa-tRNA in the preA site is very likely to dissociate after GTP hydrolysis and EF-Tu:GDP dissociation.[5,6] Peptide bond formation occurs rapidly between peptidyl-tRNA in the P site and the aa-tRNA planted in the A site. After peptide bond formation, the elongation cycle must be reinitiated, which is catalyzed by the GTP-binding translocase, elongation factor G (EF-G (now EF2), EF-2 in eukaryotes (eEF2) and archaebacteria (aEF2)). Thus, EF-G:GTP binding and GTP hydrolysis on the pre-translocated ribosome lead to the concerted translocation of the elongated peptidyl-tRNA from the A site to the P site, of the deacylated tRNA from the P site to the E site and of mRNA by one codon in the 3'-direction. The inactive EF-Tu:GDP is catalytically reactivated to the active GTP-bound form by the nucleotide exchange factor, Elongation factor Ts (EF-Ts (now EF1B), EF-1β in eukaryotes (eEF1B) and archaebacteria (aEF1B)). EF-Ts binding catalyzes the dissociation of GDP from EF-Tu and enables it to bind GTP, which is kept at high levels in the cell (~ 0.4 mM).

EF-Tu and EF-G Are G-Proteins with Switch Regions

Elongation factor EF-G and later EF-Tu were the first GTPases (G-proteins) discovered. Also, EF-Tu was the first G-protein to have its three-dimensional structure determined.[7-11] The family of G-proteins in translation also includes initiation factor IF2 and release factor RF3. Other families of G-proteins have later been recognized.[12,13] One such family includes the heterotrimeric G-proteins involved in the transduction of external signals (hormones, neurotransmitters, light, odors etc.) into intracellular chemical signals like cyclic AMP. Another family encompasses single-domain G-proteins such as Ras p21, a proto-oncogene product which is involved in the control of cell proliferation, and Ran which controls nuclear import and export. Although the basic structure of the G-domain is the same in all G-proteins, considerable variations do exist.[14]

G-proteins are sometimes referred to as molecular switches which are "on" in their GTP form and "off" in their GDP form. The G-proteins are thus switched on by interaction with nucleotide exchange factors that catalyze the binding of GTP. They are switched off by GTP hydrolysis, either by an intrinsic GTPase activity or by stimulation of the GTPase activity by other proteins (GTPase activating proteins, GAP). A given protein is easily recognized as a G-protein by consensus sequence motifs, all of which are found in loops of the G-domain that bind the GDP/GTP nucleotide together with Mg^{2+}.[14] These are described in the later section on the structure of EF-Tu domains (see also Figs. 1 and 2). The region between the first helix and the second β-strand of a G-protein is called the switch I region because its structure depends on the identity of the bound nucleotide. It is also termed the "effector loop" originating from studies of ras-p21 where this region was observed to be involved in interaction with downstream effectors. It varies substantially among the G-protein families as it is only a short loop in Ras p21 whereas in the heterotrimeric G-proteins a helical subdomain is inserted. The region encompassing the second α-helix and surrounding loops of the G-domain core structure is called the switch II region.

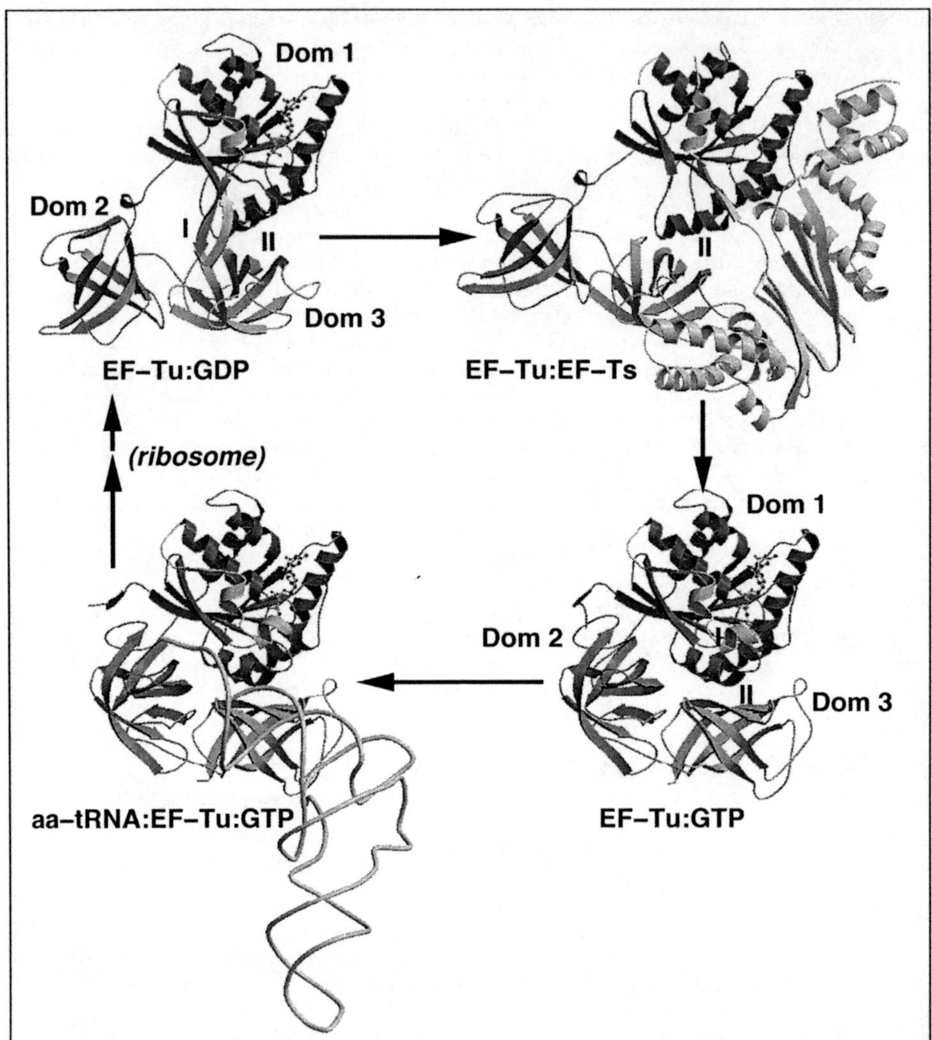

Figure 1. The elongation cycle of EF-Tu. The functional cycle of EF-Tu is illustrated by the known crystal structures.[18,23,28,30] EF-Tu domains 1 through 3 and switch regions I and II have been indicated by labels on EF-Tu:GDP and EF-Tu:GTP. Nucleotides are shown in ball-and-stick models. In the structure of EF-Tu:GDP the β-hairpin of the switch I region is seen in the centre. Just behind it is helix B of the switch II region. In the EF-Tu:EF-Ts structure, where EF-Ts is shown in grey, the switch I region is disordered and partly replaced by the C-terminal extension of EF-Ts (in *E.coli*). The pseudo two-fold symmetry of the central domain of EF-Ts is easily seen in this orientation. In the much more compact structure of EF-Tu:GTP the switch I region has an α-helical structure and the orientation of helix B of the switch II region has changed. The structure of the ternary complex of EF-Tu has tRNA in grey. This and the remaining figures have been produced using the programme MOLSCRIPT.[81]

Historical Outline of Elongation Factor Structure Determinations

The substantial progress in the structural biology of elongation factors is the outcome of a fruitful collaboration and competition among several laboratories all over the world, of which our laboratory in Aarhus has been an important contributor since 1976.

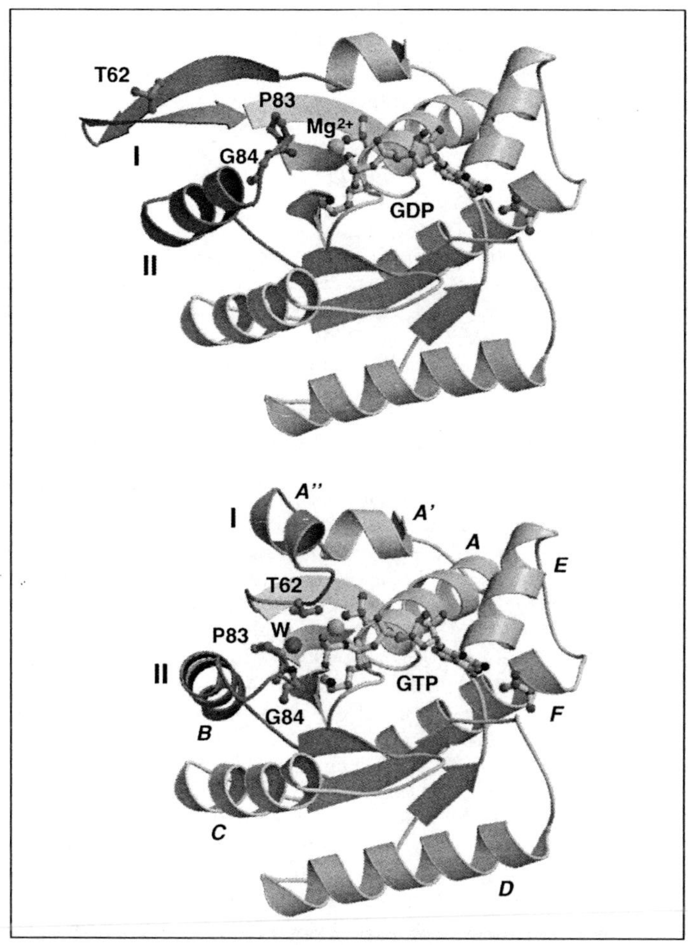

Figure 2. The structures of domain 1 of EF-Tu in the GDP and GTP forms. Most of the domain is shown in grey, but switch I is in medium grey and switch II is in dark grey. Nucleotides and amino acid residues important for the interaction with nucleotides are shown in ball-and-stick models. The Mg^{2+}-ion is shown as a sphere, as well as the proposed catalytic water molecule (W) in the GTP form. Helices are labelled A-F. Amino acid residues important for the switch mechanism are labelled. Note that the phosphates are bound by elements of the N-terminal half of the domain while the C-terminal half binds the guanine base.

The first structural details of a GTP/GDP binding domain (G-domain) were determined by standard protein crystallographic methods in 1985.[7,8] This structure, as the largest of the three domains of *E. coli* EF-Tu, ca. 200 out of 393 amino acid (although lacking a small surface peptide of 14 amino acids in the switch I region due to proteolytic cleavage), and in complex with GDP, became the model for all GDP/GTP binding proteins for several years. From these early observations on the EF-Tu G-domain structure and their comparison with the sequence of the single-domain Ras p21 protein, the G-protein motifs were established[15] with tremendous impact for the later emerging field of G-protein coupled signalling. After several years of struggling with the tracing of the entire chain of this nicked form of EF-Tu:GDP,[9,10] the refined three-domain structure at 2.6 Å resolution was published in 1992.[11] The role as a model for all G-proteins was later supported, and largely overtaken, by the later determined structure of Ras p21 in complex with a GTP analogue, which was published in 1989.[16] Before then, the EF-Tu G-domain was used quite successfully as a model for Ras p21[17] since the disposition of the secondary structural elements had been predicted correctly to be the same. More recently, the intact EF-Tu:GDP structures from *E. coli* and *Thermus aquaticus* were presented from three different crystal forms comprising a total of seven crystallographically independent models.[18,19] These structures finally confirmed that the nicked EF-Tu:GDP structure had indeed also represented the structure of the intact form. Furthermore, they completed the structure of the switch I region in the GDP-bound state. One of these crystal forms was later refined at 2.0 Å resolution.[20] Yet another version of the GDP-bound form came from the structure of bovine mitochondrial EF-Tu which has been determined at 1.9 Å resolution. It is practically identical to the eubacterial EF-Tu:GDP structure.[21]

The structure of the nicked EF-Tu:GDP form was also used in a Molecular Replacement procedure to determine the first structure of the intact GTP form of EF-Tu from *Thermus thermophilus*.[22] In parallel, the same method was used to determine the structure of EF-Tu:GTP from *T. aquaticus*.[23] In both cases, the use of a slowly hydrolyzable GTP-analogue, GDPNP, was critical for the purification and crystallization of EF-Tu in the GTP-bound state, as it had also been successfully used earlier in the determination of the GTP-bound conformation of Ras p21.[16] The EF-Tu:GDPNP structure revealed a huge conformational change in EF-Tu upon GTP binding. This was a nice surprise to the scientific community and it re-emphasized the importance of continued studies of elongation factor complexes.

The structures of nucleotide-free *T. thermophilus* EF-G[24] and of EF-G:GDP[25,26] has been determined. These structures are very similar and they reveal that the structural relationship of the G-domain and domain 2 in EF-G:GDP is surprisingly similar to that found in EF-Tu:GDPNP.[24] Despite several attempts since then, it has not yet been possible to determine the structure of EF-G:GDPNP.

After many years of experimentation in several laboratories, crystals were obtained of a ternary complex of yeast Phe-tRNA[Phe] and *T. aquaticus* EF-Tu:GDPNP.[27] The successful crystallization of a ternary complex was in fact surprising, given the fact that the aminoacyl bond is very unstable towards spontaneous hydrolysis and yet a prerequisite for the complex formation. The structure was determined soon thereafter by molecular replacement, and the refined structure was published in 1995.[28] It revealed yet a new range of completely unanticipated features in the structure and function of translation factor complexes, most notably the concept of "macromolecular mimicry" when comparing the structure of the ternary complex of EF-Tu with that of EF-G:GDP. Later, the crystal structure of *E. coli* Cys-tRNA[Cys] in complex with *T. aquaticus* EF-Tu:GDPNP[29] was presented which sustained the formulation of general rules and principles arising from the first ternary complex structure.

Excluding the ribosome-associated form(s) of EF-Tu, the final complex in the EF-Tu cycle is the EF-Tu:EF-Ts complex, which was reported as a 2.5 Å resolution structure for the *E. coli* proteins in 1996.[30] The structure revealed an elegant picture of the effective dissociation of GDP from this nucleotide exchange complex. A similar structure was later determined for the *T. thermophilus* proteins.[31] Recently, the NMR structure of a fragment of the eukaryotic equivalent of EF-Ts, human EF-1β, has also been presented.[32] Shortly after this, the complex of yeast EF-1α and the same fragment of yeast EF-1β was determined,[33] followed by structures of intermediates in the nucleotide exchange mechanism.[34] Surprisingly, the overall interactions between the elongation factors and their nucleotide exchange factors are very different in prokaryotes and in eukaryotes.

Finally, a low resolution image of the ternary complex associated with the ribosome was obtained in 1997 by electron microscopy in vitreous ice followed by image reconstruction.[35] At a resolution of approximately 20 Å, the crystal structure of the ternary complex could be recognized in the extra density compared to empty ribosomes. To optimize the fit to the EM electron density map, fitting of the individual domains was undertaken.[35] A similar image of EF-G on the ribosome has been determined by image reconstruction.[36] This could very well represent the structure of the GTP form of EF-G.

Elongation Factor Tu Structures

Primary Structure of EF-Tu

E. coli EF-Tu is 393 amino acids long, whereas *T. thermophilus* and *T. aquaticus* EF-Tu each consist of 405 amino acid residues. Ten of the extra residues are found as an extended helix motif, which may improve the stability of the thermostable *Thermus* proteins. Archaebacterial and eukaryotic EF-1α sequences contain approximately 430 and 450 residues, respectively. The extra residues are found in few kingdom-specific inserts relative to the eubacterial sequence.[3] Vast sequence homology is observed throughout the entire molecule, and the tertiary structure in all organisms must be highly conserved. The EF-Tu sequences of mitochondria and chloroplasts show homology to eubacterial sequences.[3] For reasons of simplicity, specific residue numbers in this article will refer to the *T. aquaticus* sequence (unless otherwise stated) from which tertiary structures of several functional states have been determined in our laboratory. The amino acid composition of eubacterial EF-Tu shows typical features of highly soluble bacterial proteins with a rather acidic pI around 5.5. In contrast, eukaryotic EF-1α is more typical of a nucleic acid binding protein with a basic pI.

Structure of EF-Tu Domains

Domain 1 (1-214) is the site of GTP/GDP binding and is therefore referred to as the G-domain (Fig. 1). It has a roughly spherical shape and comprises approximately one-half of the total mass of the molecule. It has the characteristic alternating α/β structure of a nucleotide-binding protein. Thus, it consists of a central six-stranded β-pleated sheet (five parallel strands and one anti-parallel) with α-helices in the crossover connections. The first and last cross-over connections occur across the top side of the sheet, the remaining crossover connections are on the back side. Typical of this structural class, the nucleotide-binding site is located in the crevice formed between the third and fourth strands, where one crossover connection comes from below and the next is over the top side of the sheet.[37] From the first structural studies of EF-Tu and its comparison with the Ras p21 sequence, the G-protein consensus sequence motifs were identified as residues at the GTP/GDP binding site. The first motif is a phosphate binding loop with the consensus sequence GxxxxGK[S/T] (residues 18-25) which is also found in many ATP binding proteins. The second motif is DxxG (residues 81-84) which is also involved in phosphate binding and which serves as a pivot point in the conformational changes in switch II. Finally, a guanine-specific NKxD motif (residues 136-139) recognizes the guanine base by hydrogen bonds. It has been proposed that the G-domain originates from the fusion of two small proteins similar to ribosomal proteins or IF3 domains.[38] This hypothesis fits well with a bifurcation that can be observed in the tertiary structure of the domain (Fig. 2). The nucleotide binding site is at the interface between these two sub-domains or modules. In fact, the phosphate groups are bound by the N-terminal module and the guanine base by the C-terminal module.

A connecting segment of about 16 amino acid residues and comprising several proline residues leads from the G-domain into domain 2. Domain 2 (215-312) and domain 3 (313-405) are β-barrel structures with a classical six-stranded "Greek key" and "Swiss roll" structural motif, respectively. Domain 2 has several large loops protruding from the domain, and its barrel rests side to end on the barrel formed by domain 3. These domains are very tightly associated. Thus, EF-Tu consists of two structural units, the G-domain and the combined domains 2 and 3.

The Structure of EF-Tu:GDP Is Open and Flat

The EF-Tu:GDP structure consists of a largely flat and triangular arrangement of the three EF-Tu domains with a hole in the middle (Fig. 1). The switch regions and helix C of domain 1 form an almost plane surface which serves as an interface to the β-barrel domain 3. The variable N-terminal part of the switch I region (residues 42-51) contains a short β-strand and a single-turn α-helix A' whereas the conserved C-terminal half adopts an extended hairpin conformation, the tip of which points towards domain 3. Helix B (switch II) and helix C are oriented almost in parallel and form contacts to the β-barrel domain 3 by amino acid side chains in their N-terminal ends. The amino acid residues of this interface between domains 1 and 3 are generally conserved. Domain 2, rigidly locked into position against domain 3, is positioned far from domain 1 and forms no contacts to this domain, except through the peptide linkages.

The EF-Tu:EF-Ts Complex Dissociates EF-Tu:GDP

The structure of the EF-Tu:EF-Ts complex (Fig. 1) was the first G-protein nucleotide exchange complex to be determined.[30,31] Thereby, rationales for the GDP dissociation were revealed. In general, GDP dissociation is catalyzed by the interaction of EF-Ts with the components of the GDP-binding site. Furthermore, an overall opening of the domain arrangement of the GDP-bound form may facilitate the transition to the GTP-bound form. As such, the enzymatic mechanism of

EF-Ts is to stabilize the transition state of the nucleotide exchange of EF-Tu, namely the nucleotide-free form of EF-Tu.

In the crystal structure of the *E. coli* proteins, two EF-Ts molecules form a dimer with each monomer binding one molecule of EF-Tu. However, this oligomerization has remained undetected in solution. The *T. thermophilus* structure shows a different mode of dimerization which is directed by intermolecular disulphide bridges in EF-Ts. The *E. coli* EF-Ts molecule comprises three domains (Fig. 1): the N-terminal domain 1 with three α-helices; the core domain 2 with two central three-stranded antiparallel β-sheets surrounded by α-helices; and a small domain 3 of three α-helices inserted into the core domain. Finally, a C-terminal extension containing one α-helix protrudes from the molecule. The core domain 2 has an internal pseudo-symmetry relating the two sheets and some of the helices. This divides the core domain into two sub-domains, 2_N and 2_C. EF-Ts makes contacts with the G-domain of EF-Tu through its N-terminal domain and through sub-domain 2_N. Sub-domain 2_C makes contacts with the tip of domain 3 of EF-Tu. These interactions dissociate the direct interactions between domains 1 and 3 seen in EF-Tu:GDP.[11] The C-terminal helix extension of EF-Ts, which is only found in *E. coli*, interacts tightly with the G-domain and in fact replaces and structurally mimics the switch I region. This could indicate a structural instability of the nucleotide-free *E. coli* EF-Tu which is then stabilized by this interaction. Neither Mg^{2+} nor GDP is observed in the nucleotide binding pocket of the EF-Tu:EF-Ts structure.

When the structure of the EF-Tu:EF-Ts complex is compared with that of EF-Tu:GDP, the Mg^{2+}-GDP-binding site is seen to be almost entirely disrupted. Residues from the N-terminal domain 1 of EF-Ts interfere with the guanine-recognition loop of EF-Tu (the NKxD motif) and prevent its interaction with the guanine base of GDP. Furthermore, residues from domains 1 and 2_N of EF-Ts interact with the phosphate binding loop of EF-Tu (with the GxxxxGK[S/T] motif) and disrupt its interaction with the β-phosphate group of GDP. Finally, the switch II region of EF-Tu:GDP, which contains the DxxG motif, is highly altered by contacts to the 2_N domain. Thus, helix B is significantly changed in its orientation.

These alterations knock out the interactions between EF-Tu and Mg^{2+}-GDP. Yet, it remains to be further studied exactly how EF-Tu escapes the EF-Ts complex by subsequent Mg^{2+}-GTP binding. However, structural studies of the yeast EF-1α:EF-1β complex with GDP and GDPNP indicate that GTP binding would precede binding of Mg^{2+}.[34]

The Structure of EF-Tu:GTP Is Compact

The major difference between the structures of EF-Tu:GDP and EF-Tu:GTP (obtained by the use of non-hydrolyzable GTP analogues) concerns the spatial arrangement of the three domains (Fig. 1). In EF-Tu:GDP, there is no interaction between domains 1 and 2, and a characteristic solvent-filled hole is found between the domains. Furthermore, only a small interface is present between domains 1 and 3. In the EF-Tu:GTP structure, however, domain 2 interacts intimately with domain 1, and domains 1 and 3 also form a different and much larger interface. In order to achieve this tight structure of EF-Tu:GTP, the switch regions in domain 1 have changed their structure upon GTP binding and domains 2 and 3 have moved and rotated by 90° as a rigid body relative to the GTP-bound domain 1. The hole seen in the EF-Tu:GDP complex disappears, and a deep cleft is formed at the interface between domains 1 and 2 which leads into a broad concave surface at the interface between domains 1 and 3. This cleft and surface form an obvious binding site for aa-tRNA.[22,23] Noteworthy, the structure of EF-Tu:GTP is unaltered upon aa-tRNA binding.

The Ribosome-Associated Form of EF-Tu

Even though a true atomic structure of EF-Tu bound to the ribosome is still to come, low-resolution electron microscopy reconstructions and docking studies are available.[35,39] An outstanding breakthrough in the attempt to approach functional states of the ribosome by electron microscopy studies was obtained by the image reconstruction of fully assembled 70S particles with a ternary complex stalled in the recognition site by the antibiotic kirromycin, which inhibits the release of EF-Tu:GDP from the ribosome after GTP hydrolysis.[35] The reconstruction, which has a nominal resolution of approximately 18 Å, allows a rough domain-by-domain interpretation starting from the crystal structure of Phe-tRNA:EF-Tu:GDPNP. This was interpreted as showing a modest conformational change, yet significant at 20 Å resolution. The adjusted domain arrangement appears as an altered GTP-bound form with domains 2 and 3 moved even further along the trajectory that brings them into the GTP-bound state from the GDP bound state. However, the specific interpretation of the domain arrangement should be treated with caution, because of the low resolution of the map.

Crystallographic studies of the ribosome have also reached remarkable new levels in recent years[40-44] (see chapter 15, by B. Wimberly). At an earlier stage, a 5 Å resolution map of the large ribosomal subunit allowed several components of the factor binding site to be fitted, namely the sarcin-ricin loop and the L11 region of the 23S RNA and the ribosomal proteins L6, L11 and L14 which make up a continuous surface onto which EF-Tu and EF-G are known to bind from several studies.[39] This site is close to the presumed position of the L7/L12 stalk although this particular region was not resolved in the map. Various sources of data were available for docking of EF-G onto this ribosomal site, most importantly hydroxyl radical cleavage data.[45]. Yet, EF-G is only known in the nucleotide-free and GDP-bound form and the structure of the switch I region in EF-G is unknown.[25,26] However, based on the structural alignment of EF-G and EF-Tu and the concept of macromolecular mimicry which imposes that the aa-tRNA:EF-Tu:GTP complex and EF-G bind to the ribosome in a similar fashion, the aa-tRNA:EF-Tu:GTP structure with its known conformation of both switch regions could replace EF-G in this docking experiment. From this model, a close interaction between the GTPase switch regions and the ribosome was postulated, most notably between the sarcin-ricin loop of 23S rRNA plus L14 and the switch I region of EF-Tu.[39]

Conformational Changes in EF-Tu Are Mediated by the Switch Regions

The two switch regions of EF-Tu undergo a spectacular rearrangement of their secondary and tertiary structures when switching from the GDP-bound to the GTP-bound state via the EF-Ts complex (Fig. 2).

In the switch II region, the conserved residue Gly84 is close to and directly influenced by the nature of the nucleotide. In EF-Tu:GDP, helix B in this region is formed by residues 85-95.[18] When GTP is introduced into the nucleotide binding site, the peptide bond Pro83-Gly84 is flipped by about 150° so that the amide is now pointing towards the γ-phosphate and a water molecule positioned so that it can make an in-line nucleophilic attack on the γ-phosphate group. This has the effect that the first turn of helix B is disrupted and its direction shifted by approximately 45°. Also, the helix is rotated around its axis by almost 40°. This allows the formation of an extra turn in the C-terminal end of the helix, which then encompasses residues 89-98 in the GTP-form.[22,23] The flexibility of the peptide chain at residue Gly95 is important for this transformation.[46] The peptide flip in Pro83-Gly84 also transmits into conformational changes in switch I by directly influencing the structure around Thr62.

The structure of the N-terminal part of switch I is the same in the GDP- and GTP-bound forms, thus with a short β-strand (residues 42-46) and a single-turn α-helix A' (residues 47-51). Yet, this structure is dislocated in the *E. coli* EF-Tu:EF-Ts complex by the C-terminal extension of EF-Ts.[30] In EF-Tu:GDP, the C-terminal part (residues 52-64) forms a β-hairpin that merges with the following β-strand in the G-domain structure (residues 52-58 and residues 62-72).[18] However, in EF-Tu:GTP, this region contains an α-helix A'' (residues 54-59) that protrudes from the protein and a loop structure that leads into the following β-strand (residues 65-72).[22,23] Thr62 as mentioned before plays a central role in this rearrangement. It provides a ligand to Mg^{2+} in the GTP form, where space for this interaction has become available because of the peptide flip in Pro83-Gly84 in switch II. In the GDP-bound form, this interaction is sterically blocked by the Pro83 side chain. Along with Thr62, Ile63 forms tight packing over the γ-phosphate moiety. These interactions with the Mg^{2+}-GTP binding site tie up the switch I region and drive the transition from the β-hairpin structure to the protruding loop-helix-loop structure. Gly60 introduces flexibility into the region, allowing it to switch between the two very different structures.[14]

The large changes in the switch regions lead to even more dramatic changes in the domain arrangement in EF-Tu. The tip of the hairpin of switch I and helix B of switch II form a small interface to domain 3 in the GDP-bound form. The change to the GTP-bound conformation creates a much larger surface in the G-domain that forms an interface to a different region of domain 3. Thereby, the tightly associated pair of domains 2 and 3 rotates and moves to form a tight packing against the G-domain. Thus, the local structures of the GTPase switch regions determine the domain arrangement in EF-Tu. Keeping the positions of domains 2 and 3 fixed, the rearrangement can

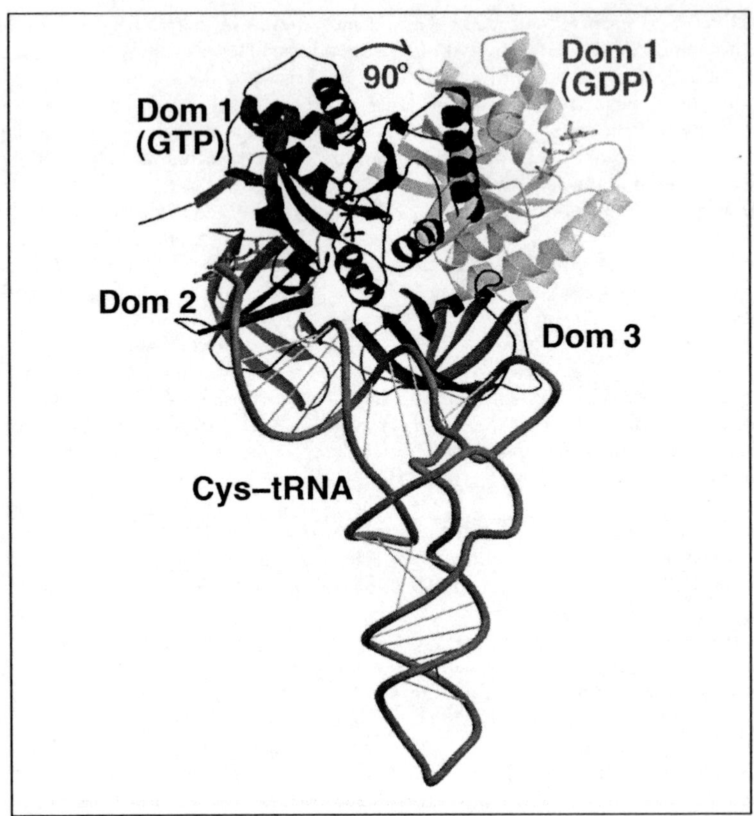

Figure 3. The structure of the ternary complex of EF-Tu and the conformational change upon GTP hydrolysis. The tRNA molecule is shown in grey with base pairs represented by white sticks. The EF-Tu:GDPNP component is shown in black with labels for the domains. The position of domain 1 rotates relative to domains 2 and 3 in the GDP-form and is shown in white. The GTP/GDP nucleotides of EF-Tu as well as the A-76 and amino acid of Cys-tRNA are shown in ball-and-stick models.[29]

be visualized as a rotation of domain 1 by approximately 90 degrees, with some regions moving over 40 Å (Fig. 3).

EF-Tu binds GDP approximately 100 times tighter than GTP, yet the isolated domain 1 of EF-Tu has an affinity for GDP that is comparable with that for GTP.[47] This suggests that the other domains of EF-Tu play an important role in the total free energy that determines the nucleotide binding affinities. Also, it hints to the mechanism of EF-Ts function: the mere separation of domains 1 and 3 and the "melting" of the switch regions in EF-Tu upon EF-Ts complex formation eliminates the preference for GDP binding and facilitates the huge conformational change associated with GTP binding.[23]

Protein-RNA Interactions Studied in the aa-tRNA:EF-Tu:GTP Complex

The tRNA Structure and Function

Each tRNA is specifically aminoacylated by a cognate synthetase (one for each of the 20 amino acids in most organisms) (see chapter 2 by D. Söll and M. Ibba). Accordingly, the 20 synthetases dictate and maintain the relations between trinucleotide codons and amino acids in the genetic code. The tRNA recognition by the synthetases involves specific identity elements in tRNA which may be direct binding to sequence motifs or which may be of more complex, structural character.[48]

Ideally the synthetases discard the cognate aa-tRNA, the product of their enzymatic function. In contrast, EF-Tu:GTP recognizes all elongator aa-tRNAs and yet discriminates by approximately three to five orders of magnitude against any deacylated tRNA in ternary complex formation. Thus, EF-Tu:GTP specifically recognizes a functional group (the aminoacyl group) and a specific RNA fold (the general tRNA structure).[49] As such, EF-Tu and the synthetases are complementary in their modes of tRNA recognition and binding. Especially for the class I synthetases, this is also clearly visible in a tRNA-based alignment of tRNA complex structures, which shows that EF-Tu and the class I synthetases (as for example Gln-tRNA synthetase) have non-overlapping binding sites on the tRNA molecule. This may even allow a direct delivery of aa-tRNA from the synthetase to EF-Tu:GTP.[49]

The tRNA molecules of any organism or organelle have been revealed to share a common L-shaped structure formed by an "acceptor helix" perpendicular to an "anticodon helix".[50,51] The spatial separation is approximately 80 Å between the anticodon loop, at the end of the anticodon helix, and the aminoacyl group attached to the single stranded 3' CCA-end that extends from the acceptor helix. This clearly emphasizes the adapter function in the structure of the tRNA molecule. Normally this structure is rooted in a clover leaf secondary structure which exhibits four base paired stem regions.[52] Thus, with a few exceptions (most noteworthy some mitochondrial tRNAs), the acceptor helix is formed by the "acceptor stem" and the "TΨC-stem" (T-stem), and the anticodon helix is formed by the "Dihydrouridine stem" (D-stem) stacked on the "anticodon stem" (see chapter 1 by R. Giegé and M. Frugier).

Overall Structure of the Ternary Complex

As mentioned earlier, the crystal structures of EF-Tu in several of its functional states have now been determined and it is clear that upon GTP binding, the overall shape is transformed from the very open and flat inactive form to a compact structure which exhibits a tRNA binding site.

The crystal structure of the ternary complex of yeast Phe-tRNA^{Phe} and *T. aquaticus* EF-Tu:GDPNP (Phe-TC) turned out to be quite different from most expectations (Fig. 3). It had been suggested from low angle scattering and hydrodynamic arguments that Phe-TC would be compact. However, the structure of the complex revealed a very elongated shape (115 Å in the longest dimension) and the EF-Tu component was observed to interact with a quite restricted region of the aa-tRNA namely one side of the acceptor helix and the aminoacylated 3' end. The recognition involves generally conserved features of EF-Tu:GTP and aa-tRNA and the structure represents a generic model of any ternary complex in any organism.

The Specific Recognition of Aminoacyl-tRNA by EF-Tu:GTP

The aa-tRNA binding site on EF-Tu:GTP is formed by four specific sites: i) a deep binding pocket between domains 1 and 2 specific for the aminoacylated 3' end (Fig. 4), ii) the junction of the three domains where a surface depression accommodates the phosphorylated 5' end, iii) the protruding switch I region which contacts the acceptor stem, and finally iv) a large surface on domain 3, which is complementary to the minor groove of the T-stem between nucleotides 63 through 67 and 50 through 54 (Fig. 5). Thus, only aminoacylated tRNA and EF-Tu:GTP can participate in ternary complex formation as both deacylated tRNA and EF-Tu:GDP are unable to match these interaction sites.

As mentioned before, EF-Tu:GDP has the domains reoriented by a relative rotation of 90°, as well as an α-helix to β-hairpin change of the switch I region.[18] These alterations destroy the tRNA binding site on EF-Tu almost completely (Fig. 3).

The binding of the 3' end is specific to the 3' aminoacyl group. The free 2' OH of the ribose, the carbonyl oxygen and the free amino group of the aminoacyl bond form hydrogen bonds and the entire group fits a tight pocket in the EF-Tu structure. Furthermore, the side chain of any amino acid will fit a large side chain binding pocket and can make various interactions with surrounding residues of EF-Tu. Thus, the ability to bind any elongator aa-tRNA as well as the discrimination against initiator N-formyl-Met-tRNA$_f^{Met}$ can readily be rationalized. However, the several orders of magnitude that separate the affinity of deacylated tRNA from, for example, Gly-tRNA cannot be explained by a trivial counting of hydrogen bonds and therefore an entropy effect has been deduced. Several observations indicate that the structure of the CCA end is influenced by aminoacylation and

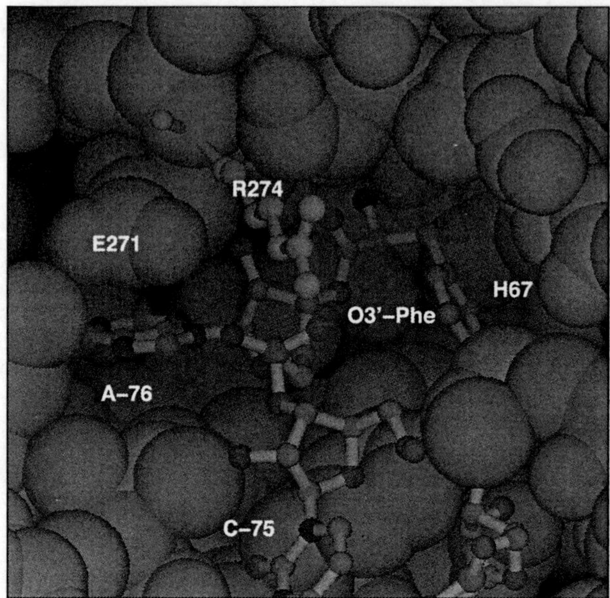

Figure 4. Details of the binding pockets for the A-76 base of tRNA and for the amino acid. Most of the surrounding amino acid residues of EF-Tu are shown in spacefill models; however, residue R274 is shown in ball-and-stick model for clarity. The 3' CCA end with attached amino acid (Phe) is also shown in ball-and-stick models.

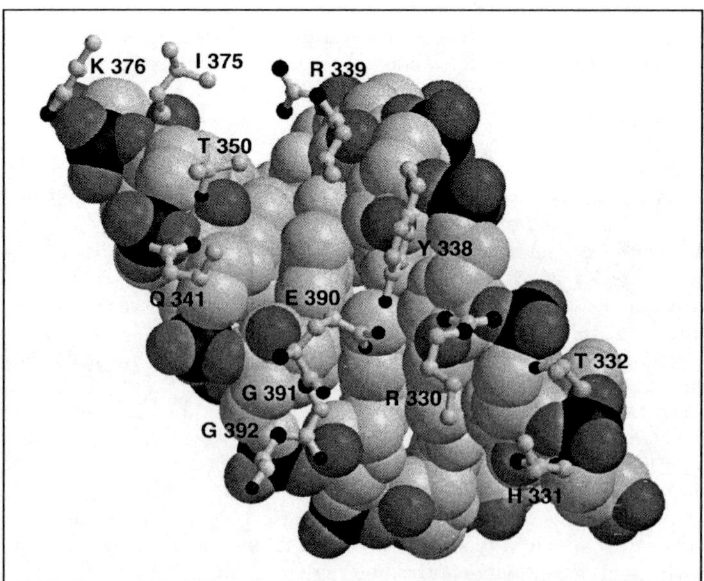

Figure 5. Interactions of EF-Tu with the minor groove of the tRNA T-stem. The hydrophobic surface of the bases and riboses in the minor groove (in grey) together with their associated phosphates (phosphorus in black and oxygens in dark grey) are shown in spacefill model. Selected amino acid residues of EF-Tu in contact with this surface are shown in ball-and-stick models.

other modifications.[53]. Thus, aminoacylation of the 3' end could yield a constrained set of conformations which lowers the entropy cost of complex formation and thereby increases the affinity considerably.

Comparison of the Structures of the Free Components and the Complex

The ternary complex can conveniently be described as the association of two rigid bodies. Small changes in both the EF-Tu and the tRNA structures are certainly observed, yet the overall structures are basically unaltered when compared with the structures of free EF-Tu:GTP and tRNA^{Phe} (Fig. 1). This is also to be expected on the basis of trivial thermodynamic considerations, as conformational changes upon complex formation (an "induced fit") necessarily represent a partial cost in free energy in contrast to a simple association of the components.

In retrospect, the pre-formed binding site for aa-tRNA can be identified in the structure of free EF-Tu:GTP. Even the GTPase switch regions which undergo very large rearrangements upon GDP to GTP exchange are located with a 1 Å precision to interact with the aa-tRNA component.[28] The side chains located in the aa-tRNA binding site in free EF-Tu:GTP form a complex network of interactions on the surface which is not maintained upon aa-tRNA binding when the phospho-ribose backbone enters and participates in the formation of intermolecular contacts.

The tRNA is slightly changed by comparison of free tRNA^{Phe} and Phe-tRNA^{Phe} of the ternary complex. The acceptor stem is rotated and twisted by approximately 10° relative to the T-stem. However, this alteration has easily been absorbed in the flexible backbone, and the two conformations may be regarded as two representatives of a coarse energy minimum in RNA structure which has the same base-stacked structure.

The Binding of aa-tRNA Is Based Upon Polar and Hydrophobic Interactions

Macromolecular complex formation is directed by changes in enthalpy and entropy. An increase in entropy is obtained by the association of complementary macromolecular surfaces which will release ordered solvent molecules into the bulk state due to the replacement of the solvation sphere by intermolecular contacts (salt bridges, hydrogen bonds, hydrophobic packing, Van der Waal contacts). These interactions may also determine the specificity of the complex formation. The interface between EF-Tu and aa-tRNA covers approximately 1560 Å^2, compared with a total surface area of 26621 Å^2 of the entire complex, and 12794 Å^2 and 16946 Å^2 of free aa-tRNA and EF-Tu:GTP, respectively (as measured by a probe with a 1.4 Å radius). The major part of this interface area is located along two stretches of the minor groove of the acceptor helix: the interaction of the GTPase switch regions with the acceptor stem, and the interface between domain 3 and the T-stem. Furthermore, the aminoacylated 3'-end is docked into a deep binding pocket and the phosphorylated 5' end is bound at the junction of the three EF-Tu domains. Together these interactions support the tRNA acceptor helix on EF-Tu:GTP as one may imagine a grip on a banister: with the palm holding on to one side (domain 3) and with the thumb left on top (the switch regions). The major groove of the acceptor helix is too deep and narrow for any protein residues to intrude. In any case, major groove recognition tends to be sequence specific and thus inappropriate for the ternary complex. As seen in Figure 5, the interface between domain 3 and the T-stem consists of large hydrophobic patches formed by the stacked bases in the minor groove and the ribose rings in tRNA that pack against aliphatic segments of polar side chains and hydrophobic side chains in EF-Tu. Polar interactions target the phospho-ribose backbone on 2' OH and phosphate groups.

Protection of the Aminoacyl Bond in the Ternary Complex

The tight binding of the aminoacyl bond by EF-Tu will sterically prevent water molecules from attacking the ester group. The several fold stronger protection in high concentrations of ammonium sulfate[27,54-56] can readily be explained by the general stabilization of the complex association by stronger hydrophobic interactions at high ionic strength.

However, by comparison of the Cys-tRNA and Phe-tRNA ternary complexes with EF-Tu,[28,29] a conserved—yet quite unanticipated—interaction with the amino group of the amino acid has been observed. It is bound in the neutral, deprotonated state due to the involvement of a hydrogen bond donor (from the main chain amide of His273) in a tetrahedral coordination sphere of the amino group. In free solution, the basic pK_a of the amino group would lead to the formation of an ammonium ion. The neutral state of the amino group imposes a lower electrophilicity of the ester carbon atom which again decreases its reactivity towards hydrolysis. Thus, this will additionally account for the high protection of the aminoacyl bond upon EF-Tu binding. Interestingly, the peptidyl transferase centre on the ribosome must also bind aa-tRNA with a neutral amino group as the free electron pair on the amino group is the nucleophile in peptide bond formation.[41]

The GTPase Switch Regions Are Involved in tRNA Binding

The GTPase switch I region of EF-Tu:GTP (residues Lys52 through Asn65) forms a loop-helix-loop structure which is positioned over a small stretch of the acceptor stem minor groove and at the beginning of the CCA end (nucleotides 1-3 and 72-73). Polar interactions to the phospho-ribose backbone as well as hydrophobic interactions and Van der Waal contacts form the thumb that supports the acceptor helix from the top (keeping an orientation in mind as in Figure 3). Introduction of a nick in the switch I region by trypsin cleavage (at position Arg59) does not hamper ternary complex formation severely, which indicates that this "thumb print" is of minor importance concerning the affinity. However, it will effectively inhibit the functional interaction with the ribosome.[57] The small helix A" is located so as to allow the side chains of Glu55 and Arg59 to point towards the minor groove. Similarly, α-helices are observed to be located across minor groove patches of tRNA molecules in synthetase-tRNA complexes,[58-60] as a general way of presenting amino acid side chains to tRNA helices. As an example, a long helix in the Ser-tRNA synthetase forms contacts with the minor groove of the extended variable arm in tRNA^{Ser} which then defines an important determinant of this system.[59]

The GTPase switch II region of EF-Tu (residues 83 through 95) also forms a loop-helix-loop structure with many conserved residues. The N-terminus of helix B points towards the major groove, with the conserved Asp87 located over the phosphates at nucleotide positions 3 and 64. The conserved Lys90 and Asn91 as well as Tyr88 (conserved as Tyr or Phe) contact the backbone of the 5' end and the following nucleotide residue. The contact to the 5' phosphate of tRNA is not obligatory for ternary complex formation, as it has been shown that dephosphorylated Tyr-tRNA can still form a ternary complex.[61] However, together with the binding of the aminoacyl group, it will clearly restrain the overall structure of the ternary complex in its elongated shape. The concerted binding of the 5' end and the aminoacylated 3' end may also serve as a control that recognizes the overhang of exactly four bases in the CCA end from the double-stranded acceptor stem, in contrast to the five base overhang of initiator tRNA which is caused by the mismatch of the 1:72 basepair.

Elongator tRNA Is Recognized on the T-Stem

A large interface is formed between the T-stem and the β-barrel domain 3, which presents a large surface complementary to the minor groove of tRNA (Fig. 5). Polar interactions of side chains of EF-Tu with the tRNA backbone as well as a close packing in the groove which also involves water molecules define a T-stem specific interaction. This intimate complementarity is probably responsible for the discrimination against prokaryotic selenocysteine specific tRNA (Sec-tRNA^{Sec}) which carries a G:U wobble pair at position 64. Besides being under influence by the side chain of Asp87 (see earlier section), the backbone of position 64 makes a close contact to a stretch of protein main chain on the surface of domain 3 (Gly391-Gly392). Furthermore, Glu390 is interacting with the base pair of tRNA residues 51 and 63. Thus, a structural alteration caused by a wobble pair in that location may result in unfavourable interactions with EF-Tu. Interestingly, the Sec-tRNA specific elongation factor, SELB, has a lysine or arginine residue corresponding to Asp87 and no apparent similarity to domain 3 in the region of Gly391-Gly392.[62] This may indicate a specific adaptation to the Sec-tRNA structure. The initiator tRNA of several eukaryotic species is phospho-ribosylated on the 2' OH of position 64,[63,64] which would lead to impaired EF-Tu binding for similar reasons.

However, despite a large interface area, the binding of the T-stem to domain 3 must be sufficiently weak to avoid effective binding of deacylated tRNA or RNA helices in general. In line with this, the affinity of EF-Tu:GTP towards aminoacylated microhelices that lack the T-stem is much lower than that of aa-tRNA.[65] Thus, this binding site must be tuned so as to accomplish sufficient discrimination against Sec-tRNA (or eukaryotic initiator Met-tRNA) and deacylated tRNA.

Recognition of Double-Stranded RNA Exploits the Spatial Separation of Two Minor Grooves Over a Major Groove

Besides specific recognition of the aminoacyl group and selection against initiator and Sec-tRNA, EF-Tu displays a very general binding site for any elongator tRNA, which recognizes the phosphorylated 5' end and two patches of minor grooves separated by a narrow major groove. In summary, the switch regions recognize the minor groove between residues 1-3 and 70-72, and domain 3 binds to the minor groove of the T-stem. This binding can be compared with the complex structure of double stranded RNA (dsRNA) and a dsRNA binding domain (dsRBD) of *Xenopus laevis* RNA binding protein A similar to the conserved core structure of a family of polynucleotidyl transferases.[66] This comparison reveals several similarities (Fig. 6). Polar interactions on the phospho-ribose backbone flank mainly hydrophobic interactions within the minor groove. Asp87 senses the presence and position of the major groove in tRNA whereas a glutamine does the same in dsRBD. The spatial separation of the acceptor stem and T-stem binding sites of EF-Tu selects for a tRNA helix which has a very narrow major groove. In contrast, two minor groove binding sites in dsRBD selects for an A-form RNA helix with a wider major groove. Both dsRBD and EF-Tu use negatively charged residues thus emphasizing that RNA binding sites are not merely positively charged regions of a protein.

Macromolecular Mimicry in Translation

Surprisingly, a comparison of the ternary complex with the ribosomal translocase, elongation factor G (EF-G),[25,26,28] revealed an odd similarity in shape, termed "macromolecular mimicry".[67-70] EF-G has a G-domain and a domain 2 homologous to EF-Tu. Interestingly these two domains in EF-G:GDP are arranged as observed in EF-Tu:GTP and not as in EF-Tu:GDP. When these domains of EF-G and EF-Tu are structurally aligned on their Cα-positions, the remaining three C-terminal domains of EF-G:GDP appear to describe the approximate position and shape of the aa-tRNA molecule of the ternary complex.[28] Thus these three domains in EF-G, which have structural folds similar to those of the ribosomal proteins S5 and S6, seem to form a tRNA mimic. This observation raises a number of new questions to the mechanisms and the origins of the translation apparatus. Furthermore, when more details become available it may reveal features of shape and surface homology between protein and RNA structure of yet unrecognized nature. Indeed, a macromolecular structure is a more complicated matter than isolated atoms displayed in a polymer structure.

The striking similarity of the overall shape of the ternary complex and EF-G most likely provides meaningful hints to the understanding of the function of the translation apparatus. Evidently, the shape similarity indicates that these macromolecular components interact with a very similar environment on the ribosome, i.e., the same binding pocket. We have strong reasons to believe that EF-G and EF-Tu interact with the same GTPase centre on the large subunit of the ribosome.[39] Both have been shown to interact with the sarcin-ricin loop of 23S rRNA, and both factors have also been mapped to the L7/L12 stalk when bound to the ribosome. However, subtle differences must exist as the GTPase activity in these factors is invoked by different signals: GTP hydrolysis in EF-Tu is under the control of codon-anticodon recognition, whereas GTP hydrolysis in EF-G is much less restrained. It is reasonable to suggest that the other G-proteins in translation, IF2 and RF3, with their G-domain and domain 2, interact in very similar ways with the ribosome and with the GTPase activating centre. This is supported by sequence alignments suggesting that they have such domains,[71] and by the recent structure determination of the eukaryotic initiation factor IF2/eIF5B.[72] The idea of a similar binding pocket on the ribosome for the ternary complex and EF-G would further implicate that EF-G must form a close contact to mRNA with the tip of domain 4 in the decoding site of the small ribosomal subunit.[73] This hypothesis is strongly supported by experiments using hydroxyl radical cleavage of rRNA.[45] Furthermore, indirect support stems from the fact that EF-G is known to be functionally linked to the effect of certain antibiotics that target the decoding centre, such as spectinomycin.[74] EF-G has been cross-linked to positions of 16S rRNA which are in close proximity to the decoding centre. Not least, cryo-EM reconstructions reveal that EF-G and aa-tRNA:EF-Tu:GTP indeed occupy very similar binding sites on the ribosome.[35,36]

GTP hydrolysis will allow both EF-Tu and EF-G to dissociate from the ribosome. EF-Tu of the ternary complex changes its three-dimensional structure dramatically upon GTP hydrolysis.

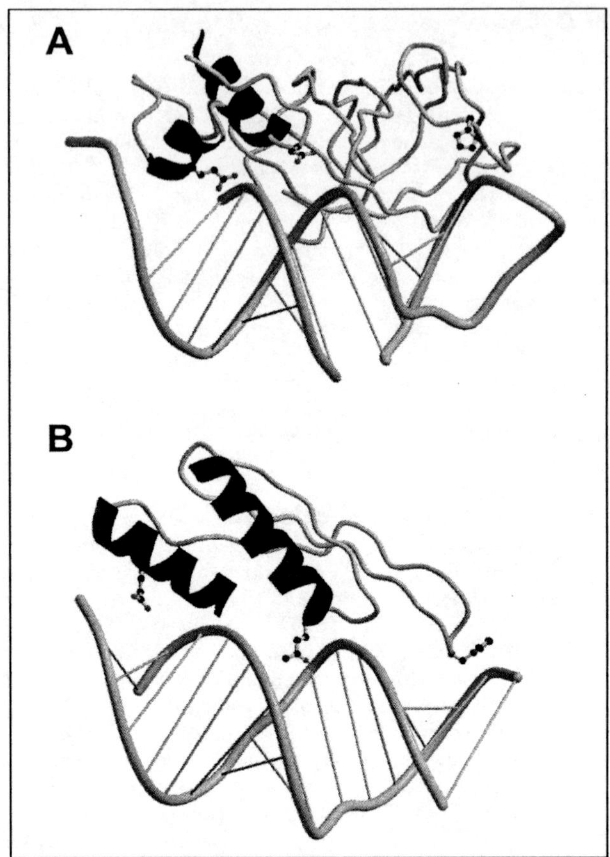

Figure 6. Structures of proteins interacting with double stranded RNA. Structures shown are in A) the ternary complex of yeast Phe-tRNA with *T. aquaticus* EF-Tu:GDPNP[28] and in B) the RNA binding domain (dsRBD) of *Xenopus laevis* double stranded RNA-binding protein A.[66] The α-helices in closest contact with RNA are shown in dark grey. Specific amino acids in contact with RNA are Glu55, Asp87 and His331 of EF-Tu shown from left to right in A) and Glu119, Gln164 and His141 of dsRBD shown similarly in B).

Unfortunately, only little is known about the structure of EF-G in the GTP conformation. Possibly, it is only slightly different from that of EF-G:GDP, with a conformational change of the G-domain relative to the tip of domain 4 reflecting the separation of two codons.

   Other protein factors of translation are known to interact with the ribosome in a "tRNA-like" manner. This is specially true for the release factors which have to react on special stop-codons and which are known to compete with suppressor tRNAs. Although there are major differences between the termination mechanisms in prokaryotes and eukaryotes, the structure of human release factor eRF1 reveals a shape somewhat similar to that of a tRNA molecule.[75] At least one domain could mimic the anticodon stem and loop. A much more obvious macromolecular mimicry of a tRNA is seen in the ribosome recycling factors from *Thermotoga maritima*[76] from *T. thermophilus*[77] and from *E. coli*.[78]

   The macromolecular mimicry seen in translation thus reflects the fact that tRNA molecules interact intimately with the functional centres of the ribosome. Such centres will have shapes and polar or hydrophobic groups which are complementary to those of functional parts of a tRNA. They will also locally adapt to the presence of a tRNA molecule, and could thereby transmit signals to other centres of the ribosome as indicated by the recent structure of an anticodon stem and loop on the small ribosomal subunit.[79] Seen in this context, it is no wonder that protein factors interacting with these same functional centres mimic the shape of tRNAs. Furthermore, specific polar and hydrophobic groups of these proteins positioned at spatially relevant areas of their surfaces would then be able to trigger ribosomal functional signals that could change the functional state of the ribosome. Future high resolution structural studies of the GTP binding translation factors in complex with the ribosome will hopefully reveal more details of how macromolecular mimicry is specifically used by the ribosome and factors to increase the speed and accuracy of the production of proteins. However, studies of the dynamic nature of the ribosome, the smooth transition between functional states of the ribosome, and the central role of tRNAs and translation factors in all of this[80] will be a major challenge for researchers for years to come.

Note added in proof: Very recent cryo-EM reconstructions of bacterial release factor RF2 bound to the ribosome by the laboratories of Joachim Frank and of Marin van Heel have shown that RF2 undergoes major conformational changes from the free crystal structure to the form bound on the ribosome. Biochemical work in the laboratory of Akira Kaji on the binding of ribosome recycling factor to the ribosome has furthermore shown that this factor does not bind the same way as tRNA does despite their very similar shape. These observations demonstrate that there is no simple shape relationship between tRNAs and protein factors acting on the A site of the ribosome.

Acknowledgements

The continued support from the Danish Natural Science Research Council especially through its Programme for Biotechnology Research and its support to the Centre for Interaction, Function, and Engineering of Macromolecules (CISFEM) is gratefully acknowledged. Poul Nissen is recipient of an Ole Rømer Fellowship from the Danish Natural Science Research Council.

References

1. Sonenberg N, Hershey JWB, Mathews MB, eds. Translational Control of Gene Expression. Cold Spring Harbor: CSHL Press, 2000.
2. Karimi R, Pavlov MY, Buckingham RH et al. Novel roles for classical factors at the interface between translation termination and initiation. Mol Cell 1999; 3(5):601-609.
3. Merrick WC, Nyborg J. The protein biosynthesis elongation cycle. In: Sonenberg N, Hershey JWB, Mathews MB, eds. Translational Control of Gene Expression. Cold Spring Harbor: CSHL Press, 2000:89-125.
4. Rodnina MV, Wintermeyer W. Fidelity of aminoacyl-tRNA selection on the sibosome: Kinetic and structural sechanisms. Annu Rev Biochem 2001; 70:415-435.
5. Ruusala T, Ehrenberg M, Kurland CG. Is there proofreading during polypeptide synthesis? EMBO J 1982; 1(6):741-745.
6. Karimi R, Ehrenberg M. Dissociation rate of cognate peptidyl-tRNA from the A-site of hyper-accurate and error-prone ribosomes. Eur J Biochem 1994; 226(2):355-360.
7. la Cour TFM, Nyborg J, Thirup S et al. Structural details of the binding of guanosine diphosphate to elongation factor Tu from *E. coli* as studied by X-ray crystallography. EMBO J 1985; 4:2385-2399.
8. Jurnak F. Structure of the GDP domain of EF-Tu and location of amino acids homologous to ras oncogen proteins. Science 1985; 230:32-36.
9. Jurnak F, Heffron S, Schick B et al. Three-dimensional models of the guanine nucleotide domain of *Escherichia coli* elongation factor Tu. Biochim Biophys Acta 1990; 1050:209-214.
10. Clark BFC, Kjeldgaard M, la Cour TFM et al. Structural determination of the functional sites of *E. coli* elongation factor Tu. Biochem Biophys Acta 1990; 1050:203-208.
11. Kjeldgaard M, Nyborg J. Refined structure of elongation factor EF-Tu from *Echerichia coli*. J Mol Biol 1992; 223:721-742.
12. Bourne HR, Sanders DA, McCormick F. The GTPase superfamily: Conserved structure and molecular mechanism. Nature 1991; 349:117-127.
13. Bourne HR, Sanders DA, McCormick F. The GTPase superfamily: A conserved switch for diverse cell functions. Nature 1990; 348:125-132.
14. Kjeldgaard M, Nyborg J, Clark BFC. The GTP-binding motif—variations on a theme. FASEB J 1996; 10:1347-1368.
15. Halliday KR. Regional homology in GTP-binding proto-oncogene products and elongation factors. J Cyc Nucl Prot Phos Res 1984; 9:435-448.
16. Pai EF, Kabsch W, Krengel U et al. Structure of the guanine-nucleotide-binding domain of the Ha-ras oncogene product p21 in the triphosphate conformation. Nature 1989; 341:209-214.
17. McCormick F, Clark BFC, la Cour TFM et al. A model for the tertiary structure of p21, the product of the ras oncogene. Science 1985; 230:78-82.
18. Polekhina G, Thirup S, Kjeldgaard M et al. Helix unwinding in the effector region of elongation factor EF-Tu:GDP. Structure 1996; 4:1141-1151.

19. Abel K, Yoder MD, Hilgenfeld R et al. An α to β conformational switch in EF-Tu. Structure 1996; 4:1153-1159.
20. Song H, Parsons MR, Rowsell S et al. Crystal structure of Intact elongation factor EF-Tu from *Escherichia coli* in GDP conformation at 2.05 Å Resolution. J Mol Biol 1999; 285:1245-1256.
21. Andersen GR, Thirup S, Spremulli LL et al. High resolution crystal structure of bovine mitochondrial EF-Tu in complex with GDP. J Mol Biol 2000; 297(2):421-436.
22. Berchtold H, Reshetnikova L, Reiser COA et al. Crystal structure of active elongation factor Tu reveals major domain rearrangements. Nature 1993; 365:126-132.
23. Kjeldgaard M, Nissen P, Thirup S et al. The crystal structure of elongation factor EF-Tu from *Thermus aquaticus* in the GTP conformation. Structure 1993; 1:35-50.
24. Ævarsson A, Brazhnikov E, Garber M et al. Three-dimensional structure of the ribosomal translocase: elongation factor G from *Thermus thermophilus*. EMBO J 1994; 13:3669-3677.
25. Czworkowski J, Wang J, Steitz TA et al. The crystal structure of elongation factor G complexed with GDP, at 2.7Å resolution. EMBO J 1994; 13:3661-3668.
26. Al-Karadaghi S, Ævarsson A, Garber M et al. The structure of elongation factor G in complex with GDP: conformational flexibility and nucleotide exchange. Structure 1996; 4:555-565.
27. Nissen P, Reshetnikova L, Siboska G et al. Purification and crystallization of the ternary complex of the elongation factor Tu:GTP and Phe-tRNA[Phe]. FEBS Lett 1994; 356:165-168.
28. Nissen P, Kjeldgaard M, Thirup S et al. Crystal Structure of the ternary complex of Phe-tRNA[Phe], EF-Tu, and a GTP analog. Science 1995; 270:1464-1472.
29. Nissen P, Thirup S, Kjeldgaard M et al. The crystal structure of Cys-tRNA[Cys]:EF-Tu:GDPNP reveals general and specific features in the ternary complex and in tRNA. Structure 1999; 7:143-156.
30. Kawashima T, Berthet-Colominas C, Wulff M et al. The structure of the *Escherichia coli* EF-Tu:EF-Ts complex at 2.5 Å resolution. Nature 1996; 379:511-518.
31. Wang Y, Jiang Y, Meyering-Voss M et al. Crystal structure of the EF-Tu:EF-Ts complex from *Thermus thermophilus*. Nature Struct Biol 1997; 4:650-656.
32. Pérez JMJ, Siegal G, Kriek J et al. The solution structure of the guanine nucleotide exchange domain of elongation factor 1βb reveals a striking resemblance to that of EF-Ts from *Escherichia coli*. Structure 1999; 7:217-226.
33. Andersen GR, Pedersen L, Valente L et al. Structural basis for the nucleotide exchange and competition with tRNA in the yeast elongation factor complex eEF1A:eEF1Bα. Mol Cell 2000; 6:1261-1266.
34. Andersen GR, Valente L, Pedersen L et al. Crystal structures of nucleotide exchange intermediates in the eEF1A:eEF1Bα complex. Nature Struct Biol 2001; 8:531-534.
35. Stark H, Rodnina MV, Rinke-Appel J et al. Visualization of elongation factor Tu on the *Escherichia coli* ribosome. Nature 1997; 389:403-406.
36. Agrawal RK, Penczek P, Grassucci RA et al. Visualization of the elongation factor G on the *Escherichia coli* 70S ribosome: The mechanism of translocation. Proc Natl Acad Sci USA 1998; 95:6134-6138.
37. Branden C, Tooze J. Introduction to Protein Structure. Second ed. New York: Garland Publishing, Inc., 1999.
38. Cousineau B, Leclerc F, Cedergren R. On the origin of protein synthesis factors: a gene duplication/fusion model. J Mol Evol 1997; 45(6):661-670.
39. Ban N, Nissen P, Hansen J et al. Placement of protein and RNA structures into a 5 Å-resolution map of the 50S ribosomal subunit. Nature 1999; 400:841-847.
40. Ban N, Nissen P, Hansen J et al. The complete atomic structure of the large ribosomal subunit at 2.4 Å resolution. Science 2000; 289:905-920.
41. Nissen P, Hansen J, Ban N et al. The structural basis of ribosome activity in peptide bond synthesis. Science 2000; 289:920-930.
42. Wimberley BT, Brodersen DE, Clemens WM et al. Structure of the 30S ribosomal subunit. Nature 2000; 407:327-339.
43. Schluenzen F, Tocilj A, Zarivach R et al. Structure of functionally activated small ribosomal subunit at 3.3 Å resolution. Cell 2000; 102:615-623.
44. Yusupov MM, Yusupova GZ, Baucom A et al. Crystal structure of the ribosome at 5.5 Å resolution. Science 2001; 292:883-896.
45. Wilson K, Noller HF. Mapping the position of translational elonagtion factor EF-G in the ribosome by directed hydrxyl radical probing. Cell 1998; 92:131-139.
46. Kjærsgaard IV, Knudsen CR, Wiborg O. Mutation of the conserved Gly83 and Gly94 in *Escherichia coli* elongation factor Tu. Indication of structural pivots. Eur J Biochem 1995; 228(1):184-190.
47. Jensen M, Cool RH, Mortensen KK et al. Structure-function relationships of elongation factor Tu. Isolation and activity of the guanine-nucleotide-binding domain. Eur J Biochem 1989; 182:247-255.
48. Arnez JG, Moras D. Structural and functional considerations of the aminoacylation reaction. Trends Biochem Sci 1997; 22:189-232.
49. Nissen P, Kjeldgaard M, Thirup S et al. The ternary complex of aminoacylated tRNA and EF-Tu:GTP. Recognition of a bond and a fold. Biochemie 1996; 78:921-933.
50. Robertus JD, Ladner JE, Finch JT et al. Structure of yeast phenylalanine tRNA at 3 Å resolution. Nature 1974; 250:546-551.
51. Kim S-H, Suddath FL, Quigley GJ et al. Three-dimensional tertiary structure of yeast phenylalanine transfer RNA. Science 1974; 185:435-440.
52. Holley RW. Structure of an alanine transfer ribonucleic acid. JAMA 1965; 194(8):868-871.

53. Nawrot B, Milius W, Ejchart A et al. The structure of 3'-O-anthraniloyladenosine, an analogue of the 3'-end of aminoacyl-tRNA. Nucleic Acids Res 1997; 25(5):948-954.
54. Antonsson B, Leberman R. Stabilization of the ternary complex EF-Tu.GTP.valyl-tRNAval by ammonium salts. Biochimie 1982; 64(11-12):1035-1040.
55. Delaria K, Guillen M, Louie A et al. Stabilization of the Escherichia coli elongation factor Tu-GTP-aminoacyl-tRNA complex. Arch Biochem Biophys 1991; 286(1):207-211.
56. Abrahams JP, Kraal B, Clark BFC et al. Isolation and stability of ternary complexes of elongation factor Tu, GTP and aminoacyl-tRNA. Nucl Acids Res 1991; 19(3):553-557.
57. Zeidler W, Schirmer NK, Egle C et al. Limited hydrolysis and amino acid replacements in the effector region of *Thermus thermophilus* elongation factor Tu. Eur J Biochem 1996; 239:265-271.
58. Cavarelli J, Moras D. Recognition of tRNAs by aminoacyl-tRNA synthetases. FASEB J 1993; 7(1):79-86.
59. Biou V, Yaremchuk A, Tukalo M et al. The 2.9 Å Crystal structure of *T. thermophilus* Seryl-tRNA Synthetase complexed with tRNA^{Ser}. Science 1994; 263:1404-1410.
60. Sekine S, Nureki O, Shimada A et al. Structural basis for anticodon recognition by discriminating glutamyl-tRNA synthetase. Nat Struct Biol 2001; 8(3):203-206.
61. Faulhammer HG, Joshi RL. Structural features in aminoacyl-tRNAs required for recognition by elongation factor Tu. FEBS Lett 1987; 217:203-211.
62. Hilgenfeld R, Bock A, Wilting R. Structural model for the selenocysteine-specific elongation factor SelB. Biochimie 1996; 78(11-12):971-978.
63. Kiesewetter S, Ott G, Sprinzl M. The role of modified purine 64 in initiator/elongator discrimination of tRNA$_i^{Met}$ from yeast and wheat germ. Nucleic Acids Res 1990; 18:4677-4682.
64. Basavappa R, Sigler PB. The 3 Å crystal structure of yeast initiator tRNA: functional implications in initiator/elongator discrimination. EMBO J 1991; 10:3105-3111.
65. Ott G, Schiesswohl M, Kiesewetter S et al. Ternary complexes of *Escherichia coli* aminoacyl-tRNAs with elongation factor EF-Tu and GTP: thermodynamic and structural studies. Biochem Biophys Acta 1990; 1050:222-225.
66. Ryter JM, Schultz SC. Molecular basis of double-stranded RNA-protein interactions: structure of a dsRNA-binding domain complexed with dsRNA. EMBO J 1998; 17(24):7505-7513.
67. Nyborg J, Nissen P, Kjeldgaard M et al. Structure of the ternary complex of EF-Tu: macromolecular mimicry in translation. Trends Biochem Sci 1996; 21:81-82.
68. Liljas A. Protein synthesis: Imprinting through molecular mimicry. Curr Biol 1996; 6:247-249.
69. Nissen P, Kjeldgaard M, Nyborg J. Macromolecular mimicry. EMBO J 2000; 19(4):489-495.
70. Pedersen GN, Nyborg J, Clark BFC. Macromolecular mimicry of nucleic acid and protein. IUBMB Life 1999; 48:1-6.
71. Ævarsson A. Structure-based sequence alignment of elongation factors Tu and G with related GTPases involved in translation. J Mol Evol 1995; 41:1096-1104.
72. Roll-Mecak A, Cao C, Dever TE et al. X-Ray structures of the universal translation initiation factor IF2/eIF5B: conformational changes on GDP and GTP binding. Cell 2000; 103:781-792.
73. Martemyanov KA, Yarunin AS, Liljas A et al. An intact conformation at the tip of elongation factor G domain IV is functionally important. FEBS-Lett 1998; 434(1-2):205-208.
74. Bilgin N, Richter AA, Ehrenberg M et al. Ribosomal RNA and protein mutants resistant to spectomycin. EMBO J 1990; 9(3):735-739.
75. Song H, Mugnier P, Das AK et al. The crystal structure of human eukaryotic release factor eRF1: Mechanism of stop codon recognition and peptidyl-tRNA hydrolysis. Cell 2000; 100:311-321.
76. Selmer M, Al-Karadaghi S, Hirokawa G et al. Crystal Structure of *Thermotoga maritima* Ribosome recycling factor: a tRNA mimic. Science 1999; 286:2349-2352.
77. Toyoda T, Tin OF, Ito K et al. Crystal structure combined with genetic analysis of the *Thermus thermophilus* ribosome recycling factor shows that a flexible hinge may act as a functional switch. RNA 2000; 6(10):1432-1444.
78. Kim KK, Min K, Suh SW. Crystal structure of the ribosome recycling factor from *Escherichia coli*. EMBO J 2000; 19:2362-2370.
79. Ogle JM, Brodersen DE, Clemons Jr. WM et al. Recognition of Cognate Transfer RNA by the 30S Ribosomal Subunit. Science 2001; 292:897-902.
80. Woese CR. Translation: in retrospect and prospect. RNA 2001; 7(8):1055-1067.
81. Kraulis PJ. MOLSCRIPT: a program to produce both detailed and schematic plots of protein structures. J Appl Cryst 1991; 24:946-950.

CHAPTER 21

Translational Termination, Ribosome Recycling and tmRNA Function

Richard Buckingham and Måns Ehrenberg

Summary

The function of peptide release factors has been greatly clarified in recent years by experiments using improved in vitro systems, analysis of mutant factors and new structural information. In class I release factors, which carry stop codon specificity, important domains have been identified and tentatively assigned to defined functions. Crystal structures of both a mammalian and a bacterial class I factor have been determined. The bacterial GTPase class II release factors stimulate recycling of the class I factors by a mechanism with unexpected features, using the ribosome-class I factor complex as guanine nucleotide exchange factor. Ribosome recycling factor in bacteria functions by dissociating the ribosomal subunits and resembles a tRNA molecule in overall form. Ribosomes stalled on messenger RNA for lack of a termination signal or inability to read a sense codon may be released by tmRNA acting both as tRNA and mRNA.

Introduction

Translation termination allows new proteins to be released from the ribosome when their synthesis is complete. In normal protein synthesis, base triplets in mRNA are decoded successively as they enter the A site of the ribosome. During this process, sense codons are recognized by tRNA molecules, partially on the basis of Watson-Crick base pairing between mRNA and the anticodon nucleotides in tRNA and, according to recent crystallographic evidence, partially on steric constraints, before the growing polypeptide chain is donated to the incoming aminoacyl-tRNA. The accuracy of codon recognition is further enhanced by proofreading; an energy driven mechanism with repeated selections of aminoacyl-tRNA. With some exceptions, to be described below, three of the 64 triplet codons are reserved as stop codons, to specify the end of the polypeptide chain. In contrast to sense codons, stop codon recognition depends on protein factors that bind to the ribosomal A site and promote hydrolysis of the ester bond between polypeptide and tRNA. This recognition lacks proofreading and yet is remarkably precise. Its structural basis remains obscure, despite decades of effort in search of an understanding. For recent reviews of translation termination, see also refs. 1-4. A particular form of termination may intervene when a ribosome stalls on mRNA. Under these circumstances, tmRNA may come into play to provide a new termination signal and release the ribosome, by acting both as a tRNA molecule and an mRNA.

The Nature of the Translational Stop Signal, Genetic Code Variations and the Importance of Codon Context

In most organisms, the three codons, UAG, UAA and UGA all signify stop. However, in several instances among ciliate species stop codons have become reassigned as sense codons (see ref.5 for a review). Thus, in *Tetrahymena* and the ciliates *Oxytricha* and *Stylonychia,* both UAG and UAA encode glutamine, and only UGA is used as a stop codon. In the ciliates *Euplotes aediculatus,* UAG and UAA are stop codons and UGA signifies cysteine. In *Mycoplasma* and in the mitochondria of most species, both UGA and UGG encode tryptophan. Apart from reassignments of this type, the

Translation Mechanisms, edited by Jacques Lapointe and Léa Brakier-Gingras.
©2003 Eurekah.com and Kluwer Academic / Plenum Publishers.

efficiency with which termination occurs at stop codons can vary considerably, and such variation often depends on the nucleotide sequence that surrounds them. The next nucleotide downstream of the stop codon is particularly important, and it has even been suggested that the stop signals should be defined as tetra- rather than trinucleotides.[6] The influence that this codon context effect has on the evolution of coding sequences can be seen in at least two ways. Firstly, in microorganisms such as *E. coli, B. subtilis* and *S. cerevisiae* a bias is seen in the frequency of occurrence of the first base downstream of stop codons, in favour of those contexts that are the most efficient for termination of protein synthesis. This bias is particularly significant in highly expressed genes. Secondly, particularly inefficient contexts for stop signals are usually seen in conjunction with sequence elements of mRNA that signal other events than termination of protein synthesis. One such example is the competition between programmed frameshifting and termination, where an unfavorable context for termination can often be understood as fine tuning of control mechanisms for gene expression that depend on frameshifting.

The nucleotide sequence upstream of stop codons in mRNA also affects termination efficiency. However, in this case, the influence is related to the sense of the translated upstream triplets, i.e., which tRNA isoacceptor and amino acid they encode. Thus, a significant bias is seen in favour of certain codons immediately upstream of stop signals both in bacteria[7-9] and in humans.[10] This may reflect the influence of the tRNA isoacceptor present in the ribosomal P-site on the efficiency of the termination process, but the nature of the amino acid itself appears also to be of importance.[9]

Codon-Specific Release Factors (Class I Factors)

Class I Release Factors in Bacteria, Eukarya and Archaea

In the late 1960s, three protein factors that were important for peptide release from terminating ribosomes were discovered. Two of these factors carried specificity for stop codon recognition, and have become known as class I release factors. Thus, RF1 recognises specifically UAG and RF2 recognises specifically UGA, whereas both RF recognise UAA. More than 50 sequences have so far been determined for class I release factors in bacteria. The alignment of these sequences shows both considerable sequence similarity, encompassing the entire group of RF1 and RF2 sequences, and other similarities specific to the RF1 or RF2 families. Proteins of the RF1 family lack about 20 amino acids at the N-terminus but are somewhat extended at the C-terminus compared to those of the RF2 family. Class I release factors from eukarya and archaea are closely related to each other but quite distinct from the bacterial class I factors.[11,12] They are larger, about 440 amino acids in length compared to about 355-365 for the bacterial factors. Comparisons of primary and secondary structures first suggested that the bacterial factors are organised very differently from their eukaryotic and archaeal counterparts. This has now been confirmed by the determination of crystallographic structures for human eRF1[13] and bacterial RF2.[14]

Stop Codon Recognition—Direct or Indirect?

By analogy with tRNA, it is commonly supposed that class I release factors recognize the stop signal in the A-site of the ribosome by interacting directly with the mRNA. However, for the moment, there is no direct proof of this hypothesis. The most direct evidence comes from the demonstration that mRNA containing S^4U can be covalently crosslinked to RF on both bacterial[15] and eukaryotic[16] ribosomes. Nevertheless, alternative mechanisms for the recognition of stop signals remain possible. It has repeatedly been suggested that the initial recognition of the stop signal may be made by rRNA, of either the large or the small subunit. Both the crosslinking data and the hypothesis that a specific region of the bacterial class I factors (the SPF/PXT region; see below) is crucial to stop codon recognition are compatible with the interaction of RF with an mRNA:rRNA complex, or even with a less direct response according to which the presence of a cognate mRNA:rRNA interaction might be transmitted to RF by an alteration in rRNA conformation.

The Unique Universal Sequence Motif GGQ

In spite of the very different organisation of bacterial RF as compared to the eukaryotic and archaeal factors, one short region is totally conserved in RF of all three kingdoms: a tripeptide sequence Gly-Gly-Gln (GGQ).[17] In the crystal structure of human eRF1 (see below, and Fig. 1), the

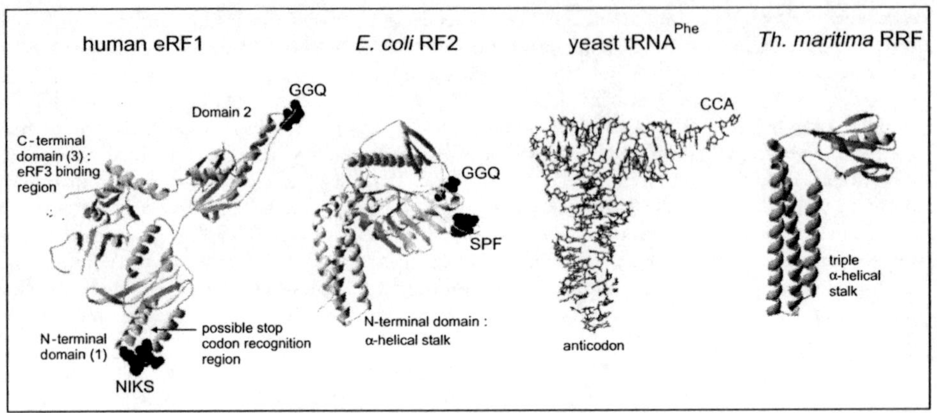

Figure 1. 3D structures of human and bacterial RF and bacterial RRF in comparison to the structure of tRNA.

tripeptide occupies a striking position in the molecule at the extremity of an extended mini-domain.[13] This is consistent with suggestions that the tripeptide plays an important role in triggering the hydrolysis of the ester bond between a tRNA and its polypeptide chain.[17] In the bacterial RF2 structure, the loop containing the GGQ tripeptide extends less from the molecule (see below), but this loop is disordered in the crystal and may adopt a different conformation on the ribosome.[14] A very precise role for the Gln residue has been proposed, namely, that its side chain amide group is responsible for coordinating a water molecule in the peptidyl transferase centre of the ribosome.[13] However, mutational studies of the human eRF1 molecule have shown that several other amino acids can be substituted for the Gln residue with retention of a substantial degree of enzymatic activity, and this is difficult to reconcile with the proposed mechanism.[18]

The Importance of Residue 246 in RF2 and the Methylation of E. coli RF

Genetic studies of RF2 in *E. coli* K12 strains showed the factor to be curiously low in activity compared to RF1, and biochemical experiments with the purified factor revealed that the activity was even lower when the protein was overproduced in vivo from plasmids. Two parameters were eventually found to account for this behaviour. A residue four positions towards the N-terminus from the GGQ motif (residue 246 in *E. coli* RF2) is always Ala or Ser in all known bacterial RF with the single exception of *E. coli* K12 RF2, where it is Thr. Other *E. coli* strains show Ala246 in RF2. The importance for release activity of residue 246 was first demonstrated by Uno et al[19] in a study of the *Salmonella typhimurium* and *E. coli* K12 factors. Restoring an Ala or Ser residue at position 246 to the K12 protein[19-21] increases the release activity of the factor. The reason why overproduced RF2 is less active than factor prepared from normal cells is because the plasmid-expressed protein lacks a post-translational modification of Gln252 in the GGQ motif: both RF1 and RF2 in *E. coli* are N5-methylated at this residue.[20] In vitro, the methylation has little or no effect on RF1 release activity but stimulates considerably the release activity of RF2. The RF methyltransferase is encoded by the *hemK* gene, immediately downstream of and translationaly coupled to *prfA*, encoding RF1.[22] HemK-deficient *E. coli* K12 mutants fail to grow on minimal medium and barely grow on rich media. Spontaneous revertants show mutations that change Thr246 in RF2 to either Ser or Ala.[22]

The Accuracy of Codon Recognition by Class I Peptide Release Factors

As mentioned above, bacterial RF1 recognizes specifically UAG stop codons, while RF2 recognizes specifically UGA. This phenomenon gives a functional identity to each of these factors, which has been used to identify amino acids involved in codon recognition as described in the next section. However, if RF1 would terminate at UGA or, vice versa, RF2 would terminate at UAG, little harm would be created for the cell, since the erroneous and correct reactions would both lead to the same termination event. There is, however, a much more important recognition problem if RF1 and RF2 induce ester bond hydrolysis and terminate protein synthesis when there is a sense, rather than a stop, codon in the ribosomal A-site. If such events were frequent, the processivity of ribosomes, i.e., the probability that a started polypeptide chain is eventually completed, would be drastically reduced and this would lead to growth impairment and possibly to cell death. Recently, Freistroffer et al[23] studied in vitro the efficiency by which RF1 and RF2 terminate at their cognate stop codons compared to the efficiency by which they terminate at closely related sense codons. This work revealed that class I release factors often are very accurate, that their remarkable ability to avoid erroneous termination at sense codons does not depend on energy driven proofreading (see below) and that a few sense codons stand out as "hot-spots", particularly vulnerable to premature termination. For instance, the efficiency by which RF1 terminates at CAA or CAG codons is close to six orders of magnitude smaller than for termination at the cognate stop codons UAA or UAG. RF1 is particularly prone to terminate at UAU (Tyr) and RF2 at UGG (Trp) codons so that the frequency of these codons in an mRNA will to a large extent determine the drop in ribosomal processivity due to termination at sense codons by bacterial class I release factors.

The Conserved Motifs PXT in RF1 and SPF in RF2 and Their Relation to Specific Stop-Codon Recognition

The different specificity of recognition of RF1 and RF2 for stop signals has allowed an ingenious approach to the search of a region of the factors that may mimic the anticodon of a tRNA molecule. By comparing amino acid sequence, Ito et al defined seven domains in *E. coli* RF1 and RF2 that displayed above average levels of conservation.[24] From this starting point, a set of 128 chimeric molecules was constructed and a pair of chimeric factors was sought in which the swap of one domain reversed the specificity of UAG and UGA recognition. Such a pair was found, and further refinement of the discriminatory element suggested that only three consecutive amino acids are needed for specific stop codon recognition.[24] The tripeptide is PXT at positions 187-189 in RF1, and SPF at positions 205-207 in RF2. The interpretation of these experiments is, however, complicated by the fact that a hybrid RF1/RF2 background has to be employed to observe the exchange of specificity. The authors propose a direct interaction between the tripeptide and the stop codon. This work marks a major step forward in understanding the relationship between the specificity of stop codon recognition and release factor structure.

The 3D Structure of Human eRF1

The 3D structure of one eRF1 is currently known, that of the human factor, determined by X-ray diffraction to a resolution of 2.8 Å.[13] Three domains may be distinguished in the structure (Fig. 1). An N-terminal domain (domain 1) contains at its extremity in a loop between two α-helices a conserved motif, NIKS, of unknown function. Mutational studies (see below) suggest that this domain interacts with the stop signal, but several strands may be implicated in the interaction.[25] Domain 2 is notable for an extended α-helix preceded by a loop containing the conserved sequence GGQ, at the extremity of the molecule opposed to the NIKS motif. According to the current hypothesis, this motif interacts with the peptidyl-transferase of the ribosome and triggers the peptidyl hydrolase activity that releases the polypeptide chain. Domain 3, the C-terminal domain, appears not to be strictly necessary for the release activity of the eukaryotic factor, but is required for the interaction with the class II factor eRF3 (see below).

The Search for the Anticodon Region of eRF1

A genetic search has been made for a part of yeast eRF1 that might be responsible for direct recognition of the stop codon.[25] The strategy was to select for mutants with disabled recognition of only one of the three stop codons. The majority of such mutants fall within a pocket on the surface of the N-terminal domain (Fig. 1), and it was suggested that this forms a stop codon recognition site. Alternative suggestions have been made. By analogy to the SPF/PXT region of bacterial RF, a tripeptide region (residues 58-60 in human eRF1; TAS) just N-terminal of the strongly conserved tetrapeptide sequence (N_{60}IKS, see Fig. 1) has been suggested to interact with the stop codon.[2] Comparisons of eRF1 sequences from organisms with canonical and non-canonical stop codon recognition implicated other residues in this region, stretching from E_{55} to N_{60}.[26] Another attempt to interpret eRF1 sequence alignments suggested that I_{35} and L_{126} are involved in stop codon recognition.[27] However, some recently determined eRF1 sequences fit poorly these hypotheses and emphasise the need for further experimental work.[28]

The 3D Structure of E. coli RF2

The crystal structure of *E. coli* RF2 is strikingly different from that of its functional analogue in man.[14] The molecule is composed of four domains (Fig. 1). An N-terminal domain consisting of four α helices contains a three-stranded coiled coil such as found in RRF. In the latter case, the domain has been proposed to mimic the anticodon stem of tRNA[29] (Fig. 1). Domain 2, composed of a five-stranded antiparallel β sheet and two α helices, shows structural homology to ribosomal proteins S6 and S10, and contains the SPF tripeptide thought to confer stop codon specificity[24], exposed in a loop on the surface of the molecule. The GGQ motif is found in another loop, in domain 3 which packs onto domain 2 and possesses homology to ribosomal protein S5. In this structure the GGQ and SPF motifs are only 23Å apart. If the orientation of domains 2 and 3 is the same when RF2 is on the ribosome as it is in the crystal, then it is clear that the SPF motif and the GGQ motif cannot interact simultaneously with the stop codon and the peptidyl transferase centre, respectively, as they have been suggested to do.

Docking the RF2 structure on to the A site of the 5.5 A model of the *Thermus thermophilus* 70S ribosome[30] so as to superimpose the SPF motif on the anticodon of A-site-bound tRNA leads to serious clashes between the protein and rRNA.[14] Vestergaard et al favor an alternative orientation of the molecule in which the GGQ loop contacts the A-loop of the peptidyl transferase centre, the SPF docks on to helix 44 of 16S rRNA and domain 1 is superimposed on the anticodon stem-loop of A-site-bound tRNA.[14] How codon-specific reading might be achieved by this model is not yet clear. This orientation of RF2 and tRNA is shown in Fig. 1.

A different model for RF2 binding to the ribosome would suppose that domains 2 and 3 change orientation when the factor binds to the ribosome, allowing the SPF/PXT and GGQ regions to interact simultaneously with the stop codon and peptidyl transferase centre respectively. To decide which, if either, of these models is correct will require further experimental work.

The Ribosome in Termination—Ribosomal Mutations Affecting Termination

Numerous studies have implicated rRNA in the overall process of translation termination. Firstly, it is now very likely that the peptidyl transferase centre itself, the activity of which in a modified form is supposed to lead to peptidyl-tRNA hydrolysis during termination, is a ribozyme. The strongest argument for this comes from the crystallographic reconstruction of an isolated 50S ribosomal subunit from *Haloarcula marismortui*, which reveals that no ribosomal proteins are present in the precise region of the ribosome where peptidyl transfer is known to take place.[31] Two regions of 16S rRNA are strongly implicated in termination. Several nucleotides in helix 34 affect termination: changes at 1054 and 1200 affect RF2 but not RF1 function, whereas changes at 1192 affect both, though the defects were larger for RF2 than RF1.[3] Part of a second helix, helix 44, is situated very close to the A-site and is clearly important to the termination process. In 23S rRNA, a region termed the GTPase center, known to be involved in GTP hydrolysis by elongation factors EF-Tu and EF-G, has also been clearly implicated in termination. Thus, a G1093A mutation in this region causes UGA-specific suppression in vivo[32] and severely reduces the efficiency (k_{cat}/K_M) for RF2-dependent termination in vitro.[33] The specificity of this change to RF2 function may indicate differences in the way that RF1 and RF2 interact with the GTPase center of the 50S ribosomal subunit.[3]

Class II Release Factors

Class II factors are involved in the termination process but do not improve the specificity of stop codon recognition.[23] The factors are ribosome-dependent GTPases, RF3 in bacteria and eRF3 in higher systems. The GTPase activity of eRF3 is also dependent on the presence of eRF1 on the ribosome, and contrary to earlier conclusions, it has recently been shown that ribosome-dependent hydrolysis of bacterial RF3 is strongly stimulated by RF1 or RF2 in a codon specific fashion. The essential role of RF3 is to accelerate the dissociation of RF1 and RF2 from the ribosome following polypeptide release. The factor is not required for cell viability, but mutants lacking the factor grow less well than wild-type strains. The role and mechanism of action of RF3 have been elucidated largely as a result of the development of efficient in vitro systems for protein synthesis based on purified components. Although the function of eRF3 may be similar to that of RF3, the absence of such in vitro systems for eukaryotic organisms has so far prevented comparable studies of eRF3.

Bacterial RF3—Recycling of Class I Release Factors and the Role of GTP Hydrolysis

Pre-termination complexes containing short synthetic mRNAs, encoding oligopeptides, and P-site-bound peptidyl-tRNA can be readily prepared in vitro and used as substrate for RF in order to study the kinetic parameters of reactions involved in translation termination. Such experiments revealed that the presence of RF3 does not affect the rate by which class I RF associate with the ribosome (k_{cat}/K_m) nor the catalytic rate (k_c) for peptide release in vitro.[34] Instead, the presence of RF3 and GTP greatly accelerates the rate at which RF1 and RF2 recycle between ribosomal termination complexes, by increasing drastically the dissociation rate constant for RF1/2 after peptide release in a way that depends on GTP hydrolysis. RF3 thus relieves the sequestering of RF1 or RF2 as well as ribosomes in a post-termination state. This serves to maintain adequate levels of available RF1/2 in the cell, to prevent queuing of ribosomes at stop signals and explains why RF3 is important for fast growth of bacterial populations.

Recent experiments have clarified the detailed mechanism by which RF3 rapidly removes RF1 and RF2 from the ribosome after, but not before, dissociation of the peptide from tRNA (Fig. 2). Unexpected findings in this work were that GDP binds three orders of magnitude more strongly to free RF3 than GTP and that the time for spontaneous dissociation of GDP from the factor is as long as half a minute. Rapid exchange of GDP on RF3 is only promoted by ribosomes in complex with a class I release factor. Dissociation of GDP from RF3 leads to a high affinity ribosomal complex containing RF1 or RF2 and RF3 in the guanine nucleotide free state. The existence of this type of high affinity complex explains why RF3 reduces, rather than enhances, the specificity of class I release factor codon recognition.[23] Not until the ester bond in peptidyl-tRNA has been hydrolyzed and the peptide dissociated from the ribosome, can GTP bind to RF3 on the ribosome and change the structure of RF3 to the GTP form. This conformation excludes RF1 or RF2 from the ribosome and catalyzes their rapid dissociation in a reaction driven forward by the high affinity of RF3·GTP for the ribosome. Removal and recycling of RF3 itself requires GTP hydrolysis and switch of the RF3 structure to its GDP form with intermediate affinity for the ribosome. This means that RF3

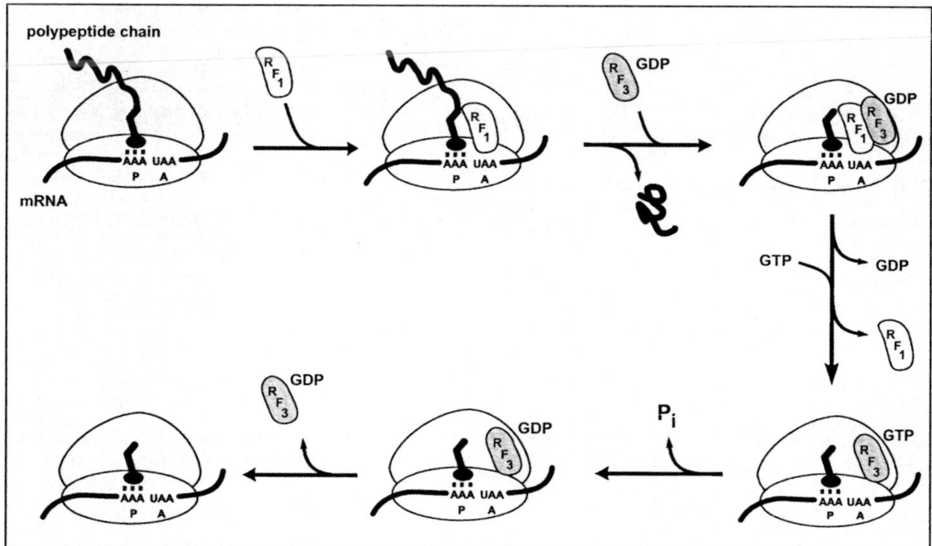

Figure 2. The mechanism of peptide release and release factor recycling.

returns to its free state in the GDP conformation and remains in this form until it returns to a new ribosome in its next cycle. This scenario also implies that RF3 and EF-Tu play opposite roles in bacterial protein synthesis. The elongation factor EF-Tu facilitates the entry of aminoacyl-tRNAs to the ribosomal A-site, and allows the accuracy of codon recognition to be enhanced by making the free energy associated with hydrolysis of its GTP available for ribosomal proofreading of aminoacyl-tRNAs.[35] Thus, discrimination between cognate and near-cognate aminoacyl-tRNAs is possible in a proofreading step that occurs after EF-Tu:GDP has left the ribosome, in addition to the discrimination at the initial selection step when aminoacyl-tRNA binds to the ribosome in complex with EF-Tu:GTP.

In contrast to EF-Tu, the class II peptide release factor RF3 removes its substrate, RF1 or RF2, from the ribosome after termination and reduces, rather than enhances, the accuracy of codon reading by these class I release factors. This occurs because nucleotide-free RF3 stabilizes the interaction of the class I factor with ribosomes containing peptidyl-tRNA, as described above. In spite of their diametrically opposite functions, EF-Tu and RF3 both belong to the same class of small G-proteins that require a guanine nucleotide exchange factor and they have a number of functional properties in common.[36]

Eukaryotic eRF3

Unlike its bacterial counterpart, eRF3 from yeast is encoded by an essential gene. A further feature that distinguishes the eukaryotic factors is that eRF1 and eRF3, in contrast to RF1 or RF2 and RF3, have a strong affinity for each other in the absence of ribosomes. Parts of the C-terminal domain of eRF1 are needed for this interaction with eRF3.[37] The last 22 amino acids of the domain are not required, consistent with the high variability of this part of the sequence of the C-terminal domain among different organisms. Two regions of eRF3 are required for the interaction, including the last ten amino acids. However, the functional significance of the interaction is unclear. Indeed, in fission yeast, it appears to be possible to largely or completely abolish the interaction between the two factors by C-terminal truncation of eRF1 without affecting cell viability.[38]

The Prion-Like Region of Yeast eRF3

Compared to bacterial RF3, eRF3 contains a considerable N-terminal extension that displays large variability between species, and may be deleted without loss of the biochemical activity of the factor.[39,40] The main function of this region may be to bind the poly(A)-binding protein PABP1.[41] In some strains of *Saccharomyces cerevisiae*, the N-terminal domain appears also to give rise to the genetic element [*PSI*+], associated with increased stop codon suppression[42] and inherited in a non-Mendelian manner like prions in mammalian cells.[43] In [*PSI*+] cells, eRF3 molecules interact with each other through their N-terminal domains.[44,45] A similar phenomenon occurs in some other yeasts, and in one case it has been shown that the [*PSI*+] state can be transmitted across an interspecies barrier.[46]

Ribosome Recycling

Following release of the polypeptide chain from the ribosome, further steps are necessary before bacterial ribosomal subunits can initiate and participate in further rounds of protein synthesis. The pioneering work of Kaji identified two proteins in *E. coli* essential to this process, RRF (Ribosome recycling factor) and EF-G, the elongation factor better known for its essential role in mRNA translocation from A- to P-site during the ribosomal elongation cycle (see ref. 47 for a review). RRF is an essential protein and, in its absence, ribosomes tend to reinitiate translation at any codon and in any frame, downstream of stop signals.[48] Though universal in all studied eubacteria, and present also in cell organelles such as mitochondria and chloroplasts, no homologue of RRF has been found in the cytoplasm of the eukaryotic cell or in archaea. How ribosomes recycle in eukarya or archaea, and whether any functional equivalent of RRF is necessary, is currently unknown.

The 3D Structure of RRF

Crystal structures of RRF from three species of bacteria have so far been determined: *Thermotoga maritima*[29], *E. coli*[49] and *T. thermophilus*.[50] These show a two-domain structure, that resembles a tRNA molecule in overall dimensions and shape. Domain I is a long three helix bundle that may mimic the stacked D-stem and anticodon-stem of the tRNA molecule. Domain II, composed of a β/α/β sandwich, may mimic the rest of the tRNA, except for the CCA-3' end. The connection between the two domains appears to be flexible and the angle between them is different in the three structures. Mutational studies of residues in this hinge region suggest that the flexibility is vital to the function of RRF.[50]

The Mechanism of Bacterial Ribosome Recycling—the Role of RRF, EF-G and IF3

RRF was first isolated as a factor that, in conjunction with EF-G and GTP, was able to catalyze polysome breakdown.[51] A more detailed understanding of how the factor functions has been provided by studies using post-termination ribosomal complexes resulting from termination reactions in vitro[52] (Fig. 3). In the absence of proteins catalyzing their breakdown, these complexes are quite stable, which is incompatible with the physiological rate of ribosome recycling. Four proteins were found to be necessary for the maximum rate of complex breakdown in vitro: RRF, EF-G, the initiation factor IF3 and RF3. The latter factor seems to be necessary only to promote the dissociation of RF1/2 from the ribosome, as described above. Consistent with its structure as a tRNA mimic, RRF is thought to bind to the ribosomal A-site and, together with EF-G, promote dissociation of the ribosomal subunits in a reaction requiring GTP hydrolysis.[52] Some authors propose that the role of EF-G is to translocate RRF from the A-site to the P-site, in analogy to the translocation of peptidyl-tRNA during elongation. However, as discussed previously,[52] several features of the RRF-dependent reaction fail to support this idea, including the fact that the deacylated tRNA remains firmly associated with the 30S ribosome following subunit dissociation. The need for IF3 in the overall reaction is to destabilize the 30S:mRNA:tRNA complex, displacing the tRNA and allowing the 30S subunit to form a new initiation complex on the same or another mRNA. This function is thus very similar to the important role of IF3 in destabilizing incorrect initiation complexes.[53] However, in this cases the tRNA is not aminoacylated, and the substrate of the reaction arises at each release of a polypeptide chain during protein synthesis.

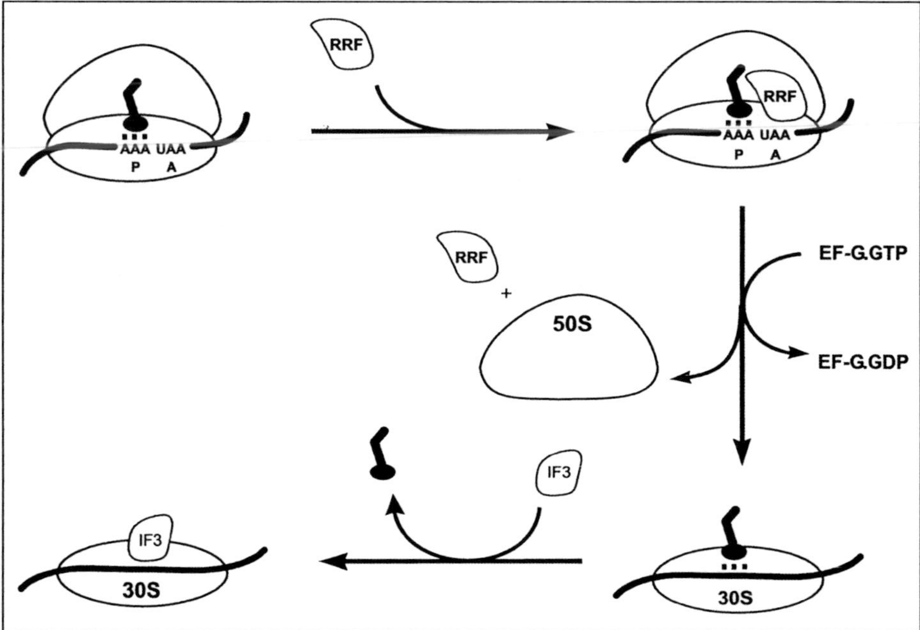

Figure 3. RRF action and ribosome subunit recycling.

tmRNA Function and Mechanism

Transfer-messenger RNA (tmRNA) is a metabolically stable RNA of about 350-410 nucleotides in length that exists in most and perhaps all eubacteria. [54-56] No tmRNA-like sequence has yet been found in archaea or eukarya. As its name implies, tmRNA can function both as a tRNA and an mRNA on the ribosome. The 3' part of the molecule shows homologies with tRNA[Ala], including the amino acid acceptor-stem and the TΨC-stem, and can be charged with Ala by Ala-tRNA synthetase. [57,58] In two sets of circumstances, tmRNA is brought into play. The first arises when ribosomes, due to lack of normal termination, reach the 3' end of an mRNA with a peptidyl-tRNA in P-site, and the A-site partially or completely lacking mRNA. This situation may for instance arise when the mRNA has been truncated and therefore lack an in-phase termination codon, or when the normal stop codon has been translated in a nonsense suppression event. The second circumstance is when ribosomes stall during mRNA translation, generally as a result of rare codon clusters and cognate tRNA shortage. [59] Translation termination can also give rise to a pause leading to the recruitment of tmRNA. [60] Tagging at pause sites is probably distinct from that occuring at the 3' end of an mRNA, though it has been suggested that recruitment of tmRNA at pause sites during mRNA translation might be the indirect result of ribosome-directed mRNA cleavage subsequent to ribosome pausing. [61] In either case, the action of tmRNA is to liberate the stalled ribosome by allowing translation to continue and to complete the nascent polypeptide chain by the addition of a tag encoded by tmRNA that promotes degradation of the tagged protein. [61]

Structure of tmRNA

At the moment of writing, the tmRNA website contains 165 sequences of tmRNA. [55] This includes the special case presented by the α-subdividion of the proteobacteria, in which the *SsrA* gene is differently organised and maturation of the transcript results in a tmRNA molecule in two fragments. [56] The primary structure and predicted secondary structure of the *E. coli* tmRNA are shown in Fig. 4. The 5' and 3' ends of the molecule come together to form a domain with strong similarity to parts of the tRNA[Ala] molecule. The domain includes the acceptor-stem and TΨC-stem, and contains notably the G:U base pair in the acceptor stem that is a determinant for recognition by Ala-tRNA synthetase. The remaining part of the sequence can be considered as an insertion in the anticodon region of the tRNA-like domain and includes the coding region for the ANDENYALAA tag, followed by a UAA stop codon. Four pseudoknots are predicted in the structure, one of which has been studied in some detail and found to be important to tmRNA function. [62]

The Mechanism of Action of tmRNA

The observation of a common C-terminal extension to a set of aberrant truncated translation products led to the discovery that SsrA, later called tmRNA, could function both as a tRNA and an mRNA. [63] Shortly afterwards, it was recognized that the extension encoded by tmRNA directed the tagged proteins towards degradation by the *E. coli* ClpXP, ClpAP and FtsH proteases. This provided a rational framework for tmRNA function and a coherent model [64] for its mechanism of action. According to this model (see Fig. 5), alanyl-tmRNA binds to a translating ribosome with an empty A-site in a step requiring a protein factor, SmpB, [65] and the elongation factor EF-Tu in its GTP form. SmpB is a protein of about 160 residues that binds strongly to tmRNA in a way that does not interfere with the interactions of tmRNA with either Ala-tRNA synthetase or EF-Tu. As in normal EF-Tu function, GTP hydrolysis occurs, except that no codon:anticodon interaction is required when tmRNA replaces tRNA, and EF-Tu:GDP dissociates from the ribosome. Peptidyl transfer then occurs from the peptidyl-tRNA in the P-site to alanyl-tmRNA in the A-site. Following a translocation step, tmRNA then acts as mRNA, encoding the addition of a further 10 aminoacids on the C-terminus of the polypeptide chain, and finally directs a normal translation termination event at the UAA stop codon.

Degradation of Tagged Proteins

The C-terminal tag added by tmRNA in different species of eubacteria varies considerably both in length (from about 10 to 27 amino acids) and in sequence. However, the last five amino acids of the tag are much more highly conserved and are similar to sequences known to target proteins for degradation. [66] Both cytoplasmic and periplasmic degradation systems for such tagged proteins exist

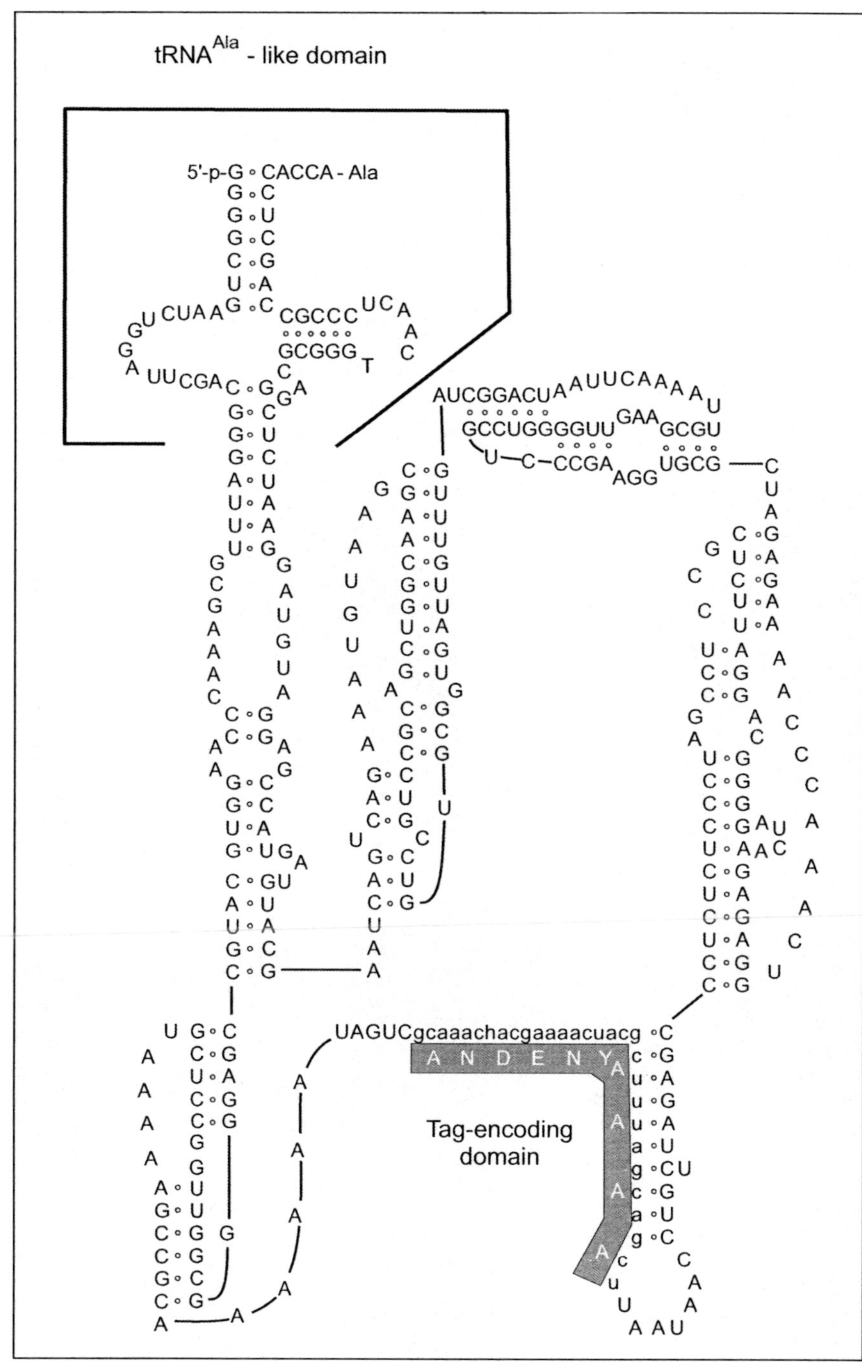

Figure 4. Domains in tmRNA.

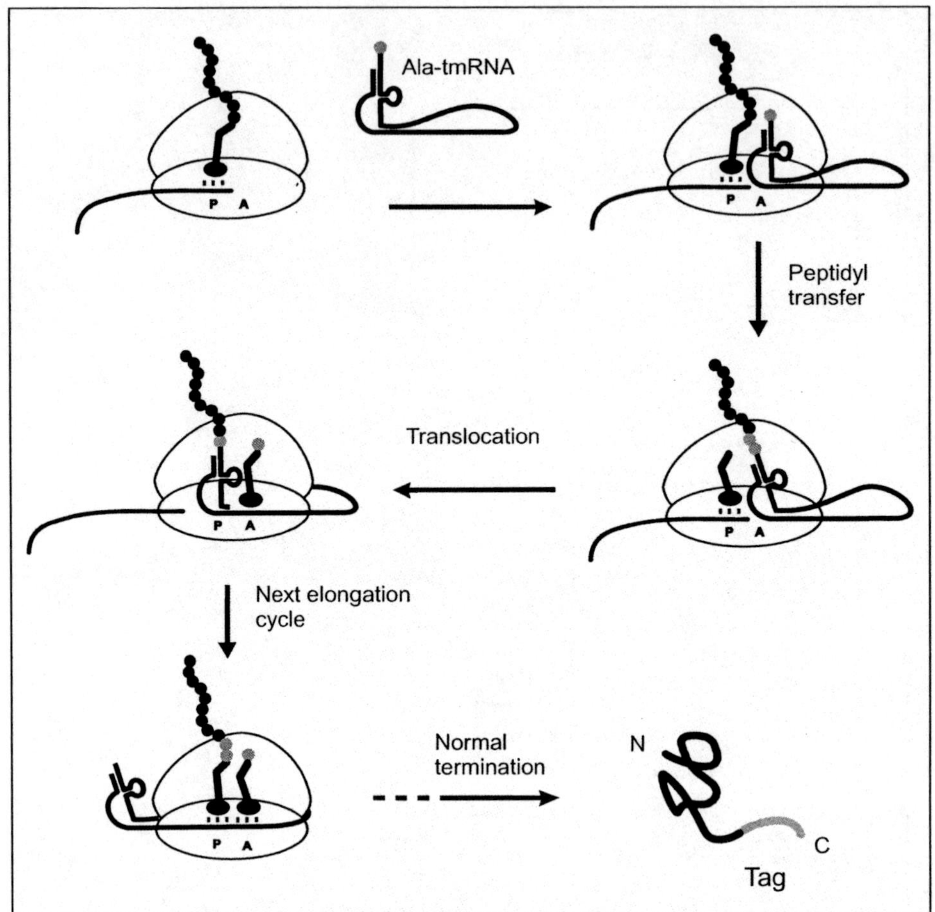

Figure 5. The mechanism of action of tmRNA.

in *E. coli* and many other eubacteria. Two energy-dependent systems have been identified in the cytoplasm in *E. coli*. One implicates the tetradecameric serine protease ClpP, able to form stacked annular complexes with the hexameric ATPases ClpX or ClpA,[67] which denature the proteins and translocate the polypeptide to ClpP for degradation. A further ATPase protease in *E. coli*, FtsH, is also able to degrade the tagged proteins,[68] and, unlike the ClpP system, seems so far to be universally present in eubacteria. The periplasm of *E. coli* possesses at least one protease, Tsp,[64] able to degrade tmRNA-tagged proteins. Other periplasmic proteases such as DegP, DegQ and DegS, which, like Tsp, have PDZ domains that bind the C-terminal domains of certain proteins, may also be involved in the process.[61]

Biological Role of tmRNA

The need for tmRNA and SmpB varies greatly between different species of bacteria and also depends on the conditions of growth. tmRNA is not essential in *E. coli*, though a variety of phenotypes are seen in cells lacking *ssrA*, including a growth defect particularly marked at elevated temperatures.[58,69] In contrast, *N. gonorrhoeae*, *M. genetalium* and *M. pneumoniae* do require tmRNA for viability.[70,71] The case of *S. typhimurium* is interesting because this bacterium survives without tmRNA when growing normally but requires the gene for growth within macrophages.[72,73] This is consistent with the idea that some growth conditions are propitious for accidents of translation and increase the dependence on tmRNA for survival.

The fact that tmRNA-encoded tags are recognized by a family of proteases leaves no doubt that this aspect of tmRNA action is of physiological importance. However, experiments in which the tag-encoding sequence has been changed to one no longer recognized by proteases and thus inactivating the targeting of tagged proteins for degradation suggest that, at least under certain environmental conditions, the ability of the SsrA-SmpB system to recycle stalled ribosomes may be more important to the cell than the targeting of truncated proteins for proteolysis.[74]

References

1. Kisselev LL, Buckingham RH. Translational termination comes of age. Trends Biochem Sci 2000; 25:561-566.
2. Nakamura Y, Ito K, Ehrenberg M. Mimicry grasps reality in translation termination. Cell 2000; 101:349-352.
3. Arkov AL, Murgola EJ. Ribosomal RNAs in translation termination: facts and hypotheses. Biochemistry (Mosc) 1999; 64:1354-1359.
4. Ehrenberg M, Tenson T. A new beginning of the end of translation. Nature Structural Biology 2002; 9:85-87.
5. Lozupone CA, Knight RD, Landweber LF. The molecular basis of nuclear genetic code change in ciliates. Curr Biol 2001; 11:65-74.
6. Tate WP, Poole ES, Dalphin ME et al. The translational stop signal: Codon with a context, or extended factor recognition element. Biochimie 1996; 78:945-952.
7. Björnsson A, Mottagui-Tabar S, Isaksson LA. Structure of the C-terminal end of the nascent peptide influences translation termination. EMBO J 1996; 15:1696-1704.
8. Arkov AL, Korolev SV, Kisselev LL. Termination of translation in bacteria may be modulated by interaction between peptide chain release factor 2 and the last peptidyl-tRNA$^{Ser/Phe}$. Nucleic Acids Res 1993; 21:2891-2897.
9. Mottagui-Tabar S, Isaksson LA. The influence of the 5' codon context on translation termination in *Bacillus subtilis* and *Escherichia coli* is similar but different from *Salmonella typhimurium*. Gene 1998; 212:189-196.
10. Arkov AL, Korolev SV, Kisselev LL. 5' contexts of *Escherichia coli* and human termination codons are similar. Nucleic Acids Res 1995; 21:2891-2897.
11. Frolova L, Le Goff X, Rasmussen HH et al. A highly conserved eukaryotic protein family possessing properties of polypeptide chain release factor. Nature 1994; 372:701-703.
12. Kisselev LL, Oparina NY, Frolova LY. Class-1 translation termination factors are structurally and functionally similar to suppressor tRNAs and belong to distinct structural/functional families (prokaryotes/mitochondria and eukaryotes/archaea). Mol Biol (Moscow) 2000; 34:427-442.
13. Song H, Mugnier P, Das AK et al. The crystal structure of human eukaryotic release factor eRF1—mechanism of stop codon recognition and peptidyl-tRNA hydrolysis. Cell 2000; 100:311-321.
14. Vestergaard B, Bich Van L, Andersen GR et al. Bacterial polypeptide release factor RF2 is structurally distinct from eukaryotic eRF1. Molec Cell 2001; 8:1375-1382.
15. Poole ES, Major LL, Mannering SA et al. Translational termination in *Escherichia coli*—3 bases following the stop codon cross-link to release factor-2 and affect the decoding efficiency of UGA-containing signals. Nucleic Acids Res 1998; 26:954-960.
16. Chavatte L, Frolova L, Kisselev L et al. The polypeptide chain release factor eRF1 specifically contacts the sUGA stop codon located in the A site of eukaryotic ribosomes. Eur J Biochem 2001; 268:2896-2904.
17. Frolova LY, Tsivkovskii RY, Sivolobova GF et al. Mutations in the highly conserved GGQ motif of class 1 polypeptide release factors abolish ability of human eRF1 to trigger peptidyl-tRNA hydrolysis. RNA 1999; 5:1014-1020.
18. Seit Nebi A, Frolova LY, Ivanova N et al. Substitutions of the glutamine residue in the ubiquitous GGQ tripeptide in human eRF1 do not entirely abolish the release factor activity. Mol Biol (Moscow) 2000; 34:899-900.
19. Uno M, Ito K, Nakamura Y. Functional specificity of amino acid at position 246 in the tRNA mimicry domain of bacterial release factor 2. Biochimie 1996; 78:935-944.
20. Dinçbas-Renqvist V, Engström Å, Mora L et al. A post-translational modification in the GGQ motif of RF2 from *Escherichia coli* stimulates termination of translation. EMBO J 2000; 19:6900-6907.
21. Wilson DN, Guevremont D, Tate WP. The ribosomal binding and peptidyl-tRNA hydrolysis functions of *Escherichia coli* release factor 2 are linked through residue 246. RNA 2000; 6:1704-1713.
22. Heurgué-Hamard V, Champ S, Engstrom A et al. The N5-glutamine methyltransferase that modifies peptide release factors in *Escherichia coli* is encoded by *hemK*. EMBO J 2002; 21:769-778.
23. Freistroffer DV, Kwiatkowski M, Buckingham RH et al. The accuracy of codon recognition by ribosome release factors. Proc Natl Acad Sci USA 2000; 97:2046-2051.
24. Ito K, Uno M, Nakamura Y. A tripepeptide anticodon deciphers stop codons in messenger RNA. Nature 2000; 403:680-684.
25. Bertram G, Bell HA, Ritchie DW et al. Terminating eukaryote translation: domain 1 of release factor eRF1 functions in stop codon recognition. RNA 2000; 6:1236-1247.

26. Muramatsu T, Heckmann K, Kitanaka C et al. Molecular mechanism of stop codon recognition by eRF1: a wobble hypothesis for peptide anticodons. FEBS Lett 2001; 488:105-109.

27. Lehman N. Molecular evolution: Please release me, genetic code. Curr Biol 2001; 11:R63-66.

28. Kervestin S, Frolova L, Kisselev L et al. Stop codon recognition in ciliates: Euplotes release factor does not respond to reassigned UGA codon. EMBO Rep 2001; 2:680-684.

29. Selmer M, Al-Karadaghi S, Hirokawa G et al. Crystal structure of *Thermotoga maritima* ribosome recycling factor: a tRNA mimic. Science 1999; 286:2349-2352.

30. Yusupov MM, Yusupova GZ, Baucom A et al. Crystal structure of the ribosome at 5.5 A resolution. Science 2001; 292:883-896.

31. Nissen P, Hansen J, Ban N et al. The structural basis of ribosome activity in peptide bond synthesis. Science 2000; 289:920-930.

32. Jemiolo DK, Pagel FT, Murgola EJ. UGA suppression by a mutant RNA of the large ribosomal subunit. Proc Natl Acad Sci USA 1995; 92:12309-12313.

33. Arkov AL, Freistroffer DV, Pavlov MY et al. Mutations in conserved regions of ribosomal RNAs decrease the productive association of peptide-chain release factors with the ribosome during translation termination. Biochimie 2000; 82:671-682.

34. Freistroffer DV, Pavlov MY, MacDougall J et al. Release factor RF3 in *E. coli* accelerates the dissociation of release factors RF1 and RF2 from the ribosome in a GTP-dependent manner. EMBO J 1997; 16:4126-4133.

35. Ruusala TEM, Ehrenberg M, Kurland CG. Is there proofreading during polypeptide synthesis? EMBO J 1982; 1:741.

36. Zavialov AV, Buckingham RH, Ehrenberg M. A post-termination ribosomal complex is the guanine nucleotide exchange factor for peptide release factor RF3. Cell 2001; 107:1-20.

37. Merkulova TI, Frolova LY, Lazar M et al. C-terminal domains of human translation termination factors eRF1 and eRF3 mediate their in vivo interaction. FEBS Lett 1999; 443:41-47.

38. Ito K, Ebihara K, Nakamura Y. The stretch of C-terminal acidic amino acids of translational release factor eRF1 is a primary binding site for eRF3 of fission yeast. RNA 1998; 4:958-972.

39. Zhouravleva G, Frolova L, Le Goff X et al. Termination of translation in eukaryotes is governed by two interacting polypeptide chain release factors, eRF1 and eRF3. EMBO J 1995; 14:4065-4072.

40. Frolova L, Le Goff X, Zhouravleva G et al. Eukaryotic polypeptide chain release factor eRF3 is an eRF1- and ribosome-dependent guanosine triphosphatase. RNA 1996; 2:334-341.

41. Hoshino S, Imai M, Mizutani M et al. Molecular cloning of a novel member of the eukaryotic polypeptide chain-releasing factors (eRF). Its identification as eRF3 interacting with eRF1. J Biol Chem 1998; 273:22254-22259.

42. Cox BS. *psi*, a cytoplasmic suppressor of super-suppressor in yeast. Heredity 1965; 20:505-521.

43. Ter-Avanesyan MD, Dagkesamanskaya AR, Kushnirov VV et al. The SUP35 omnipotent suppressor gene is involved in the maintenance of the non-Mendelian determinant [psi+] in the yeast *Saccharomyces cerevisiae*. Genetics 1994; 137:671-676.

44. Paushkin SV, Kushnirov VV, Smirnov VN et al. Propagation of the yeast prion-like *psi* ⁺ determinant is mediated by oligmerisation of the *SUP35*-encoded polypeptide chain release factor. EMBO J 1996; 15:3127-3133.

45. Patino MM, Liu JJ, Glover JR et al. Support for the prion hypothesis for inheritance of a phenotypic trait in yeast. Science 1996; 273:622-626.

46. Nakayashiki T, Ebihara K, Bannai H et al. Yeast [PSI+] «prions» that are crosstransmissible and susceptible beyond a species barrier through a quasi-prion state. Mol Cell 2001; 7:1121-1130.

47. Kaji A, Hirokawa G. In: Garrett R et al, eds. The ribosome, structure, function, antibiotics and cellular interactions. Washington DC: A.S.M. Press, 2000:527-539.

48. Janosi L, Mottagui-Tabar S, Isaksson LA et al. Evidence for in vivo ribosome recycling, the fourth step in protein biosynthesis. EMBO J 1998; 17:1141-1151.

49. Kim KK, Min K, Suh SW. Crystal structure of the ribosome recycling factor from *Escherichia coli*. EMBO J 2000; 19:2362-2370.

50. Toyoda T, Tin OF, Ito K et al. Crystal structure combined with genetic analysis of the *Thermus thermophilus* ribosome recycling factor shows that a flexible hinge may act as a functional switch. RNA 2000; 6:1432-1444.

51. Hirashima A, Kaji A. Factor-dependent release of ribosomes from messenger RNA: Requirement for two heat-stable factors. J Mol Biol 1972; 65:43-58.

52. Karimi R, Pavlov M, Buckingham RH et al. Novel roles for classical factors at the interface between translation termination and initiation. Mol Cell 1999; 3:601-609.

53. Risuleo G, Gualerzi C, Pon C. Specificity and properties of the destabilization, induced by initiation factor IF-3, of ternary complexes of the 30-S ribosomal subunit, aminoacyl-tRNA and polynucleotides. Eur J Biochem 1976; 67:603-613.

54. Ray BK, Apirion D. Characterization of 10S RNA: a new stable RNA molecule from *Escherichia coli*. Molec Gen Genet 1979; 174:25-32.

55. Williams KP. The tmRNA website. Nucleic Acids Res 2000; 28:168.

56. Keiler KC, Shapiro L, Williams KP. tmRNAs that encode proteolysis-inducing tags are found in all known bacterial genomes: A two-piece tmRNA functions in Caulobacter. Proc Natl Acad Sci USA 2000; 97:7778-7783.

57. Ushida C, Himeno H, Watanabe T et al. tRNA-like structures in 10Sa RNAs of *Mycoplasma capricolum* and *Bacillus subtilis*. Nucleic Acids Res 1994; 22:3392-3396.
58. Komine Y, Kitabatake M, Yokogawa T et al. A tRNA-like structure is present in 10Sa RNA, a small stable RNA from *Escherichia coli*. Proc Natl Acad Sci USA 1994; 91:9223-9227.
59. Roche ED, Sauer RT. SsrA-mediated peptide tagging caused by rare codons and tRNA scarcity. EMBO J 1999; 18:4579-4589.
60. Roche ED, Sauer RT. Identification of endogenous SsrA-tagged proteins reveals tagging at positions corresponding to stop codons. J Biol Chem 2001; 276:28509-28515.
61. Karzai AW, Roche ED, Sauer RT. The SsrA-SmpB system for protein tagging, directed degradation and ribosome rescue. Nat Struct Biol 2000; 7:449-455.
62. Nameki N, Felden B, Atkins JF et al. Functional and structural analysis of a pseudoknot upstream of the tag-encoded sequence in *E. coli* tmRNA. J Mol Biol 1999; 286:733-744.
63. Tu GF, Reid GE, Zhang JG et al. C-terminal extension of truncated recombinant proteins in *Escherichia coli* with a 10Sa RNA decapeptide. J Biol Chem 1995; 270:9322-9326.
64. Keiler KC, Waller PR, Sauer RT. Role of a peptide tagging system in degradation of proteins synthesized from damaged messenger RNA. Science 1996; 271:990-993.
65. Karzai AW, Susskind MM, Sauer RT. SmpB, a unique RNA-binding protein essential for the peptide-tagging activity of SsrA (tmRNA). EMBO J 1999; 18:3793-3799.
66. Parsell DA, Silber KR, Sauer RT. Carboxy-terminal determinants of intracellular protein degradation. Genes Dev 1990; 4:277-286.
67. Gottesman S, Roche E, Zhou Y et al. The ClpXP and ClpAP proteases degrade proteins with carboxy-terminal peptide tails added by the SsrA-tagging system. Genes Dev 1998; 12:1338-1347.
68. Herman C, Thevenet D, Bouloc P et al. Degradation of carboxy-terminal-tagged cytoplasmic proteins by the *Escherichia coli* protease HflB (FtsH). Genes Dev 1998; 12:1348-1355.
69. Oh BK, Apirion D. 10Sa RNA, a small stable RNA of *Escherichia coli*, is functional. Molec Gen Genet 1991; 229:52-56.
70. Huang C, Wolfgang MC, Withey J et al. Charged tmRNA but not tmRNA-mediated proteolysis is essential for *Neisseria gonorrhoeae* viability. EMBO J 2000; 19:1098-1107.
71. Hutchison CA, Peterson SN, Gill SR et al. Global transposon mutagenesis and a minimal *Mycoplasma* genome. Science 1999; 286:2165-2169.
72. Julio SM, Heithoff DM, Mahan MJ. ssrA (tmRNA) plays a role in *Salmonella enterica serovar Typhimurium* pathogenesis. J Bacteriol 2000; 182:1558-1563.
73. Baumler AJ, Kusters JG, Stojiljkovic I et al. *Salmonella typhimurium* loci involved in survival within macrophages. Infect Immun 1994; 62:1623-1630.
74. Withey J, Friedman D. Analysis of the role of trans-translation in the requirement of tmRNA for lambdaimmP22 growth in *Escherichia coli*. J Bacteriol 1999; 181:2148-2157.

CHAPTER 22

Recoding: Site- or mRNA-Specific Alteration of Genetic Readout Utilized for Gene Expression

Ivaylo P. Ivanov, Olga L. Gurvich, Raymond F. Gesteland and John F. Atkins

Abstract

A minority of genes in probably all organisms rely on "recoding" for translation of their mRNAs. Recoding can involve a proportion of ribosomes changing frame at a specific site in response to signals in mRNA, or some ribosomes reading through a stop codon to insert a standard amino acid or the 21st amino acid, selenocysteine. In other cases, ribosomes bypass a block of noncoding nucleotides present within an open reading frame. In several cases, recoding serves a regulatory function. Often there are distinct roles for both the product of standard decoding and the recoding product with which it shares amino terminal sequence. This review of the current state of the field includes a reassessment of the variety of mechanisms involved.

Since the genetic code and the mechanism of its readout were first elucidated, molecular biologists have found more and more evidence that "the code" and decoding are not written "in stone". Some organisms have branched into an alternative meaning of particular codons. Even more fascinating is the ability of certain mRNA sequences to subvert the resident translational machinery. A number of genes, from viruses to higher eukaryotes, have evolved to exploit this latter type of translational plasticity in order to regulate their own expression. To distinguish these events from "standard" decoding, they are often referred to as "recoding". Recoding events can be further subdivided into several distinctive categories: frameshifting, redefinition and hopping.

Frameshifting

At defined shift sites in mRNA, ribosomes can be programmed to efficiently change to one of the two alternative reading frames for gene expression purposes. Commonly, there are stimulatory signals in mRNA distinct from the shift site that greatly increase the level of frameshifting at the shift site. As a result, ribosomes initiating at the same start codon produce two different protein products, one being the product of standard decoding and the other the product of a recoding event. Depending on the configuration of the open reading frames (ORFs) in an mRNA relative to the site of frameshifting, two outcomes are possible. If the stop codon of the new ORF (ORF2) is 3' of the termination codon of the original ORF (ORF1), the frameshifting product is longer (sometimes much longer) than the product resulting from standard decoding. Alternatively, if the stop codon of ORF2 is 5' of the end of ORF1, the frameshift protein product is shorter. Many more examples are known where the frameshift product is longer but that may be because of inherent biases associated with the discovery of frameshifting examples.

Redefinition

In redefinition, codon meaning is changed in an mRNA specific manner (as opposed to reassignments of the "universal" genetic code that are species- or organelle-specific). Specification of an amino acid by a "stop" codon results in the production of protein product, which is longer than the product of standard decoding. Most known cases of redefinition fall into this category, perhaps only due to

Translation Mechanisms, edited by Jacques Lapointe and Léa Brakier-Gingras.

the fact that the products of standard decoding and recoding, in this case, have usually very different sizes and therefore can easily be detected and distinguished by standard laboratory techniques. Encoding of selenocysteine by special UGA codons allows incorporation of this 21st amino acid thus extending the capabilities of the genetic code. Encoding an alternative amino acid by a sense codon could, in theory, result in the production of two proteins with the same number of amino acids but differing biochemical activities. Perhaps because of the technical difficulties of identifying such cases, no examples of this kind have been discovered.

Bypassing

In translational bypassing (hopping), a fraction of the translating ribosomes "skip" a portion of the mRNA without inserting any amino acid. Regular decoding is resumed some distance downstream. As with programmed frameshifting and redefinition, translational hopping occurs at defined sequences within an mRNA. Conceptually, bypassing can be described as translational splicing. Most known cases of bypassing involve heterologous expression or synthetic constructs where the ribosome hops over a single triplet codon and lands in-frame with the original reading frame. If the bypassed triplet is a sense codon, the protein synthesized is two amino acids shorter than the product of conventional decoding. When the bypassed triplet is a stop codon, the bypassing links two open reading frames to yield a longer polypeptide than the product of standard decoding. Bypassing needs not involve three, or multiples of three, nucleotides and so is frame independent. In the best-studied case of ribosomal bypassing, that is seen in the translation of T4 gene 60, translating ribosomes skip 50 nucleotides before resuming regular decoding. Because this links two separate ORFs, the resulting polypeptide is much longer that the product of standard decoding of gene 60 mRNA.

Examples of Recoding Events in Gene Regulation

As can be seen from the brief descriptions above, recoding presents great regulatory potential that is actively exploited in diverse organisms (reviewed in ref. 1). The ability to make two or more proteins from the same mRNA is sometimes very useful, especially for organisms that put a high premium on compact genomes. Not surprisingly, most examples of recoding events have been found in viruses and transposable elements. In addition, producing two proteins from the same mRNA allows the setting up of an exact stoicheometric ratio between them. In fact, in the majority of recoding events, the proportion of ribosomes that complete the nonstandard event is tightly set.

In most viruses and retrotransposable elements, the recoding event serves to link structural (e.g., retroviral Gag) and catalytic polypeptides (e.g., retroviral Pol or Pro-Pol) (reviewed in refs. 2 and 3). Standard decoding of retroviral genomic RNA typically produces only Gag protein (whose ORF is located 5' on the mRNA) but through recoding, Gag-Pol, or Gag-Pro-Pol, is produced as a single polypeptide.[4] Viruses need many more molecules of structural proteins than catalytic ones. For this reason, in most of these cases, recoding frequencies are set in the range of 1 to 10% relative to standard decoding. The result is that most ribosomes synthesize just the structural subunits and only the minority, which undergo a recoding event, synthesize the polyproteins which are processed to yield the catalytic proteins. The frequency with which recoding events occur varies from one virus to another. So, in *S. cerevisiae* L-A virus, the -1 frameshifting required for production of its Gag-Pol counterpart occurs at 1.9%,[5,6] the +1 frameshifting of *S. cerevisiae* Ty3 retrotransposon Gag-Pol mRNA occurs at 11%,[7] and the human immunodeficiency virus -1 frameshifting required for the synthesis of Gag-Pol occurs at 0.7-7.3%.[8,9] In each case, the efficiency of recoding has presumably evolved to suit the lifestyle of the virus or retrotransposable element. Experiments have shown that the ratio between Gag and Gag-Pol and therefore the efficiency of recoding can be crucial for retroviral and retrotransposon propagation. An increased or decreased ratio due to alterations in recoding efficiency hampers virus-particle assembly and infectivity of L-A and HIV viruses.[10,11] It also leads to reduced transposition of Ty3.[12]

Another example where a recoding event (-1 translational frameshifting) sets up a fixed ratio between two polypeptide products is decoding of the *E. coli dnaX* gene,[13-15] which encodes two subunits of DNA polymerase III, τ and γ (Fig. 1). These two subunits are present in a ratio of 1:1 in the polymerase holoenzyme. The long form, τ, is synthesized by standard decoding. The short form, γ, shares its amino acid sequence with τ for the first two-thirds of the latter but has one additional amino acid at the carboxyl end. With the help of stimulatory signals in the mRNA,[16,17] 50% of

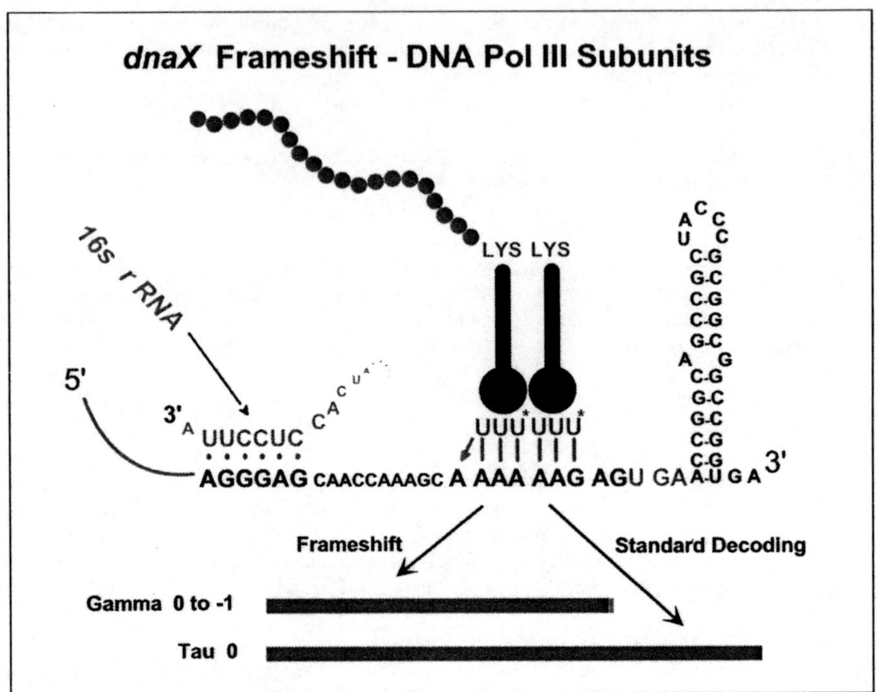

Figure 1. Schematic representation of the −1 frameshifting in the expression of *E. coli dnaX* gene.

translating ribosomes switch to the −1 frame at the "slippery sequence" A-AAA-AAG, two thirds of the way into ORF1. The ribosomes that shift frame decode one codon in the new frame before terminating at a now in-frame UGA stop codon to synthesize γ.[13-15] Because γ lacks two protein domains of τ that bind to a replication protein called DnaB and the αεθ polymerase III core subunit respectively,[18] the two polypeptides have different biochemical activities and, thus, the two subunits play different roles in DNA polymerase III.

Many physiological conditions can alter the activity of the translational apparatus. Consequently, some genes have evolved to finely tune a recoding event to an autoregulatory pathway that affects the physiology of translation. Perhaps the best-studied example of the regulatory potential of recoding is the +1 frameshifting required for expression of *E. coli* release factor-2 (RF2).[19-24] RF2 is a translational release factor that mediates termination at the stop codons UGA and UAA (see also Ch.21 by M. Ehrenberg et al). RF2 protein is encoded by two partially overlapping reading frames. The first encodes only 25 amino acids. This peptide has no known biochemical function and is rapidly degraded. +1 ribosomal frameshifting at the last sense codon of the first ORF (ORF1), at the sequence CUU-U, results in the production of full-length functional RF2. Significantly, the stop codon of ORF1 is UGA, which is recognized only by RF2. A pause of translation at the shift site caused by low levels of RF2, contributes to +1 ribosomal frameshifting, thereby increasing synthesis of RF2 and closing an autoregulatory loop (Fig. 2). In this way, the frameshifting site "senses" the levels of RF2 in the cell.

Ornithine decarboxylase antizyme (antizyme) was first defined as a biochemical activity that inhibits ornithine decarboxylase and is induced in the presence of high concentrations of polyamines. Decarboxylation of ornithine is the first and usually rate-limiting step in the biosynthesis of polyamines in the cell. Polyamines have many documented biological activities, one of which is to affect the rate and accuracy of translation.[25] Cloning of the first antizyme gene in mammals (rats) revealed that the ORF that encodes the biochemically functional portion of the antizyme protein (ORF2) does not contain an appropriate translation initiation codon.[26] Instead, translation is initiated at an upstream

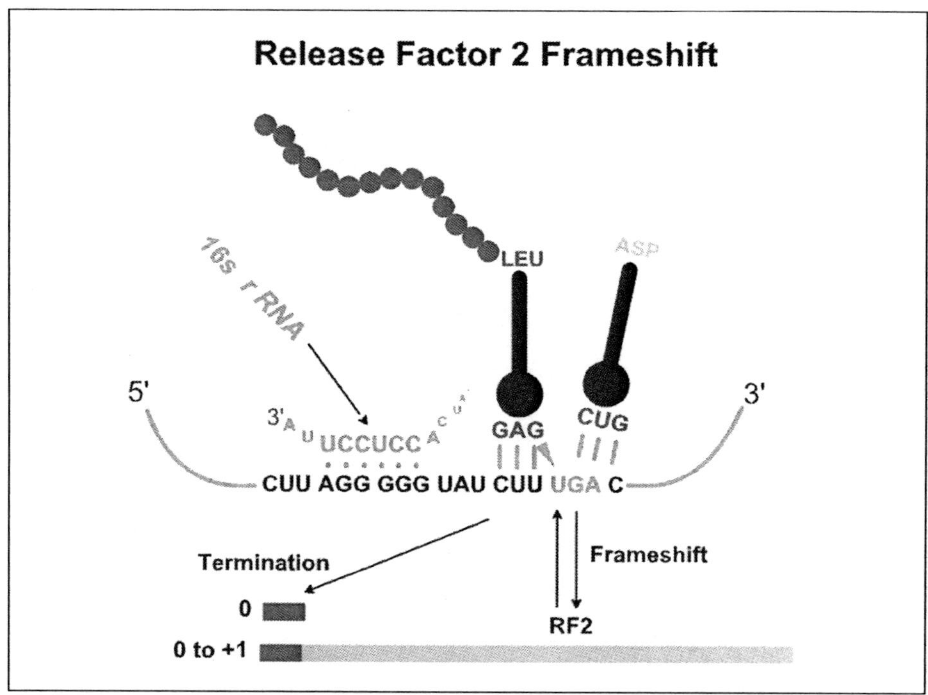

Figure 2. The regulatory +1 frameshifting involved in the expression of *E. coli* RF2.

partially overlapping ORF (ORF1) in such a way that +1 translational frameshifting in the overlap would result in the production of fully functional antizyme protein. Further analysis demonstrated that a ribosomal frameshifting event is involved and occurs at the very last sense codon of ORF1 on the sequence UCC-U.[27] Importantly, in vitro and later in vivo experiments showed that elevated levels of polyamines increase the level of +1 translational frameshifting required for synthesis of functional antizyme,[27-29] thus closing an autoregulatory loop. In this manner, the translational frameshifting on antizyme mRNA is a key component of the antizyme regulatory loop, serving as its sensor (Fig. 3). This mechanism is deduced to have evolved more than a billion years ago in the ancestor of modern day fungi, echinoderms, nematodes, insects, and vertebrates. It is amazingly conserved in all descendents that are known to possess antizyme homologs.[30]

Mechanisms of Recoding

Like all biological processes, translation is not error-free. However, translational errors due to misincorporation of the wrong amino acid or to a switch of translational reading frame are relatively rare; the most widely accepted estimate being 10^{-4} per codon.[31,32] Although conceptually and often mechanistically similar, except for specification of selenocysteine, translational errors and recoding events are separated by a quantitative degree. The stimulatory signals for recoding have been evolutionarily selected and the product is utilized. These features put errors and recoding in qualitatively different categories. All recoding events share one mechanistic feature. In every case, the recoding event is in direct competition with standard decoding. Consequently, mRNA signals that perturb normal decoding stimulate the recoding event and vice versa. Recoding signals consist of several components. The first and most important signal is the actual site, within the mRNA, where recoding is initiated. These sites are made of three to seven nucleotides, and are necessary and sufficient for induction of the recoding event (except in the incorporation of selenocysteine). However, often they are insufficient for achieving the needed efficiency, and accessory cis-acting elements are utilized. In

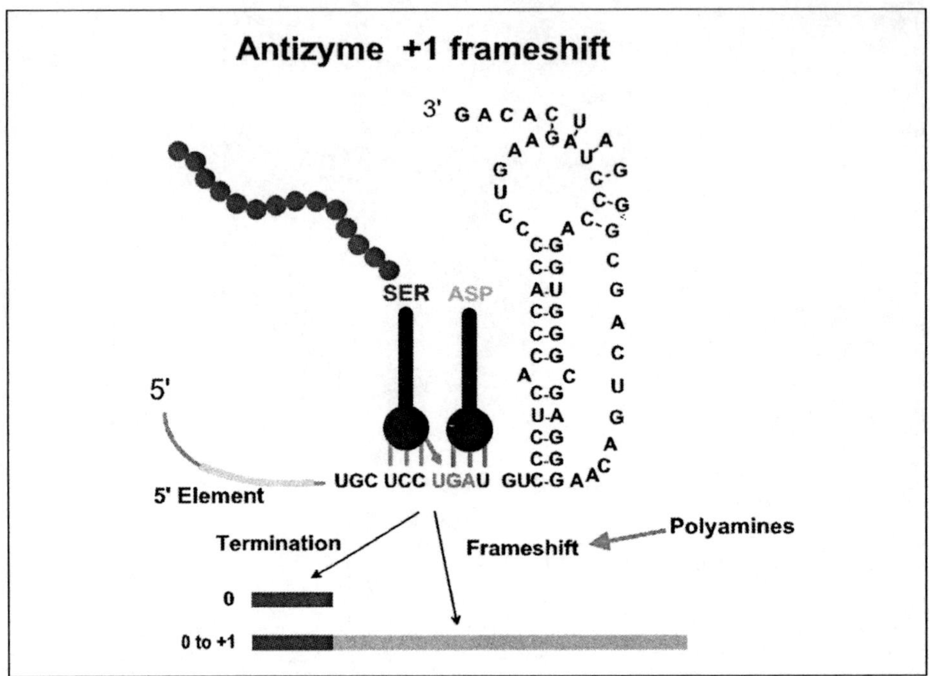

Figure 3. The +1 frameshifting event involved in the autoregulatory expression of mammalian antizyme 1.

several cases, a single recoding site uses two or more different cis-acting stimulatory elements, placed both 5' and 3'.

Classification of cis-acting stimulators can also be made on the basis of the nature of the mRNA signal. In some cases, the cis-acting elements act through their primary sequence, in others, the stimulators act through a secondary structure they form. Of the limited number known, most that work through their primary sequence are 5' of the recoding site, although exceptions do exist. Such sequences can function either directly through their nucleotide sequence or through the nascent peptide they encode. Numerous cis-acting elements have been identified, however, the exact mechanism of their action is not well understood.

Despite their wide diversity, the great majority of recoding events can be grouped in only two major classes: "tandem slippage" frameshifting and ribosomal "P-site" events. To date, most examples of programmed translational frameshifting are of the -1 tandem slippage variety. Examples are known from retroviruses, coronaviruses, plant viruses, a yeast virus-like element, bacterial insertion sequences, bacteriophages and at least one bacterial cellular gene, *E. coli dnaX*.[33] The tandem slippage model was first proposed by Jacks et al to explain the −1 frameshifting in Rous sarcoma virus.[34] According to it, frameshifting happens at a slippery heptamer with the zero frame sequence, X-XXY-YYZ (where X could be any nucleotide, Y is a weakly base-pairing nucleotide and Z is species specific). Upon reaching this sequence, the tRNAs in the P and A sites of the ribosome (reading XXY and YYZ respectively) simultaneously slip to the −1 frame (XXX-YYY). In the new frame, at least two base-pairs are preserved between each of the two tRNAs and the mRNA. (In a few cases, the base-pairing in the new frame is stronger than the pairing in the original frame). The most important feature of this sequence is its repetitive nature. In fact, the sequence AAA-AAA-G is the shiftiest in *E. coli* and can support as much as 2% -1 frameshifting.[17,35] Mutations that disrupt the runs of XXX or YYY severely decrease frameshifting. Strong base-pairing in the YYY sequence, for example by introducing runs of three C-s or G-s, can also lead to reduced frameshifting.[5,36]

A refinement of the simultaneous slippage model proposed by Weiss et al postulated that the slip occurs after peptidyl transfer (explaining why the "A-site" tRNA inserts its amino acid in the final polypeptide product) but before, or more likely during, ribosomal translocation.[35] Presumably, when the ribosome is in this state and when base-pairing between the A-site tRNAs and the message is weak, some kind of inherent special stereochemical configuration in the decoding site of the ribosome allows the A- and P-site tRNAs to simultaneously slip back by one nucleotide relative to the mRNA. Favorable base-pairing in the new position "locks" the ribosome in the new frame.

Cis-Acting mRNA Sequences Stimulate Simultaneous Slippage

Several cis-acting sequences are known to stimulate frameshifting at slippery heptamers. The most common is an RNA secondary structure downstream of the shift site. In almost all cases in eukaryotes, this downstream structure is an RNA pseudoknot, on average six nucleotides 3' of the heptamer.[37] Although in individual cases the distance between the pseudoknot and shift site is optimized (i.e., it varies from one example to another), changing it by as little as two nucleotides in either direction reduces or completely eliminates the stimulatory effect.[36] A number of different types of RNA pseudoknots are known to stimulate simultaneous slip events.[37] Instead, in a few cases, mostly in prokaryotes, the frameshift is stimulated by a stem-loop structure; either a simple stem-loop or in some cases a more complex branching structure.[38] In one case, the 3' stimulator is more than 3,800 nucleotides downstream of the frameshift site.[39] One model is that all of these RNA structures stimulate the frameshift event by stalling the ribosome at the shift site, thus allowing more time for the recoding event to take place. However, their function is not as simple as this. Some RNA structures, both pseudoknots and stem-loops, are perfectly good ribosome "stallers" but cannot stimulate frameshifting.[40-42]

In several prokaryotic cases, slippery -1 frameshift events are enhanced by 5' stimulators. Such stimulators have been described in a number of insertion sequences and in the chromosomal *dnaX* gene. In all known cases, the stimulator is a Shine-Dalgarno (SD) type sequence that acts by pairing with its complementary sequence near the 3' end of 16S rRNA of translating ribosomes. The optimal distance between the shift site and the SD sequence is between 9 and 14 nucleotides. Exact spacing is not as critical as the position of the downstream stimulators, but bringing the sequences closer, 2 to 4 nucleotides, leads to inhibition of frameshifting. One explanation for the SD-like stimulator is that it may help "pull" the ribosome backwards to facilitate the -1 frameshift. The spacing requirement for optimal stimulatory effect is consistent with this explanation. In addition, the SD-like interaction may help "stall" the ribosome at the recoding site long enough for the recoding event to occur.

P-Site Events

"P-site" mechanisms can be invoked to explain a large number of recoding events including +1 and −1 frameshifting, and ribosomal hopping. What all of these events and stop codon readthrough have in common is that they occur when the P site is occupied by a peptidyl-tRNA and the adjacent A site is unoccupied. The vacant A site provides a slowdown in standard decoding, facilitating the recoding event to take place. Simplest to describe is the mechanism of translational readthrough.

By the time the first examples of programmed translational readthrough became known, studies of nonsense suppression had "paved the way" by showing competition at stop codons between termination and chain propagation . Nonsense suppressors are mutants that allow the translation of mRNAs containing premature, in-frame nonsense codons. Such mutants, broadly, fall into two categories. Most nonsense suppressors are mutations in tRNAs that alter the anticodon loop so that the mutant tRNA can base-pair with a termination codon.[43-46] The base-pairing tRNA inserts an amino acid in place of one of the three stop codons, thus extending translation past the internal nonsense codon. Most non-tRNA suppressors are mutations in the genes encoding release factor proteins, *prfA*[47-49] and *prfB*[50,51] in *E. coli* , SUP35 and SUP45 in *S. cerevisiae*.[52] These mutations lead to reduced translational termination efficiency. By hampering termination, they allow effective competition of a near-match tRNA for the corresponding nonsense codons, thus stimulating the nonstandard decoding event.

Cis-Acting Sequences Stimulate Stop Codon Readthrough

At sites of programmed readthrough, cis-acting mRNA signals partially "disable" termination of translation. The well-known cis-acting stimulators of readthrough are located downstream of the recoding site. Certain sequences immediately 3' of a termination codon are preferred for efficient termination.[53] This phenomenon known as "3' context" is exploited by some programmed readthrough site for achieving optimum efficiency of recoding.[54] Such sites have evolved 3' contexts that are suboptimal for termination.

A 3' pseudoknot is very important for stop codon readthrough of the *gag* gene terminator of Murine Leukemia Virus (MLV) and a minority of other retroviruses.[37,55] The naturally occurring pseudoknot in MLV cannot be substituted with the pseudoknot that stimulates frameshift in Mouse Mammary Tumor Virus (MMTV).[55] The special features of the readthrough stimulatory pseudoknot presumably mean that causing a pause is not the main feature of the pseudoknot and interference with termination may be critically important. Another 3' readthrough stimulator of uncertain nature is present in Barley Dwarf Yellow Virus PAV coat protein gene. Amazingly, in this case the stimulatory element is situated nearly 700 bases 3' of the readthrough stop codon.[56]

+1 Frameshifting

Most of the small number of known cases of recoding in chromosomal genes involve +1 frameshifting events, which can be explained by a P-site mechanism. This includes the genes *prfB* (encoding RF2) in *E. coli*, EST3 and ABP140 in *S. cerevisiae*, and antizyme in metazoans. Initial data came from analysis of the frameshifting event in decoding *prfB*. As discussed above, the analysis showed that frameshifting occurs on the last sense codon of a short ORF1 at the sequence CUU-U. The last U of this sequence is part of the in-frame stop codon UGA. Frameshifting occurs when the peptidyl tRNA slips from CUU to UUU in the +1 frame. The stability of base-pairing in the new position is crucial.[23] A pause at the 0 frame A-site is also crucial as the levels of release factor can directly modulate frameshifting efficiency. High levels of RF2, which increase the efficiency of termination and thus reduce pausing at the UGA stop codon, lead to reduced levels of frameshifting. The reverse is also true.

The frameshift event occurring during translation of yeast Ty1 Gag-Pol mRNA provides an important variation in the P-site theme,[57] which was also revealed in studies of the consequence of amino acid starvation.[58-60] With Ty1, a +1 frameshift event occurs on the sequence CUU-AGG-C. The tRNA decoding CUU is in the P site and the A site of the ribosome is empty. Again, the peptidyl tRNA slips in the +1 direction relative to the mRNA to form two base pairs with the codon UUA. Unlike *prfB*, the pause is provided by a "hungry" sense codon rather than a termination codon. The tRNA that decodes AGG in *S. cerevisiae* is rare, and the AGG codon itself is also rare. This leads to a slower rate of decoding while the P-site tRNA in the ribosome is primed for slippage. Consistent with this, overexpression of the tRNA recognizing AGG codons significantly reduces the frameshifting potential of the Ty1 frameshift site,[57] while deleting this tRNA gene from the genome increases frameshifting on the Ty1 site even further.[12]

A different type of mechanism has been proposed for Ty3 frameshifting.[7] In this case, +1 frameshifting occurs on the sequence GCG-AGU-U, and, again, the first codon (GCG) is in the P site and the A-site AGU codon is vacant. As with all P-site events, ribosomal pausing at the rare AGU codon is important; however, the mechanism of shifting frames has been interpreted differently. The cognate tRNA that decodes the GCG codon cannot form standard base-pairs with the +1 codon CGA and it was proposed that tRNA slippage is not involved. Instead, it was proposed that the P-site tRNA somehow interferes with normal decoding in the adjacent A site, which allows incoming tRNAs to contact the +1 GUU codon out-of-frame. Consistent with this hypothesis, overexpression of the tRNA for GUU increases the levels of frameshifting.[61]

However, several recent results cast doubt on the necessity for invoking two separate mechanisms to explain the known cases of +1 frameshifting. Initial analysis of the frameshift site of mammalian antizyme 1 revealed that the tRNA, which is in the P-site during recoding, could not form good base-pairing with the next available +1 codon (though substantially better than in Ty3 frameshifting). This was even more obvious with several P-site mutants, which could otherwise support efficient frameshifting. Therefore, a Ty3-like mechanism was proposed.[27] However, when the same sequence was tested in *S. cerevisiae* for its frameshift potential, it supported a ribosomal shift to the +1 frame but mostly via -2 shifting (90% of the time) rather than the +1 shift (10% of the time) seen in mammals.[62] This happens even though the UCC P-site codon is recognized by tRNAs that have the same anticodon in mammals and yeast. Something similar was observed when the antizyme frameshift cassette was tested in *S. pombe*.[63] In *S. pombe*, 80% of the shift product result from a +1 shift and 20% result from a -2 shift. It is clear that the -2 shift can occur only as a result of re-pairing. It seems improbable that the same sequence can induce two rare and mechanistically different recoding events. More likely, the two antizyme mRNA decoding events are related through a single P-site re-pairing mechanism.

More recent experiments with the two yeast +1 frameshift sites, those of Ty1 and Ty3, suggest that in both cases the P-site codon:anticodon pairing is sub-standard.[64] Consistent with this, overexpressing the genes for P-site tRNAs that gave standard pairing with the codons involved, dramatically reduced +1 frameshifting on both Ty1 and a Ty3-like recoding sites. Deleting the cognate P-site tRNAs for a modified Ty1 site significantly increased frameshifting. For Ty1, the explanation for these results is that unstable tRNA: mRNA contacts combined with ribosome stalling on "hungry" A-site codons leads to a slip of the tRNA relative to the mRNA. From their earlier experiments Farabaugh and colleagues reasonably proposed that the P-site tRNA at the Ty3 frameshift site does not dissociate but that the nature of the P-site codon:anticodon interaction influences the immediate 3' codon base so that it is unavailable for pairing with an incoming tRNA.[7,64] Instead A-site bases 2,3 and 4 pair with an incoming tRNA so that the frame is shifted +1. Recent results have shown that, at least for bypassing, the requirements for successful "re-pairing" of mRNA to P-site tRNA in the shifted position may be surprisingly minimal (A. J. Herr, personal communication). While bypassing may differ from single nucleotide frameshifting in ribosome contacts with mRNA, it is tempting to consider that minimal "re-pairing" may suffice for programmed frameshifting of Ty3 also – if the shift tRNA presumed to be involved by Farabaugh and colleagues is the correct one.[64] If so, then a single type of dissociation/re-pairing model may apply to both Ty1 and Ty3, and all other known cases of +1 frameshifting. For Ty3 frameshifting, it is not possible, by simple mutagenesis experiments, to distinguish between a model that involves "once-only pairing and occlusion of the first A-site codon base" and a re-pairing model. Currently both models seem viable.

Cis-Acting mRNA Sequences Stimulating +1 Frameshifting

Cis-acting stimulators of P-site frameshfting events are noted for their variety. A 5' stimulator of P-site recoding is an SD-like sequence just upstream of the frameshift site of *prfB*, the first example of a 5' stimulator in recoding.[21] This sequence is placed only three nucleotides 5' of the CUU-U shift site (i.e., much closer than its location for stimulating –1 frameshifting in *dnaX*). The distance is crucial. Moving the SD sequence even by one base in either direction greatly reduces frameshifting. In this case, the close distance between the SD and the recoding site is thought to physically "push" the ribosome sitting on the frameshift site in the +1 direction.

Another 5' stimulator is a 40-50 nucleotides sequence placed upstream of the mammalian antizyme 1 and 2 genes, which stimulates frameshifting 2.5-5 fold.[30,62] Little is known about its mechanism of action, but it appears to exert its activity through primary nucleotide sequence and not the peptide it encodes (S. Matsufuji, personal communication). Several potential interactions have been proposed between this stimulator and rRNAs (O. Matveeva, personal communication) but none of them have been tested. It is clear that this stimulator is modular. It appears to have evolved in three separate stages over 1 billion years, with the modules evolving in order of their proximity to the shift site.[30] Antizyme genes have also evolved at least two 3' stimulators. The best studied case is an RNA pseudoknot several nucleotides downstream of the frameshift site.[27] This pseudoknot exists in two related versions, one in the vertebrate orthologs of antizyme 1 and the other in the vertebrate orthologs of antizyme 2. As with pseudoknots in -1 simultaneous slippage, spacing to the recoding site is important. Furthermore, in *S. cerevisiae*, where the mammalian antizyme sequence supports mostly -2 frameshifting, moving the pseudoknot by three nucleotides downstream results in a dramatic increase of the proportion of ribosomes shifting in the +1 direction.[62] This again demonstrates that RNA pseudoknots do more than just stall ribosomes on the recoding sites by performing additional function(s). The endogenous *S. pombe* antizyme 3' stimulator is not a recognizable pseudoknot and, though its nature is obscure, it is likely to be a new type of stimulatory element.[65]

Another interesting 3' stimulator is a short 7-15 nucleotide sequence immediately downstream of the frameshift site of Ty3 that stimulates frameshifting about 7.5 fold.[7] This sequence is thought to interfere with A-site decoding by forming base-pairing with helix 18 of 18S rRNA.[66] Finally, just like readthrough, P-site frameshifting events that are stimulated by an A-site stop codons have evolved

3' sequences that provide a poor 3' termination context. This is most obvious in a phylogenetic analysis of the known antizyme sequences. In every case where sufficient data is available, the 3' context of the antizyme gene in question is least favorable for termination.[30] This is especially striking because different taxonomic groups have different 3' context requirements.

Ribosomal Hopping

Ribosomal hopping can be seen as an extreme case of translational frameshifting. Therefore, its mechanism can be viewed as a variation of P-site slippage resulting in +1 frameshifting. Most of our knowledge about ribosomal hopping comes from work on bacteriophage T4 gene 60, which encodes a subunit of phage topoisomerase.[67,68] At the end of gene 60 ORF1, up to half of all translating ribosomes bypass 50 nucleotides and then resume normal translation on a second downstream ORF2.[69] Three stages define this remarkable event. During the first stage (take-off), the P-site tRNA dissociates from the P-site GGA codon. In the next stage (scanning), the ribosome traverses the coding gap checking for a matching P-site codon. In the last stage (landing), the P-site tRNA-mRNA pairing is re-established and regular translocation resumes. The original A-site codon is a (slowly decoded) UAG stop codon. For efficient landing, matching take-off and landing codons are essential. Several mutants that reduce gene 60 bypassing have been isolated and all of them map to the tRNA gene tRNA$_2$^{Gly} that decodes the P-site GGA codon.[70] These mutant tRNAs seem perfectly capable of dissociating from the P-site codon but are unable to find the landing site.[71,72]

Cis-Acting mRNA Sequences Stimulate Ribosome Hopping

Several cis-acting elements stimulate the ribosomal hopping in gene 60. One is a downstream stem/loop structure that partially overlaps the take-off site.[68] It appears that this structure interferes with normal decoding of the in-frame A-site termination codon.[71] How it does this is not clear. When the wild type structure is substituted with a more stable counterpart, the efficiency of recoding drops indicating that it is not just the energy of melting the structure that is important. The sequence GAG 5' of the landing site may function as an SD-like element but this has not yet been established (C. Rettberg and F. Adamski, personal communication). The length of the coding gap also seems to effect the efficiency of bypassing but the reason for that is not clear.[72] Finally, an unusual 5' stimulator element exists in gene 60. Interestingly, this stimulator works through the nascent peptide it encodes rather than its primary nucleotide sequence.[68] The most important property of this nascent peptide is its positive charge provided by several arginines.[73] It appears that the role of this element is to induce dissociation between the P-site tRNA and the mRNA.[71,74]

Selenocysteine Incorporation

A remarkable and unique example of codon redefinition is the encoding of selenocysteine. Selenocysteine is a 21st amino acid that is incorporated directly into polypeptide chains during translation in a number of bacteria, archaea and eukaryotes. The majority of studied selenoproteins are enzymes involved in oxidation-reduction reactions and contain selenocysteine in the active site. Selenocysteine is encoded by a UGA codon, which commonly specifies termination. In all three taxa, translation of UGA as selenocysteine requires distinct signals in the mRNA (termed SECIS elements for Selenocysteine Insertion Sequence), a unique tRNA that has a UCA anticodon and is charged with selenocysteine, an elongation factor which is specific for this tRNA and several enzymes essential for Sec-tRNA^{Sec} biogenesis (for recent reviews, see refs. 75 - 78 and references therein and Ch. 4 by Blanquet et al).

The mechanism has been extensively studied in *E. coli* and it appears to be similar in a number of other bacteria (Fig. 4A). The UGA codon specifying selenocysteine is followed by a stem/loop structure in the selenoprotein mRNA.[79] Aside from a few conserved nucleotides in the apical loop and in the bulge of the stem-loop, only the secondary structure is important (Fig. 5A).[80] The conserved nucleotides are responsible for interaction with the special elongation factor, SelB, which in turn binds Sec-tRNA^{Sec}. SelB is a homologue of elongation factor Tu,[81] which delivers all other tRNAs to the ribosome, but not the Sec-tRNA^{Sec}.[82] Unlike EF-Tu, SelB has a C-terminal extension domain,[83] which binds the stem-loop structure in the mRNA. The binding of the SelB in complex with Sec-tRNA^{Sec} and GTP to the RNA hairpin places the complex in the vicinity of the A-site of the ribosome and facilitates incorporation of selenocysteine into polypeptide chain.[84]

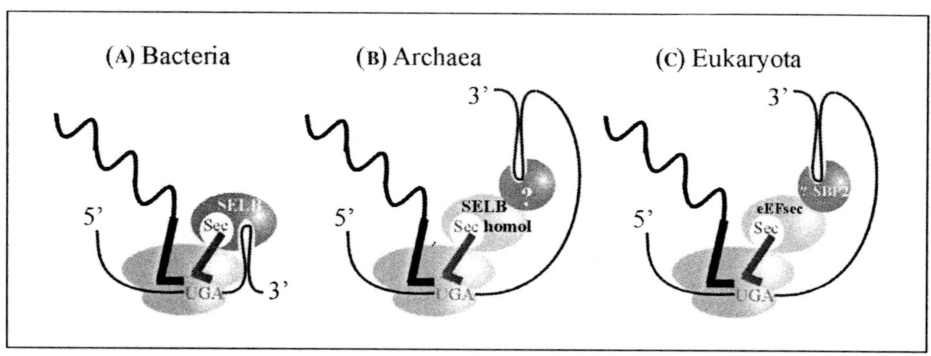

Figure 4. Models for selenocysteine incorporation into proteins. A) in bacteria; B) in archaea ; C) eukaryotes.

Archaeal and animal mechanisms (Fig. 4B,C) of selenocysteine incorporation are more complex. Although the SECIS elements have different secondary structures and conserved elements between archaea and eukaryotes (Fig. 5B-D), they do share a common feature. Unlike in *E. coli*, these SECIS elements are located in the 3' UTRs.[85-87]

How the SECIS element in the 3'UTR dictates UGAs, sometimes kilobases upstream, to specify selenocysteine has been a major question for the past decade. The last couple of years has been fruitful in this regard and yet has raised even more questions. The long-sought elongation factor homologue EF[Sec] was finally identified in the archaeon *Methanococcus jannaschii*[88] and subsequently throughout the animal kingdom from *C. elegans* to humans.[89,90] Just like its bacterial counterpart, EF[Sec] binds Sec-tRNA[Sec], has a GTP binding domain and possesses GTPase activity. Nevertheless, it is unable to bind the SECIS element.

Several proteins that bind SECIS element have been reported.[91,92] However, only one of them, SBP2, identified in rats, binds specifically to the wild-type SECIS RNA,[93] coprecipitates with EF[Sec] from cotransfected cells and stimulates selenocysteine incorporation in rabbit reticulocyte lysates and transfected cells. Gel-filtration experiments show that SBP2 is a part of a large supramolecular complex. Consequently, it was proposed that SBP2 binds to SECIS elements in the 3'UTRs and recruits other components of the selenocysteine insertion machinery to translating or initiating ribosomes. The nature of these components, aside form EF[Sec] and the tRNA, remains a mystery.

Eukaryotic SECIS elements are long stem-loop structures, which have certain important conserved features.[94] First are the sequences (A/G)UGA and GA in the 5' and 3' sides respectively of the SECIS stem (Fig. 5C,D). It was proposed that these sequences form a quartet of non-Watson-Crick base pairs, with G-A, A-G tandem pairs in the center.[95] Another conserved feature in SECIS is a stretch of two or three adenosines in the apical loop or bulge. The distance from the A-G, G-A tandem sequence to the stretch of adenosines is fixed at 9-11 base pairs, which is approximately one helical turn of A-form RNA.

Recently, analysis of ribosomal RNA structure revealed a number of G-A, A-G base pairs similar to the ones observed in SECIS elements. They play a key role in formation of a common RNA structure, which is termed a kink-turn, or K-turn, that interacts with a number of proteins with L7Ae RNA-binding motif.[96] SBP2 has the same motif, and mutations in it abrogate binding to the SECIS element and/or function in selenocysteine insertion.[97] Therefore, it is possible that interactions of SBP2 with SECIS are similar to interactions between a number of ribosomal proteins and kink-turn motifs in rRNA.

Another intriguing question is whether translation of selenoprotein mRNAs is efficient and whether or not it is processive. The efficiency of reading a single UGA as selenocysteine has been measured at about 3-7%,[98] and introduction of second UGA was reported to drop it to a marginal level.[99] However, selenoprotein P (SelP) mRNA contains 10 UGA codons in humans and rats,[100,101] 12 in cattle and 17 in zebrafish,[102,103] and all of them seem to be translated as selenocysteines in vivo. If the selenocysteine incorporation at each UGA is inefficient, then mRNAs with multiple UGAs, like SelP, are unlikely to be synthesized at detectable level. Therefore, either incorporation

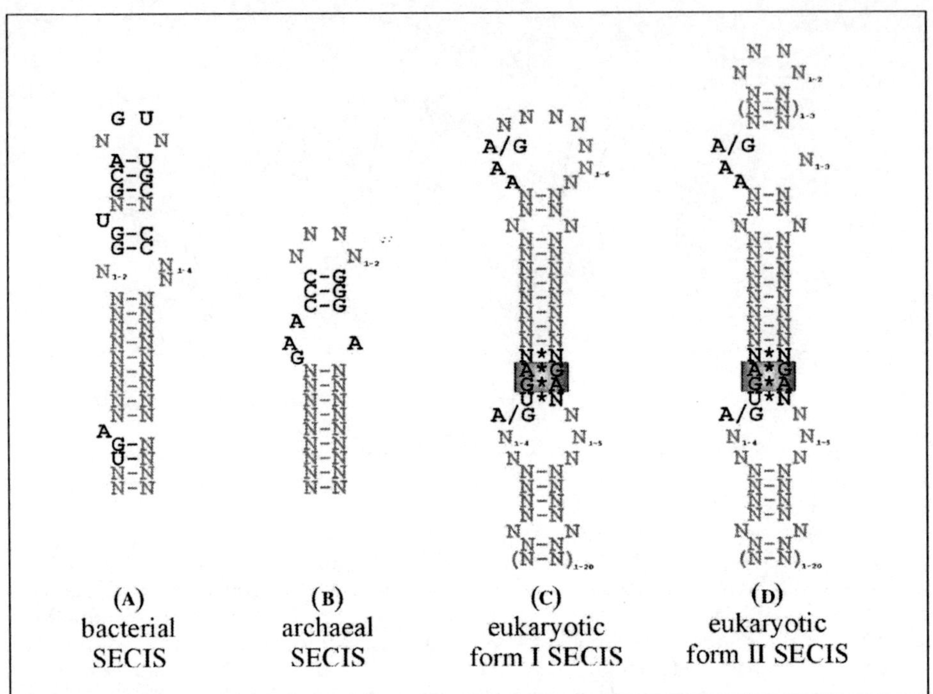

Figure 5. Secondary structure of Selenocysteine Insertion Sequences. Conserved nucleotides are in bold. A) bacterial; B) archaeal; C) and D) eukaryotic SECIS elements. Form II differs from Form I by the presence of an additional small hairpin on top of the conserved stretch of adenosines. G-A, A-G tandem, which presumably forms kink-turn motif, is boxed.

can be efficient at certain UGAs, or ribosomes translating selenoprotein's mRNA are somehow modified at translation initiation to read UGA as selenocysteine instead of ceasing translation. Nevertheless, the ability of such putatively modified ribosomes to insert selenocysteine is likely to depend on the context of the UGA since multiple isoforms of selenoprotein P have been found and these result from termination at some UGAs.[101]

The sequences surrounding UGA codon are extremely important for efficient insertion of selenocysteine. It has been shown that the identity of the two 5' codons, as well as of the 3' base dramatically influences the ratio between selenocysteine incorporation and termination.[54,104,105] It is possible that some other yet unknown elements in the mRNA affect selenocysteine incorporation. While these elements are tuned in endogenous mRNA to achieve maximum reading, they could be absent in experimental constructs. In accord with this, Tujebajeva et al did succeed in expressing full-length recombinant selenoprotein P.[106] But it should be noted that the expression was achieved only in one of the three cell lines tried. Therefore, it is possible that experiments in transfected cells do not always reflect the correct situation in vivo and their success can depend on a number of variables.

Early attempts to express selenoproteins in vitro failed. Recently, expression was achieved in a rabbit reticulocyte lysate with a single modification of the standard procedure: a 10-fold decrease of mRNA levels.[93] It is likely that, under limited conditions, different components necessary for selenoprotein mRNA translation are distributed between different mRNAs and form unproductive complexes. A similar situation could occur in transfection experiments when cells are overloaded with selenoprotein mRNA. In fact, it was observed that the ratio of termination product to full-length selenoprotein increases when the amounts of transfected DNA increase.[107] Supplementation of selenium, co-introduction of selenocysteine tRNA and sometimes SelD, stimulates incorporation several fold in transfected cells. Interestingly, while supplementation of SBP2 enhances expression of iodothyronine deiodinase 1, phospholipid glutathione peroxidase and thioredoxin reductase (all containing one selenocysteine codon) a few fold, supplementation increases expression of SelP up to 22 fold! Moreover, SBP2 doesn't readily exchange between SECIS elements.[108] Thus, during transfection experiments, SBP2 is probably predominantly bound to endogenous mRNAs, with the majority of transfected mRNAs being left unbound.

The nonsense-mediated decay (NMD) pathway can serve as a key regulator of the amounts of endogenous selenoprotein mRNA and establish stoichiometry between the levels of selenoprotein mRNA and ligands necessary for selenocysteine insertion. The NMD pathway serves to degrade mRNA containing premature termination codons, which otherwise could produce detrimental truncated proteins. A stop codon is recognized as premature if it is located more than 50-55 nucleotides upstream from the 3'-most exon-exon junction.[109] Under limited selenium, the UGA codon of selenoprotein glutathione peroxidase 1 is recognized as nonsense and triggers mRNA degradation.[110,111] This suggests that if other components of the selenocysteine insertion pathway are limited, selenocysteine codons might be recognized as premature stop codons and mRNAs would be degraded, provided that the spacing between UGA and last exon-exon junction is met. Remarkably, in most selenoprotein genes the selenocysteine-encoding UGA is located in mRNA so that it can be recognized as premature and trigger mRNA degradation if translation is terminated at it. On the contrary, cDNA based expression constructs used in transfection experiments lack exon-exon junctions, and therefore produce mRNAs that are not subject to NMD, even if translation is terminated at UGA.

SBP2 has different affinities for different SECIS elements.[108] It is proposed that SBP2 establishes an expression hierarchy between mRNAs of different selenoproteins, which becomes crucial under limitation of selenium. The other function of SBP2 could be the preselection of ribosomes for selenoprotein translation as, in the absence of mRNA, it was found to be tightly associated with ribosomes.[97] It could also be that binding of SBP2 alters the competence of the ribosome for termination and/or NMD.

Conclusion—Overview of the Field

A small but steady stream of new recoding examples is discovered each year. The majority are found in the compact genomes of viruses and transposable elements. Nevertheless, a growing number are being discovered in chromosomal genes. It is clear that only a minority of chromosomal genes employ recoding for their expression. However, recoding events are difficult to find and undoubtedly only a fraction of them is known. Currently, recoding events are discovered when the synthesis of a protein product with a known biochemical activity cannot be explained by standard decoding. An example is the recently described +1/-2 frameshifting in a hepatitis C virus gene.[112] A polypeptide shorter than that resulting from standard decoding was discovered. Initially, this polypeptide was dismissed as a product of posttranslational processing but the researchers tested for potential frameshifting. While this particular example needs further study, it does highlight that potential cases of recoding are easy to overlook. This presents a significant limitation for the rate of discovery of novel recoding examples. It has long been known that protein synthesis of a single mRNA leads to the production of a number of products (appearing as faint bands on SDS gels), sometimes bigger, sometimes shorter, than the main product. It is usually assumed that these result from preferential ribosome drop-off or from posttranslational modifications (degradation, phosphorylation, etc.). It is conceivable that at least a fraction of these products are the result of recoding events. In most cases a minor product is never investigated if it constitutes less than 10% of the main product. Our knowledge from viral recoding events shows that frameshifting/readthrough at as little as 1-2% can be physiologically significant. It is therefore possible that a large number of recoding events are currently hidden from us by limitations in technology or by our biased assumptions.

The years of incremental research has made our understanding of recoding more detailed to the point where the first attempts at general searching have been made. This approach is still in its early stages and no new recoding examples have been discovered with its application. Computer modeling using empirical data is another approach in the attempt to discover new cases of recoding. This latter approach is less comprehensive (and more biased) but promises more immediate results. Several candidate genes are being currently investigated as a result (I.P. Ivanov, unpublished). The availability of whole genome sequences greatly improves the opportunities. Database searches utilizing unique features of selenocysteine insertion machinery proved to be successful. Developed programs allow scanning of entire genomes for the presence of SECIS elements and analyze their position relative to predicted ORF-s. Candidate genes can be submitted to experimental verification. This approach has helped in discovering several mammalian[113,114] and Drosophila[115,116] selenoproteins and in identifying selenoproteins in zebrafish.[102] Remarkably, mutations in one such in silico identified novel selenoprotein, selenoprotein N, were just shown to cause muscular dystrophy in humans.[117]

Recent advances in proteomics, especially combining peptide detection with sophisticated computational algorithms, that allow assignment of individual peptide to a particular ORF, offer great promise for recoding research. This technique has the advantage of being least biased, since it does not rely on knowledge of previously identified recoding sites. The technical challenges, however, are great.

The final and potentially greatest hope comes for the recent discoveries in the structure and function of the ribosome. It is difficult to overstate the significance of these developments and likely future understanding of the conformational changes that take place during protein synthesis. Extension to recoding will throw light on the role of stimulatory elements and how events at the recoding sites exemplify the richness of decoding.

Acknowledgements

We greatly appreciate the comments of Pavel V. Baranov. We are supported by NIH (grants GM61200 to R.F.G. and GM 48152 to J.F.A.) and DOE grant DE-FG03-01ER63132 to R.F.G.

References

1. Gesteland RF, Atkins JF. Recoding: Dynamic reprogramming of translation. Annu Rev Biochem 1996; 65:741-768.
2. Brierley I. Ribosomal frameshifting viral RNAs. Gen Virol 1995; 76:1885-1892.
3. Farabaugh PJ. Programmed translational frameshifting. Annu Rev Genet 1996; 30:507-528.
4. Jacks T, Townsley K, Varmus HE et al. Two efficient ribosomal frameshifting events are required for synthesis of mouse mammary tumor virus gag-related polyproteins. Proc Natl Acad Sci USA 1987; 84:4298-4302.
5. Dinamn JD, Icho T, Wickner RB. A −1 ribosomal frameshift in a double-stranded RNA virus of yeast forms a gag-pol fusion protein. Proc Natl Acad Sci USA 1991; 88:174-178.
6. Tzeng T-H, Tu C-L, Bruenn JA. Ribosomal frameshifting requires a pseudoknot in the *Saccharomyces cerevisiae* double-stranded RNA virus. J Virol 1992; 66:999-1006.
7. Farabaugh PJ, Zhao H, Vimaladithan A. A novel programed frameshift expresses the POL3 gene of retrotransposon Ty3 of yeast: frameshifting without tRNAslippage. Cell 1993; 74:93-103.
8. Parkin NT, Chamorro M, Varmus HE. Human immunodeficiency virus type 1 *gag-pol* frameshifting is dependent on downstream mRNA secondary structure: Demonstration by expression in vivo. J Virol 1992; 66:5147-5151.
9. Bidou L, Stahl G, Grima B et al. In vivo HIV-1 frameshifting efficiency is directly related to the stability of the stem-loop stimulatory signal. RNA 1997; 3:1153-1158.
10. Dinman JD, Wickner RB. Ribosomal frameshifting efficiency and *gag/gag-pol* ratio are critical for yeast M1 double-stranded RNA virus propagation. J Virol 1992; 66:3669-3676.
11. Shehu-Xhilaga M., Crowe SM, Mak J. Maintenance of the Gag/Gag-Pol ratio is important for human immunodeficiency virus type 1 RNA dimerization and viral infectivity. J Virol 2001; 75:1834-1841.
12. Kawakami K, Pande S, Faiola B et al. A rare tRNA-Arg(CCU) that regulates Ty1 element ribosomal frameshifting is essential for Ty1 retrotransposition in *Saccharomyces cerevisiae*. Genetics 1993; 135:309-320.
13. Tsuchihashi Z, Kornberg A. Translational frameshifting generates the γ subunit of DNA polymerase III holoenzyme. Proc Natl Acad Sci USA 1990; 87:2516-2520.
14. Blinkowa AL, Walker JR. Programmed ribosomal frameshifting generates the *Escherichia coli* DNA polymerase III γ subunit from within the τ subunit reading frame. Nucleic Acids Res 1990; 18:1725-1729.
15. Flower AM, McHenry CS. The γ subunit of DNA polymerase III holoenzyme of *Escherichia coli* is produced by ribosomal frameshifting. Proc Natl Acad Sci USA 1990; 87:3713-3717.
16. Larsen B, Wills NM, Gesteland RF et al. rRNA-mRNA base pairing stimulates a programmed −1 ribosomal frameshift. J Bacteriol 1994; 176:6842-6851.
17. Larsen B, Gesteland RF, Atkins JF. Structural probing and mutagenic analysis of the stem-loop required for *E. coli dnaX* ribosomal frameshifting: programmed efficiency of 50%. J Mol Biol 1997; 271:47-60.
18. Gao D, McHenry CS. τ binds and organizes *Escherichia coli* replication proteins through distinct domains. J Biol Chem 2001; 276:4433-4440.
19. Craigen WJ, Caskey CT. Expression of peptide chain release factor 2 requires high-efficiency frameshift. Nature 1986; 322:273-275.
20. Weiss RB, Dunn DM, Atkins JF et al. 1987. Slippery runs, shifty stops, backward steps, and forward hops: -2, -1, +1, +2, +5, and +6 ribosomal frameshifting. Cold Spring Harb Symp Quant Biol 1987; 52:687-693.
21. Weiss RB, Dunn DM, Dahlberg AE et al. Reading frame switch caused by base-pair formation between the 3' end of 16S rRNA and the mRNA during elongation of protein synthesis in *Escherichia coli*. EMBO J 1988; 7:1503-1507.

22. Curran JF, Yarus M. Use of tRNA suppressors to probe regulation of *Escherichia coli* release factor 2. J Mol Biol 1988; 203:75-83.
23. Curran JF. Analysis of effects of tRNA:message stability on frameshift frequency at the *Escherichia coli* RF2 programmed frameshift site. Nucleic Acids Res 1993; 21:1837-1843.
24. Major LL, Poole ES, Dalphin ME et al. Is the in-frame termination signal of the *Escherichia coli* release factor-2 frameshift site weakened by a particularly poor context? Nucleic Acids Res 1996; 24:2673-2678.
25. Atkins JF, Lewis JB, Anderson CW et al. Enhanced differential synthesis of proteins in a mammalian cell-free system by addition of polyamines. J Biol Chem 1975; 250:5688-95.
26. Miyazaki Y, Matsufuji S, Hayashi S. Cloning and characterization of a rat gene encoding ornithine decarboxylase antizyme. Gene 1992; 113:191-197.
27. Matsufuji S, Matsufuji T, Miyazaki Y et al. 1995. Autoregulatory frameshifting in decoding mammalian ornithine decarboxylase antizyme. Cell 1995; 80:51-60.
28. Rom E, Kahana C. Polyamines regulate the expression of ornithine decarboxylase antizyme in vitro by inducing ribosomal frameshifting. Proc Natl Acad Sci USA 1994; 91:3959-3963.
29. Howard MT, Shirts BH, Zhou J et al. Cell culture analysis of the regulatory frameshift event required for the expression of mammalian antizyme. Genes to Cells 2001; 6:331-341.
30. Ivanov IP, Gesteland RF, Atkins JF. Antizyme expression: a subversion of triplet decoding, which is remarkably conserved by evolution, is a sensor for an autoregulatory circuit. Nucleic Acids Res 2000; 17:3185-3196.
31. Kurland CG. Reading-frame errors on ribosomes. In: Celis JE, Smith JD, eds. Nonsense Mutations and tRNA Suppressors. New York: Academic, 1979:97-108.
32. Kurland CG. Translational accuracy and the fitness of bacteria. Annu Rev Genet 1992; 26:29-50.
33. Baranov PV, Gurvich OL, Fayet O et al. RECODE: a database of frameshifting, bypassing and codon redefinition utilized for gene expression. Nucleic Acids Res 2001; 29:264-267.
34. Jacks T, Madhani HD, Masiarz FR et al. Signals for ribosomal frameshifting in the Rous sarcoma virus *gag-pol* region. Cell 1988; 55:447-458.
35. Weiss RB, Dunn DM, Shuh M et al. *E. coli* ribosomes re-phase on retroviral frameshift signals at rates ranging from 2 to 50 percent. New Biol 1989; 1:159-169.
36. Brierley I, Jenner AJ, Inglis SC. Mutational analysis of the "slippery-sequence" component of a coronavirus ribosomal frameshifting signal. J Mol Biol 1992; 227:463-479.
37. ten Dam EB, Pleij CW, Bosch L. RNA pseudoknots: translational frameshifting and readthrough on viral RNAs. Virus Genes 1990; 4:121-136.
38. Rettberg CC, Prere MF, Gesteland RF et al. A three-way junction and constituent stem-loops as the stimulator for programmed -1 frameshifting in bacterial insertion sequence IS911. J Mol Biol 1999; 286:1365-1378.
39. Paul CP, Barry JK, Dinesh-Kumar SP et al. A sequence required for -1 ribosomal frameshifting located four kilobases downstream of the frameshift site. J Mol Biol 2001; 310:987-999.
40. Somogyi P, Jenner AJ, Brierley I et al. Ribosomal pausing during translation of an RNA pseudoknot. Mol Cell Biol 1993; 13:6931-6940.
41. Somogyi P, Jenner AJ, Brierley I et al. Ribosomal pausing during translation of an RNA pseudoknot. Mol Cell Biol 1993; 13:6931-6940.
42. Lopinski JD, Dinman JD, Bruenn JA. Kinetics of ribosomal pausing during programmed -1 translational frameshifting. Mol Cell Biol 2000; 20:1095-1103.
43. Eggertsson G, Soll D. Transfer ribonucleic acid-mediated suppression of termination codons in *Escherichia coli*. Microbiol Rev 1988; 52:354-374.
44. Kohli J, Kwong T, Altruda F et al. Characterization of a UGA-suppressing serine tRNA from *Schizosaccharomyces pombe* with the help of a new in vitro system for eukaryotic suppressor tRNAs. J Biol Chem 1979; 254:1546-1551.
45. Kondo K, Hodgkin J, Waterston RH. Differential expression of five tRNA(UAGTrp) amber suppressors in *Caenorhabditis elegans*. Mol Cell Biol 1988; 8:3627-35.
46. Hatfield DL, Smith DW, Lee BJ et al. Structure and function of suppressor tRNAs in higher eukaryotes. Crit Rev Biochem Mol Biol. 1990; 25:71-96.
47. Oeschger MP, Oeschger NS, Wiprud et al. High efficiency temperature-sensitive amber suppressor strains of *Escherichia coli* K12: isolation of strains with suppressor-enhancing mutations. Mol Gen Genet 1980; 177:545-52.
48. Ryden SM, Isaksson LA. A temperature-sensitive mutant of *Escherichia coli* that shows enhanced misreading of UAG/A and increased efficiency for some tRNA nonsense suppressors. Mol Gen Genet 1984; 193:38-45.
49. Zhang S, Ryden-Aulin M, Isaksson LA. Functional interaction between tRNAGly2 at the ribosomal P-site and RF1 during termination at UAG. J Mol Biol 1998; 284:1243-1246.
50. Kawakami K, Inada T, Nakamura Y. Conditionally lethal and recessive UGA-suppressor mutations in the *prfB* gene encoding peptide chain release factor 2 of *Escherichia coli*. J Bacteriol 1988; 170:5378-81.
51. Karow ML, Rogers EJ, Lovett PS et al. Suppression of TGA mutations in the *Bacillus subtilis* spoIIR gene by *prfB* mutations. J Bacteriol 1998; 180:4166-4170.
52. Stansfield I, Tuite MF. Polypeptide chain termination in *Saccharomyces cerevisiae*. Curr Genet 1994; 25:385-95.
53. Poole ES, Major LL, Mannering SA et al. Translational termination in *Escherichia coli*: three bases following the stop codon crosslink to release factor 2 and affect the decoding efficiency of UGA-containing signals. Nucl Acids Res 1998; 26:954-960.

54. McCaughan KK, Brown CM, Dalphin ME et al. Translational termination efficiency in mammals is influenced by the base following the stop codon. Proc Natl Acad Sci USA 1995; 92:5431-5435.
55. Wills NM, Gesteland RF, Atkins JF. Pseudoknot-dependent read-through of retroviral gag termination codons: importance of sequences in the spacer and loop2. EMBO J 1994; 13:4137-4144.
56. Brown CM, Dinesh-Kumar SP, Miller WA. Local and distant sequences are required for efficient readthrough of the barley yellow dwarf virus PAV coat protein gene stop codon. J Virol 1996; 70:5884-5892.
57. Belcourt MF, Farabaugh PJ. Ribosomal frameshifting in the yeast retrotransposon Ty: tRNAs induce slippage on a 7 nucleotide minimal site. Cell 1990; 62:339-352.
58. Weiss R, Gallant J. Mechanism of ribosome frameshifting during translation of the genetic code. Nature 1983; 302:389-393.
59. Lindsley D, Gallant J. On the directional specificity of ribosome frameshifting at a "hungry" codon. Proc Natl Acad Sci USA 1993; 90:5469-5473.
60. Gallant J, Lindsley D, Masucci J. The unbearable lightness of peptidyl-tRNA. In: Garrett R, Douthwaite SR, Liljas A et al. eds. The Ribosome. Washington, DC: ASM Press, 2000:385-396.
61. Pande S, Vimaladithan A, Zhao H et al. Pulling the ribosome out of frame by +1 at a programmed frameshift site by cognate binding of aminoacyl-tRNA. Mol Cell Biol 1995; 15:298-304.
62. Matsufuji S, Matsufuji T, Wills NM et al. Reading two bases twice: mammalian antizyme frameshifting in yeast. EMBO J 1996; 15:1360-1370.
63. Ivanov IP, Gesteland RF, Matsufuji S et al. Programmed frameshifting in the synthesis of mammalian antizyme is +1 in mammals, predominantly +1 in fission yeast, but -2 in budding yeast. RNA 1998; 4:1230-1238.
64. Sundararajan A, Michaud WA, Qian Q et al. Near-cognate peptidyl-tRNAs promote +1 programmed translational frameshifting in yeast. Mol Cell 1999; 4:1005-1015.
65. Ivanov IP, Matsufuji S, Murakami Y et al. Conservation of polyamine regulation by translational frameshifting from yeast to mammals. EMBO J 2000; 19:1907-1917.
66. Li Z, Stahl G, Farabaugh PJ. Programmed +1 frameshifting stimulated by complementarity between a downstream mRNA sequence and an error-correcting region of rRNA. RNA 2001; 7:275-284.
67. Huang WM, Ao SZ, Casjens S et al. A persistent untranslated sequence within bacteriophage T4 DNA topoisomerase gene 60. Science 1988; 239:1005-1012.
68. Weiss RB, Huang WM, Dunn DM. A nascent peptide is required for ribosomal bypass of the coding gap in bacteriophage T4 gene 60. Cell 1990; 62:117-26.
69. Maldonado R, Herr AJ. Efficiency of T4 gene 60 translational bypassing. J Bacteriol 1998; 180:1822-1830.
70. Herr AJ, Atkins JF, Gesteland RF. Mutations which alter the elbow region of tRNA2Gly reduce T4 gene 60 translational bypassing efficiency. EMBO J 1999; 18:2886-2896.
71. Herr AJ, Gesteland RF, Atkins JF. One protein from two open reading frames: mechanism of a 50 nt translational bypass. EMBO J 2000; 19:2671-2680.
72. Herr AJ, Wills NM, Nelson CC et al. Drop-off during ribosome hopping. J Mol Biol 2001; 311:445-452.
73. Larsen B, Peden J, Matsufuji S et al. Upstream stimulators for recoding. Biochem Cell Biol 1995; 73:1123-1129.
74. Herr AJ, Nelson CC, Wills NM et al. Analysis of the roles of tRNA structure, ribosomal protein L9, and the bacteriophage T4 gene 60 bypassing signals during ribosome slippage on mRNA. J Mol Biol 2001; 309:1029-1048.
75. Berry MJ. Recoding UGA as Selenocysteine In: Sonenberg N, Hershey JWB, Mathews MB, eds. Translational Control of Gene Expression. Cold Spring Harbor: Cold Spring Harbor Laboratory Press, 2000: 763-783.
76. Kohrl J, Brigelius-Flohe R, Bock A. Selenium in biology: facts and medical perspectives. Biol Chem 2000; 381:849-864.
77. Huttenhofer A, Bock A. RNA structures involved in selenoprotein synthesis. In: Simons R, Grunberg-Manago M, eds. RNA structure and Function. Cold Spring Harbor: Cold Spring Harbor Laboratory Press, 1998:603-639.
78. Atkins JF, Bock A, Matsufuji S et al. Dynamics of the genetic code. In: Gesteland RF, Cech TR, Atkins JF, eds. The RNA World. 2nd ed. Cold Spring Harbor: Cold Spring Harbor Laboratory Press, 1999:637-673.
79. Zinoni F, Heider J, Bock A. Features of the formate dehydrogenase mRNA necessary for decoding of the UGA codon as selenocysteine. Proc Natl Acad Sci USA 1990; 87:4660-4664.
80. Heider J, Baron C, Bock A. Coding from a distance: dissection of the mRNA determinants required for the incorporation of selenocysteine into protein. EMBO J 1992; 11:3759-3766.
81. Forchhammer K, Leinfelder W, Bock A. Identification of a novel translation factor necessary for the incorporation of selenocysteine into protein. Nature 1989; 342:453-456.
82. Forster C, Ott G, Forchhammer K et al. Interaction of a selenocysteine-incorporating tRNA with elongation factor Tu from *E. coli* Nucleic Acids Res 1990; 18:487-491.
83. Kromayer M, Wilting R, Tormay P et al. Domain structure of the prokaryotic selenocysteine-specific elongation factor SelB. J Mol Biol 1996; 262:413-420.
84. Ringquist S, Schneider D, Gibson T et al. Recognition of the mRNA selenocysteine insertion sequence by the specialized translational elongation factor SELB. Genes Dev 1994; 8:376-385.
85. Berry MJ, Banu L, Chen YY et al. Recognition of UGA as a selenocysteine codon in type I deiodinase requires sequences in the 3' untranslated region. Nature 1991; 353:273-276.
86. Wilting R, Schorling S, Persson BC et al. Selenoprotein synthesis in archaea: identification of an mRNA element of *Methanococcus jannaschii* probably directing selenocysteine insertion. J Mol Biol 1997; 266:637-641.

87. Rother M, Resch A, Gardner WL et al. Heterologous expression of archaeal selenoprotein genes directed by the SECIS element located in the 3' non-translated region. Mol Microbiol 2001; 40:900-908.
88. Rother M, Wilting R, Commans S et al. Identification and characterisation of the selenocysteine- specific translation factor SelB from the archaeon *Methanococcus jannaschii.* J Mol Biol 2000; 299:351-358.
89. Tujebajeva RM, Copeland PR, Xu XM et al. Decoding apparatus for eukaryotic selenocysteine insertion. EMBO Rep 2000; 1:158-163.
90. Fagegaltier D, Hubert N, Yamada K et al. Characterization of mSelB, a novel mammalian elongation factor for selenoprotein translation. EMBO J 2000; 19:4796-805.
91. Hubert N, Walczak R, Carbon P et al. A protein binds the selenocysteine insertion element in the 3'-UTR of mammalian selenoprotein mRNAs. Nucleic Acids Res 1996; 24:464-469.
92. Fujiwara T, Busch K, Gross HJ et al. A SECIS binding protein (SBP) is distinct from selenocysteyl-tRNA protecting factor (SePF). Biochimie 1999; 81:213-218.
93. Copeland PR, Fletcher JE, Carlson BA et al. A novel RNA binding protein, SBP2, is required for the translation of mammalian selenoprotein mRNAs. EMBO J 2000; 19:306-314.
94. Berry MJ, Banu L, Harney JW et al. Functional characterization of the eukaryotic SECIS elements which direct selenocysteine insertion at UGA codons. EMBO J 1993; 12:3315-3322.
95. Walczak R, Westhof E, Carbon P et al. A novel RNA structural motif in the selenocysteine insertion element of eukaryotic selenoprotein mRNAs. RNA 1996; 2:367-379.
96. Klein DJ, Schmeing TM, Moore PB et al. The kink-turn: a new RNA secondary structure motif. EMBO J 2001; 20:4214-4221.
97. Copeland PR, Stepanik VA, Driscoll DM. Insight into mammalian selenocysteine insertion: domain structure and ribosome binding properties of Sec insertion sequence binding protein 2. Mol Cell Biol 2001; 21:1491-1498.
98. Berry MJ, Maia AL, Kieffer JD et al. Substitution of cysteine for selenocysteine in type I iodothyronine deiodinase reduces the catalytic efficiency of the protein but enhances its translation. Endocrinology 1992; 131:1848-52.
99. Nasim MT, Jaenecke S, Belduz A, et al. Eukaryotic selenocysteine incorporation follows a nonprocessive mechanism that competes with translational termination. J Biol Chem 2000; 275:14846-14852.
100. Hill KE, Lloyd RS, Yang JG et al. The cDNA for rat selenoprotein P contains 10 TGA codons in the open reading frame. J Biol Chem 1991; 266:10050-3.
101. Himeno S, Chittum HS, Burk RF. Isoforms of selenoprotein P in rat plasma. Evidence for a full-length form and another form that terminates at the second UGA in the open reading frame. J Biol Chem 1996; 271:15769-75.
102. Kryukov GV, Gladyshev VN. Selenium metabolism in zebrafish: multiplicity of selenoprotein genes and expression of a protein containing 17 selenocysteine residues. Genes Cells 2000; 5:1049-1060.
103. Tujebajeva RM, Ransom DG, Harney JW et al. Expression and characterization of nonmammalian selenoprotein P in the zebrafish, Danio rerio. Genes Cells 2000; 5:897-903.
104. Liu Z, Reches M, Engelberg-Kulka H. A sequence in the *Escherichia coli* fdhF "selenocysteine insertion sequence" (SECIS) operates in the absence of selenium. J Mol Biol 1999; 294:1073-1086.
105. Grundner-Culemann E, Martin GW 3rd, Tujebajeva R et al. Interplay between termination and translation machinery in eukaryotic selenoprotein synthesis. J Mol Biol 2001; 310:699-707.
106. Tujebajeva RM, Harney JW, Berry MJ. Selenoprotein P expression, purification, and immunochemical characterization. J Biol Chem 2000; 275:6288-6294.
107. Berry MJ, Harney JW, Ohama T. Selenocysteine insertion or termination: factors affecting UGA codon fate and complementary anticodon:codon mutations. Nucleic Acids Res 1994; 22:3753-9.
108. Low SC, Grundner-Culemann E, Harney JW et al. SECIS-SBP2 interactions dictate selenocysteine incorporation efficiency and selenoprotein hierarchy. EMBO J 2000; 19:6882-6890.
109. Zhang J, Sun X, Qian Y et al. Intron function in the nonsense-mediated decay of beta-globin mRNA: indications that pre-mRNA splicing in the nucleus can influence mRNA translation in the cytoplasm. RNA 1998; 4:801-815.
110. Moriarty PM, Reddy CC, Maquat LE. Selenium deficiency reduces the abundance of mRNA for Se-dependent glutathione peroxidase 1 by a UGA-dependent mechanism likely to be nonsense codon-mediated decay of cytoplasmic mRNA. Mol Cell Biol 1998; 18:2932-2939.
111. Weiss SL, Sunde RA. Cis-acting elements are required for selenium regulation of glutathione peroxidase-1 mRNA levels. RNA 1998; 4:816-827.
112. Xu Z, Choi J, Yen TS et al. Synthesis of a novel hepatitis C virus protein by ribosomal frameshift. EMBO J 2001; 20:3840-3848.
113. Lescure A, Gautheret D, Carbon P et al. Novel selenoproteins identified in silico and in vivo by using a conserved RNA structural motif. J Biol Chem 1999; 274:38147-38154.
114. Kryukov GV, Kryukov VM, Gladyshev VN. New mammalian selenocysteine-containing proteins identified with an algorithm that searches for selenocysteine insertion sequence elements. J Biol Chem. 1999; 274:33888-33897.
115. Martin-Romero FJ, Kryukov GV, Lobanov AV et al. Selenium metabolism in Drosophila: selenoproteins, selenoprotein mRNA expression, fertility, and mortality. J Biol Chem 2001; 276:29798-804.
116. Castellano S, Morozova N, Morey M et al. In silico identification of novel selenoproteins in the *Drosophila melanogaster* genome. EMBO Rep 2001; 2:697-702.
117. Moghadaszadeh B, Petit N, Jaillard C et al. Mutations in SEPN1 cause congenital muscular dystrophy with spinal rigidity and restrictive respiratory syndrome. Nat Genet 2001; 29:17-18.

CHAPTER 23

Control of Stable RNA Synthesis

Melanie M. Barker and Richard L. Gourse

Abstract

Transcription of rRNA is the rate-limiting step in ribosome production. In rapidly dividing *Escherichia coli*, ribosome synthesis represents the single largest expenditure of biosynthetic energy. At the fastest growth rates, over 70% of all transcription derives from the seven rRNA operons even though there are over 2000 other operons in the cell. Since rRNA production is such a crucial and costly enterprise, bacteria have evolved a number of regulatory mechanisms to ensure that the appropriate amount of rRNA is made under different environmental and nutritional conditions. Translation requires not only rRNA, but also ribosomal proteins (r-proteins), charged tRNAs, and translational factors. Therefore, the production of all of these components must be coordinated.

The first section of this chapter describes several of the inputs that contribute to the high rate of transcription of rRNA in *Escherichia coli*. The second section outlines proposed regulatory mechanisms that coordinate the rates of rRNA synthesis with the translational needs of the cell. The last section very briefly describes the regulation of production of tRNAs and r-proteins.

Acronyms and Abbreviations

αCTD	carboxy-terminal domain of the α subunit of RNAP
FIS	Factor for Inversion Stimulation
H-NS	Histone-like Nucleoid Structuring protein
NTP	nucleoside triphosphate
ppGpp	guanosine-5'-disphosphate-3' disphosphate
RNAP	RNA polymerase
r-protein	ribosomal protein

Contributors to the High Rate of Transcription of rRNA Genes

Structure of rRNA Operons and Promoters

Each of the seven rRNA operons in *E. coli* (*rrnA, rrnB, rrnC, rrnD, rrnE, rrnG, rrnH*) encode a large precursor RNA transcript that generates (in order from the 5' end): 16S rRNA, one or two "spacer" tRNAs, 23S rRNA, and 5S rRNA (reviewed in refs. 1-3). In some operons, one or two "distal" tRNAs are found after the 5S rRNA, and in the *rrnD* operon, a second 5S rRNA is encoded after the distal tRNAs . Each operon is transcribed from tandem promoters, P1 and P2, separated by about 120 bp, with the upstream promoters (*rrn* P1) predominating at medium and fast growth rates. This chapter focuses almost entirely on the P1 promoters in *Escherichia coli*, and *rrnB* P1 in particular, which has been the most intensively studied. However, the *rrn* P2 promoters share many of the regulatory properties that characterize the P1 promoters, as described briefly toward the end of this chapter. The structure of the *rrnB* operon and promoter region is shown in Figure 1.

Most transcription in the cell is performed by the RNA polymerase holoenzyme (RNAP; $\alpha_2\beta\beta'\omega\sigma$) containing the specificity factor σ^{70} ($E\sigma^{70}$).[4] The σ^{70} subunit recognizes two hexamer sequences, positioned approximately 10 and 35 bp upstream of the transcription start site (+1), separated by an optimal length of 17 bp.[4] rRNA genes are transcribed primarily by $E\sigma^{70}$, although rRNA genes may

Translation Mechanisms, edited by Jacques Lapointe and Léa Brakier-Gingras.
©2003 Eurekah.com and Kluwer Academic / Plenum Publishers.

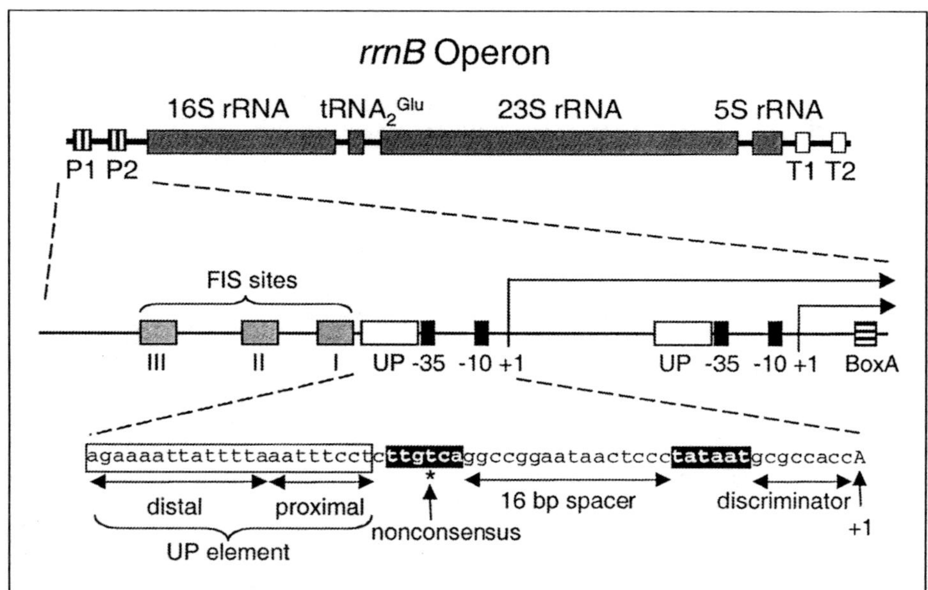

Figure 1. Schematic diagram of the *E. coli rrnB* operon and promoter region. The regions coding for the mature rRNAs (16S, 23S, and 5S) and spacer tRNA (tRNA$_2$Glu) are indicated (filled rectangles). The two promoters, P1 and P2 (vertically striped rectangles), and the two terminators, T1 and T2 (open rectangles), are also shown. The P1-P2 promoter region is expanded to illustrate the three FIS binding sites (I, II, and III; gray rectangles) upstream of P1, the UP elements (open rectangles), -35 and -10 hexamers (black rectangles), and transcription start sites (+1; lines with arrows) for both P1 and P2. The BoxA sequence (horizontally striped rectangle) is also shown. The P1 promoter sequence from -59 to +1 is shown below. The distal and proximal subsites of the UP element, the nonconsensus bp at -33, the 16 bp spacer, and the GC-rich discriminator are indicated.

also be transcribed by at least one RNAP containing an alternative σ factor.[5] The *rrn* P1 promoters contain perfect matches to the consensus -10 hexamer (TATAAT), close matches to the consensus -35 hexamer (TTGACA), and 16 bp spacers (Fig. 1).[1]

However, the excellent match of *rrn* P1 promoters to the Eσ^{70} consensus recognition sequences contributes only a small part to their extraordinary strength. In all seven rRNA operons, the "core" P1 promoters (defined as sequences from ~ -40 to +1) are over 100-fold weaker than the "full-length" P1 promoters (defined as ~ -200 to +1).[6,7] The full-length P1 promoters derive their strength primarily from two components acting upstream of the core promoter: cis-acting sequences called UP elements and the trans-acting factor FIS. In addition, sequences downstream of the P2 promoter guarantee that each initiated rRNA transcript is elongated efficiently. Lastly, the arrangement and high copy number of rRNA operons likely contribute to efficient rRNA production. The following sections describe these inputs to the high rates of transcription of rRNA genes.

UP Elements

UP elements are A+T-rich sequences positioned upstream (~ -60 to -40) from the -35 hexamers in certain promoters (reviewed in ref. 8).[9-11] UP elements interact with the C-terminal domain(s) of the α subunit(s) [αCTD(s)] of RNAP, stimulating transcription in the absence of proteins other than RNAP. UP elements can also increase transcription when appropriately positioned upstream from other core promoters in chimeric constructs. Therefore, the UP element is considered a third RNAP recognition motif (besides the -10 and -35 hexamers).

The optimal (consensus) UP element sequence was determined by in vitro selection.[12,13] UP elements consist of one or two subsites (proximal or distal to the -35 hexamer), each of which can interact with one αCTD (see Fig. 1).[13] The consensus "full" UP element sequence, composed of both proximal and distal subsites, consists primarily of alternating A- and T-tracts [AAA(a/t)(a/t)T(a/t)TTTT--AAAA] and can increase transcription in vivo more than 300-fold. Good matches to the consensus UP element sequence are found in a variety of mRNA promoters, but in a higher percentage of stable RNA promoters.[13] Indeed, all seven *rrn* P1 promoters are activated between 20 and 50-fold by UP elements.[7,9,11,14,15]

Phased A-tracts, which look similar to the consensus UP element sequence, are found upstream from the -35 hexamers of a number of strong promoters (reviewed in ref. 16). Although it was proposed that the intrinsic curvature caused by A-tracts is responsible for their stimulatory effect on transcription,[16] stimulation of the *rrnB* P1 or *lac* promoters by phased A-tracts in synthetic constructs required the αCTD, suggesting that, in many cases, A-tracts function as UP elements.[17] The unusual structure (e.g., narrow minor groove width or curvature) of A-tracts might facilitate αCTD interactions. However, a high degree of bending is not essential for UP element function, since the *rrnB* P1 UP element is not detectably bent.[18]

A large body of genetic and biochemical data has provided a very detailed picture of the interaction between α and UP element DNA. α consists of two independently folded domains, separated by a flexible linker.[19,20] High resolution structures of both domains have been determined.[21,22] RNAP lacking the α carboxy-terminal domain (αCTD) is unable to bind to UP element DNA and is unable to utilize the UP element for stimulating transcription.[9] Single substitutions in a surface-exposed patch of αCTD consisting of amino acids L262, R265, N268, C269, G296, K298, and S299 decrease UP element-dependent transcription.[23,24] The residues in α that are required for UP element function are conserved in most bacterial species sequenced to date, suggesting that UP elements are conserved determinants of bacterial promoter strength.[8,23] High resolution footprinting and structure prediction of α-DNA interactions suggested that αCTD interacts with the phosphodiester backbone and with bases in the minor groove [25,26] using two helix-hairpin-helix motifs (see Fig. 2 legend and refs. 26-27). These and other genetic data have led to a model for how αCTD interacts with a proximal UP element subsite (Fig. 2, panel A).[26] This model has recently been confirmed by the solution of an αCTD-DNA structure by x-ray crystallography.[27a]

The kinetic basis for UP element function has been investigated. Like translation initiation, transcription initiation is a multi-step process, providing many opportunities for modulation (reviewed in ref. 4). RNAP first binds to a promoter to form a closed complex. This closed complex can then isomerize through a series of intermediates to form an open complex in which the promoter DNA is unwound in the vicinity of the -10 hexamer and transcription start site. The open complex binds NTPs and sometimes produces small "abortive" RNAs before RNAP escapes the promoter and forms an elongation complex. Kinetic analysis of *rrnB* P1 suggests that the UP element increases the initial equilibrium binding constant to form the closed complex and possibly increases the rate of a subsequent isomerization step.[10,28]

UP elements increase the number of RNAP-promoter contacts and thus, in most cases, activate transcription. However, in some cases, α-DNA interactions have no effect on transcription[11] and a mispositioned UP element can even lead to formation of a nonproductive initiation complex.[29] Furthermore, an UP element fused to a promoter that is rate-limited for promoter clearance can actually inhibit transcription.[30]

FIS-Dependent Activation

rrn P1 promoters derive additional strength from activation by a transcription factor, the highly abundant nucleoid protein FIS (Factor for Inversion Stimulation)[31] (reviewed in ref. 32). FIS binds as a dimer to 3 to 5 specific sites upstream of -61 at each *rrn* P1 promoter and activates transcription 3- to 8-fold in vivo.[7,14,15,31] FIS bound at the promoter proximal site ("site I"; centered at -71) is responsible for the majority of activation at *rrnB* P1.[31]

Evidence for the identity of the interacting surfaces between FIS and RNAP was obtained from genetic studies. "Positive control" FIS mutants (i.e., mutants that bind and bend DNA normally but fail to activate *rrnB* P1 transcription) were identified and localized to a surface-exposed loop, the FIS "B-C turn" (amino acids Q68, R71, G72, and Q74).[33-35] RNAP mutants lacking the αCTD are defective for FIS-mediated activation, and alanine scanning mutagenesis of the entire αCTD identified two discrete patches on αCTD that are critical for activation by FIS.[35,36] One patch, the "265

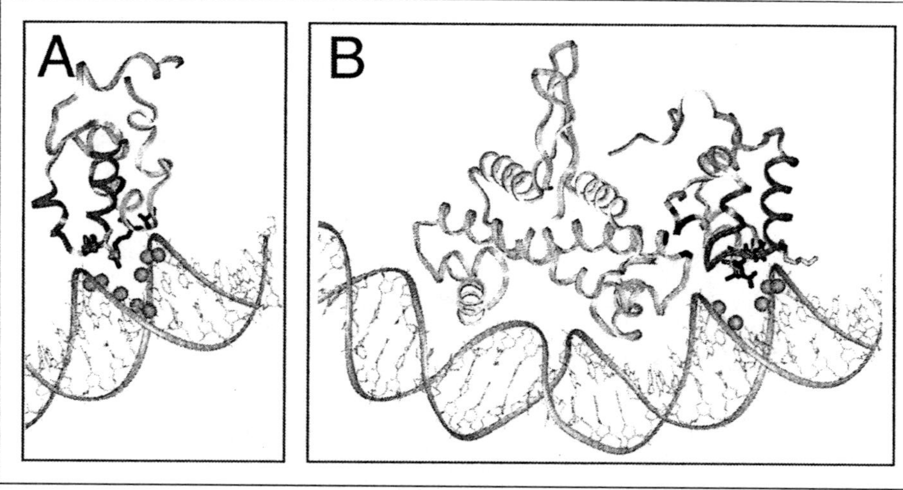

Figure 2. Model for αCTD interactions with DNA[26] and with FIS and DNA in the FIS-αCTD-DNA complex.[36] (A) Model for the αCTD interaction with the minor groove at an UP element proximal subsite.[26] The αCTD (from the NMR structure)[22,23] is shown in ribbon form in light gray and binds within the minor groove. The αCTD contains two helix-hairpin-helix (HhH) motifs [an (HhH)$_2$ domain].[27] Each HhH motif consists of two antiparallel helices separated by a hairpin loop. HhH motifs are found in a variety of proteins involved in DNA replication, recombination, and repair. In contrast to the case with αCTD, the HhH motifs found in other proteins mediate sequence-nonspecific binding.[26,27] HhH motif 2 of αCTD is shaded in medium gray in panels A and B to illustrate how the orientation of αCTD is different in the binary complex vs. the ternary complex. The ribbon at αCTD residues K271, A272, and E273 (αCTD activation patch) is dark gray. αCTD residues T263, R265, G296, K298 side chains (αCTD DNA-binding patch) are black and shown in stick form. The four deoxyribose positions on each DNA strand of the UP element proximal subsite that are protected from hydroxyl radical attack by αCTD are shown as spheres. (B) The model of the FIS dimer (from the x-ray crystal structure) is shown in ribbon form in light gray and modeled on the DNA as in ref. 152. FIS residues Q68 and R71 (FIS activation patch) are shown in a darker gray on the ribbon. The three deoxyribose positions on the nontemplate strand (-59, -60, -61) and the two positions on the template strand (-62, -63) that are protected in hydroxyl radical footprints of the FIS-RNAP-DNA ternary complex are shown as spheres. The αCTD is rotated on the DNA relative to its position in the model in (A). αCTD side chains required in the activation complex (T263, V264, R265, G296) are shown in black in stick form. K298, shown in light gray stick form, is not required in the ternary complex, although it is important for UP element recognition.

determinant," comprises the surface of αCTD that interacts with DNA.[36] The second patch, the "273 determinant" (including amino acids K271, A272, and E273), is the proposed target on αCTD for direct contact with FIS.[36] FIS-dependent activation requires contacts between one FIS subunit and one αCTD, both of which are bound to DNA.[36] Thus, the activation complex is stabilized by interactions between FIS and DNA, FIS and αCTD, and αCTD and DNA (Fig. 2, panel B).[36] Interestingly, to accommodate interactions between the FIS B-C turn and the αCTD 273 determinant, it is proposed that the αCTD rotates on the DNA relative to its position when bound at the proximal subsite of the UP element (Fig. 2, compare panels A and B).

Thus far, the majority of information about FIS-mediated activation of rRNA promoters has been obtained with *rrnB* P1. While FIS site I accounts for the majority of activation at *rrnB* P1,[31] upstream FIS sites are required for the majority of activation at *rrnA* P1 and *rrnE* P1.[7] Future studies will be required to understand the more complex architecture of these activation complexes and whether or not contacts to parts of RNAP in addition to αCTD play a role in FIS-dependent activation. Furthermore, FIS increases RNAP binding (closed complex formation), and it also increases the rate of a later isomerization step.[37-40] The magnitude and mechanism of FIS-dependent stimulation appears to depend on the particular promoter context as well as on the positions and sequences of the FIS sites.

Antitermination

In order to sustain high rates of rRNA synthesis, each rRNA transcript that is initiated must be completed. Normally transcription and translation are coupled in bacteria, and translating ribosomes prevent access of the rho transcription termination factor to an mRNA, thereby blocking premature termination (reviewed in ref. 41). In order to produce the long, highly structured, untranslated RNA from the rRNA operons, RNAP must transcribe several kilobases without being susceptible to rho-dependent termination. rRNA operons contain antitermination systems that avert this problem by converting RNAP into a termination-resistant form, allowing equal amounts of each mature rRNA (16S, 23S, and 5S) to be produced (reviewed in refs. 2,42).

Assembly of the *rrn* antitermination complex involves several trans-acting factors and cis-acting sequences, including many of those utilized in the well-characterized bacteriophage lambda antitermination system.[41] BoxA sequences, found upstream of the 16S and 23S RNA coding regions, are essential for antitermination, while the BoxB and BoxC sequences, sometimes found adjacent to BoxA, likely play accessory roles.[2,43] NusA, NusB, NusG, and several r-proteins (S10/NusE, S4, L3, L4, and L13) support efficient antitermination, but additional unidentified proteins might also be involved.[2,43,44]

It is not fully understood how the *rrn* antitermination system allows RNAP to read through terminators. The nascent BoxA RNA recruits the Nus/r-protein complex to RNAP, and this complex may remain bound to the RNAP during elongation of transcription through the rRNA operons.[2,41] *rrn* antitermination strongly inhibits many rho-dependent terminators but only weakly inhibits rho-independent terminators.[45] The *rrn* antitermination system results in an approximately two-fold increase in the rate of rRNA elongation relative to the rate of mRNA elongation, presumably by reducing pausing by RNAP,[46,47] and it has been proposed that rho factor, which translocates along the RNA, is unable to catch up to the RNAPs rapidly transcribing the rRNA operons.[2]

If the *rrn* antitermination system prevents premature rho-dependent termination, how is transcription terminated efficiently at the true end of the rRNA operons? One answer to this question might be simply that the *rrn* operons contain strong rho-independent terminators at their distal ends.[2] However, at least some *rrn* termination regions also contain rho-dependent terminators that are not subject to antitermination.[48]

The *rrn* antitermination system might play other roles in addition to reducing pausing and preventing premature termination. For example, the high rate of transcription elongation might be required for proper kinetics of rRNA folding and thus for ribosome assembly.[43,49] Assembly of the antitermination machinery might also facilitate processing and folding of the precursor RNAs by forming kinetic intermediate complexes that are not present in the final assembled ribosome.[2,50] In addition, the involvement of several r-proteins in the antitermination process suggests the possibility that some r-proteins might be delivered directly from the antitermination complex to the rRNA as it is being transcribed, potentially enhancing the rate and efficiency of ribosome assembly.[42,44] Lastly, it is possible that some r-proteins might regulate rRNA synthesis rates by coordinating the efficiency of rRNA antitermination with r-protein pools.[42,44]

Location and Copy Number of rRNA Operons

Unlike most gene products in *E. coli* which are produced from only one gene, rRNA is unusual in that it is produced from multiple operons.[1,2] All seven rRNA operons are located in the half of the genome that contains the origin of replication, and the direction of transcription in all seven rRNA operons is away from the origin. This arrangement likely contributes to the high transcription rate of rRNA genes by increasing rRNA gene copy number in fast growing cells (containing multiple origins) while at the same time avoiding head-on collisions between the transcription and DNA replication machinery.[51]

Although the seven rRNA operons are dispersed over half of the chromosome, it has been postulated that transcription of the rRNA genes might be localized to one region of the nucleoid, in a prokaryotic version of the nucleolus.[52] Recent fluorescence microscopy of green fluorescent protein-tagged RNAP in *Bacillus subtilis* suggested that RNAP localizes to a few clusters in the cell when the majority of RNAP molecules are devoted to transcribing rRNA operons (i.e., at fast growth rates).[53] The RNAP clustering dispersed when transcription from the rRNA promoters was inhibited, suggesting that rRNA expression might be organized at a specific subcellular location.[53] It will be interesting to determine whether different rRNA operons do in fact colocalize.

Originally it was postulated that multiple copies of rRNA genes were required in order to produce the enormous amounts of rRNA required for rapid growth. However, because of homeostatic regulation of rRNA promoter activity (see below), deletion of one or even several rRNA operons results in only small decreases in rRNA production in steady-state growth.[54,55] Instead, high rRNA gene copy number may allow bacteria to respond more quickly to changes in resource availability.[56,57] Because the sequences of the seven rRNA structural genes and regulatory regions differ slightly, multiple rRNA operons theoretically could be beneficial if different rRNA operons played distinct roles under different environmental conditions. The recent construction of strains in which all rRNA comes from one rRNA operon (in multicopy) could provide a means for testing whether there are functional differences for the seven rRNA operons.[58]

Regulation of Transcription of rRNA Genes

Although rRNA promoter output can be extremely high (primarily as a result of the presence of UP elements, FIS, and the antitermination system), rRNA promoters are tightly regulated to prevent overinvestment of biosynthetic energy in synthesis of the translational machinery under less than optimal conditions. There are multiple ways in which regulation of rRNA promoter activity has been assayed: responses of the promoters to amino acid starvation (stringent control), to changes in nucleotide concentrations (NTP-sensing), to different steady-state growth rates (growth rate-dependent control), to conditions such as extra rRNA gene dosage that elicit a homeostatic response (homeostatic/feedback regulation), to changes in growth phase (growth phase-dependent regulation), and to changes in nutritional or environmental conditions (upshifts or downshifts). Models for how rRNA promoter activity might be regulated in several of these situations are discussed below. A recurring theme of this section is that rRNA regulation is determined primarily by features intrinsic to the promoter-RNAP interaction.

Unusual Kinetic Properties of rRNA Promoters

The kinetic properties of rRNA promoters differ from most promoters in two important ways. First, the rRNA promoters are among the strongest in the cell.[1,3,59] As discussed above, the high binding affinity of the rRNA promoters for RNAP is conferred primarily by FIS and the UP element.[7,9,31] Second, whereas most promoters form long-lived open complexes ($t_{1/2}$ = hours), rRNA promoters form open complexes that are extraordinarily short-lived ($t_{1/2} \leq$ seconds or minutes).[4,60-63] Highly permissive conditions (e.g., supercoiled template and low ionic strength) are required for determination of *rrnB* P1 open complex half-life by standard methods.[28,60]

As a consequence of its high binding affinity for RNAP and the short lifetime of its open complex, *rrnB* P1 is fully saturated with RNAP, but the majority of the complexes are closed under standard solution conditions in vitro.[61,64-66] Because of these unusual intrinsic properties, rRNA promoters are susceptible to modulation by various types of regulatory molecules that alter open complex half-life or occupancy (see below) (reviewed in ref. 67).

Stringent Control

In response to amino acid starvation, bacteria produce high levels of the unusual nucleotide, guanosine 5'-diphosphate 3'-diphosphate (ppGpp) (reviewed in ref. 68). High levels of ppGpp rapidly shut down rRNA and tRNA transcription, a judicious response given that large numbers of ribosomes are unnecessary when the substrates for translation are unavailable. ppGpp also stimulates the expression of numerous gene products, including those from several amino acid biosynthesis/transport operons. This global adjustment of cellular metabolism to amino acid starvation is referred to as the stringent response and is dependent on the *relA* gene, which encodes the primary ppGpp synthetase.

The effect of ppGpp on transcription of rRNA operons has been studied intensely since the discovery of ppGpp over 30 years ago (reviewed in refs. 68,69). An *rrnB* P1 core promoter lacking *rrn*-specific transcribed sequences is sufficient for regulation by ppGpp, suggesting that ppGpp acts primarily by inhibiting the rate of transcription initiation and that RNAP interactions with FIS, the UP element, and the antitermination machinery are not required.[70] ppGpp also nonspecifically slows elongation and enhances pausing,[71] but these effects are mitigated in rRNA genes by the effects of the *rrn* antitermination system.[47] Because ppGpp inhibits transcription from rRNA promoters in vitro in the absence of proteins other than RNAP, ppGpp likely interacts directly with the RNAP-promoter complex to inhibit initiation.[60,62,72]

Recently, a model for inhibition of rRNA promoters by ppGpp based on the short half-life of the *rrn* open complex has been proposed. Kinetic analysis of the effects of ppGpp on various steps in transcription initiation revealed that ppGpp has minimal effects on the binding affinity of the promoter for RNAP or on the rate of promoter escape.[60] Rather, ppGpp shortens the half-lives of open complexes formed on all promoters, even promoters whose activities are unaffected by ppGpp in vivo.[60] The kinetic properties of a particular promoter determine whether or not the effect on half-life results in inhibition.[60] Promoters that form unusually short-lived open complexes, like *rrnB* P1, are inhibited, because the half-life contributes to transcriptional output. ppGpp does not inhibit promoters where the rate of open complex collapse is slow enough that, even in the presence of ppGpp, it does not contribute to the overall rate of transcription initiation.

Several observations are consistent with this model. First, footprinting studies indicate that ppGpp does not prevent binding of RNAP to *rrnB* P1, but rather decreases the fraction of complexes that are open.[64,65,73] Second, ppGpp is unable to inhibit transcription from *rrnB* P1 under solution conditions that greatly stabilize open complexes (e.g., on a supercoiled template at low ionic strength).[60,73] Third, RNAP mutants, selected for growth of ppGpp null strains in the absence of amino acids, mimic the effects of ppGpp on wild-type RNAP.[74-78] These RNAP mutants form significantly shorter-lived open complexes than wild-type RNAP on all promoters and are specifically defective for transcription of ppGpp-inhibited promoters.

Two *rrnB* P1 promoter mutants resistant to inhibition by ppGpp have been identified, and as predicted by this model, form much longer-lived open complexes than the wild-type promoter.[60,66,70] One mutant, *rrnB* P1 (CGC-5-7ATA), has a triple mutation that decreases the G+C-content of the region between the -10 hexamer and +1 start site. It was proposed long ago that the high G+C-content of this region, termed the discriminator, might make strand opening difficult.[79] The other mutant, *rrnB* P1 (T-33A, Ains-22), contains two mutations that generate a consensus -35 hexamer and spacer length. Thus, regions outside of the discriminator can also affect regulation by ppGpp.[70]

ppGpp likely reduces open complex half-life by directly binding to RNAP and inducing a conformational change (reviewed in ref. 69). Consistent with this model, cross-linkable ppGpp analogs interact with the β and β' subunits of RNAP.[80,81] The exact location of the ppGpp-binding site on RNAP is unknown, but is likely to be distinct from the catalytic site since NTPs do not appear to compete for ppGpp binding.[71,80,81] It has been proposed that ppGpp binding induces a conformation change in RNAP, because ppGpp binding changes the trypsin digestion patterns and fluorescence anisotropy of RNAP.[80,82] There are currently two general models for how ppGpp might decrease open complex lifetime. In one model, ppGpp destabilizes σ^{70}-core RNAP interactions.[78,82] Alternatively, ppGpp might alter RNAP-DNA interactions (e.g., inducing strand closure).[74]

NTP-Sensing

In addition to regulation by high levels of the unusual nucleotide ppGpp, it has been proposed that rRNA promoters might be regulated by the concentrations of their initiating nucleotides (NTP-sensing).[76] *rrn* P1 core promoters require exceptionally high concentrations of the initiating NTP (ATP or GTP) for maximal transcription in vitro.[66,76,77] Moreover, in strains that have unusually high ATP and GTP pools because of partial pyrimidine starvation, *rrn* P1 promoter activities are higher than normal.[76] This result can be interpreted in two ways. The increase in cellular ATP and GTP pools might (1) directly affect rRNA promoter activity or (2) indirectly affect promoter activity by altering the level or activity of some other factor that affects transcription from rRNA promoters (e.g., by decreasing ppGpp levels).

To distinguish between a direct or indirect effect of changing NTP pools, a *purK* mutant that results in a defect in purine biosynthesis was used to monitor the promoter activities of *rrnB* P1 and *rrnD* P1 (which initiate with ATP and GTP, respectively).[82a] ATP and GTP pools in a *purK* mutant were altered divergently when cells were grown on media containing either adenine or guanine but not both.[83] When the *purK* mutant was grown on adenine, ATP pools and *rrnB* P1 activity were high, but GTP pools and *rrnD* P1 activity were low.[82a] Conversely, when the *purK* mutant was grown on guanine, ATP pools and *rrnB* P1 activity were low, but GTP pools and *rrnD* P1 activity were high. Since all seven rRNA promoters are normally regulated in parallel,[7,84] indirect effects of changing NTP concentrations on other rRNA regulators would have been predicted to alter the activities of *rrnB* P1 and *rrnD* P1 in parallel. Since *rrnB* P1 and *rrnD* P1 responded divergently (i.e., with the concentration of their initiating NTP), this result strongly supports the model that *rrn* P1 promoters respond *directly* to the initiating NTP concentration in vivo.[82a] As further support for the direct role of NTPs, the activities of rRNA promoters vary with the concentrations of their initiating NTPs even in strains lacking ppGpp.[82a]

It has been argued that since the total ATP concentration in vivo in wild-type strains is higher than the concentration of ATP required for efficient *rrnB* P1 transcription in vitro, *rrnB* P1 should be fully saturated with ATP in vivo.[85] However, several observations suggest that this conclusion is incorrect. First, studies with pyrimidine and purine mutant strains indicate that increases in NTP pools do alter *rrnB* P1 activity in vivo.[76,82a] Second, the absolute ATP concentration required for transcription in vitro is a strong function of the solution conditions, and the ATP concentration required for transcription from *rrnB* P1 in vitro, when the salt concentration is high or when the template is not supercoiled, approaches the total cellular ATP concentrations reported in the literature.[76,86] Third, an increase in ATP concentration elicited by inhibition of protein synthesis by chloramphenicol or spectinomycin increases *rrn* P1 transcription in vivo.[82a] Thus, although the fraction of the total ATP pool available for binding to RNAP is unclear, the evidence indicates that the free NTP pools are not saturating for transcription at rRNA promoters in vivo.

Athough all promoters bind NTPs to initiate transcription, and although the concentration of the NTP needed for transcription initiation is higher at all promoters than that needed for transcription elongation (reviewed in ref. 87), the NTP pools present in vivo are likely only limiting for initiation at those promoters whose open complexes are short-lived.[66,76] High concentrations of the initiating NTP likely prevent open complex collapse by simply driving the initiation process forward (by mass-action).[66,76] Consistent with this model, *rrnB* P1 promoter mutants that form long-lived open complexes require much lower concentrations of the initiating NTP than wild-type *rrnB* P1.[66,76] Similarly, mutant RNAPs that form short-lived open complexes require high concentrations of ATP for efficient transcription.[76,77]

rrn P1 promoters respond to changes in NTP pools in strains with defects in nucleotide metabolism.[82a] There are nutritional and environmental conditions that result in changing NTP concentrations and corresponding changes in rRNA promoter activity. For example, NTP pools and rRNA promoter activity decrease upon entry into stationary phase (see below). However, given the multiplicity of factors that influence transcription of rRNA genes and the redundancy in rRNA regulation, it is unknown whether these changes in NTP concentrations are essential for the corresponding changes in *rrn* P1 promoter activity. In cases when NTP pools change but rRNA promoter activity remains unperturbed, there must be compensating changes in the other systems that affect rRNA synthesis rates.

Growth Rate-Dependent Regulation

Cells growing rapidly in nutrient-rich conditions must synthesize much more protein per unit time than cells growing slowly in nutrient-poor environments. Since the rate of translation per ribosome does not change much with growth rate, ribosome production (and the synthesis of rRNA) must increase with growth rate in order to accommodate the increasing need for protein synthesis (reviewed in refs. 1-3).

Promoter-*lacZ* fusions containing *rrnB* P1 sequences from only -41 to +1 (core promoter) are growth rate-regulated.[88] Mutations within this core promoter region that generate either a consensus -35 hexamer or a 17 bp spacer, or that decrease the G+C-content of the discriminator, disrupt regulation.[70,86,88,89] Thus, mutations that perturb growth rate regulation are found in several regions of the core promoter, and correlate well with the regions of the promoters that interact directly with RNAP. There is no evidence that these mutations define a binding site for a hypothetical regulatory protein responsible for growth rate control. Although the cis-acting sequences required for growth rate-dependent regulation of *rrn* promoters are well-defined, the identity of the trans-acting factor(s), whose activity or concentration would presumably change with growth rate, remains unclear.

Mutations in *rpoB* or *rpoC* (coding for the β or β' subunits of RNAP, respectively) have been isolated that alter the response of *rrn* promoters to growth rate.[76,77] These growth rate regulation-defective mutant RNAPs form open complexes with much shorter half-lives than wild-type RNAP.[76,77] In addition, all of the *rrnB* P1 promoter mutants that are defective for growth rate-dependent regulation form longer-lived open complexes than wild-type *rrnB* P1.[66,86] Since RNAP and promoter mutants result in both altered open complex half-life and altered regulation, the lifetime of the open complex is likely a critical determinant for growth rate-dependent regulation.

Models have been proposed in which changes in ppGpp, NTPs, FIS, and/or RNAP pools are responsible for growth rate-dependent regulation (see below).[66,76,77,85,86,91,92] If NTP-sensing were responsible for growth rate-dependent regulation, then free NTP pools should increase with the growth rate. Unfortunately, measurements of NTP concentrations as a function of growth rate vary with the extraction method used.[6,82a,90]

As stated above, the promoter sequence requirements for growth rate-dependent regulation correlate very well with the sequences required for high ATP-dependence in vitro.[86] Although this observation is consistent with the NTP-sensing model for growth rate-dependent regulation, this observation is also consistent with a model in which one or more unidentified regulatory signals are elicited in response to changes in growth rate, and these signals work on the same kinetic steps that make the promoter dependent on high concentrations of the initiating NTP for maximal transcription.[66,86]

Since ppGpp inhibits rRNA promoters during a stringent response, it has been proposed that ppGpp is the sole regulator responsible for growth rate-dependent regulation.[91,92] Consistent with this model, ppGpp levels, although 10- to 100-fold lower than in a stringent response, vary inversely with growth rate.[91] A critical prediction of this model was that rRNA promoter activity should stay constant with respect to growth rate in strains deleted for the ppGpp synthesis machinery. However, it was found that rRNA promoter activity increases with growth rate almost normally in strains devoid of ppGpp.[88,93] Lastly, if stringent control and growth rate-dependent control operate through the same mechanism, the sequence determinants for the two responses should be identical. However, several *rrnB* P1 promoter variants are stringently controlled but not growth rate regulated (see also ref. 94).[70] The data therefore do not support the model that ppGpp is the sole regulator responsible for growth rate-dependent regulation. Nevertheless, the data do not rule out the possibility that ppGpp (when present) contributes to growth rate regulation.

FIS activates *rrn* P1 transcription and FIS levels increase with growth rate, suggesting that FIS might also be a contributor to growth rate regulation.[31,95,96] Although placement of FIS sites upstream of a nonregulated promoter can confer growth rate-dependent regulation on the promoter,[97] *rrn* P1 constructs lacking FIS sites are regulated normally.[88] In addition, growth rate-dependent regulation of rRNA promoters persists in strains deleted for *fis*.[31] Although FIS is not essential for growth rate control, changing FIS levels can provide a back-up mechanism when the primary mechanism for growth rate-dependent regulation is impaired.[38]

It has been suggested that increases in rRNA promoter activity with growth rate are mediated by increases in the free RNAP pool.[85,98] One assumption of this model, that rRNA promoters compete poorly for limiting amounts of RNAP, conflicts with several observations. First, *rrnB* P1 requires the same or lower concentrations of RNAP for efficient transcription in vitro than other promoters.[74] Second, when RNAP pools are reduced in vivo by restricting RNAP synthesis, mRNA transcription decreases much more than does rRNA transcription.[74,99,100] Lastly, introduction of a multicopy plasmid containing a highly transcribed operon encoding a long nonfunctional RNA would be expected to intensify competition for RNAP. While diversion of RNAP to synthesizing the plasmid-encoded RNA diminished transcription from several mRNA promoters tested, rRNA promoter activity was not reduced.[74,101]

Thus, rRNA promoters have a competitive advantage for binding RNAP over most mRNA promoters. In fact, since the total RNAP concentration increases very little with growth rate while rRNA promoter activity increases dramatically, it has been proposed that the free RNAP concentration decreases with growth rate and might be responsible for the inverse growth rate-dependent regulation characteristic of a number of promoters required for amino acid biosynthesis.[74]

Homeostatic Regulation

It has been observed that total rRNA synthesis remains relatively constant under conditions that would be expected to perturb it; that is, genetic or biochemical alterations that would lead to under- or overproduction of ribosomes result in compensating (homeostatic) changes in rRNA promoter activity in order to minimize changes in ribosome synthesis. These observations led to the proposal that rRNA promoters are feedback regulated because they sense and respond to the translational needs of the cell[101] (reviewed in refs. 1-3).

Gene dose experiments demonstrated the capacity of rRNA homeostatic control to keep rRNA synthesis constant unless nutritional conditions change. Increased *rrn* copy number (by addition of *rrn* operons on multicopy plasmids) reduced the activity of individual *rrn* P1 promoters in inverse proportion to the gene dose.[101] Conversely, decreased *rrn* gene dose (from deletion of one or several *rrn* operons) increased the activities of the remaining *rrn* P1 promoters to keep total rRNA synthesis constant.[54,55,58,102]

Homeostatic regulation has been observed in a number of other situations. For example, deletion of *fis* or mutation of the residues in *rpoA* coding for amino acids critical for UP element binding would be expected to reduce overall rRNA output. However, *rrn* P1 core promoter activity increased in these situations, at least partially compensating for the loss of upstream activation.[9,31,86] Similarly, disruption of the *rrn* antitermination machinery also led to an increase in rRNA promoter activity, partially compensating for the loss in elongation processivity.[103] *rrn* P1 promoter activity was also stimulated by translation inhibitors (e.g., chloramphenicol and spectinomycin) in a futile attempt to compensate for the resulting reduction in protein synthesis.[82a,104]

The identity of the effector molecule(s) or the mechanism(s) responsible for feedback regulation have yet to be elucidated fully. Since homeostatic regulation is elicited by an increase in translationally-competent ribosomes but not by an increase in translationally-defective ribosomes, it has been proposed that the feedback signal is generated or consumed during the act of protein synthesis.[105,106]

Since translation is a major consumer of ATP and GTP, NTP-sensing by rRNA promoters would provide a potential mechanism for homeostatic regulation.[76] That is, overproduction of ribosomes would be expected to result in increased translation, increased consumption of NTPs, and correspondingly lower NTP pools. Conversely, inactivation or underproduction of ribosomes would be expected to result in reduced translation and higher NTP pools. Consistent with this model, total ATP pools increased in parallel with the increase in *rrn* P1 activity observed in strains treated with inhibitors of protein synthesis.[82a] This result indicates that total NTP pool size is responsive to changes in the translation capacity of the cell and is consistent with the model that NTP concentration is a signal that links rRNA synthesis to ribosome function.

ATP pools increase in *nusB* mutants, suggesting that NTP-sensing might be a molecular mechanism responsible at least in part for the increase in rRNA promoter activity observed upon disruption of *rrn* antitermination.[103] In addition, changing NTP pools might be responsible for the increase in *rrn* P1 activity in *fis* deletion mutants, since the cis-acting sequences required for homeostatic regulation elicited by deletion of *fis* correlated roughly with the sequences required for *rrnB* P1's dependence for high concentrations of ATP for maximal transcription in vitro.[86]

Feedback elicited by other methods might be mediated by factors other than NTPs. Since high levels of ppGpp inhibit *rrn* transcription and since the ppGpp synthetase RelA is ribosome-associated and generates ppGpp during the act of translation,[68] it is conceivable that ppGpp might be responsible for homeostatic regulation in certain situations. For example, transient induction of rRNA synthesis from a plasmid-borne rRNA operon under the control of an inducible promoter led to the inhibition of transcription of chromosomally encoded rRNA genes.[107] In this case, inhibition of rRNA synthesis was accompanied by a small increase in ppGpp pools, while no changes in total ATP or GTP pools were detected, suggesting that ppGpp might be the homeostatic regulator in this case.[107]

Several observations suggest that the homeostatic response to changes in steady-state rRNA gene dose might not involve either NTP-sensing or ppGpp. First, homeostatic regulation elicited by increased gene dose occurs in strains devoid of ppGpp.[93] Second, changes in NTP or ppGpp pools were not detected when rRNA gene dose was increased or decreased.[54,108] Third, it was reported that the rRNA promoter sequence requirements for response to changes in gene dose do not correlate with the sequence determinants for response to ppGpp.[108]

In summary, homeostatic regulation of rRNA promoter activity likely involves multiple mechanisms that act at the level of transcription initiation, and the regulatory factors responsible might vary with the method employed to induce the response. In addition, rRNA elongation rates increased in response to decreases in gene dose,[54] suggesting that mechanisms for altering rRNA synthesis rates in response to changes in translational capacity could include events subsequent to transcription initiation (see also refs. 44,102,109).

Shut-Down of rRNA Promoter Activity in Stationary Phase

Upon entry into stationary phase, rRNA promoter activity is markedly reduced.[110] Concurrently, the levels of many of the known inputs to transcription of the rRNA genes change in the expected direction: FIS levels plummet,[95,96,111] ppGpp levels increase,[110] NTP pools shrink,[112] and supercoiling levels decrease.[113] rRNA promoter activity decreases normally in strains deleted for *fis* but is delayed in strains lacking ppGpp.[110] It is unknown whether the reduction in NTP pools and negative supercoiling levels play a role in the decrease in rRNA promoter activity in stationary phase.

The transition to stationary phase is marked by an increase in the activity of promoters recognized by RNAP containing the σ^s subunit.[114] Since *rrn* P1 promoters are transcribed very poorly by $E\sigma^s$,[115] they are not negatively impacted by deletion of *rpoS* (the gene encoding σ^s), and in fact, the shut-off of rRNA promoters in stationary phase is delayed in a strain lacking σ^s.[110] The increase in $E\sigma^s$-dependent transcription could directly or indirectly reduce rRNA promoter activity in stationary phase in a variety of ways. For example, σ^s could indirectly reduce σ^{70}-dependent transcription by competing for binding to limiting amounts of core RNAP.[116] Alternatively, $E\sigma^s$ could increase transcription of an inhibitor of rRNA promoters.

H-NS (<u>H</u>istone-like <u>N</u>ucleoid <u>S</u>tructuring protein), a highly abundant DNA-binding nucleoid protein that modulates transcription of a variety of genes (reviewed in ref. 32), has been proposed to play a role in growth phase regulation by directly inhibiting rRNA promoter activity in stationary phase.[117] Consistent with this model, *rrnB* P1 activity is 2-fold higher in late log and stationary phase in an *hns* mutant compared to a wild-type strain.[117]

The mechanism of repression of *rrnB* P1 by H-NS has been explored in some detail. H-NS binds within and upstream of the core promoter, altering the DNA conformation and potentially antagonizing FIS-dependent activation.[118,119] H-NS also inhibits *rrnB* P1 in vitro and in vivo in the absence of FIS, potentially by inhibiting promoter escape.[117,118,120] However, additional experiments may be required to clarify the role of H-NS in rRNA regulation, since in a different study, it was reported that an *hns* mutation had no effect on rRNA growth phase regulation.[110] It is unknown whether the H-NS homolog StpA plays a role in regulation of rRNA promoter activity.[121]

Interconnections between Inputs

Now that many of the inputs to regulation have been studied individually, an important task for the future will be understanding how these contributors influence transcription of rRNA genes in combination. Many regulators of rRNA transcription influence the levels or activity of other contributors. For example, FIS not only recruits RNAP to *rrn* P1 promoters,[39] but it also decreases the concentration of ATP required for *rrn* P1 transcription.[38] In addition to their roles as transcription factors that directly regulate *rrn* P1 transcription,[36,120] FIS and H-NS could potentially indirectly regulate *rrn* P1 transcription by altering chromosome superhelicity.[122,123] Similarly, changes in ATP levels could potentially influence *rrn* P1 transcription both directly (as described above), and indirectly by altering the activity of DNA gyrase, thereby altering supercoiling.[113] Lastly, ATP and GTP pools decrease when high levels of ppGpp are produced,[124] possibly contributing to the shut-down of rRNA promoter activity during a stringent response.

rrn P2 Promoters

Although not as well-characterized as the *rrn* P1 promoters, the *rrn* P2 promoters are regulated by many of the same inputs and mechanisms.[125,125a] Like the *rrn* P1 promoters, the *rrn* P2 promoters contain A+T-rich sequences upstream of near-consensus -10 and -35 hexamers that are separated by 16 bp spacers. However, the *rrnB* P2 UP element has a much smaller effect on transcription than the *rrnB* P1 UP element [11] and FIS does not activate *rrnB* P2.[117,126] Because transcription originating from P1 can "occlude" P2, *rrnB* P2 is less active in its native context at fast growth rates when P1 activity is high than when isolated from its normal context downstream of P1.[125a,127] *rrn* P2 promoters are more active than P1 promoters at slow growth rates,[125a,128] possibly accounting for the excess ribosomes produced under these conditions that might contribute to rapid adaptation to increases in nutrient availability.[129]

Despite previous reports suggesting that *rrn* P2 promoters are constitutive,[6,130] recent studies indicate that *rrn* P2 promoters are regulated similarly to the P1 promoters, albeit with a smaller range of activities.[62,70,125,127] *rrnB* P2 forms a short-lived open complex, is sensitive to its initiating NTP concentration (CTP), and is growth rate-regulated.[125a] Furthermore, *rrn* P2 promoters are inhibited by ppGpp,[62,70,125,127] homeostatically regulated,[125,125a] and inhibited in stationary phase.[125]

tRNA Promoters

Coordination of tRNA and rRNA synthesis is achieved in part because several tRNAs genes are located within the rRNA operons. However, most tRNAs are located in other operons that encode one or more tRNAs and/or proteins.[131] Nevertheless, the inputs to strength and regulation of transcription of various tRNA genes are often similar to those used for transcription of the rRNA genes. In general, tRNA promoters have good matches to the -10 and -35 consensus hexamers and these RNAP recognition determinants are separated by 16-18 bp spacers.[1] UP elements and FIS activate some tRNA promoters [9,31,132-134] and potentially activate numerous others as well.[1,13,135,136]

The most abundant tRNAs are highly regulated. In order to optimize translation efficiency, major tRNA species, which recognize preferred codons, are stringently controlled and increase with growth rate.[137,138] Not surprisingly, many tRNA promoters contain GC-rich discriminators and form short-lived open complexes.[94,139-142] Growth rate-regulation of some tRNA promoters is entirely dependent on FIS, whereas regulation of other tRNA promoters might be mediated at least in part through NTP-sensing and/or ppGpp.[94,132,136] Many tRNA promoters are inhibited by overproduction of rRNA.[94,101,103,107,132,143]

Ribosomal Protein Synthesis

The ribosome is a complex machine composed of 3 rRNAs and more than 50 r-proteins. In order to achieve the correct stoichiometry between rRNAs and r-proteins, r-protein synthesis rates are coordinated with each other and balanced with the rate of rRNA synthesis (reviewed in refs. 1,144,145). Unlike rRNA regulation which occurs mostly at the level of transcription initiation, r-protein regulation occurs primarily through translational feedback mechanisms. Each operon encodes an r-protein (e.g., S4, S7, L1, L4) that, in addition to serving as a structural component of the ribosome, binds to a site within its own mRNA and directly or indirectly prevents synthesis of most or all the proteins encoded by the mRNA. Since most repressor r-proteins have higher affinity for their rRNA targets than for their own mRNAs, represssion occurs only when more of the repressor r-protein accumulates than can be incorporated into assembling ribosomes. However, in some cases, the repressor r-proteins have the same affinity for their rRNA targets as for their mRNA,[146,147] and the preferential incorporation of regulatory ribosomal proteins into ribosomes likely results from the high cooperativity of ribosome assembly. As might be expected, the mRNA binding sites usually show structural and sequence homologies to their rRNA binding sites.

By binding to their own mRNAs, most repressor r-proteins regulate their own synthesis at the level of translation initiation, using a variety of mechanisms to do so. For example, it has been proposed that some r-proteins bind to a site on their mRNA that overlaps the Shine-Dalgarno region, thereby sterically blocking ribosome binding. In other cases, it has been proposed that the repressor r-protein induces a conformational change in the mRNA that either prevents the ribosome from binding or locks the translation initiation complex in an inactive intermediate. The regulatory r-protein binds to a single operator, directly preventing translation of the proximal gene (which is sometimes, but not always, the first gene in the operon) and indirectly preventing translation of downstream genes through "translational coupling" (e.g., see refs. 148,149). In other words, the translation of multiple r-proteins from a single mRNA is usually coupled so that downstream cistrons are translated efficiently only when the upstream cistron is translated. In addition, cistrons upstream of the translational operator are sometimes "retroregulated" by induction of mRNA decay.[150]

Although the repressor r-proteins typically regulate expression by inhibiting translation, autogenous control is achieved in the S10 operon by regulation of mRNA elongation.[144,151] In this case, excess L4 induces RNAP to terminate prematurely 30 bp upstream from the first gene of the operon (in addition to inhibiting translation of its mRNA).

Conclusions and Future Directions

The multiplicity and redundancy of signals makes rRNA regulation incredibly robust. Although the inputs and regulatory mechanisms contributing to stable RNA synthesis in *Escherichia coli* are only partially understood, the control of rRNA transcription serves as a paradigm for other complex regulatory circuits. Figure 3 is a cartoon that illustrates the numerous regulatory inputs (e.g., presence of amino acids, nucleotides, nutrients, etc.) that are integrated to determine the appropriate amount of rRNA output for a given condition.

Despite significant progress in understanding the strength and regulation of transcription of rRNA genes, several important questions remain. For example, the architecture and mechanism of activation of the *rrn* P1 promoters containing multiple FIS binding sites is unclear. The effector molecules responsible for growth rate, growth phase, and homeostatic regulation require further definition, and how they work together remains to be determined. In addition, the precise location of the ppGpp binding site on RNAP and the allosteric mechanism by which ppGpp alters the structure of the open complex await discovery. Furthermore, whether the operons are located at a specific subcellular location remains to be determined, and lastly, it will be of great interest to determine whether the inputs and regulatory mechanisms defined for *E. coli* are conserved in other bacteria.

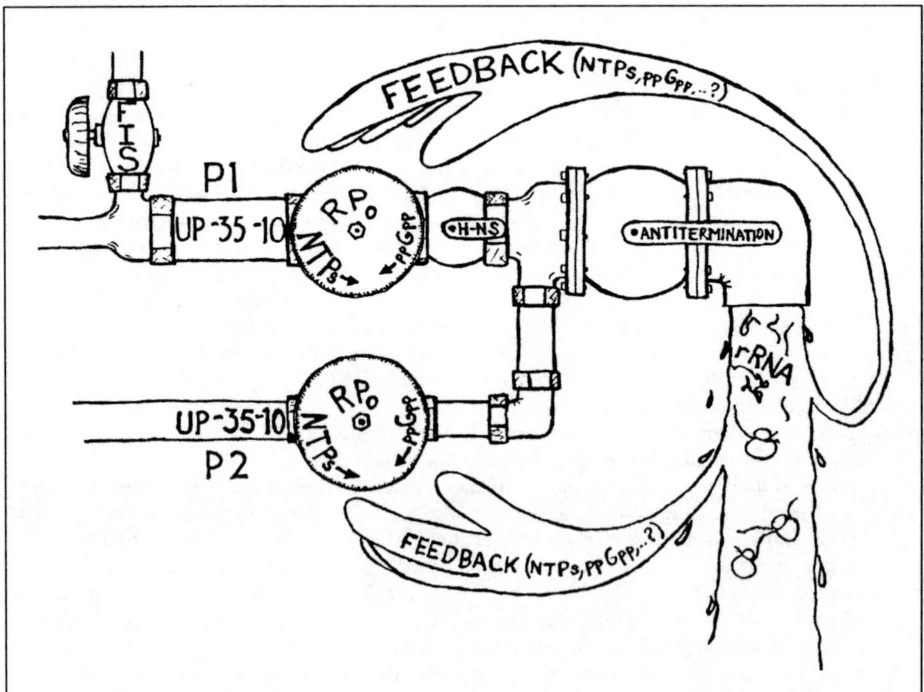

Figure 3. Schematic representation ("plumber's model") of the control of rRNA synthesis. Multiple cis-acting sequences (depicted as pipes) and trans-acting factors (depicted as valves) determine the amount of rRNA produced (outflow from the pipe). Homeostatic regulation ultimately monitors the final flow rate and adjusts it accordingly. Illustrated by J.A. Appleman (University of California-Davis).

Acknowledgements

We thank Wilma Ross, Tamas Gaal, and members of our laboratory for helpful discussions. We also thank Alex Appleman for generously illustrating Figure 3. This work was supported by N.I.H. grant GM37048 to R.L.G. and by a fellowship from Pfizer Biotechnology to M.M.B.

References

1. Keener J, Nomura M. Regulation of ribosome biosynthesis. In: Neidhardt FC, ed. *Escherichia coli* and *Salmonella* Cellular and Molecular Biology. Washington, D.C.: ASM Press, 1996:1417-1431.

2. Condon C, Squires C, Squires CL. Control of rRNA transcription in *Escherichia coli*. Microbiol Rev 1995; 59(4):623-645.

3. Gourse RL, Gaal T, Bartlett MS et al. rRNA transcription and growth rate-dependent regulation of ribosome synthesis in *Escherichia coli*. Ann Rev Microbiol 1996; 50:645-677.

4. Record MT, Jr., Reznikoff WS, Craig ML et al. *Escherichia coli* RNA polymerase ($E\sigma^{70}$), promoters, and the kinetics of the steps of transcription initiation. In: Neidhardt FC, ed. *Escherichia coli* and *Salmonella* Cellular and Molecular Biology. Washington, D.C.: ASM Press, 1996:792-820.

5. Newlands JT, Gaal T, Mecsas J et al. Transcription of the *Escherichia coli* rrnB P1 promoter by the heat shock RNA polymerase ($E\sigma^{32}$) in vitro. J Bacteriol 1993; 175(3):661-668.

6. Gourse RL, de Boer HA, Nomura M. DNA determinants of rRNA synthesis in *E. coli*: growth rate dependent regulation, feedback inhibition, upstream activation, antitermination. Cell 1986; 44(1):197-205.

7. Hirvonen CA, Ross W, Wozniak CE et al. Contributions of UP elements and the Transcription Factor FIS to Expression from the seven rrn P1 Promoters in *Escherichia coli*. J Bacteriol 2001; 183(21):6305-6314.

8. Gourse RL, Ross W, Gaal T. UPs and downs in bacterial transcription initiation: the role of the a subunit of RNA polymerase in promoter recognition. Mol Microbiol 2000; 37(4):687-695.

9. Ross W, Gosink KK, Salomon J et al. A third recognition element in bacterial promoters: DNA binding by the a subunit of RNA polymerase. Science 1993; 262(5138):1407-1413.

10. Rao L, Ross W, Appleman JA et al. Factor independent activation of rrnB P1. An "extended" promoter with an upstream element that dramatically increases promoter strength. J Mol Biol 1994; 235(5):1421-1435.

11. Ross W, Aiyar SE, Salomon J et al. *Escherichia coli* promoters with UP elements of different strength: Modular structure of bacterial promoters. J Bacteriol 1998; 180(20):5375-5383.

12. Estrem ST, Gaal T, Ross W et al. Identification of an UP element consensus sequence for bacterial promoters. Proc Natl Acad Sci USA 1998; 95(17):9761-9766.

13. Estrem ST, Ross W, Gaal T et al. Bacterial promoter architecture: subsite structure of UP elements and interactions with the carboxy-terminal domain of the RNA polymerase a subunit. Genes Dev 1999; 13(16):2134-2147.

14. Sander P, Langert W, Mueller K. Mechanisms of upstream activation of the rrnD promoter P1 of *Escherichia coli*. J Biol Chem 1993; 268(23):16907-16916.

15. Zacharias M, Goringer HU, Wagner R. Analysis of the Fis-dependent and Fis-independent transcription activation mechanisms of the *Escherichia coli* ribosomal RNA P1 promoter. Biochemistry 1992; 31(9):2621-2628.

16. Perez-Martin J, Rojo F, de Lorenzo V. Promoters responsive to DNA bending: a common theme in prokaryotic gene expression. Microbiol Rev 1994; 58(2):268-290.

17. Aiyar SE, Gourse RL, Ross W. Upstream A-tracts increase bacterial promoter activity through interactions with the RNA polymerase α subunit. Proc Natl Acad Sci USA 1998; 95(25):14652-14657.

18. Gaal T, Rao L, Estrem ST et al. Localization of the intrinsically bent DNA region upstream of the *E. coli* rrnB P1 promoter. Nucleic Acids Res 1994; 22(12):2344-2350.

19. Jeon YH, Yamazaki T, Otomo T et al. Flexible linker in the RNA polymerase α subunit facilitates the independent motion of the C-terminal activator contact domain. J Mol Biol 1997; 267(4):953-962.

20. Blatter EE, Ross W, Tang H et al. Domain organization of RNA polymerase a subunit: C-terminal 85 amino acids constitute α domain capable of dimerization and DNA binding. Cell 1994; 78(5):889-896.

21. Zhang G, Darst SA. Structure of the *Escherichia coli* RNA polymerase α subunit amino-terminal domain. Science 1998; 281(5374):262-266.

22. Jeon YH, Negishi T, Shirakawa M et al. Solution structure of the activator contact domain of the RNA polymerase α subunit. Science 1995; 270(5241):1495-1497.

23. Gaal T, Ross W, Blatter EE et al. DNA-binding determinants of the α subunit of RNA polymerase: novel DNA-binding domain architecture. Genes Dev 1996; 10(1):16-26.

24. Murakami K, Fujita N, Ishihama A. Transcription factor recognition surface on the RNA polymerase α subunit is involved in contact with the DNA enhancer element. EMBO J 1996; 15(16):4358-4367.

25. Newlands JT, Ross W, Gosink KK et al. Factor-independent activation of *Escherichia coli* rRNA transcription. II. characterization of complexes of rrnB P1 promoters containing or lacking the upstream activator region with *Escherichia coli* RNA polymerase. J Mol Biol 1991; 220(3):569-583.

26. Ross W, Ernst A, Gourse RL. Fine structure of *E. coli* RNA polymerase-promoter interactions: α subunit binding to the UP element minor groove. Genes Dev 2001; 15(5):491-506.

27. Shao X, Grishin NV. Common fold in helix-hairpin-helix proteins. Nucleic Acids Res 2000; 28(14):2643-2650.

27a. Benoff B, Yang H, Lawson CL et al. Structural basis of transcription activation: the CAP-alpha CTD-DNA complex. Science 2002; 297(5586):1562-6.

28. Leirmo S, Gourse RL. Factor-independent activation of *Escherichia coli* rRNA transcription. I. Kinetic analysis of the roles of the upstream activator region and supercoiling on transcription of the *rrnB* P1 promoter in vitro. J Mol Biol 1991; 220(3):555-568.

29. Tagami H, Aiba H. An inactive open complex mediated by an UP element at *Escherichia coli* promoters. Proc Natl Acad Sci USA 1999; 96(13):7202-7207.

30. Ellinger T, Behnke D, Knaus R et al. Context-dependent effects of upstream A-tracts. Stimulation or inhibition of *Escherichia coli* promoter function. J Mol Biol 1994; 239(4):466-475.

31. Ross W, Thompson JF, Newlands JT et al. *E. coli* Fis protein activates ribosomal RNA transcription in vitro and in vivo. EMBO J 1990; 9(11):3733-3742.

32. McLeod SM, Johnson RC. Control of transcription by nucleoid proteins. Curr Opin Microbiol 2001; 4:152-159.

33. Gosink KK, Ross W, Leirmo S et al. DNA binding and bending are necessary but not sufficient for Fis-dependent activation of *rrnB* P1. J Bacteriol 1993; 175(6):1580-1589.

34. Gosink KK, Gaal T, Bokal AJ et al. A positive control mutant of the transcription activator protein FIS. J Bacteriol 1996; 178(17):5182-5187.

35. Bokal AJ, Ross W, Gaal T et al. Molecular anatomy of a transcription activation patch: FIS-RNA polymerase interactions at the *Escherichia coli rrnB* P1 promoter. EMBO J 1997; 16(1):154-162.

36. Aiyar SE, McLeod SM, Ross W et al. Architecture of Fis-activated transcription complexes at the *Escherichia coli rrnB* P1 and *rrnE* P1 promoters. J Mol Biol 2002; in press.37. Aiyar SA. Ph.D. thesis: University of Wisconsin-Madison; 2001.

38. Bartlett MS, Gaal T, Ross W et al. Regulation of rRNA transcription is remarkably robust: FIS compensates for altered nucleoside triphosphate sensing by mutant RNA polymerases at *Escherichia coli rrn* P1 promoters. J Bacteriol 2000; 182(7):1969-1977.

39. Bokal AJ, Ross W, Gourse RL. The transcriptional activator protein FIS: DNA interactions and cooperative interactions with RNA polymerase at the *Escherichia coli rrnB* P1 promoter. J Mol Biol 1995; 245(3):197-207.

40. Zacharias M, Theissen G, Bradaczek C et al. Analysis of sequence elements important for the synthesis and control of ribosomal RNA in *E. coli*. Biochimie 1991; 73(6):699-712.

41. Richardson JP, Greenblatt J. Control of RNA chain elongation and termination. In: Neidhardt FC, ed. *Escherichia coli* and *Salmonella* Cellular and Molecular Biology. Washington, D.C.: ASM Press, 1996:822-848.

42. Squires CL, Zaporojets D. Proteins shared by the transcription and translation machines. Ann Rev Microbiol 2000; 54:775-798.

43. Squires CL, Greenblatt J, Li J et al. Ribosomal RNA antitermination in vitro: requirement for Nus factors and one or more unidentified cellular components. Proc Natl Acad Sci USA 1993; 90(3):970-974.

44. Torres M, Condon C, Balada JM et al. Ribosomal protein S4 is a transcription factor with properties remarkably similar to NusA, a protein involved in both nonribosomal and ribosomal RNA antitermination. EMBO J 2001; 20(14):3811-3820.

45. Albrechtsen B, Squires CL, Li S et al. Antitermination of characterized transcriptional terminators by the *Escherichia coli rrnG* leader region. J Mol Biol 1990; 213(1):123-134.

46. Zellars M, Squires CL. Antiterminator-dependent modulation of transcription elongation rates by NusB and NusG. Mol Microbiol 1999; 32(6):1296-1304.

47. Vogel U, Jensen KF. Effects of the antiterminator BoxA on transcription elongation kinetics and ppGpp inhibition of transcription elongation in *Escherichia coli*. J Biol Chem 1995; 270(31):18335-18340.

48. Ghosh B, Grzadzielska E, Bhattacharya P et al. Specificity of antitermination mechanisms. Suppression of the terminator cluster T1-T2 of *Escherichia coli* ribosomal RNA operon, *rrnB*, by phage lambda antiterminators. J Mol Biol 1991; 222(1):59-66.

49. Lewicki BT, Margus T, Remme J et al. Coupling of rRNA transcription and ribosomal assembly in vivo. Formation of active ribosomal subunits in *Escherichia coli* requires transcription of rRNA genes by host RNA polymerase which cannot be replaced by bacteriophage T7 RNA polymerase. J Mol Biol 1993; 231(3):581-593.

50. Pardon B, Wagner R. The *Escherichia coli* ribosomal RNA leader nut region interacts specifically with mature 16S RNA. Nucleic Acids Res 1995; 23(6):932-941.

51. French S. Consequences of replication fork movement through transcription units in vivo. Science 1992; 258(5086):1362-1365.

52. Nomura M. Engineering of bacterial ribosomes: replacement of all seven *Escherichia coli* rRNA operons by a single plasmid-encoded operon. Proc Natl Acad Sci USA 1999; 96(5):1820-1822.

53. Lewis PJ, Thaker SD, Errington J. Compartmentalization of transcription and translation in *Bacillus subtilis*. EMBO J 2000; 19(4):710-718.

54. Condon C, French S, Squires C et al. Depletion of functional ribosomal RNA operons in *Escherichia coli* causes increased expression of the remaining intact copies. EMBO J 1993; 12(11):4305-4315.

55. Asai T, Condon C, Voulgaris J et al. Construction and initial characterization of *Escherichia coli* strains with few or no intact chromosomal rRNA operons. J Bacteriol 1999; 181(12):3803-3809.

56. Condon C, Liveris D, Squires C et al. rRNA operon multiplicity in *Escherichia coli* and the physiological implications of *rrn* inactivation. J Bacteriol 1995; 177(14):4152-4156.

57. Klappenbach JA, Dunbar JM, Schmidt TM. rRNA operon copy number reflects ecological strategies of bacteria. Appl Environ Mircrobiol 2000; 66(4):1328-1333.

58. Asai T, Zaporojets D, Squires C et al. An *Escherichia coli* strain with all chromosomal rRNA operons inactivated: complete exchange of rRNA genes between bacteria. Proc Natl Acad Sci USA 1999; 96(5):1971-1976.

59. Bremer H, Dennis PP. Modulation of chemical composition and other parameters of the cell by growth rate. In: Neidhardt FC, ed. *Escherichia coli* and *Salmonella* Cellular and Molecular Biology. Washington, D.C.: ASM Press, 1996:1553-1569.

60. Barker MM, Gaal T, Josaitis CA et al. Mechanism of regulation of transcription initiation by ppGpp. I. Effects of ppGpp on transcription initiation in vivo and in vitro. J Mol Biol 2001; 305(4):673-688.

61. Gourse RL. Visualization and quantitative analysis of complex formation between *E. coli* RNA polymerase and an rRNA promoter in vitro. Nucleic Acids Res 1988; 16(20):9789-9809.

62. Heinemann M, Wagner R. Guanosine 3',5'-bis(diphosphate) (ppGpp)-dependent inhibition of transcription from stringently controlled *Escherichia coli* promoters can be explained by an altered initiation pathway that traps RNA polymerase. Eur J Biochem 1997; 247(3):990-999.

63. Langert W, Meuthen M, Mueller K. Functional characteristics of the *rrnD* promoters of *Escherichia coli*. J Biol Chem 1991; 266(32):21608-21615.

64. Ohlsen KL, Gralla JD. Melting during steady-state transcription of the *rrnB* P1 promoter in vivo and in vitro. J Bacteriol 1992; 174(19):6071-6075.

65. Ohlsen KL, Gralla JD. DNA melting within stable closed complexes at the *Escherichia coli rrnB* P1 promoter. J Biol Chem 1992; 267(28):19813-19818.

66. Barker MM, Gaal T, Ross W et al. Intrinsic properties of *Escherichia coli* rRNA promoters leading to regulation by the initiating NTP concentration. in preparation.

67. Gourse RL, Gaal T, Aiyar SE et al. Strength and regulation without transcription factors: lessons from bacterial rRNA promoters. Cold Spring Harb Symp Quant Biol 1998; 63:131-139.

68. Cashel M, Gentry DR, Hernandez VH et al. The stringent response. In: Neidhardt FC, ed. *Escherichia coli* and *Salmonella* Cellular and Molecular Biology. Washington, D.C.: ASM Press, 1996:1458-1496.

69. Chatterji D, Ojha AK. Revisiting the stringent response, ppGpp and starvation signaling. Curr Opin Microbiol 2001; 4(2):160-165.

70. Josaitis CA, Gaal T, Gourse RL. Stringent control and growth-rate-dependent control have nonidentical promoter sequence requirements. Proc Natl Acad Sci USA 1995; 92(4):1117-1121.

71. Kingston RE, Nierman WC, Chamberlin MJ. A direct effect of guanosine tetraphosphate on pausing of *Escherichia coli* RNA polymerase during RNA chain elongation. J Biol Chem 1981; 256(6):2787-2797.

72. Raghavan A, Kameshwari DB, Chatterji D. The differential effects of guanosine tetraphosphate on open complex formation at the *Escherichia coli* ribosomal protein promoters *rplJ* and *rpsA* P1. Biophys Chem 1998; 75(1):7-19.

73. Ohlsen KL, Gralla JD. Interrelated effects of DNA supercoiling, ppGpp, and low salt on melting within the *Escherichia coli* ribosomal RNA *rrnB* P1 promoter. Mol Microbiol 1992; 6(16):2243-2251.

74. Barker MM, Gaal T, Gourse RL. Mechanism of regulation of transcription initiation by ppGpp. II. Models for positive control based on properties of RNAP mutants and competition for RNAP. J Mol Biol 2001; 305(4):689-702.

75. Zhou YN, Jin DJ. The *rpoB* mutants destabilizing initiation complexes at stringently controlled promoters behave like "stringent" RNA polymerases in *Escherichia coli*. Proc Natl Acad Sci USA 1998; 95(6):2908-2913.

76. Gaal T, Bartlett MS, Ross W et al. Transcription regulation by initiating NTP concentration: rRNA synthesis in bacteria. Science 1997; 278(5346):2092-2097.

77. Bartlett MS, Gaal T, Ross W et al. RNA polymerase mutants that destabilize RNA polymerase-promoter complexes alter NTP-sensing by *rrn* P1 promoters. J Mol Biol 1998; 279(2):331-345.

78. Hernandez VJ, Cashel M. Changes in conserved region 3 of *Escherichia coli* σ^{70} mediate ppGpp-dependent functions in vivo. J Mol Biol 1995; 252(5):536-549.

79. Travers AA. Conserved features of coordinately regulated *E. coli* promoters. Nucleic Acids Res 1984; 12(6):2605-2618.

80. Toulokhonov II, Shulgina I, Hernandez VJ. Binding of the transcription effector ppGpp to *Escherichia coli* RNA polymerase is allosteric, modular, and occurs near the N terminus of the β' subunit. J Biol Chem 2001; 276(2):1220-1225.

81. Chatterji D, Fujita N, Ishihama A. The mediator for stringent control, ppGpp, binds to the β subunit of *Escherichia coli* RNA polymerase. Genes Cells 1998; 3(5):279-287.

82. Raghavan A, Chatterji D. Guanosine tetraphosphate-induced dissociation of open complexes at the *Escherichia coli* ribosomal protein promoters *rplJ* and *rpsA* P1: nanosecond depolarization spectroscopic studies. Biophys Chem 1998; 75(1):21-32.

82a. Schneider DA, Gaal T, Gourse RL. NTP-sensing by rRNA promoters in *Escherichia coli* is direct. Proc Natl Acad Sci USA 2002; 99(13):8602-7.

83. Jensen KF. Regulation of *Salmonella typhimurium pyr* gene expression: effect of changing both purine and pyrimidine nucleotide pools. J Gen Microbiol 1989; 135(Pt 4):805-815.

84. Condon C, Philips J, Fu ZY et al. Comparison of the expression of the seven ribosomal RNA operons in *Escherichia coli*. EMBO J 1992; 11(11):4175-4185.

85. Liang S, Bipatnath M, Xu Y et al. Activities of constitutive promoters in *Escherichia coli*. J Mol Biol 1999; 292(1):19-37.

86. Barker MM, Gourse RL. Regulation of rRNA transcription correlates with nucleoside triphosphate sensing. J Bacteriol 2001; 183(21):6315-6323.

87. Chamberlin MJ. RNA polymerase—An overview. In: Losick R, Chamberlin MJ, eds. RNA polymerase. Cold Spring Harbor: Cold Spring Harbor Laboratory, 1976:17-68.

88. Bartlett MS, Gourse RL. Growth rate-dependent control of the *rrnB* P1 core promoter in *Escherichia coli*. J Bacteriol 1994; 176(17):5560-5564.

89. Dickson RR, Gaal T, deBoer HA et al. Identification of promoter mutants defective in growth-rate-dependent regulation of rRNA transcription in *Escherichia coli*. J Bacteriol 1989; 171(9):4862-4870.
90. Petersen C, Moller LB. Invariance of the nucleoside triphosphate pools of *Escherichia coli* with growth rate. J Biol Chem 2000; 275(6):3931-3935.
91. Ryals J, Little R, Bremer H. Control of rRNA and tRNA syntheses in *Escherichia coli* by guanosine tetraphosphate. J Bacteriol 1982; 151(3):1261-1268.
92. Hernandez VJ, Bremer H. Guanosine tetraphosphate (ppGpp) dependence of the growth rate control of *rrnB* P1 promoter activity in *Escherichia coli*. J Biol Chem 1990; 265(20):11605-11614.
93. Gaal T, Gourse RL. Guanosine 3'-diphosphate 5'-diphosphate is not required for growth rate-dependent control of rRNA synthesis in *Escherichia coli*. Proc Natl Acad Sci USA 1990; 87(14):5533-5537.
94. Pokholok DK, Redlak M, Turnbough CL Jr. et al. Multiple mechanisms are used for growth rate and stringent control of *leuV* transcriptional initiation in *Escherichia coli*. J Bacteriol 1999; 181(18):5771-5782.
95. Ball CA, Osuna R, Ferguson KC et al. Dramatic changes in Fis levels upon nutrient upshift in *Escherichia coli*. J Bacteriol 1992; 174(24):8043-8056.
96. Nilsson L, Verbeek H, Vijgenboom E et al. FIS-dependent trans activation of stable RNA operons of *Escherichia coli* under various growth conditions. J Bacteriol 1992; 174(3):921-929.
97. Appleman JA. Ph.D. thesis: University of Wisconsin-Madison; 1998.
98. Jensen KF, Pedersen S. Metabolic growth rate control in *Escherichia coli* may be a consequence of subsaturation of the macromolecular biosynthetic apparatus with substrates and catalytic components. Microbiol Rev 1990; 54(2):89-100.
99. Bedwell DM, Nomura M. Feedback regulation of RNA polymerase subunit synthesis after the conditional overproduction of RNA polymerase in *Escherichia coli*. Mol Gen Genet 1986; 204(1):17-23.
100. Nomura M, Bedwell DM, Yamagishi M et al. RNA polymerase and regulation of RNA synthesis in *Escherichia coli*: RNA polymerase concentration, stringent control, and ribosome feedback regulation. In: Reznikoff WS, ed. RNA polymerase and the Regulation of Transcription. New York: Elsevier Science Publishing Co., Inc., 1987:137-149.
101. Jinks-Robertson S, Gourse RL, Nomura M. Expression of rRNA and tRNA genes in *Escherichia coli*: evidence for feedback regulation by products of rRNA operons. Cell 1983; 33(3):865-876.
102. Voulgaris J, French S, Gourse RL et al. Increased *rrn* gene dosage causes intermittent transcription of rRNA in *Escherichia coli*. J Bacteriol 1999; 181(14):4170-4175.
103. Sharrock RA, Gourse RL, Nomura M. Defective antitermination of rRNA transcription and derepression of rRNA and tRNA synthesis in the *nusB5* mutant of *Escherichia coli*. Proc Natl Acad Sci USA 1985; 82(16):5275-5279.
104. Shen V, Bremer H. Chloramphenicol-induced changes in the synthesis of ribosomal, transfer, and messenger ribonucleic acids in *Escherichia coli* B/r. J Bacteriol 1977; 130(3):1098-1108.
105. Cole JR, Olsson CL, Hershey JW et al. Feedback regulation of rRNA synthesis in *Escherichia coli*. Requirement for initiation factor IF2. J Mol Biol 1987; 198(3):383-392.
106. Yamagishi M, de Boer HA, Nomura M. Feedback regulation of rRNA synthesis. A mutational alteration in the anti-Shine-Dalgarno region of the 16 S rRNA gene abolishes regulation. J Mol Biol 1987; 198(3):547-550.
107. Gourse RL, Takebe Y, Sharrock RA et al. Feedback regulation of rRNA and tRNA synthesis and accumulation of free ribosomes after conditional expression of rRNA genes. Proc Natl Acad Sci USA 1985; 82(4):1069-1073.
108. Voulgaris J, Pokholok D, Holmes WM et al. The feedback response of *Escherichia coli* rRNA synthesis is not identical to the mechanism of growth rate-dependent control. J Bacteriol 2000; 182(2):536-539.
109. Nodwell JR, Greenblatt J. Recognition of boxA antiterminator RNA by the *E. coli* antitermination factors NusB and ribosomal protein S10. Cell 1993; 72(2):261-268.
110. Aviv M, Giladi H, Oppenheim AB et al. Analysis of the shut-off of ribosomal RNA promoters in *Escherichia coli* upon entering the stationary phase of growth. FEMS Microbiol Lett 1996; 140(1):71-76.
111. Talukder AA, Iwata A, Nishimura A et al. Growth phase-dependent variation in protein composition of the *Escherichia coli* nucleoid. J Bacteriol 1999; 181(20):6361-6370.
112. Kahru A, Vilu R. On characterization of the growth of *Escherichia coli* in batch culture. Arch Microbiol 1983; 135(1):12-15.
113. Balke VL, Gralla JD. Changes in the linking number of supercoiled DNA accompany growth transitions in *Escherichia coli*. J Bacteriol 1987; 169(10):4499-4506.
114. Gross CA, Chan C, Dombroski A et al. The functional and regulatory roles of σ factors in transcription. Cold Spring Harb Symp Quant Biol 1998; 63:141-155.
115. Gaal T, Ross W, Estrem ST et al. Promoter recognition and discrimination by Eσ^S RNA polymerase. Mol Microbiol 2001; 42(4):939-954.
116. Farewell A, Kvint K, Nystrom T. Negative regulation by RpoS: a case of s factor competition. Mol Microbiol 1998; 29(4):1039-1051.
117. Afflerbach H, Schroder O, Wagner R. Effects of the *Escherichia coli* DNA-binding protein H-NS on rRNA synthesis in vivo. Mol Microbiol 1998; 28(3):641-653.
118. Tippner D, Afflerbach H, Bradaczek C et al. Evidence for a regulatory function of the histone-like *Escherichia coli* protein H-NS in ribosomal RNA synthesis. Mol Microbiol 1994; 11(3):589-604.
119. Afflerbach H, Schroder O, Wagner R. Conformational changes of the upstream DNA mediated by H-NS and FIS regulate *E. coli rrnB* P1 promoter activity. J Mol Biol 1999; 286:339-353.
120. Schroder O, Wagner R. The bacterial DNA-binding protein H-NS represses ribosomal RNA transcription by trapping RNA polymerase in the initiation complex. J Mol Biol 2000; 298(5):737-748.

121. Zhang A, Rimsky S, Reaban ME et al. *Escherichia coli* protein analogs StpA and H-NS: regulatory loops, similar and disparate effects on nucleic acid dynamics. EMBO J 1996; 15(6):1340-1349.
122. Mojica FJ, Higgins CF. In vivo supercoiling of plasmid and chromosomal DNA in an *Escherichia coli hns* mutant. J Bacteriol 1997; 179(11):3528-3533.
123. Schneider R, Travers A, Muskhelishvili G. FIS modulates growth phase-dependent topological transitions of DNA in *Escherichia coli*. Mol Microbiol 1997; 26(3):519-530.
124. Gallant J, Harada B. The control of ribonucleic acid synthesis in *Escherichia coli*. 3. The functional relationship between purine ribonucleoside triphosphate pool sizes and the rate of ribonucleic acid accumulation. J Biol Chem 1969; 244(12):3125-3132.
125. Liebig B, Wagner R. Effects of different growth conditions on the in vivo activity of the tandem *Escherichia coli* ribosomal RNA promoters P1 and P2. Mol Gen Genet 1995; 249(3):328-335.
125a. Murray HD, Appleman JA, Gourse RL. Regulation of the *Escherichia coli rrnB* P2 promoter. J Bacteriol 2003; 185(1): in press.
126. Aiyar SE, Gaal T, Gourse RL. rRNA promoter activity in the fast-growing bacterium *Vibrio natriegens*. J Bacteriol 2002; 184(5): in press.
127. Gafny RS, Cohen S, Nachaliel N et al. Isolated P2 rRNA promoters of *E. coli* are strong promoters that are subject to stringent control. J Mol Biol 1994; 243:152-156.
128. Sarmientos P, Cashel M. Carbon starvation and growth rate-dependent regulation of the *Escherichia coli* ribosomal RNA promoters: differential control of dual promoters. Proc Natl Acad Sci USA 1983; 80(22):7010-7013.
129. Koch AL, Deppe CS. In vivo assay of protein synthesizing capacity of *Escherichia coli* from slowly growing chemostat cultures. J Mol Biol 1971; 55(3):549-562.
130. Zacharias M, Goringer HU, Wagner R. Influence of the GCGC discriminator motif introduced into the ribosomal RNA P2- and *tac* promoter on growth-rate control and stringent sensitivity. EMBO J 1989; 8(11):3357-3363.
131. Komine Y, Adachi T, Inokuchi H et al. Genomic organization and physical mapping of the transfer RNA genes in *Escherichia coli* K12. J Mol Biol 1990; 212(4):579-598.
132. Ross W, Salomon J, Holmes WM et al. Activation of *Escherichia coli leuV* transcription by FIS. J Bacteriol 1999; 181:3864-3868.
133. Muskhelishvili G, Buckle M, Heumann H et al. FIS activates sequential steps during transcription initiation at a stable RNA promoter. EMBO J 1997; 16(12):3655-3665.
134. Nilsson L, Vanet A, Vijgenboom E et al. The role of FIS in trans activation of stable RNA operons of *E. coli*. EMBO J 1990; 9(3):727-734.
135. Nilsson L, Emilsson V. Factor for inversion stimulation-dependent growth rate regulation of individual tRNA species in *Escherichia coli*. J Biol Chem 1994; 269(13):9460-9465.
136. Emilsson V, Nilsson L. Factor for inversion stimulation-dependent growth rate regulation of serine and threonine tRNA species. J Biol Chem 1995; 270(28):16610-16614.
137. Dong H, Nilsson L, Kurland CG. Co-variation of tRNA abundance and codon usage in *Escherichia coli* at different growth rates. J Mol Biol 1996; 260(5):649-663.
138. Ikemura T, Dahlberg JE. Small ribonucleic acids of *Escherichia coli*. II. Noncoordinate accumulation during stringent control. J Biol Chem 1973; 248(14):5033-5041.
139. Travers AA, Lamond AI, Weeks JR. Alteration of the growth-rate-dependent regulation of *Escherichia coli tyrT* expression by promoter mutations. J Mol Biol 1986; 189(1):251-255.
140. Kupper H, Contreras R, Khorana HG et al. The tyrosine tRNA promoter. In: Losick R, Chamberlin M, eds. RNA polymerase. Cold Spring Harbor: Cold Spring Harbor Press, 1976:473-484.
141. Figueroa-Bossi N, Guerin M, Rahmouni R et al. The supercoiling sensitivity of a bacterial tRNA promoter parallels its responsiveness to stringent control. EMBO J 1998; 17(8):2359-2367.
142. Pemberton IK, Muskhelishvili G, Travers AA et al. The G+C-rich discriminator region of the *tyrT* promoter antagonises the formation of stable preinitiation complexes. J Mol Biol 2000; 299(4):859-864.
143. Gourse RL, Nomura M. Level of rRNA, not tRNA, synthesis controls transcription of rRNA and tRNA operons in *Escherichia coli*. J Bacteriol 1984; 160(3):1022-1026.
144. Zengel JM, Lindahl L. Diverse mechanisms for regulating ribosomal protein synthesis in *Escherichia coli*. Prog Nucleic Acid Res Mol Biol 1994; 47:331-370.
145. Springer M. Translational control of gene expression in *E. coli* and bacteriophage. In: Lin ECC, Lynch AS, eds. Regulation of Gene Expression in *Escherichia coli*. Austin: R.G. Landes Co., 1996:85-126.
146. Deckman IC, Draper DE. Specific interaction between ribosomal protein S4 and the α operon messenger RNA. Biochemisty 1985; 24:7860-7865.
147. Robert F, Brakier-Gingras L. Ribosomal protein S7 from *Escherichia coli* uses the same determinants to bind 16S ribosomal RNA and its messenger RNA. Nucleic Acids Research 2001; 29:677-682.
148. Saito K, Mattheakis LC, Nomura M. Post-transcriptional regulation of the *str* operon in *Escherichia coli*. Ribosomal protein S7 inhibits coupled translation of S7 but not its independent translation. J Mol Biol 1994; 235(1):111-124.
149. Saito K, Nomura M. Post-transcriptional regulation of the *str* operon in *Escherichia coli*. Structural and mutational analysis of the target site for translational repressor S7. J Mol Biol 1994; 235(1):125-139.
150. Mattheakis L, Vu L, Sor F et al. Retroregulation of the synthesis of ribosomal proteins L14 and L24 by feedback repressor S8 in *Escherichia coli*. Proc Natl Acad Sci USA 1989; 86(2):448-452.
151. Lindahl L, Archer R, Zengel JM. Transcription of the S10 ribosomal protein operon is regulated by an attenuator in the leader. Cell 1983; 33(1):241-248.
152. Pan CQ, Finkel SE, Cramton SE et al. Variable structures of Fis-DNA complexes determined by flanking DNA-protein contacts. J Mol Biol 1996; 264(4):675-695.

Regulation of the Expression of Aminoacyl-tRNA Synthetases and Translation Factors

Harald Putzer and Soumaya Laalami

Abstract

Aminoacyl-tRNA synthetases and translation factors are key enzymes required for pro tein biosynthesis. *Escherichia coli* and *Bacillus subtilis* often use different strategies to regulate the expression of the genes encoding these enzymes. Synthesis of several *E. coli* aminoacyl-tRNA synthetases is controlled by different mechanisms acting at the transcriptional or translational level. By contrast, in *B. subtilis*, expression of the majority of these proteins is regulated by a common, yet specific transcriptional antitermination mechanism. However, all of these controls share a common effector, the tRNA. In the case of the *E. coli* translation factors, the primary enzyme function is often exploited for autoregulating their own synthesis at the translational level. Here we will focus on the gene organization and the multiple types of gene regulation governing prokaryotic aminoacyl-tRNA synthetase and translation factor expression.

Introduction

Decoding the genetic message is the major and most energy consuming process in the cell. In addition to ribosomes, mRNA and tRNA, translation requires amino acids, nucleotides and special-ized proteins. The latter include aminoacyl-tRNA synthetases and translation factors transiently associated with the ribosomes. These proteins catalyze sequential steps during translation, starting with the charging of tRNA, ribosome-dependent polypeptide synthesis, final release of the protein and ribosome recycling. Here, we will consider our current knowledge of the gene organization and the expression of aminoacyl-tRNA synthetases and translation factors in prokaryotes. We focus essentially on two bacterial systems, the Gram negative bacterium *Escherichia coli* and *Bacillus subtilis*, the best-studied Gram positive organism. Translation factors are only considered for *E. coli*. Structural and mechanistic aspects of these enzymes are treated in accompanying chapters.

Aminoacyl-tRNA Synthetases

Aminoacyl-tRNA synthetases (aaRS) play a central role in protein biosynthesis by catalyzing the attachment of a given amino acid to the 3' end of its cognate tRNA. They do this by forming an energy-rich aminoacyl-adenylate intermediate of the cognate amino acid, which serves to transfer the amino acid to the tRNA. The intrinsic proofreading capacities of the aaRS and their balanced expression contribute greatly to the accuracy of translation of the genetic code. In addition to their crucial role in protein biosynthesis, aaRS are involved in a number of regulatory processes via their product, the charged tRNA. The control of the expression of amino acid biosynthetic operons by the level of tRNA aminoacylation in vivo is well documented. Another phenomenon, the pleiotro-pic stringent response, operates when tRNA aminoacylation is diminished in wild-type (relA$^+$) strains and inhibits rRNA and tRNA synthesis as well as the synthesis of some other macromolecules involved in translation.[1] As a by-product of the amino acid activation step aaRS can form diadenosine 5'-5'''-P1,P4-tetraphosphate (AppppA), considered as a pleiotropically acting alarmone

Translation Mechanisms, edited by Jacques Lapointe and Léa Brakier-Gingras.
©2003 Eurekah.com and Kluwer Academic / Plenum Publishers.

that has been associated with oxidative stress or timing of cell division.[2] Eucaryotes exploit yet other functions of aaRS, e.g., some aaRS are essential factors in certain splicing activities.[3] It is obvious that control of the cellular levels of the aaRS is important for any organism.

E. coli *Aminoacyl-tRNA Synthetase Genes*

General Regulatory Phenomena

In *E. coli* a single aaRS is found for each amino acid, with the exception of lysyl-tRNA synthetase for which two species have been characterized. The cellular abundance of the different aaRS is quite similar;[4] at a doubling time of 40 minutes, the cell contains 10 to 20 times less molecules of each enzyme than ribosomes. This amounts to between 1300 and 2600 molecules per cell. In contrast, the concentrations of different charged tRNA isoacceptor families vary more than ten-fold, from about 700 (tRNA^{Gln}) to 8000 (tRNA^{Val}) at a doubling time of 60 min.[5] Under the same conditions the individual aminoacyl-tRNA/synthetase ratios vary by a factor of 15 and the turnover rate of the aminoacyl-tRNAs is between two and eight per second. Surprising differences in the in vivo activity of individual aaRS result from these measurements; while glutamyl-tRNA synthetase charges only two tRNAs per second, threonyl-tRNA synthetase (ThrS) charges as many as 48 per second. These values are much higher than what can be obtained in vitro with purified enzymes, up to 240-fold higher in the case of ThrS.[5]

A quantitative analysis of O'Farrel 2D gels in which 18 out of the 21 synthetases were identified shows that in most cases their cellular concentration increases with growth rate.[6] Overall, the level of enzyme (normalized to total cellular proteins) increases two- to threefold for each fivefold increase in growth rate. This increase is less than that observed with ribosomes but corresponds to that of elongation factor EF-Tu and the initiation factors. Growth rate-dependent regulation affects aaRS expression rather rapidly; new steady-state levels are reached within 2 to 3 minutes after a upshift. There is no general rule as to whether growth rate control of the aaRS is transcriptionally or translationally regulated. Recent DNA array data (LaRossa, personal communication) show a decrease of mRNA levels for several synthetases with increasing growth rate while the earlier proteome data clearly found increased protein concentrations under similar conditions. Even an increased mRNA level is not necessarily an indication of transcriptional regulation and may reflect translation enhanced mRNA stability.

Superimposed on growth rate-dependent regulation is a regulatory response of individual synthetase genes to limiting amounts of the cognate amino acid. At least half of the aaRS are synthesized more rapidly under these conditions.[7] Derepression can be transient (*hisS, leuS, metG, proS, serS, thrS* and *valS*) or permanent (*argS, ileS* and *pheST*). In all cases, addition of the amino acid after starvation reverses the effect. However, this leaves open the question of whether it is the free amino acid or the cognate tRNA, which is involved in regulation. For *E. coli metG*, or *hisS* from the closely related organism *Salmonella typhimurium*, it appears clear that regulation depends on the level of aminoacylation of the corresponding tRNA. The specific control mechanisms governing *pheST* and *thrS* expression (see below) also respond directly to the tRNA aminoacylation level. In contrast, there is no clear evidence so far that aaRS genes are coregulated with those of their cognate tRNAs. In both enteric and Gram-positive organisms amino acid deprivation provokes an immediate adjustment of cellular metabolism, the stringent response, by inhibiting the synthesis of rRNA and tRNAs. The extent of aminoacylation of tRNAs plays a central role in this regulatory response probably by controlling the level of guanosine-5'-diphosphate-3'-diphosphate (ppGpp) synthesized. However, the stringent response has only a minor effect, if any, on the expression of at least 10 of the 21 aaRS.[8] At present, no mutation is known which affects the control of all or even a small number of aaRS genes.

Genetic Organization in *E. coli*

The molecular weights and subunit structures are surprisingly variable among the different aaRS (see chapter 2 by D. Söll and M. Ibba). With the exception of the threonyl- and phenylalanyl-tRNA synthetase genes (*thrS* and *pheST*) which are in close vicinity, all aaRS genes are scattered throughout the chromosome (Fig. 1). The orientation of transcription of different aaRS genes is not correlated with the direction of replication. There is also no correlation between the positions of the aaRS

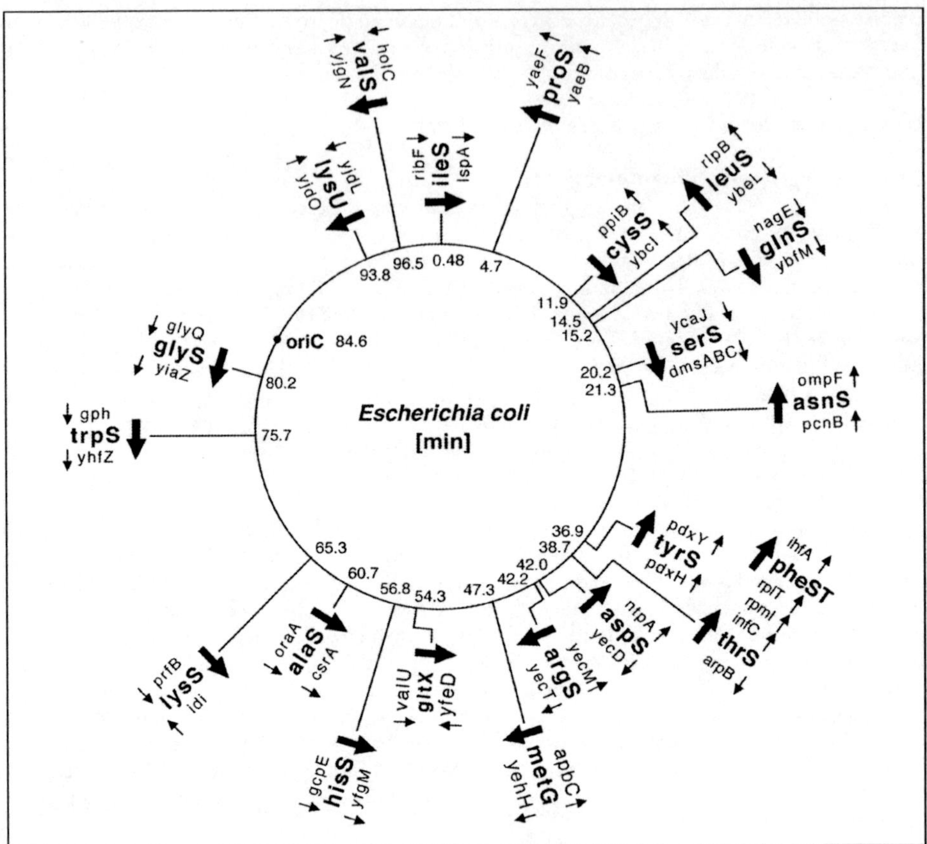

Figure 1. *E. coli* aminoacyl-tRNA synthetase genes. The aaRS genes are shown in boldface type together with adjacent genes. Positions are given in min and are calculated values from the complete genome sequence (Colibri database). Arrows indicate the directions of transcription. The replication origin is shown as oriC at 84.6 min.

genes and tRNA genes except for the gene *gltX* encoding glutamyl-tRNA synthetase (see below). All aaRS genes are expressed from ς^{70} promoters with weak consensus sequences, including *lysU*, which is heat-shock induced. In most cases the transcription initiation sites are known.[9]

The glycyl- and phenylalanyl-tRNA synthetases are the only synthetases composed of two different subunits. Both are tetramers of the $\alpha^2\beta^2$ structural type. The genes for the two different subunits are grouped on the chromosome and are expressed as an operon in which the promoter-proximal gene codes for the small subunit. Finally, there are a few aaRS that are co-expressed with other proteins, generally also involved in translation (see below).

Specific Control Mechanisms

Many aaRS genes are, often only transiently, derepressed between two and fourfold following starvation for the cognate amino acid. This is much less than what is observed for the corresponding biosynthetic operons under the same conditions (generally 10-fold and more). In the following section we will describe several studies at the molecular level. They shed considerable light on the ingenuity of the cell to regulate the expression of individual aaRS genes at all possible steps of genetic expression.

Autoregulation by Transcriptional Repression (alaS)

Alanyl-tRNA synthetase (AlaS), encoded by the *alaS* gene, is a tetramer of the α^4 type. In an in vitro system, the synthetase is able to repress its own transcription when present in micromolar quantities.[10] On the other hand, when present in amounts corresponding to the intracellular concentration, autogenous repression is only observed in the presence of alanine. Experiments analyzing protection against DNase I digestion show that AlaS can actually bind to the −10 region of its promoter, suggesting that transcription initiation of *alaS* might be controlled by the alanine concentration in the cell. Unfortunately, the derepression of AlaS synthesis expected upon alanine starvation was never shown to occur, and no in vivo data have been presented to confirm and further investigate the model.

Regulation by Transcriptional Attenuation (pheST)

Phenylalanyl-tRNA synthetase (PheST) is a tetrameric enzyme of the $\alpha^2\beta^2$ type. The *pheS* and *pheT* genes, encoding the small and the large subunits, respectively, are cotranscribed, with *pheS* being the first gene in the operon. The major promoter initiates transcription 368 bp upstream of the *pheS* initiation codon. However, about 30% of the *pheST* transcripts originate from the upstream *thrS-infC-rpmI-rplT* operon. The *pheST* genes are derepressed 2.5-fold under conditions in which the cellular concentration of phenylalanyl-charged tRNA is decreased. This was achieved, for example, by using a host strain carrying a mutated PheST with a high K_m for phenylalanine. This strain is bradytrophic (a leaky auxotroph) for phenylalanine. The key elements of *pheST* regulation are a Rho-independent transcription terminator, located just in front of *pheS*, preceded by an open reading frame coding for a 14-residue peptide (the leader peptide) containing five phenylalanine residues, of which three are consecutive. A number of in vitro and in vivo experiments, including a detailed mutational analysis and the use of transcriptional and translational fusions of the *pheST* leader to *lacZ*, established the regulatory model presented in Figure 2.[8,11,12] This regulatory mechanism, described in the legend of Figure 2, depends on differential translation of the leader peptide in the presence or absence of phenylalanine. This translation determines the local position of the ribosome, which in turn influences the formation of alternative RNA secondary structures, among them, the terminator. The default state for this mechanism, that is in the absence of any leader peptide translation, is termination (Fig. 2A). The *pheST* operon is different from similarly controlled amino acid biosynthetic operons, since its expression is essential for bacterial growth; absence of its expression cannot be compensated for by supplementation of the growth medium. Thus, some readthrough of the terminator must occur even in the repressed state. There is also a necessity for tight coupling between transcription and leader mRNA translation (Fig. 2C). Only when the ribosome immediately follows the RNA polymerase can its influence on RNA folding have an immediate effect on the RNA polymerase. Accordingly, increasing the distance between the leader mRNA stop codon and segment 2 leads to derepression; the ribosome, although having reached the stop codon due to the availability of phenylalanine, cannot hinder the formation of 2/3 antiterminator structure.

The *pheST* operon is also derepressed in strains carrying a mutant allele of the *miaA* gene whose product modifies tRNA^{Phe} and tRNA^{Trp}, among other tRNAs. The absence of the isopentenyl modification of the adenine adjacent to the anticodon of these tRNAs alters their translational properties. A concomitant slow-down in leader peptide translation favors formation of the antiterminator structure (Fig. 2C) and is thus the likely cause for the observed derepression.[11]

Autoregulation at the Translational Step (thrS)

Threonyl-tRNA synthetase (ThrS) is a homodimeric enzyme whose transcription initiates 162 bp upstream of its structural gene. Transcripts generally extend at least beyond the first downstream gene, *infC*, which is separated from *thrS* by only three nucleotides. ThrS expression is negatively feedback regulated at the translational level. When ThrS is overproduced from a plasmid, expression of translational *thrS-lacZ* fusions (producing a fused ThrS-LacZ polypeptide) is repressed, but that of transcriptional fusions (translation of *thrS* and *lacZ* is initiated at distinct sites) is not.[13] In vitro, addition of exogenous ThrS to a cell-free protein synthesizing system decreases the initial rate of ThrS synthesis by 90% without affecting *thrS* mRNA synthesis.[14] Constitutive mutations in the *thrS* leader, i.e., mutations that abolish negative feedback control, mostly occur in a GU sequence located at positions −31 and −32 relative to the first nucleotide £è the initiation codon (Fig. 3A). Not by

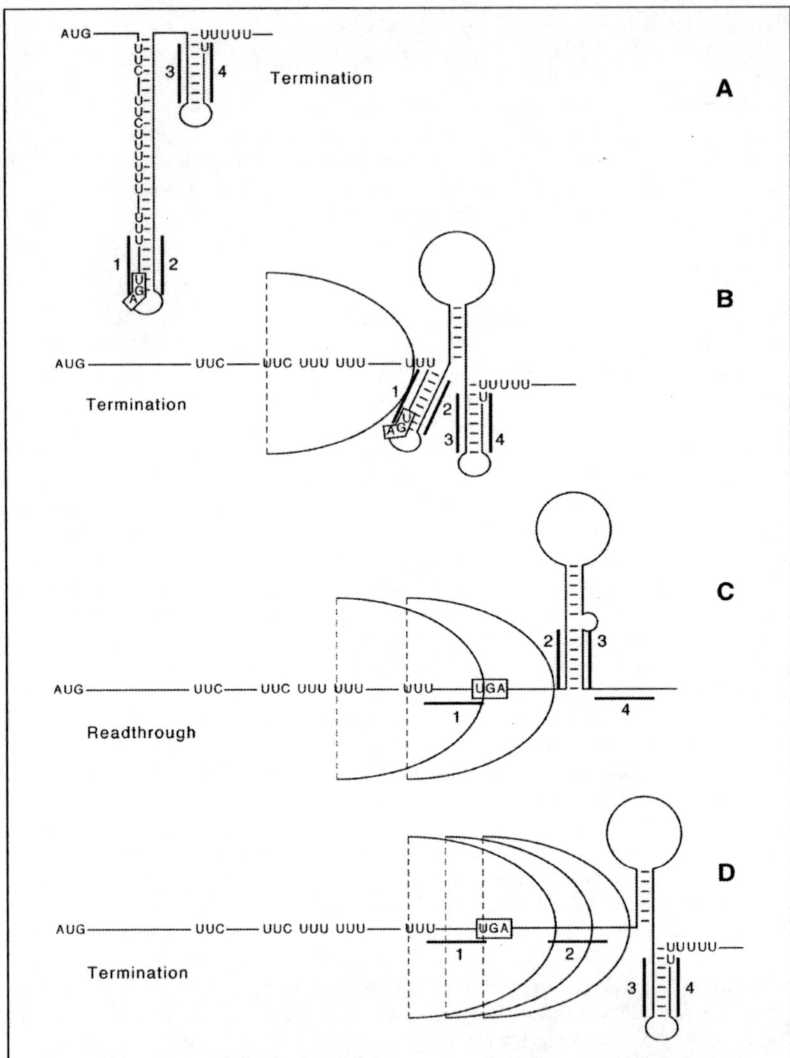

Figure 2. Alternative structures of the *pheST* leader mRNA. The initiation codon, the five phenylalanine codons of the leader peptide and the U stretch of the terminator are written out, the stop codon is boxed. Important regions 1, 2, 3, and 4 are indicated. A) mRNA structure in the absence of translation initiation, formation of the terminator 3/4 provokes premature termination. B) A ribosome stalled at the first or second Phe codon does not hinder the formation of the anti-antiterminator 1/2, termination occurs. C) A ribosome stalling at the last two Phe codons sequesters region 1 which favors formation of the antiterminator 2/3, readthrough occurs. D) Efficient translation of all Phe codons prevents formation of the antiterminator causing premature termination. Dashes indicate Watson-Crick base pairs.

chance, this dinucleotide also occurs as the second and third bases of the anticodons of all tRNA[Thr] isoacceptors (Fig. 3B) and represents the major recognition site for the synthetase. It is also present in the loops of domains 2 and 4 which makes them structurally similar to the anticodon stem of tRNA[Thr]. Both domains are part of the translational operator of *thrS*, which is confined to the first 110 bases upstream of the *thrS* structural gene[15] (Fig. 3A). Threonyl-tRNA synthetase can bind to its operator in vitro with an affinity close to that of the tRNA[Thr]-ThrS complex. In doing so, domains

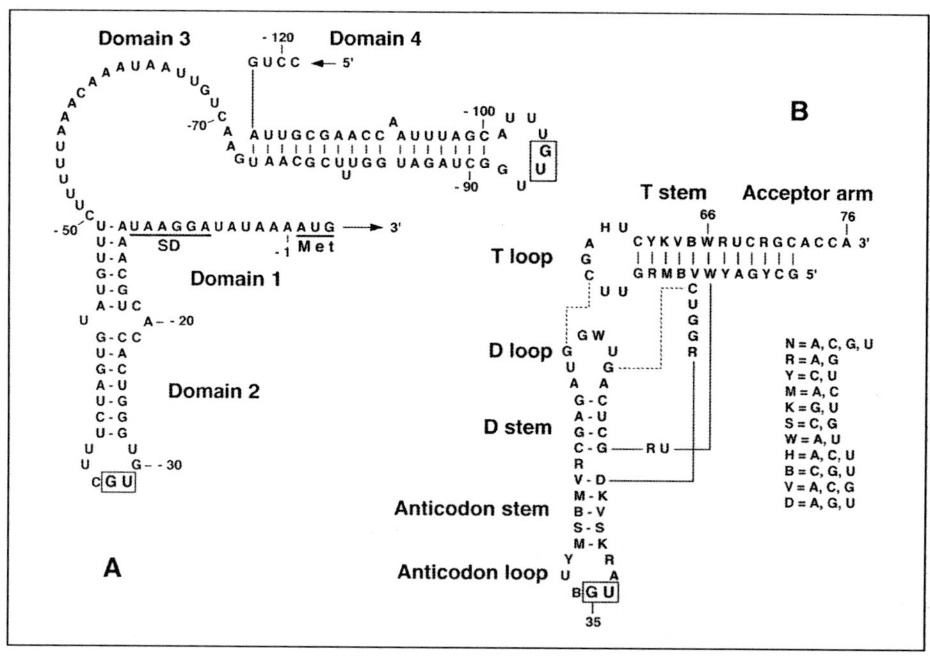

Figure 3. The *thrS* leader mRNA and the composite structure of tRNA^{Thr} isoacceptors in the L-shaped representation. The *thrS* leader structure between nucleotides −121 and +3 is shown in A. The Shine-Dalgarno sequence and the initiation codon are underlined and marked SD and Met, respectively. The A of the AUG codon is nucleotide 1. The nucleotides involved in identity are boxed.

2 and 4 are specifically protected from chemical modification or ribonuclease digestion. In addition, ribosome binding to a *thrS* leader transcript is completely inhibited by the addition of ThrS in physiological concentrations but a 10-fold excess of tRNA^{Thr} over synthetase completely relieves this inhibitory effect. Together with similar results obtained in vivo, this shows that tRNA^{Thr} acts as an antirepressor and that the intracellular concentration of uncharged tRNA^{Thr} modulates the repressor activity of ThrS. Since the tRNA^{Thr} concentration in the cell increases with growth rate, specific and growth rate-dependent regulation may be achieved via a single mechanism in the case of *thrS*. Surprisingly, operator mutations, which increase *thrS* expression also, increase the steady-state mRNA concentration, without altering mRNA stability. This is best explained if one assumes that *thrS* mRNA molecules that escape repression are fully translated and stable while those where the ribosome is blocked from initiating translation are immediately degraded.[16] The operator-ThrS interaction is now extremely well characterized, practically down to the atomic level. Domain 2, which contains the anticodon-like arm, is the major recognition site of the synthetase. The crystal structure of a ThrS-domain 2 complex clearly indicates that domain 2 and the anticodon domain of tRNA^{Thr} are recognized by the same C-terminal site of the synthetase (A.C. Dock-Brégeon, pers. comm.). This is a perfect example of molecular mimicry where the structure of an RNA molecule has evolved to fit a binding site on a protein that normally interacts with a different RNA. Final proof for the authentic functional similarity comes from experiments based on tRNA identity rules. In both *E. coli* tRNA^{Thr} and tRNA^{Met}, the anticodon is the major determinant of aminoacylation specificity. The replacement of CGU in the anticodon-like sequence of the *thrS* domain 2 by CAU is sufficient to abolish control by ThrS and establish control by methionyl-tRNA synthetase in vivo and in vitro.[17,18] Domain 3 is a single-stranded linker region between domains 2 and 4 but also important for efficient translation of *thrS*.[19] In fact, the 3' single-stranded end of domain 3 is part of a split ribosomal binding site whose two regions are brought into close proximity by the hairpin structure of domain 2 (Fig. 3A). The RNA domains involved in ribosome and ThrS binding are thus not

strictly overlapping, but interspaced. Domain 4 behaves similarly to domain 2 in all aspects but with a lower efficiency. ThrS is a homodimeric enzyme and binds two tRNA^{Thr} molecules. Domains 2 and 4 bind to ThrS with the same stoichiometry as tRNA^{Thr} but only one intact operator binds per dimeric enzyme.[20] Recent data show that it is the N-terminal domain of the synthetase which is responsible for the competition with the ribosome. ThrS thus displays a truly modular structure reminiscent of transcriptional regulators (M. Springer, pers. comm.). Finally, based on sequence and structural comparisons this molecular mimicry mechanism appears to be well conserved in other Gram-negative organisms such as Salmonella, Yersinia, Vibrio and Haemophilus (J. Caillet, pers. comm.).

Two aaRS with the Same Specificity (lysS *and* lysU)

In *E. coli*, there are two lysyl-tRNA synthetases encoded by the genes *lysS* and *lysU*. They are located at 65.3 and 93.8 min, respectively, on the *E. coli* chromosome (Fig. 1). This is the only case known, in this organism, of two synthetases being able to activate the same tRNA isoacceptor family. Individually, each of the two synthetases is dispensable for growth but disruption of *lysS* causes a cold-sensitive phenotype. The *lysS* gene encodes the housekeeping synthetase; it is constitutively expressed and is subject to growth rate dependent control. The dicistronic transcripts of *lysS* originate upstream of the *prfB* gene, encoding peptide chain release factor RF2, which is translationally autoregulated (see below). Due to the proximity (9 nucleotides) of the stop codon of *prfB* to the downstream initiator codon of *lysS*, it is likely that *lysS* expression is influenced by the translation of *prfB*.

The *lysU* gene, on the other hand, is normally silent at low temperatures which explains the cold-sensitive phenotype of a *lysS* mutant strain. Its expression is induced by a variety of stimuli such as high temperature, anaerobiosis, low external pH, the stationary phase, or the presence of specific metabolites including leucine or leucine-containing dipeptides in the growth medium.[9] Expression of *lysU* involves several genes including *lrp*, *rpoH*, and *hns*. The leucine-response regulatory protein Lrp is a transcriptional regulator which acts as a repressor or activator for numerous genes.[21] The *lysU* gene is part of the leucine regulon and normally repressed by Lrp [22] which binds just upstream of the promoter.[23] Repression is relieved upon addition of L-leucine. Independent of Lrp, heat shock induction of *lysU* occurs only in the presence of a functional rpoH gene, coding for the heat shock sigma factor ς^{32}. However, *lysU* does not have a σ^{32} dependent promoter and induction of RpoH synthesis, without temperature shift, does not cause *lysU* derepression, indicating that its effect is indirect.[24] The abundant, histone-like protein H-NS, is potentially involved in *lysU* regulation by binding to a curved DNA region upstream of *lysU*. Disruption of *hns* increases *lysU* expression at low temperatures 2.5-fold. Since H-NS also interferes with the expression of *lrp*[25] its role in *lysU* expression might be more complex.

Why does *E. coli* have two lysyl-tRNA synthetases in the first place ? The most noticeable difference between the two enzymes is an affinity of LysU for lysine nearly 8-fold higher than that of LysS. In agreement with this observation, LysU is relatively less sensitive to the presence of cadaverine, an inhibitor of lysine binding, than LysS. Remarkably, this polyamine is produced at the expense of lysine under several stress conditions that also induce *lysU* expression.[26] Increasing the synthesis of LysU under these conditions could thus be useful for the cell and explain the conservation of two lysine specific synthetases.

Glutamyl-tRNA Synthetase (gltX)

The gene for glutamyl-tRNA synthetase, *gltX*, shares an intergenic regulatory region with the divergent *valU* tRNA operon.[27] This is the only aminoacyl-tRNA synthetase gene that is adjacent to a tRNA gene in *E. coli*. The factor for inversion stimulation (FIS), which is one of the major positive regulators of rRNA operons, binds to three sites within the promoter region of *gltX* and *valU*. FIS stimulates *valU* transcription about two-fold during steady-state exponential growth, but not during growth acceleration. In contrast, *gltX* transcription is repressed two-fold by the presence of FIS, but only during growth acceleration. This leads to a concomitant decrease of the glutamyl-tRNA synthetase level. This lack of synchrony between the influence of FIS on *gltX* and *valU* transcription indicates that their expression is not coordinated.[28] Most *gltX* transcripts originate from the *gltX* proximal promoter which does not have a directly overlapping FIS binding site. Interestingly, FIS mediated repression of this major *gltX* promoter depends on the presence of the *valU* promoter which lies 120 bp upstream. FIS can bend DNA by 40° to 90° and due to the topology of its three binding sites can bring both promoters closer to each other. The influence of FIS on *gltX* transcription during growth acceleration is thus of modulatory nature. On the other hand, since FIS does not influence expression during steady-state growth it has been concluded that it is not involved in growth rate control of *gltX* expression.[28]

Methionyl-tRNA Synthetase (metG)

In its native form methionyl-tRNA synthetase is a homodimeric enzyme encoded by the *metG* gene. Expression of the synthetase is regulated by the level of aminoacylation of tRNA^{Met}. Two promoters mediate metG transcription.[29] The upstream promoter lies within the coding region of the divergently transcribed apbC gene (initially called *mrp*) which is involved in thiamine synthesis by the alternative pyrimidine biosynthetic pathway.[30] The downstream promoter lies in the *apbC-metG* intergenic region so that the −35 promoter regions of *apbC* and *metG* overlap. In addition, a transcription terminator is located between the two *metG* promoters. S1 nuclease mapping experiments show that transcription from the upstream promoter is attenuated at the terminator. The presence of a potential tRNA-like secondary structure with a CAU anticodon-like sequence upstream of the terminator led to the proposal of an autoregulatory model. In this model, an excess of methionyl-tRNA synthetase could bind to this structure and somehow affect transcription termination.[29] Further experiments are needed to prove this theory.

Aminoacyl-tRNA Synthetases in Gram-Positive Bacteria

Progress made during the last 15 years such as the completion of genome sequences from several Gram-positive organisms now gives a broader view on how eubacteria regulate and coordinate the expression of the aminoacyl-tRNA synthetases (aaRS). *Bacillus subtilis*, due to its highly developed genetics, has become the best studied representative of the Gram-positive bacteria.

There are strong structural similarities between the aaRS from *Bacillus* and *E. coli* and many *Bacillus* synthetases are functional in *E. coli* as judged by the successful complementation of diverse *E. coli* synthetase mutants. However, there are significant differences between *B. subtilis* and *E. coli* in the organization and expression of aaRS genes.

1) The chromosomal locations and arrangements of the *B. subtilis* genes are completely different from those of its Gram-negative counterpart.

2) *B. subtilis* has no glutaminyl-tRNA synthetase ; instead, a single glutamyl-tRNA synthetase aminoacylates both tRNA^{Glu} and tRNA^{Gln} with glutamate.[31] The mischarged tRNA^{Gln} is subsequently converted to Gln-tRNA^{Gln} by an amidotransferase.[32, 33]

3) In *B. subtilis* there are two cases of aaRS gene redundancy : there are two threonyl- and two tyrosyl-tRNA synthetase genes (*thrS/thrZ* and *tyrS/tyrZ*, respectively).

4) In contrast to the diversity of regulatory mechanisms thus far identified in *E. coli*, the expression of 14 out of the 21 aaRS genes in *B. subtilis* is regulated in a common manner: a transcriptional antitermination mechanism that does not exist in Gram-negative bacteria.[34,35,36]

Genetic Organization

B. subtilis has 21 genes (excluding genes for subunits) coding aaRS specific for 19 amino acids (see Fig. 1). Two of them, glycyl- and phenylalanyl-tRNA synthetase are composed of two subunits α and β, encoded by distinct, but co-transcribed genes (*glyQ, glyS* and *pheS, pheT*). As mentioned above, duplicate genes exist for the threonyl- and tyrosyl-tRNA synthetases. However, this rather appears to be an exception; other fully sequenced Gram-positive genomes such as those of *Staphylococcus aureus, Bacillus halodurans, Lactococcus lactis* or *Deinococcus radiodurans* do not display such a genotype. A gene initially thought to encode a second *B. subtilis* histidyl-tRNA synthetase (*hisZ*) actually encodes an aaRS paralog which lacks aminoacylation activity. It has been characterized as an essential component of the first enzyme in histidine biosynthesis.[37] None of the *B. subtilis* aaRS genes is in a chromosomal context similar to that of *E. coli* (compare Figs. 1 and 4). In addition, most of them are grouped in three regions of the chromosome (Fig. 4). With the exception of *trpS* and *tyrZ*, transcription of all of the *B. subtilis* aaRS genes is in the same orientation as replication. This is typical of highly expressed genes and probably prevents transcription from interfering with genome replication.[38] There is only one documented case of cotranscribed aaRS genes in *B. subtilis*,

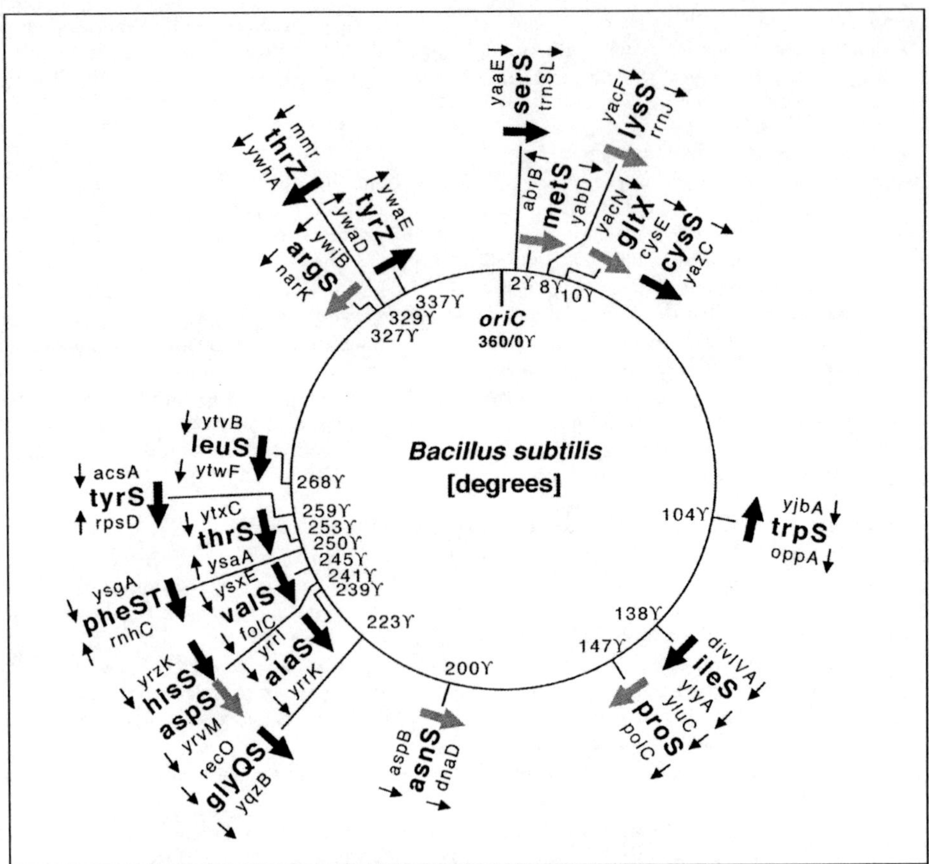

Figure 4. *B. subtilis* aminoacyl-tRNA synthetase genes. The aaRS genes are shown in boldface type together with adjacent genes. Positions are given in degrees and are calculated values from the complete genome sequence (Subtilist database). Arrows indicate the directions of transcription. Genes belonging to the T-box family are indicated by the large black arrows.

the *gltX-cysE-cysS* operon. It contains genes for both the glutamyl- and cysteinyl-tRNA synthetases separated by a gene encoding serine acetyl-transferase, the first enzyme of cysteine biosynthesis. Transcription of this operon originates 45 nucleotides upstream of *gltX* from a ς^A promoter.[39] The genes encoding histidyl- and aspartyl-tRNA synthetases (*hisS* and *aspS*) are probably also cotranscribed. Only 13 nucleotides separate aspS from the upstream *hisS* gene, putting it potentially under control of the *hisS* regulatory sequences, although this has yet to be shown. In both cases, the genetic organization of the described genes is conserved in closely related organisms like *Bacillus halodurans* or *Staphylococcus aureus* but different in more distantly related Gram-positive bacteria.

A Specific but Conserved Control Mechanism

Nineteen genes in *B. subtilis* have been identified as members of the so-called T-box family, on the basis of conservation of sequence and structural elements in their leader regions. They include fourteen aaRS genes, three biosynthetic operons (*ilvB-leu*, *proBA*, *proI*) one potential aminoacid permease (*yvbW*) and one gene whose product intervenes in the regulation of the *trp* operon (*yczA*). Their untranslated leader sequences are about 300 nucleotides long and include a factor-independent transcription terminator, just upstream of the translation initiation site, which attenuates transcription

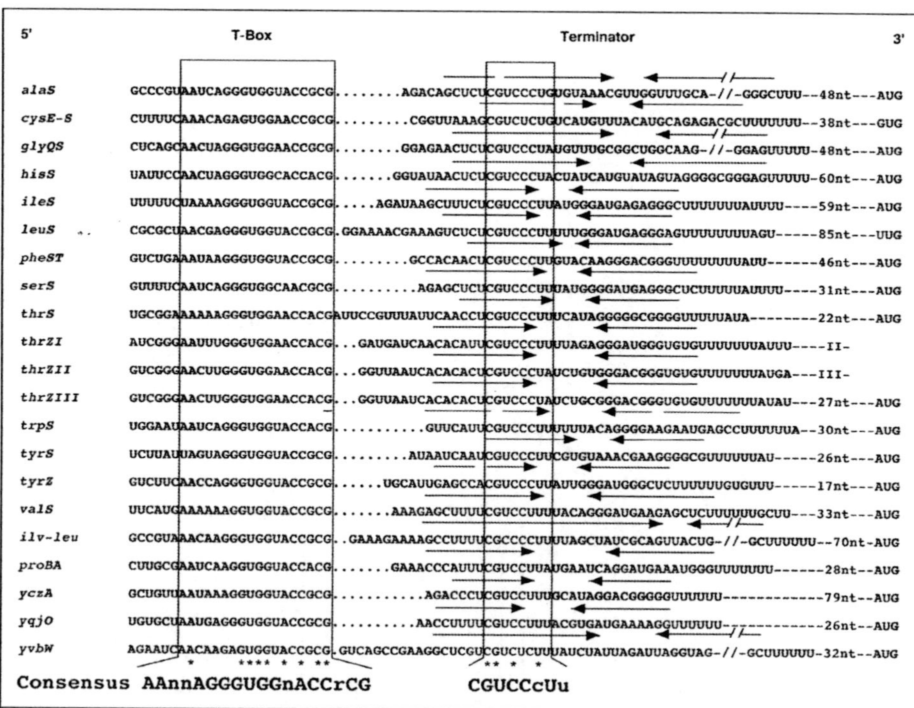

Figure 5. T-box and downstream complementary sequences from 14 aaRS and 5 other genes (see text) in *B. subtilis*. Proposed factor-independent transcription terminators are depicted by inverted arrows. The distances to the initiation codon are indicated. Uppercase nucleotides in the consensus sequence are at least 90% conserved. Lowercase nucleotides are at least 50% conserved. Residues marked with an asterisk are 100% conserved (r = purine ; n = any nucleotide).

of the structural gene.[40-42] Deletion of the leader terminator leads to constitutively high levels of expression. A particular case is the normally silent *thrZ* gene coding the second threonyl-tRNA synthetase.[43] Its leader spans over 800 nucleotides and can be considered as containing the *thrS* leader-terminator motif repeated three times. The presence of three strong transcription terminators upstream of the structural gene probably explains why this gene is normally not expressed.[40]

The aaRS genes of this family are specifically derepressed in response to starvation for their cognate amino acid by increasing transcriptional readthrough at the terminator. Induction ratios vary from 2.5 to about 30. The primary leader sequences of the various genes are very different but can be folded into similar secondary structures.[44] A scheme of the *thrS* leader based on an experimental determination of the RNA structure is shown in Figure 6.[45] Despite great sequence diversity, several elements crucial for regulation are conserved in both sequence and position.[34,35,44] The most prominent is the T-box element, located immediately upstream of the leader terminator (Fig. 5). A sequence complementary to the central part of the T-box in the 5' half of the terminator stem allows the formation of a mutually exclusive but unstable antiterminator structure. The model for specific modulation of terminator read-through is based on the presence of a triplet, known as the specifier codon, which specifies the cognate amino acid for the synthetase in question (Table 1). The specifier codon is always found in the same strategical position within the so-called specifier domain (Fig. 6). The effector molecule which signals limitation for the specific amino acid is the cognate uncharged tRNA which can interact directly with the leader by making at least two contacts.[44] The first is a codon-anticodon interaction with the specifier codon. The second occurs by basepairing between the 5'-UGGN-3' of the T-box in the antiterminator side bulge and the 5'-N'CCA-3' acceptor end

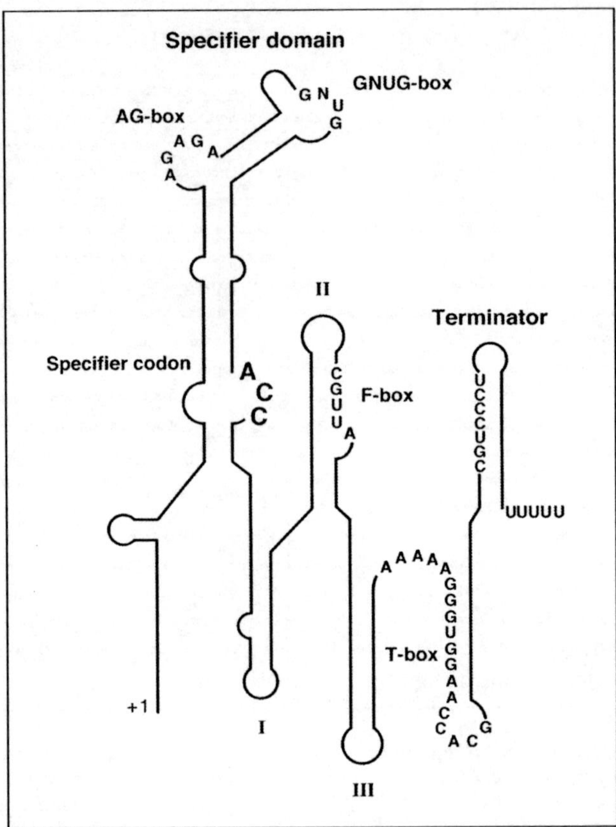

Figure 6. Scheme of the *B. subtilis thrS* leader. Locations of conserved sequences (see text) are shown. The threonine specifier codon ACC is also marked.

of uncharged tRNA (Fig. 7). The discriminator base (N') of the tRNA acceptor stem and the variable position within the T-box (N) co-vary. Thus, the first interaction is responsible for the specificity of induction while the second is thought to stabilize the antiterminator structure allowing the RNA-polymerase to escape termination. Both interactions have been well documented by genetic experiments. In several cases, altering the specifier codon affords a switch in the specificity of induction, generally with lower efficiency.[44,46,47] However, adaptation of both points of interaction in the leader to optimally accommodate a different tRNA does not always permit a switch in induction specificity.[46,48] At present, no specific tRNA determinants have been identified that can explain why an attempt to change the specificity of control was successful or not. Nevertheless, the tRNA-mRNA interaction is highly constrained and probably takes into account the overall tertiary structure of the tRNA.[49] Additional contacts between tRNA and mRNA might also be important for regulation. An exceptionally high complementarity exists between the D- and T-arms of tRNA[Trp] and the leader of the T-box controlled biosynthetic *trp* operon in *Lactococcus lactis*. In this system overexpression of wild-type tRNA[Trp] induces expression even in the absence of the codon-anticodon interaction, albeit at a 20-fold lower level. This suggests that the extensive complementarity between the tRNA and the leader actually helps to recruit the tRNA for antitermination.[50] This phenomena cannot be generalized to other T-box leaders, notably in *B. subtilis* where potential complementarities are much less convincing. However, they would be a good starting point to look for additional tRNA-leader interactions.

Table 1. Aminoacyl-ARNt synthetase genes in Bacillus subtilis

Amino Acid	Gene	Mol. Mass of Monomer (x10³)	Specifier Codon
Ala	alaS	97,2	GCU
Arg	argS	62,5	-
Asn	asnS	49,2	-
Asp	aspS	65,9	-
Cys	cysS	53,9	UGC
Glu	gltX	55.7	-
Gly	glyQ	33,9	GGC
	glyS	76,2	-
His	hisS	48,1	AAC or CAC
Ile	ileS	104,6	AUC
Leu	leuS	92	CUC*
Lys	lysS	57,5	-
Met	metS	76,1	-
Phe	pheS	38,9	UUC*
	pheT	87,9	-
Pro	proS	63,1	-
Ser	serS	48,8	UCC
Thr	thrS	73,4	ACC*
	thrZ	73,3	ACC (3x)
Trp	trpS	37,2	UGG
Tyr	tyrS	47,7	UAC*
	tyrZ	47,1	UAC
Val	valS	101,7	GUC*

Table 1. Specifier codons marked with an asterisk have been verified experimentally, the others are deduced from RNA folding and sequence.

Currently, only four of the 14 nucleotides of the T-box itself have yet been assigned a role and other short conserved leader sequences depicted in Figure 6 are also crucial for control (unpublished results).[51] It has been suggested that some of the T-box nucleotides are conserved because they are involved in base pairing interactions with nucleotides which are also used by the terminator.[51A] Most substitutions in the AG-, GNUG and F-box completely abolish regulation. At present their function is totally unknown. Structural probing of the *thrS* leader in vitro and in vivo clearly shows that the specifier domain is thermodynamically more stable and better defined in vivo than in vitro.[45] This difference might be explained by the presence of interacting proteins in vivo. Binding of these factors, if required for the regulation of all T-box genes, would presumably occur at conserved sequence or structural elements such as the AG- or GNUG boxes. The resulting stabilization of the specifier domain could be important for a productive interaction with the tRNA. The importance of proteins in the T-box system has recently been highlighted by the successful reconstitution of this antitermination system (*thrS*) in vitro, using the wild type regulatory tRNA[Thr] isoacceptor and a partially purified protein fraction. As predicted by the model, aminoacylation of the tRNA with threonine completely abolishes its ability to act as an effector.[51B] A functional in vitro system should greatly increase the chances of identifying the missing protein factors likely to be common to all T-box regulated genes.

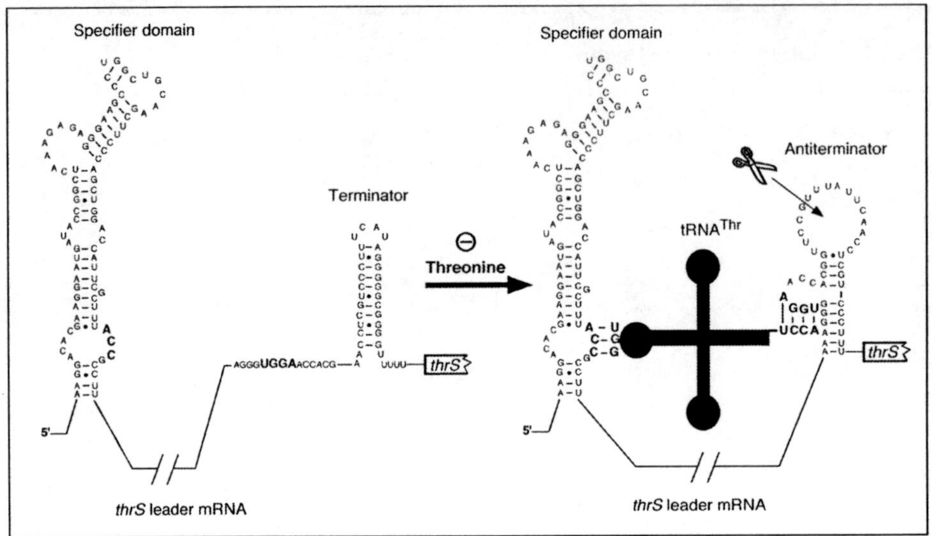

Figure 7. Model of transcriptional antitermination and processing of the *thrS* leader. The endonucleolytic cleavage site is depicted by a scissors symbol.

Uncharged tRNA is the principal effector of antitermination in this system. However, it is the ratio between charged and uncharged tRNA that is sensed by the system. An increase in expression of the biosynthetic *ilv-leu* operon caused by a less chargeable leucine tRNA mutant can essentially be suppressed in a strain diploid for mutant and wild-type alleles.[52] This suggests that both charged and uncharged tRNA may compete for interaction with the *ilv-leu* leader. Moreover, a 10-fold increase in the absolute amount of uncharged tRNA[Trp] does not increase expression of the *L. lactis trp* operon, if the amount of charged tRNA increases in parallel. On the other hand, a low level of uncharged tRNA can be sufficient for induction if the ratio to charged tRNA is increased.[53] Based on this observation, overexpression of any T-box regulated aaRS should repress its own expression. This is true for the threonyl-tRNA synthetase ThrS; a 10-fold overproduction of the synthetase causes a 10-fold repression of a *thrS-lacZ* transcriptional fusion. Moreover, a progressive reduction of ThrS synthesis leads to induction of the normally silent *thrZ* gene in a dose-compensatory manner. Overproduction of either gene on multicopy plasmids causes various degrees of auto- and cross-repression, with ThrS being the much more potent repressor.[54] Similar results are obtained with *leuS*[55] *cysS*[39] and *valS*[56] but curiously overexpression of the PheST synthetase has no effect on the expression of a *pheS-lacZ* fusion.[46] The most likely scenario is that the negative role played by the synthetases is indirect, and occurs by altering the ratio of charged to uncharged tRNA in the cell.

Growth Rate Dependent Control

Production of tRNA increases with growth rate and, logically, so too should the synthesis of the enzymes responsible for charging them. In *E. coli* the expression of most aaRS is induced two- to three-fold for a five-fold increase in growth rate. Few data are available in *B. subtilis*. Proteome and transcriptome data on the global expression pattern of *B. subtilis* (including the aaRS) have recently become available for a limited set of conditions. Decreasing the doubling time from 45 min to 25 min through addition of amino acids to a minimal medium does not significantly alter either the protein or the mRNA concentrations of 18 aaRS (U. Mäder and M. Hecker, pers. communication). On the other hand, expression of a *thrS-lacZ* transcriptional fusion is increased 3.5-fold when bacteria are grown in complex rather than minimal medium. This regulation does not occur at the level of transcription initiation but does depend on leader terminator read-through. Various mutants in the specifier codon and other regions of the leader apart from the terminator are still regulated by growth rate suggesting that this type of control is independent of the tRNA-leader mRNA interaction and tRNA charging levels.[46] One possible explanation is that growth in complex medium increases the transcription elongation rate which is known to favor readthrough of intrinsic terminators.[57]

Leader mRNA Processing

The leader regions of several aaRS genes in *B. subtilis* are cleaved between the T-box and the terminator.[58] The efficiency of the cleavage of the *thrS* leader is specifically increased by threonine starvation, suggesting that it occurs in the antiterminator conformation in close proximity to the tRNA: :T-box interaction (Fig. 7). Cleavage of the *thrS* leader at this site creates a functional *thrS* mRNA starting with a « terminator » (i.e., a putative 5' stabilizer motif, see chapters 9 and 10 by Beran et al and Dreyfus and Joyce, respectively) and ending with a terminator. This configuration of the processed transcript makes it up to five times more stable than the full-length mRNA. During threonine starvation, almost 90% of *thrS* expression comes from the more stable processed transcript. This cleavage event can thus be considered as an amplifier of antitermination by increasing the half-life and hence the steady state concentration of the read-through mRNA transcripts.[58]

When transferred on a plasmid, processing of the *B. subtilis thrS* leader can occur at the same site in *E. coli*. This cleavage is catalyzed by *E. coli* RNase E, both in vivo and in vitro, suggesting that a functional homologue of RNase E is responsible for *thrS* processing in *B. subtilis*.[59] The *Bacillus* counterpart to RNase E with respect to maturation of the 5S ribosomal RNA is RNase M5. In contrast to the single strand cleavages by RNase E, the latter cleaves twice in a double-helical region of a precursor of 5S rRNA to yield mature 5S rRNA in a single step.[60] Moreover, RNase M5 activity depends on the presence of ribosomal protein L18 which probably locks the precursor rRNA into a conformation recognizable by the nuclease.[61] It is therefore not surprising that inactivation of its gene (*rnmV*) does not affect *thrS* leader processing.[62] The identity of the nuclease responsible for this cleavage is still unknown.

Another cleavage, which differs in both position and mechanism from the *thrS* processing event occurs in the *gltX-cysE-cysS* operon. Expression of *cysE/cysS* is partially uncoupled from that of *gltX* by the presence of a tRNA^{Cys} dependent transcriptional attenuator of the T-box type between *gltX* and *cysE*.[39] Transcripts escaping termination at this site are cleaved just 3' to the terminator structure in vivo and in vitro, which explains why one finds mainly monocistronic *gltX* mRNA in the cell. The fact that cleavage occurs in vitro in the absence of proteins other than RNA polymerase suggests that it is due either to the latter or to self-cleavage. This processing leaves a stable secondary structure (the terminator) at the 3' end of the *gltX* transcript and also allows the formation of a putative hairpin structure at the 5' end of the *cysE/cysS* transcript. Cleavage also removes any single stranded residue 3' of the *gltX* terminator which probably renders the upstream mRNA particularly resistant to 3' exonucleases. Accordingly, the steady-state mRNA level of *gltX* is much greater than that of *cysE/cysS*.[36,63]

Structure and Expression of *E. coli* Translation Factor Genes

Decoding the genetic message is an active process requiring ribosomes, translation factors and the amino acids which are brought to the ribosome by specific tRNAs. Protein synthesis can be divided into four steps: peptide chain initiation, elongation, termination and the last step, disassembly of the termination complex to provide a pool of ribosomal subunits ready for new initiation events. Each step requires specific proteins that interact with the tRNAs, mRNA, and/or ribosomes. Expression of these proteins has to be regulated and adapted in order to obtain optimal protein synthesis rates under various conditions. Table 2 gives an overview of the principal functions of *E. coli* translation factors.

Initiation Factors

Translation initiation comprises a series of events ensuring the correct recognition of the initiation signals on the mRNA. It is an important control point and probably the rate-limiting step in protein biosynthesis. A key intermediate of initiation is the 30S initiation complex consisting of the 30S ribosomal subunit, mRNA, the initiator tRNA (fMet-tRNA$_f^{Met}$) and the three initiation factors IF1, IF2-GTP, and IF3 [64]. IF3 assisted by IF1 promotes the dissociation of vacant 70S ribosomes thereby providing the pool of free ribosomal subunits ready for initiation. IF2 specifically recruits fMet-tRNA$_f^{Met}$ and positions it correctly in the ribosomal P site. IF3 ensures the exclusive usage of AUG, GUG or UUG as start codons through specific recognition of the anticodon stem of the

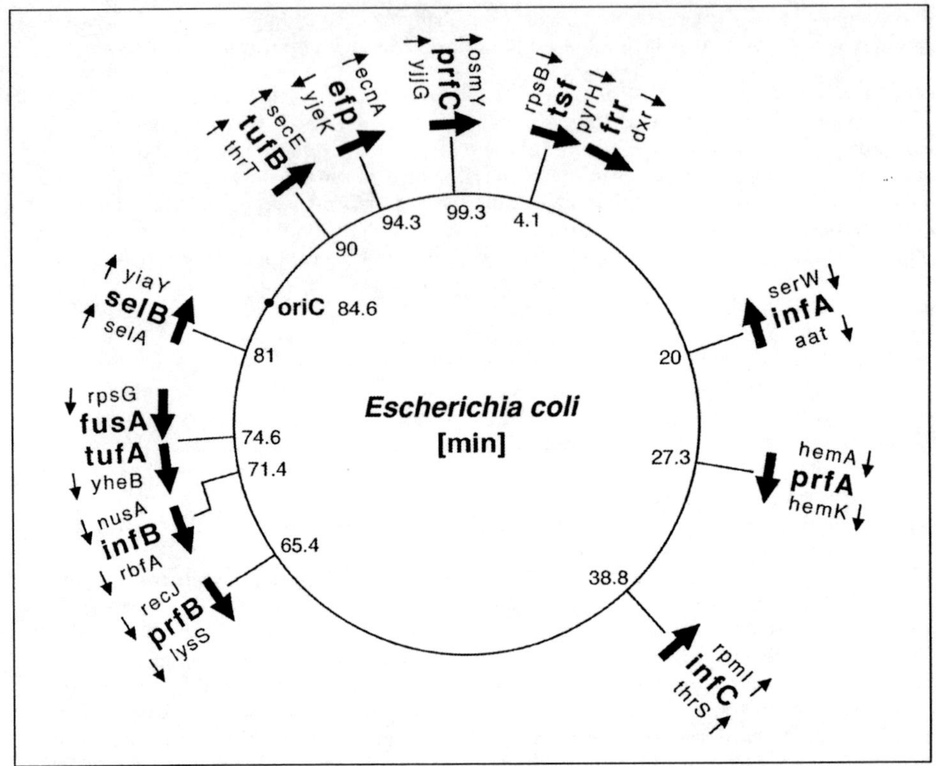

Figure 8. *E. coli* translation factor genes. The translation factor genes are shown in boldface type together with adjacent genes. Positions are given in min and are calculated values from the complete genome sequence (Colibri database). Arrows indicate the directions of transcription. The replication origin is shown as oriC at 84.6 min.

initiator tRNA. IF1, by binding to the ribosomal A site, confers specificity on the formation of the 30S initiation complex by occluding premature access of an elongator tRNA to the A site. Binding of the 50S subunit with the concomitant release of IF1 and IF3 gives rise to the 70S initiation complex with fMet-tRNA$_f^{Met}$ in the ribosomal P site. The GTP carried by IF2 is hydrolyzed, IF2-GDP is released and the ribosomal A site is now ready to accept the first elongator aminoacyl-tRNA carried by elongation factor Tu (for details, see chapter 19 by de Smit and van Duin).

The three *E. coli* initiation factors are present in approximately equimolar amounts in the cell at about 0.2 to 0.3 molecule of each factor per ribosome.[4] This is enough to saturate free 30S particles.[64] The steady-state levels of the initiation factors are coordinately regulated relative to one another and relative to ribosome levels in exponentially growing bacteria.[65] The genes for the three initiation factors (*infA*/IF1, *infB*/IF2, *infC*/IF3) are dispersed on the *E. coli* chromosome. With the exception of *infA*, they are associated with other components of the translational apparatus (Fig. 8).

Initiation Factor 1 (*infA*)

IF1 is the smallest initiation factor (8.1 kD) and is an essential protein.[66] When IF1 levels in the cell are reduced, polysomes become smaller and cell growth decreases. The *infA* gene is located at 20 min on the *E. coli* chromosome. It is transcribed counterclockwise from two ς^{70} promoters, P$_1$ and P$_2$, yielding monocistronic mRNAs of 525 and 330 nucleotides, respectively. The smaller transcript is about two-fold more abundant than the larger one, but both end at the same Rho-independent terminator located immediately downstream of the *infA* coding region.[67] Measurements of transcriptional fusions to *lacZ* show that P$_2$ is indeed twice as active as P$_1$. Growth rate-dependent control occurs exclusively at the P_2 promoter. P_2 directed β-galactosidase activity increases 1.7-fold with a doubling of the growth rate. Therefore, the entire increase in IF1 levels under these conditions can be attributed to an increased rate of IF1 synthesis. The *infA* gene is not autogenously regulated; neither transcription nor translation of *infA* is affected by high cellular levels of IF1.[67]

Initiation Factor 2 (*infB*)

IF2, the largest initiation factor is an essential GTP binding protein.[68,69] In *E. coli* three natural forms of IF2 exist in the cell, IF2α (97,2 kD), IF2β₁ (79,7 kD) and IF2β₂ (7 amino acids shorter than IF2β₁). The shorter IF2β species result from translation at in-frame start codons in the *E. coli infB* gene.[70,71] The N-terminal part of the protein accounts for the affinity of IF2 to the 30S subunit[72] but each of the three forms of IF2 can assure cell survival. However, expression of only one form slows growth at 37°C and results in a cold-sensitive phenotype.[71] Even truncation of the entire N-terminal domain of IF2 still permits cell survival provided the protein is overexpressed.[69] The *infB* gene is located at 69 min. on the *E. coli* chromosome and is part of the complex, multi-functional *metY-nusA-infB* operon (Fig. 9). The first gene of this operon, *metY*, encodes a minor form of the initiator tRNA$_{f2}$^{Met}. The rest of the operon consists of protein-coding genes including *yhbC* (p15A, a protein of unknown function), *nusA* (NusA, involved in the modulation of transcription termination), *infB*, *rbfA* (RbfA, a ribosome binding factor) and *truB* (a tRNA-modifying pseudouridine synthase). The *nusA-infB* core of the operon is conserved in evolutionarily very distant organisms such as *B. subtilis*, *B. stearothermophilus*, *M. xanthus* and *T. thermophilus*.

Three promoters (P_{-1}, P_0 and P_2) direct transcription of the operon[8] (Fig. 9). Transcripts can extend into the downstream *rpsO-pnp* operon but the latter is essentially expressed from its own strong promoters. P_{-1} and P_0, located upstream of *metY*, are the principal promoters of the operon.[73,74] An upstream activating sequence (UAS) is recognized by the protein FIS, albeit with a ten-fold lower efficiency than that of the *tufB* operon[75] (see below); The effect of FIS binding on the expression of the *nusA-infB* operon in vivo is not known. Most of the transcripts initiating at P_{-1} and P_0 are terminated at the Rho-independent terminators t₁ and t₂ located between *metY* and *yhbC* producing short initiator tRNA precursors. Readthrough of t₁ and t₂ allows for the expression of *nusA* and *infB*. Most transcripts further extend past the weak Rho-independent terminator t₃ downstream of *infB* allowing expression of *rbfA* and *truB* (Fig. 9). In addition, the various readthrough transcripts can be cleaved by RNase III immediately downstream of *metY*. This has two consequences: it separates the initiator tRNA from the coding polycistronic mRNA and, by removing the

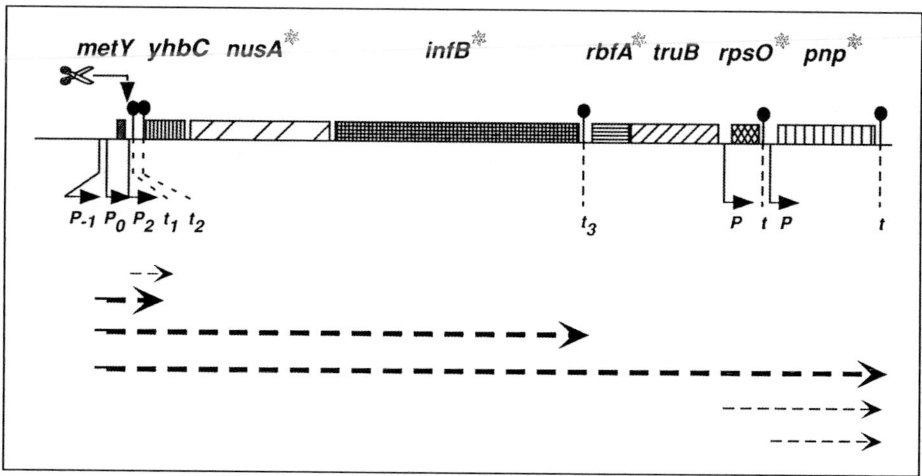

Figure 9. Gene organisation and transcripts of the *metY-infB* and the adjacent *rpsO-pnp* operons. The gene names are given above the boxes. Promoters (P) and terminators (t) are indicated, the transcripts are shown by dashed arrows. Genes known to be induced in the cold shock response are indicated by the *. The RNase III processing site between metY and yhbC is indicated by a scissors symbol.

Table 2.   Function and nomenclature of translation factors

Former Name	New Name IUBM[125]	Mass (kDa)	Gene	Main Function(s) in Translation
E. coli Initiation factors				
IF1	IF-1	8.1	*infA*	Occludes ribosomal A site, thus prevents premature access of elongator tRNAs Stimulates IF2 and IF3 functions
IF2α	IF2-1	97.2	*infB*	Binds fMet-tRNA$_f^{Met}$ to ribosomal P site, GTPase
IF2β	IF2-2	79.7		Binds fMet-tRNA$_f^{Met}$ to ribosomal P site, GTPase
IF3	IF3	20.5	*infC*	Prevents ribosome association; monitors correct fMet-tRNA$_f^{Met}$ / initiation codon interaction
E. coli Elongation factors				
EF-Tu	EF1A	43.1	*tufA & tufB*	Forms ternary complex with aa-tRNA and GTP; Binds aa-tRNA to ribosomal A site ; GTPase
EF-Ts	EF1B	30.3	*tsf*	Promotes guanine nucleotide exchange on EF1A (EF-Tu)
EF-G	EF2	77.4	*fusA*	Promotes translocation reaction; GTPase
EF-P	-	20.4	*efp*	Stimulates peptidyl transferase reaction for certain amino acids
SelB	-	68.7	*selb*	Promotes selenocysteine incorporation (specific for selenocysteyl-tRNA)
E. coli Release Factors				
RF-1	RF1	40.4	*prfa*	Promotes termination at UAA, UAG
RF-2	RF2	41	*prfb*	Promotes termination at UAA, UGA
RF-3	RF3	59.4	*prfc*	Promotes action of RF1 and RF2; GTPase
E. coli Ribosome recycling factor				
RRF	RF4	20.5	*frr*	Dissociates ribosomes from mRNA after termination of translation (needs EF-G or RF3)

5' terminal structural motif (the tRNA) from the mRNA causes a more rapid decay of its 5' end. However, this has no effect on the expression of NusA and IF2 since they are not overproduced in a RNase III mutant strain.[76] Dissociation of the tRNA from the mRNA is also achieved directly by terminating *metY* transcripts at t$_1$ or t$_2$ and using the minor P$_2$ promoter downstream of *metY*.

The NusA protein negatively controls the expression of its operon at the transcriptional level. This was shown by overexpressing NusA from a multicopy plasmid. Under these conditions the expression of *lacZ*-fusions containing various regions of the operon is repressed two-fold. Protein and gene fusions behave the same way indicating that NusA acts at the transcriptional level.[77] Similarly, in a NusA mutant strain expressing only 30% of the wild-type NusA level, IF2 synthesis is increased five-fold.[78] Overexpression of IF2 does not affect the regulation of the operon.

NusA, IF2, RbfA and the products of the downstream *rpsO-pnp* operon, the ribosomal protein S15 (*rpsO*) and polynucleotide phosphorylase (*pnp*) are all cold shock-induced proteins.[79,80] At 37°C, the *rpsO-pnp* operon is essentially expressed from its own transcriptional signals. However, a cold shock results in a dramatic change in the spectrum and length of transcripts originating upstream of *metY*. Several major cold shock proteins, namely CspA, CspC and CspE, which are induced upon temperature decrease, increase readthrough of the t₁ and t₂ terminators and allow efficient cotranscription of both the *nusA-infB* and *rpsO-pnp* operons.[81]

Initiation Factor 3 (*infC*)

Like the other initiation factors, IF3 is an essential protein in *E. coli*.[82] The *infC* gene, located at 37 min on the *E. coli* chromosome, is part of the *thrS-infC-rpmI-rplT* gene cluster encoding threonyl-tRNA synthetase, IF3, and the ribosomal proteins L35 and L20, respectively.[83] Only three base pairs separate the *thrS* stop codon from the atypical AUU initiation codon of *infC*.[83,84] Transcription of *infC* is initiated from three promoters, P$_{thrS}$, P$_0$ and P$_0$'[85] (Fig. 10). P$_{thrS}$, located upstream of *thrS*, is the source of bicistronic *thrS-infC* and tetracistronic *thrS-infC-rpmI-rplT* transcripts. P$_0$ and P$_0$' lie within *thrS* and the latter is the major promoter for the expression of *infC*, *rpmI* and *rplT*. The t₁ terminator located immediately downstream of *infC* is not very efficient and roughly 50% of the transcripts generated from the upstream promoters are continued through *rplT*. Finally, transcription of *rpmI* and *rplT* can also initiate at the weak P₁ promoter within the *infC* reading frame (Fig. 10). Thus, initiation and termination signals are arranged to yield a set of overlapping transcripts from this operon.

Expression of *infC* is controlled independently from that of *thrS*. There is five times more IF3 protein and mRNA in the cell than threonyl-tRNA synthetase and *thrS* mRNA. Moreover, the steady state level of *infC* mRNA does not vary with cell growth indicating that growth rate dependent synthesis of IF3 is not under transcriptional control nor due to differential mRNA stability.[86]

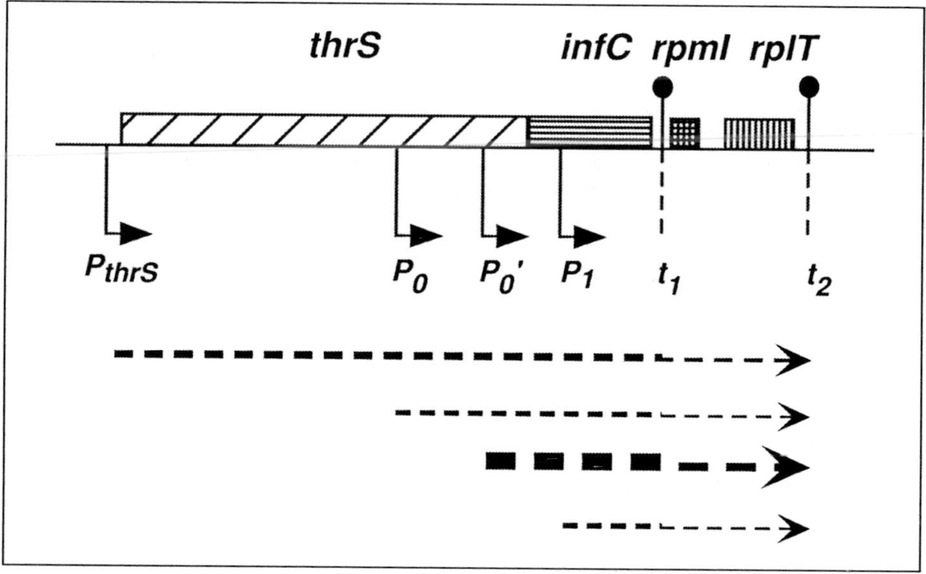

Figure 10. Gene organisation and transcripts of the *thrS-infC* operon. The gene names are given above the boxes. Promoters (P) and terminators (t) are indicated, and the corresponding transcripts are shown by dashed arrows whose thickness is roughly proportional to transcript strength.

IF3 uses its ability to differentiate between 'normal' and 'abnormal' initiation codons to autoregulate its translation, which begins with an AUU initiation codon. Such an initiation codon is unique in *E. coli* but well conserved in other IF3 genes even in distantly related organisms. An exception is *Myxococcus xanthus* which initiates *infC* translation at an equally unusual AUC codon. The AUU start codon is not only essential,[87] but also sufficient,[88] for autoregulation. Changing the AUG initiation codons of the *thrS* or *rpsO* genes to AUU increases their expression in an *infC* mutant background. However, under conditions of IF3 excess, repression of these mutant genes is weaker than that observed for *infC*, suggesting that the *infC* message has specific features that render its expression particularly sensitive to regulation.[88] One of them could be a very rare second potential ribosomal binding site between the probable Shine-Dalgarno region and the AUU start codon. Finally, parts of the *infC* mRNA can be folded into a long range secondary structure occluding the ribosomal binding site of the downstream *rpmI* gene. This structure is specifically recognized and stabilized by the L20 protein (*rplT*) and allows for translational repression of the downstream *rpmI* and *rplT* genes.[89]

Elongation Factors

After formation of the 70S initiation complex, three major protein factors, EF-Tu, EF-Ts and EF-G, catalyze the next step in translation, peptide chain elongation. EF-Tu transports the aminoacylated tRNAs to the A-site of the ribosome in the form of the ternary complex EF-Tu•GTP•aa-tRNA. Subsequent GTP hydrolysis causes its release in the form of a binary complex, EF-Tu•GDP. EF-Ts catalyses the GDP/GTP nucleotide exchange on EF-Tu in order to reactivate the factor. An analog of EF-Tu, SelB, brings the only elongator tRNA not recognized by EF-Tu, selenocysteyl-tRNA^{Sec}, to the ribosome. After peptide bond formation, EF-G complexed with GTP promotes the translocation of the mRNA bound peptidyl-tRNA from the A to the P site of the ribosome, a process requiring hydrolysis of the GTP molecule (for details, see chapter 20 by P. Nissen et al). A less well-characterized translation factor, EF-P, probably allows peptide bond formation to occur more efficiently with some aminoacyl-tRNAs that are poor acceptors for the ribosomal peptidyltransferase.

Elongation Factor Tu (*tufA, tufB*)

Gene Organization and Structure

Two unlinked genes *tufA* and *tufB* encode the elongation factor EF-Tu, a monomeric protein of 43.1 kD. The two genes are located in different operons and differ at only 13 nucleotides. EF-TuA and EF-TuB are identical except for the C-terminal residue.[90,91] The tufA gene located at 74.6 min on the E. coli chromosome is part of the streptomycin (*str*) operon *rpsL-rpsG-fusA-tufA* encoding ribosomal proteins S12 and S7, EF-G and EF-TuA, respectively (Fig. 11). This operon is one of the most conserved in prokaryotic evolution. The *tufB* gene maps at 90 min and is preceded by the four tRNA genes *thrU, tyrU, glyT* and *thrT* coding for tRNA$_4^{Thr}$, tRNA$_2^{Tyr}$, tRNA$_2^{Gly}$ and tRNA$_3^{Thr}$, respectively (Fig. 11).

Regulation of Expression

EF-Tu is one of the most abundant proteins in the cell, representing up to 10% of the total protein content. There is about one EF-Tu molecule for every tRNA and this ratio is constant under different growth conditions.[92] Normally, 75-90% of the tRNAs are charged and bound to EF-Tu•GTP, forming a reactive ternary complex.[93] Two gene copies are probably maintained to achieve efficient expression of EF-Tu. Either of the *tuf* genes can be deleted without loss of viability. However, deletion of *tufA* increases the generation time by one third.[94] At least one half of the cellular EF-Tu is derived from *tufA*[95] and this proportion remains constant under different growth conditions.

Three to four times more EF-TuA is synthesized than S12, S7 and EF-G whose genes are cotranscribed with *tufA* from a common upstream promoter (P$_1$, Fig. 11A). Two observations account for this differential expression. First, there are two additional promoters (P$_2$ and P$_3$) located within the upstream *fusA* gene (EF-G) which are functional in vivo and together contribute up to 25% to *tufA* transcription.[96] Second, *tufA* expression is not affected by S7 which acts as a translational repressor for the first three genes of the operon (see below, EF-G). Thus, increased transcription and uncoupled translation allows for a selectively enhanced expression of EF-TuA. Depletion or overexpression

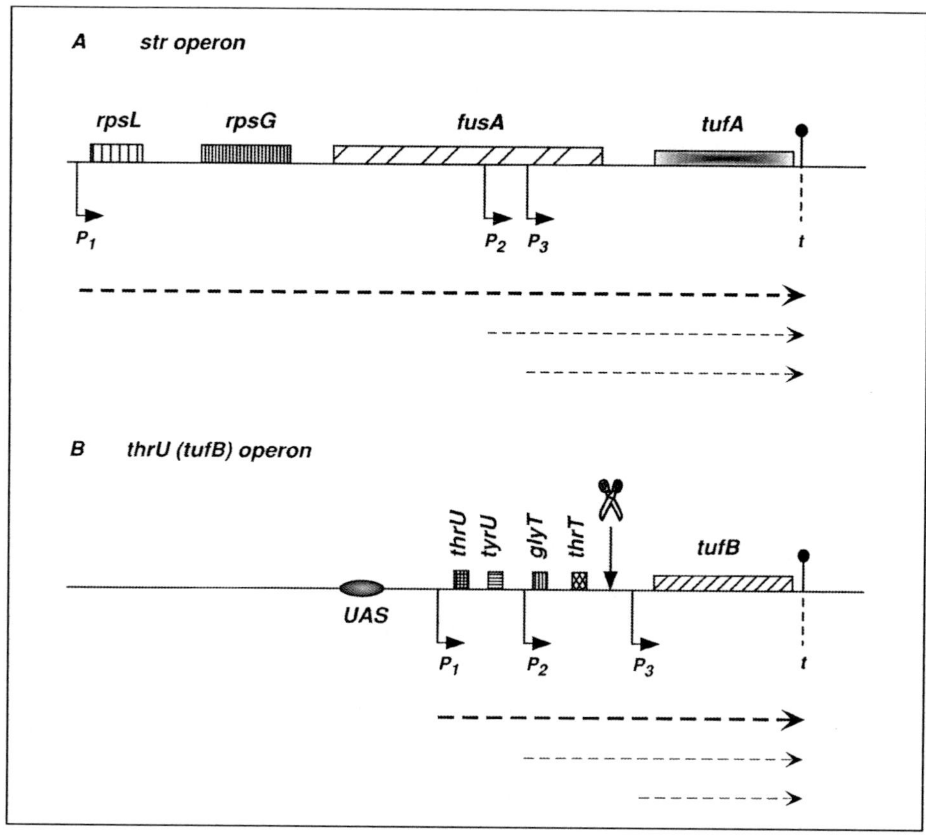

Figure 11. Gene organisation and transcripts of the two operons carrying a gene for EF-Tu. (A) *str* operon comprising *rpsL, rpsG, fusA* and *tufA* encoding ribosomal proteins S12, S7 and elongation factors EF-G and EF-TuA, respectively. (B) thrU-tufB operon showing the UAS regulatory element , the four tRNA genes *thrU, tyrU glyT* and *thrT* and the gene *tufB* (EF-TuB). The processing site between the tRNA gene cluster and tufB mRNA is indicated by a scissors symbol. The gene names are given above the boxes. The promoters (P) and terminators (t) are indicated, and the corresponding transcripts are shown by dashed arrows whose thickness is roughly proportional to transcript strength.

of EF-Tu has no effect on *tufA* expression. However, some results suggest that overproduction of EF-Tu negatively autoregulates EF-TuB expression.[97]

The *tufB* gene is transcribed with four upstream tRNA genes yielding a multicistronic mRNA of about 1.8 kb. This transcript can be processed in the *thrT-tufB* intergenic region separating the structural genes from the *tufB* mRNA (Fig. 11B).[98] A partial deletion of the four tRNA genes has no significant effect on *tufB* expression. The major control element of this operon is a cis-acting AT-rich upstream activator sequence (UAS), 5' to the major P_1 promoter.[75] Deletion of this sequence results in a 10- to 15-fold drop in transcription. The UAS constitutes a binding site for the pleiotropic regulator, FIS,[99] which also recognizes similar sequence elements upstream of rRNA and tRNA operons (see chapter 23 by M.M. Barker and R.L. Gourse). FIS binds to two sites in the UAS of the *tufB* operon and activates transcription by facilitating the binding of RNA polymerase to the promoter. Activation of the *tufB* operon upon a nutritional upshift is drastically reduced in *fis* mutant strains.[100] In addition, expression of this operon is subject to stringent control. In vitro, selective inhibition of *tufB* transcription by ppGpp depends on the presence of a G/C rich discriminator sequence between the –10 region of the promoter and the first base of *thrU*. Similarly located G/C elements are crucial for many stringently controlled promoters. Thus, the *tufB* operon is very well adapted to respond quickly and efficiently to nutritional changes.

Elongation Factor SelB (*selB*)

The elongation factor SelB is the key molecule for the specific incorporation of the amino acid selenocysteine into polypeptides. It specifically recognizes the selenocysteine charged tRNA^{Sec}, which has a UCA anticodon, in an EF-Tu like manner. This allows insertion of selenocysteine at in-frame UGA stop codons. In *E. coli* SelB binds GTP, selenocysteyl-tRNA^{Sec} and a stem-loop structure immediately downstream of the UGA codon (the SECIS sequence).[101] The absence of active SelB prevents the participation of selenocysteyl-tRNA^{Sec} in translation.[102]

The *selB* gene is cotranscribed with *selA* encoding the selenocysteine synthase, which converts the serine attached to tRNA^{Sec} to selenocysteine. The *selAB* operon maps at 81 min on the *E. coli* chromosome and is expressed from a rather weak ς^{70} promoter affording transcription initiation 48 bases upstream of *selA*.[103] The *selA* termination codon overlaps the selB initiation codon suggesting tight translational coupling and hence coordinate synthesis of the two proteins. Consistent with this idea, both SelA and SelB are present in the cell at 1200 to 2000 copies. Expression of *selAB* and other *sel* genes is constitutive in both aerobically and anaerobically grown *E. coli* cells. This reflects the need to synthesize selenoproteins under both conditions even though some of them are differentially expressed. Weak transcription and low stability of the transcripts account for the low level of full-length *selAB* mRNAs in the cell. These features probably obviate the requirement for regulation, for example, in response to the intracellular selenium concentration.[103]

Elongation Factor G (*fusA*)

EF-G in complex with GTP stimulates the translocation of the mRNA/tRNA complex from the A to the P site of the ribosome. It is an essential monomeric protein of 77.4 kD and is present in the cell at about one molecule per ribosome.[4]

The *fusA* gene encoding EF-G is located upstream of the *tufA* gene within the str operon (see Fig. 11). All four genes of this operon are cotranscribed from a common promoter. Expression of the first three, *rpsL* (S12), rpsG (S7) and *fusA* is translationally coupled. The ribosomal protein S7 acts as a translational repressor by binding to a complex RNA stem-loop structure in the intergenic region between *rpsL* and *rpsG*, probably inducing a conformational change in this structure.[104,105] A crucial element required for repression is an anti-SD sequence that sequesters the ribosomal binding site of *rpsG* in a double stranded structure thereby inhibiting ribosome binding. This translational feedback regulation by S7 inhibits translation of its own gene and that of *fusA* whose start codon is only 28 bp downstream of *rpsG*. Interestingly, the stem-loop structure is also important for the coupled translation of S12 and S7. It has thus been proposed that ribosomes skip the intercistronic loop in order to reinitiate at the *rpsG* initiation site which is brought right next to the *rpsL* stop codon by RNA folding. In addition, S7 represses only coupled translation of its gene allowing for a small basal level of unregulated expression of S7 and presumably also EF-G.[106] All three proteins S12, S7 and EF-G are thus coordinately expressed. This is consistent with the fact that the cellular level of EF-G corresponds to that of ribosomes. EF-Tu expression from the distal *tufA* gene is not under control of the translational repressor S7 and is translationally uncoupled. However, synthesis of both EF-Tu and EF-G proteins is stringently controlled and subject to growth rate dependent regulation (see above, EF-Tu).

Elongation Factor Ts (*tsf*)

The EF-Tu nucleotide exchange factor EF-Ts is present in the cell at about 0.2 molecule per ribosome which is 30 times less than its « substrate » EF-Tu.[4] EF-Ts is expressed from a bicistronic operon, located at 4.1 min on the *E. coli* map, together with ribosomal protein S2 (*rpsB*).[107] A single major promoter immediately upstream of *rpsB* directs transcription of the *rpsB-tsf* operon. Two potential Rho-independent transcription terminators are present in the operon, one in the intergenic region between the two genes and one distal to *tsf*. Like EF-Tu and EF-G, EF-Ts synthesis is under stringent control and coordinated with ribosomal protein synthesis in a growth rate dependent manner.[108]

Elongation Factor P (*efp*)

The role of the EF-P protein in the elongation process has been little considered thus far. Yet, it is an essential well-conserved factor required for protein synthesis and has a eukaryotic counterpart, eIF5A. EF-P stimulates the peptidyltransferase activity of fully assembled 70S ribosomes,[109,110] preferentially promoting first peptide bond synthesis. EF-P activity requires the ribosomal protein L16, suggesting that the binding site of EF-P may overlap the peptidyltransferase center. In contrast to the other elongation factors it does not require GTP for its action. This 20.5 kD protein is encoded by the *efp* gene located at 94.3 min on the *E. coli* genome. There are 800-900 EF-P molecules in the cell (0.1-0.2 copy per ribosome) and this ratio is independent of the stage of cell growth.[111]

Release and Ribosome Recycling Factors

The elongation cycle of protein biosynthesis on the ribosome is brought to an end when a stop codon appears in the A (decoding) site of the ribosome. In contrast to translation initiation and elongation, stop codon recognition is not achieved by a tRNA, but by two specialized proteins called release factors (RF). RF1 recognizes UAG and UAA, and RF2 UGA and UAA stop codons. Binding of either one of the two factors triggers hydrolysis of peptidyl-tRNA. A third release factor, the GTP binding protein RF3, catalyses the release of RF1/RF2 and thereby accelerates the transition from termination to ribosome recycling. The latter step requires the ribosome recycling factor (RRF) which together with EF-G or RF3 dissociates the post-termination complex, a process involving GTP hydrolysis. Finally, the deacylated tRNA still present on the 30S particle is displaced by IF3 allowing recycling of the 30S subunit (for details, see chapter 21 by R.H. Buckingham and M. Ehrenberg).

Release Factors: RF1 (*prfA*) RF2 (*prfB*) and RF3 (*prfC*)

Of the three release factors, only RF3 is non-essential for cell growth. However, it is necessary for optimal translational activity, particularly under environmental stress conditions.[112,113] The genes for RF1 and RF2 can both be knocked out if a mutant form of RF2, capable of terminating translation at all three stop codons, is present.[114] The number of RF1 molecules per cell increases from 1200 to 4900, and that of RF-2 from 5900 to 24900 as growth rates increases from 0.3 to 2.4 doublings per hour. The strict one to five ratio between RF1 and RF2 is maintained independently of growth rate and corresponds well to the one to four ratio of UAG and UGA stop codons in *E. coli*. The concentration of RF1/2 in the cell is thus similar to that of the initiation factors. This seems logical since they are required at the same rate at opposite ends of the translational process. On the other hand, RF3 is present at a roughly 60-fold lower level than RF1/2 and its concentration can vary significantly from one strain to the other.[115] Thus different strains can accommodate low RF3 concentrations very well. Some species, such as Mycoplasma have simply dispensed with the RF3 gene altogether.[116]

RF1 (40.5 kD) is encoded by the *prfA* gene, located at 27.3 min on the *E. coli* genome. Transcription probably initiates at a ς^{70} promoter 65 base pairs upstream of the initiation codon and results in a monocistronic transcript ending at a terminator distal to *prfA*. The termination codon for RF1 translation is UGA, recognized by RF2.[117]

RF2 (41.2 kD) is encoded by the *prfB* gene located at 65.4 min. A strong ς^{70} promoter with a stringent discriminator, immediately upstream of *prfB*, directs transcription of a 2.8 kb bicistronic transcript, including *lysS* and which ends at the *lysS* distal terminator.[118] RF2 translation is terminated with a UGA codon recognized by itself. A premature in-frame UGA codon at position 26 is crucial to RF2 expression. A +1 frameshift at this position is necessary for complete translation of RF2. Two elements are required to achieve the remarkably high frameshifting rate of 50% : a correctly placed Shine-Dalgarno-like sequence and a particular codon context (Fig. 12). A CUU (leucine) codon preceding the UGA provides for a uracil-rich stretch, favorable for frameshifting. Furthermore, tRNA^{Leu}GAG is a « shifty » tRNA with the ability to cause a 4-base translocation.[119] Pairing of the Shine-Dalgarno-like sequence with the 16S rRNA leads to stalling or slowing down of the elongating ribosome which probably favors and orients the shift to the leucine codon (Fig. 12, see also chapter 22 by I.P. Ivanov et al). This provides a natural mechanism of autogenous control at the translational level. When RF2 levels in the cell are low, the U of the UGA codon is made more

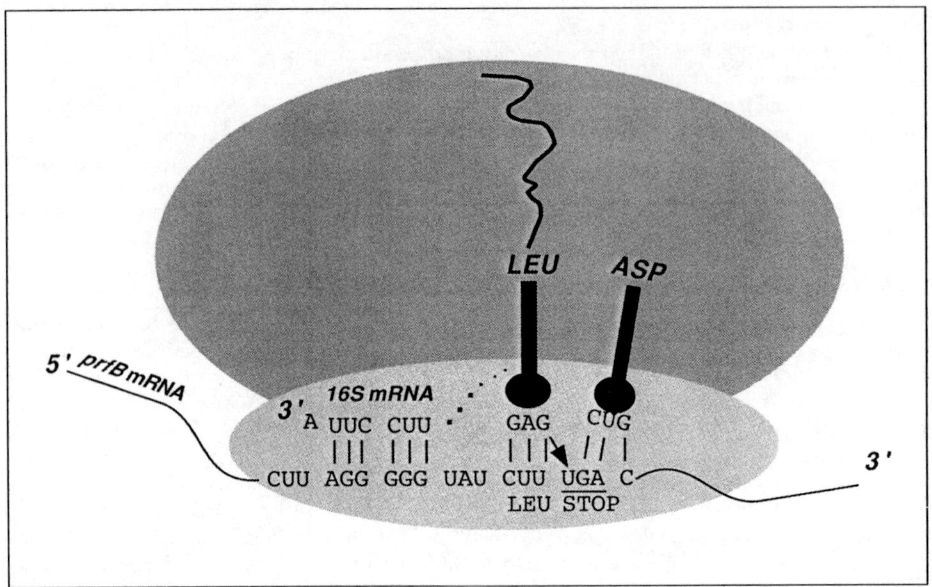

Figure 12. Frameshift region in the *prfB* gene. Shown are the Shine-Dalgarno interaction and the CUU UGA C sequence where the +1 frameshift occurs. Shifting of tRNA[Leu] from CUU to UUU is indicated by an arrow.

available for re-pairing by tRNA[Leu]$_{GAG}$ and frameshifting is favored over termination and vice versa. Autoregulation has been demonstrated to occur in vitro[120] and in vivo.[121]

Release factor 3 (59.6 kD) is encoded by the *prfC* gene located at 99.3 min on the *E. coli* map. Transcription probably originates about 60 base pairs upstream of the initiation codon at a ς^{70} promoter.[112] The presence of a terminator downstream of *prfC* suggests monocistronic transcription. However, nothing is known about the regulation of *prfC* as yet.

Ribosome Fecycling Factor (*frr*)

Ribosome recycling factor (RRF) is an essential protein.[122] It acts after RF1/RF2 mediated peptidyl-tRNA hydrolysis to participate in the disassembly of the post-termination complex. In addition, it maintains translational fidelity during chain elongation. The molar amount of RRF in the cell is about 50% of that of the ribosomes and approximately 30% of total RRF is bound to ribosomes.[123]

The *frr* gene, coding for RRF (20.5 kD), is separated by only one gene from *tsf*, encoding EF-Ts. Albeit transcribed in the same orientation (Fig. 8) they do not form an operon. A very atypical ς^{70} type promoter directs transcription initiation 58 bp upstream of the initiation codon. Spacing between the weak –35 and the –10 region is 20 nucleotides which is extremely rare and, in fact, only one other case is known in *E. coli*. Deletion of the –35 region still allows for RNA polymerase binding and reduced promoter activity.[124] A potential regulatory role for this promoter configuration remains to be analyzed. Finally, overexpression of RRF only slightly affects growth, suggesting that it is not harmful for elongating ribosomes.

Conclusions and Perspectives

Expression of the genes described in this chapter is regulated by many different mechanisms and yet represents only a portion of the regulatory capabilities of the prokaryotic cell. It has become evident that the highly developed organisms that we study today dedicate appreciable resources to regulatory functions. By doing so, they ensure survival in various natural habitats and acquire the flexibility to respond efficiently to the many challenges they experience. Different organisms often use different strategies to regulate corresponding genes, as is the case for the aminoacyl-tRNA synthetases. In *E. coli* expression of all aaRS is regulated in a growth rate dependent manner, but specific mechanisms, such as those employed by the *pheST* and *thrS* genes, allow induction in response to starvation for the cognate amino acid. In *B. subtilis* 14 aaRS genes (the T-box family) are specifically induced by tRNA-mediated transcription antitermination. To our current knowledge, this mechanism is essentially confined to Gram positive bacteria. As different and ingenious these controls may be, they share a common effector, the tRNA. In *E. coli* it provokes the crucial ribosome stalling in *pheST* attenuation, it titrates ThrS away from binding to its operator on the mRNA and it directly favors formation of the antiterminator in *B. subtilis*. The conservation of many different regulatory mechanisms allows for both gene-specific regulation and more global coordinate control of genes with related functions. Some control mechanisms that evolved early, have possibly persisted because they performed well and there has been no selective pressure to improve or replace them.

The high induction ratios observed for some *B. subtilis* aaRS genes in the absence of one of their substrates appear counterintuitive. However, the cost of increased aaRS synthesis is probably outweighed by the increased efficiency acquired to scavenge for diminishing amino acid pools. This option might be advantageous to *B. subtilis*, which is often confronted with poor nutritional conditions, thereby avoiding premature commitment to the even more costly process of spore formation. The same control mechanism also allows a very limited adaptation of gene expression suggested by the moderate derepression of several other aaRS genes under similar conditions.

Many aspects of aaRS biosynthesis are still unresolved in both organisms. The mechanism underlying growth rate-dependent regulation of virtually all aaRS genes in *E. coli* is not understood; for *pheST* and *thrS* this regulation can be explained at least partially by the specific control mechanisms. In B. subtilis a metabolic upshift can induce thrS expression by increasing read-through of the leader terminator in the absence of a tRNA: :mRNA interaction. This might occur via a faster transcription elongation rate in rich medium which is known to reduce terminator efficiency. The tRNA alone is not sufficient to promote specific antitermination in vitro. The additional protein factors which are required to make this system work still remain to be identified.

The regulatory mechanisms described support the notion that, in many cases, the primary function of particular components of the translation apparatus has been exploited for the purposes of regulation of their synthesis. The mechanisms used to control translation factor expression clearly support this view. For example, IF3 autoregulates its own translation using its capacity to discriminate between "good" and "bad" initiation codons. It does this, not by binding to its mRNA, but to the ribosome. RF2 normally brings translation to an end when a final UGA stop codon is reached. However, when in excess it blocks its own translation at an in-frame stop codon on its mRNA. In order to make this mechanism work, an efficient alternate reading of the genetic code was designed: a + 1 frameshift allowing synthesis of the functional protein. These very specific mechanisms permit a perfect coordination of the synthesis of these translation factors to that of the ribosomes. The synthesis of several other factors involved in translation is controlled by mechanisms that are not fully understood yet. A great number of overlapping regulatory signals exists in different operons, i.e., multiple promoters, transcription terminators and RNA structures mediating translational coupling. It is likely that they all serve the ultimate purpose of coordinating the expression of the different components of the translational apparatus and rendering it as efficient as possible. A good example are the normally independently transcribed *nusA-infB* and *rpsO-pnp* operons. During a cold shock, when these proteins are all required at elevated levels, specific proteins promote transcription attenuation at the multiple terminators present and transform the two operons into a single coordinately regulated transcription unit. Studies in other organisms stimulated by newly available genome sequences will allow us to compare how different bacteria cope with similar challenges.

Acknowledgements

We thank Drs. C. Condon, J. Lapointe and M. Springer for critical reading of the manuscript. This work was supported by funds from the Centre National de la Recherche Scientifique (UPR9073 and UMR) and PRFMMIP from the Ministère de l'Education Nationale.

References

1. Chatterji D, Ojha AK. Revisiting the stringent response, ppGpp and starvation signaling. Curr Opin Microbiol 2001; 4(2):160-5.
2. Kisselev LL, Justesen J, Wolfson AD et al. Diadenosine oligophosphates (Ap(n)A), a novel class of signalling molecules? FEBS Lett 1998; 427(2):157-63.
3. Martinis SA, Plateau P, Cavarelli J et al. Aminoacyl-tRNA synthetases: a new image for a classical family. Biochimie 1999; 81(7):683-700.
4. Bremer H, Dennis PP. Modulation of chemical composition and other parameters of the cell by growth rate. In: Neidhardt FC, Curtiss R, Ingraham JL, Lin EC, Low KB, Magasanik B, Reznikoff WS, Riley M, Schaechter M, Umbarger HE, eds. *Escherichia coli* and *Salmonella*: cellular and molecular biology. Washington, D.C.: American Society for Microbiology; 1996: 1553-1569.
5. Jakubowski H, Goldman E. Quantities of individual aminoacyl-tRNA families and their turnover in *Escherichia coli*. J Bacteriol 1984; 158(3):769-76.
6. VanBogelen RA, Abshire KZ, Moldover B et al. *Escherichia coli* proteome analysis using the gene-protein database. Electrophoresis 1997; 18(8):1243-51.
7. Morgan S, Söll D. Regulation of the biosynthesis of amino acid:tRNA ligases and of tRNA. Prog Nucleic Acid Res Mol Biol 1978; 21:181-207.
8. Grunberg-Manago M. Regulation of the expression of aminoacyl-tRNA synthetases and translation factors. In: Neidhardt FC, Curtiss R, Ingraham JL, Lin EC, Low KB, Magasanik B, Reznikoff WS, Riley M, Schaechter M, Umbarger HE, eds. *Escherichia coli* and *Salmonella*: cellular and molecular biology. Washington, D.C.: American Society for Microbiology; 1996: 1432-1457.
9. Putzer H, Grunberg-Manago M, Springer M. Bacterial aminoacyl-tRNA synthetases: Genes and regulation of expression. In: Söll D, RajBhandary, eds. tRNA: Structure, Biosynthesis, and Function. Washington, D.C.; 1995: 293-333.
10. Putney SD, Schimmel P. An aminoacyl-tRNA synthetase binds to a specific DNA sequence and regulates its gene transcription. Nature 1981; 291:632-635.
11. Springer M, Mayaux JF, Fayat G et al. Attenuation control of the *Escherichia coli* phenylalanyl-tRNA synthetase operon. J Mol Biol 1985; 181:467-478.
12. Fayat G, Mayaux JF, Sacerdot C et al. *Escherichia coli* phenylalanyl-tRNA synthetase operon region. Evidence for an attenuation mechanism. Identification of the gene for the ribosomal protein L20. J Mol Biol 1983; 171:239-261.
13. Springer M, Plumbridge JA, Butler JS et al. Autogenous control of *Escherichia coli* threonyl-tRNA synthetase expression in vivo. J Mol Biol 1985; 185:93-104.
14. Lestienne P, Plumbridge JA, Grunberg-Manago M et al. Autogenous repression of *E. coli* threonyl-tRNA synthetase expression in vitro. J Biol Chem 1984; 259:5232-5237.
15. Moine H, Romby P, Springer M et al. Messenger RNA structure and gene regulation at the translational level in *Escherichia coli*: the case of threonine:tRNA^{Thr} ligase. Proc Natl Acad Sci USA 1988; 85:7892-7896.
16. Nogueira T, de Smit M, Graffe M et al. The Relationship Between Translational Control and mRNA Degradation for the *Escherichia coli* Threonyl-tRNA synthetase Gene. J Mol Biol 2001; 310(4):709-22.
17. Graffe M, Dondon J, Caillet J et al. The specificity of translational control switched using tRNA identity rules. Science 1992; 225:994-996.
18. Romby P, Brunel C, Caillet J et al. Molecular mimicry in translational control of *E. coli* threonyl-tRNA synthetase gene. Competitive inhibition in tRNA aminoacylation and operator-repressor recognition switch using tRNA identity rules. Nucleic Acids Res 1992; 20:5633-5640.
19. Sacerdot C, Caillet J, Graffe M et al. The *Escherichia coli* threonyl-tRNA synthetase gene contains a split ribosomal binding site interrupted by a hairpin structure that is essential for autoregulation. Mol Microbiol 1998; 29(4):1077-90.
20. Romby P, Caillet J, Ebel C et al. The expression of *E. coli* threonyl-tRNA synthetase is regulated at the translational level by symmetrical operator-repressor interactions. EMBO J 1996; 15:5976-5987.
21. Newman EB, Lin R. Leucine-responsive regulatory protein: a global regulator of gene expression in *E. coli*. Annu Rev Microbiol 1995; 49:747-75.
22. Lin R, Ernting B, Hirshfield IN et al. The *lrp* product regulates expression of *lysU* in *E. coli* K12. J Bacteriol 1992; 174:2779-2784.
23. Gazeau M, Delort F, Dessen P et al. *E. coli* leucine-responsive regulatory protein (Lrp) controls lysyl-tRNA synthetase expression. FEBS letters 1992; 300:254-258.
24. VanBogelen RA, Acton MA, Neidhardt FC. Induction of the heat shock regulon does not produce thermotolerance in *E. coli*. Genes and Develop 1987; 1:525-531.
25. Oshima T, Ito K, Kabayama H et al. Regulation of lrp gene expression by H-NS and Lrp proteins in *Escherichia coli*: dominant negative mutations in *lrp*. Mol Gen Genet 1995; 247:521-528.
26. Brevet A, Chen J, Leveque F et al. Comparison of the enzymatic properties of the two *Escherichia coli* lysyl-tRNA synthetase species. J Biol Chem 1995; 270(24):14439-44.
27. Brun YV, Sanfaçon H, Breton R et al. Closely spaced and divergent promoters for an aminoacyl-tRNA synthetase gene and a tRNA operon in *E. coli*. Transcriptional and post-transcriptional regulation of *glutX*, *valU* and *alaW*. J Mol Biol 1990; 214:845-864.
28. Champagne N, Lapointe J. Influence of FIS on the transcription from closely spaced and non- overlapping divergent promoters for an aminoacyl-tRNA synthetase gene (*gltX*) and a tRNA operon (valU) in *Escherichia coli*. Mol Microbiol 1998; 27(6):1141-56.
29. Dardel F, Panvert M, Fayat G. Transcription and regulation of expression of the *E. coli* methionyl-tRNA synthetase gene. Mol Gen Genet 1990; 223:121-133.
30. Petersen L, Downs DM. Mutations in *apbC* (*mrp*) prevent function of the alternative pyrimidine biosynthetic pathway in Salmonella typhimurium. J Bacteriol 1996; 178(19):5676-82.

31. Lapointe J, Duplain L, Proulx M. A single glutamyl-tRNA synthetase aminoacylates tRNA^{Glu} and tRNA^{Gln} in *Bacillus subtilis* and efficiently misacylates *Escherichia coli* tRNA$_1^{Gln}$ in vitro. J Bacteriol 1986; 165(1):88-93.

32. Wilcox M. g-Phosphoryl ester of Glu-tRNA^{Gln} as an intermediate in *Bacillus subtilis* glutaminyl-tRNA synthesis. Cold Spring Harbor Symp Quant Biol 1969; 34:521-528.

33. Curnow AW, Hong K, Yuan R et al. Glu-tRNAGln amidotransferase: a novel heterotrimeric enzyme required for correct decoding of glutamine codons during translation. Proc Natl Acad Sci U S A 1997; 94(22):11819-26.

34. Henkin TM. tRNA-directed transcription antitermination. Mol Microbiol 1994; 13:381-387.

35. Condon C, Grunberg-Manago M, Putzer H. Aminoacyl-tRNA synthetase gene regulation in Bacillus subtilis. Biochimie 1996; 78(6):381-9.

36. Pelchat M, Lapointe J. Aminoacyl-tRNA synthetase genes of *Bacillus subtilis*: organization and regulation. Biochem Cell Biol 1999; 77(4):343-7.

37. Sissler M, Delorme C, Bond J et al. An aminoacyl-tRNA synthetase paralog with a catalytic role in histidine biosynthesis. Proc Natl Acad Sci U S A 1999; 96(16):8985-90.

38. French S. Consequences of replication fork movement through transcription units in vivo. Science 1992; 258:1362-1365.

39. Gagnon Y, Breton R, Putzer H et al. Clustering and co-transcription of the *Bacillus subtilis* genes encoding the aminoacyl-tRNA synthetases specific for glutamate and for cysteine and the first enzyme for cysteine biosynthesis. J Biol Chem 1994; 269:7473-7482.

40. Putzer H, Gendron N, Grunberg-Manago M. Co-ordinate expression of the two threonyl-tRNA synthetase genes in *Bacillus subtilis*: control by transcriptional antitermination involving a conserved regulatory sequence. EMBO J 1992; 11:3117-3127.

41. Henkin TM, Glass BL, Grundy FJ. Analysis of the *Bacillus subtilis tyrS* gene: Conservation of a regulatory sequence in multiple tRNA synthetase genes. J Bacteriol 1992; 174:1299-1306.

42. Grandoni JA, Zahler SA, Calvo JM. Transcriptional regulation of the ilv-leu operon of Bacillus subtilis. J Bacteriol 1992; 174(10):3212-3219.

43. Putzer H, Brackhage AA, Grunberg-Manago M. Independent genes for two threonyl-tRNA synthetases in *Bacillus subtilis*. J Bacteriol 1990; 172(8):4593-4602.

44. Grundy FJ, Henkin TM. Transfer RNA as a positive regulator of transcription antitermination in *B. subtilis*. Cell 1993; 74(3):475-482.

45. Luo D, Condon C, Grunberg-Manago M et al. In vitro and in vivo secondary structure probing of the *thrS* leader in *Bacillus subtilis*. Nucleic Acids Res 1998; 26(23):5379-87.

46. Putzer H, Laalami S, Brakhage AA et al. Aminoacyl-tRNA synthetase gene regulation in *Bacillus subtilis*: induction, repression and growth-rate regulation. Mol Microbiol 1995; 16:709-718.

47. Marta PT, Ladner RD, Grandoni JA. A CUC triplet confers leucine-dependent regulation of the *Bacillus subtilis ilv-leu* operon. J Bact 1996; 178:2150-2153.

48. Grundy FJ, Hodil SE, Rollins SM et al. Specificity of tRNA-mRNA interactions in *Bacillus subtilis tyrS* antitermination. J Bacteriol 1997; 179(8):2587-94.

49. Grundy FJ, Collins JA, Rollins SM et al. tRNA determinants for transcription antitermination of the *Bacillus subtilis tyrS* gene. RNA 2000; 6(8):1131-41.

50. van De Guchte M, Ehrlich SD, Chopin A. Identity elements in tRNA-mediated transcription antitermination: implication of tRNA D- and T-arms in mRNA recognition. Microbiology 2001; 147(Pt 5):1223-33.

51. Rollins SM, Grundy FJ, Henkin TM. Analysis of cis-acting sequence and structural elements required for antitermination of the *Bacillus subtilis tyrS* gene. Mol Microbiol 1997; 25(2):411-21.

51A. Grundy FJ, Moir TR, Haldeman MT et al. Sequence requirements for terminators and antiterminators in the T box transcription antitermination system: disparity between conservation and functional requirements. Nucleic Acids Res 2002; 30(7):1646-55.

51B. Putzer H, Condon C, Brechemier-Baey D et al. Transfer RNA-mediated antitermination in vitro. Nucleic Acids Res 2002; 30(14):3016-3023.

52. Garrity DB, Zahler SA. Mutations in the gene for a tRNA that functions as a regulator of a transcriptional attenuator in *Bacillus subtilis*. Genetics 1994; 137:627-636.

53. van de Guchte M, Ehrlich DS, Chopin A. tRNATrp as a key element of antitermination in the *Lactococcus lactis trp* operon. Mol Microbiol 1998; 29(1):61-74.

54. Gendron N, Putzer H, Grunberg-Manago M. Expression of both *Bacillus subtilis* threonyl-tRNA synthetase genes is autogenously regulated. J Bacteriol 1994; 176:486-494.

55. Vanderhorn PB, Zahler SA. Cloning and nucleotide sequence of the leucyl-transfer RNA synthetase gene of *Bacillus subtilis*. J Bacteriol 1992; 174(12):3928-3935.

56. Luo D, Leautey J, Grunberg-Manago M et al. Structure and regulation of expression of the *Bacillus subtilis* valyl-tRNA synthetase gene. J Bacteriol 1997; 179(8):2472-2478.

57. McDowell JC, Roberts JW, Jin DJ et al. Determination of intrinsic transcription termination efficiency by RNA polymerase elongation rate. Science 1994; 266(5186):822-5.

58. Condon C, Putzer H, Grunberg-Manago M. Processing of the leader mRNA plays a major role in the induction of thrS expression following threonine starvation in *B. subtilis*. Proc Natl Acad Sci USA 1996; 93:6992-6997.

59. Condon C, Putzer H, Luo D et al. Processing of the *Bacillus subtilis thrS* leader mRNA is RNase E-dependent in *Escherichia coli*. J Mol Biol 1997; 268(2):235-42.

60. Sogin ML, Pace B, Pace NR. Partial purification and properties of a ribosomal RNA maturation endonuclease from *Bacillus subtilis*. J Biol Chem 1977; 252(4):1350-7.

61. Pace B, Stahl DA, Pace NR. The catalytic element of a ribosomal RNA-processing complex. J Biol Chem 1984; 259(18):11454-8.
62. Condon C, Brechemier-Baey D, Beltchev B et al. Identification of the gene encoding the 5S ribosomal RNA maturase in *Bacillus subtilis*: Mature 5S rRNA is dispensable for ribosome function. RNA 2001; 7:242-253.
63. Pelchat M, Lapointe J. In vivo and in vitro processing of the *Bacillus subtilis* transcript coding for glutamyl-tRNA synthetase, serine acetyltransferase, and cysteinyl-tRNA synthetase. RNA 1999; 5(2):281-9.
64. Hershey JWB. Protein synthesis. In: Ingraham JL, Low KB, Magasanik B, Schaechter M, Umbarger HE, eds. *Escherichia coli* and *Salmonella typhimurium*. Washington, D.C.: American Society for Microbiology.; 1987: 614-647.
65. Howe JG, Hershey JWB. Initiation factor levels are coordonately controlled in *Escherichia coli* growing at different rates. J Biol Chem 1983; 258:1954-1959.
66. Cummings HS, Hershey JW. Translation initiation factor IF1 is essential for cell viability in *Escherichia coli*. J Bacteriol 1994; 176(1):198-205.
67. Cummings HS, Sands JF, Foreman PC et al. Structure and expression of the *infA* operon encoding translational initiation factor IF1. Transcriptional control by growth rate. J Biol Chem 1991; 266(25):16491-8.
68. Cole JR, Olsson CL, Hershey JW et al. Feedback regulation of rRNA synthesis in *Escherichia coli*. Requirement for initiation factor IF2. J Mol Biol 1987; 198(3):383-92.
69. Laalami S, Putzer H, Plumbridge JA et al. A severely truncated form of translational initiation factor 2 supports growth of *Escherichia coli*. J Mol Biol 1991; 220(2):335-49.
70. Plumbridge JA, Deville F, Sacerdot C et al. Two translational initiation sites in the *infB* gene are used to express initiation factor IF2 alpha and IF2 beta in *Escherichia coli*. EMBO J 1985; 4(1):223-9.
71. Sacerdot C, Vachon G, Laalami S et al. Both forms of translational initiation factor IF2 (alpha and beta) are required for maximal growth of *Escherichia coli*. Evidence for two translational initiation codons for IF2 beta. J Mol Biol 1992; 225(1):67-80.
72. Moreno JM, Drskjotersen L, Kristensen JE et al. Characterization of the domains of *E. coli* initiation factor IF2 responsible for recognition of the ribosome. FEBS Lett 1999; 455(1-2):130-4.
73. Granston AE, Thompson DL, Friedman DI. Identification of a second promoter for the *metY-nusA-infB* operon of *Escherichia coli*. J Bacteriol 1990; 172(5):2336-42.
74. Ishii S, Kuroki K, Imamoto F. tRNA$_{f2}$^Met gene in the leader region of the *nusA* operon in *E.coli*. Proc Natl Acad Sci USA 1984; 81:409-413.
75. Verbeek H, Nilsson L, Baliko G et al. Potential binding sites of the trans-activator FIS are present upstream of all rRNA operons and of many but not all tRNA operons. Biochim Biophys Acta 1990; 1050(1-3):302-6.
76. Régnier P, Grunberg-Manago M. Cleavage by RNAse III in the transcripts of the *metY-nusA-infB* operon of *E. coli* releases the tRNA and initiates decay of the downstream mRNA. J Mol Biol 1989; 210:293-302.
77. Plumbridge JA, Dondon J, Nakamura Y et al. Effect of NusA protein on expression of the nusA,infB operon in *E. coli*. Nucleic Acids Res 1985; 13(9):3371-88.
78. Nakamura Y, Plumbridge J, Dondon J et al. Evidence for autoregulation of the nusA-infB operon of *Escherichia coli*. Gene 1985; 36(1-2):189-93.
79. Jones MD, Lowe DM, Borgford T et al. Natural variation of tyrosyl-tRNA synthetase and comparison with engineered mutants. Biochemistry 1986; 25:1887-1891.
80. Jones PG, Inouye M. The cold-shock response—a hot topic. Mol Microbiol 1994; 11(5):811-8.
81. Bae W, Xia B, Inouye M et al. *Escherichia coli* CspA-family RNA chaperones are transcription antiterminators. Proc Natl Acad Sci U S A 2000; 97(14):7784-9.
82. Olsson CL, Graffe M, Springer M et al. Physiological effects of translation initiation factor IF3 and ribosomal protein L20 limitation in *Escherichia coli*. Mol Gen Genet 1996; 250(6):705-14.
83. Sacerdot C, Fayat G, Dessen P et al. Sequence of a 1.26-kb DNA fragment containing the structural gene for initiation factor IF3: presence of an AUU initiator codon. EMBO J 1982; 1:311-315.
84. Mayaux JF, Fayat G, Fromant M et al. Structural and transcriptional evidence for related *thrS* and *infC* expression. Proc Natl Acad Sci U S A 1983; 80(20):6152-6.
85. Wertheimer SJ, Klotsky R-A, Schwartz I. Transcriptional patterns for the *thrS-infC-rplT* operon of *Escherichia coli*. Gene 1988; 63:309-320.
86. Liveris D, Klotsky RA, Schwartz I. Growth rate regulation of translation initiation factor IF3 biosynthesis in *Escherichia coli*. J Bacteriol 1991; 173(12):3888-93.
87. Butler JS, Springer M, Grunberg-Manago M. AUU to AUG mutation in the initiator codon of the translation initiation factor IF3 abolishes translational autocontrol of its own gene (*infC*) in vivo. Proc Natl Acad Sci USA 1987; 84:4022-4025.
88. Sacerdot C, Chiaruttini C, Engst K et al. The role of the AUU initiation codon in the negative feedback regulation of the gene for translation initiation factor IF3 in *Escherichia coli*. Mol Microbiol 1996; 21(2):331-46.
89. Lesage P, Chiaruttini C, Graffe M et al. Messenger RNA secondary structure and translational coupling in the *Escherichia coli* operon encoding translation initiation factor IF3 and the ribosomal proteins, L35 and L20. J Mol Biol 1992; 228(2):366-86.
90. Arai K, Clark BF, Duffy L et al. Primary structure of elongation factor Tu from *Escherichia coli*. Proc Natl Acad Sci U S A 1980; 77(3):1326-30.
91. Jones MD, Petersen TE, Nielsen KM et al. The complete amino-acid sequence of elongation factor Tu from *Escherichia coli*. Eur J Biochem 1980; 108(2):507-26.
92. Gouy M, Grantham R. Polypeptide elongation and tRNA cycling in *Escherichia coli*: a dynamic approach. FEBS Lett 1980; 115(2):151-5.

93. Louie A, Ribeiro NS, Reid BR et al. Relative affinities of all *Escherichia coli* aminoacyl-tRNAs for elongation factor Tu-GTP. J Biol Chem 1984; 259(8):5010-6.

94. Zuurmond AM, Rundlof AK, Kraal B. Either of the chromosomal tuf genes of *E. coli* K-12 can be deleted without loss of cell viability. Mol Gen Genet 1999; 260(6):603-7.

95. van der Meide PH, Vijgenboom E, Talens A et al. The role of EF-Tu in the expression of *tufA* and *tufB* genes. Eur J Biochem 1983; 130(2):397-407.

96. Zengel JM, Lindahl L. Mapping of two promoters for elongation factor Tu within the structural gene for elongation factor G. Biochim Biophys Acta 1990; 1050(1-3):317-22.

97. Van Delft JH, Talens A, De Jong PJ et al. Control of the tRNA-tufB operon in *Escherichia coli*. 2. Mechanisms of the feedback inhibition of tufB expression studied in vivo and in vitro. Eur J Biochem 1988; 175(2):363-74.

98. Van Delft JH, Schmidt DS, Bosch L. The tRNA-*tufB* operon transcription termination and processing upstream from *tufB*. J Mol Biol 1987; 197(4):647-57.

99. Travers A, Schneider R, Muskhelishvili G. DNA supercoiling and transcription in *Escherichia coli*: The FIS connection. Biochimie 2001; 83(2):213-7.

100. Nilsson L, Vanet A, Vijgenboom E et al. The role of FIS in trans activation of stable RNA operons of *E. coli*. EMBO J 1990; 9(3):727-34.

101. Bock A. Biosynthesis of selenoproteins—an overview. Biofactors 2000; 11(1-2):77-8.

102. Leinfelder W, Stadtman TC, Bock A. Occurrence in vivo of selenocysteyl-tRNA(SERUCA) in *Escherichia coli*. Effect of sel mutations. J Biol Chem 1989; 264(17):9720-3.

103. Sawers G, Heider J, Zehelein E et al. Expression and operon structure of the sel genes of *Escherichia coli* and identification of a third selenium-containing formate dehydrogenase isoenzyme. J Bacteriol 1991; 173(16):4983-93.

104. Saito K, Nomura M. Post-transcriptional regulation of the *str* operon in *Escherichia coli*. Structural and mutational analysis of the target site for translational repressor S7. J Mol Biol 1994; 235(1):125-39.

105. Spiridonova VA, Rozhdestvensky TS, Kopylov AM. A study of the thermophilic ribosomal protein S7 binding to the truncated S12-S7 intercistronic region provides more insight into the mechanism of regulation of the str operon of *E. coli*(1). FEBS Lett 1999; 460(2):353-6.

106. Saito K, Mattheakis LC, Nomura M. Post-transcriptional regulation of the str operon in *Escherichia coli*. Ribosomal protein S7 inhibits coupled translation of S7 but not its independent translation. J Mol Biol 1994; 235(1):111-24.

107. An G, Lee JS, Friesen JD. Evidence for an internal promoter preceding tufA in the *str* operon of *Escherichia coli*. J Bacteriol 1982; 149(2):548-53.

108. Miyajima A, Kaziro Y. Coordination of levels of elongation factors Tu, Ts, and G, and ribosomal protein SI in *Escherichia coli*. J Biochem (Tokyo) 1978; 83(2):453-62.

109. Aoki H, Adams SL, Chung DG et al. Cloning, sequencing and overexpression of the gene for prokaryotic factor EF-P involved in peptide bond synthesis. Nucleic Acids Res 1991; 19(22):6215-20.

110. Aoki H, Dekany K, Adams SL et al. The gene encoding the elongation factor P protein is essential for viability and is required for protein synthesis. J Biol Chem 1997; 272(51):32254-9.

111. An G, Glick BR, Friesen JD et al. Identification and quantitation of elongation factor EF-P in Escherichia coli cell-free extracts. Can J Biochem 1980; 58(11):1312-4.

112. Grentzmann G, Brechemier-Baey D, Heurgue V et al. Localization and characterization of the gene encoding release factor RF3 in *Escherichia coli*. Proc Natl Acad Sci U S A 1994; 91(13):5848-52.

113. Mikuni O, Ito K, Moffat J et al. Identification of the *prfC* gene, which encodes peptide-chain-release factor 3 of Escherichia coli. Proc Natl Acad Sci U S A 1994; 91(13):5798-802.

114. Ito K, Uno M, Nakamura Y. Single amino acid substitution in prokaryote polypeptide release factor 2 permits it to terminate translation at all three stop codons. Proc Natl Acad Sci USA 1998; 95(14):8165-9.

115. Holst-Hansen P, Kildsgaard J, MacDougall J et al. Immunochemical determination of the cellular content of polypeptide chain release factor RF3 in *Escherichia coli*. Biochimie 1997; 79(12):725-9.

116. Fraser CM, Gocayne JD, White O et al. The minimal gene complement of Mycoplasma genitalium. Science 1995; 270(5235):397-403.

117. Craigen WJ, Cook RG, Tate WP et al. Bacterial peptide chain release factors : conserved primary structure and possible frameshift regulation of release factor 2. Proc Natl Acad Sci, USA 1985; 82:3616-3620.

118. Kawakami K, Jönsson YH, Björk GR et al. Chromosomal location and structure of the operon encoding peptide-chain-release factor 2 of *E. coli*. Proc Natl Acad Sci USA 1988; 85:5620-5624.

119. Weiss R, Gallant J. Mechanism of ribosome frameshifting during translation of the genetic code. Nature 1983; 302(5907):389-93.

120. Craigen WJ, Caskey CT. Expression of peptide chain release factor 2 requires high-efficiency frameshift. Nature 1986; 322(6076):273-5.

121. Kawakami K, Nakamura Y. Autogenous suppression of an opal mutation in the gene encoding peptide chain release factor 2. Proc Natl Acad Sci USA 1990; 87(21):8432-6.

122. Janosi L, Ricker R, Kaji A. Dual functions of ribosome recycling factor in protein biosynthesis: disassembling the termination complex and preventing translational errors. Biochimie 1996; 78(11-12):959-69.

123. Kaji A, Hirokawa G. Ribosome-recycling factor: an essential factor for protein synthesis. In: Garrett RA, Douthwaite S, Liljas A, Matheson AT, Moore PB, Noller HF, eds. The Ribosome: Structure, function, antibiotics, and cellular interactions. Washington, DC: ASM Press; 2000: 527-539.

124. Shimizu I, Kaji A. Identification of the promoter region of the ribosome-releasing factor cistron (frr). J Bacteriol 1991; 173(16):5181-7.

125. International Union of Biochemistry and Molecular Biology (IUBMB). Prokaryotic and eukaryotic translation factors. Biochimie 1996; 78:1119-1122.

Inhibitors of Aminoacyl-tRNA Synthetases as Antibiotics and Tools for Structural and Mechanistic Studies

Robert Chênevert, Stéphane Bernier and Jacques Lapointe

Abstract

Aminoacyl-tRNA synthetases (aaRS) catalyze the esterification of a particular tRNA with its corresponding amino acid. In the first reaction step, the appropriate amino acid is recognized by the enzyme and reacts with ATP to form an enzyme-bound mixed anhydride; in the second step, this activated amino acid is esterified with one of the two hydroxyl groups of the tRNA. AaRSs are classified into two main groups of ten enzymes each, on the basis of common structural and functional features.

The design of aaRS inhibitors has three main objectives: 1° to facilitate the crystallization and X-ray structure determination of these enzymes, 2° to gain mechanistic informations about them, and 3° to discover new antibiotics. Several natural products including pseudomonic acid, SB-203207, SB-219383, indolmycin, capsaicin and ascamycin are selective inhibitors of aaRSs. Pseudomonic acid is a potent inhibitor of bacterial IleRS and is the sole aaRS inhibitor currently marketed as an antibacterial agent. Synthetic inhibitors are usually stable analogues of the mixed anhydride intermediate. The stability is achieved by replacement of the labile anhydride function by nonhydrolyzable bioisosteres. Several aminoalkyl adenylates (replacement of the anhydride by a phosphate ester) and aminoacylsulfamoyl adenosines (replacement of the phosphate by a sulfamoyl group) have been synthesized and shown to be potent inhibitors of aaRSs.

Abbreviations

aaRS	aminoacyl-tRNA synthetase
AlaRS	alanyl-tRNA synthetase
aa-AMP	aminoacyl adenylate
aa-AMS	aminoacylsulfamoyl adenosine

Introduction

In the first stage of protein biosynthesis, the standard amino acids (aa) are esterified to their corresponding transfer ribonucleic acid (tRNA) by aminoacyl-tRNA synthetases (aaRS). The overall reaction is a two-step event[1-4] (Fig. 1).

The first step is the activation of the α-carboxyl group of the amino acid. A carboxylic acid does not react with an alcohol to form an ester unless a strong acid is used as a catalyst. The formation of an acid chloride or an anhydride is an alternative way to activate the carboxyl group. In the cells, activation compatible with biochemical materials results from the reaction of the amino acid with adenosine triphosphate (ATP) to form a mixed anhydride (carboxylic-phosphoric). In this intermediate, the high-energy anhydride bond activates the carboxyl group of the amino acid. The role of the Mg^{2+} ion is to stabilize the conformation of ATP and withdraw electrons from the β-phosphate, thus facilitating its displacement by the amino acid.

Translation Mechanisms, edited by Jacques Lapointe and Léa Brakier-Gingras.
©2003 Eurekah.com and Kluwer Academic / Plenum Publishers.

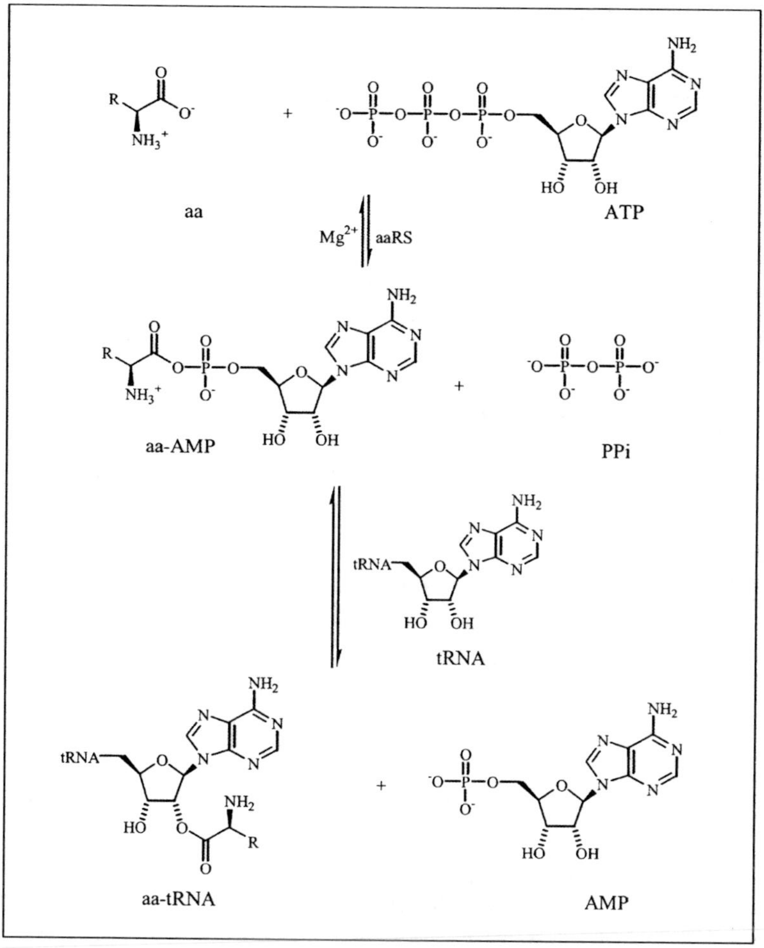

Figure 1. The activation and transfer steps of tRNA aminoacylation.

In the second step, the activated amino acid is transferred to the CCA end of the corresponding tRNA to form aminoacyl-tRNA (aa-tRNA) and adenosine monophosphate (AMP). This transfer (esterification) is a nucleophilic attack by the 2' or 3' ribose hydroxyl group of the terminal AMP residue at the 3' end of the tRNA on the activated carboxyl group of the mixed anhydride intermediate. Both reaction steps are readily reversible and the overall reaction is driven to completion by the pyrophosphatase-catalyzed hydrolysis of PPi generated in the first step.

Aminoacyl-tRNA synthetases are classified into two main groups (classes I and II) of ten enzymes each, on the basis of common structural features[5-7] (for a review, see Chapter 2 by M. Ibba and D. Söll). Class I enzymes share two sequence motifs called HIGH and KMSKS (one letter amino acid nomenclature) forming part of the ATP-binding site. Their active sites are based on a parallel β-sheet nucleotide-binding fold called Rossmann fold. On the functional level, the main common features are: 1° ATP is in an extended conformation, 2° the aminoacylation site is the 2'-OH of the terminal adenosine of the tRNA and then the aminoacyl group moves to the 3'-OH by an intramolecular transesterification, 3° the activated carbonyl group of the aminoacyl adenylate (aa-AMP) intermediate is exposed to the solvent and has a minimal interaction with the protein.

Class II enzymes share conserved sequence motifs called motif 1 (in a long α-helix linked to a β-strand), motif 2 (in two antiparallel β-strands connected by a long loop) and motif 3 (in a β-strand followed immediately by an α helix). Motifs 2 and 3 are part of the active site built on an antiparallel β-sheet surrounded by α-helices. The common functional characteristics are: 1° ATP is in a bent conformation, 2° the aminoacylation site is the 3'-OH (exception: PheRS), 3° the activated carbonyl group of aa-AMP is involved into a network of hydrogen bonds with the protein and the adjacent amino group.

The design of aaRS inhibitors has three main objectives: 1° to facilitate the crystallization and X-ray structure determination of these enzymes, 2° to gain mechanistic informations about them, and 3° to discover new antibiotics.

The dissemination of antibiotic resistance has become a major problem in clinical medicine and there is a critical need to develop anti-bacterial agents with novel modes of action. Aminoacyl-tRNA synthetases have been subjected to significant evolutionary divergence and selective inhibition of bacterial enzymes has been observed.[8-10]

Natural Products and Analogues

Inhibitors of IleRS, LeuRS or ValRS

Pseudomonic acid **1**, isolated from *Pseudomonas fluorescens*, is a highly potent and selective inhibitor of bacterial IleRS (K_i^{Ile} = 6 nM for *E. coli* IleRS). This compound inhibits bacterial IleRS about 10^4-fold more than the corresponding mammalian enzymes.[11] This natural product is the sole aaRS inhibitor currently marketed as an antibacterial agent (generic name: mupirocin). This drug is widely used as a topical antibiotic and is marketed by Glaxo SmithKline under the tradename of Bactroban®. Mupirocin is only used for topical applications because of its rapid hydrolysis in blood and tissues.

Synthesis of stable analogues of mupirocin is an important area of research and several structural modifications have been reported.[12,13] Analogue **2** is an exquisitely potent inhibitor (K_i < 0.001 nM) of bacterial IleRS. This synthetic compound was designed rationally from the binding model for pseudomonic acid together with a detailed understanding of the reaction cycle of IleRS and the characterization of the mode of binding of the reaction intermediate isoleucyl adenylate.[14]

A part of this compound **2** is identical to the side chain of isoleucine. It is noteworthy that several naturally occurring analogues or derivatives of standard amino acids are specific inhibitors of the corresponding aminoacyl-tRNA synthetases.

SB-203207 (**3**) was recently isolated from a *Streptomyces* strain and shown to inhibit bacterial IleRS (IC_{50} = 1.7 nM) and mammalian IleRS (IC_{50} < 2 nM).[15,16] In SB-203207, isoleucine is bonded to altemicidin (**4**), another natural product, via a sulfamoyl linkage. Substitution of isoleucine residue of SB-203207 with leucine and valine produced potent inhibitors (**5** and **6**) of LeuRS (IC_{50} = 16 nM) and ValRS (IC_{50} = 30 nM) respectively.[17] These compounds displayed low level antibacterial activity against any of the microorganisms tested.

Granaticin[18] and furanomycin[19] inhibit respectively LeuRS and IleRS.

TrpRS Inhibitors

Indolmycin (**7**), produced by *Streptomyces griseus*,[20,21] inhibits the growth of both gram-negative and gram-positive microorganisms by selective and competitive inhibition of TrpRS (K_i^{Trp} = 8 to 9 μM with *Escherichia coli* TrpRS) whereas rat liver TrpRS is only slightly affected.[21] Again, the inhibition implies a structural resemblance between this natural product and the enzyme's substrate, tryptophan. Structural modifications did not improve the antibacterial and enzyme inhibitory potency.[22,23]

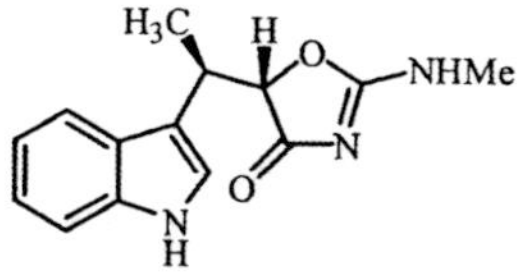

7 (indolmycin)

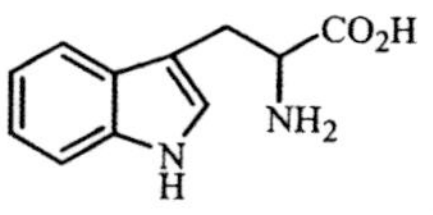

tryptophan

PheRS and ValRS Inhibitors

The mycotoxin ochratoxin A (**8**), an inhibitor of PheRS, contains a phenylalanine moiety linked by its amino group to a chlorinated dihydroisocoumarin acid.[24,25] When phenylalanine is replaced by valine, the resulting val-ochratoxin compound is a specific inhibitor of ValRS indicating that the specificity is due to the amino acid and not to the dihydroisocoumarin moiety.[26] Ochratoxin A inhibits the growth of gram-positive bacteria but no inhibition was found with gram-negative species. This mycotoxin, produced by ubiquitous strains of *Aspergillus* and *Penicillium*, is highly toxic to mammals.

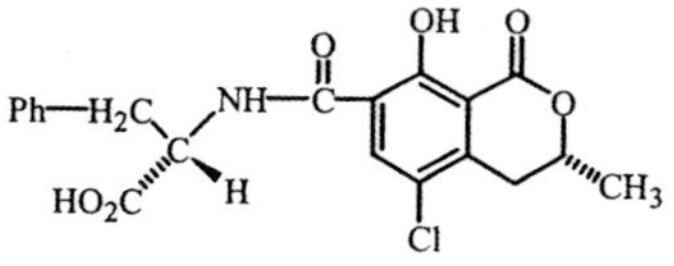

8 (ochratoxin)

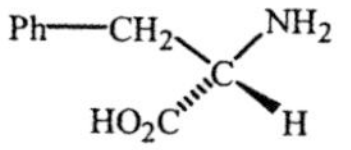

phenylalanine

TyrRS Inhibitors

Capsaicin (**9**) is the pungent ingredient found in the fruit of the genus *Capsicum* (paprika, cayenne). This natural structural analogue of tyrosine is a competitive inhibitor (K_i = 42 μM) of mammalian TyrRS.[27-28]

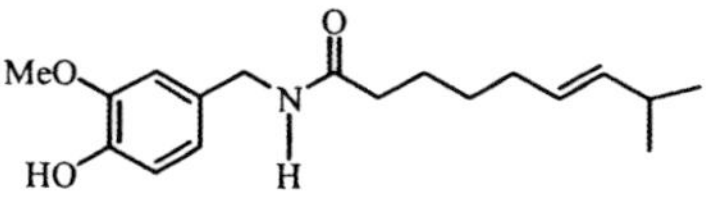

9 (capsaicin)

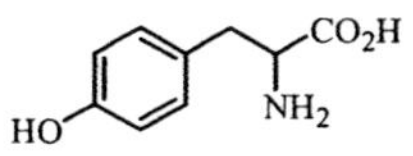

tyrosine

SB-219383 (**10**) isolated from *Micromonospora* sp., is a dipeptide with an N-terminal L-tyrosine coupled with a complex α-amino acid bearing a C-glycoside side chain.[29-30] This natural tyrosine derivative was found to be a potent inhibitor of bacterial TyrRS (IC_{50} = 1.2 nM) with a good selectivity over the mammalian enzyme (IC_{50} = 22 μM). Several synthetic analogues exhibit nanomolar levels of inhibition of TyrRS.[31-34] These compounds, however, show only weak antibacterial activity.

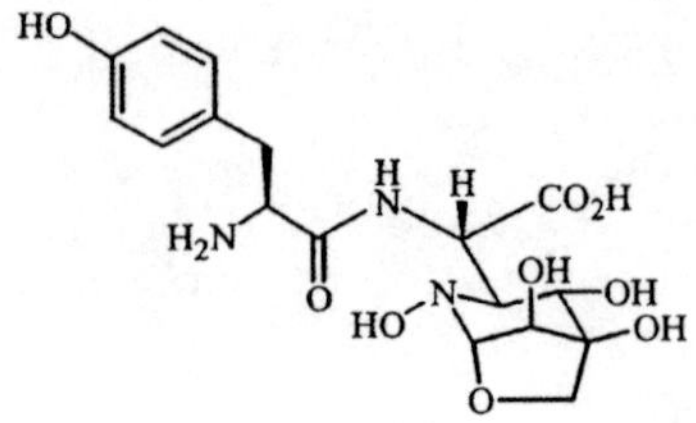

10 (SB-219383)

AlaRS Inhibitor

Ascamycin (**11**) is a nucleoside antibiotic isolated from a *Streptomyces* and an inhibitor of AlaRS. This natural product is alanine linked by a sulfamoyl bridge to adenosine bearing a chlorine atom at position 2 of the purine base.[35]

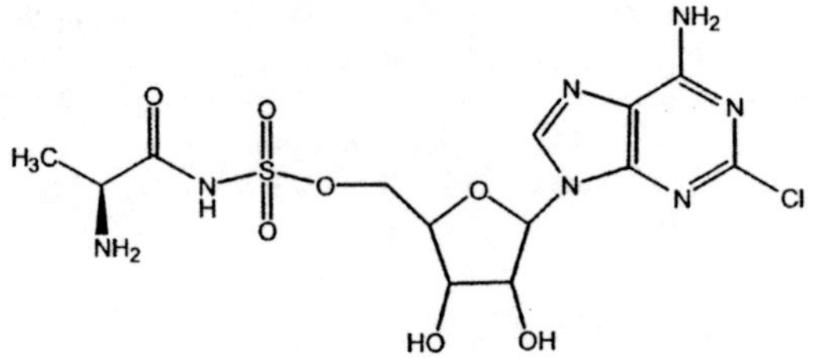

11 (ascamycin)

ThrRS Inhibitor

Borrelidin is a macrolide-type antibiotic produced by a variety of *Streptomyces* species. Its antibiotic effect is caused by the specific inhibition of ThrRS.[36-38] It has a certain structural similarity to AMP carrying a threonyl residue in 3' position. The analysis of borrelidin-resistant mutants of *Saccharomyces cerevisiae*, *E. coli* and Chinese hamster ovary cells revealed either constitutively increased levels of wild-type ThrRS, or structurally altered ThrRS with lowered K_m value for threonine and a higher K_i value for borrelidin. There was no therapeutic use of borrelidin so far because of its high toxicity.[39]

HisRS Inhibitor

Histidinol is an intermediate in the biosynthesis of histidine. It is a competitive inhibitor of *E. coli* HisRS, with a K_i = 35 μM.[40] A comparison of the structures of the HisRS:histidyl-adenylate and ATP:HisRS:histidinol complexes provided information at the atomic level on the mechanism of histidine activation by this enzyme.[41] Histidinol-resistant Chinese hamster ovary cells isolated by stepwise adaptation to this protein biosynthesis inhibitor had up to a 30-fold increase in HisRS activity and increased levels of translatable mRNA encoding this enzyme.[42]

ProRS Inhibitors

The proline analogue cispentacin[43] produced by *Bacillus cereus* and *Streptomyces setonii*, is a weak inhibitor of *Candida albicans* ProRS and has modest in vitro antifungal activity, but it protects effectively mice against systemic *C. albicans* infections.

Inhibitor of All Aminoacylation Reactions

Purpuromycin, an antibiotic produced by *Actinoplanes ianthinogenes*, binds fairly tightly to all tRNAs, inhibiting specifically their acceptor activity; none of the tested functions of aa-tRNA are affected.[44]

Synthetic Inhibitors

Potent synthetic inhibitors are usually stable analogues of aa-AMP, the mixed anhydride intermediate shown on Fig. 1. Aminoacyl adenylate (aa-AMP) represents the product of the first step and the substrate for the second step. This intermediate is a high energy molecule, very sensitive to nucleophiles including water. It is more tightly bound to the enzymes than substrates (amino acid and ATP) by two or three orders of magnitude, and stabilized analogues have the potential to be tight binding inhibitors. The stability is achieved by replacement of the labile anhydride function of aa-AMP by nonhydrolyzable bioisosteres. Enzymes are highly stereoselective and L-amino acid derivatives (the natural configuration) are much more potent inhibitors than D-amino acid derivatives.

Aminoalkyl Adenylates (aa-ol-AMP)

The replacement of the anhydride functionality by a phosphate ester was first proposed by Cassio et al[45] in 1967. Following their pioneering work, several aminoalkyl adenylates (general structure **12**, R represents the side chains of standard amino acids) have been described as inhibitors of the corresponding aaRS[46-52] and ligands for affinity chromatography[53] or crystal structure determination.[54-55] Met-ol-AMP is bacteriostatic for *E. coli*.[56]

12   aaAMP

Representative inhibition studies are shown in Table 1. Aminoalkyl adenylates are potent inhibitors of class I enzymes but seem to have less affinity for class II enzymes.[46] Aminoalkyl adenylates lack the carbonyl group (C=O) found in aa-AMP. In class I enzymes, this carbonyl group has a minimal interaction with the protein whereas in class II complexes, it plays an important role and is involved in a network of hydrogen bonds to the protein.

Aminoacylsulfamoyl Adenosine (aa-AMS)

The replacement of the phosphate by a sulfamoyl group provides aminoacylsulfamoyl adenosines (**13**), stable non hydrolyzable analogues of aminoacyl-AMP. These analogues were modelled after the natural product ascamycin **11**. The sulfamoyl linkage is also found in synthetic compound **2**.

13   aaAMP

The molecular dimensions of the sulfamoyl group are nearly the same as those of the phosphate group. Also, the sulfamoyl can exist in solution and in solid state in the anionic form due to acidity of the NH function; the negative charge is delocalized over several atoms and the anion is a good mimic of the phosphate anion.

Several aminoacylsulfamoyl adenosine derivatives (also called aminoacylsulfamates) have been synthesized and shown to be potent inhibitors of both class I and class II aaRSs (Table 2). They

Table 1.   *Inhibition of aminoacyl-tRNA synthetases by the corresponding aminoalkyl adenylates*

Class	Source[1]	Inhibition[2] by aa-ol-AMP	References
Class I			
Subclass Ia			
ArgRS	*S. aureus*	IC_{50} = 7.5 nM	46
CysRS			
IleRS	*S. aureus*	IC_{50} = 780 nM (comp)	47
IleRS	*E. coli*	K_i^{Ile} = 7.4 nM (comp) (exch)	45
LeuRS			
MetRS	*E. coli*	K_i^{Met} = 8.6 nM (comp) (exch)	45
MetRS	*E. coli*	K_i^{Met} = 4.7 nM (comp) (aacyl)	45
ValRS	*E. coli*	K_i^{Val} = 29 nM (comp) (exch)	45
LysRS I			
Subclass Ib			
GlnRS	*E. coli*	K_i^{Gln} = 280 nM (comp) (aacyl)	50
GlnRS	*E. coli*	K_i^{ATP} = 860 nM (comp) (aacyl)	50
GluRS	*E. coli*	K_i^{Glu} = 3 µM (comp) (aacyl)	49
Subclass Ic			
TrpRS			
TyrRS	*S. aureus*	IC_{50} = 11 nM	48
TyrRS	*E. coli*	K_i^{Tyr} = 29 nM (comp) (exch)	45
Class II			
Subclass IIa			
GlyRS			
HisRS	*S. aureus*	20% at 300 µM	46
ProRS			
ThrRS	*S. aureus*	60% at 300 µM	46
SerRS			
Subclass IIB			
AsnRS			
AspRS			
LysRS II			
Subclass IIc			
AlaRS	*E. coli*	K_i^{Ala} = 9.1 µM (comp) (exch)	45
GlyRS			
PheRS	*E. coli*	$K_i^{Phe,\ ATP}$ = 1 to 2 µM (comp) (aacyl) (exch)	51
PheRS	*E. coli*	K_i^{Phe} = 2.5 µM (comp) (exch)	45

1. *S. aureus* = *Staphylococcus aureus*; *E. coli* = *Escherichia coli*.
2. comp = competitive inhibition; aacyl = measured in the aminoacylation reaction; exch = measured in the ATP=PPi exchange reaction.

retain the carbonyl of the amino acid moiety which plays an important role in the interaction with the class II enzymes. The crystal structures of several aaRS complexed with the corresponding aminoacylsulfamates and a cognate tRNA (Table 2) provided insights into the catalytic mechanisms of several aaRSs: binding of Lys-AMS induces significant conformational changes in the vicinity of the active site of *T. thermophilus* LysRS:tRNA complex,[58] whereas the ternary complex of Ser-AMS with *Thermus thermophilus* SerRS and tRNA^{Ser} revealed a mechanism which may promote an ordered passage through the activation and transfer steps.[59] The comparison of *E. coli* LeuRS with its complexes with leucine and Leu-AMS revealed conformational changes in the active site required for amino acid activation and tight binding of the adenylate.[60]

Table 2. **Aminoacylsulfamoyl adenosines:aaRS interaction revealed by inhibition or structural studies**

Class	Source[1]	Inhibition[2] and/or Structure	References
Class I			
Subclass Ia			
ArgRS	*S. aureus*	IC_{50} = 4.5 nM	46
CysRS			
IleRS	*S. aureus*	IC_{50} = 4 nM	47
LeuRS	*E. coli*	Structure	60
MetRS			
ValRS	*T. thermophilus*	Structure	61
LysRS I			
Subclass Ib			
GlnRS	*E. coli*	K_i^{Gln} = 1.32 µM (comp) (aacyl)	64
GluRS			
Subclass Ic			
TrpRS			
TyrRS	*S. aureus*	IC_{50} = 26 nM	48
Class II			
Subclass IIa			
GlyRS			
HisRS	*S. aureus*	IC_{50} = 130 nM	46
	E. coli	tRNA binding	66
ProRS	*E. coli*	K_i^{Pro} = 4.3 nM (NC) (exch)	65
	human	K_i^{ATP} = 1.7 nM (M) (exch)	65
ThrRS	*S. aureus*	IC_{50} = 15 nM	46
SerRS	*E. coli*	K_i^{Ser} in the nM range	62
Subclass IIb			
AsnRS	*T. thermophilus*	Structure	63
AspRS			
LysRS II	*T. thermophilus*	Structure	58
Subclass IIc			
AlaRS		Structure of Ala-AMS	57
GlyRS			
PheRS			

1. *S. aureus = Staphylococcus aureus; E. coli = Escherichia coli; T. thermophilus = Thermus thermophilus.*
2. *See footnote of Table 1; NC = noncompetitive inhibition; M = mixed type of inhibition.*

The mechanism of cognate amino acid selection by some aaRSs was better understood from the structure of aa-AMS bound to the corresponding aaRS or aaRS:tRNA complex: that of *T. thermophilus* tRNA^Val:ValRS:Val-AMS revealed the structural basis for double-sieve discrimination of L-valine from L-isoleucine and L-threonine by ValRS;[61] the quality of the electron density map of the *T. thermophilus* AsnRS:Asn-AMS complex was much better than that of the AsnRS:Asn-AMP complex; this allowed a clearer definition of the interactions between the active site residues and the intermediate complex, and revealed differences with AspRS which explain how these similar class IIb aaRSs discriminate their respective and very similar amino acid substrates.[63] A similar reward came from the structure of Gln-AMS cocrystallized with *E. coli* GlnRS and tRNA$_2$^Gln which showed for the first time the interactions between glutamine and its binding site, and revealed how this enzyme discriminates against glutamate.[64]

Other Synthetic Inhibitors

So far, we have considered analogues of aminoacyl adenylates. Other aaRS inhibitors are designed by introducing chemical modifications in the substrates: amino acid, ATP, and tRNA.

Hydroxamate derivatives of several amino acids (aa-Hdx) inhibit the corresponding aaRS, but only a few of them inhibit bacterial growth. L-Met-Hdx (**14**) inhibits *E. coli* MetRS in a

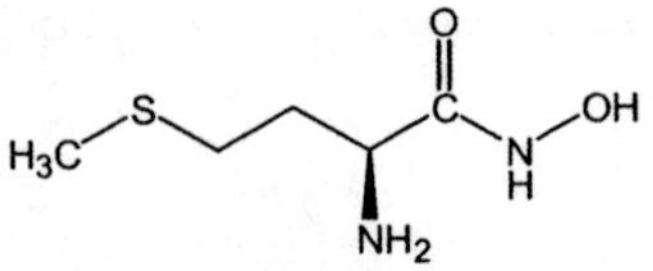

competitive manner (K_i^{Met} = 20 μM)[67] and is active against some gram-negative bacteria. L-Ser-Hdx is a competitive inhibitor of *E. coli* SerRS (K_i^{Ser} = 30 μM) and is bacteriostatic for *E. coli*;[68,69] some resistant mutants have SerRSs with increased K_i values. On the other hand, hydroxamate derivatives of Ala, Gly, Leu, Thr and Trp did not affect the growth of *E. coli* at the concentrations used (0.6 mM) and L-Lys-Hdx had only a small effect[68,69] in spite of the fact that Leu-Hdx, Thr-Hdx and Gly-Hdx inhibited the corresponding aaRSs; competition with the intracellular amino acid pool might explain their lack of bacteriocidal effect. Lys-Hdx is a strong inhibitor of *Bacillus stearothermophilus* LysRS in the L-lysine-dependent ATP-PPi exchange reaction (K_i = 0.6 μM);[70] it produced a stable complex with the enzyme and ATP, LysRS:Lys-Hdx-AMP probably being formed with a P-O-N(H)- linkage between lysine and AMP[71], as found for the Ser-Hdx-AMP complex with *T. thermophilus* SerRS.[62] It is not yet known if the inhibition of other aaRSs by the corresponding aa-Hdx is due to the formation of tightly bound aa-Hdx-AMP.

Thiosine (also called thialysine or thiolysine), an analogue of lysine where a sulfur atom replaces the gamma carbon, is the only lysine analogue reported to inhibit cell growth.[72]

Geminal biphosphonates (**15**) are nonhydrolyzable analogues of pyrophosphate (PPi, Fig. 1) and are important therapeutic drugs for the treatment of bone disorders such as osteoporosis. The effects of biphosphonates towards bone-resorbing osteoclasts may be due to interaction with aminoacyl-tRNA synthetases.[73]

2'-O-Methyl oligonucleotides complementary to the major nucleotide determinants for aminoacylation of *Escherichia coli* tRNA^{Cys} inhibits aminoacylation of the tRNA by CysRS.[74] The K_d of the interaction between a 10-mer and tRNA^{Cys} is several orders of magnitude lower than that of the CysRS:tRNA^{Cys} complex, such that a stoichiometric amount of this 10-mer is sufficient to completely inhibit tRNA aminoacylation.

Ester and hydroxamate analogues of methionyl and isoleucyl adenylate (**16**) are micromolar inhibitors of the corresponding aaRSs. The level of inhibition did not change significantly when adenine was replaced by surrogate heterocycles bearing carbon linkers to the ribose.[75,76] The decrease of affinity (nanomolar to micromolar range) displayed by compounds **16** compared to aminoalkyl or aminoacylsulfamoyl adenylates (**12** and **13**) can be explained by the absence of the phosphate or a mimic of this bridge.

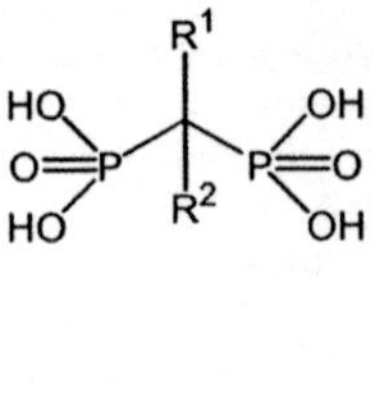

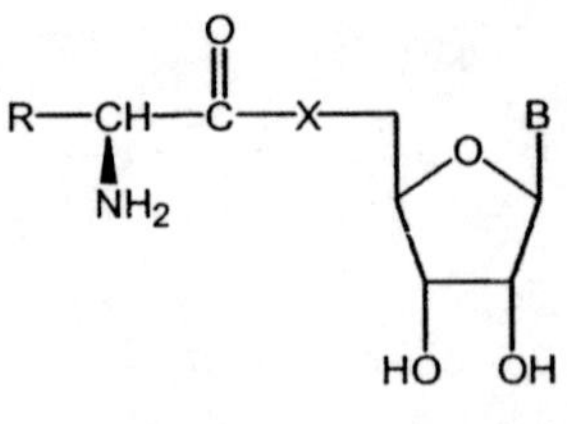

X = O, NH, N-OH, NH-O
B = adenine or surrogates

The transition state theory suggests that enzymes can enhance the rate of a reaction only to the extent that they bind the substrate more tightly in the transition state than in the ground state. This theory has led to a rational basis for drug design based on the understanding of reaction mechanisms.[77-79] Therefore, potent inhibitors would be stable compounds whose structures resemble that of the substrates at the transition state (or transient intermediates) of the reaction. Many transition state analogues inhibitors for various enzymatic reactions have been reported.

Phosphonic anhydrides (**17**) are competitive and specific inhibitors ($K_i \sim 0.1$ to 0.5 μM) of aaRSs.[80,81] The phosphonic group is an analogue of the transition state (**18**) leading to the tetrahedral intermediate of the addition at the activated carboxyl of aminoacyl adenylates (Fig. 1, step 2).

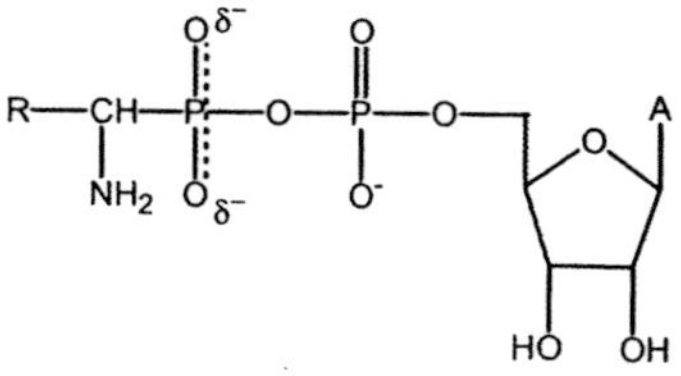

17 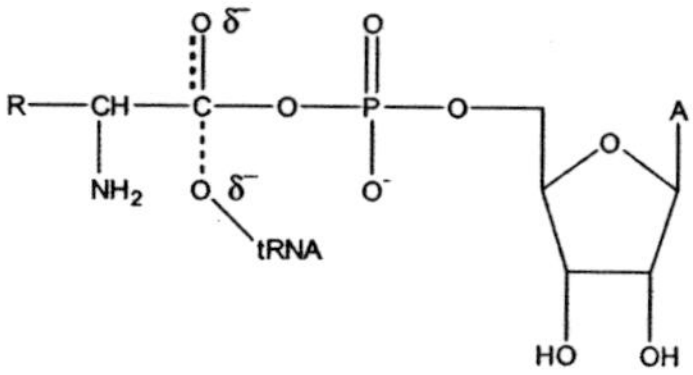								**18**

Conclusion and Perspectives

The structural diversity of the microbial secondary metabolites makes up a valuable source of aaRS inhibitors. Also, natural products can be used as lead compounds for chemical synthesis. The recent finding of natural products such as SB-203207 (**3**) and SB-219383 (**10**) and the design of bioactive analogues are good examples of this approach.

So far, many inhibitors were designed to serve as probes for mechanistic studies or ligands for single crystal X-ray structure determination. These inhibitors were complex and polar molecules and they exhibited little growth inhibition of microorganisms. Recently, aminoacyl adenylate mimics, in which the phosphate was replaced by a sulfamate and the adenine by various heterocycles, were synthesized during a program to develop novel antibiotics.[82,83]

Combinatorial chemistry, linked to automated high-throughput screening, has become an important part of drug discovery. Combinatorial synthesis is a set of techniques for generating large libraries of organic compounds (molecular diversity) and rapidly identifying the bioactive ones. The availability of many aaRSs for high-throughput screening provides an opportunity for developing new aaRS inhibitors.[84,85] For instance, the quinolinecarboxylic acid **19** was identified from high-throughput screening of a large library.[84] Compound **19** inhibits *Candida albicans* ProRS (IC_{50} = 0.5 μM). A structure-activity relationship study[84] revealed that analogue **20** is a more potent and selective inhibitor of ProRS (*C. albicans*, IC_{50} = 5 nM; human IC_{50} > 20 μM).

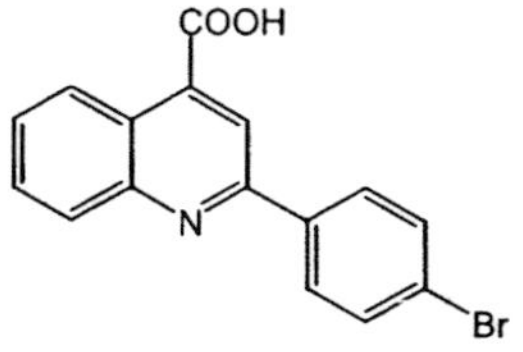

19 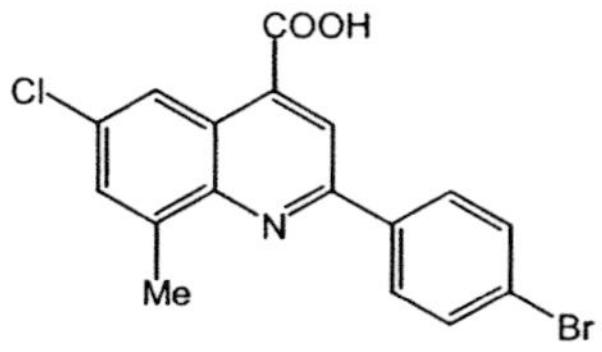								**20**

Finally, many bacteria use a different pathway to form glutamine-tRNA^{Gln} and asparaginyl-tRNA^{Asn}. In the initial step of the so-called transamidation pathway, a nondiscriminatory GluRS aminoacylates tRNA^{Gln} with glutamate or a nondiscriminatory AspRS aminoacylates tRNA^{Asn} with aspartate; in the final step, the incorrectly aminoacylated tRNAs are transformed respectively into Gln-tRNA^{Gln} and

Asn-tRNA^{Asn} by specific amidotransferases[86-88] which transform the side chain carboxyl function of Glu or Asp fixed on tRNA into an amide group (**21** to **22**) (see also Chapter 4 by S. Blanquet et al). The widespread use of the transamidation pathway among bacteria[89] and its absence in the mammalian cytoplasm, identifies these amidotransferases as targets for antibiotic chemotherapy.

21 X= OH
22 X= NH$_2$

Acknowledgments

Work from the authors' laboratories was supported by grant 97-ER-2481 from the "Fonds pour la Formation de Chercheurs et l'Aide à la Recherche" (FCAR) to RC and JL, and by grant OGP0009597 from the Natural Sciences and Engineering Research Council of Canada (NSERC) to JL.

References

1. Ibba M, Söll D. Aminoacyl-tRNA synthesis. Annu Rev Biochem 2000; 69:617-650.
2. Delarue M. Aminoacyl-tRNA synthetases. Curr Opin Struct Biol 1995; 5:48-55.
3. Nomanbhoy TK, Hendrickson TL, Schimmel P. Transfer RNA-dependent translocation of misactivated amino acids to prevent errors in protein synthesis. Molecular Cell 1999; 4:519-528.
4. Kluger R, Loo RW, Mazza V. Biomimetically activated amino acids. Catalysis in the hydrolysis of alanyl ethyl phosphate. J Am Chem Soc 1997; 119:12089-12094.
5. Cramer F, Freist W. Aminoacyl-tRNA synthetases: the division into two classes predicted by the chemistry of substrates and enzymes. Angew Chem Int Ed Engl 1993; 32:190-200.
6. Eriani G, Delarue M, Poch O et al. Partition of tRNA synthetases into two classes based on mutually exclusive sets of sequence motifs. Nature 1990; 347:203-206.
7. Cusak S, Berthet-Colominas C, Hartlein M et al. A second class of synthetase structure revealed by X-ray analysis of *Escherichia coli* seryl-tRNA synthetase at 2.5 Å. Nature 1990; 347:249-255.
8. von der Haar F, Gabius HJ, Cramer F. Target directed drug synthesis: the aminoacyl-tRNA synthetases as possible targets. Angew Chem Int Ed Engl 1981; 20:217-223.
9a. Beaulieu D, Ohemeng KA. Patents on bacterial tRNA synthetase inhibitors: January 1996 to March 1999. Exp Opin Ther Patents 1999; 9:1021-1028.
9b. Tao J, Schimmel P. Inhibitors of aminoacyl-tRNA synthetases as novel anti-infectives. Exp Opin Invest Drugs 2000; 9:1767-1775.
10. Schimmel P, Tao J, Hill J. Aminoacyl-tRNA synthetases as targets for new anti-infectives. FASEB J 1998; 12:1599-1609.
11. Hugues J, Mellows G. Interaction of pseudomonic acid A with *Escherichia coli* B isoleucyl-tRNA synthetase. Biochem J 1980; 191:209-219.
12. Class YJ, DeShong P. The pseudomonic acids. Chem Rev 1995; 95:1843-1857.
13. Broom NJP, Cassels R, Cheng HY et al. The chemistry of pseudomonic acid. 17. Dual-action C-1 oxazole derivatives of pseudomonic acid having an extended-spectrum of antibacterial activity. J Med Chem 1996; 39:3596-3600.
14. Brown MJB, Mensah LM, Doyle ML et al. Rational design of femtomolar inhibitors of isoleucyl tRNA synthetase from a binding model for pseudomonic acid A. Biochemistry 2000; 39:6003-6011.
15. Stefanska AL, Cassels R, Ready SJ et al. SB-203207 and SB-203208, two novel isoleucyl tRNA synthetase inhibitors from a *Streptomyces* sp I. Fermentation, isolation and properties. J Antibiotics 2000; 53:357-363.
16. Houge Frydrych CSV, Gilpin ML, Skett PW et al. SB-203207 and SB-203208, two novel isoleucyl tRNA synthetase inhibitors from a *Streptomyces* sp II. Structure determination. J Antibiotics 2000; 53:364-372.
17. Banwell MG, Crasto CF, Easton CJ et al. Analogues of SB-203207 as inhibitors of tRNA synthetases. Bioorg Med Chem Lett 2000; 10:2263-2266.
18. Ogilvie A, Wiebauer K, Kersten W. Inhibition of leucyl-transfer ribonucleic acid synthetase in *Bacillus subtilis* by granaticin. Biochem J 1975; 152:511-515.
19. Tanaka K, Tamaki M, Watanabe S. Effects of furanomycin on the synthesis of isoleucyl-tRNA. Biochem Biophys Acta 1969; 195:244-245.
20. Werner RG. Uptake of indolmycin in gram-positive bacteria. Antimicrob Agents Chemother 1980; 18:858-862.
21. Werner RG, Thorpe LF, Reuter W et al. Indolmycin inhibits prokariotic tryptophanyl-tRNA ligase. Eur J Biochem 1976; 68:1-3.

22. Witty DR, Walker G, Bateson JH et al. Synthesis of conformationally restricted analogues of the tryptophanyl tRNA synthetase inhibitor indolmycin. Tetrahedron Lett 1996; 37:3067-3070.

23 Witty DR, Walker G, Bateson JH et al. Structure-activity dependency of new bacterial tryptophanyl tRNA synthetase inhibitors. Bioorg Med Chem Lett 1996; 6:1375-1380.

24. Konrad I, Roschenthaler R. Inhibition of phenylalanine tRNA synthetase from *Bacillus subtilis* by ochratoxin A. FEBS Lett 1977; 83:341-347.

25. Roth A, Eriani G, Dirheimer G et al. Kinetic properties of pure overproduced *Bacillus subtilis* phenylalanyl-tRNA synthetase do not favour its in vivo inhibition by ochratoxin A. FEBS Lett 1993; 326:87-91.

26. Creppy EE, Mayer M, Kern D et al. In vivo inhibition of yeast valyl-tRNA synthetase by the valine homologue of ochratoxin A. Biochim Biophys Acta 1981; 656:265-268.

27. Cochereau C, Sanchez D, Bourhaoui A et al. Capsaicin, a structural analog of tyrosine, inhibits the aminoacylation of tRNA^{Tyr}. Toxicol Appl Pharmacol 1996; 141:133-137.

28. Cochereau C, Sanchez D, Creppy EE. Tyrosine prevents capsaicin-induced protein-synthesis inhibition in cultured-cells. Toxicology 1997; 117:133-139.

29. Stefanska AL, Coates NJ, Mensah LM et al. SB-219383, a novel tyrosyl tRNA synthetase inhibitor from a *Micromonospora* sp I. Fermentation, isolation and properties. J Antibiotics 2000; 53:345-350.

30. Houge-Frydrych CSV, Readshaw SA, Bell DJ. SB-219383, a novel tyrosyl tRNA synthetase inhibitor from a *Micromonospora* sp II. Structure determination. J Antibiotics 2000; 53:351-356.

31. Berge JM, Broom NJP, Houge-Frydrych CSV et al. Synthesis and activity of analogues of SB-219383: novel potent inhibitors of bacterial tyrosyl tRNA synthetase. J Antibiotics 2000; 53:1282-1292.

32. Berge JM, Copley RCB, Eggleston DS et al. Inhibitors of bacterial tyrosyl tRNA synthetase: synthesis of four stereoisomeric analogues of the natural product SB-219383. Bioorg Med Chem Lett 2000; 10:1811-1814.

33. Brown P, Eggleston DS, Haltiwanger RC et al. Synthetic analogues of SB-219383. Novel C-glycosyl peptides as inhibitors of tyrosyl-tRNA synthetase. Bioorg Med Chem Lett 2001; 11:711-714.

34. Jarvest RL, Berge JM, Brown P et al. Potent synthetic inhibitors of tyrosyl-tRNA synthetase derived from C-pyranosyl analogues of SB-219383. Bioorg Med Chem Lett 2001; 11:715-718.

35. Isono K, Uramoto M, Kusakabe H et al. Ascamycin and dealanylascamycin, nucleoside antibiotics from *Streptomyces* sp. J Antibiotics 1984; 37:670-672.

36. Paetz W, Nass G. Biochemical and immunological characterization of threonyl-tRNA synthetase of two borrelidin-resistant mutants of *Escherichia coli* K-12. Eur J Biochem 1973; 35:331-337.

37. Nass G, Thomale J. Alteration of structure or level of threonyl-tRNA synthetase in borrelidin-resistant mutants of *E. coli*. FEBS Letters 1974; 39:182-186.

38. Freist W, Gauss DH. Threonyl-tRNA synthetase. Biol Chem Hoppe-Seyler 1995; 376:213-224.

39. Poralla K. Borrelidin. In: Corcoran JW, Hahn FE, eds. Antibiotics. Vol 3. Springer Verlag, 1975:365-369.

40. Lepore GC, Di Natale P, Guarini L et al. Histidyl-tRNA synthetase from *Salmonella typhimurium*: specificity in the binding of histidine analogues. Eur J Biochem 1975; 56:369-374.

41. Arnez JG, Augustine JG, Moras D et al. The first step of aminoacylation at the atomic level in histidyl-tRNA synthetase. Proc Natl Acad Sci USA 1997; 94:7144-7149.

42. Tsui FW, Andrulis IL, Murialdo H et al. Amplification of the gene for histidyl-tRNA synthetase in histidinol-resistant Chinese hamster ovary cells. Mol Cell Biol 1985; 5:2381-2388.

43. Konishi M, Nishio M, Saitoh K et al. Cispentacin, a new antifungal antibiotic. J Antibiotics 1989; 42:1749-1755.

44. Kirillov S, Vitali LA, Goldstein BP et al. Purpuromycin—an antibiotic inhibiting transfer-RNA aminoacylation. RNA 1997; 3:905-913.

45. Cassio D, Lemoine F, Waller JP et al. Selective inhibition of aminoacyl ribonucleic acid synthetases by aminoalkyl adenylates. Biochemistry 1967; 6:827-835.

46. Forrest AK, Jarvest RL, Mensah LM et al. Aminoalkyl adenylate and aminoacyl sulfamate intermediate analogues differing greatly in affinity for their cognate *Staphylococcus aureus* aminoacyl tRNA synthetases. Bioorg Med Chem Lett 2000; 10:1871-1874.

47. Pope AJ, Moore KJ, McVey M et al. Characterization of isoleucyl-tRNA synthetase from *Staphylococcus aureus* II. Mechanism of inhibition by reaction intermediate and pseudomonic acid analogues studied using transient and steady-state kinetics. J Biol Chem 1998; 273:31691-31701.

48. Brown P, Richardson CM, Mensah LM et al. Molecular recognition of tyrosinyl adenylate analogues by prokaryotic tyrosyl tRNA synthetases. Bioorg Med Chem Lett 1999; 7:2473-2485.

49. Desjardins M, Garneau S, Desgagnés J et al. Glutamyl adenylate analogues are inhibitors of glutamyl-tRNA synthetase. Bioorg Chem 1998; 26:1-13.

50. Bernier S, Dubois DY, Therrien M et al. Synthesis of glutaminyl adenylate analogues that are inhibitors of glutaminyl-tRNA synthetase. Bioorg Med Chem Lett 2000; 10:2441-2444.

51. Santi DV, Danenberg PV, Satterly P. Phenylalanyl transfer ribonucleic acid synthetase from *Escherichia coli*. Reaction parameters and order of substrate addition. Biochemistry 1971; 10:4804-4812.

52. Santi DV, Pefia VA. Tyrosyl transfer ribonucleic acid synthetase from *Escherichia coli* B. Analysis of tyrosine and adenosine 5'-triphosphate binding sites. J Med Chem 1973; 16:273-280.

53. Clarke CM, Knowles JR. Affinity chromatography of aminoacyl-transfer ribonucleic acid synthetases. Biochem J 1977; 167:405-417.

54. Reshetnikova L, Moor N, Lavrik O et al. Crystal structures of phenylalanyl-tRNA synthetase complexed with phenylalanine and a phenylalanyl-adenylate analogue. J Mol Biol 1999; 267:555-568.

55. Brick P, Bhat TN, Blow DM. Structure of tyrosyl-tRNA synthetase refined at 2.3 Å resolution. Interaction of the enzyme with tyrosyl adenylate intermediate. J Mol Biol 1988; 208:83-98.

56. Cassio D, Robert-Gero M, Shire DJ et al. Effect of methioninyl adenylate on the growth of *E. coli* K12. FEBS Letters 1973; 35:112-116.

57. Ueda H, Shoku Y, Hayashi N et al. X-ray crystallographic conformational study of 5'-O-[N-(L-alanyl)-sulfamoyl]adenosine, a substrate analogue for alanyl-tRNA synthetase. Biochim Biophys Acta 1991; 1080:126-134.
58. Cusack S, Yaremchuk A, Tukalo M. The crystal structure of *T. thermophilus* lysyl-tRNA synthetase complexed with *E. coli* tRNA^{Lys} and a *T. thermophilus* tRNA^{Lys} transcript: anticodon recognition and conformational changes upon binding of a lysyl-adenylate analogue. EMBO J 1996; 15:6321-6334.
59. Cusack S, Yaremchuk A, Tukalo M. The crystal structure of the ternary complex of *T. thermophilus* seryl-tRNA synthetase with tRNA^{Ser} and a seryl-adenylate analogue reveals a conformational switch in the active site. EMBO J 1996; 15:2834-2842.
60. Cusack S, Yaremchuk A, Tukalo M. The 2 Å crystal structure of leucyl-tRNA synthetase and its complex with a leucyl-adenylate analogue. EMBO J 2000; 19:2351-2361.
61. Fukai S, Nureki O, Sekine S et al. Structural basis for double-sieve discrimination of L-valine from L-isoleucine and L-threonine by the complex of tRNA^{Val} and valyl-tRNA synthetase. Cell 2000; 103:793-803.
62. Belrhali H, Yaremchuk A, Tukalo M et al. Crystal structures at 2.5 angstrom resolution of seryl-tRNA synthetase complexed with two analogs of seryl adenylate. Science 1994; 263:1432-1436.
63. Berthet-Colominas C, Seignovert L, Härtlein M et al. The crystal structure of asparaginyl-tRNA synthetase from *Thermus thermophilus* and its complexes with ATP and asparaginyl-adenylate: the mechanism of discrimination between asparagine and aspartic acid. EMBO J 1998; 17:2947-2960.
64. Rath VL, Silvian LF, Beijer B et al. How glutaminyl-tRNA synthetase selects glutamine. Structure 1998; 6:439-449.
65. Heacock D, Forsyth CJ, Shiba K et al. Synthesis and aminoacyl-tRNA synthetase inhibitory activity of prolyl adenylate analogs. Bioorg Chem 1996; 24:273-289.
66. Bovee ML, Yan W, Sproat BS et al. tRNA discrimination at the binding step by a class II aminoacyl-tRNA synthetase. Biochemistry 1999; 38:13725-13735.
67. Lee J, Kang MK, Chun MW et al. Methionine analogs as inhibitors of methionyl-tRNA synthetase. Bioorg Med Chem Lett 1998; 8:3511-3514.
68. Tosa T, Pizer LI. Effect of serine hydroxamate on the growth of *Escherichia coli*. J Bacteriol 1971; 106:966-971.
69. Tosa T, Pizer LI. Biochemical bases for the antimetabolite action of L-serine hydroxamate. J Bacteriol 1971; 106:972-982.
70. Takita T, Ohkubo Y, Shima H et al. Lysyl-tRNA synthetase from *Bacillus stearothermophilus*. Purification and fluorometric and kinetic analysis of the binding of substrates, L-lysine and ATP. J Biochem 1996; 119:680-689.
71. Takita T, Hashimoto S, Ohkubo Y et al. Lysyl-tRNA synthetase from *Bacillus stearothermophilus*. Formation and isolation of an enzyme-lysyl adenylate complex and its analogue. J Biochem 1997; 121:244-250.
72. Freist W, Gauss DH. Lysyl-tRNA synthetase. Biol Chem Hoppe-Seyler 1995; 376:451-472.
73. Rogers MJ, Brown RJ, Hodkin V et al. Bisphosphonates are incorporated into adenine nucleotides by human aminoacyl-tRNA synthetase enzymes. Biochem Biophys Res Commun 1996; 224:863-869.
74. Hou YM, Gamper HB. Inhibition of tRNA aminoacylation by 2'-O-methyl oligonucleotides. Biochemistry 1996; 35:15340-15348.
75. Lee J, Kang SU, Kang MK et al. Methionyl adenylate analogues as inhibitors of methionyl-tRNA synthetase. Bioorg Med Chem Lett 1999; 9:1365-1370.
76. Lee J, Kang SU, Kim SY et al. Ester and hydroxamate analogues of methionyl and isoleucyl adenylates as inhibitors of methionyl-tRNA and isoleucyl-tRNA synthetases. Bioorg Med Chem Lett 2001; 11:961-964.
77. Wolfenden R. Conformational aspects of inhibitor design: enzyme-substrate interactions in the transition state. Bioorg Med Chem 1999; 7:647-652.
78. Schramm VL, Horenstein BA, Kline PC. Transition state analysis and inhibitor design for enzymatic reactions. J Biol Chem 1994; 269:18259-18262.
79. Mader MM, Bartlett PA. Binding energy and catalysis: the implications for transition-state analogs and catalytic antibodies. Chem Rev 1997; 97:1281-1301.
80. Biryukov AI, Zhukov YN, Lavrik OI. Influence of the aminoacyl-tRNA synthetase inhibitors and the diadenosine-5'-tetraphosphate phosphonate analogues on the catalysis of diadenosyl oligophosphates formation. FEBS 1990; 273:208-210.
81. Biryukov AI, Ishmuratov BK, Khomutov RM. Transition-state analogues of aminoacyl adenylates. FEBS Lett 1978; 91:249-252.
82. Yu XY, Hill JM, Yu G et al. Synthesis and structure-activity relationships of a series of novel thiazoles as inhibitors of aminoacyl-tRNA synthetases. Bioorg Med Chem Lett 1999; 9: 375-380.
83. Zydowsky TM, Yu G, Hill JM et al. Aminoacyl adenylate mimics as novel antimicrobial and antiparasitic agents. US Patent 5, 726, 195.
84. Yu XY, Hill JM, Yu G et al. A series of quinoline analogs as potent inhibitors of *C albicans* prolyl tRNA synthetase. Bioorg Med Chem Lett 2001; 11:541-544.
85. Leeman AH, Hammond ML, Maletic M et al. Novel catechols as antimicrobial agents. Patent WO 0066120.
86. Wilcox M, Nirenberg M. Transfer RNA as a cofactor coupling amino acid synthesis with that of protein. Proc Natl Acad Sci USA 1968; 61:229-236.
87. Curnow AW, Hong KW, Yuan R et al. Glu-tRNA^{Gln} amidotransferase: a novel heterotrimeric enzyme required for correct decoding of glutamine codons during translation. Proc Natl Acad Sci USA 1997; 94:11819-11826.
88. Ibba M, Becker HD, Stathopoulos C et al. The adaptor hypothesis revisited. Trends Biochem Sci 2000; 25:311-316.
89. Gagnon Y, Lacoste L, Champagne N et al. Widespread use of the Glu-tRNA^{Gln} transamidation pathway among bacteria. J Biol Chem 1996; 271:14856-14863.

Antibiotics as Indicators of the Functional Components of the Ribosome

Dominique Fourmy, Satoko Yoshizawa and Stephen Douthwaite

Abstract

The inhibitory action of many antibiotics is to block directly the synthesis of proteins on the bacterial ribosome. How these antibiotics interact with their targets on the bacterial ribosome has been the focus of considerable scrutiny over the last four decades. It was envisaged that elucidation of the mechanisms of action of these drugs would lead not only to a deeper understanding of ribosome structure and function, but would also indicate how the drugs might be altered to improve their clinical efficacy. These ideas are beginning to be fulfilled with the advent of the recent crystallographic models of the ribosome at atomic resolution. These models put into perspective a myriad of previous data on ribosome-drug interactions and give us a glimpse of how more effective antimicrobial agents might be rationally designed.

Introduction

As the preceding chapters bear witness, the translation of mRNA into proteins by the ribosome is a complex process involving the coordinated interactions of over one hundred macromolecules. Many of these interactions occur within and between the RNA and protein components of the ribosome; other interactions occur between the ribosome and the stream of aminoacylated tRNAs and factors that bind and dissociate during the succession of steps of translation. While all of the interactions have a function, some of them are more central in the translation process than others. Many interactions are between relatively constant structures that are concerned with maintaining the overall shape of the translational machinery. These for the most part lead a quiet existence in the suburban areas of the ribosome, and will not be considered further in this review. Instead we will concentrate on the functional hot spots on the ribosome that are directly involved in steering the translation process.

The structures of the components at the functionally active sites of the ribosome will generally be dynamic, engaging in conformational shifts and/or in rapidly changing interactions with other translational components. An important part of the early work undertaken by ribosomologists was concerned with identifying the locations of these functional sites within the ribosome structure. One successful approach was to inhibit the translation process at well-defined steps, and then investigate what parts of the ribosome were jammed. For this purpose, antibiotics proved to be excellent jamming devices.

Many antibiotics that occur in nature and that are used in the clinical treatment of disease inhibit the growth of bacteria by blocking mRNA translation by the ribosome.[1,2] Despite the structural complexity of the ribosome, antibiotics target only a few select ribosomal regions that are invariably composed of rRNA.[3-5] Pioneers in the use of antibiotics to determine which ribosomal components are responsible for what functions included David Vázquez, Julian Davies and Eric Cundliffe (for a review, see ref. 1). Using a variety of techniques, some involving radiolabelled drugs, it was shown that macrolides, lincosamides, streptogramins (MLS antibiotics), as well as chloramphenicol and puromycin are specific for the 50S ribosomal subunit. It was correctly concluded that these drugs bind at, or near to, the site of peptide bond formation as they either directly inhibit the peptidyl transferase reaction, or at least specifically displace drugs that can inhibit this reaction. This clearly localized the peptidyl transferase centre to the 50S subunit, and showed it to be a target for several chemically different classes of drugs that interacted with mutually exclusive, and thus presumably physically overlapping, binding sites (Fig. 1).

A physically distinct site on the 50S subunit was shown to be targeted by thiostrepton, micrococcin and related thiopeptide drugs. These drugs interfere with the GTPase- and factor-related steps of translation,[4] linking the 50S subunit with energy requiring processes including EF-Tu-directed A-site binding of aminoacyl-tRNA and EF-G-catalysed translocation of peptidyl-tRNA. These latter functions were also shown to involve the 30S subunit, as the binding site of tetracycline, an inhibitor of aminoacyl-tRNA binding at the A site, as well as the binding site of the translocation inhibitor spectinomycin were localized to distinct sites on the small subunit (Fig. 1). The process of decoding, whereby the ribosome checks the validity of cognate codon-anticodon interactions, is located on the 30S subunit. Decoding is disturbed by aminoglycoside antibiotics which target a specific site on the 16S rRNA (Fig. 1).

While these earlier studies of the 1960's and 70's gave ribosomologists a rough idea of which component was doing what, the 1980's saw the introduction of chemical footprinting which pinpointed the sites of drug interaction on the rRNA.[6,7] The drug footprinting sites were gratifyingly consistent with the sites of rRNA mutations and methylation that confer drug resistance, and established a clear link between these regions of the rRNA and specific ribosome functions.[4,5] The last few years research has seen the application of physical chemical methods including cryo-electron microscopy,[8,9] nuclear magnetic resonance (NMR)[10] and X-ray crystallography[11-14] to the challenge of understanding ribosome structure and antibiotic interactions. In particular, NMR and X-ray crystallography (for reviews, see Chapter 15 by B. Wimberly and refs. 15, 16 and 17) have taken our view of the ribosome to previously unseen levels of resolution, and many regions of the structure can now be viewed at the atomic scale. In this review, we summarize what is presently known about the ribosomal sites that are targeted by antibiotics. A close look is taken at one group of drugs, the aminoglycosides, which have furthered our knowledge of antibiotic interactions and mechanisms of antibiotic resistance, and have provided us with a deeper understanding of how the ribosome decodes mRNA.

Aminoglycoside Antibiotics

It has been known for more than thirty years that aminoglycosides interfere with translation by decreasing the accuracy of amino acid incorporation, and in some cases by inhibiting the subsequent step of tRNA translocation.[18,19] Aminoglycosides decrease the accuracy of translation by stabilizing the interaction of non-cognate (as well as cognate) tRNAs with mRNA in the aminoacyl-tRNA site of the ribosome (the ribosomal A site, also termed the decoding site). This in turn decreases the

Figure 1. Interfaces of the *Thermus thermophilus* 30S and 50S subunits drawn from the coordinates of the model at 5.5 Å resolution.[13] The rRNAs are shown in dark blue and the r-proteins in light blue. (These are dark and light, respectively, in the black and white version. The color version can be viewed at http://www.eurekah.com/chapter.php?chapid=887&bookid=59&catid=54). Interaction sites of ribosome targeting drugs are indicated. On the 50S subunit, the peptidyl transferase centre is the target for many chemically dissimilar compounds including macrolide, lincosamide and streptogramin (MLS) antibiotics,[1,2,64] in addition to chloramphenicol,[1,2] puromycin,[1,2,65] amicetin[66] and sparsomycin.[67] More recently, the synthetic oxazolidinone drug, linezolid, has also been shown to interact at the peptidyl transferase centre immediately adjacent to the MLS site[64,68] (although an independent study places other oxazolidinones elsewhere in the 50S).[69] In addition, the pleuromutilin drugs, tiamulin and valnemulin, have been footprinted at the peptidyl transferase centre.[70] Thiostrepton (Thiostrep) and related thiopeptide drugs bind at the GTPase-associated centre of the large subunit, which is situated at the r-protein L11 binding site within domain II of the 23S rRNA.[4,71-74] The structure of this region of rRNA complexed with L11 has been solved at higher resolution by X-ray crystallography.[75,76] The oligosaccharide antibiotic, evernimicin (Evern), binds to a novel site in the 50S subunit,[77] and this is consistent with its supposed function in obstructing access to initiation factor IF-2 at the same site.[78]

The drug sites on the 30S subunit have been mapped at atomic resolution by X-ray crystallography.[26,27] This has brought into perspective a wealth of previous data on tetracycline (Tet),[6] spectinomycin (Spec),[6,79] streptomycin (Strep),[6,80,81] hygromycin B (Hygro B),[6,23] and aminoglycosides of the neomycin/paromomycin and gentamicin/kanamycin classes (Paromo)[6,10,18,19,23] (see text). The primary and secondary sites of tetracycline are indicated (Tet 1° and Tet 2°); the secondary site lies within the body of the subunit.[27]

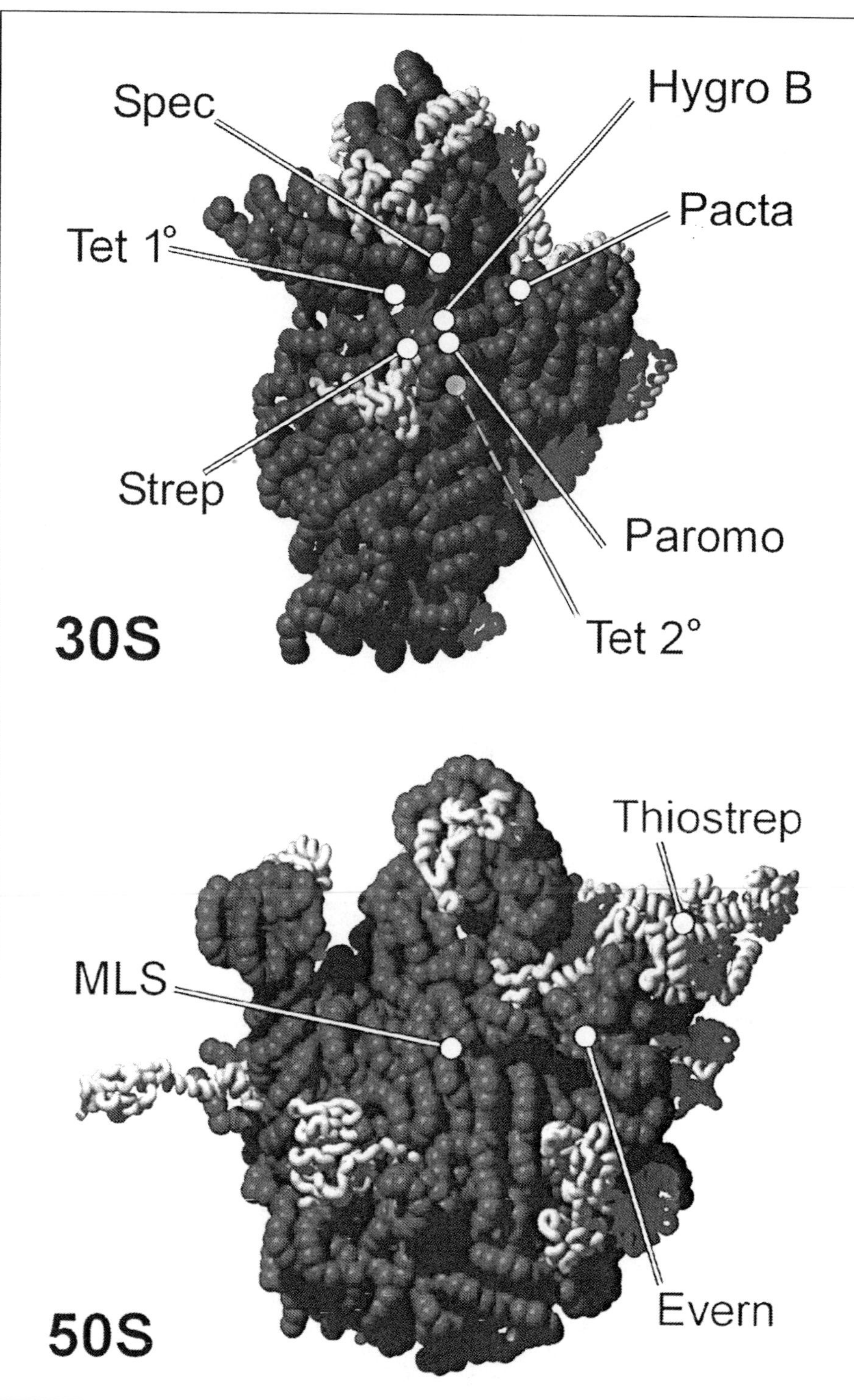
Spec
Hygro B
Tet 1°
Pacta
Strep
Paromo
Tet 2°
30S
Thiostrep
MLS
Evern
50S

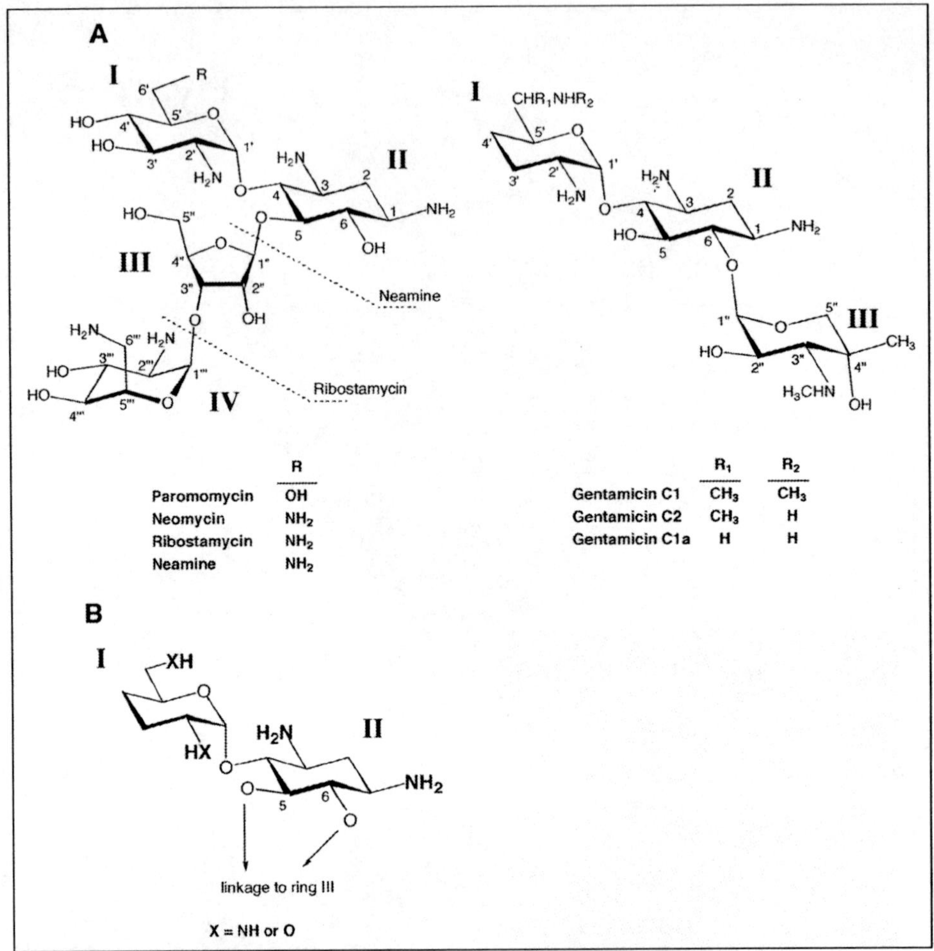

Figure 2. A) Structures of the neomycin/paromomycin (left) and gentamicin/kanamycin (right) classes of aminoglycoside antibiotics. Ribostamycin is a neomycin/paromomycin-class drug that lacks ring IV; neamine lacks both rings III and IV. B) Rings I and II showing the common structural elements (bold) of all aminoglycoside antibiotics that bind to the ribosomal A site.

dissociation rates of tRNAs from the A site and thereby interferes with the proofreading steps that ensure translational fidelity.[20] The aminoglycoside antibiotics shown in Figure 2 are related in parts of their structure, while other parts are chemically distinct. Likewise, the interactions of these drugs with the ribosome display common as well as drug-specific rRNA interactions.

The aminoglycoside antibiotics are composed of two to four rings. The rings are positively charged at biological pH, which contributes to rRNA binding.[21] The chemical groups on aminoglycosides that are essential for specific binding to the A site are presented in ring I (an amino sugar) and ring II (2-deoxystreptamine). Rings I and II are common to all these aminoglycosides[22] and conserved features include the N1 and N3 amino groups of ring II and the hydrogen bond donors at the 2' and 6' positions of ring I (Fig. 2B). However, the substitution patterns of the rings as well as the linkage of ring III to 2-deoxystreptamine can vary. In the neomycin class aminoglycosides such as ribostamycin and paromomycin, ring III is connected to position 5 of ring II. In comparison, in the gentamicin/kanamycin class, ring III is linked to position 6 of ring II. Rings III of the gentamicin/kanamycin class aminoglycosides share common chemical groups at the 2" and 3" positions.

Aminoglycosides Bind at the Ribosomal A Site

Aminoglycoside antibiotics bind to a conserved sequence in 16S rRNA adjacent to where codon-anticodon recognition takes place in the A site of the 30S subunit (Fig. 3). The location of the site of aminoglycoside interaction has been determined by a number of techniques mentioned above including chemical footprinting,[6] analyses of resistant ribosomes,[23,24] NMR[10,25] and X-ray crystallography.[26,27]

Chemical modification experiments showed that specific nucleotides in the A site are protected from chemical probes by neomycin-class aminoglycosides.[6,28] The protected nucleotides are clustered in the A site at A1408, G1491 and G1494 (Fig. 3B), implying that these nucleotides directly contact the drugs and are probably central in forming the drug interaction site. Substitutions and methylations at these nucleotides confer aminoglycoside resistance, presumably by distorting the drug site and thereby preventing interaction of the drug. The exact nucleotide-drug interactions that are perturbed by base substitutions and methylations can now be explained by the NMR and X-ray structures, and these are considered in detail below.

Structural Studies of Aminoglycosides Bound to Their Target

NMR Studies of Drug-RNA Complexes

Aminoglycoside antibiotics bind to small RNA fragments that encompass the A site of 16S rRNA with similar affinity and specificity as to 30S subunits.[29-31] This is fortuitous for the determination of the aminoglycoside-RNA interaction by NMR spectroscopy, which is most effective for solving small RNA structures. A 27-nucleotide fragment of RNA that mimics the A site (Fig. 3C) forms a specific and stable interaction with aminoglycosides and thus contains the essential elements of the ribosomal binding site for these drugs.[30,31] The solution structures of the 27-nucleotide RNA on its own[32] and in complexes with each of two aminoglycosides, paromomycin and gentamicin C1a, were solved by NMR spectroscopy.[10,25] The secondary structure of the RNA fragment (Fig. 3C) comprises an upper and a lower helix stabilized by a UUCG tetraloop. The NMR data indicate that in addition to the secondary structure elements shown in Figure 3C, C1407 and G1494 form a Watson-Crick base pair and close the asymmetrical internal loop together with non-canonical U1406-U1495 and A1408-A1493 base pairs.

Figure 4 shows the RNA fragment complexed with either paromomycin or gentamicin. In both aminoglycoside-rRNA complexes, the antibiotic binds in the distorted major groove of the RNA within a pocket created by the A1408-A1493 base pair and a single bulged adenine (A1492). The structures of rings I and II of paromomycin and gentamicin are well-defined, and engage in specific interactions with conserved nucleotides in the RNA. Ring I is positioned near the A1408-A1493 pair and stacks above the base of G1491, while ring II spans the U1406-U1495 and C1407-G1494 base pairs. Additional interactions are made from the amino groups at positions 1 and 3 of ring II via hydrogen bonds to the O4 of U1495 and the N7 of G1494, respectively. The amino group at position 3 may also make contact with the phosphate between A1493 and G1494.[31]

Superimposition of the positions of rings I and II in both structures reveals how similar hydrogen bond donor groups of two distinct class aminoglycosides specifically contact the RNA (Fig. 4). This agrees with previous results showing that rings I and II direct the binding of the truncated aminoglycosides ribostamycin and neamine to the A-site RNA.[33] Furthermore, the model suggests a structural rationale for how rings I and II of all aminoglycosides facilitate a common mode of binding to the decoding site of 16S rRNA.[25]

While the aminoglycosides described here (Fig. 2) have a common binding site on the rRNA, rings II and III are connected differently in the neomycin-class and gentamicin/kanamycin aminoglycosides. Thus, because rings I and II direct all these drugs into their binding site, ring III (and ring IV) of the two classes become positioned so that they make different interactions with the rRNA (Fig. 4). As a consequence of this, ring III of gentamicin interacts with the upper stem region (U1406-U1495 and G1405-C1496 base pairs), whereas rings III and IV of paromomycin interact with the lower stem of the A-site RNA. The gentamicin-rRNA interaction explains why no antibiotics of the gentamicin/kanamycin class has more than three rings, as this would be sterically difficult to accommodate within the major groove of the rRNA target. In contrast, rings III and IV in the neomycin class are directed along the major groove of the lower stem of the rRNA target site,

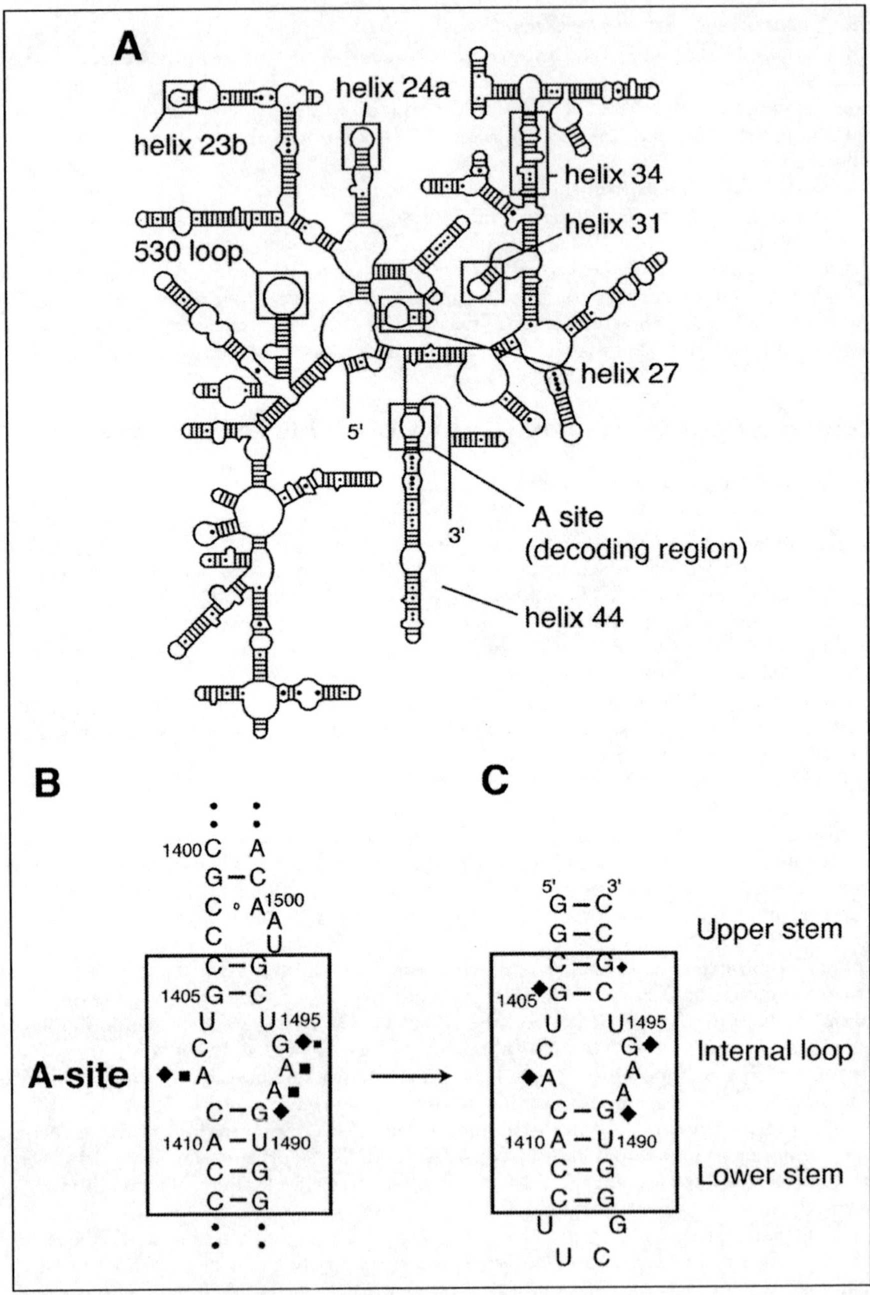

Figure 3. A) Secondary structure of 16S rRNA from *E. coli* showing the locations of the structures with which antibiotics interact (see text). B) Structure of the decoding region; nucleotides protected from dimethyl sulphate modification by A-site tRNA (▪)[82] and neomycin-class aminoglycoside antibiotics (◆)[6] are indicated. C) The 27 nucleotide RNA fragment with the 16S rRNA A site sequence boxed. Nucleotides in the RNA fragment that were protected from DMS modification by neomycin-class aminoglycoside antibiotics are indicated (◆).[30]

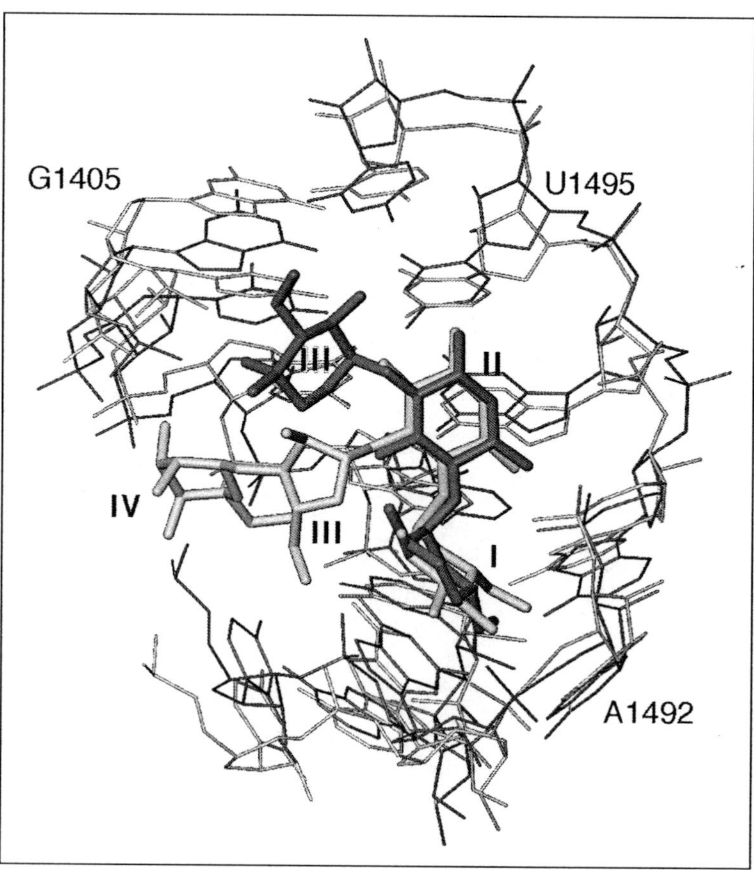

Figure 4. Best-fit superimposition of the paromomycin-RNA and gentamicin C1a-RNA complexes, viewed from the major groove side of the RNA. The heavy atoms of the core (nucleotides U1406 to A1410, and U1490 to U1495) of the RNA are superimposed. Only the core is represented. For the paromomycin-RNA complex, the RNA is represented in brown and the antibiotic in yellow (a color version of this figure can be viewed at http://www.eurekah.com/chapter.php?chapid=887&bookid=59&catid=54). For the gentamicin C1a-RNA complex, the RNA is in tan and the gentamicin in red.

where they are readily accommodated into the structure and the positive charges in ring IV can interact with the backbone of the RNA.[33,34] While all these interactions contribute to the binding affinities of the two aminoglycoside drug classes, ring III of gentamicin interacts specifically with universally conserved bases in the binding site (Fig. 4).[25] This appears to facilitate the placement of gentamicin/kanamycin drugs at their ribosomal target and has undoubtedly contributed to the clinical success of gentamicin.

Crystal Structure of the Aminoglycoside-30S Complex

The crystal structure of the 30S ribosomal subunit simultaneously complexed with paromomycin, streptomycin and spectinomycin has been solved at 3 Å resolution.[26] The crystal structure of 30S with paromomycin (Fig. 5) shows the binding site to lie in the major groove of the decoding site of 16S rRNA, and thus validates the NMR data for this drug bound to the 27-nucleotide RNA fragment. In both sets of data, ring I of paromomycin stacks against G1491 and makes contacts with the backbone of A1492 and A1493; the amino groups at positions 1 and 3 of ring II make hydrogen bonds to the O4 of U1495 and the N7 of G1494, respectively; and the amino group at position 3

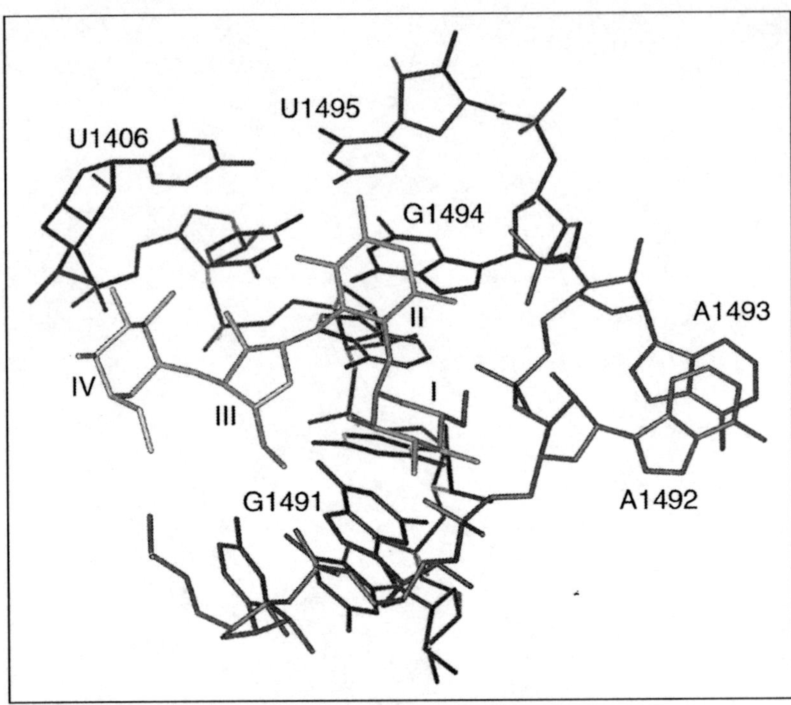

Figure 5. The paromomycin binding site on the *Thermus thermophilus* 30S subunit drawn from the coordinates of the model obtained by X-ray crystallography at 3 Å resolution.[26] The view is from from the major groove as in Figure 4. The rRNA is represented in blue and the antibiotic in yellow (a color version of this figure can be viewed at http://www.eurekah.com/chapter.php?chapid=887&bookid=59&catid=54).

(ring II) also makes contact with the phosphate of G1494. The antibiotic-rRNA interaction is extended along the major groove by rings III and IV, and the 5" hydroxyl group (ring III) makes a hydrogen bond with the N7 of G1491. The amino and hydroxyl groups of ring IV establish electrostatic and/or hydrogen bond contacts with the phosphate backbone of the rRNA.

There are, however, some subtle differences between the NMR and X-ray structures that are worthy of note. Ring IV of paromomycin, which is dynamically disordered in the NMR structure, adopts a well-defined conformation in the X-ray structure. Also, the two universally conserved residues A1492 and A1493, that were shown by NMR to be displaced towards the minor groove in presence of an antibiotic,[32] become flipped out to a far greater extent in the X-ray structure.[26] These differences could reflect the experimental conditions used: the X-ray data were collected on samples co-crystallized with spectinomycin and streptomycin and frozen with cryoprotectant, whereas the NMR data were collected at 35°C in solution with a single aminoglycoside. Alternatively, they could arise from differences in the drug target on the 30S subunit versus that on an RNA fragment. Possibly the flipped-out conformation of A1492 and A1493 is stabilized at the low temperature, and the displacement of the adenines towards the minor groove is therefore less pronounced in the NMR structure. The conformation of these two adenosines is a key feature in a model of how aminoglycosides interfere with the decoding process[35] (see below, and see also note added in proof).

One of the essential new contributions of the paromomycin-30S crystal structure is to set the drug binding site in relation to the other small subunit structures that are involved in the decoding process. Recognition of a cognate codon-anticodon interaction at the decoding site is followed by a relay of events that transmit this information through the ribosome to the peptidyl transferase centre, where the amino acid presented at the 3′-end of the A-site tRNA is added to the growing peptide chain. Crucial in this chain of events is 16S rRNA helix 27, which can exist in two alternative conformations depending on how it is base paired, and this governs whether transmission of the decoding site signal is restricted or permitted.[36] The relay of information also involves other 16S rRNA components including the 530 loop[37] and helix 34[38,39] (Fig. 3), and r-proteins S4, S5 and S12, which have long been implicated in translational fidelity by genetic studies of drug resistance mutants.[40]

Resistance to Aminoglycosides

The two main forms of aminoglycoside resistance are conferred by alterations in the 16S rRNA target site and enzymatic inactivation of the drug. These mechanisms have been observed in drug-producing and pathogenic bacteria, as well as in organisms used as laboratory models. The molecular details of these forms of resistance can now be explained by the NMR and X-ray structures. Both resistance mechanisms destroy essential drug-RNA contacts either by removing the potential for hydrogen bonds or by creating a steric clash between the drug and its RNA target.

Alterations in the Drug Target Site

Base substitution at the aminoglycoside binding site have been observed to confer resistance in several laboratory strains. For instance, in yeast, mutation of C to G in the mitochondrial rRNA at the position corresponding to 1409 of *Escherichia coli* 16S rRNA confers resistance to paromomycin.[41] Resistance to the same drug is conferred in *Tetrahymena thermophila* by a G to A mutation at the position corresponding to 1491,[23] suggesting that the base pair between positions 1409 and 1491 (Fig. 3B) is essential for paromomycin binding. Systematic disruption of this base pair in *E. coli* 16S rRNA by site-directed mutagenesis confers resistance to a variety of aminoglycosides.[24,42] From the structural models, it can be seen that stacking of the aminoglycoside ring I onto G1491 and the hydrogen bond between the ring III 5″ hydroxyl group and the N7 of G1491 would be marred by substitutions in the 1409-1491 pair. Substitution of a G at the neighboring nucleotide A1408 alters the noncanonical A1408-A1493 pair and confers resistance in both *E. coli* and *Mycobacterium* species to aminoglycosides containing a 6′ amino group.[43,44]

In aminoglycoside-producing bacteria, nature has devised an elegant means to protect the ribosomes by enzymatic modification of bases within the A-site of the 16S rRNA.[45] Methylation at the N1 of A1408 confers aminoglycoside resistance[4] presumably in a manner similar to substitution of this base. Methylation of G1405-N7 confers resistance to kanamycin and gentamicin but not to paromomycin and neomycin.[45,46] The structure of the gentamicin-RNA complex (Fig. 4) illustrates how methylation of the N7 of G1405 prevents the formation of a hydrogen bond with ring III of gentamicin while at the same time creating a steric clash with this ring.[25] Rings III and IV in the neomycin class of drugs are directed away from contact with G1405 (Fig. 4) and are thus indifferent to methylation of this nucleotide.

Reduced Binding of Aminoglycosides to Eukaryal Ribosomes

A prerequisite for the therapeutic use of aminoglycosides is that they interact more avidly with bacterial ribosomes than with those of the host. Bacterial ribosomes are sensitive to a 10- to 15-fold lower antibiotic concentration than are eukaryal ribosomes.[47] A crucial difference between these ribosomes is that bacterial rRNA sequences have an A at position 1408, whereas this position is a G in eukaryal sequences. As can be seen from the mutagenesis studies reviewed above, the A1408-A1493 pair is essential for antibiotic binding, and leads to formation of the specific binding pocket for ring I.[10,25] A variant of the 27 nucleotide RNA fragment demonstrated that the eukaryal G1408-A1493 configuration does not have an equivalent geometry, and binds paromomycin with decreased affinity.[48,49] In the higher Eukarya, cytoplasmic rRNA sequences contain G1408 in addition to a mispair at the 1409-1491 positions. In bacteria, the C1409-G1991 pair provides the floor of the antibiotic binding pocket, and its absence in the eukaryal structures prevents them from binding aminoglycoside antibiotics with high affinity.[48]

Inactivation of the Drug

The predominant form of resistance found in bacterial pathogens involves enzymatic modification of aminoglycosides.[50] Modification enzymes primarily target rings I and II, disrupting specific interactions with the A site. Acetylation of the conserved amino group on ring II (position 3) obstructs the hydrogen bond between this group and the N7 of G1494, as well as specific electrostatic and hydrogen bonding contacts with the phosphate of A1493. In the neomycin class antibiotics, the 6'-amino group on ring is also subject to acetylation, sterically hindering binding and resulting in aminoglycoside resistance. Although the affinity of the 6'-acetyl gentamicin-rRNA interaction is not known, it is most likely weaker than that of gentamicin C1 whose binding is affected by the steric clash of the 6'-N-methyl group with the RNA.

In ring I of the neomycin class drugs, the 3'-OH can be phosphorylated and 4'-OH adenylated, which in both cases sterically and electrostatically impedes complex formation. These two hydroxyl groups are absent in the gentamicins and other later-generation aminoglycoside antibiotics; this has little consequence for RNA binding while providing the advantage that the drugs are impervious to these types of modifying enzymes.

Deciphering the Genetic Code

As reviewed by Kurland,[51] it has long been appreciated that the differences in binding energy between cognate and noncognate interactions of mRNA codons with tRNA anticodons at the ribosomal A site are not enough to account for the relatively high level of translational fidelity exhibited by the ribosome. The codon-anticodon interaction obviously plays the primary role in the decoding process, but it must also function in combination with other components that provide a proof-reading step upon GTP hydrolysis and the release of EF-Tu. A number of models have been proposed for how the ribosome attains its high levels of accuracy,[52-55] but this subject remains controversial. It seems that a set of the 30S ribosomal subunit components switch between (at least) two distinct conformational states, alternatively determining whether formation of the next peptide bond should be permitted or restricted. As outlined above (and described in depth in chapter 16 by L. Brakier-Gingras et al and in Chapter 17 by J.S. Lodmell and S.P. Hennelly), helix 27 is central in the conformation switch and is part of a relay of events involving the 530 loop, helix 34 and several proteins in the 30S subunit before the information is passed across the subunit interface to the 50S subunit peptidyl transferase center. The components that initiate this relay are situated in the 30S decoding site at the end of helix 44 where the codon-anticodon interaction takes place and where aminoglycosides bind.

Aminoglycosides reduce the fidelity of translation by inducing the 30S subunit to attain the permissive conformation and signal to the 50S subunit that peptide bond formation may proceed. Earlier biophysical measurements suggested that the rRNA conformation in the aminoglycoside complex is a high affinity state for mRNA-tRNA recognition,[20] and the recent structural models give a more accurate picture of what the essential elements of this conformation might be.[10,26] The two universally conserved residues in the A site of 16S rRNA, A1492 and A1493, are displaced towards the minor groove of the RNA helix when aminoglycosides are bound.[10,25,26] There is experimental evidence for A1492 and A1493 being essential for A-site tRNA binding.[56] This, as well as theoretical support from modeling,[57] suggests that hydrogen bonding occurs between A1492 and A1493 and the A-site codon. A model was proposed in which A1492 and A1493 monitor the minor groove of the three base pairs formed between the codon and anticodon, sensing the base pair geometry and detecting any distortions arising from mispairing.[56] This model has recently been revised using the X-ray structure of the 30S subunit complexed to anticodon stem loops and mRNA.[58] The data explain how the highly flipped-out A1492 and A1493 residues of the crystal model could stringently check the fidelity of the first two base pairs of the codon-anticodon interaction while showing greater flexibility at the third, or 'wobble', position. Aminoglycosides disturb this process by holding nucleotides A1492 and A1493 in the flipped-out conformation, despite any mispairs in the codon-anticodon interaction. This conformation is presumably the high codon-anticodon affinity state that initiates the relay of the signal for peptide bond formation and, when stabilized by aminoglycosides, results in the incorporation of noncognate amino acids into the new peptide chain.

Additional Perspectives

Co-crystals of the 30S ribosomal subunit with several other drugs have recently defined exactly where these antibiotics bind and what interactions are involved in determining their specific inhibitory mechanisms. Streptomycin and spectinomycin have now been accurately mapped to their sites on the 30S,[26] which satisfactorily puts into perspective the large body of data that has accumulated on these drugs over the past few decades. More recently, the binding sites of hygromycin B, pactamycin and tetracycline were reported.[27] Biochemical[6,59] and molecular genetic experiments[42] had previously provided accurate indications of where hygromycin B and pactamycin interact with the 16S rRNA. Hygromycin B binds at the upper stem region of the portion of helix 44 shown in Figure 3B, and this fits well with previous footprinting and genetic data. The hygromycin B site is immediately adjacent to, but distinct from, the neomycin/gentamicin site described above (Fig. 1). The site of interaction of pactamycin in the crystal structure also fits well with previous data,[59,60] and places this drug close to the initiation factor 3 site[61] at helices 23b and 24a (Fig. 3A), where it can interfere with rRNA-mRNA contacts during translational initiation and block the formation of the 70S ribosome.

Due to its wide clinical use, tetracycline is one of the most intensely studied of the 30S specific antibiotic.[62] However, tetracycline remained uncooperative in attempts to map its location by footprinting or by selection of strains with resistance conferred by mutations in their ribosomal components. From the crystal structure,[27] it can now be seen that the previous scarcity of binding data was partly due to the manner in which tetracycline interacts with the 30S, and partly because of difficulty in distinguishing between the two sites with which this drug interacts on the 30S. The primary binding site (Fig. 1) is formed by the irregular minor groove of helix 34 with an additional drug contact to the phosphate of G966 in helix 31 (Fig. 3A). The binding pocket is composed largely of contacts with the phosphates and riboses in the backbone of helix 34 rather than with the bases, and this explains why base substitutions conferring resistance had been notoriously hard to come by. One notable exception, which does confer tetracycline resistance, is the single nucleotide substitution of G1058 to C.[63] The location of this mutation fits well with the crystal data, as it would disturb the G1058-U1199 pair in helix 34 and thereby distorts the tetracycline binding pocket. Early chemical footprinting of 30S subunits had revealed an enhancement in the accessibility of nucleotides U1052 and C1054 on binding of tetracycline,[6] fitting well with interaction of the drug at its primary target. At the same time, footprinting protection of A892 in helix 27 was observed,[6] and this can now be attributed to binding at the secondary tetracycline site (Fig. 1).[27] Binding of tetracycline to its primary target in helix 34 explains how the drug prevents access of aminoacyl-tRNA to the A site on the 30S subunit.[1,2] Tetracycline binding at the secondary site in helix 27 would further complicate the decoding process, but it is not presently clear whether this site is of physiological relevance for the action of the drug.

Data on 50S subunits co-crystallized with drugs specific for this subunit will soon become available (see note added in proof), and will hopefully enlighten us about the nature of the binding interactions of MLS and other antibiotics with levels of resolution approaching those which the 30S subunit crystallographers have shown us. Many questions remain to be answered, including how many chemically dissimilar drugs manage to interact with the same region of the 50S at the peptidyl transferase centre (see Fig. 1 and legend). With the enormous advance in the structural information that has become available to us in the last couple of years, the question that many ribosomologists are beginning to ask is no longer what antibiotics can tell us about the whereabouts of functional sites on the ribosome, but rather what the ribosome can now tell us about the existence of potential novel targets against which new synthetic antibiotics can be designed.

Acknowledgements

We wish to thank Jacob Pøhlsgaard for making Figure 1. The authors have received financial support from the European Commission's Fifth Framework Program (contract number QLK2-2000-00935), the Danish Biotechnology Instrument Centre, the Carlsberg Foundation, the Nucleic Acid Centre of the Danish Grundforskningsfond and the Danish Medical Research Council.

Note Added in Proof

Since this manuscript was submitted, several important crystallographic studies have been published that are relevant to this review. A small RNA, mimicking the A site in Figure 3, was crystallized together with the aminoglycosides paromomycin[83] and tobramycin.[84] The manner in which rings I and II of the aminoglycosides (Fig. 2) support the bulged conformations of A1492 and A1493 is clearly demonstrated in these structures. The 50S subunit of the bacterium *Deinococcus radiodurans* has been crystallized with macrolide, lincosamide and chloramphenicol antibiotics.[85] This study was rapidly followed by the crystal structure of the *Haloarcula marismortui* 50S subunit complexed with a range of macrolides.[86] These studies now provide us with a clear picture of the drug-target interactions at atomic resolution.

References

1. Gale EF, Cundliffe E, Reynolds PE et al. The Molecular Basis of Antibiotic Action. London: John Wiley and Sons, 1981.
2. Vázquez D. Inhibitors of Protein synthesis. New York: Springer Verlag, 1979.
3. Spahn CMT, Prescott CD. Throwing a spanner in the works:antibiotics and the translational apparatus. J Mol Med 1996; 74:423-439.
4. Cundliffe E. Recognition sites for antibiotics within rRNA. In: Hill WE, Dahlberg A, Garret RA et al. The Ribosome: Structure, Function and Evolution. American Society for Microbiology, Washington DC 1990; 479-490.
5. Green R, Noller HF. Ribosomes and translation. Annu Rev Biochem 1997; 66:679-716.
6. Moazed D, Noller HF. Interaction of antibiotics with functional sites in 16S ribosomal RNA. Nature 1987; 327:389-394.
7. Moazed D, Noller HF. Chloramphenicol, erythromycin, carbomycin and vernamycin B protect overlapping sites in the peptidyl transferase region of 23S ribosomal RNA. Biochimie 1987; 69:879-884.
8. Frank J. Cryo-electron microscopy as an investigative tool:the ribosome as an example. Bioessays 2001; 23:725-732.
9. van Heel M. Unveiling ribosomal structures:the final phases. Current Opinion Structural Biology 2000; 10 :259-264
10. Fourmy D, Recht MI, Blanchard SC et al. Structure of the A site of *E. coli* 16S rRNA complexed with an aminoglycoside antibiotic. Science 1996; 274:1367-1371.
11. Wimberly BT, Brodersen DE, Clemons WM et al. Structure of the 30S ribosomal subunit. Nature 2000; 407:327-339.
12. Schluenzen F, Tocilj A, Zarivach R et al. Structure of functionally activated small ribosomal subunits at 3.3 Å resolution. Cell 2000; 102:615-623.
13. Yusupov MM, Yusupova GZ, Baucom A et al. Crystal structure of the ribosome at 5.5 Å resolution. Science 2001; 292:883-896.
14. Ban N, Nissen M, Hansen J et al. The complete atomic structure of the large ribosomal subunit at 2.4 Å resolution. Science 2000; 289:905-920.
15. Ramakrishnan V, Moore PB. Atomic structures at last:the ribosome in 2000. Curr Opin Struct Biol 2001; 11:144-154.
16. Moore PB. The ribosome at atomic resolution. Biochemistry 2001; 40:3243-3250.
17. Puglisi JD, Blanchard SC, Green R. Approaching translation at atomic resolution. Nat Struct Biol 2000; 7:855-861
18. Davies J, Davis BD. Misreading of ribonucleic acid code words induced by aminoglycoside antibiotics. J Biol Chem 1968; 243:3312-3316.
19. Davies J, Gorini L, Davis BD. Misreading of RNA codewords induced by aminoglycoside antibiotics. Mol Pharmacol 1965; 1:93-106.
20. Karimi R, Ehrenberg M. Dissociation rate of cognate peptidyl-tRNA from the A-site of hyper-accurate and error-prone ribosomes. Eur J Biochem 1994; 226:355-360.
21. Botto RE, Coxon B. Nitrogen-15 nuclear magnetic resonance spectroscopy of neomycin B and related aminoglycosides. J Am Chem Soc 1983; 105:1021-1028.
22. Benveniste R, Davies J. Structure-activity relationships among the aminoglycoside antibiotics:Role of hydroxyl and amino groups. Antimicrob. Ag. Chemother 1973; 4:402-409.
23. Spangler EA, Blackburn EH. The nucleotide sequence of the 17S ribosomal RNA gene of Tetrahymena thermophila and the identification of point mutations resulting in resistance to the antibiotics paromomycin and hygromycin. J Biol Chem 1985; 260:6334-40.
24. De Stasio EA, Moazed D, Noller HF et al. Mutations in 16S ribosomal RNA disrupt antibiotic-RNA interactions. EMBO J 1989; 8:1213-6.
25. Yoshizawa S, Fourmy D, Puglisi JD. Structural origins of gentamicin antibiotic action. EMBO J 1998; 17:6437-6448.
26. Carter AP, Clemons Jr WM, Brodersen DE et al. Functional insights from the structure of the 30S ribosomal subunit and its interactions with antibiotics. Nature 2000; 407:340-348.
27. Brodersen DE, Clemons WM, Carter AP et al. The structural basis for the action of the antibiotics tetracycline, pactamycin and hygromycin B on the 30S ribosomal subunit. Cell 2000; 103:1143-1154.
28. Woodcock J, Moazed D, Cannon M et al. Interaction of antibiotics with A- and P-site-specific bases in 16S ribosomal RNA. EMBO J 1991; 10:3099-3103.
29. Purohit P, Stern S. Interaction of a small RNA with antibiotic and RNA ligands of the 30S subunit. Nature, 1994; 370:659-662.
30. Recht MI, Fourmy D, Blanchard SC et al. RNA sequence determinants for aminoglycoside binding to an A-site rRNA model oligonucleotide. J Mol Biol 1996; 262:421-436.
31. Blanchard SC, Fourmy D, Eason RG et al. rRNA chemical groups required for aminoglycoside binding. Biochemistry 1998; 37:7716-7724.
32. Fourmy D, Yoshizawa S, Puglisi J D. Paromomycin binding induces a local conformational change in the A site of 16S rRNA. J Mol Biol 1998; 277:333-345.

33. Fourmy D, Recht MI, Puglisi JD. Binding of neomycin-class aminoglycoside antibiotics to the A Site of 16S rRNA. J Mol Biol 1998; 277:347-362.
34. Alper PB, Hendrix M, Sears P et al. Probing the specificity of aminoglycoside-ribosomal RNA interactions with designed synthetic analogs. J Am Chem Soc 1998; 120:1965-1978.
35. Puglisi JD, Blanchard SC, Dahlquist KD et al. Aminoglycosides antibiotics and decoding. In: Garrett RA, Douthwaite SR, Liljas A et al, eds. The Ribosome. Structure, Function, Antibiotics and Cellular Interactions. Washington DC: ASM Press, 2000:419-430.
36. Lodmell JS, Dahlberg AE. A conformational switch in *Escherichia coli* 16S ribosomal RNA during decoding of messenger RNA. Science 1997; 277:1262-1267.
37. Powers T, Noller HF. Evidence for functional interaction between elongation factor EF-Tu and 16S ribosomal RNA. Proc Natl Acad Sci USA 1993; 90:1364-1368.
38. Murgola EJ, Arkov AL, Chernyaeva NS et al. rRNA functional sites and structures for peptide chain termination. In: Garrett RA, Douthwaite SR, Liljas A et al, eds. The Ribosome. Structure, Function, Antibiotics and Cellular Interactions. Washington DC: ASM Press, 2000:509-518.
39. Murgola EJ, Pagel FT, Hijazi KA et al. Variety of nonsense suppressor phenotypes associated with mutational changes at conserved sites in *Escherichia coli* ribosomal RNA. Biochem Cell Biol 1995; 73:925-931.
40. Kurland CG, Hughes D, Ehrenberg M. In: Neidhardt FC et al, eds. Limitations of Translational accuracy. Washington DC: ASM, 1996:979-1003.
41. Li M, Tzagoloff A, Underbrink-Lyon K et al. Identification of the paromomycin-resistance mutation in the 15S rRNA gene of yeast mitochondria. J Biol Chem 1982; 257:5921-5928.
42. DeStasio EA, Dahlberg AE. Effects of mutagenesis of a conserved base-paired site near the decoding region of *Escherichia coli* 16S ribosomal RNA. J Mol Biol 1990; 212:127-133.
43. Recht MI, Douthwaite S, Dahlquist KD et al. Effect of mutations in the A site of 16S rRNA on aminoglycoside antibiotic-ribosome interaction. J Mol Biol 1999; 286:33-43.
44. Prammanaman T, Sander P, Brown BA et al. A single 16S ribosomal RNA substitution is responsible for resistance to amikacin and other 2-deoxystreptamine aminoglycosides in *Mycobacterium abscessus* and *Mycobacterium chelonae*. J Infect Dis 1998; 177:1573-1581.
45. Beauclerk AA, Cundliffe E. Sites of action of two ribosomal RNA methylases responsible for resistance to aminoglycosides. J Mol Biol 1987; 193:661-671.
46. Thompson J, Skeggs A, Cundliffe E. Methylation of 16S ribosomal RNA and resistance to the aminoglycoside antibiotics gentamicin and kanamycin determined by DNA from the gentamicin-producer, Micromonospora purpurea. Mol Gen Genet 1985; 201:168-173.
47. Wilhem JM, Pettit SE, Jessop JJ. Aminoglycoside antibiotics and eukaryotic protein synthesis:structure-function relationships in the stimulation of misreading with a wheat embryo system. Biochemistry 1978; 17:1143-1149.
48. Recht MI, Douthwaite S, Puglisi JD. Basis of prokaryotic specificity of action of aminoglycoside antibiotics. EMBO J 1999; 18:3133-3138.
49. Lynch SR, Puglisi JD. Structural origins of aminoglycoside specificity for prokaryotic ribosomes. J Mol Biol 2001; 306:1037-1058.
50. Shaw K J, Rather PN, Hare RS et al. Molecular genetics of aminoglycoside resistance genes and familial relationships of the aminoglycoside-modifying enzymes. Microbiol Rev 1993; 57:138-163.
51. Kurland CG. Translational accuracy and the fitness of bacteria. Annu Rev Genet 1992; 26:9-50.
52. Pape T, Wintermeyer W, Rodnina MV. Conformational switch in the decoding region of 16S rRNA during aminoacyl-tRNA selection on the ribosome. Nature Struct Biol 2000; 7:104-110.
53. Rodnina MV, Wintermeyer W. Fidelity of aminoacyl-tRNA selection on the ribosome:Kinetics and structural mechanisms. Annu Rev Biochem 2001; 70:415-435.
54. Gabashvili IS, Agrawal RK, Grassucci R et al. Major rearrangements in the 70S ribosomal 3D structure caused by a conformational switch in 16S ribosomal RNA. EMBO J 1999; 18:6501-6507.
55. Nierhaus KH. The allosteric three-site model for the ribosomal elongation cycle:features and future. Biochemistry 1990; 29:4997-5008.
56. Yoshizawa S, Fourmy D, Puglisi JD. Recognition of the codon-anticodon helix by ribosomal RNA. Science 1999; 285:1722-1725.
57. VanLoock MS, Easterwood TR, Harvey SC. Major groove binding of the tRNA/mRNA complex to the 16S ribosomal RNA decoding site. J Mol Biol 1999; 285:2069-2078.
58. Ogle JM, Brodersen DE, Clemons WM Jr et al. Recognition of cognate transfer RNA by the 30S ribosomal subunit. Science 2001; 292:897-902.
59. Egebjerg J, Garrett RA. Binding sites of the antibiotics pactamycin and celesticetin on ribosomal RNA. Biochimie 1991; 73:1145-1149.
60. Mankin AS. Pactamycin resistance mutations in functional sites of 16S rRNA. J Mol Biol 1997; 274:8-15.
61. Moazed D, Samaha RR, Gualerzi C et al. Specific protection of 16S rRNA by translational initiation factors. J Mol Biol 1995; 248:207-210.
62. Chopra I, Hawkey PM, Hinton M. Tetracyclines, molecular and clinical aspects. J Antimicrob Chemother 1992; 29:245-277.
63. Ross JI, Eady EA, Cove JH et al. 16S rRNA mutation associated with tetracycline resistance in a gram-positive bacterium. Antimicrob. Agents Chemother 1998; 42:1702-1705.
64. Mankin AS. Ribosomal antibiotics. Mol Biol 2001; 35:509-520.
65. Nissen P, Hansen J, Ban N et al. The structural basis of ribosome activity in peptide bond synthesis. Science 2000; 289:920–30.

66. Leviev IG, Rodrigez-Fonseca C, Phan H et al. A conserved secondary structural motif in 23S rRNA defines the site of interaction of amicetin, a universal inhibitor of peptide bond formation. EMBO J 1994; 13:1682-1686.
67. Lázaro E, Rodrigrez-Fonseca C, Porse B et al. A sparsomycin-resistant mutant of *Halobacterium salinarium* lacks a modification at nucleotide U2603 in the peptidyl transferase centre of 23S rRNA. J Mol Biol 1996; 261:231-238.
68. Xiong L, Kloss P, Douthwaite S et al. Oxazolidinone resistance mutations in 23S ribosomal RNA of *Escherichia coli* reveal the central region of domain V as the primary site of drug action. J Bacteriol 2000; 182:5325-5331.
69. Matassova N B, Rodnina M V, Endermann R et al. Ribosomal RNA is the target for oxazolidinones, a novel class of translational inhibitors. RNA 1999; 5:939-946.
70. Poulsen SM, Karlsson M, Johansson LB et al. The pleuromutilin drugs tiamulin and valnemulin bind to the RNA at the peptidyl transferase centre on the ribosome. Mol Microbiol 2001; 41:1091-1100.
71. Cundliffe E, Thompson J. Concerning the mode of action of micrococcin upon bacterial protein synthesis. Eur J Biochem 1981; 118:47-52.
72. Egebjerg J, Douthwaite S, Garrett RA. Antibiotic interactions at the GTPase-associated centre within *Escherichia coli* 23S rRNA. EMBO J 1989; 8:607-608.
73. Rosendahl G, Douthwaite S. Ribosomal proteins L11 and L10.(L12)$_4$ and the antibiotic thiostrepton interact with overlapping regions of the 23S rRNA backbone in the ribosomal GTPase centre. J Mol Biol 1993; 234:1013-1020.
74. Ryan PC, Lu M, Draper DE. Recognition of the highly conserved GTPase center of 23S ribosomal RNA by ribosomal protein L11 and the antibiotic thiostrepton. J Mol Biol 1991; 221:1257-1268.
75. Conn GL, Draper DE, Lattman EE et al. Crystal structure of a conserved ribosomal protein-RNA complex. Science 1999; 284:1171-1174.
76. Wimberly BT, Guymon R, McCutcheon JP et al. A detailed view of a ribosomal active site: the structure of the L11-RNA complex. Cell 1999; 97:491-502.
77. Belova L, Tenson T, Xiong L et al. A novel site of antibiotic action in the ribosome:Interaction of evernimicin with the large ribosomal subunit. Proc Natl Acad Sci USA 2001; 98:3726-3731.
78. La Teana A, Gualerzi CO, Dahlberg AE. Initiation factor IF 2 binds to the alpha-sarcin loop and helix 89 of *Escherichia coli* 23S ribosomal RNA. RNA 2001; 7:1173-1179
79. Bilgin N, Richter AA, Ehrenberg M et al. Ribosomal RNA and protein mutants resistant to spectinomycin. EMBO J 1990; 9:735-739.
80. Leclerc D, Melançon P, Brakier-Gingras L. The interaction between streptomycin and ribosomal RNA. Biochimie 1991; 73:1431-1438.
81. Leclerc D, Melançon P, Brakier-Gingras L. Mutations in the 915 region of *Escherichia coli* 16S ribosomal RNA reduce the binding of streptomycin to the ribosome. Nucleic Acids Research 1991; 14:3973-3977.
82. Moazed D, Noller HF. Transfer RNA shields specific nucleotides in 16S ribosomal RNA from attack by chemical probes. Cell 1986; 47: 985-994.
83. Vicens Q, Westhof E. Crystal structure of paromomycin docked into the eubacterial ribosomal decoding A site. Structure 2001; 9:647-658.
84. Vicens Q, Westhof E. Crystal structure of a complex between the aminoglycoside tobramycin and an oligonucleotide containing the ribosomal decoding A site. Chem Biol 2002; 9:747-755.
85. Schlunzen F, Zarivach R, Harms J et al. Structural basis for the interactoin of antibiotics with the peptidyl transferase centre in eubacteria. Nature 2001; 413:814-821.
86. Hansen JL, Ippolito JA, Ban N et al. The structures of four macrolide antibiotics bound to the large ribosomal subunit. Mol Cell 2002; 10:117-128.

Index